Advances in

CHEMICAL ENGINEERING

MATHEMATICS IN CHEMICAL KINETICS AND ENGINEERING

VOLUME **34**

ADVANCES IN

CHEMICAL ENGINEERING

Editor-in-Chief

GUY B. MARIN
Department of Chemical Engineering
Ghent University
Ghent, Belgium

Editorial Board

DAVID H. WEST
Research and Development
The Dow Chemical Company
Freeport, Texas, U.S.A.

PRATIM BISWAS
Department of Chemical and Civil Engineering
Washington University
St. Louis, Missouri, U.S.A.

JINGHAI LI
Institute of Process Engineering
Chinese Academy of Sciences
Beijing, P.R. China

SHANKAR NARASIMHAN
Department of Chemical Engineering
Indian Institute of Technology
Chennai, India

Advances in

CHEMICAL ENGINEERING

MATHEMATICS IN CHEMICAL KINETICS AND ENGINEERING

VOLUME **34**

Edited by

GUY B. MARIN

Department of Chemical Engineering
Ghent University, Ghent, Belgium

DAVID WEST

The Dow Chemical Company
Freeport, Texas, USA

GREGORY S. YABLONSKY

Parks College, Department of Chemistry
Saint Louis University
Missouri, USA

ELSEVIER

Amsterdam • Boston • Heidelberg • London • New York • Oxford
Paris • San Diego • San Francisco • Singapore • Sydney • Tokyo
Academic Press is an imprint of Elsevier

Academic Press is an imprint of Elsevier
Radarweg 29, PO Box 211, 1000 AE Amsterdam, The Netherlands
32 Jamestown Road, London NW1 7BY, UK
30 Corporate Drive, Suite 400, Burlington, MA 01803, USA
525 B Street, Suite 1900, San Diego, CA 92101-4495, USA

First edition 2008

Library of Congress Cataloging-in-Publication Data
A catalog record for this book is available from the Library of Congress

British Library Cataloguing in Publication Data
A catalogue record for this book is available from the British Library

ISBN: 978-0-12-374506-4
ISSN: 0065-2377

For information on all Academic Press publications
visit our website at elsevierdirect.com

Printed and bound in USA

08 09 10 11 12 10 9 8 7 6 5 4 3 2 1

CONTENTS

CONTRIBUTORS

Numbers in parenthesis indicate the pages on which the authors' contributions begins.

C.J. van Duijn, *Department of Mathematics and Computer Science, TU Eindhoven, P.O. Box 513, 5600 MB Eindhoven, The Netherlands* (1)

A.N. Gorban, *Department of Mathematics, University of Leicester, LE1 7RH, UK* (103)

Mark Z. Lazman, *Independent Consultant, Calgary AB, Canada* (47)

Andro Mikelić, *Université de Lyon, Lyon, F-69003, France; Université Lyon 1, Institut Camille Jordan, Site de Gerland, 50, avenue Tony Garnier 69367 Lyon Cedex 07, France* (1)

I.S. Pop, *Department of Mathematics and Computer Science, TU Eindhoven, P.O. Box 513, 5600 MB Eindhoven, The Netherlands* (1)

O. Radulescu, *IRMAR, UMR 6625, University of Rennes 1, Campus de Beaulieu, 35042 Rennes, France* (103)

Carole Rosier, *L.M.P.A., Université du Littoral; 50 rue F. Buisson, B.P. 699, 62228 Calais Cedex, France* (1)

Liqiu Wang, *Department of Mechanical Engineering, The University of Hong Kong, Pokfulam Road, Hong Kong, P. R. China* (175)

Xiaohao Wei, *Department of Mechanical Engineering, The University of Hong Kong, Pokfulam Road, Hong Kong, P. R. China* (175)

Mingtian Xu, *Institute of Thermal Science and Technology, Shandong University, Jinan, P. R. China* (175)

Gregory S. Yablonsky, *Parks College, Department of Chemistry, 3450 Lindell Blvd, Saint Louis University, Missouri, MO 63103, USA; Department of Energy, Environmental & Chemical Engineering, One Brookings Dr., Campus Box 1180, Washington Univ. in St. Louis, St. Louis, MO 63130-4899, USA* (47)

PREFACE

The cross fertilization of physico-chemical and mathematical ideas has a long historical tradition. A conference "Mathematics in Chemical Kinetics and Engineering" (MaCKiE-2007) was held in Houston, Texas in 2007 bringing together about 40 mathematicians, chemists and chemical engineers from 10 countries (Australia, Belgium, France, Germany, Sweden, Russia, Switzerland, Ukraine, the United Kingdom, and the United States) to discuss the application and development of mathematical tools in their respective fields. More attention was paid to biological applications at this conference than at the first conference at Ghent University (Belgium) in 2002. In order to make sure that the main messages of this conference would not be lost, it was decided to publish some of the important contributions in an archival journal. The series *Advances in Chemical Engineering* was considered to serve best this purpose, among other things because there would be no page limitations on the different contributions. Volume 34 of Advances in Chemical Engineering is therefore mainly dedicated to MaCKiE-2007.

The problem of dispersion of a solute in a flow through a small diameter channel, with or without adsorption or reaction at the wall, has stimulated a great deal of research since the classical work of G.I. Taylor in 1953. The popularity of this topic arises from at least two properties of the problem: it is a canonical version of technologically important flows (e.g. transport in a pipeline or porous medium), and it is amenable to analysis. Consequently, the Taylor dispersion problem is a good setting for the development and testing of *effective* (low dimensional) descriptions of multiscale phenomena. The contribution by C.J. van Duijn, A. Mikelic, I.S. Pop and C. Rosier, entitled "Effective dispersion equations for reactive flows with dominant Peclet and Damkohler numbers" discusses the use of anisotropic singular perturbation methods to obtain an effective equation for the solute concentration similar to what has been obtained before by other methods, but which also obtain a rigorous estimate of the error.

Non-linearity is one of the main problems of theoretical chemical kinetics.

The paper by M. Lazman and G. Yablonsky "Overall Reaction Rate Equation of Single-Route Complex Catalytic Reaction in Terms of Hypergeometric Series" presents a further development of the approach formulated by the authors in the early 1980s using contemporary algebraic methods. For a single-route catalytic reaction with a non-linear mechanism, the authors previously showed that under assumptions of the steady state or quasi-steady state the non-linear kinetic model corresponding to the complex mechanism can be transformed into just one algebraic equation, so called "kinetic polynomial"(K.P.), that is an overall (steady state or quasi-steady state) rate equation. However, in contrast to traditional overall equations, in a general case the K.P. is not an explicit function of reaction conditions, $r = f(c, T)$, but an implicit function $F(r, c, T) = 0$. Many properties of the K.P. which are interesting from the physico-chemical point of view have been described before:

(a) K.P. may have several physically meaningful solutions (roots), and this fact can be used for understanding of the multiplicity of steady-states.
(b) A free term of the K.P. relates to the overall reaction, so despite the non-linearity under equilibrium conditions K.P. is transformed into a simple equilibrium relationship which does not depend on the details of the complex mechanism.

In this contribution the authors posed and solved a new problem: how to find an explicit reaction rate expression r at least for the thermodynamic branch of the K.P. Such rate equation was obtained in terms of hypergeometric series, "the four-term rate equation". The mentioned terms are following: (1) *the kinetic apparent coefficient*; (2) *the potential term* or *driving force* related to the thermodynamics of the overall reaction; (3) *the resistance term*, i.e. the denominator, which reflects the complexity of reaction, both its multistep character and its non-linearity; finally, (4) *the non-linear term* which is caused exclusively by non-linear steps. Known rate equations, such as Langmuir–Hinshelwood, Hougen–Watson and Horiuti–Boreskov ones, are particular cases of the presented "four-term equation". Distinguishing the fourth, "non-linear", term is the original result of this chapter. In classical theoretical kinetics of heterogeneous catalysis (Langmuir–Hinshelwood and Hougen–Watson equations, and equations for linear mechanisms), such term is absent.

Chemists have always tried to overcome the mathematical difficulties of analyzing detailed kinetic models introducing simplifying hypotheses such as a hypothesis on the existence of a single rate-limiting step. Reaction rate equations grounded on these assumptions found a wide area of practical application including situations far beyond the assumptions of the original theory. However the validity of the popular assumption on the limiting step is not trivial. In their contribution, "Dynamic and static limitation in multiscale reaction networks, revisited" A. N. Gorban and O. Radulescu analyzed the mathematical status of this classical hypothesis and found many non-trivial cases and misinterpretations as well. They rigorously studied differences between transient and steady-state (quasi-steady-state) regimes and classified different scenarios in chemical kinetics. Studying networks of linear reactions, the authors introduce the concept of the "dominant system", i.e. the system that produces the main asymptotic terms of the steady-state and relaxation in the limit when rate constants are well separated, i.e. of clear separation of time scales. In the simplest case, the dominant system is determined by the ordering of the rate constants. For distinguishing dominant systems of complex networks, a special mathematical technique was developed. It is defined as a "surgery of cycles", because its main operations are gluing and cutting cycles in graphs of auxiliary discrete dynamic systems. The interesting fact is that the limiting step for relaxation of complex chemical reaction is not the slowest reaction, nor the second slowest reaction in the hierarchy of the full spectrum of kinetic constants, but the slowest reaction of the dominant system. Moreover, the limiting step constant is not obligatory a reaction rate constant of the initial system. It can be represented by a function (monomial) of kinetic constants as well. The obtained results are illustrated by concrete examples of complex mechanisms. Computationally, the proposed methods are quite cheap. It is worth mentioning that the concept of "dominant systems" goes back to classical ideas proposed by Newton.

The last contribution to this volume "Multiscale theorems" by L. Wang, M. Xu and X. Wei is exceptional in two ways. It does not correspond to a contribution to MaCKiE-2007 and it is longer. It comes from the Department of Mechanical Engineering of the University of Hong Kong and provides a comprehensive overview of the mathematical foundations of multiscale modeling. This overview extends way beyond the "classics" of multiscale theorems such as the Gauss divergence theorem, the Reynolds transport theorem and the Leibnitz rule. The authors not only present but prove a selected set of 71 (!) mathematical theorems. They emphasise the role of so-called indicator functions. These are an extension of Heaviside step functions and can be used to identify a region of interest by taking the value of one in the interior and zero exterior to the region. In this way they allow to transform line, surface and volume integrals into integrals over all space while they and/or their derivatives are appearing in the integrand. In doing so, "the indicator functions are mathematical catalysts: they facilitate the derivations but do not appear in the end product." It is their strong belief that "without detailed proofs, users have little structured access to the physics, application conditions and the ways of improving the theorems" and little chances to address successfully what they consider to be the key questions to develop a good scaling approach: "(1) When is simple aggregation (i.e. linear addition of elements) sufficiently accurate for upscaling? (2) Are processes which are observed or models which are formulated, at points or small spatial scales transferable to larger scales? (3) Where such scaling is possible, how should it be done? (4) How do these means change with scale? (5) How does the variability change with scale? (6) How does the sensitivity change with scale? (7) Under what circumstances would non-linear responses be either amplified or dampened as scales change? (8) How does heterogeneity change with scale? (9) How does predictability change with space and time scales? (10) What types of conceptual errors are involved, wittingly or unwittingly, by scientists in their up- or downscaling assumptions? and (11) How can observations made at two scales be reconciled?" Applications in the fields of modeling of single-phase turbulent flow, heat conduction in two-phase systems and transport in porous and multiphase systems are discussed.

Of course the axiom of separability of length scales, more precisely of the molecular scale, the microscale, the macroscale and the megascale, is taken as the starting point. The molecular scale is characterized by the mean free path between molecular collisions, the microscale by the smallest scale at which the laws of continuum mechanics apply, the macroscale by the smallest scale at which a set of averaged properties of concern can be defined and the megascale by the length scale corresponding to the domain of interest.

Certainly, the rigorous mathematical nature of this contribution may deter part of the readership. It is our conviction, however, that it will become a reference text for those active in the field of multiscale modeling. It provides the foundations for techniques such as mathematical homogenization, mixture and hybrid mixture theory, spatial averaging, filtering techniques, moment methods, central limit or Martingale methods, stochastic-convective approaches, projection operators, renormalization group techniques, variational approaches, space transformation methods, continuous time random walks, Eulerian and Lagrangian perturbation schemes, heterogeneous multiscale methods, serial methods, onion-type hybrid methods, coarse-graining methods, multigrid-type hybrid methods, parallel approaches, dynamic methods and concurrent methods.

Guy B. Marin
Ghent University

David West
The Dow Chemical Company

Gregory S. Yablonsky
Saint Louis University
December 12, 2007

CHAPTER 1

Effective Dispersion Equations for Reactive Flows with Dominant Péclet and Damkohler Numbers

C.J. van Duijn[1], **Andro Mikelić**[2,*], **I.S. Pop**[1] and **Carole Rosier**[3]

Contents

[1] Department of Mathematics and Computer Science, TU Eindhoven, P.O. Box 513, 5600 MB Eindhoven, The Netherlands

[2] Université de Lyon, Lyon, F-69003, France; Université Lyon 1, Institut Camille Jordan, Site de Gerland, 50, avenue Tony Garnier 69367 Lyon Cedex 07, France

[3] L.M.P.A., Université du Littoral; 50 rue F.Buisson, B.P. 699, 62228 Calais Cedex, France

*Corresponding author.
E-mail address: mikelic@univ-lyon1.fr

Advances in Chemical Engineering, Volume 34
ISSN 0065-2377, DOI 10.1016/S0065-2377(08)00001-X

Abstract In this chapter we study a reactive flow through a capillary tube. The solute particles are transported and diffused by the fluid. At the tube lateral boundary they undergo an adsorption–desorption process. The transport and reaction parameters are such that we have large, dominant Péclet and Damkohler numbers with respect to the ratio of characteristic transversal and longitudinal lengths (the small parameter ε). Using the anisotropic singular perturbation technique we derive the effective equations. In the absence of the chemical reactions they coincide with Taylor's dispersion model. The result is compared with the turbulence closure modeling and with the center manifold approach. Furthermore, we present a numerical justification of the model by a direct simulation.

1. INTRODUCTION

In many processes involving reactive flows different phenomena are present at different order of magnitude. It is fairly common that transport dominates diffusion and that chemical reaction happen at different timescales than convection/diffusion. Such processes are of importance in chemical engineering, pollution studies, etc.

In bringing the models to a non-dimensional form, the presence of dominant Péclet and Damkohler numbers in reactive flows is observed. The problems of interest arise in complex geometries-like porous media or systems of capillary tubes.

Taylor's dispersion is one of the most well-known examples of the role of transport in dispersing a flow carrying a dissolved solute. The simplest setting for observing it is the injection of a solute into a slit channel. The solute is transported by Poiseuille's flow. In fact this problem could be studied in three distinct regimes: (a) *diffusion-dominated mixing*, (b) *Taylor dispersion-mediated mixing* and (c) *chaotic advection*.

In the first flow regime, the velocity is small and Péclet number is of order one or smaller. Molecular diffusion plays the dominant role in solute dispersion. This case is well-understood even for reactive flows (see e.g. Conca et al., 2004, 2003; van Duijn and Knabner, 1997; van Duijn and Pop, 2004; van Duijn et al., 1998; Hornung and Jäger, 1991; Knabner et al., 1995; Mikelić and Primicerio, 2006 and references therein).

If the flow rate is increased so that Péclet number $Pe \gg 1$, then there is a timescale at which transversal molecular diffusion smears the contact discontinuity into a plug. In Taylor (1993), Taylor found an effective long-time axial diffusivity, proportional to the square of the transversal Péclet number and occurring in addition to the molecular diffusivity. After this pioneering work of Taylor, a vast literature on the subject developed, with over 2000 citations to date. The most notable references are the article (Aris, 1956) by Aris, where Taylor's intuitive approach was explained through moments expansion and the lecture notes (Caflisch and Rubinstein, 1984), where a probabilistic justification of Taylor's dispersion is given. In addition to these results, addressing the tube flow with a dominant Péclet number and in the absence of chemical reactions, there is a huge literature on mechanical dispersion for flows through porous media. Since this is not the scope of our paper, we refer to the book (Bear and Verruijt, 1987) for more details about the modeling. For the derivation of Taylor's dispersion in porous media using formal two-scale expansions, we refer to Auriault and Adler (1995) and the references therein.

In the third regime, we observe the *turbulent mixing*.

Our goal is the study of *reactive flows* through slit channels in the regime of Taylor dispersion-mediated mixing and in this chapter we will develop new effective models using the technique of anisotropic singular perturbations.

As already said, Taylor's effective model contains a contribution in the effective diffusion coefficient, which is proportional to the square of the transversal Péclet number. Frequently this term is *more important than the original molecular diffusion*. After his work, it is called *Taylor's dispersion coefficient* and it is generally accepted and used in chemical engineering numerical simulations. For the practical applications we refer to the classical paper (Rubin, 1983) by Rubin. The mathematical study of the models from Rubin (1983) was undertaken in Friedman and Knabner (1992).

Even with this enormous number of scientific papers on the subject, mathematically rigorous results on the subject are rare. Let us mention just ones aiming toward a rigorous justification of Taylor's dispersion model and its generalization to reactive flows. We could distinguish them by their approach

- The averaging of the equations over the section leads to an infinite system of equations for the moments. A parallel could be drawn with the turbulence and in the article Paine et al. (1983), Paine, Carbonell and Whitaker used an ad hoc closure approach borrowed from Launder's "single point" closure schemes of turbulence modeling, for obtaining an effective model for reactive flows in capillary tubes. We will see that this approach leads to correct general form of the effective equations, but it does not give the effective coefficients. Furthermore, let us remark that it is important to distinguish between the turbulent transport, arising for very high Péclet numbers, and the Taylor dispersion arising for dominant Péclet number, but smaller than some threshold value.
- The center manifold approach of Mercer and Roberts (see the article Mercer and Roberts, 1990 and the subsequent article by Rosencrans, 1997) allowed to calculate approximations at any order for the original Taylor's model. Even if the error estimate was not obtained, it gives a very plausible argument for the validity of the effective model. This approach was applied to reactive flows in the article by Balakotaiah and Chang (1995). A number of effective models for different Damkohler numbers were obtained. Some generalizations to reactive flows through porous media are in Mauri (1991) and the preliminary results on their mathematical justification are in Allaire and Raphael (2007).
- Another approach consisting of the Liapounov–Schmidt reduction coupled with a perturbation argument is developed in the articles Balakotaiah and Chang (2003), Balakotaiah (2004) and Chakraborty and Balakotaiah (2005). It allows developing multi-mode hyperbolic upscaled models.

- Recent approach using the anisotropic singular perturbation is the article by Mikelić et al. (2006). This approach gives the error estimate for the approximation and, consequently, the rigorous justification of the proposed effective models. It uses the strategy introduced by Rubinstein and Mauri (1986) for obtaining the effective models.

We continue by applying the later approach for reactive transport with adsorption–desorption through a capillary tube.

2. NON-DIMENSIONAL FORM OF THE PROBLEM AND STATEMENT OF THE RESULTS

We study the diffusion of the solute particles transported by the Poiseuille velocity profile in a semi-infinite two-dimensional (2D) channel. Solute particles are participants in a chemical reaction with the boundary of the channel. They do not interact between them. The simplest example is described by the following model for the solute concentration c^*:

$$\frac{\partial c^*}{\partial t^*} + q(z)\frac{\partial c^*}{\partial x^*} - D^*\Delta_{x^*,z}c^* = 0 \quad \text{in } \mathbb{R}_+ \times (-H,H) \tag{1}$$

where $q(z) = Q^*(1-(z/H)^2)$ and Q^* (*velocity*) and D^* (*molecular diffusion*) are positive constants. At the lateral boundaries $z = \pm H$

$$-D^*\partial_z c^* = \frac{\partial \hat{c}}{\partial t^*} = \hat{k}^*(c^* - \hat{c}/K_e) \qquad \text{on } z = \pm H \tag{2}$$

where $\hat{k}^*$ represents the rate constant for adsorption and K_e the linear adsorption equilibrium constant.

The natural way of analyzing this problem is to introduce the appropriate scales. They would come from the characteristic concentration $\hat{c}$, the characteristic length L_R, the characteristic velocity Q_R, the characteristic diffusivity D_R and the characteristic time T_R. The characteristic length L_R coincides in fact with the "observation distance". Setting

$$c_F = \frac{c^*}{c_R}, \quad x = \frac{x^*}{L_R}, \quad y = \frac{z}{H}, \quad t = \frac{t^*}{T_R}, \quad Q = \frac{Q^*}{Q_R}, \quad D = \frac{D^*}{D_R},$$

$$k = \frac{\hat{k}^*}{k_R}, \quad c_s = \frac{\hat{c}}{\hat{c}_R}, \quad K = \frac{K_e}{K_{eR}}$$

we obtain the dimensionless equations

$$\frac{\partial c_F}{\partial t} + \frac{Q_R T_R}{L_R}Q(1-y^2)\frac{\partial c_F}{\partial x} - \frac{D_R T_R}{L_R^2}D\frac{\partial^2 c_F}{\partial x^2} - \frac{D_R T_R}{H^2}D\frac{\partial^2 c_F}{\partial y^2} = 0 \qquad \text{in } \Omega \tag{3}$$

and

$$-\frac{DD_R}{H}c_R\frac{\partial c_F}{\partial y} = \frac{\hat{c}_R}{T_R}\frac{\partial c_s}{\partial t} = k_R k(c_R c_F - \frac{\hat{c}_R c_s}{KK_{eR}}) \qquad \text{at } y = 1 \tag{4}$$

where

$$\Omega = (0, +\infty) \times (-1, 1), \quad \Gamma^+ = (0, +\infty) \times \{1\} \text{ and } \Gamma = (0, +\infty) \times \{-1, 1\} \tag{5}$$

This problem involves the following timescales:

T_L = characteristic longitudinal timescale = L_R/Q_R
T_T = characteristic transversal timescale = H^2/D_R
$T_{DE} = K_{eR}/k_R$ (characteristic desorption time)
$T_A = \hat{c}_R/(c_R k_R)$ (characteristic adsorption time)
T_{react} = superficial chemical reaction timescale $= H/k_R$

and the following characteristic non-dimensional numbers

$$Pe = \frac{L_R Q_R}{D_R} \text{(Péclet number);} \qquad Da = \frac{L_R}{T_A Q_R} \text{(Damkohler number)}$$

Further we set $\varepsilon = H/L_R \ll 1$ and choose $T_R = T_L$.

Solving the full problem for arbitrary values of coefficients is costly and one would like to find the effective (or averaged) values of the dispersion coefficient and the transport velocity and an effective corresponding $1D$ parabolic equation for the effective concentration.

We consider the case when $K_{eR} = H$, $T_A \approx T_L \approx T_{DE}$. We choose $Q = Q^*/Q_R = \mathcal{O}(1)$, and

$$\frac{T_T}{T_L} = \frac{HQ_R}{D_R}\varepsilon = \mathcal{O}(\varepsilon^{2-\alpha}) = \varepsilon^2 Pe$$

Then the situation from Taylor's article corresponds to the case when $0 \leqslant \alpha < 2$, i.e. the transversal Péclet number is equal to $(1/\varepsilon)^{\alpha-1}$ and $\hat{k}^* = 0$ (no chemistry). It is interesting to remark that in his paper Taylor has $\alpha = 1.6$ and $\alpha = 1.9$.

Our domain is now the infinite strip $Z^+ = \mathbb{R}_+ \times (0, 1)$. Then using the anti-symmetry of $c^\varepsilon = c_F$, our equations in their non-dimensional form are

$$\frac{\partial c^\varepsilon}{\partial t} + Q(1 - y^2)\frac{\partial c^\varepsilon}{\partial x} = D\varepsilon^\alpha \partial_{xx} c^\varepsilon + D\varepsilon^{\alpha-2}\partial_{yy} c^\varepsilon \qquad \text{in } Z^+ \tag{6}$$

$$c^\varepsilon(x, y, 0) = 1, \qquad (x, y) \in \mathbb{R}_+ \times (0, 1) \tag{7}$$

$$-D\varepsilon^{\alpha-2}\frac{\partial c^\varepsilon}{dy} = \frac{T_A}{T_{DE}}\frac{\partial c_s^\varepsilon}{\partial t} = \frac{T_L}{T_{DE}} k\left(c^\varepsilon - \frac{(T_A/T_{DE})c_s}{K}\right)\Bigg|_{y=1} \quad on\ \Gamma^+ \times (0, T) \tag{8}$$

$$\partial_y c^\varepsilon(x, 0, t) = 0, \qquad (x, t) \in \mathbb{R}_+ \times (0, T) \tag{9}$$

$$c_s^\varepsilon(0, t) = 0 \text{ and } c^\varepsilon(0, y, t) = 0, \qquad (y, t) \in (0, 1) \times (0, T) \tag{10}$$

We study the behavior of the solution to Equations (6)–(10), with square integrable gradient in x and y, when $\varepsilon \to 0$ and try to obtain an *effective problem*.

In the paragraph from Section 3.1, we will give a detailed derivation of the effective equations. Our technique is motivated by the paper by Rubinstein and Mauri, 1986, where the analysis is based on the hierarchy of timescales and a corresponding two-scale expansion. For $\hat{k}^* = 0$, our approach gives the effective problem from Taylor's (1993) paper :

$$\begin{cases} \partial_t c^{\mathrm{Tay}} + \frac{2Q}{3}\partial_x c^{\mathrm{Tay}} = \left(\frac{D}{Pe} + \frac{8}{945}\frac{Q^2}{D}\frac{T_{\mathrm{T}}}{T_{\mathrm{L}}}\right)\partial_{xx} c^{\mathrm{Tay}}, \\ \text{in } \mathbb{R}_+ \times (0,T), \qquad c^{\mathrm{Tay}}|_{x=0} = 1, \\ c^{\mathrm{Tay}}|_{t=0} = 0, \qquad \partial_x c^{\mathrm{Tay}} \in L^2(\mathbb{R}_+ \times (0,T)) \end{cases} \tag{11}$$

What is known concerning derivation of effective equations? Our approach and calculations performed in Section 3, gives the following non-dimensional effective equations in $(0, +\infty) \times (0, T)$:

$$\partial_t\left(c + \frac{T_{\mathrm{A}}}{T_{\mathrm{DE}}}c_{\mathrm{s}}\right) + \left(\frac{2Q}{3} + \frac{2Qk}{45D}\frac{T_{\mathrm{T}}}{T_{\mathrm{DE}}}\right)\partial_x c - \left(D\varepsilon^{\alpha} + \frac{8}{945}\frac{Q^2}{D}\varepsilon^{2-\alpha}\right)\partial_{xx} c$$
$$= \frac{2Qk}{45DK}\frac{T_{\mathrm{A}}}{T_{\mathrm{DE}}}\frac{T_{\mathrm{T}}}{T_{\mathrm{DE}}}\partial_x c_{\mathrm{s}} \tag{12}$$

$$\left(1 + \frac{k}{3D}\frac{T_{\mathrm{T}}}{T_{\mathrm{DE}}}\right)\partial_t c_{\mathrm{s}} = k\frac{T_{\mathrm{L}}}{T_{\mathrm{A}}}\left(c + \frac{2Q}{45D}\varepsilon^{2-\alpha}\partial_x c - \frac{(T_{\mathrm{A}}/T_{\mathrm{DE}})c_{\mathrm{s}}}{K}\right) \tag{13}$$

The system (12)–(13) could be compared with the corresponding non-dimensional effective equations obtained by Paine et al. (1983). After substituting the Equation (13) at the place of $\partial_t c_{\mathrm{s}}$ in Equation (12), we see that our effective Equations (12) and (13) coincide with the effective non-dimensional system (39)–(40), Paine et al. (1983, p. 1784). There is however a notable difference: the system (39)–(40) from Paine et al. (1983) contains the parameters A_1, A_2, K^* and Sh which depend non-locally on c and c_{s}. Instead we give explicit values of the effective coefficients.

This case cannot be compared with the results from Balakotaiah and Chang (1995), since they have a different timescale on the pages 61–73. Nevertheless, a comparison will be possible in the case $K_{\mathrm{e}} = +\infty$.

Balakotaiah and Chakraborty introduce a four-mode hyperbolic model but with non-linear reactions, in three-dimensional geometry and with much bigger Damkohler number (Balakotaiah, 2004; Chakraborty and Balakotaiah, 2005). The effective model cannot be directly compared with our system (12)–(13). Nevertheless, in Section 3.1 we derive a four-mode hyperbolic model, analogous to the models from Balakotaiah (2004) and Chakraborty and Balakotaiah (2005). We show that it is formally equivalent to our model at the order $\mathcal{O}(\varepsilon^{2(2-\alpha)})$. This shows the relationship between the upscaled models developed by Balakotaiah and Chakraborty and our results (Balakotaiah, 2004; Chakraborty and Balakotaiah, 2005).

In its dimensional form our effective problem reads

$$\partial_{t^*}\left(c^* + \frac{\hat{c}}{H}\right) + \left(\frac{2Q^*}{3} + \frac{2Q^* Da_{\mathrm{T}}}{45}\right)\partial_{x^*} c^* - D^*\left(1 + \frac{8}{945}Pe_{\mathrm{T}}^2\right)\partial_{x^* x^*} c^*$$
$$= \frac{2Q^* Da_{\mathrm{T}}}{45 K_{\mathrm{e}}}\partial_{x^*}\hat{c} \tag{14}$$

$$\left(1+\frac{1}{3}Da_{\mathrm{T}}\right)\partial_{t^*}\hat{c} = \hat{k}^*\left(c^* + \frac{2HPe_{\mathrm{T}}}{45}\partial_{x^*}c^* - \frac{\hat{c}}{K_{\mathrm{e}}}\right) \tag{15}$$

where $Pe_{\mathrm{T}} = Q^*H/D^*$ is the transversal Péclet number and $Da_{\mathrm{T}} = \hat{k}^*H/D^*$ the transversal Damkohler number.

Taking the transversal section mean gives

$$\partial_{t^*}\left(c^{\mathrm{moy}} + \frac{\hat{c}^{\mathrm{moy}}}{H}\right) + \frac{2Q^*}{3}\partial_{x^*}c^{\mathrm{moy}} - D^*\partial_{x^*x^*}c^{\mathrm{moy}} = 0 \tag{16}$$

$$\partial_{t^*}\hat{c} = \hat{k}^*\left(c^{\mathrm{moy}} - \frac{\hat{c}}{K_{\mathrm{e}}}\right) \tag{17}$$

We will compare numerically our effective Equations (14) and (15) with the system (16)–(17), but we have even stronger arguments in our favor.

Why we prefer our model to other models from the literature? Because we are able to prove the error estimate. They were established in Mikelić et al. (2006) for the particular case when $K_{\mathrm{e}} = +\infty$ (the case of an irreversible, first order, heterogeneous reaction).

In this case the effective non-dimensional problem is

$$\begin{aligned}\partial_t c + Q\left(\frac{2}{3} + \frac{4k}{45D}\varepsilon^{2-\alpha}\right)\partial_x c + k\left(1 - \frac{k}{3D}\varepsilon^{2-\alpha}\right)c \\ = \left(D\varepsilon^{\alpha} + \frac{8}{945}\frac{Q^2}{D}\varepsilon^{2-\alpha}\right)\partial_{xx}c\end{aligned} \tag{18}$$

Our result could be stated in dimensional form: Let us suppose that $L_{\mathrm{R}} > \max\{D_{\mathrm{R}}/Q_{\mathrm{R}}, Q_{\mathrm{R}}H^2/D_{\mathrm{R}}, H\}$. Then the upscaled dimensional problem corresponding to the case $K_{\mathrm{e}} = +\infty$ reads

$$\begin{aligned}\frac{\partial c^{*,\mathrm{eff}}}{\partial t^*} + \left(\frac{2}{3} + \frac{4}{45}Da_{\mathrm{T}}\right)Q^*\frac{\partial c^{*,\mathrm{eff}}}{\partial x^*} + \frac{k^*}{H}\left(1 - \frac{1}{3}Da_{\mathrm{T}}\right)c^{*,\mathrm{eff}} \\ = D^*\left(1 + \frac{8}{945}Pe_{\mathrm{T}}^2\right)\frac{\partial^2 c^{*,\mathrm{eff}}}{\partial (x^*)^2}\end{aligned} \tag{19}$$

Let us now compare the physical concentration c^{ε} with the effective concentration c. $H(x)$ denotes Heaviside's function.

Theorem 1. Let c be the unique solution of (18) and let $\Omega_K = (0, K) \times (0, 1)$, $K > 0$. Then we have

$$\max_{0 \le t \le T} t^3 \int_{\Omega_K} \left|c^{\varepsilon}(x, y, t) - c(x, t)\right| \mathrm{d}x\mathrm{d}y \le C\varepsilon^{2-\alpha} \tag{20}$$

$$\left(\int_0^T \int_{\Omega_K} t^6 \left|\partial_y c^{\varepsilon}(x, y, t)\right|^2 \mathrm{d}x\mathrm{d}y\mathrm{d}t\right)^{1/2} \le C\left(\varepsilon^{2-5\alpha/4}H(1-\alpha) + \varepsilon^{3/2-3\alpha/4}H(\alpha - 1)\right) \tag{21}$$

$$\left(\int_0^T \int_{\Omega_K} t^6 \left|\partial_x \left(c^\varepsilon(x,y,t) - c(x,t)\right)\right|^2\right)^{1/2} \leq C\left(\varepsilon^{2-7\alpha/4} H(1-\alpha) + \varepsilon^{3/2-5\alpha/4} H(\alpha-1)\right) \tag{22}$$

Furthermore, there exists a linear combination $C_{\mathrm{cor}}(x,y,t,\varepsilon)$ of products between polynomials in y and derivatives of c up to order 3, such that for all $\delta>0$, we have

$$\max_{0\leq t\leq T} \max_{(x,y)\in\Omega^+} \left|t^3\left(c^\varepsilon(x,\ y,\ t) - c(x,\ t) - C_{\mathrm{cor}}(x,\ y,\ t)\right)\right| \leq \begin{cases} C\varepsilon^{4-7\alpha/2-\delta}, & \text{if } \alpha<1, \\ C\varepsilon^{(3/2)-\alpha-\delta} & \text{if } \alpha \geq 1 \end{cases} \tag{23}$$

For details of the proof we refer to Mikelić et al. (2006).

If we compare the non-dimensional effective equation (18) with the corresponding equation (57), from Paine et al. (1983, p. 1786), we find out that they have the same form. Contrary to Paine et al. (1983), we have calculated the effective coefficients and we find them independent of the time and of the moments of c.

In the article Balakotaiah and Chang (1995) the surface reactions are much faster and do not correspond to our problem. In order to compare two approaches we will present in the paragraph from Section 3.4 computations with our technique for the timescale chosen in Balakotaiah and Chang (1995) and we will see that one gets identical results. This shows that our approach through the anisotropic singular perturbation reproduces exactly the results obtained using the center manifold technique.

2.1 Statement of the results in the case of non-linear reactions

At sufficiently high concentrations of the transported solute particles, the surface coverage becomes important and non-linear laws for the rate of adsorption should be used.

Now we study some of non-linear cases. First, the condition (36) is replaced by

$$-D^*\partial_z c^* = \frac{\partial \hat{c}}{\partial t^*} = \hat{\Phi}\left(c^*\right) - \hat{k}_{\mathrm{d}}^* \hat{c} \ \text{ on } z = \pm H \tag{24}$$

where $\hat{k}_{\mathrm{d}}^*$ represents the constant desorption rate. For simplicity we suppose $\hat{\Phi}(0) = 0$. Examples of $\hat{\Phi}$ are

$$\begin{cases} \hat{\Phi}(c) = \dfrac{k_1^* c}{1 + k_2^* c}, & \text{(Langmuir's adsorption);} \\ \hat{\Phi}(c) = k_1^* c^{k_2}, & \text{(Freundlich's adsorption)} \end{cases} \tag{25}$$

Let us write non-dimensional forms for both non-linear adsorption laws.

We start with *Langmuir's isotherm*. In this case the adsorption speed is k_1^*, having the characteristic size $k_{1\mathrm{R}}$ and $k_1^* = k_{1\mathrm{R}} k_1$. For the second parameter we set $k_2^* c_{\mathrm{R}} = k_2$, where k_2 is a dimensionless positive constant.

Let $\Phi(u) = k_1 u/(1 + k_2 u)$. The characteristic times linked with the surface reactions are now:

$$T_{\mathrm{A}} = \hat{c}_{\mathrm{R}}/(c_{\mathrm{R}} k_{1\mathrm{R}}) \text{ (characteristic adsorption time)}$$
$$T_{\mathrm{react}} = \text{superficial chemical reaction time scale} = H/k_{1\mathrm{R}}$$

Then after a short calculation we get the non-dimensional form of Equation (24):

$$-D\varepsilon^{\alpha-2}\frac{\partial c^{\varepsilon}}{\partial y} = \frac{T_{\mathrm{A}}}{T_{\mathrm{react}}}\frac{\partial c_{\mathrm{s}}^{\varepsilon}}{\partial t} = \frac{T_{\mathrm{L}}}{T_{\mathrm{react}}}\left(\Phi(c^{\varepsilon}) - k_{\mathrm{d}}^{*} T_{\mathrm{A}} c_{\mathrm{s}}^{\varepsilon}\right)\Big|_{y=1} \text{ on } \Gamma^{+} \times (0, T) \qquad (26)$$

We suppose $T_{\mathrm{L}} \approx T_{\mathrm{A}} \approx 1/k_{\mathrm{d}}^{*}$ and k_1 and k_2 of order 1.

Next we consider *Freundlich's isotherm*. In this case it makes sense to suppose that $k_1^{*} = k_1 k_{1\mathrm{R}} c_{\mathrm{R}}^{1-k_2}$ and k_1 and k_2 of order 1. Then we get once more Equation (26) but with $\Phi(u) = k_1 u^{k_2}$.

After the calculations from Section 3.2, we find out that the effective equations in $(0, +\infty) \times (0, T)$:

$$\partial_t\left(c_{\mathrm{FN}}^{0} + \frac{T_{\mathrm{A}}}{T_{\mathrm{react}}} c_{\mathrm{sN}}^{\mathrm{eff}}\right) + \frac{2Q}{3}\partial_x\left(c_{\mathrm{FN}}^{0} + \frac{1}{15D}\frac{T_{\mathrm{T}}}{T_{\mathrm{react}}}\Phi\left(c_{\mathrm{FN}}^{0}\right)\right)$$
$$= \varepsilon^{\alpha}\left(D + \frac{8}{945}\frac{Q^2}{D}\varepsilon^{2(1-\alpha)}\right)\partial_{xx} c_{\mathrm{FN}}^{0} + \frac{2Q}{45D}\frac{T_{\mathrm{A}} T_{\mathrm{T}} k_{\mathrm{d}}^{*}}{T_{\mathrm{react}}}\partial_x c_{\mathrm{sN}}^{\mathrm{eff}} \qquad (27)$$

$$\partial_t c_{\mathrm{sN}}^{\mathrm{eff}} = \frac{T_{\mathrm{L}}}{T_{\mathrm{A}}}\left(\Phi\left(c_{\mathrm{FN}}^{0} + \varepsilon^{2-\alpha} c_{\mathrm{FN}}^{1}\big|_{y=1}\right) - k_{\mathrm{d}}^{*} T_{\mathrm{A}} c_{\mathrm{sN}}^{0}\right) \qquad (28)$$

$$c_{\mathrm{FN}}^{1}\big|_{y=1} = \frac{2}{45}\frac{Q}{D}\partial_x c_{\mathrm{FN}}^{0} - \frac{T_{\mathrm{A}}}{3DT_{\mathrm{react}}}\partial_t c_{\mathrm{sN}}^{\mathrm{eff}} \qquad (29)$$

$$c_{\mathrm{FN}}^{0}\big|_{x=0} = 0, \qquad c_{\mathrm{FN}}^{0}\big|_{t=0} = 1, \qquad c_{\mathrm{sN}}^{\mathrm{eff}}\big|_{t=0} = c_{\mathrm{s0}} \qquad (30)$$

In its dimensional form our effective problem for the volume and surface solute concentrations $\{c_{\mathrm{N}}^{*}, \hat{c}_{\mathrm{N}}\}$ reads

$$\partial_{t^*}\left(c_{\mathrm{N}}^{*} + \frac{\hat{c}_{\mathrm{N}}}{H}\right) + \partial_{x^*}\left(\frac{2Q^*}{3}c_{\mathrm{N}}^{*} + \frac{Pe_{\mathrm{T}}}{15}\hat{\Phi}\left(c_{\mathrm{N}}^{*}\right)\right)$$
$$= D^{*}\left(1 + \frac{8}{945}Pe_{\mathrm{T}}^{2}\right)\partial_{x^*x^*} c_{\mathrm{N}}^{*} + \frac{2k_{\mathrm{d}}^{*} Pe_{\mathrm{T}}}{45}\partial_{x^*}\hat{c}_{\mathrm{N}} \qquad (31)$$

$$\partial_{t^*}\hat{c}_{\mathrm{N}} = \hat{\Phi}\left(c_{\mathrm{N}}^{*} + Pe_{\mathrm{T}}\tilde{c}_{\mathrm{N}}^{1}\right) - k_{\mathrm{d}}^{*}\hat{c}_{\mathrm{N}} \qquad (32)$$

$$\tilde{c}_{\mathrm{N}}^{1} = \frac{2H}{45}\partial_{x^*} c_{\mathrm{N}}^{*} - \frac{1}{3}\partial_{t^*}\hat{c}_{\mathrm{N}} \qquad (33)$$

where $Pe_{\mathrm{T}} = Q^{*}H/D^{*}$ is the transversal Péclet number.

Similar to the linear case, taking the mean over the transversal section gives

$$\partial_{t^*}\left(c_{\mathrm{N}}^{\mathrm{moy}} + \frac{\hat{c}_{\mathrm{N}}^{\mathrm{moy}}}{H}\right) + \frac{2Q^*}{3}\partial_{x^*} c_{\mathrm{N}}^{\mathrm{moy}} - D^{*}\partial_{x^*x^*} c_{\mathrm{N}}^{\mathrm{moy}} = 0 \qquad (34)$$

$$\partial_{t^*}\hat{c}_{\mathrm{N}} = \hat{\Phi}\left(c_{\mathrm{N}}^{\mathrm{moy}}\right) - k_{\mathrm{d}}^{*}\hat{c}_{\mathrm{N}} \qquad (35)$$

We point out that for the non-negligible local Péclet number, taking the simple mean over the section does not lead to a good approximation.

Here also we could propose four-mode models in the sense of Balakotaiah (2004) and Chakraborty and Balakotaiah (2005).

2.2 Statement of the results in the case of an infinite adsorption rate

Here we concentrate our attention to the case when the adsorption rate constant $\hat{k}^*$ is infinitely large.

This means that the reaction at channel wall $\Gamma^* = \{(x^*, z) : 0 < x^* < +\infty, |z| = H\}$ is described by the following flux equation

$$-D^* \partial_z c^* = K_e \frac{\partial c^*}{\partial t^*} \quad \text{on } \Gamma^* \tag{36}$$

where K_e is, as before, the linear adsorption equilibrium constant. Now we see that Equation (2) is replaced by Equation (36), which corresponds to taking the limit $\hat{k}^* \to \infty$.

The characteristic times T_A and T_{DE} cannot be used anymore and we introduce the new characteristic time $T_C = K_{eR}/\varepsilon Q_R$, which has a meaning of the superficial chemical reaction timescale. As before, we set $\varepsilon = H/L_R \ll 1$ and choose $T_R = T_L$.

Introducing the dimensionless numbers into the starting and considering constant initial/boundary conditions yields the problem:

$$\frac{\partial c^\varepsilon}{\partial t} + Q(1 - y^2)\frac{\partial c^\varepsilon}{\partial x} = D\varepsilon^\alpha \frac{\partial^2 c^\varepsilon}{\partial x^2} + D\varepsilon^{\alpha-2}\frac{\partial^2 c^\varepsilon}{\partial y^2} \quad \text{in } \Omega^+ \times (0, T) \tag{37}$$

$$-D\varepsilon^{\alpha-2}\frac{\partial c^\varepsilon}{\partial y} = -D\frac{1}{\varepsilon^2 Pe}\frac{\partial c^\varepsilon}{\partial y} = \frac{T_C}{T_L} K \frac{\partial c^\varepsilon}{\partial t} \quad \text{on } \Gamma^+ \times (0, T) \tag{38}$$

$$c^\varepsilon(x, y, 0) = 1 \quad \text{for } (x, y) \in \Omega^+ \tag{39}$$

$$c^\varepsilon(0, y, t) = 0 \quad \text{for } \quad (y, t) \in (0, 1) \times (0, T) \tag{40}$$

$$\frac{\partial c^\varepsilon}{\partial y}(x, 0, t) = 0, \quad \text{for } (x, t) \in (0, +\infty) \times (0, T) \tag{41}$$

Further, we suppose that $T_C \approx T_L$.

After the calculations from Section 3.3 we find that the effective problem for the concentration $c_K^{*,\text{eff}}$ in its dimensional form reads

$$\begin{aligned} &(1 + Da_K)\frac{\partial c_K^{*,\text{eff}}}{\partial t^*} + \frac{2Q^*}{3}\frac{\partial c_K^{*,\text{eff}}}{\partial x^*} \\ &= D^*\left(1 + \frac{4}{135}Pe_T^2\left[\frac{2}{7} + \frac{Da_K(2 + 7Da_K)}{(1 + Da_K)^2}\right]\right)\frac{\partial^2 c_K^{*,\text{eff}}}{\partial (x^*)^2} \end{aligned} \tag{42}$$

In Equation (42) $Pe_T = Q^* H/D^*$ is the transversal Péclet number and $Da_K = K_e/H$ is the transversal Damkohler number.

The transversal section mean gives

$$(1 + Da_K)\partial_{t^*} c_K^{\text{moy}} + \frac{2Q^*}{3}\partial_{x^*} c_K^{\text{moy}} - D^* \partial_{x^*x^*} c_K^{\text{moy}} = 0 \tag{43}$$

Once more, for the non-negligible local Péclet and Damkohler numbers, taking the simple mean over the section does not lead to a good approximation and our numerical simulations, presented in the last section, will confirm these theoretical results. For an error estimate analogous to Theorem 1, we refer to the articles Mikelić and Rosier (2007) and Choquet and Mikelić (2008).

We note the possible similarities of the effective model (42) with Golay's theory as presented in Paine et al. (1983). In the effective dispersion term this theory predicts a rational function of K_e and we confirm it. Nevertheless, there is a difference in particular coefficients.

3. DERIVATION OF THE EFFECTIVE MODELS IN THE NON-DIMENSIONAL FORM

In this section, we will obtain the non-dimensional *effective or upscaled* equations using a two-scale expansion with respect to the *transversal Péclet number* $\varepsilon^{2-\alpha}$. Note that the transversal Péclet number is equal to the ratio between the characteristic transversal timescale and longitudinal timescale. Then we use Fredholm's alternative[1] to obtain the effective equations. However, they do not follow immediately. Direct application of Fredholm's alternative gives hyperbolic equations which are not satisfactory for our model. To obtain a better approximation, we use the strategy from Rubinstein and Mauri (1986) and embed the hyperbolic equation to the next order equations. This approach leads to the effective equations containing Taylor's dispersion type terms. Since we are in the presence of chemical reactions, dispersion is not caused only by the important Péclet number, but also by the effects of the chemical reactions, entering through Damkohler number.

3.1 Full linear model with adsorption–desorption

We start with the problem (6)–(10) and search for c^ε in the form

$$c^\varepsilon = c_{\text{F}}^0(x,t;\varepsilon) + \varepsilon^{2-\alpha} c_{\text{F}}^1(x,y,t) + \varepsilon^{2(2-\alpha)} c_{\text{F}}^2(x,y,t) + \cdots \tag{44}$$

$$c_{\text{s}}^\varepsilon = c_{\text{s}}^0(x,t;\varepsilon) + \varepsilon^{2-\alpha} c_{\text{s}}^1(x,y,t) + \varepsilon^{2(2-\alpha)} c_{\text{s}}^2(x,y,t) + \cdots \tag{45}$$

[1] Comment for a non-mathematical reader: Fredholm's alternative gives a necessary and sufficient criteria for solvability of an equation, in the critical situation when we are in a spectrum. For linear algebraic system $A\mathbf{x} = \mathbf{b}$, it says that if 0 is an eigenvalue of the matrix A, the system has a solution if and only if $\mathbf{b}$ is orthogonal to the eigenvectors of A that correspond to the eigenvalue 0. Except the last example, that is borrowed from Balakotaiah and Chang (1995), in all examples considered here 0 is a simple eigenvalue. Therefore the corresponding boundary value problem in y-variable admits a solution if and only if the mean of the right-hand side with respect to the transversal variable y is equal to the value of the flux at $y = 1$. We refer to the textbooks as Wloka (1987), for the Fredholm theory the partial differential equations.

After introducing Equation (44) into Equation (6) we get

$$\varepsilon^0\{\partial_t c_F^0 + Q(1-y^2)\partial_x c_F^0 - D\partial_{yy}c_F^1\} + \varepsilon^{2-\alpha}\{\partial_t c_F^1 + Q(1-y^2)\partial_x c_F^1 \\ -D\varepsilon^{2(\alpha-1)}\partial_{xx}c_F^0 - D\varepsilon^{\alpha}\partial_{xx}c_F^1 - D\partial_{yy}c_F^2\} = \mathcal{O}(\varepsilon^{2(2-\alpha)}) = \mathcal{O}\left(\left(\frac{T_T}{T_L}\right)^2\right) \tag{46}$$

At the lateral boundary $y = 1$, after introducing Equation (45) into Equation (8) we get:

$$\left(-D\partial_y c_F^1 - \frac{T_A}{T_{DE}}\frac{\partial c_S^0}{\partial t}\right) + \varepsilon^{2-\alpha}\left(-D\partial_y c_F^2 - \frac{T_A}{T_{DE}}\frac{\partial c_S^1}{\partial t}\right) + \cdots = 0 \tag{47}$$

$$\left(\frac{T_A}{T_{DE}}\frac{\partial c_S^0}{\partial t} - \frac{T_L}{T_{DE}}k\left(c_F^0 - \frac{T_A}{T_{DE}}\frac{c_S^0}{K}\right)\right) \\ + \varepsilon^{2-\alpha}\left(\frac{T_A}{T_{DE}}\frac{\partial c_S^1}{\partial t} - \frac{T_L}{T_{\text{react}}}k\left(c_F^1 - \frac{T_A}{T_{DE}}\frac{c_S^1}{K}\right)\right) + \cdots = 0 \tag{48}$$

To satisfy Equations (46)–(48) for every $\varepsilon \in (0, \varepsilon_0)$, all coefficients in front of the powers of $\varepsilon^{2-\alpha}$ should be zero.

Equating the ε^0 terms gives the problem

$$\begin{cases} -D\partial_{yy}c_F^1 = Q\left(y^2 - \frac{1}{3}\right)\partial_x c_F^0 - \left(\partial_t c_F^0 + \frac{2Q\partial_x c_F^0}{3}\right) \quad \text{on } (0,1), \\ -D\partial_y c_F^1 = \frac{T_A}{T_{DE}}\frac{\partial c_S^0}{\partial t} = \frac{T_L}{T_{DE}}k\left(c_F^0 - \frac{T_A}{T_{DE}}\frac{c_S^0}{K}\right) \quad \text{on } y = 1 \\ \text{and } \partial_y c_F^1 = 0 \text{ on } y = 0 \end{cases} \tag{49}$$

for every $(x, t) \in (0, +\infty) \times (0, T)$. By Fredholm's alternative, this problem has a solution if and only if

$$\partial_t c_F^0 + \frac{2Q\partial_x c_F^0}{3} + \frac{T_A}{T_{DE}}\frac{\partial c_S^0}{\partial t} = 0 \tag{50}$$

$$\text{and } \frac{\partial c_S^0}{\partial t} = \frac{T_L}{T_A}k\left(c_F^0 - \frac{T_A}{T_{DE}}\frac{c_S^0}{K}\right) \tag{51}$$

in $(0, \infty) \times (0, T)$. Unfortunately our initial and boundary data are incompatible and therefore the solution to this hyperbolic equation with a memory is discontinuous. Since the asymptotic expansion for c^{ε} involves derivatives of c_F^0, system (50)–(51) does not suit our needs. In the case $k = 0$, considered in Bourgeat et al. (2003), this difficulty was overcome by assuming compatible initial and boundary data. Such an assumption does not always suit the experimental data and we proceed by following an idea from Rubinstein and Mauri (1986). More precisely, we suppose that expression (50) is of the next order in our asymptotic expansion, i.e.

$$\partial_t c_F^0 + \frac{2Q\partial_x c_F^0}{3} + \frac{T_A}{T_{DE}}\frac{\partial c_S^0}{\partial t} = \mathcal{O}(\varepsilon^{2-\alpha}) \qquad \text{in } (0, +\infty) \times (0, T) \tag{52}$$

This hypothesis will be justified *a posteriori*, after getting an equation for c_F^0 and c_S^0.

Combining Equations (49) and (50) and using hypothesis (52) gives

$$\begin{cases} -D\partial_{yy}c_F^1 = -Q\left(\frac{1}{3} - y^2\right)\partial_x c_F^0 + \frac{T_A}{T_{DE}}\frac{\partial c_s^0}{\partial t} \text{ on } (0,1), \\ -D\partial_y c_F^1 = \frac{T_A}{T_{DE}}\frac{\partial c_s^0}{\partial t} = \frac{T_L}{T_{DE}}k\left(c_F^0 - \frac{T_A}{T_{DE}}\frac{c_s^0}{K}\right) \text{ on } y = 1, \\ \text{and } \partial_y c_F^1 = 0 \text{ on } y = 0 \end{cases} \tag{53}$$

for every $(x,t) \in (0,+\infty) \times (0,T)$. Consequently

$$c_F^1(x,y,t) = \frac{Q}{D}\left(\frac{y^2}{6} - \frac{y^4}{12} - \frac{7}{180}\right)\partial_x c_F^0 + \frac{1}{D}\left(\frac{1}{6} - \frac{y^2}{2}\right)\frac{T_A}{T_{DE}}\frac{\partial c_s^0}{\partial t} + A(x,\ t) \tag{54}$$

where $A(x,\ t)$ is an arbitrary function.

The problem corresponding to the order $\varepsilon^{2-\alpha}$ is

$$\begin{cases} -D\partial_{yy}c_F^2 = \varepsilon^\alpha D\partial_{xx}c_F^1 - Q(1-y^2)\partial_x c_F^1 + D\varepsilon^{2(\alpha-1)}\partial_{xx}c_F^0 \\ \quad -\partial_t c_F^1 - \varepsilon^{\alpha-2}\left(\partial_t c_F^0 + \frac{2Q\partial_x c_F^0}{3} + \frac{T_A}{T_{DE}}\frac{\partial c_s^0}{\partial t}\right) \text{ on } (0,1), \\ -D\partial_y c_F^2 = \frac{T_A}{T_{DE}}\frac{\partial c_s^1}{\partial t} = \frac{T_L}{T_{DE}}k\left(c_F^1 - \frac{T_A}{T_{DE}}\frac{c_s^1}{K}\right) \text{ on } y = 1 \\ \quad \text{and } \partial_y c_F^2 = 0 \text{ on } y = 0 \end{cases} \tag{55}$$

for every $(x,t) \in (0,+\infty) \times (0,T)$. Note that in order have an expression for c_s^1 compatible with Equation (54), when adding an arbitrary function $A(x,\ t)$ to c_F^1 in Equation (54), we should also add to c_s^1 a function $B(x,\ t)$ satisfying

$$\partial_t B = \frac{T_L k}{T_A}\left(A - \frac{T_A}{T_{DE}}\frac{B}{K}\right) \tag{56}$$

The problem (55) has a solution if and only if

$$\begin{aligned} &\partial_t c_F^0 + \frac{2Q\partial_x c_F^0}{3} + \frac{T_A}{T_{DE}}\frac{\partial c_s^0}{\partial t} + \varepsilon^{2-\alpha}\frac{T_A}{T_{DE}}\frac{\partial c_s^1}{\partial t} + \varepsilon^{2-\alpha}\partial_t\left(\int_0^1 c_F^1\,dy\right) \\ &-\varepsilon^\alpha D\partial_{xx}c_F^0 + Q\varepsilon^{2-\alpha}\partial_x\left(\int_0^1 (1-y^2)c_F^1\,dy\right) \\ &-D\varepsilon^2\partial_{xx}\left(\int_0^1 c_F^1\,dy\right) = 0 \end{aligned} \tag{57}$$

in $(0,+\infty) \times (0,T)$. Note that this is the equation for c_F^0 and c_s^0. Next let us remark that

$$\int_0^1 c_F^1\,dy = A(x,t) \tag{58}$$

$$\int_0^1 (1-y^2)\,c_F^1\,dy = \frac{2}{3}A(x,t) - \frac{Q}{D}\frac{8}{945}\partial_x c_F^0 + \frac{2}{45D}\frac{T_A}{T_{DE}}\frac{\partial c_s^0}{\partial t} \tag{59}$$

and Equation (57) becomes

$$\partial_t\left(c_F^0+\frac{T_A}{T_{DE}}\left(c_s^0+\varepsilon^{2-\alpha}c_s^1\right)\right)+\frac{2Q}{3}\partial_x c_F^0-\varepsilon^\alpha\tilde{D}\partial_{xx}c_F^0$$
$$=-\varepsilon^{2-\alpha}\frac{T_A}{T_{DE}}\frac{2Q}{45D}\partial_{xt}c_s^0-\varepsilon^{2-\alpha}\left\{\frac{T_A}{T_{DE}}\partial_t B+\partial_t A+\frac{2Q}{3}\partial_x A-\mathrm{D}\varepsilon^\alpha\partial_{xx}A\right\} \tag{60}$$

in $(0,+\infty)\times(0,T)$, with

$$\tilde{D}=D+\frac{8}{945}\frac{Q^2}{D}\varepsilon^{2(1-\alpha)} \tag{61}$$

Let

$$\mathcal{L}_1\{A,B\}=\frac{T_A}{T_{DE}}\partial_t B+\partial_t A+\frac{2Q}{3}\partial_x A-\mathrm{D}\varepsilon^\alpha\partial_{xx}A \tag{62}$$

$$\mathcal{L}_2\{A,B\}=\partial_t B-\frac{T_L}{T_A}\left(A-\frac{T_A}{T_{DE}}\frac{B}{K}\right) \tag{63}$$

There is no clear criterion for choosing the functions A and B. Nevertheless, if $\mathcal{L}_2\{A,B\}=0$, it is possible to introduce the change of unknown functions $c_F^0\to c_F^0+\varepsilon^{2-\alpha}A$ and $c_s^1\to c_s^1+B$. Then Equation (60) differs only by the term $\varepsilon^{2(2-\alpha)}(2Q/45D)\partial_{xt}B$ from its variant with $A=B=0$. Hence $\{c_F^0, c_s^0+\varepsilon^{2-\alpha}c_s^1\}$ would change at order $\mathcal{O}(\varepsilon^{2(2-\alpha)})$ and the contribution appears at the next order in the expansion for c^ε. Optimal choice of $\{A,\ B\}$ could come only from higher order calculations. For simplicity we choose $A=B=0$. This choice simplifies Equation (60) and the boundary condition at $y=1$ to the following system of partial differential equations on $(0,+\infty)\times(0,T)$:

$$\partial_t\left(c_F^0+\frac{T_A}{T_{DE}}c_s^{\mathrm{eff}}\right)+\left(\frac{2Q}{3}+\frac{2}{45}\frac{T_T}{T_{DE}}\frac{Qk}{D}\right)\partial_x c_F^0$$
$$=\varepsilon^\alpha\left(D+\frac{8}{945}\frac{Q^2}{D}\varepsilon^{2(1-\alpha)}\right)\partial_{xx}c_F^0+\frac{2}{45}\frac{T_A T_T}{(T_{DE})^2}\frac{Qk}{DK}\partial_x c_s^{\mathrm{eff}} \tag{64}$$

$$\left(1+\frac{T_T}{T_{DE}}\frac{k}{3D}\right)\partial_t c_s^{\mathrm{eff}}=\frac{T_L k}{T_A}\left(c_F^0+\frac{2}{45}\frac{Q}{D}\varepsilon^{2-\alpha}\partial_x c_F^0-\frac{T_A}{KT_{DE}}c_s^{\mathrm{eff}}\right) \tag{65}$$

where $c_s^{\mathrm{eff}}=c_s^0+\varepsilon^{2-\alpha}c_s^1$.

In fact it is possible to proceed differently and to "hyperbolize" the effective model. Following Balakotaiah and Chang (2003), Balakotaiah (2004) and Chakraborty and Balakotaiah (2005) we set

$$C_m(x,t)=\int_0^1(1-y^2)(c_F^0+\varepsilon^{2-\alpha}c_F^1)\,\mathrm{d}y=\frac{2}{3}c_F^0(x,t)+\varepsilon^{2-\alpha}\int_0^1(1-y^2)c_F^1\,\mathrm{d}y \tag{66}$$

(the mixing-cup concentration)

$$C_w(x,t)=c_F^0(x,t)+\varepsilon^{2-\alpha}c_F^1(x,1,t) \tag{67}$$

(the effective solute concentration at the wall)

Then if $1<\alpha<2$ we can drop the axial diffusion term $\varepsilon^\alpha \tilde{D}\partial_{xx}c_F^0$ and write Equation (57) in the form

$$\frac{\partial c_F^0}{\partial t}+Q\frac{\partial C_m}{\partial x}+\frac{T_A}{T_{DE}}\frac{\partial c_s^{eff}}{\partial t}=0 \tag{68}$$

Next we have

$$\frac{T_A}{T_{DE}}\frac{\partial c_s^{eff}}{\partial t}=-\frac{\partial c_F^0}{\partial t}-\frac{2Q}{3}\frac{\partial C_F^0}{\partial x}+\mathcal{O}(\varepsilon^{2(2-\alpha)}) \tag{69}$$

and after replacing $\partial_t c_s^0$ by the right hand side of Equation (69) we get

$$c_F^1(x,y,t)=\frac{Q}{D}\left(\frac{y^2}{2}-\frac{y^4}{12}-\frac{3}{20}\right)\partial_x c_F^0-\frac{1}{D}\left(\frac{1}{6}-\frac{y^2}{2}\right)\partial_t c_F^0 \tag{70}$$

$$\int_0^1 (1-y^2)c_F^1\,dy=-\frac{Q}{D}\frac{4}{105}\partial_x c_F^0-\frac{2}{45D}\partial_t c_F^0 \tag{71}$$

$$C_m(x,t)=\frac{2}{3}c_F^0(x,t)-\frac{1}{D}\varepsilon^{2-\alpha}\left\{\frac{4Q}{105}\partial_x c_F^0+\frac{2}{45}\partial_t c_F^0\right\} \tag{72}$$

$$C_w(x,t)=c_F^0(x,t)+\frac{1}{D}\varepsilon^{2-\alpha}\left\{\frac{4Q}{15}\partial_x c_F^0+\frac{1}{3}\partial_t c_F^0\right\} \tag{73}$$

Equation (65) now reads

$$\frac{T_A}{T_{DE}}\partial_t c_s^{eff}=\frac{T_L k}{T_{DE}}\left(C_w-\frac{T_A}{KT_{DE}}c_s^{eff}\right) \tag{74}$$

The system (68), (74), (72) and (73) is analogous to the four-mode hyperbolic model (60)–(63), from Balakotaiah (2004, p. 324) and to the four-mode model (90)–(93), from Chakraborty and Balakotaiah (2005, p. 233).

In this chapter our goal is to have a generalization of Taylor's dispersion and we search for a parabolic operator for c_F^0.

3.2 Non-linear reactions

Now we study some non-linear surface reactions.

We start with the problem (6)–(10), but with (8) replaced by (26) (i.e. we have a non-linear adsorption). As before we search for c^ε in the form

$$c^\varepsilon=c_{FN}^0(x,t;\varepsilon)+\varepsilon^{2-\alpha}c_{FN}^1(x,y,t)+\varepsilon^{2(2-\alpha)}c_{FN}^2(x,y,t)+\ldots \tag{75}$$

$$c_s^\varepsilon=c_{sN}^0(x,t;\varepsilon)+\varepsilon^{2-\alpha}c_{sN}^1(x,y,t)+\varepsilon^{2(2-\alpha)}c_{sN}^2(x,y,t)+\ldots \tag{76}$$

After introducing Equations (75)–(76) into Equation (6) we get once more Equation (46). To satisfy it for every $\varepsilon\in(0,\varepsilon_0)$, all coefficients in front of the powers of ε should be zero.

In addition we have the following equations for the boundary reactions at $y = 1$:

$$\left(-D\partial_y c^1_{\mathrm{FN}} - \frac{T_{\mathrm{A}}}{T_{\mathrm{react}}}\frac{\partial c^0_{\mathrm{sN}}}{\partial t}\right) + \varepsilon^{2-\alpha}\left(-D\partial_y c^2_{\mathrm{FN}} - \frac{T_{\mathrm{A}}}{T_{\mathrm{react}}}\frac{\partial c^1_{\mathrm{sN}}}{\partial t}\right) + \cdots = 0 \tag{77}$$

$$\begin{aligned}&\left(\frac{T_{\mathrm{A}}}{T_{\mathrm{react}}}\frac{\partial c^0_{\mathrm{sN}}}{\partial t} - \frac{T_{\mathrm{L}}}{T_{\mathrm{react}}}\left(\Phi(c^0_{\mathrm{FN}}) - k^*_{\mathrm{d}} T_{\mathrm{A}} c^0_{\mathrm{sN}}\right)\right)\\ &+\varepsilon^{2-\alpha}\left(\frac{T_{\mathrm{A}}}{T_{\mathrm{react}}}\frac{\partial c^1_{\mathrm{sN}}}{\partial t} - \frac{T_{\mathrm{L}}}{T_{\mathrm{react}}}\left(\Phi'(c^0_{\mathrm{FN}})c^1_{\mathrm{FN}} - k^*_{\mathrm{d}} T_{\mathrm{A}} c^1_{\mathrm{sN}}\right)\right) + \cdots = 0\end{aligned} \tag{78}$$

As before, the ε^0 terms give the problem

$$\begin{cases} -D\partial_{yy} c^1_{\mathrm{FN}} = Q\left(y^2 - \frac{1}{3}\right)\partial_x c^0_{\mathrm{FN}} - \left(\partial_t c^0_{\mathrm{FN}} + \frac{2Q\partial_x c^0_{\mathrm{FN}}}{3}\right) \text{ on } (0,1),\\ -D\partial_y c^1_{\mathrm{FN}} = \frac{T_{\mathrm{A}}}{T_{\mathrm{react}}}\frac{\partial c^0_{\mathrm{sN}}}{\partial t} = \frac{T_{\mathrm{L}}}{T_{\mathrm{react}}}\left(\Phi(c^0_{\mathrm{FN}}) - k^*_{\mathrm{d}} T_{\mathrm{A}} c^0_{\mathrm{sN}}\right) \text{ on } y=1,\\ \text{and } \partial_y c^1_{\mathrm{FN}} = 0 \text{ on } y = 0 \end{cases} \tag{79}$$

for every $(x, t) \in (0, +\infty) \times (0, T)$. By Fredholm's alternative, this problem has a solution if and only if

$$\partial_t c^0_{\mathrm{FN}} + \frac{2Q\partial_x c^0_{\mathrm{FN}}}{3} + \frac{T_{\mathrm{A}}}{T_{\mathrm{react}}}\frac{\partial c^0_{\mathrm{sN}}}{\partial t} = 0 \tag{80}$$

$$\text{and } \frac{T_{\mathrm{A}}}{T_{\mathrm{react}}}\frac{\partial c^0_{\mathrm{sN}}}{\partial t} = \frac{T_{\mathrm{L}}}{T_{\mathrm{react}}}\left(\Phi(c^0_{\mathrm{FN}}) - k^*_{\mathrm{d}} T_{\mathrm{A}} c^0_{\mathrm{sN}}\right) \tag{81}$$

in $(0, \infty) \times (0, T)$. Unfortunately our initial and boundary data are incompatible and therefore the solution to this hyperbolic equation with a memory is discontinuous. Since the asymptotic expansion for c^ε involves derivatives of c^0_{FN}, system (80)–(81) does not suit our needs and, as in the previous subsection, we proceed by following an idea from Rubinstein and Mauri (1986). More precisely, we suppose that expression (80) is of the next order in our asymptotic expansion, i.e.

$$\partial_t c^0_{\mathrm{FN}} + \frac{2Q\partial_x c^0_{\mathrm{FN}}}{3} + \frac{T_{\mathrm{A}}}{T_{\mathrm{react}}}\frac{\partial c^0_{\mathrm{sN}}}{\partial t} = \mathcal{O}(\varepsilon^{2-\alpha}) \text{ in } (0, +\infty) \times (0, T) \tag{82}$$

This hypothesis will be justified *a posteriori*, after getting an equation for c^0_{FN} and c^0_{sN}.

Combining (49) and (80) and using hypothesis (82) gives

$$\begin{cases} -D\partial_{yy} c^1_{\mathrm{FN}} = -Q\left(\frac{1}{3} - y^2\right)\partial_x c^0_{\mathrm{FN}} + \frac{T_{\mathrm{A}}}{T_{\mathrm{react}}}\frac{\partial c^0_{\mathrm{sN}}}{\partial t} \text{ on } (0,1),\\ -D\partial_y c^1_{\mathrm{FN}} = \frac{T_{\mathrm{A}}}{T_{\mathrm{react}}}\frac{\partial c^0_{\mathrm{sN}}}{\partial t} = \frac{T_{\mathrm{L}}}{T_{\mathrm{react}}}\left(\Phi(c^0_{\mathrm{FN}}) - k^*_{\mathrm{d}} T_{\mathrm{A}} c^0_{\mathrm{sN}}\right) \text{ on } y=1\\ \text{and } \partial_y c^1_{\mathrm{FN}} = 0 \text{ on } y = 0 \end{cases} \tag{83}$$

for every $(x,t) \in (0,+\infty) \times (0,T)$. Consequently

$$c^1_{\mathrm{FN}}(x,y,t) = \frac{Q}{D}\left(\frac{y^2}{6} - \frac{y^4}{12} - \frac{7}{180}\right)\partial_x c^0_{\mathrm{FN}} + \frac{1}{D}\left(\frac{1}{6} - \frac{y^2}{2}\right)\frac{T_{\mathrm{A}}}{T_{\mathrm{react}}}\frac{\partial c^0_{\mathrm{sN}}}{\partial t} + A_{\mathrm{N}}(x,t) \quad (84)$$

where $A_{\mathrm{N}}(x,t)$ is an arbitrary function.

The problem corresponding to the order $\varepsilon^{2-\alpha}$ is

$$\begin{cases} -D\partial_{yy}c^2_{\mathrm{FN}} = \varepsilon^\alpha D\partial_{xx}c^1_{\mathrm{FN}} - Q(1-y^2)\partial_x c^1_{\mathrm{FN}} + D\varepsilon^{2(\alpha-1)}\partial_{xx}c^0_{\mathrm{FN}} \\ \quad -\partial_t c^1_{\mathrm{FN}} - \varepsilon^{\alpha-2}\left(\partial_t c^0_{\mathrm{FN}} + \dfrac{2Q\partial_x c^0_{\mathrm{FN}}}{3} + \dfrac{T_{\mathrm{A}}}{T_{\mathrm{react}}}\dfrac{\partial c^0_{\mathrm{sN}}}{\partial t}\right) \text{ on } (0,1), \\ -D\partial_y c^2_{\mathrm{FN}} = \dfrac{T_{\mathrm{A}}}{T_{\mathrm{react}}}\dfrac{\partial c^1_{\mathrm{sN}}}{\partial t} = \dfrac{T_{\mathrm{L}}}{T_{\mathrm{react}}}\left(\Phi'(c^0_{\mathrm{FN}})c^1_{\mathrm{FN}} - k^*_{\mathrm{d}}T_{\mathrm{A}}c^1_{\mathrm{sN}}\right) \text{ on } y=1, \\ \text{and } \partial_y c^2_{\mathrm{FN}} = 0 \text{ on } y=0 \end{cases} \quad (85)$$

for every $(x,\ t) \in (0,\ +\infty) \times (0,\ T)$. Note that in order to have an expression for c^1_{sN} that is compatible with Equation (84), when adding an arbitrary function $A_{\mathrm{N}}(x,t)$ to c^1_{FN} in Equation (54), we should also add to c^1_{sN} a function $B_{\mathrm{N}}(x,t)$ satisfying

$$\partial_t B_{\mathrm{N}} = \frac{T_{\mathrm{L}}}{T_{\mathrm{A}}}\left(\Phi(A_{\mathrm{N}}) - k^*_{\mathrm{d}}T_{\mathrm{A}}B_{\mathrm{N}}\right) \quad (86)$$

The problem (85) has a solution if and only if

$$\partial_t c^0_{\mathrm{FN}} + \frac{2Q\partial_x c^0_{\mathrm{FN}}}{3} + \frac{T_{\mathrm{A}}}{T_{\mathrm{react}}}\frac{\partial c^0_{\mathrm{sN}}}{\partial t} + \varepsilon^{2-\alpha}\frac{T_{\mathrm{A}}}{T_{\mathrm{react}}}\frac{\partial c^1_{\mathrm{sN}}}{\partial t} + \varepsilon^{2-\alpha}\partial_t\left(\int_0^1 c^1_{\mathrm{FN}}\,\mathrm{d}y\right)$$
$$-\varepsilon^\alpha D\partial_{xx}c^0_{\mathrm{FN}} + Q\varepsilon^{2-\alpha}\partial_x\left(\int_0^1 (1-y^2)c^1_{\mathrm{FN}}\,\mathrm{d}y\right) - D\varepsilon^2\partial_{xx}\left(\int_0^1 c^1_{\mathrm{FN}}\,\mathrm{d}y\right) = 0 \quad (87)$$

in $(0,+\infty) \times (0,T)$.[2] Note that this is the equation for c^0_{FN} and c^0_{sN}.

Next let us remark that

$$\int_0^1 c^1_{\mathrm{FN}}\,\mathrm{d}y = A_{\mathrm{N}}(x,\ t) \quad (88)$$

$$\int_0^1 (1-y^2)\,c^1_{\mathrm{FN}}\,\mathrm{d}y = \frac{2}{3}A_{\mathrm{N}}(x,t) - \frac{Q}{D}\frac{8}{945}\partial_x c^0_{\mathrm{FN}} + \frac{2}{45D}\frac{T_{\mathrm{A}}}{T_{\mathrm{react}}}\frac{\partial c^0_{\mathrm{sN}}}{\partial t} \quad (89)$$

and Equation (87) becomes

$$\partial_t\left(c^0_{\mathrm{FN}} + \frac{T_{\mathrm{A}}}{T_{\mathrm{react}}}\left(c^0_{\mathrm{sN}} + \varepsilon^{2-\alpha}c^1_{\mathrm{sN}}\right)\right) + \frac{2Q}{3}\partial_x c^0_{\mathrm{FN}} - \varepsilon^\alpha\tilde{D}\partial_{xx}c^0_{\mathrm{FN}}$$
$$= -\varepsilon^{2-\alpha}\frac{T_{\mathrm{A}}}{T_{\mathrm{react}}}\frac{2Q}{45D}\partial_{xt}c^0_{\mathrm{sN}}$$
$$-\varepsilon^{2-\alpha}\left\{\frac{T_{\mathrm{A}}}{T_{\mathrm{react}}}\partial_t B_{\mathrm{N}} + \partial_t A_{\mathrm{N}} + \frac{2Q}{3}\partial_x A_{\mathrm{N}} - D\varepsilon^\alpha\partial_{xx}A_{\mathrm{N}}\right\} \quad (90)$$

[2] Note that Freundlich's adsorption non-linearity is not differentiable since in most applications $0<k_2<1$. Nevertheless at the end we will get expressions which do not involve derivative of Φ. Hence in manipulations we can use a smooth regularization of Φ. Clearly, a lacking smoothness of Φ would deteriorate precision of the approximation.

in $(0, +\infty) \times (0, T)$, with

$$\tilde{D} = D + \frac{8}{945}\frac{Q^2}{D}\varepsilon^{2(1-\alpha)} \tag{91}$$

Let

$$\mathscr{L}_1\{A, B\} = \frac{T_A}{T_{\text{react}}}\partial_t B + \partial_t A + \frac{2Q}{3}\partial_x A - D\varepsilon^{\alpha}\partial_{xx} A \tag{92}$$

$$\mathscr{L}_2\{A, B\} = \partial_t B - \frac{T_L}{T_A}(\Phi(A) - k_d^* T_A B) \tag{93}$$

There is no clear criterion for choosing the functions A_N and B_N. With the same arguing as in Section 3.1 we choose $A_N = B_N = 0$. This choice simplifies (90). The next simplification is to eliminate the term $\partial_{xt} c_{sN}^0$ using Equation (81), i.e. $\frac{T_A}{T_{\text{react}}}\frac{\partial c_{sN}^0}{\partial t} = \frac{T_L}{T_{\text{react}}}\left(\Phi(c_{FN}^0) - k_d^* T_A c_{sN}^0\right)$. Then

$$\varepsilon^{2-\alpha}\frac{T_A}{T_{\text{react}}}\frac{2Q}{45D}\partial_{xt} c_{sN}^0 = \frac{2Q}{45D}\frac{T_T}{T_{\text{react}}}\partial_x \Phi(c_{FN}^0) - \frac{2Q}{45D}\frac{T_T T_A k_d^*}{T_{\text{react}}}\partial_x c_{sN}^0 \tag{94}$$

and Equation (90) reads

$$\begin{aligned}
&\partial_t\left(c_{FN}^0 + \frac{T_A}{T_{\text{react}}}c_{sN}^{\text{eff}}\right) + \frac{2Q}{3}\partial_x\left(c_{FN}^0 + \frac{1}{15D}\frac{T_T}{T_{\text{react}}}\Phi(c_{FN}^0)\right) \\
&= \varepsilon^{\alpha}\left(D + \frac{8}{945}\frac{Q^2}{D}\varepsilon^{2(1-\alpha)}\right)\partial_{xx} c_{FN}^0 \\
&+ \frac{2Q}{45D}\frac{T_A T_T k_d^*}{T_{\text{react}}}\partial_x c_{sN}^{\text{eff}} \text{ in } (0, +\infty) \times (0, T)
\end{aligned} \tag{95}$$

where $c_{sN}^{\text{eff}} = c_s^0 + \varepsilon^{2-\alpha} c_s^1$. Next, after putting together the expansions for the ordinary differential equations from Equations (83)–(85) for surface concentration at $y = 1$, we obtain

$$\partial_t c_{sN}^{\text{eff}} = \frac{T_L}{T_A}\left(\Phi\left(c_{FN}^0 + \varepsilon^{2-a} c_{FN}^1|_{y=1}\right) - k_d^* T_A c_{sN}^{\text{eff}}\right) \tag{96}$$

$$c_{FN}^1|_{y=1} = \frac{2}{45}\frac{Q}{D}\partial_x c_{FN}^0 - \frac{T_A}{3DT_{\text{react}}}\partial_t c_{sN}^{\text{eff}} \tag{97}$$

in $(0, +\infty) \times (0, T)$

The effective problem is now

$$\begin{aligned}
&\partial_t\left(c_{FN}^0 + \frac{T_A}{T_{\text{react}}}c_{sN}^{\text{eff}}\right) + \frac{2Q}{3}\partial_x\left(c_{FN}^0 + \frac{1}{15D}\frac{T_T}{T_{\text{react}}}\Phi\left(c_{FN}^0\right)\right) \\
&= \varepsilon^{\alpha}\left(D + \frac{8}{945}\frac{Q^2}{D}\varepsilon^{2(1-\alpha)}\right)\partial_{xx} c_{FN}^0 + \frac{2Q}{45D}\frac{T_A T_T k_d^*}{T_{\text{react}}}\partial_x c_{sN}^{\text{eff}} \text{ in } (0, +\infty) \times (0, T), \\
&\partial_t c_{sN}^{\text{eff}} = \frac{T_L}{T_A}\left(\Phi(c_{FN}^0 + \varepsilon^{2-\alpha} c_{FN}^1|_{y=1}) - k_d^* T_A c_{sN}^{\text{eff}}\right) \text{ in } (0, +\infty) \times (0, T),
\end{aligned}$$

$$c_{\mathrm{FN}}^{1}|_{y=1} = \frac{2}{45}\frac{Q}{D}\partial_x c_{\mathrm{FN}}^{0} - \frac{T_{\mathrm{A}}}{3DT_{\mathrm{react}}}\partial_t c_{\mathrm{sN}}^{\mathrm{eff}} \text{ in } (0,+\infty)\times(0,T),$$
$$c_{\mathrm{FN}}^{0}|_{x=0} = 0,\ c_{\mathrm{FN}}^{0}|_{t=0} = 1,\ c_{\mathrm{sN}}^{\mathrm{eff}}|_{t=0} = c_{s0},\ \partial_x c \in L^2((0,+\infty)\times(0,T)) \tag{98}$$

3.3 Infinite adsorption rate

We start with Equations (37) and (38) and search for c^{ε} in the form

$$c^{\varepsilon} = c_K^0(x,t;\varepsilon) + \varepsilon^{2-\alpha}c_K^1(x,y,t) + \varepsilon^{2(2-\alpha)}c_K^2(x,y,t) + \ldots \tag{99}$$

After introducing (44) into Equation (37) we get

$$\varepsilon^0\{\partial_t c_K^0 + Q(1-y^2)\partial_x c_K^0 - D\partial_{yy}c_K^1\} + \varepsilon^{\alpha-2}\{\partial_t c_K^1 + Q(1-y^2)\partial_x c_K^1 - D\varepsilon^{2(\alpha-1)}\partial_{xx}c_K^0 - D\varepsilon^{\alpha}\partial_{xx}c_K^1 - D\partial_{yy}c_K^2\} = \mathcal{O}(\varepsilon^{2(2-\alpha)}) \tag{100}$$

In order to have Equation (100) for every $\varepsilon \in (0,\varepsilon_0)$, all coefficients in front of the powers of ε should be zero.

The problem corresponding to the order ε^0 is

$$\begin{cases} -D\partial_{yy}c_K^1 = -Q\left(\dfrac{1}{3} - y^2\right)\partial_x c_K^0 - \left(\partial_t c_K^0 + \dfrac{2Q\partial_x c_K^0}{3}\right) \text{ on } (0,1), \\ \partial_y c_K^1 = 0 \text{ on } y=0 \text{ and } -D\partial_y c_K^1 = K\dfrac{T_{\mathrm{C}}}{T_{\mathrm{L}}}\partial_t c_K^0 \text{ on } y=1 \end{cases} \tag{101}$$

for every $(x,t)\in(0,+\infty)\times(0,T)$. By the Fredholm's alternative, the problem (101) has a solution if and only if

$$\left(1 + K\frac{T_{\mathrm{C}}}{T_{\mathrm{L}}}\right)\partial_t c_K^0 + \frac{2Q\partial_x c_K^0}{3} = 0 \text{ in } (0,L)\times(0,T) \tag{102}$$

Unfortunately our initial and boundary data are incompatible and the hyperbolic Equation (102) has a discontinuous solution. Since the asymptotic expansion for c^{ε} involves derivatives of c_K^0, Equation (102) does not suit our needs. As before, we proceed by following an idea from Rubinstein and Mauri (1986) and suppose that

$$\left(1 + K\frac{T_{\mathrm{C}}}{T_{\mathrm{L}}}\right)\partial_t c_K^0 + \frac{2Q\partial_x c_K^0}{3} = \mathcal{O}(\varepsilon^{2-\alpha}) \text{ in } (0,+\infty)\times(0,T) \tag{103}$$

The hypothesis (103) will be justified *a posteriori*, after getting an equation for c_K^0.

Hence Equation (101) reduces to

$$\begin{cases} -D\partial_{yy}c_K^1 = -Q\left(\dfrac{1}{3} - y^2\right)\partial_x c_K^0 + K\dfrac{T_{\mathrm{C}}}{T_{\mathrm{L}}}\partial_t c_K^0 \text{ on } (0,1) \\ \partial_y c_K^1 = 0 \text{ on } y=0 \text{ and } -D\partial_y c_K^1 = K\dfrac{T_{\mathrm{C}}}{T_{\mathrm{L}}}\partial_t c_K^0 \text{ on } y=1 \end{cases} \tag{104}$$

for every $(x,t)\in(0,+\infty)\times(0,T)$, and we have

$$c_K^1(x,y,t) = \frac{Q}{D}\left(\frac{y^2}{6} - \frac{y^4}{12} - \frac{7}{180}\right)\partial_x c_K^0 + \frac{K\,T_{\mathrm{C}}}{D\,T_{\mathrm{L}}}\left(\frac{1}{6} - \frac{y^2}{2}\right)\partial_t c_K^0 + C_{0K}(x,t) \tag{105}$$

where C_{0K} is an arbitrary function.

Let us go to the next order. Then we have

$$\begin{cases} -D\partial_{yy}c_K^2 = -Q(1-y^2)\partial_x c_K^1 + D\varepsilon^{2(\alpha-1)}\partial_{xx}c_K^0 - \partial_t c_K^1 \\ \quad +D\varepsilon^\alpha \partial_{xx}c_K^1 - \varepsilon^{\alpha-2}\left(\left(1+K\dfrac{T_C}{T_L}\right)\partial_t c_K^0 + \dfrac{2Q\partial_x c_K^0}{3}\right) \text{ on } (0,1), \\ \partial_y c_K^2 = 0 \text{ on } y=0 \text{ and } -D\partial_y c_K^2 = K\dfrac{T_C}{T_L}\partial_t c_K^1 \text{ on } y=1 \end{cases} \tag{106}$$

for every $(x, t) \in (0, +\infty) \times (0, T)$. The problem (106) has a solution if and only if

$$\partial_t c_K^0 + \frac{2Q\partial_x c_K^0}{3} + K\frac{T_C}{T_L}(\partial_t c_K^0 + \varepsilon^{2-\alpha}\partial_t c_K^1|_{y=1}) + \varepsilon^{2-\alpha}\partial_t\left(\int_0^1 c_K^1\,dy\right)$$
$$- \varepsilon^\alpha D\partial_{xx}c_K^0 + Q\varepsilon^{2-\alpha}\partial_x\left(\int_0^1 (1-y^2)\,c_K^1\,dy\right)$$
$$= D\varepsilon^2\partial_{xx}\left(\int_0^1 c_K^1\,dy\right) \text{ in } (0,+\infty)\times(0,T) \tag{107}$$

Equation (107) is the equation for c_K^0. Next let us remark that

$$\int_0^1 c_K^1\,dy = C_{0K}(x,t) \tag{108}$$

$$\int_0^1 (1-y^2)\,c_K^1\,dy = \frac{2}{3}C_{0K}(x,t) - \frac{Q}{D}\frac{8}{945}\partial_x c_K^0 + \frac{2K}{45D}\frac{T_C}{T_L}\frac{\partial c_K^0}{\partial t} \tag{109}$$

$$\left.\frac{\partial c_K^1}{\partial t}\right|_{y=1} = \frac{2Q}{45D}\partial_{xt}c_K^0 - \frac{K}{3D}\frac{T_C}{T_L}\partial_{tt}c_K^0 + \partial_t C_{0K} \tag{110}$$

In order to get a parabolic equation for c_K^0 we choose C_{0K} such that $\partial_{tt}c_K^0$ and $\partial_{xt}c_K^0$ do not appear in the effective equation.[3] Then C_{0K} is of the form $C_{0K} = a\partial_t c_K^0 + b\partial_x c_K^0$ and after a short calculation we find that

$$C_{0K}(x,t) = \frac{1}{3D}\left(\frac{T_C}{T_L}\right)^2 \frac{K^2}{1+KT_C/T_L}\partial_t c_K^0 - \frac{2Q}{45D}\frac{T_C}{T_L}\frac{K(2+7KT_C/T_L)}{(1+KT_C/T_L)^2}\partial_x c_K^0 \tag{111}$$

Now c_K^1 takes the form

$$c_K^1(x,y,t) = \frac{Q}{D}\left(\frac{y^2}{6} - \frac{y^4}{12} - \frac{7}{180} - \frac{2}{45}\frac{T_C}{T_L}\frac{K(2+7KT_C/T_L)}{(1+KT_C/T_L)^2}\right)\partial_x c_K^0$$
$$+\frac{K}{D}\frac{T_C}{T_L}\left(\frac{1}{6} + \frac{1}{3}\frac{T_C}{T_L}\frac{K}{1+KT_C/T_L} - \frac{y^2}{2}\right)\partial_t c_K^0 \tag{112}$$

[3] Note that this strategy differs from the approach in the previous section, and the current effective equations cannot be obtained as a limit $\tilde{k}^* \to 0$ of the effective equations obtained before. Nevertheless they are of the same order.

For $\alpha \geqslant 1, 2 \geqslant 2(2-\alpha)$ and we are allowed to drop the term of order $\mathcal{O}(\varepsilon^2)$. Now the Equation (107) becomes

$$\left(1+\frac{KT_{\rm C}}{T_{\rm L}}\right)\partial_t c_K^0+\frac{2Q}{3}\partial_x c_K^0=\varepsilon^\alpha \tilde{D}\partial_{xx}c_K^0 \text{ in } (0,+\infty)\times(0,T) \tag{113}$$

with

$$\tilde{D}=D+\frac{8}{945}\frac{Q^2}{D}\varepsilon^{2(1-\alpha)}+\frac{4Q^2}{135D}\frac{T_{\rm C}}{T_{\rm L}}\frac{K(2+7KT_{\rm C}/T_{\rm L})}{(1+KT_{\rm C}/T_{\rm L})^2}\varepsilon^{2(1-\alpha)} \tag{114}$$

Now the problem (106) becomes

$$\begin{cases} -D\partial_{yy}c_K^2=-\dfrac{Q^2}{D}\partial_{xx}c_K^0\left\{\dfrac{8}{945}+(1-y^2)\left(\dfrac{y^2}{6}-\dfrac{y^4}{12}-\dfrac{7}{180}\right)\right\} \\ +\partial_{xt}c_K^0\dfrac{Q\tilde{K}}{D}\left\{\dfrac{2}{45}-(1-y^2)\left(\dfrac{1}{6}-\dfrac{y^2}{2}\right)\right\}+\dfrac{2Q\tilde{K}}{45D}\left(1-\dfrac{\tilde{K}(7\tilde{K}+2)}{(1+\tilde{K})^2}\right)\partial_{xt}c_K^0 \\ -\left(\dfrac{\tilde{K}^2}{3D}-\dfrac{\tilde{K}^3}{3D(1+\tilde{K})}\right)\partial_{tt}c_K^0-\left(\dfrac{y^2}{6}-\dfrac{y^4}{12}-\dfrac{7}{180}\right)\partial_{xt}c_K^0\dfrac{Q}{D} \\ +\dfrac{Q\tilde{K}\left(\dfrac{1}{3}-y^2\right)}{D(1+\tilde{K})}\left(\dfrac{2Q}{45}\partial_{xx}c_K^0\dfrac{7\tilde{K}+2}{1+\tilde{K}}-\dfrac{\tilde{K}}{3}\partial_{xt}c_K^0\right)-\left(\dfrac{1}{6}-\dfrac{y^2}{2}\right)\partial_{tt}c_K^0\dfrac{\tilde{K}}{D}\Bigg\} \\ \text{on } (0,1),\ \partial_y c_K^2=0 \text{ on } y=0 \text{ and on } y=1 \\ -\mathrm{D}\partial_y c_K^2=\dfrac{2\tilde{K}Q}{45D}\left(1-\dfrac{\tilde{K}(7\tilde{K}+2)}{(1+\tilde{K})^2}\right)\partial_{xt}c_K^0-\dfrac{\tilde{K}^2}{3D}\left(1-\dfrac{\tilde{K}}{1+\tilde{K}}\right)\partial_{tt}c_K^0 \end{cases} \tag{115}$$

where $\tilde{K}=KT_{\rm C}/T_{\rm L}$.

If we choose c^2 such that $\int_0^1 c^2 dy=0$, then

$$\begin{aligned} c^2(x,y,t)=&-\frac{Q^2}{D^2}\partial_{xx}c_K^0\left(\frac{281}{453600}+\frac{23}{1512}y^2-\frac{37}{2160}y^4+\frac{1}{120}y^6-\frac{1}{672}y^8\right) \\ &+\frac{Q}{D^2}\partial_{xt}c_K^0\left(\frac{31}{7560}-\frac{7}{360}y^2+\frac{y^4}{72}-\frac{y^6}{360}\right) \\ &-\frac{Q}{D^2}\left(-\frac{y^4}{12}+\frac{y^2}{6}-\frac{7}{180}\right)\left(\frac{2Q}{45}\partial_{xx}c_K^0\frac{\tilde{K}(7\tilde{K}+2)}{(1+\tilde{K})^2}-\frac{\tilde{K}^2}{3(1+\tilde{K})}\partial_{xt}c_K^0\right) \\ &+\frac{Q\tilde{K}}{D^2}\partial_{xt}c_K^0\left(\frac{y^6}{60}-\frac{y^4}{18}+\frac{11y^2}{180}-\frac{11}{945}\right)+\frac{\tilde{K}}{2D^2}\partial_{tt}c_K^0\left(-\frac{y^4}{12}+\frac{y^2}{6}-\frac{7}{180}\right) \\ &-\left(\left(\frac{\tilde{K}Q}{45D^2}+\frac{Q\tilde{K}}{45D^2}\frac{\tilde{K}(7\tilde{K}+2)}{(1+\tilde{K})^2}\right)\partial_{xt}c_K^0\right. \\ &\left.-\left(\frac{\tilde{K}^2}{6D^2}-\frac{\tilde{K}^3}{6D^2(1+\tilde{K})}\right)\partial_{tt}c_K^0\right)\left(\frac{1}{3}-y^2\right) \end{aligned} \tag{116}$$

3.4 An irreversible very fast first order reaction

The goal of this subsection is to compare our approach with the center manifold technique from Balakotaiah and Chang (1995). We study the 2D variant of the model from Balakotaiah and Chang (1995, pp. 58–61), and we keep the molecular diffusion. Then the corresponding analog of the problem (6)–(10), with $K=+\infty$, is

$$\frac{\partial c^{\varepsilon}}{\partial t}+Q(1-y^{2})\frac{\partial c^{\varepsilon}}{\partial x}=D\varepsilon^{\alpha}\frac{\partial^{2}c^{\varepsilon}}{\partial x^{2}}+D\varepsilon^{\alpha-2}\frac{\partial^{2}c^{\varepsilon}}{\partial y^{2}} \tag{117}$$

$$-D\varepsilon^{\alpha-2}\left.\frac{\partial c^{\varepsilon}}{\partial x}\right|_{y=1}=k\varepsilon^{\alpha-2}c^{\varepsilon}\Big|_{y=1}\ \text{and}\ \left.\frac{\partial c^{\varepsilon}}{\partial y}\right|_{y=0}=0 \tag{118}$$

Owing to the very fast reaction, we expect fast decay of the solution in time. We search for c^{ε} in the form

$$c^{\varepsilon}=\mathrm{e}^{-\lambda_0\varepsilon^{\alpha-2}t}\left(c^{0}(x,t;\varepsilon)\psi_0(y)+\varepsilon^{2-\alpha}c^{1}+\varepsilon^{2(2-\alpha)}c^{2}+\ldots\right)+\mathcal{O}(\mathrm{e}^{-\lambda_1\varepsilon^{\alpha-2}t}) \tag{119}$$

After introducing Equation (119) into Equation (117) we get

$$\begin{aligned}&\varepsilon^{\alpha-2}\{-\lambda_0c^{0}\psi_0-D\partial_{yy}\psi_0c^{0}\}+\varepsilon^{0}\{\psi_0(y)(\partial_t c^{0}+Q(1-y^{2})\partial_x c^{0})\\&-D\partial_{yy}c^{1}-\lambda_0c^{1}\}+\varepsilon^{2-\alpha}\{\partial_t c^{1}+Q(1-y^{2})\partial_x c^{1}\\&-D\varepsilon^{2(\alpha-1)}\partial_{xx}c^{0}\psi_0(y)-D\varepsilon^{\alpha}\partial_{xx}c^{1}-D\partial_{yy}c^{2}-\lambda_0c^{2}\}\\&=\mathcal{O}(\varepsilon^{2(2-\alpha)})=\mathcal{O}\left(\left(\frac{T_{\mathrm{T}}}{T_{\mathrm{L}}}\right)^{2}\right)\end{aligned} \tag{120}$$

To satisfy Equation (120) for every $\varepsilon\in(0,\varepsilon_0)$, all coefficients in front of the powers of ε should be zero.

The problem corresponding to the order $\varepsilon^{\alpha-2}$ is

$$\begin{cases}-D\partial_{yy}\psi_0=\lambda_0\psi_0\ \text{on}\ (0,1),\\ \partial_y\psi_0=0\ \text{on}\ y=0\ \text{and}\ -D\partial_y\psi_0=k\psi_0\ \text{on}\ y=1\end{cases} \tag{121}$$

for every $(x,t)\in(0,+\infty)\times(0,T)$. This spectral problem[4] has one-dimensional (1D) proper space, spanned by $\psi_0(y)=\sqrt{2}\cos(\sqrt{\lambda_0/D}y)$, where the eigenvalue λ_0 is the first positive root of the equation $\sqrt{\lambda_0/D}\ \tan(\sqrt{\lambda_0/D})=k/D$.

Next, the ε^{0} problem reads

$$\begin{cases}-D\partial_{yy}c^{1}-\lambda_0c^{1}=-\psi_0(y)(Q(1-y^{2})\partial_x c^{0}+\partial_t c^{0})\ \text{on}\ (0,1),\\ \partial_y c^{1}=0\ \text{on}\ y=0\ \text{and}\ -D\partial_y c^{1}=kc^{1}\ \text{on}\ y=1\end{cases} \tag{122}$$

By Fredholm's alternative, this problem has a solution if and only if

$$\partial_t c^{0}+Q\left(\int_0^1\psi_0^2(y)(1-y^{2})\,\mathrm{d}y\right)\partial_x c^{0}=0 \tag{123}$$

in $(0,\infty)\times(0,T)$. As before, our initial and boundary data are incompatible and therefore the solution to this linear transport equation does not suit our needs.

[4] References for spectral problems for partial differential equations are e.g. Vladimirov (1996) and Wloka (1987).

We proceed by using again the idea in Rubinstein and Mauri (1986) and suppose that expression (123) is of the next order in our asymptotic expansion:

$$\psi(y)\left(\partial_t c^0 + Q\left(\int_0^1 \psi_0^2(y)(1-y^2)\,dy\right)\partial_x c^0\right) = \mathcal{O}(\varepsilon^{2-\alpha}) \text{ in } (0,+\infty)\times(0,T) \quad (124)$$

and justify it *a posteriori*, after getting an equation for c^0. Following Balakotaiah and Chang (1995) we set $\alpha_{00} = \int_0^1 \psi_0^2(y)(1-y^2)\,dy$.

Combining Equations (122) and (123) and using hypothesis (124) leads us to consider

$$\begin{cases} -D\partial_{yy}c^1 - \lambda_0 c^1 = -Q\psi_0(y)\big((1-y^2)-\alpha_{00}\big)\partial_x c^0 \text{ on } (0,1), \\ -D\partial_y c^1 = kc^1 \text{ on } y=1, \text{ and } \partial_y c^1 = 0 \text{ on } y=0 \end{cases} \quad (125)$$

for every $(x,t)\in(0,+\infty)\times(0,T)$. Consequently

$$c^1(x,y,t) = Q\partial_x c^0 q_0(y) + \psi_0(y)A(x,t) \quad (126)$$

where A is arbitrary and q_0 is the solution for Equation (125) with $Q\partial_x c$ replaced by 1, such that $\int_0^1 \psi_0(y)q_0(y)\,dy = 0$.

The problem corresponding to the order $\varepsilon^{2-\alpha}$ is

$$\begin{cases} -D\partial_{yy}c^2 - \lambda_0 c^2 = -\partial_t c^1 - Q(1-y^2)\partial_x c^1 + D\varepsilon^{2(\alpha-1)}\partial_{xx}c^0\psi_0(y) \\ +\varepsilon^\alpha D\partial_{xx}c^1 - \varepsilon^{\alpha-2}\psi_0(y)\big(\partial_t c^0 + Q\alpha_{00}\partial_x c^0\big) \text{ on } (0,1), \\ -D\partial_y c^2 = kc^2 \text{ on } y=1 \text{ and } \partial_y c^2 = 0 \text{ on } y=0 \end{cases} \quad (127)$$

for every $(x,t)\in(0,+\infty)\times(0,T)$. This problem has a solution if and only if

$$\partial_t c^0 + Q\alpha_{00}\partial_x c^0 - \left(\varepsilon^\alpha D - Q^2\left(\int_0^1 \psi_0(y)q_0(y)\,(1-y^2)\,dy\right)\right)\partial_{xx}c^0 = 0 \quad (128)$$

in $(0,+\infty)\times(0,T)$. We note that the arbitrary function A enters into Equation (128) as $\partial_t A + Q\alpha_{00}\partial_x A$ and this term is of higher order for reasonable choice of A. We take $A=0$.

Next, we note that, through Hilbert–Schmidt expansion,[5] q_0 is given by

$$q_0(y) = -\sum_{k=1}^{+\infty}\frac{\alpha_{0k}\psi_k(y)}{\lambda_k-\lambda_0} \quad (129)$$

where $\{\lambda_k,\psi_k\}_{k\geq 0}$ is the orthonormal basis defined by the spectral problem (121). Now we see that

$$\int_0^1 \psi_0(y)q_0(y)(1-y^2)\,dy = -\sum_{k=1}^{+\infty}\frac{\alpha_{0k}^2}{\lambda_k-\lambda_0}$$

and Taylor's contribution to the effective diffusion coefficient is strictly positive. We note that this result confirms the calculations from Balakotaiah and

[5] For an elementary presentation of the Hilbert–Schmidt expansion see Vladimirov (1996).

Chang (1995, pp. 58–61). Since

$$\lambda_0 = \frac{T_L}{T_{\text{react}}} k\psi_0(1)$$

in the limit when $T_{\text{react}} \gg T_T$ we obtain the effective Equation (18). In fact our calculations indicate the relationship between the center manifold approach and approach using Bloch's waves and a factorization principle for the two-scale convergence (see the recent papers by Allaire and Raphael, 2006, 2007).

4. NUMERICAL TESTS

For carrying out the numerical tests we have chosen the data from the original paper by Taylor (1993). Analogous data are taken in the presence of chemistry.

The representative case considered in Taylor (1993) is his case **(B)**, where the longitudinal transport time L/u_0 is much bigger than the transversal diffusive time a^2/D. The problem of a diffusive transport of a solute was studied experimentally and analytically. Two basically different cases were subjected to experimental verification in Taylor's paper:

Case (B1). Solute of mass M concentrated at a point $x = 0$ at time $t = 0$.

The effective concentration is given by

$$C_m(x, t) = \frac{M}{2a^2\sqrt{\pi^3 kt}} \exp\left\{ -\frac{(x - u_0 t/2)^2}{4kt} \right\} \tag{130}$$

Case (B2). Dissolved material of uniform concentration C_0 enters the pipe at $x = 0$, starting at time $t = 0$. Initially, the concentration of the solvent was zero.

Clearly, it is Taylor's case (B2) which is well suited for the numerical simulations and it dictates the choice of the initial/boundary value conditions:

$$c^*|_{x^*=0} = c_R \text{ and } c^*|_{t^*=0} = 0 \tag{131}$$

In the presence of the boundary concentration $\hat{c}$ we choose the following initial condition

$$\hat{c}|_{t^*=0} = 0 \tag{132}$$

Originally this problem is formulated in a semi-infinite channel. In our numerical computations we have considered a finite one of length $2L_R$. At the outflow we have imposed a homogeneous Neumann boundary condition

$$\partial_{x^*} c^*|_{x^*=2L_R} = 0 \tag{133}$$

In a similar fashion, taking a homogeneous Neumann condition in the z^*direction along the x^* axis $z^* = 0$, the anti-symmetry of the concentrations allows considering only the upper half of the channel.

In each of the cases we will solve the *full physical problem* numerically. Its section average will be compared with the solution the proposed effective 1D model with Taylor's dispersion. Finally, if one makes the unjustified hypothesis that the average of a product is equal to the product of averages, averaging over sections gives a 1D model which we call the "simple mean". We will make a comparison with the solution of that problem as well.

Numerical solution of the full physical problem is costly, due to dominant Péclet and Damkohler numbers. We solve it using two independent methods.

In the first approach we use the package *FreeFem++* by Pironneau, Hecht and Le Hyaric. For more information we refer to http://www.freefem.org/ff++/. For the problem (6)–(10) the method of characteristics from Pironneau (1988) is used. We present a very short description of the method:

- Discretization in time:
  The first order operator is discretized using the method of characteristics. More precisely, the Equation (6) is written as:

$$\frac{\partial c}{\partial t} + (\vec{q}.\nabla)c = D\varepsilon^{\alpha}\partial_{xx}c + D\varepsilon^{\alpha-2}\partial_{yy}c = f(x,y,t) \tag{134}$$

  Let c^m be an approximation for the solution c at a time $m\delta t$. Then the one step backward convection scheme by the method of characteristics reads as follows:

$$\frac{1}{\delta t}\left(c^{m+1}(x,y) - c^m\left(x - q(y)\delta t, y\right)\right) = f^m(x,y)$$

- Space discretization:
  One of the characteristics of our problem is the presence of a smeared front. To track it correctly, the Lagrange P1 finite elements, with adaptive mesh, are used. The mesh is adapted in the neighborhood of front after every 10 time steps.

Second method consists of a straightforward discretization method: first order (Euler) explicit in time and finite differences in space. Both the time step and the grid size are kept constant and satisfying the Courant Friedrichs Lewy (CFL) condition to ensure the stability of the calculations. To deal with the transport part we have considered the minmod slope limiting method based on the first order upwind flux and the higher order Richtmyer scheme (see, e.g. Quarteroni and Valli, 1994, Chapter 14). We call this method *SlopeLimit*.

A similar procedure is considered for the upscaled, 1D problems, obtained either by our approach or by taking the simple mean. It is refined in the situations when we have explicit formulas for the solution, using the direct numerical evaluation of the error function erf.

4.1 Examples from Taylor's article (no chemistry)

First let us note that in Taylor's article (Taylor, 1993) the problem is axially symmetric with zero flux at the lateral boundary. The solute is transported by Poiseuille velocity.

For simplicity we will consider the flow through the 2D slit $\Omega^* = (0, +\infty) \times (0, H)$. To have a 2D problem equivalent to the case (B) from Taylor's article, we reformulate the characteristic velocity and the radius. Obviously we have

$$Q^* = \frac{3}{4} u_0, \qquad H = a\sqrt{\frac{35}{32}} \tag{135}$$

Then we start with

4.1.1 CASE A: First example from Taylor's paper with the time of flow: $t^* = 11,220$ s

Here we are in absence of the chemistry, i.e. $k_R = 0$. We solve

1. The 2D problem (1), (2), (131). It is solved using the FreeFem++ package and with (SlopeLimit). On the images the solution is denoted (pbreel).
2. The effective problem

$$\partial_{t^*} c^{\mathrm{Tay}} + \frac{2Q^*}{3} \partial_{x^*} c^{\mathrm{Tay}} = D^* \left(1 + \frac{8}{945} Pe_{\mathrm{T}}^2 \right) \partial_{x^* x^*} c^{\mathrm{Tay}} \text{ for } x, t > 0 \tag{136}$$

$$c^{\mathrm{Tay}}|_{x=0} = 1 \text{ and } c^{\mathrm{Tay}}|_{t=0} = 0 \tag{137}$$

   On the images its solution is denoted by (taylor).
3. The problem obtained by taking the simple mean over the vertical section:

$$\partial_{t^*} c^{\mathrm{moy}} + \frac{2Q^*}{3} \partial_{x^*} c^{\mathrm{moy}} - D^* \partial_{x^* x^*} c^{\mathrm{moy}} = 0 \text{ in } (0, +\infty) \times (0, T) \tag{138}$$

   with initial/boundary conditions (137). On the images its solution is denoted by (moyenne).
   Parameter values are at Table 1.

We note that Table 2 is analogous to Table 2, page 196 from Taylor's (1993) article.

Note that in the absence of the chemical reactions we can solve explicitly the problems (136)–(137), respectively (138)–(137). With $\bar{Q} = 2Q^*/3$ and

Table 1 Case A. Parameter values for the longest time example ($t^* = 11{,}220$ s) from Taylor's paper

Parameters	Values
Width of the slit: H	2.635×10^{-4} m
Characteristic length: L_R	0.319 m
$\varepsilon = H/L_R$	0.826×10^{-3}
Characteristic velocity: Q^*	4.2647×10^{-5} m/s
Diffusion coefficient: D^*	1.436×10^{-10} m^2/s
Longitudinal Péclet number: $Pe = L_R Q^*/D^* =$	0.94738×10^5
$\alpha = \log Pe/\log(1/\varepsilon) =$	1.614172
Transversal Péclet number: $Pe_T = HQ^*/D^* =$	0.7825358×10^2

Table 2 Comparison between the concentrations c^{Tay}, c^{moy} and $(1/H)\int_0^H c^*\,dz$ for the Case A at the time $t^* = 11{,}220$ s

x^*	c^{Tay}	c^{moy}	$\frac{1}{H}\int_0^H c^*\text{dz}$ (SlopeLimit)	$\frac{1}{H}\int_0^H c^*\text{dz}$ (FreeFem + +)
0	1	1	1	1
0.3	0.930	0.968	0.97	0.945
0.308	0.805	0.863	0.888	0.885
0.313	0.685	0.725	0.775	0.844
0.314	0.659	0.695	0.75	0.821
0.317	0.571	0.588	0.665	0.69
0.324	0.359	0.329	0.439	0.58
0.3255	0.317	0.279	0.39	0.5625
0.33	0.206	0.155	0.256	0.427
0.3365	0.094	0.05	0.115	0.2957
0.337	0.088	0.048	0.107	0.2677
0.3385	0.070	0.035	0.085	0.2398
0.34	0.057	0.025	0.067	0.1839
0.344	0.029	0.009	0.033	0.0993
0.3475	0.016	0.003	0.016	0.04544

$\bar{D} = D^*(1 + (8/945)\, Pe_{\mathrm{T}}^2)$, the solution for (136)–(137) reads

$$c^{\text{Tay}}(x,t) = 1 - \frac{1}{\sqrt{\pi}}\left[\exp\left\{\frac{\bar{Q}x}{\bar{D}}\right\}\int_{(x+\bar{Q}t)/(2\sqrt{\bar{D}t})}^{\infty} \mathrm{e}^{-\eta^2}\,\mathrm{d}\eta + \int_{(x-\bar{Q}t)/(2\sqrt{\bar{D}t})}^{\infty} \mathrm{e}^{-\eta^2}\,\mathrm{d}\eta\right] \tag{139}$$

For the problems (138) and (137), everything is analogous.

4.1.2 CASE B: second example from Taylor's paper with the time of flow: $t^* = 240$ s

We solve the same equations as in Section 4.1.1. Since α is very close to the threshold value $\alpha^* = 2$, the difference between the solution to the effective equation obtained by taking the simple mean, at one side, and the solutions to the original problem and to our upscaled equation, are spectacular. Our model approximates fairly well with the physical solution even without adding the correctors (Table 4). Parameters are given at Table 3.

Since no chemistry is considered here, an explicit solution can be given in this case as well and it is given by Equation (139). The results are presented in Table 4 and Figure 2. Figures 1 and 2 show clearly the advantage of the upscaled model over the model obtained by taking the simple mean over the vertical section. Presence of the important enhanced diffusion is very important for numerical schemes. Note that in the case considered in Section 4.1.2, the transversal Péclet number is 10 times larger than in the case from Section 4.1.1, explaining the difference in the quality of the approximation.

Table 3 Case B. Parameter values for the characteristic time 240 s for the second example from Taylor's paper

Parameters	Values
Width of the slit: H	$2.635 \times 10^{-4}\,\text{m}$
Characteristic length: L_R	$0.632\,\text{m}$
$\varepsilon = H/L_R$	0.41693×10^{-3}
Characteristic velocity: Q^*	$0.393 \times 10^{-2}\,\text{m/s}$
Diffusion coefficient: D^*	$0.6 \times 10^{-9}\,\text{m}^2/\text{s}$
Longitudinal Péclet number: $Pe = L_R Q^*/D^* =$	4.1396×10^6
$\alpha = \log Pe/\log(1/\varepsilon) =$	1.95769
Transversal Péclet number: $Pe_T = HQ^*/D^* =$	1.72592×10^3

Table 4 Comparison between the concentrations c^{Tay}, c^{moy} and $(1/H)\int_0^H c^* dz$ for the Case B, corresponding to the second example from Taylor's paper, at the time $t^* = 240$ s

x^*	c^{Tay}	c^{moy}	$\frac{1}{H}\int_0^H c^*\,dz$ (SlopeLimit)	$\frac{1}{H}\int_0^H c^*\,dz$ (FreeFem + +)
0	1	1	1	1
0.45	0.986	1	0.99	0.98438
0.537	0.876	1	0.89	0.942785
0.58	0.741	0.993	0.758	0.751335
0.605	0.636	0.882	0.65	0.675492
0.638	0.484	0.327	0.49	0.501282
0.667	0.351	0.033	0.348	0.456008
0.68	0.296	0.007	0.288	0.323355
0.711	0.182	0.	0.166	0.20671
0.74	0.106	0.	0.086	0.116112
0.75	0.086	0.	0.065	0.0926387
0.76	0.069	0.	0.049	0.0723552
0.77	0.055	0.	0.035	0.0549984
0.795	0.029	0.	0.014	0.0407674
0.804	0.023	0.	0.009	0.0201409

4.2 Examples with the linear surface adsorption–desorption reactions

In the case of the full 2D problem with linear surface adsorption–desorption reactions (1), (2), (131) and (132), we present two tests.

4.2.1 Linear surface adsorption–desorption reactions. Case A2 with the times of flow: $t^* = 100$, 211 and 350 s

This first case is with slighty modified data of the Case A from Section 4.1.1. We just modify the width of the channel, the diffusivity and choose a shorter time of the flow (Table 5).

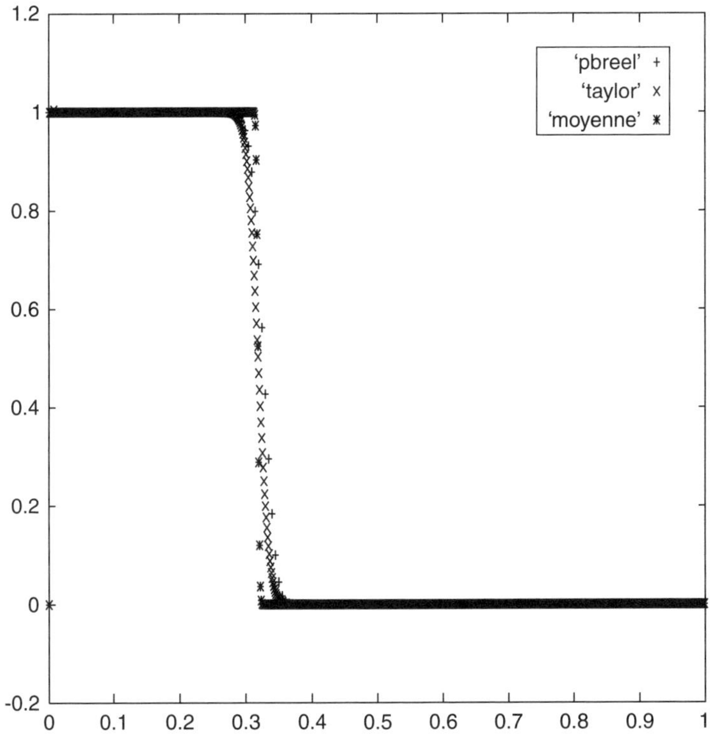

Figure 1 Comparison between concentration from Taylor's paper (taylor), from the original problem (pbreel) and the simple average (moyenne) at $t = 11{,}220$ s.

We note that our scaling impose $\hat{k}^* = \varepsilon Q^*$ and $K_e = H$. This gives $Da_T = \varepsilon Pe_T$. Now the system to solve is Equations (14) and (15):

$$\partial_{t^*}\left(c^* + \frac{\hat{c}}{H}\right) + \left(\frac{2Q^*}{3} + \frac{2Q^* Da_T}{45}\right)\partial_{x^*}c^* - D^*\left(1 + \frac{8}{945}Pe_T^2\right)\partial_{x^*x^*}c^* = \frac{2Q^* Da_T}{45K_e}\partial_{x^*}\hat{c}$$

$$\left(1 + \frac{1}{3}Da_T\right)\partial_{t^*}\hat{c} = \hat{k}^*\left(c^* + \frac{2HPe_T}{45}\partial_{x^*}c^* - \frac{\hat{c}}{K_e}\right)$$

and no explicit solution is known. We should compare between the solutions to (1)–(2) with the initial/boundary conditions (137), $\hat{c}|_{t=0} = 0$ (giving us all together (pbreel3)) and (14)–(15) (giving us (eff)) and (16)–(17) (giving us (moy)), with the same initial/boundary conditions.

The results are shown on the Tables 6, 7 and 8 and on the Figures 3, 4 and 5.

Note that the solution to the problem obtained by taking the simple section average develops a physically incorrect contact discontinuity. Also our upscaled problem gives a good approximation for the original 2D problem, which is not the case with the simple mean.

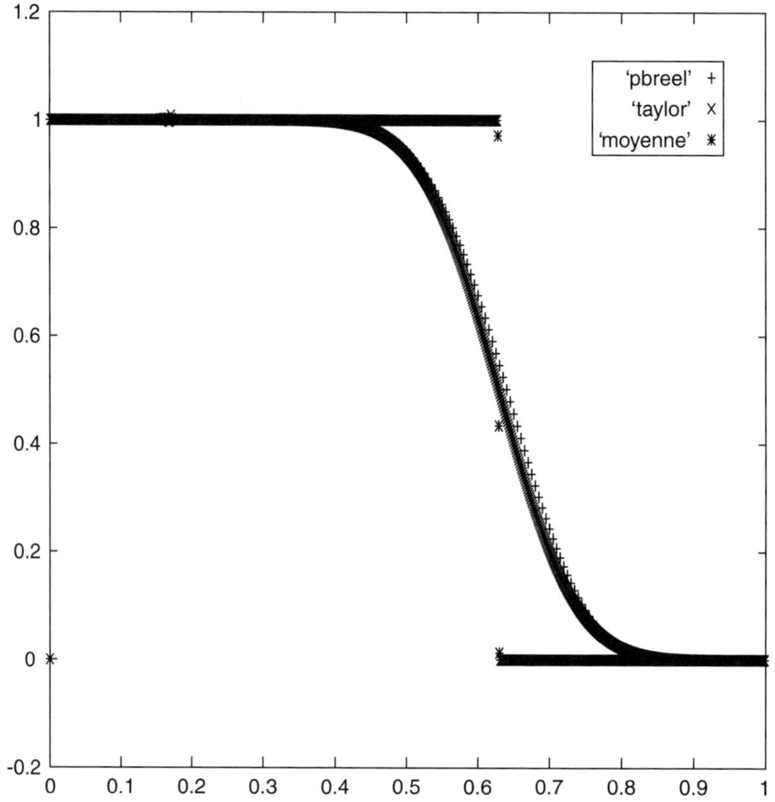

Figure 2 Case B: second case from Taylor's paper. Comparison between the solution for the original problem (pbreel), the solution to the upscaled problem (taylor) and the solution for the problem obtained by taking a simple section average (moyenne) at $t^* = 240$ s.

Table 5 Full linear surface adsorption–desorption problem: parameter values at the Case A2: diffusive transport with surface reaction

Parameters	Values
Width of the slit: H	0.5×10^{-2} m
Characteristic length: L_R	0.632 m
$\varepsilon = H/L_R$	0.7911×10^{-2}
Characteristic velocity: Q^*	0.3×10^{-2} m/s
Diffusion coefficient: D^*	0.2×10^{-6} m^2/s
Longitudinal Péclet number: $Pe = L_R Q^*/D^* =$	9.48×10^3
$\alpha = \log Pe/\log(1/\varepsilon) =$	1.670972
Transversal Peclet number: $\mathbf{Pe}_T = HQ^*/D^* =$	75
Characteristic reaction velocity: $\hat{k}^* = \varepsilon Q^* =$	0.237×10^{-4} m/s
Transversal Damkohler number: $Da_T = \varepsilon(HQ^*/D^*) =$	0.5933

Table 6 Comparison between the volume concentrations c^{Tay}, c^{moy} and $(1/H)\int_0^H c^*\,\mathrm{dz}$ for the linear surface adsorption–desorption reactions, Case A2, at the time $t^* = 100\,\mathrm{s}$

x^*	c^{Tay}	c^{moy}	$\frac{1}{H}\int_0^H c^*\mathrm{dz}$
0	1	1	1
0.01	0.98669465	0.990034274	0.97837
0.05	0.950946235	0.950663125	0.92873
0.1	0.903593771	0.896561247	0.876323
0.2	0.79700151	0.776023352	0.7669
0.225	0.759276074	0.745201145	0.728739
0.25	0.715756063	0.71148785	0.678978
0.275	0.65174438	0.696567508	0.613898
0.29	0.603878726	0.693955625	0.566586
0.3	0.567950276	0.590067563	0.532094
0.31	0.539037927	0.371543232	0.495586
0.32	0.498188037	0.213820021	0.457112
0.35	0.377225997	0.00495647031	0.333673
0.4	0.172223512	2.41496286E-07	0.134612
0.45	0.0591622065	3.07462138E-13	0.0160686

Table 7 Comparison between the volume concentrations c^{Tay}, c^{moy} and $(1/H)\int_0^H c^*\,\mathrm{dz}$ for the linear surface adsorption–desorption reactions, Case A2, at the time $t^* = 211\,\mathrm{s}$

x^*	c^{Tay}	c^{moy}	$\frac{1}{H}\int_0^H c^*\mathrm{dz}$
0	1	1	1
0.01	0.989694187	0.994090699	0.986112
0.05	0.967015027	0.971961203	0.952705
0.1	0.934075267	0.936547842	0.91569
0.2	0.861407801	0.857677963	0.836403
0.3	0.781074907	0.765463212	0.750173
0.4	0.694746658	0.662811744	0.662342
0.5	0.600404621	0.553304147	0.574491
0.55	0.544239838	0.497265165	0.521332
0.6	0.474489299	0.438951289	0.452928
0.65	0.386694802	0.318097632	0.366176
0.7	0.284796763	0.0115430139	0.269368
0.75	0.183421956	1.67295192E-05	0.172172
0.8	0.100489679	3.46962941E-09	0.088037
0.9	0.017165388	1.93051599E-19	0.00981583

Table 8 Comparison between the volume concentrations c^{Tay}, c^{moy} and $(1/H)\int_0^H c^*\,\mathrm{d}z$ for the linear surface adsorption–desorption reactions, Case A2, at the time $t^* = 350$ s

x^*	c^{Tay}	c^{moy}	$\frac{1}{H}\int_0^H c^*\mathrm{d}z$
0	1	1	1
0.1	0.95909192	0.965613038	0.9484
0.2	0.911441678	0.919474858	0.897755
0.4	0.794454955	0.793564942	0.775743
0.6	0.657701569	0.631584001	0.624061
0.7	0.583632368	0.542316066	0.545435
0.8	0.508150772	0.453470264	0.469133
0.9	0.431290446	0.363040727	0.39611
1.	0.34825939	0.276213033	0.319716
1.05	0.298816871	0.237173717	0.273235
1.1	0.247412008	0.109554202	0.224233
1.15	0.19336287	0.00589796516	0.175742
1.2	0.140469463	3.17192071E-05	0.128868
1.3	0.058066265	5.57849169E-12	0.0512471
1.4	0.0152972824	4.65348193E-21	0.0131282

Adding correctors would get us even closer to the solution for the 2D problem.
Figures 3, 4 and 5 show the simulation by FreeFm++ in the case from Section 4.2.1. Advantage of our approach is again fairly clear and the errors of the model obtained by taking a simple mean persist in time.

4.2.2 Linear surface adsorption–desorption reactions. Case B2 with the times of flow: $t^* = 240$ s

In this case we consider the data of Case B, Section 4.1.2, as are given in Table 3. The results are shown in Figure 6.

4.3 An example with the first order irreversible surface reaction

In this situation we take $K = K_e/H \to +\infty$. Equation (1) does not change but the boundary condition (2) becomes

$$-D^*\partial_z c^* = \frac{\partial \hat{c}}{\partial t^*} = \hat{k}^* c^* \quad \text{on } z = \pm H \tag{140}$$

The system (14)–(15) becomes

$$\begin{cases} \partial_{t^*} c^* + \left(\dfrac{2Q^*}{3} + \dfrac{4Q^* Da_{\mathrm{T}}}{45}\right)\partial_{x^*} c^* \\ \quad + \dfrac{\hat{k}^*}{H}\left(1 - \dfrac{Da_{\mathrm{T}}}{3}\right) c^* - D^*\left(1 + \dfrac{8}{945} Pe_{\mathrm{T}}^2\right)\partial_{x^* x^*} c^* = 0 \\ \text{in } (0, +\infty) \times (0, T) \end{cases} \tag{141}$$

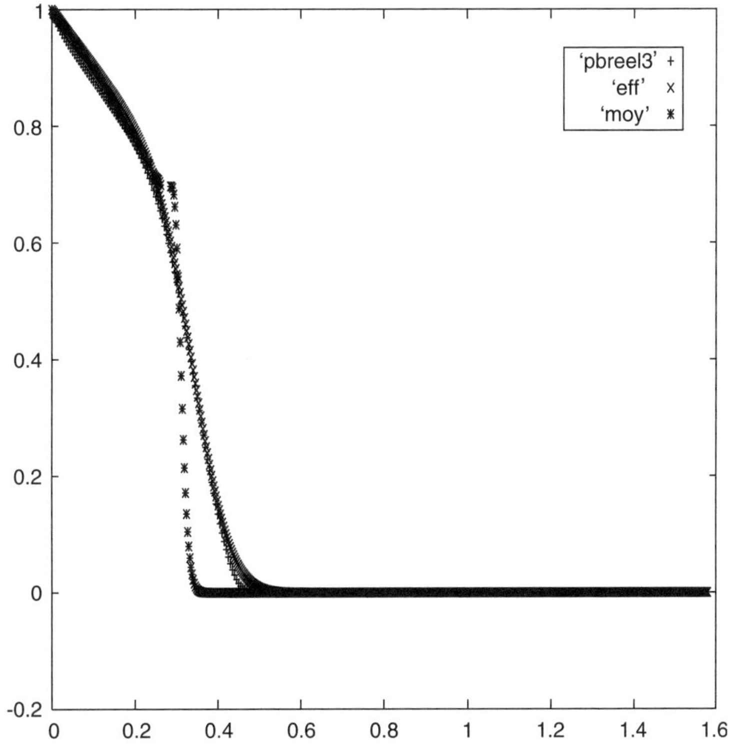

Figure 3 Comparison between the volume concentrations c^{Tay}, $(1/H)\int_0^H c^* \mathrm{d}z$ and c^{moy} for the linear surface adsorption–desorption reactions, Case A2, obtained using our effective problem (eff), average of the section of the concentration from the original problem (pbreel3) and the concentration coming from the simple average (moy) at time. $t^* = 100\,\mathrm{s}$.

and the equation corresponding to a simple mean reads

$$\begin{cases} \partial_{t^*} c^{\mathrm{moy}} + \dfrac{2Q^*}{3} \partial_{x^*} c^{\mathrm{moy}} + \dfrac{\hat{k}}{H} c^{\mathrm{moy}} - D^* \partial_{x^* x^*} c^{\mathrm{moy}} = 0 \\ \text{in } (0, +\infty) \times (0, T) \end{cases} \tag{142}$$

We impose $\hat{k}^* = Q^*/400$.

For this particular reactive flow, the problem (141) has an explicit solution for the following initial/boundary data:

$$c^*|_{x^*=0} = 0 \text{ and } c^*|_{t^*=0} = 1 \tag{143}$$

It reads

$$c^*(x^*, t^*) = \mathrm{e}^{-k_1 t^*} \left(1 - \frac{1}{\sqrt{\pi}} \left[\mathrm{e}^{\frac{2Q_1 x^*}{3D_1}} \int_{\frac{x + 2t^* Q_1/3}{2\sqrt{D_1 t^*}}}^{+\infty} \mathrm{e}^{-\eta^2}\, \mathrm{d}\eta + \int_{\frac{x - 2t^* Q_1/3}{2\sqrt{D_1 t^*}}}^{+\infty} \mathrm{e}^{-\eta^2}\, \mathrm{d}\eta \right] \right) \tag{144}$$

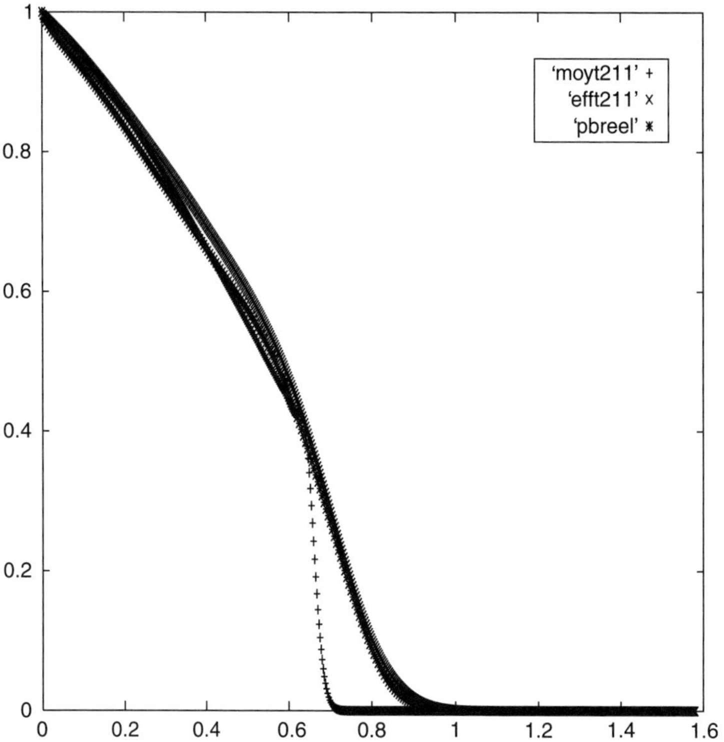

Figure 4 Comparison between the volume concentrations c^{Tay}, $(1/H)\int_0^H c^*\,dz$ and c^{moy} for the linear surface adsorption–desorption reactions, Case A2, obtained using our effective problem (eff), average of the section of the concentration from the original problem (pbreel3) and the concentration coming from the simple average (moy) at time $t^* = 211$ s.

where

$$k_1 = \frac{\hat{k}^*}{H}\left(1 - \frac{Da_{\mathrm{T}}}{3}\right), \qquad Q_1 = Q^*\left(1 + \frac{2Da_{\mathrm{T}}}{15}\right) \text{ and } D_1 = D^*\left(1 + \frac{8}{945}\, Pe_{\mathrm{T}}^2\right)$$

For problem (142) we also impose the initial/boundary condition (143) and c^{moy} is given by the formula (144) as well, but with $k_1 = \hat{k}^*/H$, $Q_1 = Q^*$ and $D_1 = D^*$.

The data are given in Table 9, whereas the results are shown in Tables 10, 11 and 12 and in Figures 7, 8 and 9, corresponding to the times $t^* = 50$, 70 and 100 s.

We see that the solution to the problem obtained by taking a simple mean over the vertical section has incorrect amplitude.

4.4 Numerical experiments in the case of an infinite adsorption rate

In this subsection we solve Equation (42)

$$(1 + Da_K)\frac{\partial c_K^{*,\mathrm{eff}}}{\partial t^*} + \frac{2Q^*}{3}\frac{\partial c_K^{*,\mathrm{eff}}}{\partial x^*} = D^*\left(1 + \frac{4}{135}Pe_{\mathrm{T}}^2\left[\frac{2}{7} + \frac{Da_K(2 + 7Da_K)}{(1 + Da_K)^2}\right]\right)\frac{\partial^2 c_K^{*,\mathrm{eff}}}{\partial (x^*)^2}$$

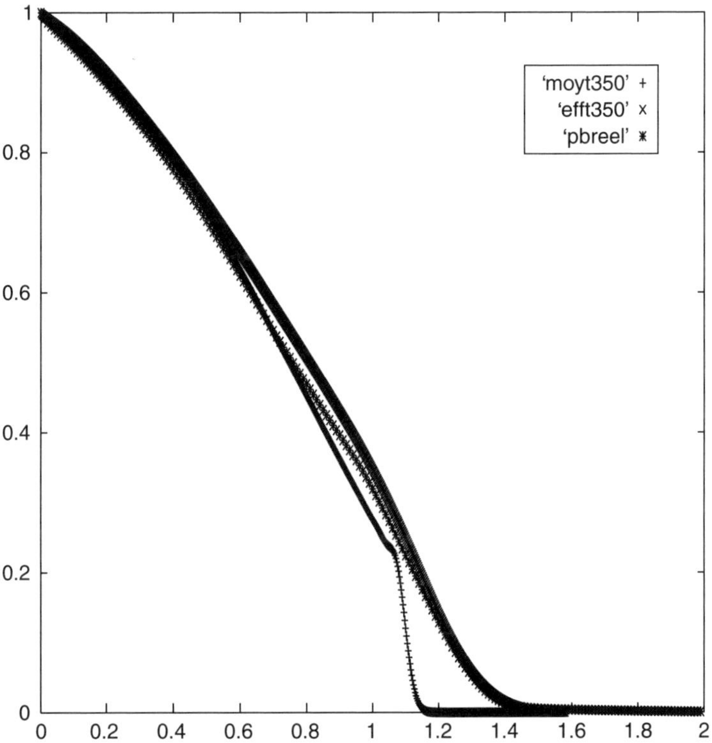

Figure 5 Comparison between the volume concentrations c^{Tay}, $(1/H)\int_0^H c^*\,\mathrm{d}z$ and c^{moy} for the linear surface adsorption–desorption reactions, Case A2, obtained using our effective problem (eff), average of the section of the concentration from the original problem (pbreel3) and the concentration coming from the simple average (moy) at time $t^* = 350\,\mathrm{s}$.

with the initial/boundary data

$$c_K^{*,\mathrm{eff}}|_{x^*=0} = 0 \text{ and } c_K^{*,\mathrm{eff}}|_{t^*=0} = 1 \tag{145}$$

Parameters are shown on the Table 13.

Results are shown at Tables 14, 15 and 16 and on corresponding Figures 10, 11 and 12, at times $t^* = 863$, 2,877 and 5,755 s.

Once more the model obtained by the simple averaging over vertical section gives an approximation which is not good and which gets worse during time evolution.

5. CONCLUSIONS AND PERSPECTIVES

In this chapter we have justified by direct numerical simulation the effective (or upscaled) equations obtained using the techniques of anisotropic singular

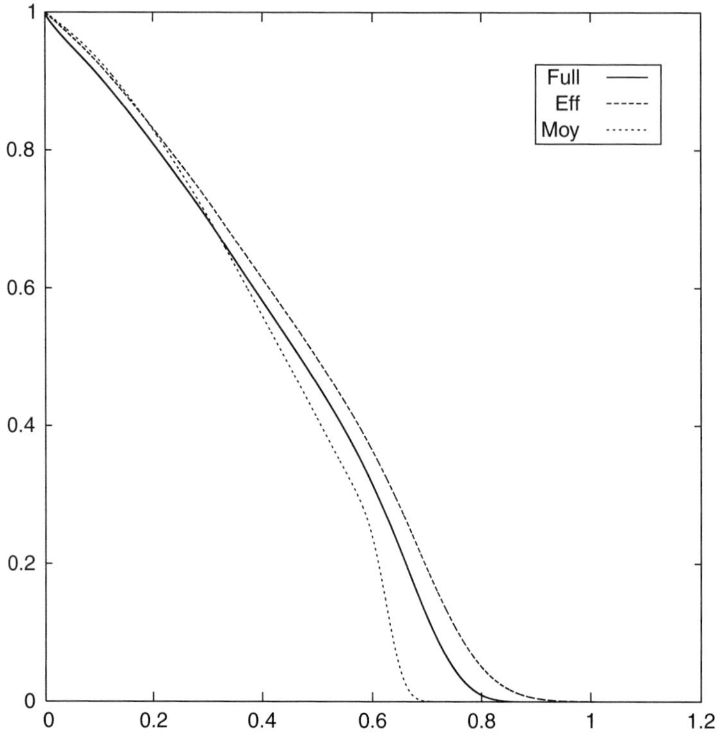

Figure 6 Volume concentrations (linear surface adsorption–desorption reactions, Case B2): Comparison between concentration obtained using our effective problem (eff), average of the section of the concentration from the original problem (full) and the concentration coming from the simple average (moy) at $t = 240$ s.

Table 9 Parameter values in the case of the first order irreversible surface reaction ($K = +\infty$)

Parameters	Values
Width of the slit: H	2.635×10^{-4} m
Characteristic length: L_R	0.632 m
$\varepsilon = H/L_R$	0.41693×10^{-3}
Characteristic velocity: Q^*	0.393×10^{-2} m/s
Diffusion coefficient: D^*	1.2×10^{-8} m^2/s
Longitudinal Péclet number: $Pe = L_R Q^*/D^* =$	2.0698×10^5
$\alpha = \log Pe/\log(1/\varepsilon) =$	1.572789
transversal Péclet number: $Pe_T = HQ^*/D^* =$	86.296

Table 10 Case of the first order irreversible surface reaction ($K = +\infty$): comparison between the volume concentrations c^{Tay}, c^{moy} and $(1/H)\int_0^H c^*\,\mathrm{d}z$ at the time $t^* = 50\,\mathrm{s}$

x^*	c^{Tay}	c^{moy}	$\frac{1}{H}\int_0^H c^*\mathrm{d}z$
0	0	0	0
0.1	1.37300401E-17	2.17207153E-17	5.17763e-05
0.11	0.000418590074	1.04498908E-16	0.00231391
0.13	0.0519752326	0.0280014326	0.0170583
0.14	0.128440421	0.155000571	0.0655227
0.145	0.153338539	0.155000571	0.0990472
0.15	0.169945407	0.155000571	0.130369
0.155	0.175667748	0.155000571	0.152722
0.16	0.176884544	0.155000571	0.165339
0.165	0.177199575	0.155000571	0.170635
0.17	0.177233822	0.155000571	0.172341
0.18	0.177238982	0.155000571	0.173227
0.19	0.177239004	0.155000571	0.173531
0.2	0.177239004	0.155000571	0.173718
0.3	0.177239004	0.155000571	0.174536

Table 11 Case of the first order irreversible surface reaction ($K = +\infty$): comparison between the volume concentrations c^{Tay}, c^{moy} and $(1/H)\int_0^H c^*\,\mathrm{d}z$ at the time $t^* = 70\,\mathrm{s}$

x^*	c^{Tay}	c^{moy}	$\frac{1}{H}\int_0^H c^*\mathrm{d}z$
0	0	0	0
0.1	4.86944849E-17	3.51572972E-18	4.27436e-06
0.16	0.000252862511	1.27275794E-17	0.000184936
0.18	0.0178727814	0.0003202901	0.00657295
0.185	0.0303102565	0.0699896992	0.0133435
0.19	0.0488626237	0.0735303344	0.0240309
0.195	0.0630697107	0.0735303475	0.0377344
0.2	0.0765167157	0.0735303475	0.0524721
0.205	0.0830143704	0.0735303475	0.0658466
0.21	0.0869467435	0.0735303475	0.0755906
0.215	0.0881363431	0.0735303475	0.081589
0.22	0.0885999673	0.0735303475	0.0845812
0.3	0.0887121329	0.0735303475	0.0869702
0.6	0.0887121329	0.0735303475	0.0875448

Table 12 Case of the first order irreversible surface reaction ($K = +\infty$): comparison between the volume concentrations c^{Tay}, c^{moy} and $(1/H)\int_0^H c^*\,dz$ at the time $t^* = 100$ s

x^*	c^{Tay}	c^{moy}	$\frac{1}{H}\int_0^H c^* dz$
0	0	0	0
0.1	7.40437514E-18	1.14872338E-18	9.51673e-07
0.2	3.20048527E-10	4.15858702E-18	1.22748e-05
0.24	0.000270157055	4.15858702E-18	0.000189098
0.26	0.00693874586	0.00236291078	0.00313599
0.28	0.025151632	0.0240251771	0.0175925
0.285	0.0278139146	0.0240251771	0.021842
0.29	0.0298614239	0.0240251771	0.0252274
0.295	0.0307142265	0.0240251771	0.0276271
0.3	0.0311937104	0.0240251771	0.0291264
0.31	0.0313963959	0.0240251771	0.0302246
0.32	0.0314129272	0.0240251771	0.0304318
0.35	0.0314136645	0.0240251771	0.030591
0.36	0.0314136645	0.0240251771	0.0306213
0.5	0.0314136645	0.0240251771	0.0308346

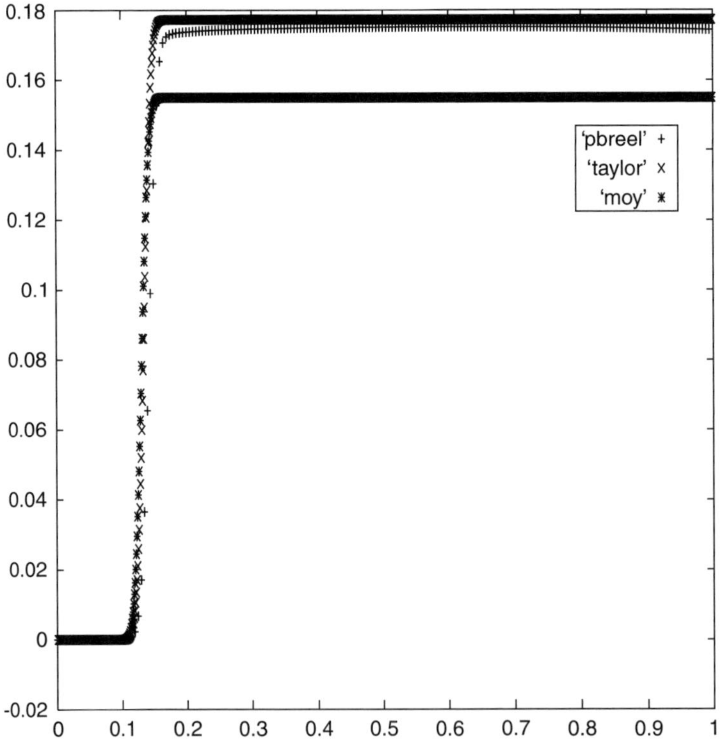

Figure 7 Case of the first order irreversible surface reaction ($K = +\infty$): Comparison between concentration obtained using our effective problem (eff), average of the section of the concentration from the original problem (pbreel3) and the concentration coming from the simple average (moy) at $t = 50$ s.

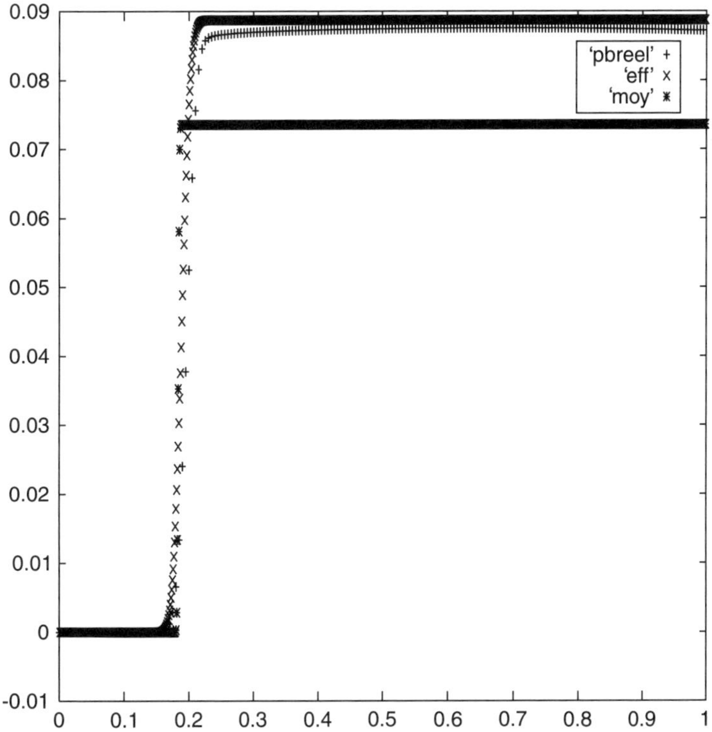

Figure 8 Case of the first order irreversible surface reaction ($K = +\infty$): Comparison between concentration obtained using our effective problem (eff), average of the section of the concentration from the original problem (pbreel) and the concentration coming from the simple average (moy) at $t = 70$ s.

perturbation for the partial differential equations describing reactive flows through a slit under dominant Péclet and Damkohler numbers.

To have a good comparison with classical Taylor's paper we were forcing our models to be parabolic, when it was possible.

Nevertheless, there is the possibility of obtaining hyperbolic models, at same order of precision, $\mathcal{O}(\varepsilon^{2(2-\alpha)})$. We note that such models where derived by Balakotaiah and Chang (2003) for a number of practical situations. In several articles, Balakotaiah et al. used the Liapounov–Schmidt reduction coupled with perturbation, to develop multi-mode models, which exhibit hyperbolic behavior (Balakotaiah, 2004; Chakraborty and Balakotaiah, 2005). Our comparison calculation from Section 3.1 shows that formally multi-mode models are of the same order as our parabolic effective equations. This was already argued in Balakotaiah and Chang (2003). It would be interesting to calculate the error estimate for the multi-mode hyperbolic models, introduced

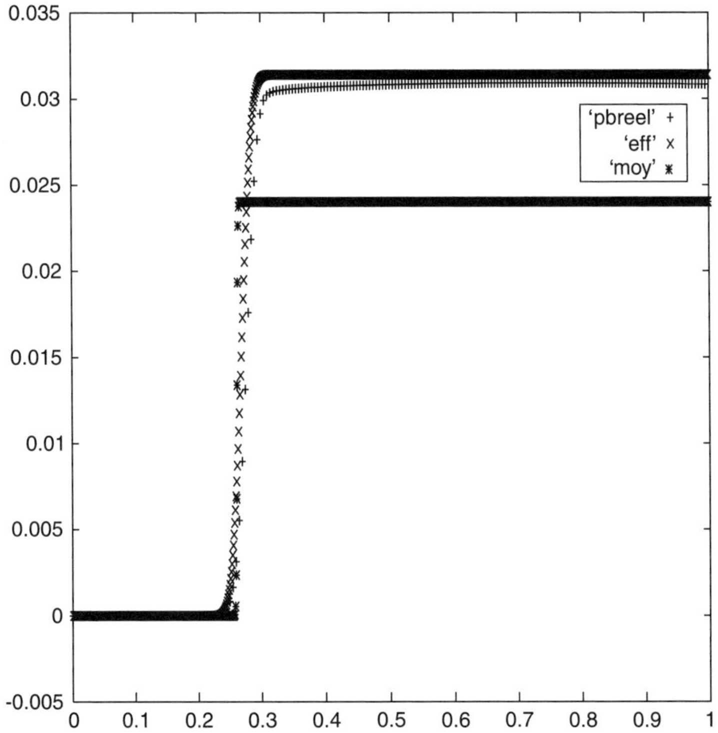

Figure 9 Case of the first order irreversible surface reaction ($K = +\infty$): Comparison between concentration obtained using our effective problem (eff), average of the section of the concentration from the original problem (pbreel) and the concentration coming from the simple average (moy) at $t = 100$ s.

Table 13 Parameter values in the case of an infinite adsorption rate $\hat{k}^* = +\infty$

Parameters	Values
Width of the slit: H	5×10^{-3} m
Characteristic length: L_R	0.8632 m
$\varepsilon = H/L_R$	5.7924001×10^{-3}
Characteristic velocity: Q^*	0.3×10^{-3} m/s
Diffusion coefficient: D^*	2×10^{-7} m^2/s
Longitudinal Péclet number: $Pe_T = L_R Q^*/D^* =$	1.2948×10^5
$\alpha = \log Pe/\log(1/\varepsilon) =$	1.83815052
Transversal Péclet number: $Pe_T = HQ^*/D^* =$	75
Transversal Damkohler number: $Da_T = K_e/H =$	1

Table 14 Comparison between the volume concentrations c^{Tay}, c^{moy} and $(1/H)\int_0^H c^*\,\text{dz}$ for the case of an infinite adsorption rate $\hat{k}^* = +\infty$ at the time $t^* = 863$ s

x^*	c^{Tay}	c^{moy}	$\frac{1}{H}\int_0^H c^*\text{dz}$
0	0	0	0
0.1	0.000476974507	1.48207153E-17	0.00019277
0.3	0.00665410189	2.36823908E-16	0.00402643
0.4	0.0169799929	4.65358482E-16	0.0127379
0.6	0.0739152145	1.9895652E-15	0.074789
0.8	0.212484001	2.14373031E-06	0.23459
1.0	0.436195692	0.5	0.474176
1.1	0.561624158	0.989232525	0.59902
1.3	0.783030278	1.	0.807166
1.5	0.920190592	1.	0.928339
1.7	0.983578518	1.	0.979062
1.9	0.996962548	1.	0.994446
2.	0.998850322	1	0.99675
3.	0.99999	1.	0.99969
4.	1.	1.	0.999917

Table 15 Comparison between the volume concentrations c^{Tay}, c^{moy} and $(1/H)\int_0^H c^*\,\text{dz}$ for the case of an infinite adsorption rate $\hat{k}^* = +\infty$ at the time $t^* = 2{,}877$ s

x^*	c^{Tay}	c^{moy}	$\frac{1}{H}\int_0^H c^*\text{dz}$
0	0	0	0
1.	2.15685873E-05	1.10453096E-16	0.00264385
2.	0.0129950594	5.15917604E-16	0.0186693
2.3	0.0422251119	6.74184016E-16	0.0401594
2.6	0.110382208	7.66476726E-16	0.095134
2.9	0.234665783	2.44409991E-08	0.205974
3.	0.288903773	1.35588449E-05	0.256387
3.2	0.411909396	0.0466182486	0.373383
3.4	0.544317915	0.799344896	0.499505
3.6	0.671938419	0.999606221	0.625723
3.8	0.782076721	0.999999998	0.742053
4.	0.867184442	1.	0.832967
4.3	0.94674831	1.	0.921783
5.	0.997306633	1.	0.983445
6.	0.99999576	1.	0.992814

Table 16 Comparison between the volume concentrations c^{Tay}, c^{moy} and $(1/H)\int_0^H c^* \,\mathrm{d}z$ for the case of an infinite adsorption rate $\hat{k}^* = +\infty$ at the time $t^* = 5{,}755$ s

x^*	c^{Tay}	c^{moy}	$\frac{1}{H}\int_0^H c^*\mathrm{d}z$
0	0	0	0
5.	0.0245430842	7.99463577E-16	0.0481293
5.5	0.0841804114	2.05329898E-15	0.102444
6.	0.21560078	1.47268882E-09	0.223168
6.3	0.332534457	0.000549169915	0.330783
6.6	0.468630165	0.276435631	0.453472
6.8	0.562546008	0.882371619	0.536562
7.	0.653050221	0.998497928	0.619671
7.2	0.735557754	0.999998971	0.700753
7.4	0.806714035	1.	0.771646
7.6	0.864767429	1.	0.830057
7.8	0.909573473	1.	0.876335
8.	0.942287957	1.	0.910953
8.5	0.984791852	1.	0.957911
9.	0.997065201	1.	0.973686

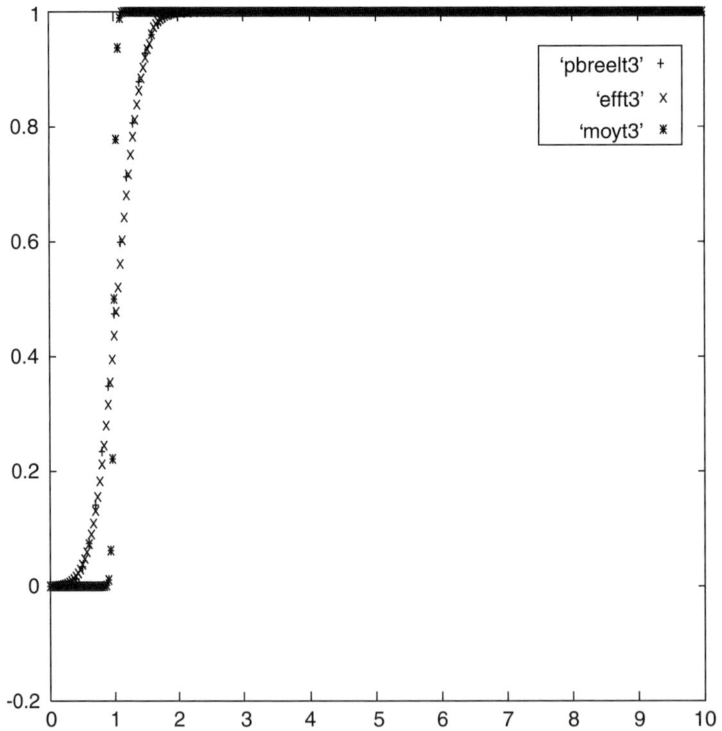

Figure 10 Case of an infinite adsorption rate $\hat{k}^* = +\infty$: Comparison between concentration obtained using our effective problem (eff), average of the section of the concentration from the original problem (pbreel3) and the concentration coming from the simple average (moy) at $t = 863$ s.

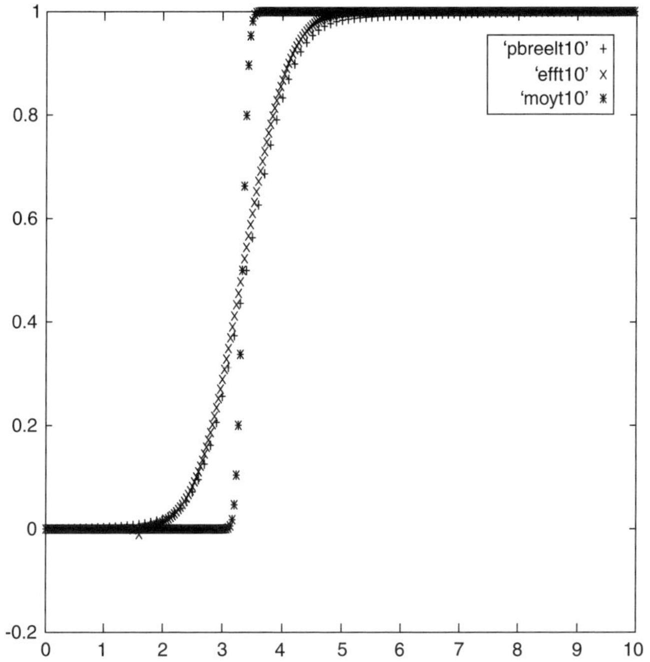

Figure 11 Case of an infinite adsorption rate $\hat{k}^* = +\infty$: Comparison between concentration obtained using our effective problem (eff), average of the section of the concentration from the original problem (pbreel3) and the concentration coming from the simple average (moy) at $t = 2{,}877$ s.

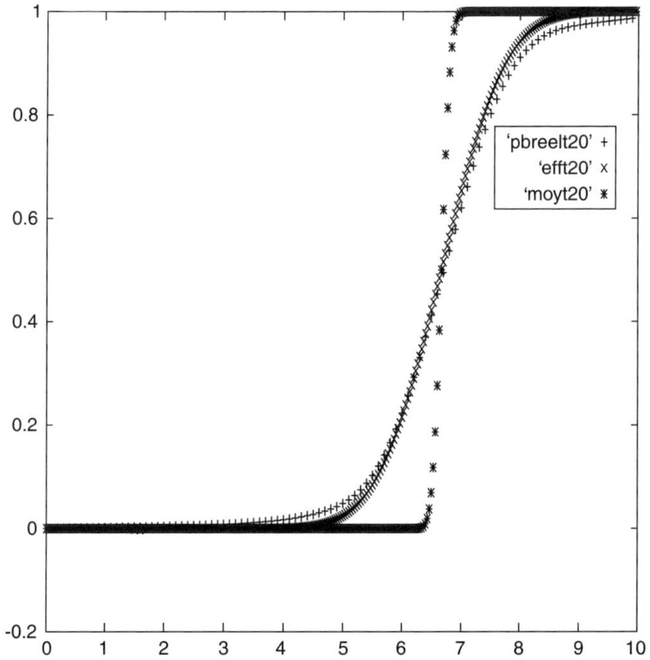

Figure 12 Case of an infinite adsorption rate $\hat{k}^* = +\infty$: Comparison between concentration obtained using our effective problem (eff), average of the section of the concentration from the original problem (pbreel3) and the concentration coming from the simple average (moy) at $t = 5{,}755$ s.

by Balakotaiah et al., and to compare the approximations on mathematically rigorous way.

Furthermore, there is approach by Camacho using a viewepoint of Irreversible Thermodynamics and leading to the Telegraph equation. For more details we refer to Camacho (1993a, b, c) and to the doctoral thesis Berentsen (2003). We plan to address this subject in the near future and extend our result in this direction.

ACKNOWLEDGEMENTS

This author acknowledge the referee for his careful reading of the article and valuable comments.

This work was initiated during the sabbatical visit of A. Mikelić to the TU Eindhoven in Spring 2006, supported by the Visitors Grant B-61-602 of the Netherlands Organisation for Scientific Research (NWO).

The research of C.J. van Duijn and I.S. Pop was supported by the Dutch government through the national program BSIK: knowledge and research capacity, in the ICT project BRICKS (http://www.bsik-bricks.nl), theme MSV1.

The research of A. Mikelić and C. Rosier was supported by the Groupement MOMAS (Modélisation Mathématique et Simulations numériques liées aux problèmes de gestion des déchets nucléaires: (PACEN/CNRS, ANDRA, BRGM, CEA, EDF, IRSN) as a part of the project "*Modèles de dispersion efficace pour des problèmes de Chimie-Transport: Changement d'échelle dans la modélisation du transport réactif en milieux poreux, en présence des nombres caractéristiques dominants*".

REFERENCES

Allaire, G., and Raphael, A.-L. Homogénéisation d'un modèle de convection-diffusion avec chimie/adsorption en milieu poreux, preprint R.I. 604, Ecole Polytechnique, Centre de Mathématiques appliquées, Paris, November 2006.

Allaire, G., and Raphael, A.-L. *Comptes rendus Mathématique* **344**(8), 523–528 (2007).

Aris, R. *Proc. R. Soc. Lond. A* **235**, 67–77 (1956).

Auriault, J. L., and Adler, P. M. *Adv. Water Resour.* **18**, 217–226 (1995).

Balakotaiah, V. *Korean J. Chem. Eng.* **21**(2), 318–328 (2004).

Balakotaiah, V., and Chang, H.-C. *Phil. Trans. R. Soc. Lond. A* **351**(1695), 39–75 (1995).

Balakotaiah, V., and Chang, H.-C. *SIAM J. Appl. Maths.* **63**, 1231–1258 (2003).

Bear, J., and Verruijt, A., "Modeling Groundwater Flow and Pollution". D. Reidel Publishing Company, Dordrecht (1987).

Berentsen, C. Upscaling of Flow in Porous Media form a Tracer Perspective, Ph. D. thesis, University of Delft (2003).

Bourgeat, A., Jurak, M., and Piatnitski, A. L. *Math. Model Method Appl. Sci.* **26**, 95–117 (2003).

Caflisch, R. E., and Rubinstein, J., "Lectures on the Mathematical Theory of Multiphase-Flow". Courant Institute of Mathematical Sciences, New York (1984).

Camacho, J. *Phys. Rev. E* **47**(2), 1049–1053 (1993a).

Camacho, J. *Phys. Rev. E* **48**(1), 310–321 (1993b).

Camacho, J. *Phys. Rev. E* **48**(3), 1844–1849 (1993c).

Chakraborty, S., and Balakotaiah, V. *Adv. Chem. Eng.* **30**, 205–297 (2005).

Choquet, C., and Mikelić, A. "Laplace Transform Approach to the Upscaling of the Reactive Flow under Dominant Péclet Number", accepted for publication in "Applicable Analysis" (2008).

Conca, C., Diaz, J. I., Liñán, A., and Timofte, C., Homogenization in Chemical Reactive Flows through Porous Media. Electron. J. Differ. Eq., paper no. 40, 22pp. (2004).

Conca, C., Diaz, J. I., and Timofte, C. *Math. Model Method Appl. Sci.* **13**, 1437–1462 (2003).

van Duijn, C. J., and Knabner, P. *Euro. J. Appl. Math.* **8**, 49–72 (1997).

van Duijn, C. J., Knabner, P., and Schotting, R. T. *Adv. Water Resour.* **22**, 1–16 (1998).
van Duijn, C. J., and Pop, I. S. *J. Reine Angew. Math.* **577**, 171–211 (2004).
Friedman, A., and Knabner, P. *J. Differ. Equ.* **98**, 328–354 (1992).
Hornung, U., and Jäger, W. *J. Differ. Equ.* **92**, 199–225 (1991).
Knabner, P., van Duijn, C. J., and Hengst, S. *Adv. Water Resour.* **18**, 171–185 (1995).
Mauri, R. *Phys. Fluids A*, **3**, 743–755 (1991).
Mercer, G. N., and Roberts, A. J. *SIAM J. Appl. Math.* **50**, 1547–1565 (1990).
Mikelić, A., and Primicerio, M. *M^3AS: Math. Model Method Appl. Sci.* **16**(11), 1751–1782 (2006).
Mikelić, A., and Rosier, C. *Ann. Univ. Ferrara. Sez. VII Sci. Mat.* **53**, 333–359 (2007).
Mikelić, A., Devigne, V., and van Duijn, C. J. *SIAM J. Math. Anal.* **38**, 1262–1287 (2006).
Paine, M. A., Carbonell, R. G., and Whitaker, S. *Chem. Eng. Sci.* **38**, 1781–1793 (1983).
Pironneau, O., "Méthodes des éléments finis pour les fluides". Masson, Paris (1988).
Quarteroni, A., and Valli, A., "Numerical Approximation of Partial Differential Equations". Springer, Berlin (1994).
Rosencrans, S. *SIAM J. Appl. Math.* **57**, 1216–1241 (1997).
Rubin, J. *Water Resour. Res.* **19**, 1231–1252 (1983).
Rubinstein, J., and Mauri, R. *SIAM J. Appl. Math.* **46**, 1018–1023 (1986).
Taylor, G. I. *Proc. R. Soc. A* **219**, 186–203 (1993).
Vladimirov, V. S., "Equations of Mathematical Physics". URSS, Moscow (1996).
Wloka, J., "Partial Differential Equations". Cambridge University Press, Cambridge (1987).

CHAPTER 2

Overall Reaction Rate Equation of Single-Route Complex Catalytic Reaction in Terms of Hypergeometric Series

Mark Z. Lazman[1] and **Gregory S. Yablonsky**[2,*]

Contents

[1] Independent Consultant, Calgary AB, Canada

[2] Parks College, Department of Chemistry, 3450 Lindell Blvd, Saint Louis University, Missouri, MO 63103, USA; Department of Energy, Environmental & Chemical Engineering, One Brookings Dr., Campus Box 1180, Washington Univ. in St. Louis, St. Louis, MO 63130-4899, USA

*Corresponding author.
E-mail address: gy@seas.wustl.edu

Advances in Chemical Engineering, Volume 34
ISSN 0065-2377, DOI 10.1016/S0065-2377(08)00002-1

Abstract

The non-linear theory of steady-steady (quasi-steady-state/pseudo-steady-state) kinetics of complex catalytic reactions is developed. It is illustrated in detail by the example of the single-route reversible catalytic reaction. The theoretical framework is based on the concept of the *kinetic polynomial* which has been proposed by authors in 1980–1990s and recent results of the algebraic theory, i.e. an approach of hypergeometric functions introduced by Gel'fand, Kapranov and Zelevinsky (1994) and more developed recently by Sturnfels (2000) and Passare and Tsikh (2004). The concept of ensemble of *equilibrium subsystems* introduced in our earlier papers (see in detail Lazman and Yablonskii, 1991) was used as a physico-chemical and mathematical tool, which generalizes the well-known concept of "equilibrium step". In each equilibrium subsystem, ($n-1$) steps are considered to be under equilibrium conditions and one step is limiting (n is a number of steps of the complex reaction). It was shown that *all solutions* of these equilibrium subsystems define coefficients of the *kinetic polynomial*.

As a result, it was obtained an analytical expression of the reaction rate in terms of hypergeometric series with *no classical simplifications about the "limiting step" or the "vicinity of the equilibrium"*. The obtained explicit equation, "four-term equation", can be presented as follows in the Equation (77):

$$R = \frac{k_+\left(f_+(c) - K_{\mathrm{eq}}^{-1} f_-(c)\right)}{\sum(k,c)} N(k,c)$$

It has four terms:

(1) *an apparent kinetic coefficient* k_+; (2) *a "potential term"* ($f_+(c) - K_{eq}^{-1} f_-(c)$) related to the net reaction ("driving force" of irreversible thermodynamics); (3) a *"resistance" term,* $\Sigma(k,c)$, denominator of the polynomial type, which reflects complexity of chemical reaction, both its non-elementarity (many-step character) and non-linearity of elementary steps as well; (4) finally, *the "fourth term"*, $N(k,c)$. This fourth term is generated exclusively by the non-linearity of reaction steps. This term is the main distinguishing feature of this equation in comparison with Langmuir–Hinshelwood–Hougen–Watson (LHHW) equations based on simplifying assumptions. In absence of non-linear steps the "fourth" term is also absent.

It was demonstrated that the hypergeometric rate representation covers descriptions, which correspond to typical simplifications ("vicinity of the equilibrium", "rate-limiting step").

Using kinetic models of typical catalytic mechanisms (Eley–Rideal and Langmuir–Hinshelwood (LH) mechanisms) as examples, we found parametric domains, in which the hypergeometric representation is an excellent approximation of the exact solution. Unexpectedly, this representation works well even very far from the equilibrium in the small rate domain (*low-rate branch*). Convergence problems of the hypergeometric representation have been discussed.

The obtained results can be used for a description of kinetic behavior of steady-state open catalytic systems as well as quasi (pseudo)-steady-state catalytic systems, both closed and open.

1. INTRODUCTION

A single-route complex catalytic reaction, steady state or quasi (pseudo) steady state, is a favorite topic in kinetics of complex chemical reactions. The practical problem is to find and analyze a steady-state or quasi (pseudo)-steady-state kinetic dependence based on the detailed mechanism or/and experimental data. In both mentioned cases, the problem is to determine the concentrations of intermediates and *overall reaction rate* (i.e. rate of change of reactants and products) as dependences on concentrations of reactants and products as well as temperature. At the same time, the problem posed and analyzed in this chapter is directly related to one of main problems of theoretical chemical kinetics, i.e. search for general law of complex chemical reactions at least for some classes of detailed mechanisms.

By definition of the steady-state regime, the steady-state rate of change of the intermediate concentrations equals zero.

As for the quasi (pseudo)-steady-state case, the basic assumption in deriving kinetic equations is the well-known Bodenshtein hypothesis according to which the rates of formation and consumption of intermediates are equal. In fact, Chapman was first who proposed this hypothesis (see in more detail in the book by Yablonskii et al., 1991). The approach based on this idea, the Quasi-Steady-State Approximation (QSSA), is a common method for eliminating intermediates from the kinetic models of complex catalytic reactions and corresponding transformation of these models. As well known, in the literature on chemical problems, another name of this approach, the Pseudo-Steady-State Approximation (PSSA) is used. However, the term "Quasi-Steady-State Approximation" is more popular. According to the Internet, the number of references on the QSSA is more than 70,000 in comparison with about 22,000, number of references on PSSA.

All our analysis is done under the assumption that chemical kinetics is the limiting factor of the complex chemical process, not transport of reactants/products.

The original model regarding surface intermediates is a system of ordinary differential equations. It corresponds to the detailed mechanism under an assumption that the surface diffusion factor can be neglected. Physico-chemical status of the QSSA is based on the presence of the 'small parameter', i.e. the total amount of the surface active sites is small in comparison with the total amount of gas molecules. Mathematically, the QSSA is a zero-order approximation of the original (singularly perturbed) system of differential equations by the system of the algebraic equations (see in detail Yablonskii et al., 1991). Then, in our analysis we simply replace the differential equations of the "fast" catalytic intermediates by the corresponding algebraic equations.

In comparison with the linear case, the non-linear one represents much greater challenge in deriving rate as no simple explicit formula like Cramer's rule exists here. Chemists always tried to overcome the mathematical difficulties of the analysis of the non-linear detailed kinetic model introducing simplifying hypotheses such as a hypothesis on existence of the single rate-limiting stage or hypothesis of the vicinity of thermodynamic equilibrium. First results here were obtained independently by Professor J. Horiuti (Japan) and Academician G.K. Boreskov (Russia). They introduced the fundamental concepts of *stoichiometric number* and *molecularity* of steps and found kinetic equations based on the limiting step assumption. Reaction rate equations grounded on these assumptions found wide area of practical application including situations far beyond the assumptions of the original theory. It has to be mentioned however that the validity of the popular assumption on the limiting step is not so trivial problem. Recently, Gorban (2005) and Gorban and Radulescu (2008) revisited the mathematical status of this classical hypothesis and found many non-trivial cases and misinterpretations as well. They rigorously studied the difference in limiting between transient and quasi-steady-state regimes and classified different scenarios of limiting in chemical kinetics. Quasi-steady-state regimes can be considered as specific invariant manifolds which general theory was developed by Gorban and Karlin (2003) and Gorban (2005).

Nevertheless, in this area, particularly in presenting explicit solutions of QSSA-kinetic models, theoretical chemical kinetics is still far from completing.

What could be done in the general non-linear case? Authors (Lazman and Yablonsky) started answering this question by applying constructive algebraic geometry. In early 1980s, we have proved that QSSA system corresponding to the single-route reaction mechanism of catalytic reaction can be reduced to a *single polynomial* in terms of reaction rate and concentrations (parameters of reactions are Arrhenius dependences of the temperature). In this case, the reaction rate of complex reaction is now *an implicit, not explicit* function of concentrations and temperature. Mathematically, this polynomial (i.e. *kinetic polynomial*) is a resultant of the QSSA algebraic system. Its vanishing is a necessary and sufficient condition for steady state. Thus, the roots of kinetic polynomial are the values of reaction rate in the steady state.

In this chapter, we will try to answer the next obvious question: can we find an *explicit* reaction rate equation for the general *non-linear* reaction mechanism, at least for its *thermodynamic branch*, which goes through the equilibrium. Applying the kinetic polynomial concept, we introduce the new *explicit* form of reaction rate equation in terms of hypergeometric series.

The second motive of this chapter is concerned with evergreen topic of interplay of chemical kinetics and thermodynamics. We analyze the generalized form of the explicit reaction rate equation of the *thermodynamic branch* within the context of relationship between forward and reverse reaction rates (we term the corresponding problem as the Horiuti–Boreskov problem). We will compare our equation with the well-known LHHW equation and demonstrate both similarities and differences.

In our approach, the concept of ensemble of equilibrium subsystems introduced in our earlier chapters (see in detail Lazman and Yablonskii, 1991) was used as a very efficient tool of mathematical analysis and physico-chemical understanding. The equilibrium subsystem is such a system that corresponds to the following assumption: $(n-1)$ steps are considered to be under equilibrium conditions, one step is limiting, where n is a number of steps. In fact, the concept of "equilibrium subsystems" is a generalization of the concept of "equilibrium step", which is well known in chemical kinetics. Then, we take n of these equilibrium subsystems (*an ensemble of equilibrium subsystems*). It was shown that solutions of these subsystems ("roots", "all roots", not just one "root") define coefficients of the *kinetic polynomial*.

We term our equation obtained for the thermodynamic branch of the reaction rate as "the four-term rate equation". This equation generalizes the known explicit forms of overall reaction rate equations. The mentioned terms are following: (1) *the kinetic apparent coefficient*; (2) *the potential term*, or *driving force* related to the thermodynamics of the net reaction; (3) *the term of resistance*, i.e. the denominator, which reflects the complexity of reaction, both its multi-step character and its non-linearity; finally, (4) *the non-linear term* which is caused exclusively by non-linear steps. In the case of linear mechanism, this term is vanishing. Distinguishing this fourth term is the original result of this chapter. In classical theoretical kinetics of heterogeneous catalysis (LH and Hougen—Watson (HW) equations) such term is absent.

Finally, we present the results of the case studies for Eley–Rideal and LH reaction mechanisms illustrating the practical aspects (i.e. convergence, relation to classic approximations) of application of this new form of reaction rate equation. One of surprising observations here is the fact that hypergeometric series provides the good fit to the exact solution not only in the vicinity of thermodynamic equilibrium but also far from equilibrium. Unlike classical approximations, the approximation with truncated series has non-local features. For instance, our examples show that approximation with the truncated hypergeometric series may supersede the conventional rate-limiting step equations. For thermodynamic branch, we may think of the domain of applicability of reaction rate series as the domain, in which the reaction rate is relatively small.

However, first we have to explain in detail a situation in the corresponding area of theoretical chemical kinetics and our chemico-mathematical framework.

1.1 Linear and non-linear mechanisms

Regarding the participation of intermediate in the steps of detailed mechanism, Temkin (1963) classified catalytic reaction mechanisms as linear and non-linear ones. For linear mechanisms, every reaction involves the participation of only one molecule of the intermediate substance. The typical linear mechanism is the two-step catalytic scheme (Temkin–Boudart mechanism), e.g. water–gas shift reaction:

$$\frac{\begin{array}{l}1.\ \ Z + H_2O \Leftrightarrow ZO + H_2 \\ 2.\ \ ZO + CO \Leftrightarrow Z + CO_2\end{array}}{\begin{array}{l}\text{Overall reaction:} \\ CO + H_2O \Leftrightarrow H_2 + CO_2\end{array}},$$

where Z and ZO are reduced and oxidized forms of catalyst, respectively, Boudart and Diega-Mariadassou (1984).

The typical non-linear mechanism is three-step adsorption mechanism (LH mechanism), e.g.

$$\frac{\begin{array}{l}1.\ \ O_2 + 2Pt \Leftrightarrow 2PtO \\ 2.\ \ CO + Pt \Leftrightarrow PtCO \\ 3.\ \ PtO + PtCO \rightarrow 2Pt + CO_2\end{array}}{\begin{array}{l}\text{Overall reaction:} \\ 2CO + O_2 \rightarrow 2CO_2\end{array}},$$

where Pt is am empty site of platinum catalyst; PtO and PtCO are surface oxygen and CO, respectively.

In many cases it is assumed that the rate of any elementary reaction is governed by the Mass-Action-Law (MAL). We will make the same assumption as well.

1.2 Rate equation for one-route linear mechanism

The overall rate equation of complex single-route reaction with the linear detailed mechanism was derived and analyzed in detail by many researchers. King and Altman (1956) derived the overall reaction rate equation for single-route enzyme reaction with an arbitrary number of intermediates

$$\frac{\begin{array}{l}1.\ \ S + E \Leftrightarrow X_1 \\ 2.\ \ X_1 \Leftrightarrow X_2 \\ \qquad \vdots \\ n.\ \ X_{n-1} \Leftrightarrow P + E.\end{array}}{\begin{array}{l}\text{Overall reaction:} \\ S \Leftrightarrow P\end{array}}$$

using the methods of graph theory. The graph of this complex reaction was presented as

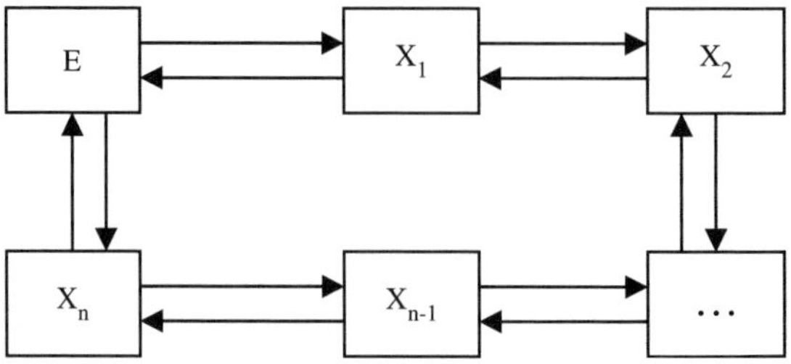

Later, in 1970s and 1980s, Evstigneev et al. (1978, 1979, 1981) systematically analyzed this equation applying methods of graph theory. They found a variety of its interesting structural properties regarding the link between kinetics of the complex reaction and structure of the reaction mechanism.

1. This equation can be always presented in the form

$$R = \frac{C}{\sum} \tag{1}$$

where R is the reaction rate and C is a cyclic characteristic;

$$C = K^{+}f^{+}(\vec{c}) - K^{-}f^{-}(\overleftarrow{c}) \tag{1a}$$

$$K^{+} = \prod_i k_i^{+}, \quad K^{-} = \prod_i k_i^{-} \tag{1b}$$

$$K^{+}/K^{-} = K_{\text{eq}} \tag{1c}$$

$\vec{c}$ and $\overleftarrow{c}$ are sets of concentrations of reactants and products, respectively; k_i^{+} and k_i^{-} are Arrhenius-type kinetic constants of the forward and reverse reactions, respectively; K_{eq} is the equilibrium constant of the overall reaction.

2. The numerator C of expression (1) does not depend on the complexity of the detailed mechanism. It always corresponds to the overall reaction.
3. The denominator $\sum$, the "resistance" term, reflects the real complexity of the detailed mechanism. We can present it as

$$\sum = \sum_j K_j \prod_l c_l^{p_{jl}} \tag{2}$$

where K_j is either the product of kinetic constants of some reactions of detailed mechanism or sum of such products, c_l the concentration of reactant (product) and p_{jl} the positive integer. The physical meaning of the denominator $\sum$ is a "resistance", i.e. the "retardation" of the overall reaction rate by the intermediates of detailed mechanism. No doubt, from the energetic point of view catalytic intermediates accelerate the whole reaction decreasing the apparent activation energy. However, at the same time the complex reaction is not a "single event". It is occurred via catalytic intermediates that take a part of reacting substances.

Equation (2) can be termed as the "Langmuir form". Equation (1) can be presented in the form

$$\left(\sum\right) R - C = 0 \tag{3}$$

which is a particular case of the *kinetic polynomial* (see below).

4. We can write the rate equation (1) as a difference of forward and reverse overall rates

$$R = R^{+} - R^{-} \tag{4}$$

where

$$R^+ = \frac{K^+ f^+(\vec{c})}{\sum}, \qquad R^- = \frac{K^- f^-(\overleftarrow{c})}{\sum} \tag{5}$$

can be termed as overall reaction rates of forward and reverse reactions, respectively, and

$$\frac{R^+}{R^-} = \frac{K^+}{K^-} \frac{f^+(\vec{c})}{f^-(\overleftarrow{c})} \tag{6}$$

5. At the thermodynamic equilibrium, when $R = 0$, we have

$$R^+ = R^- \tag{7}$$

and

$$\frac{K^+}{K^-} = K_{eq} = \frac{f^-(\overleftarrow{c})}{f^+(\vec{c})} \tag{8}$$

6. Different properties of Equation (1), particularly a number of independent parameters K_j (see Equation (2)) and relationships between them, the properties of apparent kinetic order (i.e. $\partial \ln R/\partial \ln c_l$) and apparent activation energy (i.e. $\partial \ln R/\partial(-1/RT)$) in terms of concentrations of intermediates and parameters of detailed mechanism have been found (see the monograph by Yablonskii et al., 1991).

Some examples of kinetic equations of complex catalytic reactions are presented in Appendix 1.

Equation (3) is linear with respect to the reaction rate variable, R. In the further analysis of more complex, non-linear, mechanisms and corresponding kinetic models, we will present the polynomial as an equation, which generalizes Equation (3), and term it as the *kinetic polynomial.* We will demonstrate that the overall reaction rate, in the general non-linear case, cannot generally be presented as a difference between two terms representing the forward and reverse reaction rates. This presentation is valid only at the special conditions that will be described.

The similar analysis for particular multi-route linear mechanism was done in 1960s by Vol'kenstein and Gol'dstein (1966) and Vol'kenstein (1967). In 1970s, the rigorous "structurized" equation for the rate of multi-route linear mechanism was derived by Yablonskii and Evstigneev (see monograph by Yablonskii et al., 1991). It reflects the structure of detailed mechanism, particularly coupling between different routes (cycles) of complex reaction. Some of these results were rediscovered many years later and not once (e.g. Chen and Chern, 2002; Helfferich, 2001).

1.3 Single route overall rate equation in applied kinetics

In remarkable progress of catalytic industry in 1930–1950s, a kinetic model was considered as a basis of reactor design. Langmuir and Hinshelwood compensated a lack of information about surface intermediates and detailed mechanism via two simple assumptions:

(i) Catalytic process is occurred by competition between the components of the reaction mixture for sites on the catalyst surface.
(ii) Adsorption and desorption rates are high in comparison to other steps of chemical transformation on the catalyst surface.

When the complex catalytic reaction is irreversible, a typical form of the corresponding kinetic equation, i.e. LH equation, is written as follows:

$$R = \frac{K\prod_i c_i}{1+\sum_i K_i c_i^{m_i}} \tag{9}$$

In the case of reversible reaction, Hougen and Watson proposed the similar semi-empirical equation (HW equation). For instance, for the reaction of cyclohexane dehydrogenation this equation has the form

$$R = \frac{K^+ c_{C_6H_{12}} - K^- c_{C_6H_{10}} c_{H_2}}{(K_1 c^{\alpha}_{C_6H_{10}} + K_2 c^{\beta}_{H_2} + K_3 c^{\gamma}_{C_6H_{12}})^m} \tag{10}$$

Equations (9) and (10) are called LHHW equations indicating their similarity: they belong to the class of "numerator divided by denominator" kinetic equations.

In HW equations the numerator of the expression (10) is considered to be correspondent to the overall reaction. However, within the *semi-empirical* HW approach, which was not mechanistic, the relationships between the detailed mechanisms and kinetic dependences could not be analyzed.

1.4 Relationships between forward and reverse overall reaction rates: Horiuti–Boreskov problem

Equation (10) can be presented as the difference of two terms, the forward and reverse reaction rates,

$$R = R^+ - R^- \tag{11}$$

where

$$\frac{R^+}{R^-} = K_{eq}\frac{f^+(\vec{c})}{f^-(\overleftarrow{c})} \tag{12}$$

There is a famous question posed firstly by the Japanese scientist Juro Horiuti before WWII (1939), and then, independently, by the Russian scientist Georgii Boreskov in 1945: "How to find the equation for reaction rate in back direction knowing the rate expression in a forward direction?" (detailed description of the problem is presented by Horiuti (1973), see also Boreskov's (1945) original paper).

Horiuti posed this problem and solved it only for the special example of reaction on a hydrogen electrode. Boreskov with no knowledge about the

Horiuti's results (there was a war!) analyzed the SO_2 oxidation case. Both Horiuti and Boreskov assumed that all reaction steps, except one of them, are reversible and fast. These steps are not obligatory adsorption steps. One reversible step, i.e. rate-determining one, is much slower than the rest of other steps. Using SO_2 oxidation as an example and assuming power low kinetic expressions for the reaction rates, Boreskov showed that

$$\frac{R^+}{R^-} = \left[K_{eq}\frac{f^+(\vec{c})}{f^-(\overleftarrow{c})}\right]^M \tag{13}$$

and

$$R = R^+\left\{1 - \left[\frac{f^-(\overleftarrow{c})}{K_{eq}f^+(\vec{c})}\right]^M\right\} \tag{14}$$

where $f^+(\vec{c})$ and $f^-(\overleftarrow{c})$ are kinetic dependencies corresponding to overall reaction (forward and reverse), M a number related to the rate-determining step.[1] The rate of reverse reaction was estimated on isotope exchange data in reactions of SO_2 oxidation and ammonia synthesis. Then, the power M that Boreskov called the "molecularity" was found. For SO_2 oxidation it was 1/2.

Within the Horiuti's approach, the physical meaning of the molecularity is clear. Horiuti introduced the concept of stoichiometric numbers (*Horiuti numbers*, v).[2] Horiuti numbers are the numbers such that, after multiplying the chemical equation for every reaction step by the appropriate Horiuti number v_i and subsequent adding, all reaction intermediates are cancelled. The equation obtained is the overall reaction. In the general case, the Horiuti numbers form a matrix. Each set of Horiuti numbers (i.e. matrix column) leading to elimination of intermediates corresponds to the specific reaction route.[3]

For typical one-route linear mechanisms all the Horiuti numbers can be selected to be equal to 1.[4] This is not necessarily true for non-linear reaction mechanism, e.g. for SO_2 oxidation mechanism

$$\begin{array}{lll} 1. & 2K + O_2 \Leftrightarrow 2KO & 1 \\ 2. & KO + SO_2 \Leftrightarrow K + SO_3 & 2 \\ \hline & 2SO_2 + O_2 \Leftrightarrow 2SO_3 & \end{array} \tag{15}$$

[1] Boreskov assumed the power law dependence for reaction rate, which is mathematically incorrect. Thus, strictly speaking, he did not prove Equations (13) and (14). Authors performed the analysis of the model corresponding to the single-route reaction mechanism with the rate-limiting step and proved these relations rigorously (see Lazman and Yablonskii, 1988; Lazman and Yablonskii, 1991). Mathematically, expression (12) is the first term of infinite power series by powers kinetic parameters of rate-limiting step.

[2] It is more convenient, in our opinion, to use the term "Horiuti number" instead of the "stoichiometric number" as the latter could be mistakenly identified with the term "stoichiometric coefficient" which designates the number of molecules participating in the reaction.

[3] Obviously, Horiuti numbers are defined up to non-singular linear transformation.

[4] They can be zero for "buffer" reaction steps (such step that involves the reaction intermediate participating in this stage exceptionally).

According to the Temkin's classification, this mechanism is non-linear, because in the first step oxygen reacts with two catalyst sites K. Numbers to the right of stoichiometric equations are Horiuti numbers: $\nu_1 = 1$ and $\nu_2 = 2$. An analysis of data by Boreskov showed that the power M in Equations (13) and (14) is

$$M = 1/\nu_{\text{lim}}$$

where ν_{lim} is Horiuti number of rate limiting reaction step. Experimental data demonstrated that:

$$\frac{R^+}{R^-} = \left(K_{\text{eq}} \frac{c_{SO_2} c_{O_2}}{c_{SO_3}} \right)^{1/2}$$

Thus, $M = 1/\nu_{\text{lim}} = 1/2$, $\nu_{\text{lim}} = 2$ and the second step of reaction mechanism (15) is rate limiting step.

It is not still absolutely clear if the Horiuti–Boreskov representation is valid in the general case (for instance, for single-route non-linear mechanisms without rate-limiting step) and under which conditions it could be valid. It seems both scientists considered this representation is valid in all the domain of conditions. Both tried to find the relationship between R^+ and R^- considering that such distinguishing does exist always. However it is a problem! Having respect to scientists who first started to work in this area, we term it as the "Horiuti–Boreskov problem".

We formulate this problem as follows:

Under which conditions the presentation $R = R^+ - R^-$ is valid for the complex reaction?

2. RIGOROUS ANALYSIS OF COMPLEX KINETIC MODELS: NON-LINEAR REACTION MECHANISMS

The solution of such a problem became possible since early 1980s that was concerned with the following factors:

- increasing interest in decoding the non-linear phenomena (steady-state multiplicity, self-oscillations, etc.) and
- new mathematical results in algebraic geometry, complex analysis and computer algebra; personal motivation in solving Horiuti–Boreskov problem.

2.1 Quasi-steady-state approximation

As previously mentioned, the QSSA is a common method for eliminating intermediates from the kinetic models of complex catalytic reactions and corresponding transformation of these models. Mathematically, it is a zero-order approximation of the original (singularly perturbed) system of differential equations, which describes kinetics of the complex reaction. We simply replace the differential equations corresponding to the "fast" intermediates with algebraic equations

$$\Gamma^T \bar{w} = 0 \tag{16}$$

Let n be a number of reactions and m be a number of intermediates in the reaction mechanism. The vector $\bar{w} = (w_1, \ldots, w_n)^T$ is composed of the rates of reaction steps. Assuming the MAL, we have

$$w_i = f_i \prod_{j=1}^{m} z_j^{\alpha_{ij}} - r_i \prod_{j=1}^{m} z_j^{\beta_{ij}}, \qquad i = 1, \ldots, n \tag{17}$$

where z_j is the concentration of jth intermediate; f_i, r_i are reaction weights of forward and reverse reactions of step i.

Reaction weight is a reaction rate calculated at unit concentrations of intermediates, i.e. it is either the reaction constant or the reaction constant multiplied by power product corresponding to the "slow" components (either reagents or products). Thus, the dependencies of reaction rate on temperature and concentrations are "hidden" in reaction weights.

Non-negative integers α_{ij}, β_{ij} are stoichiometric coefficients of component j in reaction i. Matrix $\Gamma = (\gamma_{ij})$ is a stoichiometric matrix of some rank (rk), with elements $\gamma_{ij} = \beta_{ij} - \alpha_{ij}$.

Let $r = rk\Gamma$. We assume that $r < n$. In this case, there exists the $n \times P$ matrix N, such that

$$\Gamma^T \mathrm{N} = 0 \tag{18}$$

The number

$$P = n - r \tag{19}$$

is the number of *reaction routes*. Reaction route corresponds to the column of matrix N. These columns are linearly independent and form the *stoichiometric basis*. Elements of matrix N are *Horiuti numbers* ν_{sp}. The stoichiometric basis is defined up to a non-singular linear transformation. We can always define the stoichiometric basis in terms of integer stoichiometric numbers. We assume below that all ν_{sp} are integers. If we multiply each chemical equation of our mechanism by the corresponding stoichiometric number from some column of matrix N and add up the results, we obtain the chemical equation free of intermediates. This equation corresponds to the *net reaction* of the selected reaction route.

Vector $\bar{w}$ solves the homogeneous linear system (16) if and only if it belongs to the space spanned by columns of the matrix N. There exists such a vector $\bar{R} = (R_1, \ldots, R_P)^T$ that

$$\bar{w} = \mathrm{N}\bar{R} \tag{20a}$$

The element of vector $\bar{R}$ is the *rate along the reaction route*. Concentrations of intermediates satisfy $B = m - r$ linear balance equations

$$\mathbf{L}(\mathbf{z}) = 0 \tag{20b}$$

Systems (20a) and (20b)[5] of $n+B$ equations in $n+B$ unknowns $z_1, \ldots z_n$ and $R_1, \ldots R_P$, is the equivalent presentation of the original problem (16)[6]. For the single-route reaction mechanism vector $\bar{R}$ has only one element R, which is the overall reaction rate (the rate of net reaction).

2.2 Non-linear mechanisms: the kinetic polynomial

2.2.1 The resultant in reaction rate

Assuming the MAL, the system (20) consists of polynomials of variables $z_1, \ldots z_n$, and $R_1, \ldots, R_P$. Powerful techniques of effective algebraic geometry can be applied to polynomial systems. In many cases, we can apply the *variable elimination* and reduce our system to a single polynomial equation of the single variable. Mathematically, we need to find the invariant of our algebraic system, i.e. the system *resultant* (see Bykov et al., 1998; Gel'fand et al., 1994; van der Waerden, 1971 for algebraic background).

In chemical kinetics, the overall reaction rate is a natural choice for the variable. The resultant in terms of the reaction rate is a generalization of the reaction rate equations of conventional explicit form (Equation (1)) which are obtained for linear reaction mechanism. Generally, the resultant is a polynomial in terms of the overall reaction rate. The roots of this polynomial are the values of reaction rate corresponding to the solutions of the system of algebraic equations corresponding to the QSSA. It was shown before that this system (i.e. QSSA system) could be written in two equivalent forms: Equations (16) and (20). System (16) is obtained from system of differential equations of material balance of intermediates by replacing the time-derivatives of intermediate concentrations (for instance, surface coverages of intermediates in heterogeneous catalysis) with zero. System (20) is equivalent form of QSSA expressed in terms of rates $\bar{R}$ along the reaction paths.

We have termed the resultant of the overall reaction rate as the *kinetic polynomial*. Equation (3) is just the particular form of kinetic polynomial for the linear mechanism.

Authors founded the kinetic polynomial theory in early 1980s (Lazman and Yablonskii, 1991; Lazman et al., 1985a, 1985b, 1987a, 1987b; Yablonskii et al., 1982, 1983). It was further developed in collaboration with mathematicians Bykov and Kytmanov (Bykov et al., 1987, 1989).[7] Later, applying computer algebra methods,

[5] In heterogeneous catalysis equations (20b) express the preservation of number of active cites of particular type, for instance we have $\mathbf{L}(\mathbf{z}) = z_1 + \ldots + z_m - 1$ for catalyst with single type of active cites when each of intermediates occupy one active cite.

[6] Introduction of stoichiometric number concept and linear transformation of the "conventional" QSSA equations (16) to the equivalent system (20) was essentially the major (and, possibly, only) result of theory of steady reactions developed independently by J. Horiuti in 1950s and M. I. Temkin in 1960s.

[7] Remarkably, the development of kinetic polynomial stimulated obtaining pure mathematical results that became the "standard references" in mathematical texts (see, for instance, WWW sources as E. W. Weisstein. "Resultant". From *MathWorld* — A Wolfram Web Resource. http://mathworld.wolfram.com/Resultant.html, Multi-dimensional logarithmic residues, Encyclopaedia of Mathematics — ISBN 1402006098 Edited by Michiel Hazewinkel CWI, Amsterdam, 2002, Springer, Berlin).

first author (Lazman) developed the kinetic polynomial software (see Bykov et al., 1993, 1998; Lazman and Yablonsky, 2004).

Kinetic polynomial found important applications including parameter estimation (Lazman et al., 1987a, Yablonskii et al., 1992), analysis of kinetic model identifiability (Lazman et al., 1987b), asymptotic analysis of bifurcations in heterogeneous catalysis (Lazman et al., 1985b; Yablonsky and Lazman, 1996, 1997; Yablonsky et al., 2003), and, of course, finding all steady states of kinetic models (Lazman, 1997, 2000, 2002, 2003a, 2003b; Lazman and Yablonsky, 2004). Detailed discussions and mathematical proofs of more technical results have been presented in the paper by Lazman and Yablonskii (1991) and book by Bykov et al. (1998).

We studied the following system (let us call it the *Basic Case*) corresponding to the single-route mechanism of catalytic reaction with the single type of active sites

$$w_s(z_1,\ldots,z_n) - \nu_s\ R = 0, \qquad s = 1,\ldots,n \tag{21a}$$

$$1 - \sum_j z_j = 0, \qquad j = 1,\ldots,n \tag{21b}$$

where

$$w_s(z_1,\ldots,z_n) = f_s z^{\alpha^s} - r_s z^{\beta^s}$$

$$z^{\alpha^s} = \prod_{j=1}^{n} z_j^{\alpha_{sj}}, \qquad z^{\beta^s} = \prod_{j=1}^{n} z_j^{\beta_{sj}}$$

$$\alpha^s = (\alpha_{s1},\ldots,\alpha_{sn}), \qquad \beta^s = (\beta_{s1},\ldots,\beta_{sn})$$

We are assuming that

1. The rank of stoichiometric matrix

$$rk\Gamma = n - 1 \tag{22}$$

It follows from Equation (19) that in this case we have a single reaction route, i.e. $P = 1$.

We are also assuming that[8]

2.

$$\|\alpha^s\| = \sum_j \alpha_{sj} = \|\beta^s\| = p_s, \qquad \alpha^s \neq \beta^s \tag{23}$$

Finally, we assume that

3. $R = 0$ is not a generic root of system (21).

Horiutis numbers $\nu_1,\ldots,\nu_n$ have the following property

$$\sum_s \nu_s(\beta^s - \alpha^s) = 0 \tag{24}$$

[8] We consider the systems in which each intermediate contains the same number of active sites, e.g. AZ, BZ (Z is the catalytic site) or PtO, PtCO include only one catalyst site. This assumption simplifies the analysis. However, our results can be generalized to systems, in which surface intermediate include more that one active site or the catalyst surface is characterized by more than one type of active sites.

Up to the scaling, they are co-factors Δ_s of elements of any column of stoichiometric matrix Γ (see, for instance, Bykov et al., 1998; Lazman and Yablonskii, 1991).[9] We can always assign the directions of elementary reactions so that all stoichiometric coefficients are non-negative and this will be assumed later.

At these assumptions, system (21) has the resultant with respect to reaction rate R (Bykov et al., 1998; Lazman and Yablonskii, 1991). This means that there exists the polynomial

$$\mathrm{Res}(R) = B_L R^L + \ldots + B_1 R + B_0 \tag{25}$$

vanishing if R is the root of system (21). Vanishing of the resultant is the necessary (and in some cases sufficient) condition of algebraic system solvability (see Gel'fand et al., 1994; van der Waerden, 1971).

The right-hand side of Equation (25) is the kinetic polynomial. Assuming $v_1 \neq 0$, we can define the resultant with respect to R as

$$\mathrm{Res}(R) = \prod_j \left(f_1 z_{(j)}^{\alpha^1} - r_1 z_{(j)}^{\beta^1} - v_1 R\right) \tag{26}$$

where $z_{(j)}$ are the roots of the system

$$w_s(z_1, \ldots, z_n) - v_s\ R = 0, \qquad s = 1, \ldots, n,\ s \neq k \tag{27a}$$

$$1 - \sum_j z_j = 0, \qquad j = 1, \ldots, n \tag{27b}$$

at fixed R and $k = 1$ (see Bykov et al., 1987, 1998; Lazman and Yablonskii, 1991 for a background and rigorous proof).

In our assumptions, system (27) has the finite number of roots (by Lemma 14.2 in Bykov et al., 1998), so that the product in Equation (26) is well defined. We can interpret formula (26) as a corollary of Poisson formula for the classic resultant of homogeneous system of forms (i.e. the *Macaulay* (or *Classic*) *resultant*, see Gel'fand et al., 1994). Moreover, the product Res(R) in Equation (26) is a polynomial of R-variable and it is a rational function of kinetic parameters f_s and r_s (see a book by Bykov et al., 1998, Chapter 14). It is the same as the classic resultant (which is an irreducible polynomial (Macaulay, 1916; van der Waerden, 1971) up to constant in R multiplier.[10] In many cases, finding resultant allows to solve the system (21) for all variables.[11]

[9] Note, that property (22) guarantees that for some i we have non-zero v_i.

[10] It is possible to prove that our system (21) has a *sparse* (or *toric*) resultant in sense of Gel'fand et al. (1994). Non-trivial moment here is the importance of the stoichiometric condition (24).

[11] Certain properties of the resultant derivatives by parameters allow finding all the coordinates of the solution (see a book by Gel'fand et al., 1994).

2.2.2 The cyclic characteristic and the thermodynamic consistency

At the thermodynamic equilibrium, all steps of the detailed mechanism should be to be at the equilibrium, i.e.

$$w_s(\mathbf{z}) = 0, \qquad s = 1, \ldots, n \tag{28}$$

Thus, we have $R = 0$ at the equilibrium. Equation (28) together with the linear balance Equation (27b) form an overdetermined system of $z_1, \ldots z_n$. As we assume that $R = 0$ is not a generic root of the system, the only possibility for system (27b)+(28) to have a solution is satisfying certain constraints on kinetic parameters. We can rewrite Equation (28) as

$$f_s z^{\alpha^s} = r_s z^{\beta^s}, \qquad s = 1, \ldots, n \tag{29}$$

After raising both sides of equation to the power of ν_s, multiplying transformed equations and applying condition (24), we conclude that the necessary condition for system (27b)+(28) to have a solution with non-zero coordinates[12] is

$$\prod_{s=1}^{n} (f_s/r_s)^{\nu_s} = 1 \tag{30}$$

Equilibrium constraint (30) can be expressed as

$$K_{\text{eq}} = \frac{f^-(\overleftarrow{c})}{f^+(\vec{c})} \tag{31}$$

where

$$K_{\text{eq}} = \prod_{s=1}^{n} (k_s/k_{-s})^{\nu_s} \tag{32}$$

is the equilibrium constant of the net reaction (k_s and k_{-s} are kinetic constants), $f^-(\overleftarrow{c})$ the product of concentrations of the net reaction products and $f^+(\vec{c})$ the product of concentrations of the net reaction reactants, respectively. Naturally, condition (31) has exactly the same form as the equilibrium condition for the linear reaction mechanism (see Equation (8)). Both conditions (8) and (31) correspond to the MAL for the net reaction. The only difference is that for non-linear mechanism, ν_s in Equation (32) can be different from values 1 (and 0) in corresponding formulas for the linear mechanism (compare with Equations (1b) and (1c)).

It is intuitively clear that the structure of resultant (25) should reflect the equilibrium constraint (30). Let us call the binomial

$$C = \prod_{s=1}^{n} f_s^{\nu_s} - \prod_{s=1}^{n} r_s^{\nu_s} \tag{33}$$

the *cyclic characteristic*. Since $R = 0$ is not (generic) zero of system (21), the constant term B_0 of kinetic polynomial is not identically zero. We have proved

[12] Owing to assumption #3, system (27b)+(28) cannot have the solution with zero z_s for generic values of reaction weights.

(Bykov et al., 1998; Lazman and Yablonskii, 1991; Lazman et al., 1985a) that the constant term B_0 in Equation (25) is the non-zero multiple of the cyclic characteristic (33).

More precisely, the following proposition is valid (see Theorem 14.1 in a monograph by Bykov et al., 1998). We are assuming that none of rational fractions $B_0, \ldots, B_L$ in Equation (25) is reducible. We write $B_0 \sim B$ if $B_0/B \neq 0$ for generic $f_1, r_1, \ldots, f_s, r_s \neq 0$.

Proposition 1. In the assumptions of the Basic Case

$$B_0 \sim C^p \tag{34}$$

The Horiuti numbers defining the cyclic characteristic in Equation (34) are relatively prime i.e. $\mathrm{GCD}(\nu_1, \ldots, \nu_n) = 1$. The exponent p in Equation (34) is the natural number. If we assume additionally that

$$\nu_s \neq 0, \qquad s = 1, \ldots, n \tag{35}$$

then,

$$p = \mathrm{GCD}(\Delta_1, \ldots, \Delta_n) \tag{36}$$

(see[13] Corollary 14.1 in the monograph by Bykov et al., 1998). Property (34) ensures the thermodynamic consistence of kinetic polynomial: if $R = 0$, the cyclic characteristic (33) should vanish.[14]

Cyclic characteristic has following property (see Bykov et al., 1998, Corollary 14.2).

Proposition 2. If $p > 1$ and property (35) is valid, then the cyclic characteristic C is contained in the coefficients B_s of kinetic polynomial (25) with an exponent equal at least to $p - s$, $s = 0, 1, \ldots, p - 1$, but it is not contained in the leading coefficient $\mathrm{B_L}$.

This means that we could have a situation when more than one root of kinetic polynomial vanishes at the thermodynamic equilibrium. However, only one of these roots would be feasible.

2.2.3 Coefficients of the kinetic polynomial

For sufficiently small $|R|$ the following formula is valid (Bykov et al., 1998; Lazman and Yablonskii, 1991)

$$\frac{d \ln \mathrm{Res}(R)}{dR} = -\sum_{k=1}^{n} \nu_k \sum_{j_k=1}^{M_k} 1/(w_k(\mathbf{z}_{j_k}(R)) - \nu_k R) \tag{37}$$

[13] GCD in Equation (36) is the *Greatest Common Divisor*, i.e. the largest positive integer that divides integer numbers in ().

[14] We have found recently the topological interpretation of property (34). The stoichiometric constraints (24) can be interpreted in terms of the topological object, the *circuit*. Existence of the circuit "explains" the appearance of the cyclic characteristic in the constant term of kinetic polynomial. Thus, we can say that in some sense the correspondence between the detailed mechanism and thermodynamics is governed by pure topology.

where $\mathbf{z}_{j_k}(R)$ is zero of the system (27). The latter has a finite number of zeros in the assumptions of the *Basic case* (see Bykov et al., 1998; Lazman and Yablonskii, 1991). Let

$$d_k = \frac{d^k \ln \mathrm{Res}(R)}{dR^k}\bigg|_{R=0} \tag{38}$$

Then,

$$B_k = \left(\frac{1}{k!}\right) \sum_{j=1}^{k} B_{k-j} d_j/(j-1)!, \qquad k = 1, \ldots, L \tag{39}$$

It follows from Equations (37)–(39) that the coefficients of the kinetic polynomial are symmetric functions of zeroes of systems (27) for $k = 1, \ldots, n$ at $R = 0$, i.e. *equilibrium subsystems.*

Formula for the first coefficient of the kinetic polynomial is

$$\frac{B_1}{B_0} = -\sum_{k=1}^{n} \nu_k \sum_{j_k=1}^{M_k} 1/w_k(\mathbf{z}_{j_k}(0)) \tag{40}$$

The symbolic algorithm based on multi-dimensional residues theory their implementation is described in Bykov et al., 1998.[15]

2.2.4 Kinetic polynomial as the generalized overall reaction rate equation

2.2.4.1 The root count. *(The reader interested first in deriving explicit reaction rate equation may omit this section and start to read Section 2.2.4.2)*

Vanishing of the resultant Res(R) is a necessary and sufficient condition for R to be the coordinate of the zero of system (21). Thus, equation

$$\mathrm{Res}(R) = 0 \tag{41}$$

defines the overall reaction rate calculated in the assumptions of QSSA. The Res(R) is a polynomial of degree L in R. Generally, we have

$$L \leq N \tag{42}$$

where N is the number of all isolated zeros (i.e. such $z_1, \ldots, z_n,\ W \in C^{n+1}$ that solve system (21)). It is important to have estimates for number of roots in various domains.

2.2.4.1.1 Bezout root count. Let

$$p_s = \max\left(\|\alpha^s\|, \|\beta^s\|\right)$$

be the *reaction order* of stage sth stage. Let us call the reaction with index μ, such that $p_\mu = \min(p_1, \ldots, p_n)$ and $\nu_\mu \neq 0$, the *minor reaction.*

[15] Note that standard procedures, such as Groebner Bases, that are built into the modern computer algebra systems (e.g. Maple) cannot handle our systems efficiently (see monograph by Bykov et al., 1998).

Proposition.

$$N \leq L_\mu = \prod_{s \neq \mu}^{n} p_s \tag{43}$$

Proof. We can write the system (21) in the *reduced* form as

$$\begin{aligned} \nu_\mu w_i(\mathbf{z}) - \nu_i w_\mu(\mathbf{z}) = 0, &\qquad i \neq \mu \\ 1 - \sum_i z_i = 0 & \\ R - \frac{1}{\nu_\mu} w_\mu(\mathbf{z}) = 0 & \end{aligned} \tag{44}$$

Then the estimate (43) follows from the classic Bezout theorem (see van der Waerden, 1971) applied to the system (44). Estimate (43) shows that overall number of zeroes of system (21) does not exceed the product of reactions order of all reactions except the minor one.[16]

2.2.4.1.2 *Number of interior roots.* System (21) may have both *interior* ($z_i \neq 0,\ i = 1, \ldots, n$) and *boundary* roots ($\exists i\colon z_i = 0$). As a rule, boundary roots correspond to the zero reaction rate. Some classes of mechanisms can be free of boundary roots at all. For instance, system (21) corresponding to the mechanism of the *Basic case* type, satisfying the condition (35) does not have boundary roots. All interior roots belong to the algebraic tore. Bernstain theorem (see Gel'fand et al., 1994) can be applied to estimate the number of roots in this case.

2.2.4.1.3 *Number of feasible roots.* Usually we are interested mostly in the feasible roots of the system (21). The feasible roots of system (21) are real roots satisfying constraints

$$z_i \geq 0 \tag{45}$$

Thus, for system (21) the feasible domain is $S_n \times \mathcal{R}$ where S_n is n-dimensional standard simplex. Generally, the number of feasible roots depends on the parameters of system (21) (i.e. the reaction weights). The domains in the space of parameters with different number of feasible roots are the semi-algebraic sets (i.e. the sets, corresponding to some system of polynomial equations and inequalities). Analysis of the semi-algebraic domains is a difficult problem in computer algebra (see a monograph by Bykov et al., 1998). Asymptotic analysis (the *critical simplification*) allows obtaining the simple description of the domains of multiplicity for important mechanisms (Yablonsky et al., 2003).

[16] Estimate (43) could be generalized to the multi-route mechanism — we can drop at least P reactions from Bezout root count.

2.2.4.2 *Kinetic polynomial for model mechanisms*

2.2.4.2.1 Eley–Rideal mechanism. Kinetic polynomial corresponding to the mechanism

$$\begin{array}{lll} (1) & A_2 + 2Z \Leftrightarrow 2AZ & [1] \\ (2) & AZ + B \Leftrightarrow AB + Z & [2] \end{array} \tag{46}$$

of heterogeneous catalytic reaction

$$A_2 + 2B \Leftrightarrow 2AB \tag{47}$$

is

$$4(f_1 - r_1)R^2 - (4r_1r_2 + 4f_1f_2 + (f_2 + r_2)^2)R + f_1f_2^2 - r_1r_2^2 = 0 \tag{48}$$

where $f_1 = k_1c_{A_2}$, $r_1 = k_{-1}$, $f_2 = k_2c_B$, and $r_2 = k_{-2}c_{AB}$. Here k_1, k_{-1}, k_2 and k_{-2} are reaction constants of corresponding elementary reactions of mechanism (46) and c_{A_2}, c_B and c_{AB} are concentrations of corresponding reactants and products. Note that coefficients of the kinetic polynomial (48) are polynomials in reaction weights (sometimes named reaction frequencies) f_1, r_1, f_2 and r_2.[17] Obviously, these coefficients are *not* constants: they depend on reaction weights that, in turn, depend on temperature and concentrations of reactants and products of the net reaction. Thus, coefficients of kinetic polynomial are functions of temperature and concentrations of net reaction reagents.

If we add up the chemical equations of mechanism (46) multiplied by corresponding stoichiometric numbers (i.e. 1 and 2), we obtain the equation of net reaction (47). Note how the constant term of Equation (48) and exponents in its terms correspond to the Proposition 1 (and property (36)). The only feasible root of polynomial (48) is

$$R = \frac{2(f_1f_2^2 - r_1r_2^2)}{4r_1r_2 + 4f_1f_2 + (f_2 + r_2)^2 + (f_2 + r_2)\sqrt{(f_2 + r_2)^2 + 8(f_1f_2 + r_1r_2) + 16f_1r_1}} \tag{49}$$

2.2.4.2.2 Langmuir–Hinshelwood mechanism. More interesting is the kinetic polynomial corresponding to the mechanism

$$\begin{array}{lll} (1) & A_2 + 2Z \Leftrightarrow 2AZ & [1] \\ (2) & B + Z \Leftrightarrow BZ & [2] \\ (3) & AZ + BZ \Leftrightarrow AB + 2Z & [3] \end{array} \tag{50}$$

of reaction (47). This mechanism is the standard "building block" of the models describing the critical phenomena in heterogeneous catalytic reactions (for instance, CO oxidation) (Bykov and Yablonskii, 1977a, b; Yablonskii and Lazman, 1996). Kinetic polynomial corresponding to the mechanism (50) is

$$B_4R^4 + B_3R^3 + B_2R^2 + B_1R + B_0 = 0 \tag{51}$$

[17] f_i are *reaction weights* (or *reaction frequencies*) of forward reactions and r_i reaction weights of reverse reactions.

where

$$B_0 := -r_2^2 r_1 (f_1 f_2^2 f_3^2 - r_2^2 r_3^2 r_1)$$

$$\begin{aligned}
B_1 :=\ & 8r_2^3 r_3^2 r_1^2 + 2r_2^4 r_1 r_3^2 + 4r_2^4 r_1^2 r_3 + f_2^2 r_2^2 f_3^2 r_1 + 2f_2^2 r_2^2 f_3 r_3 r_1 \\
& + 4f_2^2 r_2^2 r_1^2 r_3 + f_2^4 f_3^2 r_1 + 2f_2 r_2^3 f_3 r_3 r_1 + 8f_2 r_2^3 r_1^2 r_3 + 2f_2^3 r_2 f_3^2 r_1 \\
& - f_1 f_2^2 r_2^2 f_3^2 + 8f_1 f_2^2 r_2^2 r_1 f_3 + 4f_1 r_2^3 f_3 r_3 r_1 + 4f_1 r_2^4 r_3 r_1 - 4f_1 f_2^2 r_2 f_3^2 r_1 \\
& + 8f_1 f_2 r_2^3 f_3 r_1 + 4f_1 f_2 r_2^2 f_3^2 r_1
\end{aligned}$$

$$\begin{aligned}
B_2 :=\ & -4r_2^4 r_3 r_1 + 4f_1^2 r_2^2 f_3^2 + 24f_2^2 r_2^2 r_1^2 + 8f_1^2 f_3 r_2^3 - 4f_2^4 r_1 f_3 - 8f_1 r_2^4 r_1 \\
& + f_2^4 f_3^2 + 16f_2^3 r_2 r_1^2 + 16r_2^3 r_1^2 r_3 + f_2^2 r_2^2 f_3^2 + 16f_2 r_2^3 r_1^2 + 4f_1 r_2^4 r_3 \\
& - 4f_2^3 f_3^2 r_1 + 8r_2^3 r_1 r_3^2 + 2f_2^3 r_2 f_3^2 + 24r_2^2 r_3^2 r_1^2 + 4f_1^2 r_2^4 + 4r_2^4 r_1^2 + r_2^4 r_3^2 \\
& + 4f_2^4 r_1^2 - 16f_1 f_2 r_2^3 r_1 + 4f_1 f_2^2 r_2^2 f_3 - 4f_1 f_2^2 r_2 f_3^2 - 4f_1 f_2^2 f_3^2 r_1 \\
& + 4f_1 f_2 f_3 r_2^3 + 4f_1 r_2^3 f_3 r_3 - 8f_1 r_2^3 r_1 f_3 + 16f_1 r_2^3 r_3 r_1 - 4f_1 r_2^2 f_3^2 r_1 \\
& + 2f_2 r_2^3 f_3 r_3 - 4f_2 r_2^3 r_1 f_3 - 8f_2 r_2^3 r_3 r_1 + 32f_2 r_2^2 r_1^2 r_3 - 8f_1 f_2^2 r_2^2 r_1 \\
& + 2f_2^2 r_2^2 f_3 r_3 - 12f_2^2 r_2^2 r_1 f_3 - 4f_2 r_2^2 f_3^2 r_1 - 8f_2^2 r_2 f_3^2 r_1 + 16f_2^2 r_2 r_1^2 r_3 \\
& - 4f_2^2 r_2^2 r_3 r_1 - 8r_2^3 f_3 r_3 r_1 - 12f_2^3 r_2 r_1 f_3 - 8f_2 r_2^2 f_3 r_3 r_1 + 16f_1 f_2 r_2^2 r_1 f_3 \\
& + 24f_1 r_2^2 f_3 r_3 r_1 + 24f_1 f_2^2 r_2 r_1 f_3 + 16f_1 f_2 r_2 f_3^2 r_1
\end{aligned}$$

$$\begin{aligned}
B_3 :=\ & 16f_1^2 r_2^2 f_3 + 16f_1^2 f_3^2 r_2 - 4f_1 r_2^2 f_3^2 - 8f_1 f_2^2 f_3^2 + 32r_3^2 r_1^2 r_2 + 8r_2^2 r_3^2 r_1 \\
& + 16r_2^2 r_3 r_1^2 + 4r_2^2 r_1 f_3^2 - 16r_2^2 r_3 r_1 f_3 + 16f_2^2 r_3 r_1^2 + 4f_2^2 r_1 f_3^2 + 8f_2 r_2 r_1 f_3^2 \\
& - 24f_2 r_2 r_3 r_1 f_3 + 32f_2 r_2 r_3 r_1^2 - 8f_2^2 r_3 r_1 f_3 + 48f_1 r_2 r_3 r_1 f_3 + 16f_1 r_2^2 r_3 r_1 \\
& - 16f_1 r_2 r_1 f_3^2 - 16f_1 r_2^2 r_1 f_3 + 8f_1 r_2^2 f_3 r_3 + 16f_1 f_2^2 r_1 f_3 + 16f_1 f_2 r_1 f_3^2 \\
& - 8f_1 f_2 f_3^2 r_2
\end{aligned}$$

$$B_4 := 16f_1^2 f_3^2 - 16f_1 r_1 f_3^2 + 16r_3^2 r_1^2 + 32f_1 r_3 r_1 f_3$$

here $f_1 = k_1 c_{A_2}$, $r_1 = k_{-1}$, $f_2 = k_2 c_B$, $r_2 = k_{-2}$, $f_3 = k_3$, and $r_3 = k_{-3} c_{AB}$. The 4th degree polynomial in R on the left-hand side of Equation (51) may have 1 or 3 feasible roots. Figure 1 shows the dependence of all four roots of this polynomial on the parameter f_2.[18] We can interpret the overall reaction rate R as a multi-valued algebraic function of f_2. Only one branch (denoted as R1) of this function passes through the point of thermodynamic equilibrium (see Figure 2). Far from equilibrium, however, this branch disappears in the bifurcation point where the branch R1 merges with the branch R2. Beyond the bifurcation point both branches become infeasible. They correspond to the pair of complex conjugated roots. Similar metamorphosis happens to the branches R2 and R3; the branch R4

[18] Up to scale, this is the dependence of overall reaction rate on concentration C_B in the assumption of constant temperature and concentrations c_{A_2} and C_{AB}. All figures in this chapter illustrate certain qualitative features of kinetic behavior, i.e. rate-limitation, vicinity of equilibrium, steady-state multiplicity, etc. Parameter values are selected to illustrate these qualitative features. Certainly these features could be illustrated with "realistic" kinetic parameters.

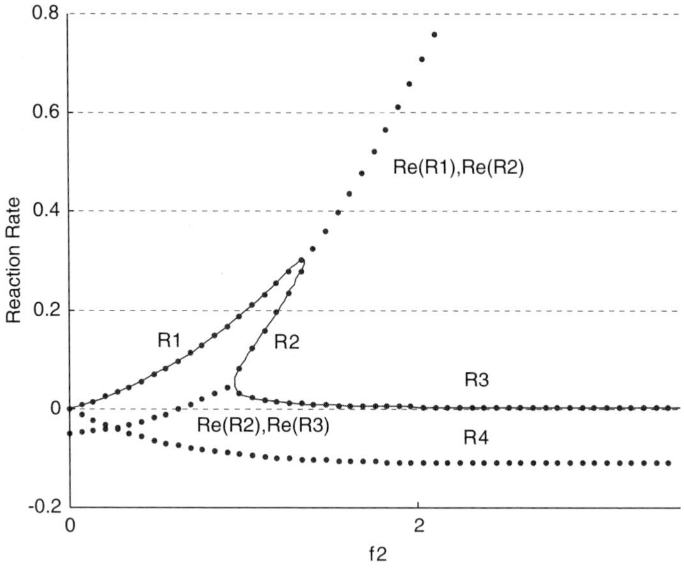

Figure 1 Dependence of overall reaction rate on the parameter f_2 (LH mechanism). Branches R1, R2, R3 and R4 represent the roots of kinetic polynomial. Solid line indicates feasible steady states. Branches Re(R1), Re(R2) and Re(R3) correspond to the real parts of conjugated complex roots of kinetic polynomial. Parameter values: $f_1 = 1.4$, $r_1 = 0.1$, $r_2 = 0.1$, $f_3 = 15$ and $r_3 = 2$.

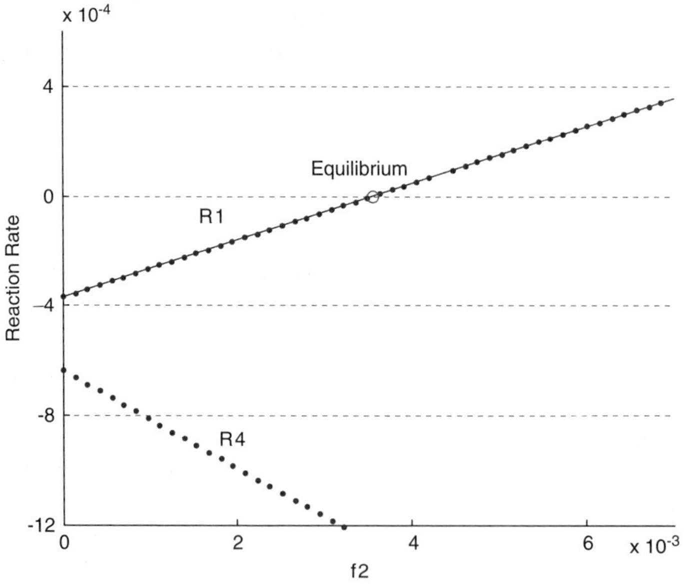

Figure 2 Overall reaction rate dependence (see Figure 1) in the vicinity of thermodynamic equilibrium.

is always infeasible. All these branches can be described implicitly by the single equation (51) in single variable R as the roots of *kinetic polynomial.*

3. REACTION RATE APPROXIMATIONS

Although it is possible to solve in radicals the 4th order polynomial corresponding to the LH mode, we will not even try to reproduce corresponding equations due to their intractability. Higher order polynomials would present even more difficulties for analytic treatment. Although analytic expressions for roots of algebraic equations are still possible (in terms of theta-functions; Mumford, 1984) one can imagine that tractability issue becomes more and more difficult for higher degree polynomials.

We consider below the possibilities for simplification of overall reaction rate equations and introduce the main result of this chapter — *the hypergeometric series for reaction rate.*

3.1 Classic approximations

3.1.1 Rate-limiting step

Let our mechanism contains a reaction step with index k such that

$$f_k, r_k \ll f_i, r_i, \qquad i \neq k$$

This step is called the rate-limiting step. Reaction rate approximation can be found in the form of power series (by f_k or r_k) (see Lazman and Yablonskii, 1988, 1991).

The first term of this series is the rate of rate-limiting step calculated at the equilibrium of the rest of reactions. This first term approximation is widely applied in heterogeneous catalysis. We have derived the following explicit formula (in the assumptions of the *Basic case*; see Lazman and Yablonskii, 1988).

$$R = \frac{f_k}{\nu_k} \frac{\left(1 - \prod_{j=1}^{n} (1/\kappa_j)^{\nu_j/\nu_k}\right)}{\left(\sum_{i=1}^{n} \prod_{j \neq k} (1/\kappa_j)^{\Delta_{ji}^k/(p_k \Delta_k)}\right)^{p_k}} \tag{52}$$

where $\kappa_j = f_j/r_j$, $p_k = ||\alpha^k|| = ||\beta^k||$ is the reaction order of the step k, Δ_k are co-factors of the element of kth row of matrix — Γ and $\Delta_{ji}^k$are co-factors of elements with indices j,i of matrix,

$$\Gamma_k' = \begin{pmatrix} \alpha_{11} - \beta_{11} & . & . & . & \alpha_{1n} - \beta_{1n} \\ . & . & . & . & . \\ \alpha_{k1} & . & . & . & \alpha_{kn} \\ . & . & . & . & . \\ \alpha_{n1} - \beta_{n1} & . & . & . & \alpha_{nn} - \beta_{nn} \end{pmatrix}$$

where

$$\prod_{j=1}^{n}\left(\frac{1}{\kappa_j}\right)^{\nu_j}=\left(\frac{1}{K_{\rm eq}}\right)\frac{F(c_{\rm pr})}{F(c_{\rm r})}$$

$K_{\rm eq}$ is the equilibrium constant, $F(c_{\rm pr})$ and $F(c_{\rm r})$ are functions of concentrations of products and reactants, respectively, which relate to the net reaction.

Example: LH mechanism, second step is the rate-limiting one
We have $\Delta_2=\nu_2=2, p_2=1$,

$$\Gamma_2'=\begin{pmatrix}2 & -2 & 0\\ 1 & 0 & 0\\ -2 & 1 & 1\end{pmatrix},\quad \Delta_{11}^2=0,\quad \Delta_{31}^2=0,\quad \Delta_{12}^2=-1,$$

$$\Delta_{32}^2=0,\quad \Delta_{13}^2=1,\quad \Delta_{33}^2=2$$

Substituting to Equation (52), we obtain

$$R=\frac{r_2}{2}\frac{\sqrt{\kappa_1\kappa_2\kappa_3}-1}{1+\sqrt{\kappa_1\kappa_3}+\kappa_1\kappa_3}$$

3.1.2 Vicinity of thermodynamic equilibrium

The cyclic characteristic C is small in the vicinity of thermodynamic equilibrium. We can find the overall reaction rate approximation in the vicinity of equilibrium either directly from kinetic polynomial or by expanding the reaction rate in power series by the small parameter C. The explicit expression for the first term is presented by Lazman and Yablonskii (1988, 1991). It is written as follows:

$$R=\frac{\prod\limits_{j=1}^{n}\kappa_j^{\nu_j}-1}{\sum\limits_{k=1}^{n}\nu_k^2 f_k^{-1}\left(\sum\limits_{i=1}^{n}\prod\limits_{j\neq k}^{n}(1/\kappa_j)^{\Delta_{ji}^k/(p_k\Delta_k)}\right)^{p_k}\Bigg|_{\rm eq}} \qquad (53)$$

where $\prod_{j=1}^{n}\kappa_j^{\nu_j}=K_{\rm eq}F(c_{\rm r})/F(c_{\rm pr})$, $K_{\rm eq}$ is the equilibrium constant, $F(c_{\rm pr})$ and $F(c_{\rm r})$ functions of concentrations of products and reactants, respectively, which relate to the net reaction.

Note that denominator of Equation (53) should be calculated at equilibrium conditions. Formula (53) corresponds to the assumption of linear relation between the reaction rate and affinity of each reaction step (see Lazman and Yablonskii (1988, 1991) for detailed discussion). We can write

Equation (53) as

$$R = \frac{\prod_{j=1}^{n} \kappa_j^{\nu_j} - 1}{\sum_{k=1}^{n} \nu_k^2 / w_k|_{eq}} \tag{53a}$$

where $w_k|_{eq}$ is the reaction rate of forward (or reverse) reaction calculated at equilibrium conditions. In this form, this formula was known (Nacamura, 1958), however it was not the explicit expression of overall reaction rate in terms of parameters of reaction mechanism presented in this chapter, see Equation (53).

This linear affinity approximation does not always correspond to the linear approximation of kinetic polynomial $R \approx -(B_0)/(B_1)$. This happens only when degree p of cyclic characteristic in Proposition 1 (see Equation (34)) is one.[19] If $p>1$, linear approximation of the kinetic polynomial does not correspond to the linear affinity relation (53). Equation $B_0 + B_1R + \cdots + B_pR^p \approx 0$ is correct approximation in this case (see Lazman and Yablonskii, 1991).

3.2 Overall reaction rate as a hypergeometric series

The goal of this chapter, generally, is to present an analytical expression of the reaction rate branch which goes through the equilibrium point (*"thermodynamical branch" of the overall reaction rate*) with no classical assumption discussed earlier, i.e. "vicinity of the equilibrium" and "limiting step". This result is a logical continuation of the previously developed theory of the kinetic polynomial: that is why we had to explain main results of this theory, which are still not widely known. Posing this problem, i.e. deriving such analytical expression, and moreover its solving became possible only on the basis of last achievements of the algebraic theory. Some next sections will be devoted to the explanation of contemporary theoretical situation and its applications to the problems of complex kinetics.

3.2.1 Roots of algebraic equations: The hypergeometric series

As well known, we cannot solve general algebraic equation (with complex coefficients)

$$a_n x^n + \cdots + a_1 x + a_0 = 0 \tag{54}$$

in terms of radicals if $n>4$ (Abel-Ruffini theorem, Galois theory, see a book by van der Waerden, 1971). For a long time mathematicians tried to find an analytic (not necessarily algebraic) solution to Equation (54). The situation was more or less clear already in the 19th century. First successes were summarized in the Felix Klein's (1888) book. A root of the quintic equation has been presented by Klein in terms of elliptic modular functions. At the same time, Klein mentioned

[19] In this case, we can derive (53) from linear approximation of kinetic polynomial (Lazman and Yablonskii, 1991). See also Appendix 3, (A3.11) for details.

that solution of his equations could also be reduced to the studying solutions of some differential equations in terms of *hypergeometric* functions.[20] Well known to mathematicians, these facts are relatively obscure outside pure mathematics (most probably, due to absence of "standard" modern texts). Moreover, certain aspects of the problem are belonging to the relatively hot topic even in modern mathematics. Our major sources of mathematical facts are relatively recent papers by Sturmfels (2000) and Passare and Tsikh (2004). We will use the Sturmfels' results as a source of our formulas and we will use the very recent Passare and Tsikh's paper as a source for convergence conditions.

Sturmfels (2000) presented the complete set of equation (54) solutions in terms of the A-hypergeometric functions introduced before by Gel'fand et al. (1994). These functions are associated with the *Newton polytope N* of the polynomial on the left-hand side of Equation (54). The latter is the convex hull of the set A of the exponents of all monomials considered as integer lattice points in the corresponding real vector space. In the case of polynomial in single variable, the set A is the configuration of $n+1$ points 0, 1, ..., n on the affine line. The Newton polytope here is just the line segment $[0,n]$. There are 2^{n-1} distinct complete sets of solutions of Equation (54) in terms of A-hypergeometric series. Each of these solutions corresponds to the particular *triangulation* of the set A that in our case is simply some subdivision

$$[0, i_1], [i_1, i_2], \ldots, [i_s, n], \qquad i_1 < i_2 < \ldots < i_s, \qquad i_k \in Z^+$$

of the segment $[0,n]$. The finest subdivision divides A into n segments of unit length whereas the coarsest one is just a single segment $[0,n]$. Sturmfels (2000) presented the general formula for the series solutions of Equation (54) and proved that there is a domain in the space of parameters of equation (54) where all n series for n roots of equation converge. For the finest subdivision $[0, 1], [1, 2], \ldots, [n-1, n]$, these series solution are

$$X_j = -\left[\frac{a_{j-1}}{a_j}\right] + \left[\frac{a_{j-2}}{a_{j-1}}\right], \qquad j = 1, 2, \ldots, n \tag{55}$$

where $\left[\frac{a_{-1}}{a_0}\right] = 0$, and[21]

$$\left[\frac{a_{j-1}}{a_j}\right] = \sum_{i_0 \geq 0, \ldots [j-1], [j], \ldots i_n \geq 0} \frac{(-1)^{i_j}}{i_{j-1}+1} \binom{i_j}{i_1 \ldots i_{j-1} i_{j+1} \ldots i_n} \frac{a_0^{i_0} a_1^{i_1} \ldots a_{j-2}^{i_{j-2}} a_{j-1}^{i_{j-1}+1} a_{j+1}^{i_{j+1}} \ldots a_n^{i_n}}{a_j^{i_j+1}} \tag{56}$$

where $i_0, i_1, \ldots, i_n$ are non-negative integers satisfying the relations

$$\begin{aligned} i_0 + i_1 + \cdots + i_{j-1} - i_j + i_{j+1} + \cdots + i_n &= 0 \\ i_1 + 2i_2 + \cdots + (j-1)i_{j-1} - ji_j + (j+1)i_{j+1} + \cdots + ni_n &= 0 \end{aligned} \tag{57}$$

Note that series (56) has integer coefficients.

[20] Klein considered this approach too cumbersome. Note that hypergeometric functions were applied to problem (54) as early as in 18th century.

[21] Here and below [] indicates that the corresponding index is omitted (see Equation (56)).

Recently Passare and Tsikh (2004) provided the detailed description of the domains of convergence of multi-dimensional hypergeometric series representing the roots of algebraic equations. They have relied on results of Birkeland obtained in 1920s. Birkeland found the Taylor series solutions to algebraic equations of the type

$$b_0 + b_1 y + \cdots + y^p + \cdots + y^q + \cdots + b_{n-1} y^{n-1} + b_n y^n = 0 \tag{58}$$

where p and q are two integers satisfying $0 \le p < q \le n$.

We can always reduce the problem (54) to the *dehomogenized* problem (58). Let

$$\begin{aligned} b_i &= \lambda_0 \lambda_1^i a_i, \quad & i = 0, \ldots, n \\ y &= x/\lambda_1, \quad & \lambda_0, \lambda_1 = C\setminus\{0\} \end{aligned} \tag{59}$$

To satisfy Equation (54) we can set $\lambda_1 = (a_p/a_q)^{1/(q-p)}$ and $\lambda_0 = a_p^{-1}\lambda_1^{-p}$ just substitute expressions (59) into Equation (58). There are exactly $(q-p)$ possible choices for parameters λ_0 and λ_1. The following Birkeland formula expresses the $(q-p)$ zeroes of Equation (58) as the Taylor series by parameters $b_0, \ldots, [p], \ldots, [q], \ldots, b_n$

$$Y_i = \sum_{k \in N^{n-1}} \frac{\varepsilon_i^{-\langle \beta_q, k\rangle + 1}}{(q-p)k!} \frac{\Gamma\left(\left(-\langle \beta_q, k\rangle + 1\right)/(q-p)\right)}{\Gamma\left(1 + \left(\langle \beta_p, k\rangle + 1\right)/(q-p)\right)} b_0^{k_0} b_1^{k_1}, \ldots, [p], \ldots, [q], \ldots, b_n^{k_n} \tag{60}$$

where ε_i is any of the radicals $(-1)^{1/(q-p)}$, $i = 1, \ldots, (q-p)$, $k! = k_0! k_1!, \ldots, [p], \ldots, [q], \ldots, k_n!$, vectors β_p and β_q are corresponding rows of $(n+1) \times (n-1)$ matrix $B_{pq} = (\beta_\mu^v)$ which column vectors β^v are defined as $(q-p)e_v + (v-q)e_p + (p-v)e_q, v \ne p, q$ and $e_0, \ldots, e_n$ are the standard basis vectors in $\mathcal{R}^{n+1}$.

Note that both Equations (56) and (60) result in the same series for the root of kinetic polynomial corresponding the "thermodynamic branch" (see Appendix 4 for the proof).

3.2.2 Examples

We have applied the computer algebra (Maple) to generate the partial sums corresponding to the "brackets" (55).

3.2.2.1 Eley–Rideal mechanism. Kinetic polynomial here is quadratic in R (see Equation (48)). There is only one feasible solution (49) here. The feasible branch should vanish at the thermodynamic equilibrium. Thus, the only candidate for the feasible branch expansion is $R = -[B_0/B_1]$ because the second branch expansion is $R' = -B_2/B_1 + [B_0/B_1]$ and it does not vanish at equilibrium. First terms of series for reaction rate generated by formula (55) at $j = 1$ are

$$\begin{aligned} R = & -16{,}796\frac{B_0^{11}B_2^{10}}{B_1^{21}} - \frac{4{,}862B_0^{10}B_2^9}{B_1^{19}} - \frac{1{,}430B_0^9B_2^8}{B_1^{17}} - \frac{429B_0^8B_2^7}{B_1^{15}} - \frac{132B_0^7B_2^6}{B_1^{13}} \\ & - \frac{42B_0^6B_2^5}{B_1^{11}} - \frac{14B_0^5B_2^4}{B_1^9} - \frac{5B_0^4B_2^3}{B_1^7} - \frac{2B_0^3B_2^2}{B_1^5} - \frac{B_0^2B_2}{B_1^3} - \frac{B_0}{B_1} \end{aligned} \tag{61}$$

Birkeland's formula (60) applied to the dehomogenized polynomial $(1+y+b_2y^2)$ gives the following series (see Passare and Tsikh, 2004).

$$y = -\sum_{k=0}^{\infty} \frac{(2k)!}{(k+1)!} \frac{b_2^k}{k!} \equiv \frac{-1+\sqrt{1-4b_2}}{2b_2} \tag{62}$$

In our case, $n = 2$, $p = 0$, $q = 1$ and relations (59) give

$$R = B_0/B_1y, \qquad b_2 = B_0B_2/B_1^2 \tag{63}$$

It is easy to test that Equations (62) and (63) results in series (61).

Figure 3 compares the exact solution (49) and its approximations obtained by truncating series (61) at first, second, etc. terms. It shows that the even first term $(-B_0/B_1)$ provides reasonable approximation in the finite neighborhood of equilibrium. Addition of higher order terms increases the domain of close approximation. This is not surprising because the condition of convergence of this series is $b_2<1/4$ i.e. series (61) converges if the root is real which is always the case (see Equation (49)) for feasible values of parameters.

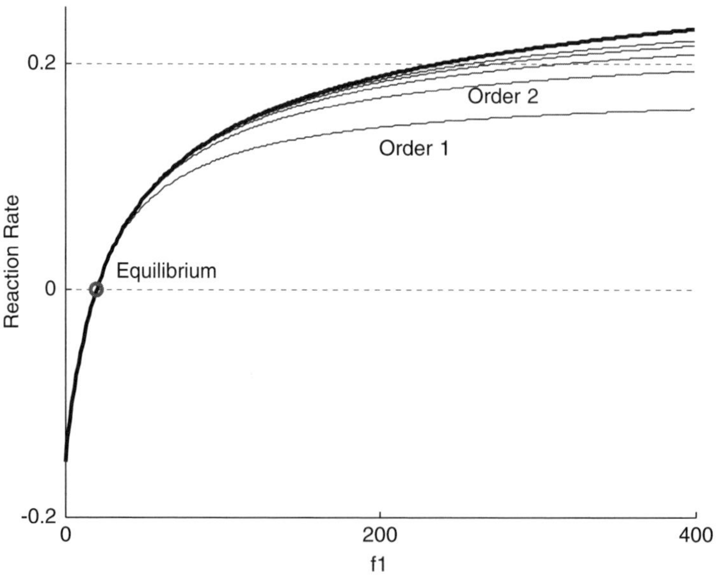

Figure 3 Overall reaction rate and its approximations by first 1, 2,..., 5 terms of hypergeometric series (Eley–Rideal mechanism). Parameters: $f_2 = 0.71$, $r_1 = 0.2$ and $r_2 = 7$.

3.2.2.2 Langmuir–Hinshelwood mechanism. Series expressions for roots of 4th order polynomial (51) are

$$
\begin{aligned}
X_1 &= -\left[\frac{B_0}{B_1}\right],\\
X_1 &= -\left[\frac{B_1}{B_2}\right] + \left[\frac{B_0}{B_1}\right],\\
X_2 &= -\left[\frac{B_2}{B_3}\right] + \left[\frac{B_1}{B_2}\right],\\
X_1 &= -\left[\frac{B_3}{B_4}\right]
\end{aligned}
\tag{64}
$$

where[22]

$$
\begin{aligned}
\left[\frac{B_0}{B_1}\right] &= \frac{B_0}{B_1} + \frac{B_0^2 B_2}{B_1^3} - \frac{B_0^3 B_3}{B_1^4} + \frac{2B_0^3 B_2^2}{B_1^5} + \frac{B_0^4 B_4}{B_1^5} - \frac{5B_0^4 B_2 B_3}{B_1^6} + \cdots,\\
\left[\frac{B_1}{B_2}\right] &= \frac{B_1}{B_2} + \frac{B_1^2 B_3}{B_2^3} + \frac{2B_0 B_1 B_4}{B_2^3} - \frac{B_1^3 B_4}{B_2^4} - \frac{3B_0 B_1 B_3^2}{B_2^4}\\
&\quad + \frac{2B_1^3 B_3^2}{B_2^5} + \frac{12 B_0 B_1^2 B_3 B_4}{B_2^5} + \frac{6B_0^2 B_1 B_4^2}{B_2^5} + \cdots,\\
\left[\frac{B_2}{B_3}\right] &= \frac{B_2}{B_3} + \frac{B_2^2 B_4}{B_3^3} - \frac{3B_1 B_2 B_4^2}{B_3^4} + \frac{2B_2^3 B_4^2}{B_3^5} + \frac{4B_0 B_2 B_4^3}{B_3^5} + \cdots\\
\left[\frac{B_3}{B_4}\right] &= \frac{B_3}{B_4}
\end{aligned}
\tag{65}
$$

Figures 4 and 5 compare the exact solutions of the kinetic polynomial (51) (i.e. the quasi-steady-state values of reaction rate) to their approximations. Even the first term of series (65) gives the satisfactory approximation of all four branches of the solution (see Figure 4). Figure 5, in which "brackets" from Appendix 2 are compared to the exact solution, illustrates the existence of the region (in this case, the interval of parameter f_2), where hypergeometric series converge for each root.

However, regions of convergence strongly depend on model parameters. Figures 6 and 7 illustrate this for the case of the steady-state multiplicity (parameters are the same as for Figure 1). Figure 7 shows that the approximation of feasible solution is satisfactory in the vicinity of the thermodynamic equilibrium. Intuitively, it is clear, because we can expect the convergence of our series for $\left[\frac{B_0}{B_1}\right]$ (see Equation (65)) at smaller values of B_0. Less obvious, it is an excellent approximation of the low-rate branch of the steady-state kinetic dependence *very far from thermodynamic equilibrium* (see Figure 6). For comparison, Figures 6 and 7 show the dependencies of the equilibrium approximation (53) as well as the approximation which corresponds to first

[22] See Appendix 2 for more detailed "bracket" expressions.

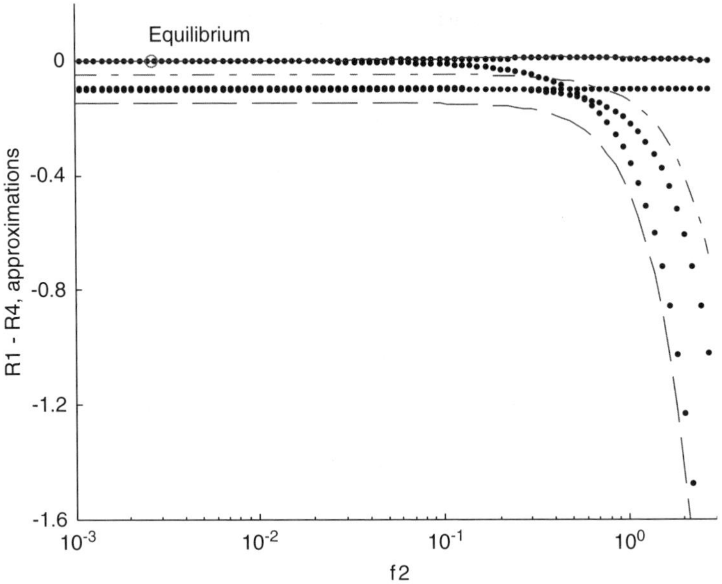

Figure 4 All roots of kinetic polynomial (dots) and their first-term (lines) approximations for LH mechanism. Parameters: $f_1 = 1.4$, $r_1 = 0.1$, $r_2 = 0.1$, $f_3 = 0.1$ and $r_3 = 0.01$.

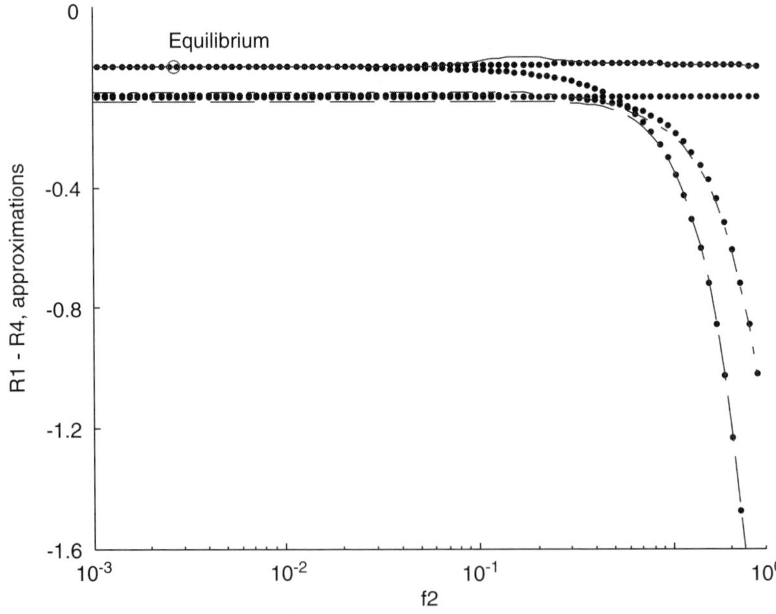

Figure 5 All roots of the kinetic polynomial from Figure 4 (dots) and their higher-order ($m = 3$) approximations (lines).

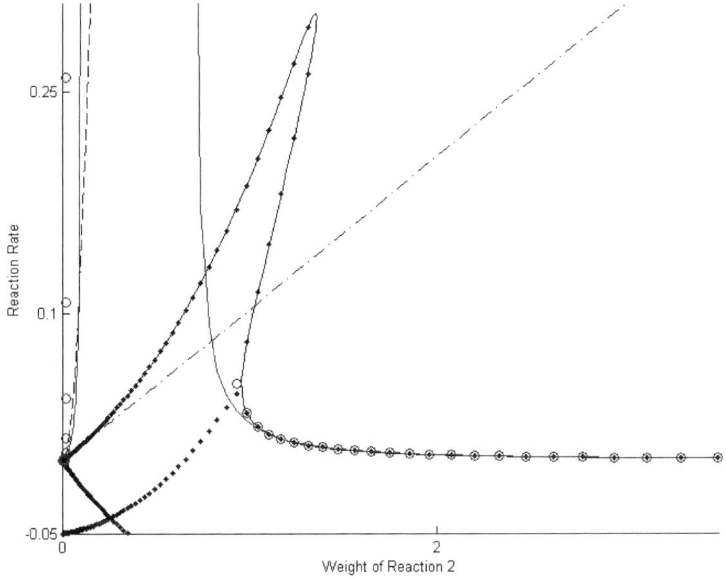

Figure 6 Approximations of the thermodynamic branch: steady-state multiplicity case (see Figure 1). Solid line is the first-term hypergeometric approximation. Circles correspond to the higher-order hypergeometric approximation ($m = 3$). Dashed line is the first-order approximation in the vicinity of thermodynamic equilibrium. Dash-dots correspond to the second-order approximation in the vicinity of thermodynamic equilibrium.

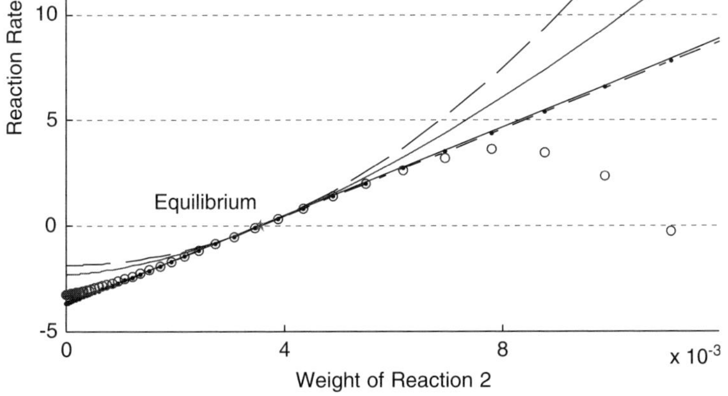

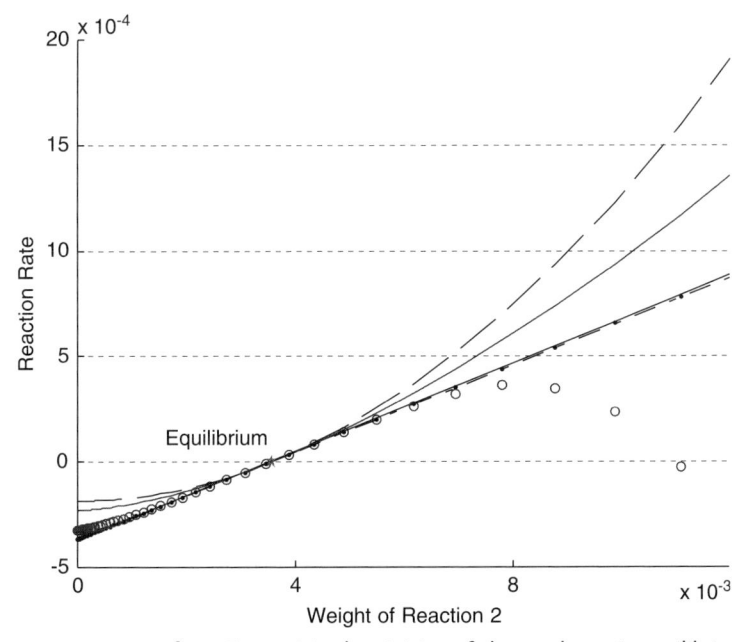

Figure 7 Approximations from Figure 6 in the vicinity of thermodynamic equilibrium.

terms of the expansion by parameter f_2 of feasible reaction rate branch in the vicinity of thermodynamic equilibrium. As expected, both formula (53) and expansion work in the vicinity of thermodynamic equilibrium. However, these approximations cannot predict the lower branch of reaction rate dependence. This example shows *the non-local nature* of hypergeometric series approximation.

At the same time, we see that practical application of this type of approximation requires the knowledge of the exact convergence conditions of these series. Surprisingly, the exact mathematical results here were obtained very recently (Passare and Tsikh, 2004) and the convergence theory still looks unfinished.

We are going to focus below on the series for the *thermodynamic branch*, i.e. the branch described by formula

$$R = -\left[\frac{B_0}{B_1}\right] \tag{66}$$

3.2.3 The thermodynamic branch

3.2.3.1 The conventional representation. We can write Equation (66) explicitly as follows

$$R = -\frac{B_0}{B_1} \cdot D \tag{67}$$

where

$$D = \sum_{i_2=0,\ldots,i_n=0}^{\infty} \frac{(-1)^{i_1}}{i_0+1} \binom{i_1}{i_2,\ldots,i_n} \cdot \frac{B_0^{i_0} \prod_{k=2}^{n} B_k^{i_k}}{B_1^{i_1}} \tag{68}$$

and

$$\begin{aligned} i_1 &= \sum_{k=2}^{n} k i_k, \\ i_0 &= \sum_{k=2}^{n} (k-1) i_k \end{aligned} \tag{69}$$

We know from Proposition 1 that the constant term $B_0 \sim C$ vanishes at the thermodynamic equilibrium. Some features of Equation (67) similar to the known LHHW-kinetic equation. There is a "potential term" B_0 responsible for thermodynamic equilibrium, there is a "denominator" B_1 of the polynomial type. However there is also a big difference. Equation (67) includes the term D, which is generated by the non-linear steps.

$$D = 1 + T_1 + T_2 + \cdots,$$

where T_i is the term of series (68). According to Propositions 1 and 2 each term T_i includes the multiplier C (i.e. the cyclic characteristic) in some integer degree d_i

$$T_i = t_i C^{d_i} \tag{70}$$

3.2.3.2 Validity of the thermodynamic branch. To show that series (67) actually represents the thermodynamic branch we have to prove that all terms of this series include the cyclic characteristic C in *positive* degree.

In the assumptions of the *Basic case*, this follows from the following facts

Proposition 3. If $p \geq 1$ and property (35) is valid, then the cyclic characteristic C is contained in the coefficient B_1 of kinetic polynomial with an exponent equal to $p-1$.

Proof. See Appendix 3.

Proposition 4. Degree d_i in Equation (70) is non-negative.

Proof. It follows from Propositions 1–3 and Equation (69) that in the case $p \leq 2$ we have

$$d_i \geq i_0 p - i_1(p-1) = \sum_{k=2}^{n} i_k(k-p) \geq 0$$

and in the case $p > 2$ we have

$$d_i \geq i_0 p - i_1(p-1) + \sum_{k=2}^{p-1} i_k(p-k) = \sum_{k=p}^{n} i_k(k-p) \geq 0$$

3.2.3.3 The four-term overall reaction rate equation. It follows from Propositions 1, 3 and the fact that the kinetic polynomial defined by formula (26) is a rational function of reaction weights f_s and r_s that we can write Equation (67) as

$$R = -\frac{B_0'}{B_1'} \cdot C(1 + t_1 C^{d_1} + \cdots + t_i C^{d_i} + \cdots) \tag{71}$$

where B_0' and B_1' are polynomials in f_s and r_s. Moreover, it follows from the theory (Bykov et al., 1998) that in the assumptions of the *Basic Case*, B_0' is a monomial and we have proved in the Appendix 3 that B_1' is not vanishing at the thermodynamic equilibrium. Note, that formula (1) for the linear mechanism is just a particular case of formula (71).

We can write (71) as a four-term overall reaction rate equation

$$R = \frac{k_+(f_+(c) - K_{\text{eq}}^{-1} f_-(c))}{-B_1'} \left(1 + \sum_{i=1}^{\infty} t_i C^{d_i}\right) \tag{72}$$

where B_1' is either coefficient B_1 or its multiplier, t_i $i = 1, 2, \ldots$ are rational functions in $k_{\pm i}$, c and integer degree d_i is non-negative. The latter property guarantees vanishing of reaction rate at the thermodynamic equilibrium. Equation (72) corresponds to Equation (14) with $M = 1$. In addition, similar to the HW equation, we have the kinetic term k_+, the potential term $f_+(c) - K_{\text{eq}}^{-1} f_-(c)$ as well as the term $-B_1'$ which can be interpreted as the adsorption resistance term.

However, unlike the known approximations, our exact expression has the fourth term $(1+\sum_{i=1}^{\infty} t_i C^{d_i})$ that represents the infinite series. Clearly, this term is generated by the non-linear steps. If all steps of the detailed mechanism are linear, such term is absent. It is important to understand the convergence of the series of fourth term.

3.2.4 The convergence domain

Detailed description of the domains of convergence of hypergeometric series in terms of *amoeba* of the *discriminant* of the polynomial has been given recently in Passare and Tsikh (2004). The discriminant $\Delta(a)$ is an irreducible polynomial with integer coefficients in terms of the coefficients a_i of polynomial (54) that vanishes if this polynomial has multiple roots. For instance, for cubic polynomial the discriminant is

$$\Delta(a) = 27a_0^2a_3^2 + 4a_1^3a_3 + 4a_0a_2^3 - 18a_0a_1a_2a_3 - a_1^2a_2^2 \tag{73}$$

The amoeba of a polynomial is the image of its zero locus under the mapping Log which relates each variable to the logarithm of its absolute value.

In the case of the quadratic equation, the convergence condition for the "thermodynamic branch" series is simply positive discriminant (Passare and Tsikh, 2004). For kinetic polynomial (48) this discriminant is always positive for feasible values of parameters (see Equation (49)). This explains the convergence pattern for this series, in which the addition of new terms extended the convergence domain.

For certain types of the series the explicit inequalities involving "mirror reflections" of the discriminant were possible (Passare and Tsikh, 2004). The situation is clearer for series depending on fewer variables. For instance, applying Birkeland approach, we can reduce to two the number of parameters in the case of cubic equation. The "thermodynamic branch" corresponds to the Birkeland series (60) for $p = 0$ and $q = 1$. The discriminant for cubic equation in Birkeland form is

$$\Delta(b_2, b_3) = 27b_3^2 + 4b_3 + 4b_2^3 - 18b_2b_3 - b_2^2 \tag{74}$$

The boundary of convergence domain is

$$\Delta(|b_2|, -|b_3|) = 0 \tag{75}$$

and the convergence domain is

$$27|b_3^2| - 4|b_3| + 4|b_2|^3 + 18|b_2||b_3| - |b_2|^2 < 0 \tag{76}$$

The following case study demonstrates the convergence behavior for the LH mechanism (50) with irreversible first stage (i.e. $r_{-1} = 0$). In this case the kinetic polynomial (51) always has (structurally unstable with respect to feasibility) zero root whereas three other roots could be found from the cubic equation

$$k_3R^3 + k_2R^2 + k_1R + k_0 = 0$$

where

$$k_3 := 16f_1^2f_3^2$$

$$\begin{aligned} k_2 := & -8f_2^2f_1f_3^2 - 8f_2f_1f_3^2r_2 + 16f_1^2f_3r_2^2 + 8f_1f_3r_2^2r_3 \\ & - 4f_1f_3^2r_2^2 + 16f_1^2f_3^2r_2 \end{aligned}$$

$$\begin{aligned} k_1 := & f_3^2f_2^4 + 2f_3^2r_2f_2^3 - 4f_1f_3^2r_2f_2^2 + r_2^2f_3^2f_2^2 + 2f_3f_2^2r_2^2r_3 \\ & + 4f_1f_3f_2^2r_2^2 + 2r_2^3f_3f_2r_3 + 4f_1r_2^3f_3f_2 + 8f_1^2f_3r_2^3 + 4f_1^2r_2^4 \\ & + 4f_1r_2^4r_3 + 4f_1f_3r_2^3r_3 + 4f_1^2f_3^2r_2^2 + r_2^4r_3^2 \end{aligned}$$

$$k_0 := -f_1f_3^2r_2^2f_2^2$$

Figure 8 shows the convergence domain (75) (i.e. the rhomboid) and coefficients $b_2 = (k_0k_2)/(k_1^2)$ and $b_3 = (k_0^2k_3)/(k_1^3)$ as parametric function of parameter $f_2 \in [0, \infty]$ at different values of f_3 (i.e. ovals). At lower values of parameter f_3 the whole loop is located within the convergence domain. This means that the series will converge for any $f_2 \in [0, \infty]$. At some value of f_3, the ovals start intersecting the rhomboid boundary. In this case we can have (at least) two convergence intervals $f_2 \in [0, f_2^*]$ and $f_2 \in [f_2^{**}, \infty]$ separated by interval $<s>$ of non-convergence.

Figure 9 shows the convergence domain as well as steady-state multiplicity domain on f_2, f_1 plane. We can see that steady-state multiplicity is not generally

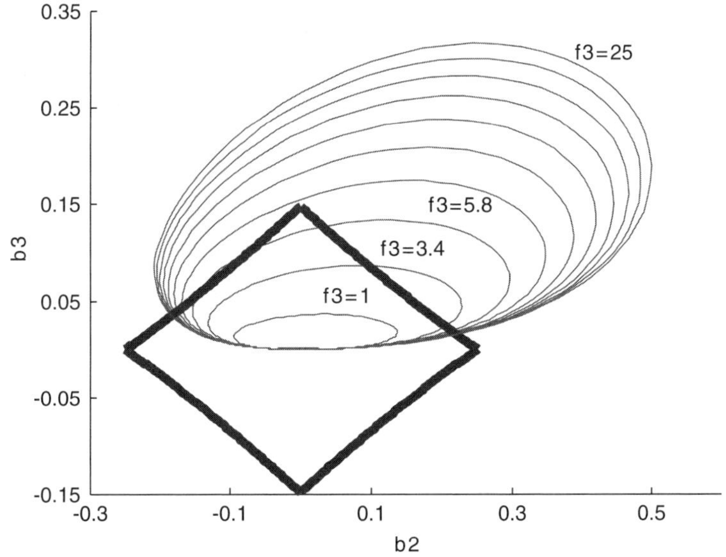

Figure 8 A convergence domain (rhomboid) and coefficients of the kinetic polynomial (ovals). The ovals represent the coefficients b2 and b3 as parametric functions of parameter f_2 at different values of parameter f_3. Parameters: $f_1 = 1.4$, $r_2 = 0.9$ and $r_3 = 0.4$.

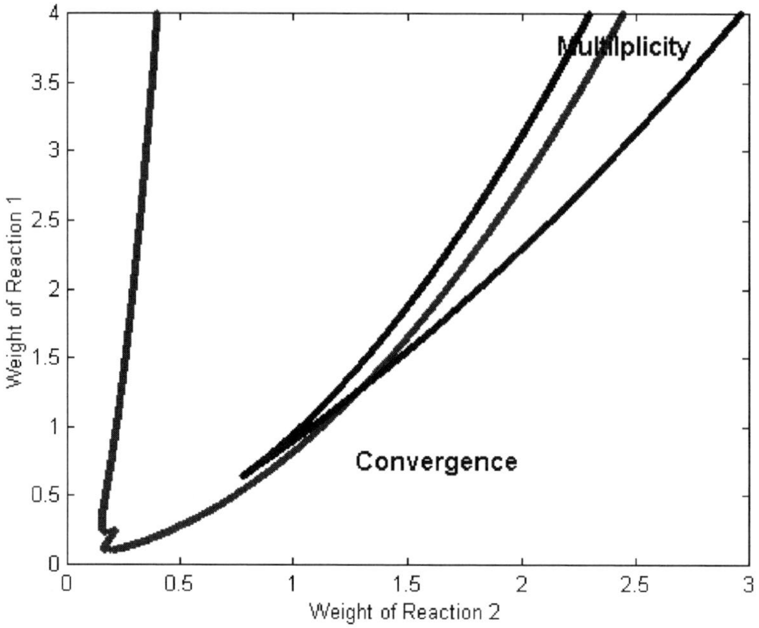

Figure 9 A convergence domain and steady-state multiplicity domain.

an obstacle to the convergence of hypergeometric series — one of the branches of the convergence domain boundary extends to the steady-state multiplicity region. Figure 9 shows also that convergence is lost before the bifurcation.

Figures 10–13 compare the exact dependencies of the (feasible) reaction rate and their first term approximation (i.e. $R = -(k_0)/(k_1)$) as well as approximation corresponding to $m = 3$:

$$R = -10{,}010\frac{k_0^{10}k_2^3k_3^3}{k_1^{16}} - \frac{1{,}430k_0^9k_2^2k_3^3}{k_1^{14}} - \frac{165k_0^8k_2k_3^3}{k_1^{12}} - \frac{990k_0^8k_2^3k_3^2}{k_1^{13}} - \frac{12k_0^7k_3^3}{k_1^{10}} + \frac{180k_0^7k_2^2k_3^2}{k_1^{11}} + \frac{28k_0^6k_2k_3^2}{k_1^9} - \frac{84k_0^6k_2^3k_3}{k_1^{10}} + \frac{3k_0^5k_3^2}{k_1^7} - \frac{21k_0^5k_2^2k_3}{k_1^8} - \frac{5k_0^4k_2k_3}{k_1^6} + \frac{5k_0^4k_2^3}{k_1^7} - \frac{k_0^3k_3}{k_1^4} + \frac{2k_0^3k_2^2}{k_1^5} + \frac{k_0^2k_2}{k_1^3} + \frac{k_0}{k_1}$$

Figure 9 shows that at smaller f_1 values the hypergeometric approximation works for the whole ray $f_2 \in [0, \infty]$ (see Figure 10).

At higher f_1 values we can have two convergence intervals (see Figure 11) or even 3 such intervals (see Figure 12, note that there is no significant difference between Figures 11 and 12, most probably due to the very small length of the middle interval). Finally, Figure 13 illustrates the convergence pattern in the region of steady-state multiplicity.

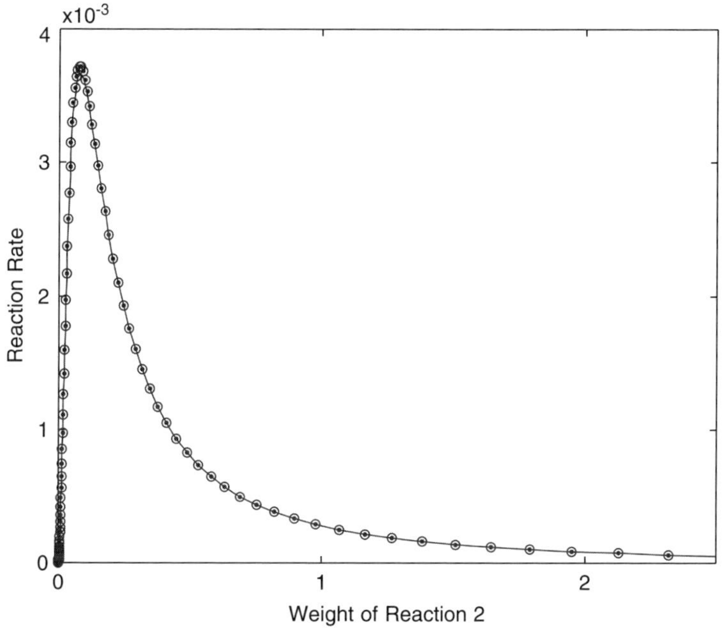

Figure 10 Exact overall reaction rate dependence (dots), its first-term hypergeometric approximation (solid line) and its higher-order ($m = 3$) hypergeometric approximation. Parameters: $f_1 = 0.01$, $r_2 = 0.2$, $r_3 = 1$ and $f_3 = 10$.

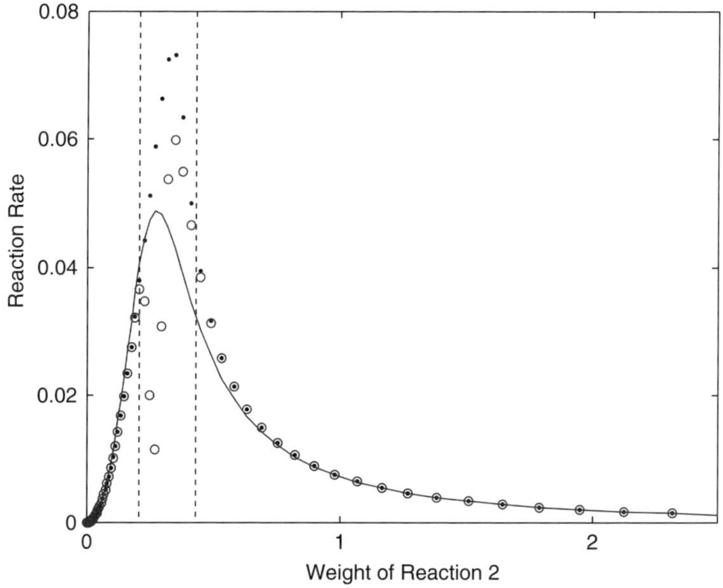

Figure 11 Dependencies from Figure 10 at $f_1 = 0.23$. Vertical dotted lines correspond to the boundaries of the convergence domain: there is no convergence in the interval between these boundaries.

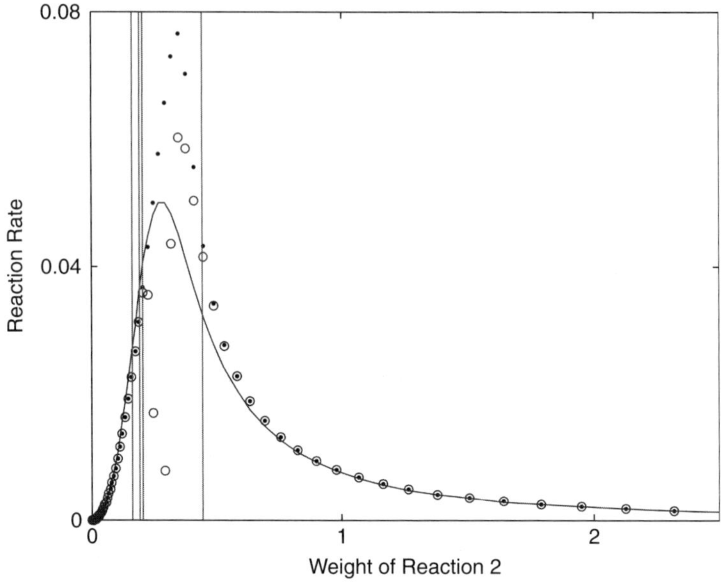

Figure 12 Dependencies from Figure 10 at $f_1 = 0.24$. There are four convergence boundaries in this case and three convergence domains (one of them is really narrow) separated by two non-convergence domains.

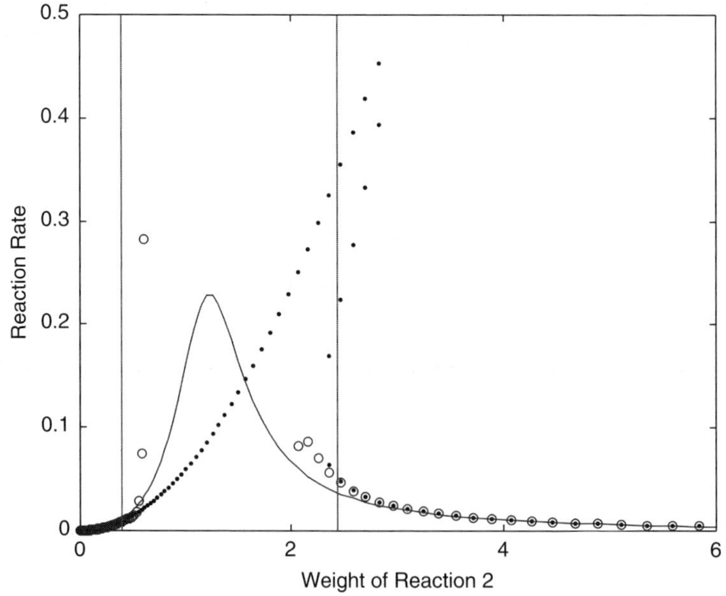

Figure 13 Dependencies from Figure 10 at $f_1 = 4$: case of the steady-state multiplicity.

3.2.5 Comparison to classic approximations

We are going to compare our hypergeometric approximation of the thermodynamic branch to the classic rate-limiting step and linear equilibrium approximations (see Sections 3.1.1 and 3.1.2).

We use the model corresponding to the LH mechanism as an example.

The case of step 1 limiting is shown in Figure 14. We can see that both classical and hypergeometric approximations fit well with the exact dependence in the vicinity of equilibrium as well as at lower values of parameter f_1. At the same time, we can see that hypergeometric approximation provides better fit at higher values of parameter f_1.

Figure 15 illustrates the case of step 2 limiting. Growth of the magnitude of kinetic parameters (in our case, 10-fold) unavoidably results in the degradation of the quality of rate-limiting type approximations (see Figure 15c). Whereas the equilibrium approximation works as expected in the vicinity of equilibrium, it does not produce good fit far from the equilibrium. The hypergeometric approximation produces uniformly good fit of the exact dependence (see Figure 15a–c).

Figure 16 demonstrates the similar features for the case of step 3 limiting.

Finally, Figure 17 compares all types of approximations and the exact reaction rate dependence. The equilibrium approximation works well at smaller values f_2 (the equilibrium point is close to the origin). Limitation of the step 1 works at higher values of parameter f_2 whereas limitation of step 2 fits the initial

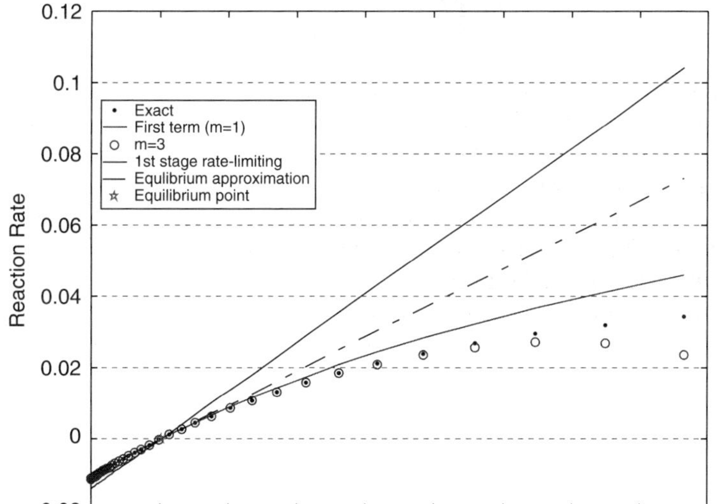

Figure 14 Overall reaction rate and its approximations: LH mechanism. Parameters: $r_1 = 0.1$, $f_2 = 14$, $r_2 = 10$, $f_3 = 1$ and $r_3 = 2$ (see Plate 1 in Color Plate Section at the end of this book).

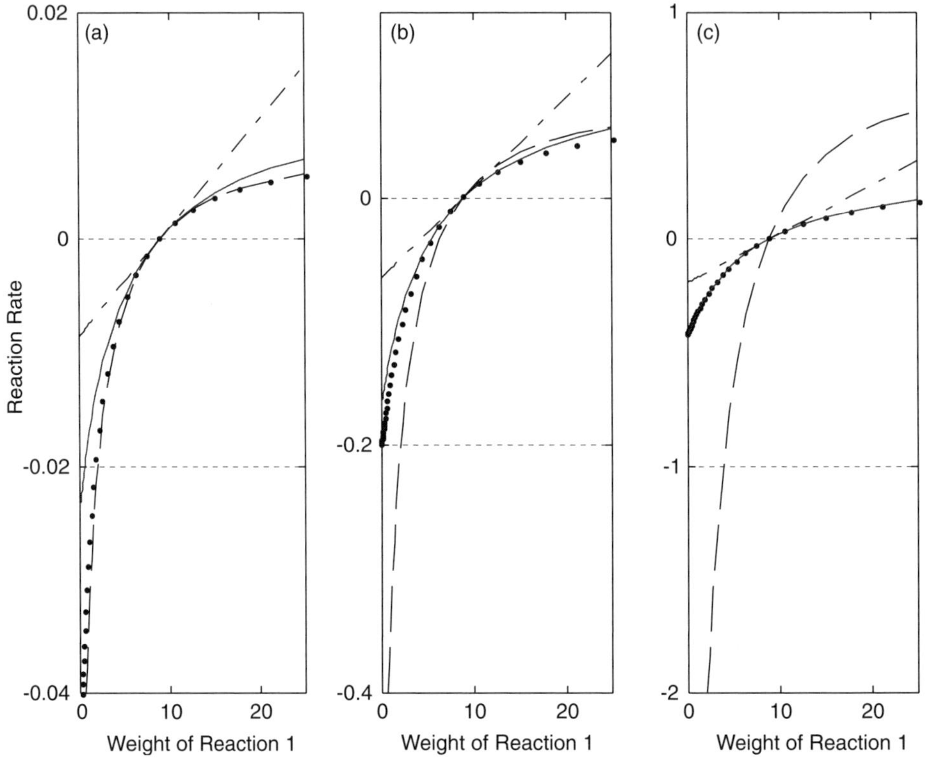

Figure 15 Overall reaction rate and its approximations: step 2 is rate-limiting. Dots represent the exact reaction rate dependence, solid line is the first-term hypergeometric approximation, dashed line corresponds to the reaction-rate equation that assumes the limitation of step 2 and dash-dots represent the equilibrium approximation. Hypergeometric approximation "survives" the 100-times increase in rate-limiting stage kinetic parameters and it works when there is no rate-limiting step at all. Parameters: $r_1 = 5$, $f_3 = 15$, $r_3 = 10$; $r_2 = 0.2$, $f_2 = 0.1$ (a); $r_2 = 2$, $f_2 = 1$ (b); $r_2 = 20$, $f_2 = 10$, (c).

increasing part of dependence. The hypergeometric approximation (we used just the first-term approximation, i.e. $R = -(B_0)/(B_1)$ works for all listed cases within the convergence boundaries.

Thus, when applicable, the hypergeometric approximation covers all classical cases. It is obvious advantage for the situations where we do not have the clear hierarchy of reaction rate parameters.

4. DISCUSSION AND CONCLUSIONS

We consider that there was a step towards to the development of the general theoretical description of kinetic behavior of complex reactions which covers many parametric domains, particularly domains on "both sides" of the chemical equilibrium, domains with step limiting, regions of the vicinity of the

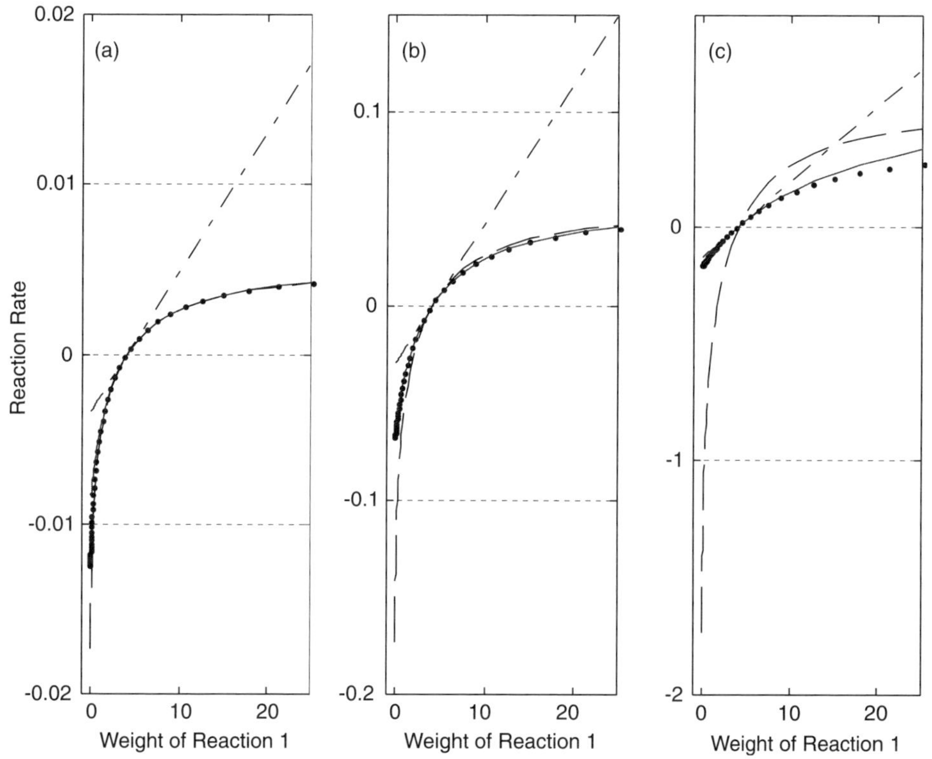

Figure 16 Overall reaction rate and its approximations: step 3 is rate-limiting. Parameters: $r_1 = 2$, $f_2 = 14$, $r_2 = 10$; $r_3 = 0.2$, $f_3 = 0.1$ (a); $r_3 = 2$, $f_3 = 1$ (b); $r_3 = 20$, $f_3 = 10$ (c). See Figure 15 for notation and comments.

equilibrium, and even domains, in which the branch of overall reaction rate is not unique ("multiplicity of steady states"). The developed mathematical constructions, i.e. *kinetic polynomial* and its analytical solution as a hypergeometric series, describe all these domains. Validity of these constructions is rigorously justified based on recent advanced results of the algebraic theory (Gel'fand–Kapranov–Zelevinsky's and Sturmfels' results). In our approach, the concept of ensemble of equilibrium subsystems introduced in our earlier papers (see in detail Lazman and Yablonskii, 1991) was used as a very efficient tool of mathematical analysis and physico-chemical understanding. The equilibrium subsystem is such a system that corresponds to the following assumption: $(n-1)$ steps are considered to be under equilibrium conditions, one step is limiting, where n is a number of steps. In fact, the concept of "equilibrium subsystems" is a generalization of the concept of "equilibrium step", which is well-known in chemical kinetics. Then, we take n of these equlibrium subsystems ("an ensemble of equilibrium subsystems"). It was shown that solutions of these subsystems ("roots", "all roots", not just one "root") define coefficients of the kinetic polynomial.

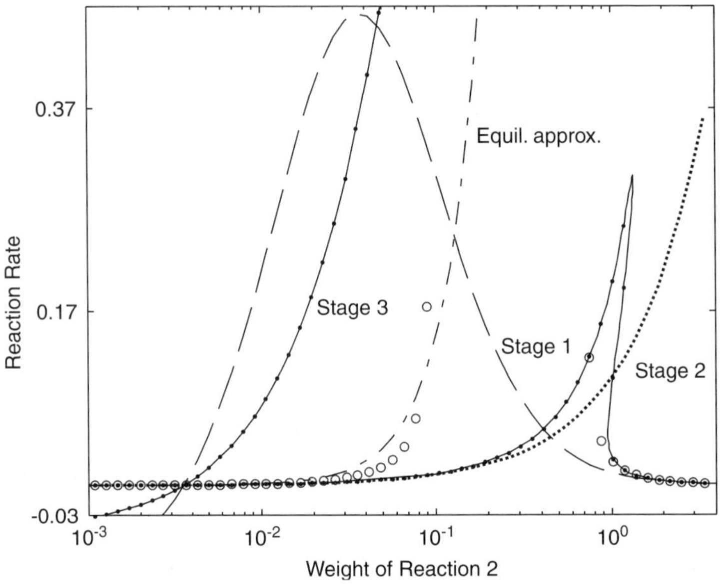

Figure 17 Overall reaction rate, its classic approximations and its hypergeometric approximation (circles): LH mechanism.

In the future, such algebraic representations can be constructed and analyzed for more complex mechanisms, e.g. two-route mechanisms, mechanisms with two sites of active centers, etc. It is quite interesting that the hypergeometric representation describes also the "low-rate" branch, which is located in the domain "very far" from the equilibrium.

Applying "kinetic polynomial" approach we found the analytical representation for the "thermodynamic branch" of the overall reaction rate of the complex reaction *with no traditional assumptions on the rate limiting and "fast" equilibrium of steps.*

The obtained explicit equation, "four-term equation", can be presented in a simple manner as follows:

$$R = \frac{k_+(f_+(c) - K_{eq}^{-1} f_-(c))}{\sum(k,c)} N(k,c) \tag{77}$$

This equation can be used for a description of kinetic behavior of steady-state open catalytic systems as well as quasi (pseudo)-steady-state catalytic systems, both closed and open.

Equation (77) has four terms:

(1) *an apparent kinetic coefficient* k_+; (2) *a "potential term"* $(f_+(c) - K_{eq}^{-1} f_-(c))$ related to the net reaction ("driving force" of irreversible thermodynamics); (3) a *"resistance" term,* $\Sigma(k,c)$, denominator of the polynomial type, which reflects complexity of chemical reaction, both its non-elementarity (many-step character) and non-linearity of elementary steps as well; (4) finally, *the "fourth term"*, $N(k,c)$, a polynomial in concentrations and kinetic parameters. This fourth term is

generated exclusively by the non-linearity of reaction steps. This term is the main distinguishing feature of the general equation (77) in comparison with LHHW equations based on simplifying assumptions.

In the absence of non-linear steps the "fourth" term is also absent.

The interesting feature of our representation is that many sub-terms of the "fourth", non-linear, term may contain the "potential term" (the cyclic characteristic C) as well. It means that even in the domain "far from equilibrium" the open system still may have a "memory" about the equilibrium. Particular forms of this general Equation (77), i.e. for the cases of step limiting and the vicinity of equilibrium, respectively, are presented.

These explicit forms do not contain the "fourth term".

The obtained result gives a desired answer regarding the validity of the Horiuti–Boreskov form. So, the presentation of the overall reaction rate of the complex reaction as a difference between two terms, overall rates of forward and backward reactions respectively, is valid, if we are able to present this rate in the form of Equation (77). We can propose a reasonable hypothesis (it has to be proven separately) that it is always possible even for the non-linear mechanism, if the "physical" branch of reaction rate is unique, i.e. multiplicity of steady states is not observed. As it has been proven for the MAL systems, the steady state is unique, if the detailed mechanism of surface catalytic reaction does not include the step of interaction between the different surface intermediates (Yablonskii et al., 1991). This hypothesis will be analyzed in further studies.

REFERENCES

Boreskov, G. K. *Zh. Fiz. Khim.* **19**, 92–94 (1945).

Boudart, M., and Diega-Mariadassou, G., "Kinetics of Heterogeneous Catalytic Reactions". Princeton University, Princeton (1984).

Bykov, V. I., Kytmanov, A. M., and Lazman, M. Z., The modified elimination methods in computer algebra of polynomials. *Abstracts of International Congress on Comp. Syst., and Appl. Math,* pp. 106–107. St. Petersburg State Univ., St. Petersburg (1993).

Bykov, V. I., Kytmanov, A. M., and Lazman, M. Z., and Passare, M. (Eds), Elimination Methods in Polynomial Computer Algebra, *in* "Mathematics and Its Applications", Vol. 448, 237pp. Kluwer Academic Publishers, Dordrecht (1998).

Bykov, V. I., Kytmanov, A. M., Lazman, M. Z., and Yablonskii, G. S. *Khim Fiz.* **6**, 1549–1554 (1987).

Bykov, V. I., Kytmanov, A. M., Lazman, M. Z., and Yablonskii, G. S., A kinetic polynomial for one-route n-step catalytic reaction, *in* "Mathematical problems of Chemical Kinetics", pp. 125–149. Novosibirsk, Nauka (1989).

Bykov, V. I., and Yablonskii, G. S. *Kinetika i Kataliz.* **18**, 1561–1567 (1977a).

Bykov, V. I., and Yablonskii, G. S. *Dokl. Akad. Nauk USSR* **233**, 642–645 (1977b).

Chen, T. S., and Chern, J.-M. *Chem. Eng. Sci.* **57**, 457–467 (2002).

Evstigneev, V. A., Yablonskii, G. S., and Bykov, V. I. *Dokl. Akad. Nauk USSR* **238**, 645 (1978).

Evstigneev, V. A., Yablonskii, G. S., and Bykov, V. I. *Dokl. Akad. Nauk USSR* **245**, 871–874 (1979).

Evstigneev, V. A., Yablonskii, G. S., Noskov, A. S., and Bykov, V. I. *Kinetika i Kataliz* **22**, 738–743 (1981).

Gel'fand, I. M., Kapranov, M. M., and Zelevinsky, A. V., "Discriminants, Resultants and Multi-dimensional Determinants". Birkhauser, Boston (1994).

Gorban, A. N., and Karlin, I. V. *Chem. Eng. Sci.* **58, 21**, 4751–4768 Preprint online: http://arxiv.org/abs/cond-mat/020731(2003).
Gorban, A. N., Invariant manifolds for physical and chemical kinetics, Lecture Notes, 660, Springer, Berlin (2005).
Gorban, A. N., and Radulescu, O. *Adv. Chem. Eng.* **34**, this volume (2008).
Helfferich, F. G., "Kinetics of Homogeneous Multi-Step Reactions". Elsevier, Amsterdam (2001).
Horiuti, J. *Ann. New York Acad. Sci.* **213**, 5–30 (1973).
King, E. L., and Altman, C. A. *J. Phys. Chem.* **60**, 1375–1378 (1956).
Klein, F., "Lectures on the Ikosahedron and the Solution of Equations of the Fifth Degree". Trubner & Co, London (1888).
Lazman, M., Finding all the roots of nonlinear algebraic equations: A global approach and application to chemical problems, *15th IMACS World Congress on Scientific Computation, Modeling and Applied Mathematics*, Berlin, August 1997, 6: Application in Modeling and Simulation, A. Sydow (ed.), Wissenschaff & Technic Verlag, 329–334 (1997).
Lazman M., Effective Process Simulation: Analytical Methods, *16th IMACS World Congress On Scientific Computation, Applied Mathematics and Simulation*, Lausanne, Switzerland, August 21–25, 2000, CD ROM of congress proceedings, ISBN 3-9522075-1-9 (2000).
Lazman M., Reaction Rate is an eigenvalue: Polynomial elimination in chemical kinetics, "MaCKiE-2002 Mathematics in Chemical Kinetics and Engineering, Book of Abstracts, Part 2", Ghent, Belgium, May 5–8, pp. 25–28 (2002).
Lazman M., Algebraic geometry methods in analysis of quasi steady state and dynamic models of catalytic reactions, Proceedings of the 4th International Conference on Unsteady-State Processes in Catalysis USPC-4, Montreal, Quebec, Canada, October 26–29, 2003, Dr. H. Sapoundjiev (Ed.), Natural Resources Canada, 92–93 (2003a).
Lazman M., Advanced Process Simulation: Models and methods, "CESA'2003 IMACS Multiconference Computational Engineering in Systems Applications" (P. Borne, E. Craye, N. Dandoumau Eds.), CD ROM, # S3-R-00-0095, Lille, France, July 9–11, ISBN: 2-9512309-5-8 (2003b).
Lazman, M. Z., Spivak, S. I., and Yablonskii, G. S. *Sov. J. Chem. Phys.* **4**, 781–789 (1987b).
Lazman M., and Yablonsky G. Computer algebra in chemical kinetics: Theory and application, Computer algebra in Scientific computing. Proceedings of CASC'2004, Muenchen, 313–324 (2004).
Lazman, M. Z., and Yablonskii, G. S. *React. Kinet. Catal. Lett.* **37, 2**, 379–384 (1988).
Lazman, M. Z., and Yablonskii, G. S., Kinetic polynomial: A new concept of chemical kinetics, *Patterns and Dynamics in Reactive Media*, The IMA Volumes in Mathematics and its Applications, pp. 117–150, Berlin, Springer (1991).
Lazman, M. Z., Yablonskii, G. S., and Bykov, V. I. *Sov. J. Chem. Phys.* **2**, 404–418 (1985a).
Lazman, M. Z., Yablonskii, G. S., and Bykov, V. I. *Sov. J. Chem. Phys.* **2**, 693–703 (1985b).
Lazman, M. Z., Yablonskii, G. S., Vinogradova, G. M., and Romanov, L. N. *Sov. J. Chem. Phys.* **4**, 1121–1134 (1987a).
Macaulay, F. S., "Algebraic Theory of Modular Systems". Cambridge University Press, Cambridge (1916).
Mumford, D., "Tata Lectures on Theta II. Jacobian Theta Functions and Differential Equations". Birkhäuser, Boston (1984).
Nacamura, T. *J. Res. Inst. Catal., Hokkaido Univ.* **6**, 20–27 (1958).
Passare, M., and Tsikh, A., Algebraic equations and hypergeometric series. *in* "Legacy of Niels Henrik Abel: The Abel Bicentennial", Oslo, Springer, June 3–8, 653–672 (2002).
Sturmfels, B. *Discrete Math.* **2101-3**, 171–181 (2000).
Temkin, M. I. *Dokl. Akad. Nauk USSR* **152**, 156–159 (1963).
Van der Waerden, B. L., "Algebra". Springer, Berlin (1971).
Vol'kenstein, M. V., "Physics of Enzymes". Nauka, Moscow (1967).
Vol'kenstein, M. V., and Gol'dstein, B. N. *Biochim. Biophys. Acta* **115**, 471–477 (1966).
Yablonskii, G. S., Bykov, V. I., Gorban, A. N., and Elokhin, V. I., Kinetic models of catalytic reactions, *in* "Comprehensive Chemical Kinetics, 32" (R. Compton Ed.), p. 392. Elsevier, Amsterdam. (1991).
Yablonskii, G. S., Spivak, S. I., and Lazman, M. Z., Modeling of complex catalytic reactions, *in* "Proceedings of the 4th International Symposium on Systems Analysis and Simulation (Berlin, Germany)" (A. Sydow Ed.), pp. 651–656. Elsevier, New York. (1992).

Yablonskii, G. S., and Lazman, M. Z. *React. Kin. Cat. Lett.* **59, 1**, 145–150 (1996).
Yablonsky, G. S., and Lazman, M. Z., Non-Linear Steady-State Kinetics of Complex Catalytic Reactions: Theory and Experiment, Dynamics of Surfaces and Reaction Kinetics in Heterogeneous Catalysis, Proceedings of the International Symposium, Antwerpen, September 371–378 (1997).
Yablonskii, G. S., Lazman, M. Z., and Bykov, V. I. *React. Kinet. Catal. Lett.* **20**(1–2), 73–77 (1982).
Yablonskii, G. S., Lazman, M. Z., and Bykov, V. I. *Dokl. Akad. Nauk USSR* **269**(1), 166–168 (1983).
Yablonsky, G. S., Mareels, M. Y., and Lazman, M. Z. *Chem. Eng. Sci.* **58**, 4833–4842 (2003) .

APPENDIX 1. REACTION OVERALL RATE EQUATIONS FOR LINEAR MECHANISMS

A. Two-stage mechanism of water–gas shift reaction

$$\begin{array}{l}1.\quad Z + H_2O \Leftrightarrow ZO + H_2\\ 2.\quad ZO + CO \Leftrightarrow Z + CO_2\\ \hline H_2O + CO \Leftrightarrow H_2 + CO_2\end{array}$$

Overall reaction rate is

$$R = \frac{K^+ c_{H_2O} c_{CO} - K^- c_{H_2} c_{CO_2}}{\sum}$$

where

$$K^+ = k_1^+ k_2^+, \qquad K^- = k_1^- k_2^-, \qquad \sum = k_1^+ c_{H_2O} + k_2^+ c_{CO} + k_1^- c_{H_2} + k_2^- c_{CO_2},$$

$$R^+ = \frac{K^+ c_{H_2O} c_{CO}}{\sum}, \quad R^- = \frac{K^- c_{H_2} c_{CO_2}}{\sum}, \quad \frac{R^+}{R^-} = K_{eq} \frac{c_{H_2O} c_{CO}}{c_{H_2} c_{CO_2}}$$

B. Three-stage mechanism of catalytic isomerization

$$\begin{array}{l}1.\quad A + Z \Leftrightarrow AZ\\ 2.\quad AZ \Leftrightarrow BZ\\ 3.\quad BZ \Leftrightarrow B + Z\\ \hline \qquad A \Leftrightarrow B\end{array}$$

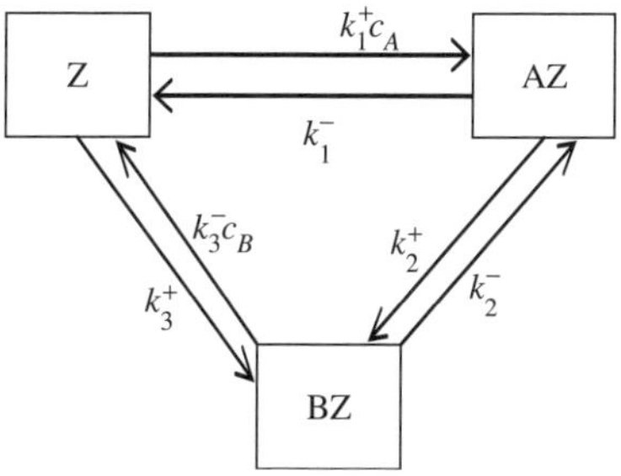

Overall reaction rate is

$$R = \frac{K^+ c_A - K^- c_B}{\sum}$$

where

$$K^+ = k_1^+ k_2^+ k_3^+, \quad K^- = k_1^- k_2^- k_3^-, \quad \sum = K_1 c_A + K_2 c_B + K_3,$$

$$K_1 = k_1^+ (k_2^+ + k_3^+ + k_2^-), \quad K_2 = k_3^- (k_2^+ + k_2^- + k_1^-), \quad K_3 = k_2^+ k_3^+ + k_1^- k_2^- + k_1^- k_3^+$$

$$R^+ = \frac{K^+ c_A}{\sum}, \quad R^- = \frac{K^- c_B}{\sum}, \quad \frac{R^+}{R^-} = K_{eq} \frac{c_A}{c_B}$$

APPENDIX 2. COMPUTER GENERATED "BRACKETS" FOR 4TH DEGREE POLYNOMIAL

Formulas correspond to the upper boundary of m for each independent index in the summation (56).

$\begin{bmatrix} \frac{B_0}{B_1} \end{bmatrix}$, $m = 3$:

$$-414{,}414{,}000 \frac{B_0^{19} B_2^3 B_3^3 B_4^3}{B_1^{28}} - \frac{33{,}649{,}000 B_0^{18} B_2^2 B_3^3 B_4^3}{B_1^{26}} - \frac{2{,}018{,}940 B_0^{17} B_2 B_3^3 B_4^3}{B_1^{24}}$$

$$+ \frac{24{,}227{,}280 B_0^{17} B_2^3 B_3^2 B_4^3}{B_1^{25}} - \frac{67{,}830 B_0^{16} B_3^3 B_4^3}{B_1^{22}} + \frac{2{,}238{,}390 B_0^{16} B_2^2 B_3^2 B_4^3}{B_1^{23}}$$

$$- \frac{17{,}160{,}990 B_0^{16} B_2^3 B_3^3 B_4^2}{B_1^{24}} + \frac{155{,}040 B_0^{15} B_2 B_3^2 B_4^3}{B_1^{21}} - \frac{1{,}085{,}280 B_0^{15} B_2^3 B_3 B_4^3}{B_1^{22}}$$

$$- \frac{1{,}627{,}920 B_0^{15} B_2^2 B_3^3 B_4^2}{B_1^{22}} + \frac{6{,}120 B_0^{14} B_3^2 B_4^3}{B_1^{19}} - \frac{116{,}280 B_0^{14} B_2^2 B_3 B_4^3}{B_1^{20}}$$

$$- \frac{116{,}280 B_0^{14} B_2 B_3^3 B_4^2}{B_1^{20}} + \frac{1{,}162{,}800 B_0^{14} B_2^3 B_3^2 B_4^2}{B_1^{21}} - \frac{9{,}520 B_0^{13} B_2 B_3 B_4^3}{B_1^{18}}$$

$$- \frac{4{,}760 B_0^{13} B_3^3 B_4^2}{B_1^{18}} + \frac{28{,}560 B_0^{13} B_2^3 B_4^3}{B_1^{19}} + \frac{128{,}520 B_0^{13} B_2^2 B_3^2 B_4^2}{B_1^{19}}$$

$$- \frac{542{,}640 B_0^{13} B_2^3 B_3^3 B_4}{B_1^{20}} - \frac{455 B_0^{12} B_3 B_4^3}{B_1^{16}} + \frac{3{,}640 B_0^{12} B_2^2 B_4^3}{B_1^{17}}$$

$$+ \frac{10{,}920 B_0^{12} B_2 B_3^2 B_4^2}{B_1^{17}} - \frac{61{,}880 B_0^{12} B_2^3 B_3 B_4^2}{B_1^{18}} - \frac{61{,}880 B_0^{12} B_2^2 B_3^3 B_4}{B_1^{18}}$$

$$+ \frac{364 B_0^{11} B_2 B_4^3}{B_1^{15}} + \frac{546 B_0^{11} B_3^2 B_4^2}{B_1^{15}} - \frac{8{,}190 B_0^{11} B_2^2 B_3 B_4^2}{B_1^{16}}$$

$$- \frac{5{,}460 B_0^{11} B_2 B_3^3 B_4}{B_1^{16}} + \frac{43{,}680 B_0^{11} B_2^3 B_3^2 B_4}{B_1^{17}} + \frac{22 B_0^{10} B_4^3}{B_1^{13}} - \frac{858 B_0^{10} B_2 B_3 B_4^2}{B_1^{14}}$$

$$- \frac{286 B_0^{10} B_3^3 B_4}{B_1^{14}} + \frac{2{,}002 B_0^{10} B_2^3 B_4^2}{B_1^{15}} + \frac{6{,}006 B_0^{10} B_2^2 B_3^2 B_4}{B_1^{15}}$$

$$
\begin{aligned}
&-\frac{10{,}010B_0^{10}B_2^3B_3^3}{B_1^{16}}-\frac{55B_0^9B_3B_4^2}{B_1^{12}}+\frac{330B_0^9B_2^2B_4^2}{B_1^{13}}+\frac{660B_0^9B_2B_3^2B_4}{B_1^{13}}\\
&-\frac{2{,}860B_0^9B_2^3B_3B_4}{B_1^{14}}-\frac{1{,}430B_0^9B_2^2B_3^3}{B_1^{14}}+\frac{45B_0^8B_2B_4^2}{B_1^{11}}+\frac{45B_0^8B_3^2B_4}{B_1^{11}}\\
&-\frac{495B_0^8B_2^2B_3B_4}{B_1^{12}}-\frac{165B_0^8B_2B_3^3}{B_1^{12}}+\frac{990B_0^8B_2^3B_3^2}{B_1^{13}}+\frac{4B_0^7B_4^2}{B_1^9}\\
&-\frac{72B_0^7B_2B_3B_4}{B_1^{10}}-\frac{12B_0^7B_3^3}{B_1^{10}}+\frac{120B_0^7B_2^3B_4}{B_1^{11}}+\frac{180B_0^7B_2^2B_3^2}{B_1^{11}}-\frac{7B_0^6B_3B_4}{B_1^8}\\
&+\frac{28B_0^6B_2^2B_4}{B_1^9}+\frac{28B_0^6B_2B_3^2}{B_1^9}-\frac{84B_0^6B_2^3B_3}{B_1^{10}}+\frac{6B_0^5B_2B_4}{B_1^7}+\frac{3B_0^5B_3^2}{B_1^7}\\
&-\frac{21B_0^5B_2^2B_3}{B_1^8}+\frac{B_0^4B_4}{B_1^5}-\frac{5B_0^4B_2B_3}{B_1^6}+\frac{5B_0^4B_2^3}{B_1^7}-\frac{B_0^3B_3}{B_1^4}+\frac{2B_0^3B_2^2}{B_1^5}+\frac{B_0^2B_2}{B_1^3}+\frac{B_0}{B_1}
\end{aligned}
$$

$\left[\frac{B_1}{B_2}\right]$, $m = 3$:

$$
\begin{aligned}
&92{,}400\frac{B_0^3B_1^4B_3^3B_4^3}{B_2^{13}}+\frac{8{,}400B_0^3B_1^3B_3^2B_4^3}{B_2^{11}}+\frac{560B_0^3B_1^2B_3B_4^3}{B_2^9}\\
&\quad-\frac{2{,}520B_0^3B_1^2B_3^3B_4^2}{B_2^{10}}+\frac{20B_0^3B_1B_4^3}{B_2^7}-\frac{210B_0^3B_1B_3^2B_4^2}{B_2^8}\\
&\quad-\frac{120{,}120B_0^2B_1^6B_3^3B_4^3}{B_2^{14}}-\frac{13{,}860B_0^2B_1^5B_3^2B_4^3}{B_2^{12}}-\frac{1{,}260B_0^2B_1^4B_3B_4^3}{B_2^{10}}\\
&\quad+\frac{6{,}300B_0^2B_1^4B_3^3B_4^2}{B_2^{11}}-\frac{70B_0^2B_1^3B_4^3}{B_2^8}+\frac{840B_0^2B_1^3B_3^2B_4^3}{B_2^9}+\frac{90B_0^2B_1^2B_3B_4^2}{B_2^7}\\
&\quad-\frac{210B_0^2B_1^2B_3^3B_4}{B_2^8}+\frac{6B_0^2B_1B_4^2}{B_2^5}-\frac{30B_0^2B_1B_3^2B_4}{B_2^6}+\frac{60{,}060B_0B_1^8B_3^3B_4^3}{B_2^{15}}\\
&\quad+\frac{7{,}920B_0B_1^7B_3^2B_4^3}{B_2^{13}}+\frac{840B_0B_1^6B_3B_4^3}{B_2^{11}}-\frac{4{,}620B_0B_1^6B_3^3B_4^2}{B_2^{12}}+\frac{56B_0B_1^5B_4^3}{B_2^9}\\
&\quad-\frac{756B_0B_1^5B_3^2B_4^2}{B_2^{10}}-\frac{105B_0B_1^4B_3B_4^2}{B_2^8}+\frac{280B_0B_1^4B_3^3B_4}{B_2^9}-\frac{10B_0B_1^3B_4^2}{B_2^6}\\
&\quad+\frac{60B_0B_1^3B_3^2B_4}{B_2^7}+\frac{12B_0B_1^2B_3B_4}{B_2^5}-\frac{10B_0B_1^2B_3^3}{B_2^6}+\frac{2B_0B_1B_4}{B_2^3}-\frac{3B_0B_1B_3^2}{B_2^4}\\
&\quad-\frac{10{,}010B_1^{10}B_3^3B_4^3}{B_2^{16}}-\frac{1{,}430B_1^9B_3^2B_4^3}{B_2^{14}}-\frac{165B_1^8B_3B_4^3}{B_2^{12}}+\frac{990B_1^8B_3^3B_4^2}{B_2^{13}}\\
&\quad-\frac{12B_1^7B_4^3}{B_2^{10}}+\frac{180B_1^7B_3^2B_4^2}{B_2^{11}}+\frac{28B_1^6B_3B_4^2}{B_2^9}-\frac{84B_1^6B_3^3B_4}{B_2^{10}}+\frac{3B_1^5B_4^2}{B_2^7}\\
&\quad-\frac{21B_1^5B_3^2B_4}{B_2^8}-\frac{5B_1^4B_3B_4}{B_2^6}+\frac{5B_1^4B_3^3}{B_2^7}-\frac{B_1^3B_4}{B_2^4}+\frac{2B_1^3B_3^2}{B_2^5}+\frac{B_1^2B_3}{B_2^3}+\frac{B_1}{B_2}
\end{aligned}
$$

$\left[\frac{B_2}{B_3}\right]$, $m = 10$:

$$
\begin{aligned}
&2{,}002\frac{B_0^3B_2^2B_4^{10}}{B_3^{15}} + \frac{220B_0^3B_2B_4^9}{B_3^{13}} + \frac{6{,}006B_0^2B_1^2B_2B_4^{10}}{B_3^{15}} - \frac{30{,}030B_0^2B_1B_2^3B_4^{10}}{B_3^{16}} \\
&- \frac{4{,}290B_0^2B_1B_2^2B_4^9}{B_3^{14}} - \frac{495B_0^2B_1B_2B_4^8}{B_3^{12}} + \frac{24{,}024B_0^2B_2^5B_4^{10}}{B_3^{17}} \\
&+ \frac{5{,}005B_0^2B_2^4B_4^9}{B_3^{15}} + \frac{990B_0^2B_2^3B_4^8}{B_3^{13}} + \frac{180B_0^2B_2^2B_4^7}{B_3^{11}} + \frac{28B_0^2B_2B_4^6}{B_3^9} \\
&- \frac{30{,}030B_0B_1^3B_2^2B_4^{10}}{B_3^{16}} - \frac{2{,}860B_0B_1^3B_2B_4^9}{B_3^{14}} + \frac{120{,}120B_0B_1^2B_2^4B_4^{10}}{B_3^{17}} \\
&+ \frac{20{,}020B_0B_1^2B_2^3B_4^9}{B_3^{15}} + \frac{2{,}970B_0B_1^2B_2^2B_4^8}{B_3^{13}} + \frac{360B_0B_1^2B_2B_4^7}{B_3^{11}} \\
&- \frac{136{,}136B_0B_1B_2^6B_4^{10}}{B_3^{18}} - \frac{30{,}030B_0B_1B_2^5B_4^9}{B_3^{16}} - \frac{6{,}435B_0B_1B_2^4B_4^8}{B_3^{14}} \\
&- \frac{1{,}320B_0B_1B_2^3B_4^7}{B_3^{12}} - \frac{252B_0B_1B_2^2B_4^6}{B_3^{10}} - \frac{42B_0B_1B_2B_4^5}{B_3^8} + \frac{43{,}758B_0B_2^8B_4^{10}}{B_3^{19}} \\
&+ \frac{11{,}440B_0B_2^7B_4^9}{B_3^{17}} + \frac{3{,}003B_0B_2^6B_4^8}{B_3^{15}} + \frac{792B_0B_1^5B_4^7}{B_3^{13}} + \frac{210B_0B_2^4B_4^6}{B_3^{11}} \\
&+ \frac{56B_0B_2^3B_4^5}{B_3^9} + \frac{15B_0B_2^2B_4^4}{B_3^7} + \frac{4B_0B_2B_4^3}{B_3^5} - \frac{3{,}003B_1^5B_2B_4^{10}}{B_3^{16}} \\
&+ \frac{40{,}040B_1^4B_2^3B_4^{10}}{B_3^{17}} + \frac{5{,}005B_1^4B_2^2B_4^9}{B_3^{15}} + \frac{495B_1^4B_2B_4^8}{B_3^{13}} - \frac{136{,}136B_1^3B_2^5B_4^{10}}{B_3^{18}} \\
&- \frac{25{,}025B_1^3B_2^4B_4^9}{B_3^{16}} - \frac{4{,}290B_1^3B_2^3B_4^8}{B_3^{14}} - \frac{660B_1^3B_2^2B_4^7}{B_3^{12}} - \frac{84B_1^3B_2B_4^6}{B_3^{10}} \\
&+ \frac{175{,}032B_1^2B_2^7B_4^{10}}{B_3^{19}} + \frac{40{,}040B_1^2B_2^6B_4^9}{B_3^{17}} + \frac{9{,}009B_1^2B_2^5B_4^8}{B_3^{15}} + \frac{1{,}980B_1^2B_2^4B_4^7}{B_3^{13}} \\
&+ \frac{420B_1^2B_2^3B_4^6}{B_3^{11}} + \frac{84B_1^2B_2^2B_4^5}{B_3^9} + \frac{15B_1^2B_2B_4^4}{B_3^7} - \frac{92{,}378B_1B_2^9B_4^{10}}{B_3^{20}} \\
&- \frac{24{,}310B_1B_2^8B_4^9}{B_3^{18}} - \frac{6{,}435B_1B_2^7B_4^8}{B_3^{16}} - \frac{1{,}716B_1B_2^6B_4^7}{B_3^{14}} - \frac{462B_1B_2^5B_4^6}{B_3^{12}} \\
&- \frac{126B_1B_2^4B_4^5}{B_3^{10}} - \frac{35B_1B_2^3B_4^4}{B_3^8} - \frac{10B_1B_2^2B_4^3}{B_3^6} - \frac{3B_1B_2B_4^2}{B_3^4} + \frac{16{,}796B_2^{11}B_4^{10}}{B_3^{21}} \\
&+ \frac{4{,}862B_2^{10}B_4^9}{B_3^{19}} + \frac{1{,}430B_2^9B_4^8}{B_3^{17}} + \frac{429B_2^8B_4^7}{B_3^{15}} + \frac{132B_2^7B_4^6}{B_3^{13}} + \frac{42B_2^6B_4^5}{B_3^{11}} \\
&+ \frac{14B_2^5B_4^4}{B_3^9} + \frac{5B_2^4B_4^3}{B_3^7} + \frac{2B_2^3B_4^2}{B_3^5} + \frac{B_2^2B_4}{B_3^3} + \frac{B_2}{B_3}
\end{aligned}
$$

APPENDIX 3. PROOF OF PROPOSITION 3.

By formula (40)

$$\frac{B_1}{B_0} = -\sum_{k=1}^{n} v_k S_k,$$

$$S_k = \sum_{j_k=1}^{M_k} \frac{1}{w_k(j_k)}$$

where $w_k(j_k)$ is the value of the rate of stage with index k calculated at zero with index j_k of the system (27) at $R = 0$. These zeros could be *boundary* (i.e. $z_l = 0$ for some $1 \le l \le n$) and *interior* (otherwise). Note, that system (27) always has the interior (toric) roots and may or may not have the boundary roots. Consider some $S_k : v_k \neq 0$. We can present it as

$$S_k = S_k^B + S_k^I \tag{A3.1}$$

where $S_k^B = \sum_{j \in B} 1/w_k(j)$, $S_k^I = \sum_{j \in I} 1/w_k(j)$ and B and I are the sets of boundary and interior zeros.

It follows from Lemma 14.3 (see Bykov et al., 1998) that in the case $v_k \neq 0$ we have $w_k(j) = f_k z^{\alpha^k} \neq 0$ or $w_k(j) = r_k z^{\beta^k} \neq 0$ i.e. w_k (j) is generically non-zero and finite in the point of equilibrium.[23] Thus, S_k^B cannot have C^m, $m \neq 0$ as a factor.

There are $|\Delta_k|$ interior zeroes (see Bykov et al., 1998). They can be subdivided into p groups, each group producing the multiplier C in the coefficient B_0 (see proof of the Theorem 14.1, Bykov et al., 1998). We can write S_k^I as

$$S_k^I = \frac{\sum\limits_{l=1}^{p} \sum\limits_{j \in I_l} \prod\limits_{i \in I[l]} w_k(i) \cdot \prod\limits_{i \in I_l\ i \neq j} w_k(i)}{\prod\limits_{j \in I} w_k(j)} \tag{A3.2}$$

where I_l is lth group. Let $c = (\prod_{i=1}^{n} (f_i/r_i)^{v_i} - 1)$, v_i is defined as in Proposition 1 and let $w_{-k}(j) = r_k z(j)^{\beta^k}$. Then

$$S_k^I = (-1)^{v_k - 1} \frac{\sum\limits_{l=1}^{p} \prod\limits_{i \in I[l]} w_{-k}(i) \prod\limits_{i \in I_l, i \neq j} w_k(i)}{c \prod\limits_{j \in I} w_{-k}(j)} \tag{A3.3}$$

There are exactly p interior zeroes $\mathbf{z}(j_l^{\text{eq}})$, $l = 1, \ldots, p$ such that $w_k(j_l^{\text{eq}})$ vanishes at the equilibrium. At the equilibrium

$$\prod_{i \neq j_l^{\text{eq}}} w_k(i) = (-1)^{v_k - 1} v_k \prod_{i \neq j_l^{\text{eq}}} w_{-k}(i)$$

[23] This follows from the formula for solution z_l (j) for non-zero z_l (see Lemma 14.1 in Bykov et al., 1998).

and at the vicinity of equilibrium (i.e. at $c \to 0$) we have

$$S_k^I\big|_{\text{eq}} \cong \frac{v_k}{c}\sum_{l=1}^{p}\frac{1}{w_{-k}(j_l^{\text{eq}})} \tag{A3.4}$$

As S_k^B is finite, in the neighborhood of equilibrium we have

$$S_k\big|_{\text{eq}} \cong S_k^I\big|_{\text{eq}}$$

Now, it is sufficient to prove that $y_k = \sum_{l=1}^{p}(1/w_{-k}(j_l^{\text{eq}}))\neq 0$ and finite at the equilibrium. This will ensure that y_k cannot produce additional c-factors, so that $(B_0)/(B_1) \sim 1/C$ which means that $B_1 \sim C^{p-1}$ (see Propositions 1 and 2). It is obvious that $y_k > 0$ if $p = 1$ (the only root of system (27) is positive at $R = 0$). Let us prove this for the case of $p \geq 1$.

Lemma $y_k > 0$ and finite

We may assume $k = 1$. Let $w_{1\text{eq}} = w_{-1}(j_1^{\text{eq}}) \equiv w_1(j_1^{\text{eq}})$. Applying the explicit formula for interior roots (see Bykov et al., 1998; Lazman and Yablonskii, 1991), we have[24]

$$\frac{1}{w_{1\text{eq}}} = \frac{1}{f_1}\left[\sum_{l=1}^{n}\prod_{j=2}^{n}\kappa_j^{-\frac{\Delta_{jl}^1}{p_1\Delta_1}}\cdot\exp\left(2\pi i\sum_{j=2}^{n}m_j\frac{\Delta_{jl}^1}{p_1\Delta_1}\right)\right]^{p_1} \tag{A3.5}$$

We can define integers m_j, $j = 2, \ldots, n$ as (see Bykov et al., 1998, p. 145)

$$\begin{aligned} m_j &= v_1 q m_j^0, \qquad q = 0,\ v_1, \ldots, (p-1)v_1, \\ m_1^0 v_1 &+ \cdots + m_n^0 v_n = -1 \end{aligned} \tag{A3.6}$$

Remind that v_1 belongs to the set of relatively prime stoichiometric numbers, so that

$$\Delta_1 = p v_1 \tag{A3.7}$$

Taking into account Equations (A3.5) and (A3.6) we can write

$$\sum_{l=1}^{p}\frac{1}{w_{1\text{eq}}} = \frac{1}{f_1}\sum_q\left[\sum_{l=1}^{n}\prod_{j=2}^{n}\kappa_j^{-\frac{\Delta_{jl}^1}{p_1 p v_1}}\cdot\exp\left(2\pi i q\sum_{j=2}^{n}m_j^0\frac{\Delta_{jl}^1}{p_1 p}\right)\right]^{p_1}, \qquad q = 0,\ v_1, \ldots, (p-1)v_1 \tag{A3.8}$$

Let us now expand the [] in Equation (A3.8) and consider one of the expansion terms

$$T = C_a t_1^{a_1}\cdot\ldots\cdot t_n^{a_n}, \ \ \|a\| = p_1$$

[24] Notation is the same as in formula (52), (see Section 3.1.1).

We have

$$T = C_a \prod_{j=2}^{n} \kappa_j^{-\frac{\mu_j}{p_1 p \nu_1}} \cdot \exp\left(2\pi i q \frac{\sum_{j=2}^{n} m_j^0 \mu_j}{p_1 p}\right) \tag{A3.9}$$

where

$$\mu_j = \begin{vmatrix} \alpha_{11} & \cdots & \alpha_{1n} \\ \vdots & \gamma & \vdots \\ a_1 & \{row\ j\} & a_n \\ & \gamma & \end{vmatrix} = p_1 \mu_j^0 \tag{A3.10}$$

The property (A3.10) becomes obvious after adding up the columns (all sums are 0 except sums in rows 1 and j where they are equal to p_1). Collecting all terms corresponding to the term T for all values of q we have

$$\sum_{q=0}^{p-1} T_q = C_a \prod_{j=2}^{n} \kappa_j^{-\frac{\mu_j^0}{p\nu_1}} s$$

where

$$s = 1 + g + \cdots + g^{p-1},$$

$$g = \exp\left(2\pi i \frac{\sum_{j=2}^{n} m_j^0 \mu_j^0}{p}\right)$$

We have

$$s = \begin{cases} p, & \text{if } p \text{ divides} \sum_{j=2}^{n} m_j^0 \mu_j^0 \\ 0, & \text{otherwise} \end{cases}$$

Thus, $s \geq 0$. However $\mu_j^0 = 0$ if $\alpha_{11} = a_1, \ldots, \alpha_{1n} = a_n$. Thus, we have strictly positive addendum that proves the Lemma.

In the vicinity of equilibrium we have

$$-\frac{B_0}{B_1} \cong \frac{\prod_{i=1}^{n} (f_i/r_i)^{\nu_i} - 1}{\sum_{k=1}^{n} \nu_k^2 \sum_{l=1}^{p} 1/w_{-k}(j_l^{eq})} \tag{A3.11}$$

and the denominator of Equation (A3.11) is strictly positive at the equilibrium.

Note that Equation (A3.11) produces formula (53) for the overall reaction rate at the linear vicinity of thermodynamic equilibrium. Formula (53) follows from Equation (A3.11) when $p = 1$.

APPENDIX 4. THERMODYNAMIC BRANCH IS FEASIBLE

We are going to show that in the vicinity of equilibrium, the thermodynamic branch (66), corresponds to the feasible solution of system (21) (i.e. solution with $z_j > 0,\ j = 1, \ldots, n$).

Proposition A4.1. System (21) can have one and only one feasible zero with $R = 0$[25] if

$$f_1^{\nu_1}, \ldots, f_n^{\nu_n} = r_1^{\nu_1}, \ldots, r_n^{\nu_n} \tag{A4.0}$$

Proof. At the equilibrium, $z_1, \ldots, z_n$ should solve equilibrium subsystem (27). Let $s = 1$.

Interior solutions of equilibrium subsystem satisfy the following relations (see Bykov et al., 1998, p. 144)

$$\begin{aligned} z_1 &= z_n \mathbf{C}^{\Delta^1/\Delta_1} e^{2\pi i < \mathbf{m}, \Delta^1/\Delta_1 >}, \\ &\vdots \\ z_{n-1} &= z_n \mathbf{C}^{\Delta^{n-1}/\Delta_1} e^{2\pi i < \mathbf{m}, \Delta^{n-1}/\Delta_1 >}, \\ z_1 &+ \ldots + z_n = 1 \end{aligned} \tag{A4.1}$$

where $\Delta_1 = \det \Gamma_1, \Gamma_1 = (\alpha_j^s - \beta_j^s)_{j=1,\ldots,n-1}^{s=2,\ldots,n}$, $\Delta^j = (\Delta_2^j, \ldots, \Delta_n^j)$ is vector of co-factors of the elements of jth column of matrix Γ_1, $\mathbf{C} = (f_2/r_2, \ldots, f_n/r_n)$ and $\mathbf{m} = (m_2, \ldots, m_n)$ is some integer vector.

The only feasible solution of Equation (A4.1) is

$$z_i = z_n K^{\Delta^i/\Delta_1}, \qquad z_n = 1/(1 + K^{\Delta^1/\Delta_1} + K^{\Delta^{n-1}/\Delta_1}) \tag{A4.2}$$

At the equilibrium, we have

$$f_1 z^{\alpha^1} = r_1 z^{\beta^1} \tag{A4.3}$$

Substituting Equation (A4.2) into Equation (A4.3) we have the following necessary and sufficient condition of the equilibrium

$$\mathrm{K_{eq}} = 1 \tag{A4.4}$$

where

$$\mathrm{K_{eq}} = (f_1/r_1)(f_2/r_2)^{\nu_2/\nu_1}, \ldots, (f_n/r_n)^{\nu_n/\nu_1} \tag{A4.5}$$

[25] Equilibrium point is unique, as expected from thermodynamics of ideal systems.

Conditions (A4.2) and (A4.4) (it is equivalent to the condition (A4.0) for feasible values of f_i, r_i) are necessary and sufficient conditions for $z_1>0, \ldots, z_n>0$ to solve the system (21).

Proposition A4.2. Thermodynamic branch $R=-[(B_0)/(B_1)]$ corresponds to the feasible solution in the vicinity of equilibrium.

(i) Let $p=1$ (see Propositions 1 and 2). Equilibrium subsystem (21), corresponding to $s=1$, has $|\Delta_1|$ interior solutions (see Lemma 14.1 in Bykov et al., 1998).

If $p=1$, $|\Delta_1|=\nu_1$. It follows from Equation (A4.4) that at the equilibrium each of these ν_1 solutions should satisfy condition

$$K_{eq}e^{2\pi i(j/\nu_1)}=1, \qquad j=0,1,\ldots,\nu_1-1 \tag{A4.6}$$

(for details of obtaining formula (A4.6) see proof of Proposition 1, Lazman and Yablonskii (1991) and Bykov et al., 1998).

For $K_{eq}>0$, condition (A4.6) could be satisfied only, if $j=0$. There is a bijection between solution (A4.1) and condition (A4.6), and the case $j=0$ corresponds to the only feasible solution (A4.2) (see Proposition A4.1). However, when $p=1$, there is only one branch of solutions of kinetic polynomial vanishing at the equilibrium.[26] As the thermodynamic branch satisfies the equilibrium condition (A4.0) and there are no other branches vanishing at the equilibrium (we proved in Appendix 3 that $B_1\neq 0$ at the equilibrium (see also Lazman and Yablonskii, 1991), this branch should be feasible. By continuity, this property should be valid in some vicinity of equilibrium.

(ii) Let now $p>1$. In this case, we have p branches of kinetic polynomial zeros vanishing at the equilibrium. Which one corresponds to the thermodynamic branch?

In the vicinity of equilibrium, the overall reaction rate satisfies Equation (53a), i.e.

$$R=\frac{\prod_{j=1}^{n}\kappa_j^{\nu_j}-1}{\sum_{k=1}^{n}\nu_k^2/w_{k_{eq}}} \tag{A4.7}$$

Formula (A4.7) is valid for $w_{k_{eq}}$ defined by formula (A3.4). They may correspond to feasible or non-feasible solutions of system (27) at the equilibrium. Let w_k^f is feasible and $w_k^{[f]}$ is non-feasible value of $w_{k_{eq}}$. For, $k=1$ we have[27]

$$1/\left|w_1^{[f]}\right|<(1/f_1)\left|\sum\prod\right|^{p_1}\leq(1/f_1)\left(\sum\left|\prod\right|\right)^{p_1}=1/w_1^f$$

[26] By definition of the resultant, every root of kinetic polynomial solves the system and vice versa.

[27] Formula (A3.4) is presented schematically here.

Then,

$$\left|\sum_{k=1}^{n} v_k^2/w_k^{[f]}\right| \le \sum_{k=1}^{n} v_k^2 \Big/ |w_k^{[f]}| < \sum_{k=1}^{n} v_k^2 \Big/ \left|w_k^f\right|.$$

It follows from (A.4.7) that

$$\left|R^{[f]}\right| \ge \left|R^f\right| \tag{A4.8}$$

Note that the equal condition in Equation (A4.8) requires the equilibrium (i.e. Equation (A.4.0)). Thus, in the vicinity of the equilibrium, the non-zero absolute value of feasible reaction rate is always smaller than absolute values of reaction rate corresponding to non-feasible solutions vanishing at the equilibrium. Near the equilibrium, these p "small" solutions are close to the roots of the equation (see Lazman and Yablonskii, 1991, in more detail)

$$B_0' C^p + B_1' C^{p-1} R + \cdots + B_p R^p = 0$$

where B_0' and B_1' are defined in Equation (71). Let $R = vC$. Then $\left|v^f\right| < \left|v^{[f]}\right|$ and we are interested in the root of polynomial

$$B_0' + B_1' v + \cdots + B_p v^p = 0 \tag{A4.9}$$

with the smallest absolute value. This root vanishes if $B_0' = 0$, which happens for the thermodynamic branch, i.e. $v = -[B_0'/B_1']$.

Example

For mechanism of para–ortho conversion of hydrogen

$$\begin{aligned}&(1)\quad H_2(\text{para}) + 2Z \Leftrightarrow 2HZ\\ &(2)\quad 2HZ \Leftrightarrow H_2(\text{ortho}) + 2Z\end{aligned}$$

we have $p = 2$. Figure 18 shows both roots of corresponding kinetic polynomial $(r_1 + f_2 - r_2 - f_1)^2 R^2 - 2(f_1 f_2 - r_1 r_2)(r_1 + f + r_2 + f_2)R + (f_1 f_2 - r_1 r_2)^2$ and their hypergeometric approximations. Note, that the thermodynamic branch converges to the feasible root which absolute value is always smaller than the non-feasible root.

APPENDIX 5. STURMFELS SERIES COINCIDES WITH THE BIRKELAND SERIES FOR THE CASE OF "THERMODYNAMIC BRANCH"

Sturmfels series. It follows from Equations (34) and (35) at $j = 1$ that

$$x = \sum_{i_2=0,\ldots,\, i_n=0}^{\infty} \frac{(-1)^{i_1+1}}{1+i_0} \binom{i_1}{i_2, \ldots, i_n} \cdot \frac{a_0^{1+i_0} \prod_{k=2}^{n} a_k^{i_k}}{a_1^{1+i_1}} \tag{A5.1}$$

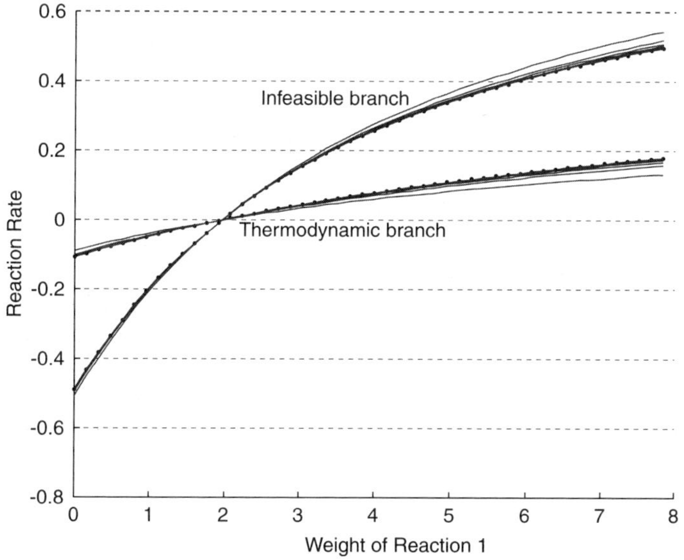

Figure 18 Overall reaction rate and its approximation for the mechanism of hydrogen para–ortho conversion.

where

$$i_0 = \sum_{k=2}^{n}(k-1)i_k, \qquad i_1 = \sum_{k=2}^{n} k i_k$$

Birkeland series. Consider formula (60) at $p = 0$ and $q = 1$. We have

$$\varepsilon = (-1)^{1/(q-p)} = -1,$$

$$-\left\langle \beta_q, k \right\rangle = \sum_{k=2}^{n} i_k k = i_1,$$

$$\left\langle \beta_p, k \right\rangle = \sum_{k=2}^{n}(k-1)i_k = i_0,$$

$$Y_1(b_2,\ldots,b_n) = \sum_{i_2=0,\ldots,\ i_n=0}^{\infty} \frac{(-1)^{1+i_1}\Gamma(1+i_1)}{i_2!\ldots i_n!\ \Gamma(1+i_0+1)} \cdot \prod_{k=2}^{n} b_k^{i_k}$$

$$\equiv \sum_{i_2=0,\ldots,\ i_n=0}^{\infty} \frac{(-1)^{1+i_1} i_1!}{i_2!\ldots i_n!\ (i_0+1)!} \cdot \prod_{k=2}^{n} b_k^{i_k}$$

$$\equiv \sum_{i_2=0,\ldots,\ i_n=0}^{\infty} \frac{(-1)^{1+i_1}\left(\sum\limits_{k=2}^{n} k i_k\right)!}{(i_0+1)\ \ i_2!\ldots i_n!\left(\sum\limits_{k=2}^{n} k i_k - \sum\limits_{k=2}^{n} i_k\right)!} \cdot \prod_{k=2}^{n} b_k^{i_k}$$

$$\equiv \sum_{i_2=0,\ldots,\ i_n=0}^{\infty} \frac{(-1)^{1+i_1}!}{(i_0+1)} \cdot \begin{pmatrix} i_1 \\ i_2\ldots i_n \end{pmatrix} \cdot \prod_{k=2}^{n} b_k^{i_k}$$

Then,

$$x = \lambda_1 Y(\lambda_0 \lambda_1^2 a_2, \ldots, \lambda_0 \lambda_1^n a_n)$$

where

$$\lambda_0 = \frac{1}{a_0}, \qquad \lambda_1 = \frac{a_0}{a_1}$$

Thus

$$x = \frac{a_0}{a_1} \sum_{i_2=0,\ldots,\, i_n=0}^{\infty} \frac{(-1)^{1+i_1}!}{(i_0+1)} \cdot \begin{pmatrix} i_1 \\ i_2 \ldots i_n \end{pmatrix} \cdot \prod_{k=2}^{n} \left(\frac{a_0^{k-1} a_k}{a_1^k} \right)^{i_k}$$

$$\equiv \sum_{i_2=0,\ldots,\, i_n=0}^{\infty} \frac{(-1)^{1+i_1}!}{(i_0+1)} \cdot \begin{pmatrix} i_1 \\ i_2 \ldots i_n \end{pmatrix} \cdot \frac{a_0^{1+i_0} \prod\limits_{k=2}^{n} a_k^{i_k}}{a_1^{1+i_1}}$$

CHAPTER 3

Dynamic and Static Limitation in Multiscale Reaction Networks, Revisited

A.N. Gorban[1,*] and **O. Radulescu**[2]

Contents

[1] Department of Mathematics, University of Leicester, LE1 7RH, UK

[2] IRMAR, UMR 6625, University of Rennes 1, Campus de Beaulieu, 35042 Rennes, France

*Corresponding author.
E-mail address: ag153@le.ac.uk

Advances in Chemical Engineering, Volume 34
ISSN 0065-2377, DOI 10.1016/S0065-2377(08)00003-3

Abstract

The concept of the limiting step gives the limit simplification: the whole network behaves as a single step. This is the most popular approach for model simplification in chemical kinetics. However, in its elementary form this idea is applicable only to the simplest linear cycles in steady states. For simple cycles the nonstationary behavior is also limited by a single step, but not the same step that limits the stationary rate. In this chapter, we develop a general theory of static and dynamic limitation for all linear multiscale networks. Our main mathematical tools are auxiliary discrete dynamical systems on finite sets and specially developed algorithms of "cycles surgery" for reaction graphs. New estimates of eigenvectors for diagonally dominant matrices are used.

Multiscale ensembles of reaction networks with well-separated constants are introduced and typical properties of such systems are studied. For any given ordering of reaction rate constants the explicit approximation of steady state, relaxation spectrum and related eigenvectors ("modes") is presented. In particular, we prove that for systems with well-separated constants eigenvalues are real (damped oscillations are improbable). For systems with modular structure, we propose the selection of such modules that it is possible to solve the kinetic equation for every module in the explicit form. All such "solvable" networks are described. The obtained multiscale approximations, that we call "dominant systems" are computationally cheap and robust. These dominant systems can be used for direct computation of steady states and relaxation dynamics, especially when kinetic information is incomplete, for design of experiments and mining of experimental data, and could serve as a robust first approximation in perturbation theory or for preconditioning.

1. INTRODUCTION

Which approach to model reduction is the most important? Population is not the ultimate judge, and popularity is not a scientific criterion, but "Vox populi, vox Dei", especially in the epoch of citation indexes, impact factors and bibliometrics. Let us ask Google. It gave on 31st December 2006:

- for "quasi-equilibrium" — 301,000 links;
- for "quasi-steady state" 347,000 and for "pseudo-steady state" 76,200, 42,3000 together;
- for our favorite "slow manifold" (Gorban and Karlin, 2003, 2005) 29,800 links only, and for "invariant manifold" slightly more, 98,100;
- for such a framework topic as "singular perturbation" Google gave 361,000 links;
- for "model reduction" even more, as we did expect, 373,000;
- but for "limiting step" almost two times more — 714,000!

Our goal is the general theory of static and dynamic limitation for multiscale networks. The concept of the limiting step gives, in some sense, the limit simplification: the whole network behaves as a single step. As the first result of our chapter we introduce further detail in this idea: the whole network behaves as a single step in statics, and as *another* single step in dynamics: even for simplest cycles the stationary rate and the relaxation time to this stationary rate are limited by different reaction steps, and we describe how to find these steps.

The concept of limitation is very attractive both for theorists and experimentalists. It is very useful to find conditions when a selected reaction step becomes the limiting step. We can change conditions and study the network experimentally, step-by-step. It is very convenient to model a system with limiting steps: the model is extremely simple and can serve as a very elementary building block for further study of more complex systems, a typical situation both in industry and in systems biology.

In the IUPAC Compendium of Chemical Terminology (2007) one can find two articles with a definition of limitation.

- Rate-determining step (rate-limiting step) (2007): "These terms are best regarded as synonymous with rate-controlling step".
- Rate-controlling step (2007): "A rate-controlling (rate-determining or rate-limiting) step in a reaction occurring by a composite reaction sequence is an elementary reaction the rate constant for which exerts a strong effect — stronger than that of any other rate constant — on the overall rate".

It is not wise to object to a definition and here we do not object, but, rather, complement the definition by additional comments. The main comment is that usually when people are talking about limitation they expect significantly more: there exists a rate constant which exerts such a strong effect on the overall rate that the effect of all other rate constants together is significantly smaller. Of course, this is not yet a formal definition, and should be complemented by a definition of "effect", for example, by "control function" identified by derivatives of the overall rate of reaction, or by other overall rate "sensitivity parameters" (Rate-controlling step, 2007).

For the IUPAC Compendium definition a rate-controlling step always exists, because among the control functions generically exists the biggest one. On the contrary, for the notion of limitation that is used in practice, there exists a difference between systems with limitation and systems without limitation.

An additional problem arises: are systems without limitation rare or should they be treated equitably with limitation cases? The arguments in favor of limitation typicality are as follows: the real chemical networks are multi-scale with very different constants and concentrations. For such systems it is improbable to meet a situation with compatible effects of all different stages. Of course, these arguments are statistical and apply to generic systems from special ensembles.

During the last century, the concept of the limiting step was revised several times. First simple idea of a "narrow place" (a least conductive step) could be applied without adaptation only to a simple cycle of irreversible steps that are of the first order (see Chapter 16 of the book Johnston (1966) or the paper of Boyd (1978)). When researchers try to apply this idea in more general situations they meet various difficulties such as:

- Some reactions have to be "pseudomonomolecular". Their constants depend on concentrations of outer components, and are constant only under condition that these outer components are present in constant concentrations, or change sufficiently slow. For example, the simplest Michaelis–Menten enzymatic reaction is E+S→ES→E+P (E here stands for enzyme, S for substrate and P for product), and the linear catalytic cycle here is S→ES→S. Hence, in general we must consider nonlinear systems.
- Even under fixed outer components concentration, the simple "narrow place" behavior could be spoiled by branching or by reverse reactions. For such reaction systems definition of a limiting step simply as a step with the smallest constant does not work. The simplest example is given by the cycle: $A_1 \leftrightarrow A_2 \rightarrow A_3 \rightarrow A_1$. Even if the constant of the last step $A_3 \rightarrow A_1$ is the smallest one, the stationary rate may be much smaller than $k_3 b$ (where b is the overall balance of concentrations, $b = c_1+c_2+c_3$), if the constant of the reverse reaction $A_2 \rightarrow A_1$ is sufficiently big.

In a series of papers, Northrop (1981, 2001) clearly explained these difficulties with many examples based on the isotope effect analysis and suggested that the concept of rate-limiting step is "outmoded". Nevertheless, the main idea of limiting is so attractive that Northrop's arguments stimulated the search for modification and improvement of the main concept.

Ray (1983) proposed the use of sensitivity analysis. He considered cycles of reversible reactions and suggested a definition: *The rate-limiting step in a reaction sequence is that forward step for which a change of its rate constant produces the largest effect on the overall rate.* In his formal definition of sensitivity functions the reciprocal reaction rate ($1/W$) and rate constants ($1/k_i$) were used and the connection between forward and reverse step constants (the equilibrium constant) was kept fixed.

Ray's approach was revised by Brown and Cooper (1993) from the system control analysis point of view (see the book of Cornish-Bowden and Cardenas, 1990). They stress again that there is no unique rate-limiting step specific for an enzyme, and this step, even if it exists, depends on substrate, product and effector concentrations. They also demonstrated that the control coefficients

$$C_{k_i}^W = \left(\frac{k_i}{W}\frac{\partial W}{\partial k_i}\right)_{[S],[P],\ldots}$$

where W is the stationary reaction rate and k_i are constants, are additive and obey the summation theorems (as concentrations do). A simple relation between control coefficients of rate constants and intermediate concentrations was reported by Kholodenko et al. (1994). This relation connects two type of experiments: measurement of intermediate levels and steady-state rate measurements.

For the analysis of nonlinear cycles the new concept of *kinetic polynomial* was developed (Lazman and Yablonskii, 1991; Yablonskii et al., 1982). It was proven that the stationary state of the single-route reaction mechanism of catalytic reaction can be described by a single polynomial equation for the reaction rate. The roots of the kinetic polynomial are the values of the reaction rate in the steady state. For a system with limiting step the kinetic polynomial can be approximately solved and the reaction rate found in the form of a series in powers of the limiting-step constant (Lazman and Yablonskii, 1988).

In our approach, we analyze not only the steady-state reaction rates, but also the relaxation dynamics of multiscale systems. We focused mostly on the case when all the elementary processes have significantly different timescales. In this case, we obtain "limit simplification" of the model: all stationary states and relaxation processes could be analyzed "to the very end", by straightforward computations, mostly analytically. Chemical kinetics is an inexhaustible source of examples of multiscale systems for analysis. It is not surprising that many ideas and methods for such analysis were first invented for chemical systems.

In Section 2 we analyze a simple example and the source of most generalizations, the catalytic cycle, and demonstrate the main notions on this example. This analysis is quite elementary, but includes many ideas elaborated in full in subsequent sections.

There exist several estimates for relaxation time in chemical reactions (developed, e.g. by Cheresiz and Yablonskii, 1983), but even for the simplest cycle with limitation the main property of relaxation time is not widely known. For a simple irreversible catalytic cycle with limiting step the stationary rate is controlled by the smallest constant, but the relaxation time is determined by the second in order constant. Hence, if in the stationary rate experiments for that cycle we mostly extract the smallest constant, in relaxation experiments another, the second in order constant will be observed.

It is also proven that for cycles with well-separated constants damped oscillations are impossible, and spectrum of the matrix of kinetic coefficients is real. For general reaction networks with well-separated constants this property is proven in Section 4.

Another general effect observed for a cycle is robustness of stationary rate and relaxation time. For multiscale systems with random constants, the standard deviation of constants that determine stationary rate (the smallest constant for a cycle) or relaxation time (the second in order constant) is approximately n times smaller than the standard deviation of the individual constants (where n is the cycle length). Here we deal with the so-called order statistics. This decrease of the deviation as n^{-1} is much faster than for the standard error summation, where it decreases with increasing n as $n^{-1/2}$.

In more general settings, robustness of the relaxation time was studied by Gorban and Radulescu (2007) for chemical kinetics models of genetic and signaling networks. Gorban and Radulescu (2007) proved that for large multiscale systems with hierarchical distribution of timescales the variance of the inverse relaxation time (as well as the variance of the stationary rate) is much lower than the variance of the separate constants. Moreover, it can tend to 0 faster than $1/n$, where n is the number of reactions. It was demonstrated that similar phenomena are valid in the nonlinear case as well. As a numerical illustration we used a model of a signaling network that can be applied to important transcription factors such as NFkB.

Each multiscale system is characterized by its structure (the system of elementary processes) and by the rate constants of these processes. To make any general statement about such systems when the structure is given but the constants are unknown it is useful to take the constant set as random and independent. But it is not obvious how to chose the random distribution. The usual idea to take normal or uniform distribution meets obvious difficulties, the timescales are not sufficiently well separated.

The statistical approach to chemical kinetics was developed by Li et al. (2001, 2002), and high-dimensional model representations (HDMR) were proposed as efficient tools to provide a fully global statistical analysis of a model. The work of Feng et al. (2004) was focused on how the network properties are affected by random rate constant changes. The rate constants were transformed to a logarithmic scale to ensure an even distribution over the large space.

The log-uniform distribution on sufficiently wide interval helps us to improve the situation, indeed, but a couple of extra parameters appears: $\alpha = \min \log k$ and $\beta = \max \log k$. We have to study the asymptotics $\alpha \to -\infty$, $\beta \to \infty$. This approach could be formalized by means of the uniform invariant distributions of $\log k$ on $\mathbb{R}^n$. These distributions are finite-additive, but not countable-additive (not σ-additive).

The probability and measure theory without countable additivity has a long history. In Euclid's time only arguments based on finite-additive properties of volume were legal. Euclid meant by equal area the scissors congruent area. Two polyhedra are scissors-congruent if one of them can be cut into finitely many polyhedral pieces which can be reassembled to yield the second. But all proofs of the formula for the volume of a pyramid involve some form of limiting process. Hilbert asked in his third problem: are two Euclidean polyhedra of the same volume scissors congruent? The answer is "no" (a review of old and recent results is presented by Neumann, 1998). There is another invariant of cutting and gluing polyhedra.

Finite-additive invariant measures on non-compact groups were studied by Birkhoff (1936) (see also the book of Hewitt and Ross, 1963, Chapter 4). The frequency-based Mises approach to probability theory foundations (von Mises, 1964), as well as logical foundations of probability by Carnap (1950) do not need σ-additivity. Non-Kolmogorov probability theories are discussed now in the context of quantum physics (Khrennikov, 2002), nonstandard analysis (Loeb, 1975) and many other problems (and we do not pretend provide here is a full review of related works).

We answer the question: What does it mean "to pick a multiscale system at random"? We introduce and analyze a notion of multiscale ensemble of reaction systems. These ensembles with well-separated variables are presented in Section 3.

The best geometric example that helps us to understand this problem is one of the Lewis Carroll's Pillow Problems published in 1883 (Carroll, 1958): "Three points are taken at random on an infinite plane. Find the chance of their being the vertices of an obtuse-angled triangle." (In an acute-angled triangle all angles are comparable, in an obtuse-angled triangle the obtuse angle is bigger than others and could be much bigger.) The solution of this problem depends significantly on the ensemble definition. What does it mean "points are taken at random on an infinite plane"? Our intuition requires translation invariance, but the normalized translation invariant measure on the plain could not be σ-additive. Nevertheless, there exist finite-additive invariant measures.

Lewis Carroll proposed a solution that did not satisfy some of modern scientists. There exists a lot of attempts to improve the problem statement (Eisenberg and Sullivan, 1996; Falk and Samuel-Cahn, 2001; Guy, 1993; Portnoy, 1994): reduction from infinite plane to a bounded set, to a compact symmetric space, etc. But the elimination of paradox destroys the essence of Carroll's problem. If we follow the paradox and try to give a meaning to "points are taken at random on an infinite plane" then we replace σ-additivity of the probability measure by finite-additivity and come to the applied probability theory for finite-additive probabilities. Of course, this theory for abstract probability spaces would be too poor, and some additional geometric and algebraic structures are necessary to build rich enough theory.

This is not just a beautiful geometrical problem, but rather an applied question about the proper definition of multiscale ensembles. We need such a definition to make any general statement about multiscale systems, and briefly analyze lessons of Carroll's problem in Section 3.

In this section, we use some mathematics to define the multiscale ensembles with well-separated constants. This is necessary background for the analysis of systems with limitation, and technical consequences are rather simple. We need only two properties of a typical system from the multiscale ensemble with well-separated constants:

(i) Every two reaction rate constants k, k', are connected by the relation $k \gg k'$ or $k \ll k'$ (with probability close to 1);
(ii) The first property persists (with probability close to 1), if we delete two constants k and k' from the list of constants, and add a number kk' or a number k/k' to that list.

If the reader can use these properties (when it is necessary) without additional clarification, it is possible to skip reading Section 3 and go directly to more applied sections. In Section 4 we study static and dynamic properties of linear multiscale reaction networks. An important instrument for that study is a hierarchy of auxiliary discrete dynamical system. Let A_i be nodes of the network ("components"), $A_i \to A_j$ be edges (reactions), and k_{ji} be the constants of these reactions (please pay attention to the inverse order of subscripts). A discrete dynamical system ϕ is a map that maps any node A_i in a node $A_{\phi(i)}$. To construct a first auxiliary dynamical system for a given network we find for each A_i the maximal constant of reactions $A_i \to A_j$: $k_{\phi(i)i} \geq k_{ji}$ for all j, and $\phi(i) = i$ if there are no reactions $A_i \to A_j$. Attractors in this discrete dynamical system are cycles and fixed points.

The fast stage of relaxation of a complex reaction network could be described as mass transfer from nodes to correspondent attractors of auxiliary dynamical system and mass distribution in the attractors. After that, a slower process of mass redistribution between attractors should play a more important role. To study the next stage of relaxation, we should glue cycles of the first auxiliary system (each cycle transforms into a point), define constants of the first derivative network on this new set of nodes, construct for this new network an (first) auxiliary discrete dynamical system, etc. The process terminates when we get a discrete dynamical system with one attractor. Then the inverse process of cycle restoration and cutting starts. As a result, we create an explicit description of the relaxation process in the reaction network, find estimates of eigenvalues and eigenvectors for the kinetic equation, and provide full analysis of steady states for systems with well-separated constants.

The problem of multiscale asymptotics of eigenvalues of nonself-adjoint matrices was studied by Vishik and Ljusternik (1960) and Lidskii (1965). Recently, some generalizations were obtained by idempotent (min-plus) algebra methods (Akian et al., 2004). These methods provide a natural language for discussion of some multiscale problems (Litvinov and Maslov, 2005). In the Vishik–Ljusternik–Lidskii theorem and its generalizations the asymptotics of eigenvalues and eigenvectors for the family of matrices $A_{ij}(\varepsilon) = a_{ij}\varepsilon^{A_{ij}} + o(\varepsilon^{A_{ij}})$ is studied for $\varepsilon > 0$, $\varepsilon \to 0$.

In the chemical reaction networks that we study, there is no small parameter ε with a given distribution of the orders $\varepsilon^{A_{ij}}$ of the matrix nodes. Instead of these powers of ε we have orderings of rate constants. Furthermore, the matrices of kinetic equations have some specific properties. The possibility to operate with the graph of reactions (cycles surgery) significantly helps in our constructions. Nevertheless, there exists some similarity between these problems and, even for general matrices, graphical representation is useful. The language of idempotent algebra (Litvinov and Maslov, 2005), as well as nonstandard analysis with infinitisemals (Albeverio et al., 1986), can be used for description of the multiscale reaction networks, but now we postpone this for later use.

We summarize results of relaxation analysis and describe the algorithm of approximation of steady state and relaxation in Section 4.3. After that, several examples of networks are analyzed. In Section 5 we illustrate the analysis of dominant systems on a simple example, the reversible triangle of reactions: $A_1 \leftrightarrow A_2 \leftrightarrow A_3 \leftrightarrow A_1$. This simplest example became very popular for the lumping analysis case study after the well-known work of Wei and Prater (1962). The most important mathematical proofs are presented in the appendices.

In multiscale asymptotic analysis of reaction network we found several very attractive *zero-one laws*. First of all, components eigenvectors are close to 0 or ± 1. This law together with two other zero-one laws are discussed in Section 6: "Three zero-one laws and nonequilibrium phase transitions in multiscale systems".

A multiscale system where every two constants have very different orders of magnitude is, of course, an idealization. In parametric families of multiscale systems there could appear systems with several constants of the same order. Hence, it is necessary to study effects that appear due to a group of constants of the same order in a multiscale network. The system can have modular structure, with different time scales in different modules, but without separation of times inside modules. We discuss systems with modular structure in Section 7. The full theory of such systems is a challenge for future work, and here we study structure of one module. The elementary modules have to be solvable. That means that the kinetic equations could be solved in explicit analytical form. We give the necessary and sufficient conditions for solvability of reaction networks. These conditions are presented constructively, by algorithm of analysis of the reaction graph.

It is necessary to repeat our study for nonlinear networks. We discuss this problem and perspective of its solution in the concluding Section 8. Here we again use the experience summarized in the IUPAC Compendium (Rate-controlling step, 2007) where the notion of controlling step is generalized onto nonlinear elementary reaction by inclusion of some concentration into "pseudo-first-order rate constant".

2. STATIC AND DYNAMIC LIMITATION IN A LINEAR CHAIN AND A SIMPLE CATALYTIC CYCLE

2.1 Linear chain

A linear chain of reactions, $A_1 \to A_2 \to \dots A_n$, with reaction rate constants k_i (for $A_i \to A_{i+1}$), gives the first example of limitation Let the reaction rate constant k_q be the smallest one. Then we expect the following behavior of the reaction chain in timescale $\sim 1/k_q$: all the components $A_1, \dots, A_{q-1}$ transform fast into A_q, and all the components $A_{q+1}, \dots, A_{n-1}$ transform fast into A_n, only two components,

A_q and A_n are present (concentrations of other components are small), and the whole dynamics in this time scale can be represented by a single reaction $A_q \rightarrow A_n$ with reaction rate constant k_q. This picture becomes more exact when k_q becomes smaller with respect to other constants.

The kinetic equation for the linear chain is

$$\dot{c}_i = k_{i-1}c_{i-1} - k_i c_i \tag{1}$$

where c_i is concentration of A_i and k_{i-1} for $i = 1$. The coefficient matrix K of this equation is very simple. It has nonzero elements only on the main diagonal, and one position below. The eigenvalues of K are $-k_i$ $(i = 1, \ldots, n-1)$ and 0. The left and right eigenvectors for 0 eigenvalue, l^0 and r^0, are:

$$l^0 = (1, 1, \ldots, 1), \;\; r^0 = (0, 0, \ldots, 0, 1) \tag{2}$$

all coordinates of l^0 are equal to 1, the only nonzero coordinate of r^0 is r_n^0 and we represent vector-column r^0 in row.

Below we use explicit form of K left and right eigenvectors. Let vector-column r^i and vector-row l^i be right and left eigenvectors of K for eigenvalue $-k_i$. For coordinates of these eigenvectors we use notation r_j^i and l_j^i. Let us choose a normalization condition $r_i^i = l_i^i = 1$. It is straightforward to check that $r_j^i = 0$ $(j<i)$ and $l_j^i = 0$ $(j>i)$, $r_{j+1}^i = k_j r_j / (k_{j+1} - k_i)$ $(j \geqslant i)$ and $l_{j-1}^i = k_{j-1} l_j / (k_{j-1} - k_j)$ $(j \leqslant i)$, and

$$r_{i+m}^i = \prod_{j=1}^{m} \frac{k_{i+j-1}}{k_{i+j} - k_i}; \qquad l_{i-m}^i = \prod_{j=1}^{m} \frac{k_{i-j}}{k_{i-j} - k_i} \tag{3}$$

It is convenient to introduce formally $k_0 = 0$. Under selected normalization condition, the inner product of eigenvectors is: $l^i r^j = \delta_{ij}$, where δ_{ij} is the Kronecker delta.

If the rate constants are well separated (i.e. any two constants, k_i and k_j are connected by relation $k_i \gg k_j$ or $k_i \ll k_j$,

$$\frac{k_{i-j}}{k_{i-j} - k_i} \approx \begin{cases} 1, & \text{if } k_i \ll k_{i-j}; \\ 0, & \text{if } k_i \gg k_{i-j} \end{cases} \tag{4}$$

Hence, $|l_{i-m}^i| \approx 1$ or $|l_{i-m}^i| \approx 0$. To demonstrate that also $|r_{i+m}^i| \approx 1$ or $|r_{i+m}^i| \approx 0$, we shift nominators in the product (3) on such a way:

$$r_{i+m}^i = \frac{k_i}{k_{i+m} - k_i} \prod_{j=1}^{m-1} \frac{k_{i+j}}{k_{i+j} - k_i}$$

Exactly as in Equation (4), each multiplier

$$\frac{k_{i+j}}{(k_{i+j} - k_i)}$$

here is either almost 1 or almost 0, and

$$\frac{k_i}{(k_{i+m} - k_i)}$$

is either almost 0 or almost −1. In this zero-one asymptotics

$$\begin{aligned}
&l^i_i = 1,\ l^i_{i-m} \approx 1 \\
&\qquad \text{if } k_{i-j} > k_i \text{ for all } j = 1, \ldots, m, \text{ else } l^i_{i-m} \approx 0; \\
&r^i_i = 1,\ r^i_{i+m} \approx -1 \\
&\qquad \text{if } k_{i+j} > k_i \text{ for all } j = 1, \ldots, m-1 \\
&\qquad \text{and } k_{i+m} < k_i, \text{ else } r^i_{i+m} \approx 0
\end{aligned} \tag{5}$$

In this asymptotic, only two coordinates of right eigenvector r^i can have nonzero values, $r^i_i = 1$ and $r^i_{i+m} \approx -1$ where m is the first such positive integer that $i+m<n$ and $k_{i+m}<k_i$. Such m always exists because $k_n = 0$. For left eigenvector $l^i, l^i_i \approx \ldots l^i_{i-q} \approx 1$ and $l^i_{i-q-j} \approx 0$ where $j>0$ and q the first such positive integer that $i-q-1>0$ and $k_{i-q-1}<k_i$. It is possible that such q does not exist. In that case, all $l^i_{i-j} \approx 1$ for $j \geqslant 0$. It is straightforward to check that in this asymptotic $l^i r^j = \delta_{ij}$.

The simplest example gives the order $k_1 \gg k_2 \gg \cdots \gg k_{n-1} : l^i_{i-j} \approx 1$ for $j \geqslant 0$, $r^i_i = 1$, $r^i_{i+1} \approx -1$ and all other coordinates of eigenvectors are close to zero. For the inverse order, $k_1 \ll k_2 \ll \cdots \ll k_{n-1}$, $l^i_i = 1$, $r^i_i = 1$, $r^i_n \approx -1$ and all other coordinates of eigenvectors are close to zero.

For less trivial example, let us find the asymptotic of left and right eigenvectors for a chain of reactions:

$$A_1 \xrightarrow{5} A_2 \xrightarrow{3} A_3 \xrightarrow{4} A_4 \xrightarrow{1} A_5 \xrightarrow{2} A_6$$

where the upper index marks the order of rate constants: $k_4 \gg k_5 \gg k_2 \gg k_3 \gg k_1$ (k_i is the rate constant of reaction $A_i \to \ldots$).

For left eigenvectors, rows l^i, we have the following asymptotics:

$$\begin{aligned}
&l^1 \approx (1,0,0,0,0,0),\ l^2 \approx (0,1,0,0,0,0), \\
&l^3 \approx (0,1,1,0,0,0),\ l^4 \approx (0,0,0,1,0,0), \\
&l^5 \approx (0,0,0,1,1,0)
\end{aligned} \tag{6}$$

For right eigenvectors, columns r^i, we have the following asymptotics (we write vector-columns in rows):

$$\begin{aligned}
&r^1 \approx (1,0,0,0,0,-1),\ r^2 \approx (0,1,-1,0,0,0), \\
&r^3 \approx (0,0,1,0,0,-1),\ r^4 \approx (0,0,0,1,-1,0), \\
&r^5 \approx (0,0,0,0,1,-1)
\end{aligned} \tag{7}$$

The correspondent approximation to the general solution of the kinetic equations is:

$$c(t) = (l^0 c(0))r^0 + \sum_{i=1}^{n-1} (l^i c(0))r^i \exp(-k_i t) \tag{8}$$

where $c(0)$ is the initial concentration vector, and for left and right eigenvectors l^i and r^i we use their zero-one asymptotic.

Asymptotic formulas allow us to transform kinetic matrix K to a matrix with value of diagonal element could not be smaller than the value of any element from the correspondent column and row.

Let us represent the kinetic matrix K in the basis of approximations to eigenvectors (7). The transformed matrix is $\tilde{K}_{ij} = l^i K r^j$ $(i, j = 0, 1, \ldots, 5)$:

$$K = \begin{bmatrix} -k_1 & 0 & 0 & 0 & 0 & 0 \\ k_1 & -k_2 & 0 & 0 & 0 & 0 \\ 0 & k_2 & -k_3 & 0 & 0 & 0 \\ 0 & 0 & k_3 & -k_4 & 0 & 0 \\ 0 & 0 & 0 & k_4 & -k_5 & 0 \\ 0 & 0 & 0 & 0 & k_5 & 0 \end{bmatrix},$$
$$\tilde{K} = \begin{bmatrix} 0 & 0 & 0 & 0 & 0 & 0 \\ 0 & -k_1 & 0 & 0 & 0 & 0 \\ 0 & k_1 & -k_2 & 0 & 0 & 0 \\ 0 & k_1 & k_3 & -k_3 & 0 & 0 \\ 0 & 0 & -k_3 & k_3 & -k_4 & 0 \\ 0 & 0 & -k_3 & k_3 & -k_5 & -k_5 \end{bmatrix} \quad (9)$$

The transformed matrix has an important property

$$|\tilde{K}_{ij}| \leq \min\{|\tilde{K}_{ii}|, |\tilde{K}_{jj}|\}$$

The initial matrix K is diagonally dominant in columns, but its rows can include elements that are much bigger than the correspondent diagonal elements.

We mention that a naive expectation $\tilde{K}_{ij} \approx \delta_{ij}$ is not realistic: some of the nondiagonal matrix elements $\tilde{K}_{ij}$ are of the same order than $\min\{\tilde{K}_{ii}, \tilde{K}_{jj}\}$. This example demonstrates that a good approximation to an eigenvector could be not an approximate eigenvector. If $Ke = \lambda e$ and $\|e - f\|$ is small then f is an approximation to eigenvector e. If $Kf \approx \lambda f$ (i.e. $\|Kf - \lambda f\|$ is small), then f is an approximate eigenvector for eigenvalue λ. Our kinetic matrix K is very ill-conditioned. Hence, nobody can guarantee that an approximation to eigenvector is an approximate eigenvector, or, inverse, an approximate eigenvector (a "quasimode") is an approximation to an eigenvector.

The question is, what do we need for approximation of the relaxation process (8). The answer is obvious: for approximation of general solution (8) with guaranteed accuracy we need approximation to the genuine eigenvectors ("modes") with the same accuracy. The zero-one asymptotic (5) gives this approximation. Below we always find the modes approximations and not quasimodes.

2.2 General properties of a cycle

The catalytic cycle is one of the most important substructures that we study in reaction networks. In the reduced form the catalytic cycle is a set of linear reactions:

$$A_1 \to A_2 \to \ldots A_n \to A_1$$

Reduced form means that in reality some of these reaction are not monomolecular and include some other components (not from the list $A_1, \ldots, A_n$). But in the study of the isolated cycle dynamics, concentrations of these components are taken as constant and are included into kinetic constants of the cycle linear reactions.

For the constant of elementary reaction $A_i \to$ we use the simplified notation k_i because the product of this elementary reaction is known, it is A_{i+1} for $i<n$ and A_1 for $i = n$. The elementary reaction rate is $w_i = k_i c_i$, where c_i is the concentration of A_i. The kinetic equation is:

$$\dot{c}_i = w_{i-1} - w_i \tag{10}$$

where by definition $w_0 = w_n$. In the stationary state ($\dot{c}_i = 0$), all the w_i are equal: $w_i = w$. This common rate w we call the cycle stationary rate, and

$$w = \frac{b}{(1/k_1) + \cdots + (1/k_n)}; \quad c_i = \frac{w}{k_i} \tag{11}$$

where $b = \sum_i c_i$ is the conserved quantity for reactions in constant volume (for general case of chemical kinetic equations see elsewhere, for example, the book by Yablonskii et al., 1991). The stationary rate w (11) is a product of the arithmetic mean of concentrations, b/n, and the harmonic mean of constants (inverse mean of inverse k_i).

2.3 Static limitation in a cycle

If one of the constants, $k_{\min}$, is much smaller than others (let it be $k_{\min} = k_n$), then

$$\begin{aligned} c_n &= b\left(1 - \sum_{i<n} \frac{k_n}{k_i} + o\left(\sum_{i<n} \frac{k_n}{k_i}\right)\right), \\ c_i &= b\left(\frac{k_n}{k_i} + o\left(\sum_{i<n} \frac{k_n}{k_i}\right)\right), \\ w &= k_n b\left(1 + O\left(\sum_{i<n} \frac{k_n}{k_i}\right)\right) \end{aligned} \tag{12}$$

or simply in linear approximation

$$c_n = b\left(1 - \sum_{i<n} \frac{k_n}{k_i}\right), \quad c_i = b\frac{k_n}{k_i}, \quad w = k_n b \tag{13}$$

where we should keep the first-order terms in c_n in order not to violate the conservation law.

The simplest zero order approximation for the steady state gives

$$c_n = b, \quad c_i = 0 \; (i \neq n) \tag{14}$$

This is trivial: all the concentration is collected at the starting point of the "narrow place", but may be useful as an origin point for various approximation procedures.

So, the stationary rate of a cycle is determined by the smallest constant, $k_{\min}$, if $k_{\min}$ is sufficiently small:

$$w = k_{\min} b \text{ if } \sum_{k_i \neq k_{\min}} \frac{k_{\min}}{k_i} \ll 1 \tag{15}$$

In that case we say that the cycle has a limiting step with constant $k_{\min}$.

2.4 Dynamical limitation in a cycle

If k_n/k_i is small for all $i<n$, then the kinetic behavior of the cycle is extremely simple: the coefficients matrix on the right-hand side of kinetic equation (10) has one simple zero eigenvalue that corresponds to the conservation law $\sum c_i = b$ and $n-1$ nonzero eigenvalues

$$\lambda_i = -k_i + \delta_i \ (i<n) \tag{16}$$

where $\delta_i \to 0$ when $\sum_{i<n}(k_n/k_i) \to 0$.

It is easy to demonstrate Equation (16): let us exclude the conservation law (the zero eigenvalue) $\sum c_i = b$ and use independent coordinates c_i $(i = 1,\ldots,n-1)$; $c_n = b - \sum_{i<n} c_i$. In these coordinates the kinetic equation (10) has the form

$$\dot{c} = Kc - k_n Ac + k_n b e^1 \tag{17}$$

where c is the vector-column with components c_i $(i<n)$, K the lower triangle matrix with nonzero elements only in two diagonals: $(K)_{ii} = -k_i (i = 1,\ldots,n-1)$, $(K)_{i+1,i} = k_i$ $(i = 1,\ldots, n-2)$ (this is the kinetic matrix for the linear chain of $n-1$ reactions $A_1 \to A_2 \to \ldots A_n$); A the matrix with nonzero elements only in the first row: $(A)_{1i} \equiv 1$, e^1 the first basis vector ($e^1_1 = 1$, $e^1_i = 0$ for $1<i<n$). After that, Equation (16) follows simply from continuous dependence of spectra on matrix.

The relaxation time of a stable linear system (17) is, by definition,

$$\tau = [\min\{Re(-\lambda_i)|i = 1,\ldots,n-1\}]^{-1}$$

For small k_n,

$$\tau \approx 1/k_\tau, \quad k_\tau = \min\{k_i|i = 1,\ldots,n-1\} \tag{18}$$

In other words, k_τ is the second slowest rate constant: $k_{\min} \leq k_\tau \leq \cdots$

2.5 Relaxation equation for a cycle rate

A definition of the cycle rate is clear for steady states because stationary rates of all elementary reactions in cycle coincide. There is no common definition of the cycle rate for nonstationary regimes. In practice, one of steps is the step of product release (the "final" step of the catalytic transformation), and we can consider its rate as the rate of the cycle. Formally, we can take any step and study relaxation of its rate to the common stationary rate. The single relaxation time approximation gives for rate w_i of any step:

$$\begin{aligned} \dot{w}_i &= k_\tau(k_{\min}b - w_i); \\ w_i(t) &= k_{\min}b + e^{-k_\tau t}(w_i(0) - k_{\min}b) \end{aligned} \tag{19}$$

where $k_{\min}$ is the limiting (the minimal) rate constant of the cycle and k_τ the second in order rate constant of the cycle.

So, for catalytic cycles with the limiting constant $k_{\min}$, the relaxation time is also determined by one constant, but another one. This is k_τ, the second in order rate constant. It should be stressed that the only smallness condition is required, $k_{\min}$ should be much smaller than other constants. The second constant, k_τ should be just smaller than others (and bigger than $k_{\min}$), but there is no $\ll$ condition for k_τ required.

One of the methods for measurement of chemical reaction constants is the relaxation spectroscopy (Eigen, 1972). Relaxation of a system after an impact gives us a relaxation time or even a spectrum of relaxation times. For catalytic cycle with limitation, the relaxation experiment gives us the second constant k_τ, whereas the measurement of stationary rate gives the smallest constant, $k_{\min}$. This simple remark may be important for relaxation spectroscopy of open system.

2.6 Ensembles of cycles and robustness of stationary rate and relaxation time

Let us consider a catalytic cycle with random rate constants. For a given sample constants $k_1, \ldots, k_n$ the ith order statistics is equal its ith smallest value. We are interested in the first order (the minimal) and the second order statistics.

For independent identically distributed constants the variance of $k_{\min} = \min\{k_1, \ldots, k_n\}$ is significantly smaller than the variance of each k_i, Var(k). The same is true for statistic of every order. For many important distributions (e.g. for uniform distribution), the variance of ith order statistic is of order $\sim$Var$(k)/n^2$. For big n it goes to zero faster than variance of the mean that is of order $\sim$Var$(k)/n$. To illustrate this, let us consider n constants distributed in interval $[a, b]$. For each set of constants, $k_1, \ldots, k_n$ we introduce "symmetric coordinates" s_i: first, we order the constants, $a \le k_{i_1} \le k_{i_2} \le \cdots k_{i_n} \le b$, then calculate $s_0 = k_{i_1} - a$, $s_j = k_{i_{j+1}} - k_{i_j}$ $(j = 1, \ldots, n-1)$, $s_n = b - k_{i_n}$. Transformation $(k_1, \ldots, k_n) \mapsto (s_0, \ldots, s_n)$ maps a cube $[a, b]^n$ onto n-dimensional simplex $\Delta_n = \{(s_0, \ldots, s_n) | \sum_i s_i = b - a\}$ and uniform distribution on a cube transforms into uniform distribution on a simplex.

For large n, almost all volume of the simplex is concentrated in a small neighborhood of its center and this effect is an example of measure concentration effects that play important role in modern geometry and analysis (Gromov, 1999). All s_i are identically distributed, and for normalized variable $s = s_i/(b - a)$

the first moments are: $\mathbf{E}(s) = 1/(n+1) = 1/n + o(1/n)$, $\mathbf{E}(s^2) = 2/[(n+1)(n+2)] = 2/n^2 + o(1/n^2)$,

$$\mathrm{Var}(s) = \mathbf{E}(s^2) - (\mathbf{E}(s))^2$$
$$= \frac{n}{(n+1)^2(n+2)} = \frac{1}{n^2} + o\left(\frac{1}{n^2}\right)$$

Hence, for example, $\mathrm{Var}(k_{\min}) = (b-a)^2/n^2 + o(1/n^2)$. The standard deviation of $k_{\min}$ goes to zero as $1/n$ when n increases. This is much faster than $1/\sqrt{n}$ prescribed to the deviation of the mean value of independent observation (the "law of errors"). The same asymptotic $\sim 1/n$ is true for the standard deviation of the second constant also. These parameters fluctuate much less than individual constants, and even less than mean constant (for more examples with applications to statistical physics we address to the paper by Gorban, 2006).

It is impossible to use this observation for cycles with limitation directly, because the inequality of limitation (15) is not true for uniform distribution. According to this inequality, ratios $k_i/k_{\min}$ should be sufficiently small (if $k_i \neq k_{\min}$). To provide this inequality we need to use at least the log-uniform distribution: $k_i = \exp \Delta_i$ and Δ_i are independent variables uniformly distributed in interval $[\alpha,\beta]$ with sufficiently big $(\beta-\alpha)/n$.

One can interpret the log-uniform distribution through the Arrhenius law: $k = A\exp(-\Delta G/kT)$, where ΔG is the change of the Gibbs free energy inreaction (it includes both energetic and entropic terms: $\Delta G = \Delta H - T\Delta S$, where ΔH the enthalpy change and ΔS the entropy change in reaction, T the temperature). The log-uniform distribution of k corresponds to the uniform distribution of ΔG.

For log-uniform distribution of constants $k_1, \ldots, k_n$, if the interval of distribution is sufficiently big (i.e. $(\beta-\alpha)/n \gg 1$), then the cycle with these constants has the limiting step with probability close to one. More precisely we can show that for any two constants k_i and k_j the probability $\mathbb{P}[k_i/k_j > r \text{ or } k_j/k_i > r] = (1 - \log(r)/(\beta - \alpha))^2$ approaches one for any fixed $r > 1$ when $\beta - \alpha \to \infty$. Relaxation time of this cycle is determined by the second constant k_τ (also with probability close to one). Standard deviations of $k_{\min}$ and k_τ are much smaller than standard deviation of single constant k_i and even smaller than standard deviation of mean constant $\sum_i k_i/n$. This effect of stationary rate and relaxation time robustness seems to be important for understanding robustness in biochemical networks: behavior of the entire system is much more stable than the parameters of its parts; even for large fluctuations of parameters, the system does not change significantly the stationary rate (statics) and the relaxation time (dynamics).

2.7 Systems with well-separated constants and monotone relaxation

The log-uniform identical distribution of independent constants $k_1, \ldots, k_n$ with sufficiently big interval of distribution ($(\beta-\alpha)/n \gg 1$) gives us the first example of ensembles with well-separated constants: any two constants are connected by relation $\gg$ or $\ll$ with probability close to one. Such systems (not only cycles, but much more complex networks too) could be studied analytically "up to the end".

Some of their properties are simpler than for general networks. For example, the damping oscillations are impossible, i.e. the eigenvalues of kinetic matrix are real (with probability close to one). If constants are not separated, damped oscillations could exist, for example, if all constants of the cycle are equal, $k_1 = k_2 = \cdots = k_n = k$, then $(1 + \lambda/k)^n = 1$ and $\lambda_m = k(\exp(2\pi im/n) - 1)$ $(m = 1,\ldots,n-1)$, the case $m = 0$ corresponds to the linear conservation law. Relaxation time of this cycle may be relatively big: $\tau = (1/k)\ (1 - \cos(2\pi/n))^{-1} \sim n^2/(2\pi k)$ (for big n).

The catalytic cycle without limitation can have relaxation time much bigger then $1/k_{\min}$, where $k_{\min}$ is the minimal reaction rate constant. For example, if all k are equal, then for $n = 11$ we get $\tau \approx 20/k$. In more detail the possible relations between τ and the slowest constant were discussed by Yablonskii and Cheresiz (1984). In that paper, a variety of cases with different relationships between the steady-state reaction rate and relaxation was presented.

For catalytic cycle, if a matrix $K - k_nA$ (17) has a pair of complex eigenvalues with nonzero imaginary part, then for some $g \in [0, 1]$ the matrix $K - gk_nA$ has a degenerate eigenvalue (we use a simple continuity argument). With probability close to one, $k_{\min} \ll |k_i - k_j|$ for any two k_i and k_j that are not minimal. Hence, the $k_{\min}$-small perturbation cannot transform matrix K with eigenvalues k_i (16) and given structure into a matrix with a degenerate eigenvalue. For proof of this statement it is sufficient to refer to diagonal dominance of K (the absolute value of each diagonal element is greater than the sum of the absolute values of the other elements in its column) and classical inequalities.

The matrix elements of A in the eigenbasis of K are $(A)_{ij} = l^iAr^j$. From obtained estimates for eigenvectors we get $|(A)_{ij}| \lesssim 1$ (with probability close to one). This estimate does not depend on values of kinetic constants. Now, we can apply the Gershgorin theorem (see, e.g. the review of Marcus and Minc (1992) and for more details the book of Varga (2004)) to the matrix $K - k_nA$ in the eigenbasis of K: the characteristic roots of $K - k_nA$ belong to discs $|z + k_i| \leq k_nR_i(A)$, where $R_i(A) = \sum_j |(A)_{ij}|$. If the discs do not intersect, then each of them contains one and only one characteristic number. For ensembles with well-separated constants these discs do not intersect (with probability close to one). Complex conjugate eigenvalues could not belong to different discs. In this case, the eigenvalues are real — there exist no damped oscillations.

2.8 Limitation by two steps with comparable constants

If we consider one-parametric families of systems, then appearance of systems with two comparable constants may be unavoidable. Let us imagine a continuous path $k_i(s)$ ($s \in [0, 1]$, s is a parameter along the path) in the space of systems, which goes from one system with well-separated constants ($s = 0$) to another such system ($s = 1$). On this path $k_i(s)$ such a point s that $k_i(s) = k_j(s)$ may exist, and this existence may be stable, that is, such a point persists under continuous perturbations. This means that on a path there may be points where not all the constants are well separated, and trajectories of some constants may intersect.

For catalytic cycle, we are interested in the following intersection only: $k_{\min}$ and the second constant are of the same order, and are much smaller than other constants. Let these constants be k_j and k_l, $j \neq l$. The limitation condition is

$$\frac{1}{k_j} + \frac{1}{k_l} \gg \sum_{i \neq j,l} \frac{1}{k_i} \tag{20}$$

The steady-state reaction rate and relaxation time are determined by these two constants. In that case their effects are coupled. For the steady state we get in first-order approximation instead of Equation (13):

$$\begin{aligned} w &= \frac{k_j k_l}{k_j + k_l} b; \quad c_i = \frac{w}{k_i} = \frac{b}{k_i} \frac{k_j k_l}{k_j + k_l} (i \neq j, l); \\ c_j &= \frac{b k_l}{k_j + k_l} \left(1 - \sum_{i \neq j,l} \frac{l}{k_i} \frac{k_j k_l}{k_j + k_l} \right); \\ c_l &= \frac{b k_j}{k_j + k_l} \left(1 - \sum_{i \neq j,l} \frac{l}{k_i} \frac{k_j k_l}{k_j + k_l} \right) \end{aligned} \tag{21}$$

Elementary analysis shows that under the limitation condition (20) the relaxation time is

$$\tau = \frac{1}{k_j + k_l} \tag{22}$$

The single relaxation time approximation for all elementary reaction rates in a cycle with two limiting reactions is

$$\begin{aligned} \dot{w}_i &= k_j k_l b - (k_j + k_l) w_i; \\ w_i(t) &= \frac{k_j k_l}{k_j + k_l} b + e^{-(k_j + k_l)t} \left(w_i(0) - \frac{k_j k_l}{k_j + k_l} b \right) \end{aligned} \tag{23}$$

The catalytic cycle with two limiting reactions has the same stationary rate w (21) and relaxation time (22) as a reversible reaction $A \leftrightarrow B$ with $k^+ = k_j$ and $k^- = k_l$.

In two-parametric families three constants can meet. If three smallest constants k_j, k_l and k_m have comparable values and are much smaller than others, then static and dynamic properties would be determined by these three constants. Stationary rate w and dynamic of relaxation for the whole cycle would be the same as for 3-reaction cycle $A \to B \to C \to A$ with constants k_j, k_l and k_m. The damped oscillation here are possible, for example, if $k_j = k_l = k_m = k$, then there are complex eigenvalues $\lambda = k(-(3/2) \pm i(\sqrt{3}/2))$. Therefore, if a cycle manifests damped oscillation, then at least three slowest constants are of the same order. The same is true, of course, for more general reaction networks.

In N-parametric families of systems $N+1$ smallest constants can meet, and near such a "meeting point" a slow auxiliary cycle of $N+1$ reactions determines behavior of the entire cycle.

2.9 Irreversible cycle with one inverse reaction

In this subsection, we represent a simple example that gives the key to most of subsequent constructions of "cycles surgery". Let us add an inverse reaction to the irreversible cycle: $A_1 \to \ldots \to A_i \leftrightarrow A_{i+1} \to \ldots \to A_n \to A_1$. We use the previous notation $k_1, \ldots, k_n$ for the cycle reactions, and k_i^- for the inverse reaction $A_i \leftarrow A_{i+1}$. For well-separated constants, influence of k_i^- on the whole reaction is determined by relations of three constants: k_i, k_i^- and k_{i+1}. First of all, if $k_i^- \ll k_{i+1}$ then in the main order there is no such influence, and dynamic of the cycle is the same as for completely irreversible cycle.

If the opposite inequality is true, $k_i^- \gg k_{i+1}$, then equilibration between A_i and A_{i+1} gives $k_i c_i x \approx k_i^- c_{i+1}$. If we introduce a lumped component A_i^1 with concentration $c_i^1 = c_i + c_{i+1}$, then $c_i \approx k_i^- c_i^1/(k_i + k_i^-)$ and $c_{i+1} \approx k_i c_i^1/(k_i + k_i^-)$. Using this component instead of the pair A_i, A_{i+1} we can consider an irreversible cycle with $n-1$ components and n reactions $A_1 \to \ldots \to A_{i-1} \to A_i^1 \to A_{i+2} \to \ldots \to A_n \to A_1$. To estimate the reaction rate constant k_i^1 for a new reaction, $A_i^1 \to A_{i+2}$, let us mention that the correspondent reaction rate should be $k_{i+1}c_{i+1} \approx k_{i+1}k_i c_i^1/(k_i + k_i^-)$. Hence,

$$k_i^1 \approx k_{i+1}k_i/(k_i + k_i^-)$$

For systems with well-separated constants this expression can be simplified: if $k_i \gg k_i^-$ then $k_i^1 \approx k_{i+1}$ and if $k_i \ll k_i^-$ then $k_i^1 \approx k_{i+1}k_i/k_i^-$. The first case, $k_i \gg k_i^-$ is limitation in the small cycle (of length two) $A_i \leftrightarrow A_{i+1}$ by the inverse reaction $A_i \leftarrow A_{i+1}$. The second case, $k_i \ll k_i^-$, means the direct reaction is the limiting step in this small cycle.

To estimate eigenvectors, we can, after identification of the limiting step in the small cycle, delete this step and reattach the outgoing reaction to the beginning of this step. For the first case, $k_i \gg k_i^-$, we get the irreversible cycle, $A_1 \to \ldots \to A_i \to A_{i+1} \to \ldots \to A_n \to A_1$, with the same reaction rate constants. For the second case, $k_i \ll k_i^-$ we get a new system of reactions: a shortened cycle $A_1 \to \ldots \to A_i \to A_{i+2} \to \ldots \to A_n \to A_1$ and an "appendix" $A_{i+1} \to A_i$. For the new elementary reaction $A_i \to A_{i+2}$ the reaction rate constant is $k_i^1 \approx k_{i+1}k_i/k_i^-$. All other elementary reactions have the same rate constants, as they have in the initial system. After deletion of the limiting step from the "big cycle" $A_1 \to \ldots \to A_i \to A_{i+2} \to \ldots \to A_n \to A_1$, we get an acyclic system that approximate relaxation of the initial system.

So, influence of a single inverse reaction on the irreversible catalytic cycle with well-separated constants is determined by relations of three constants: k_i, k_i^- and k_{i+1}. If k_i^- is much smaller than at least one of k_i, k_{i+1}, then there is no influence in the main order. If $k_i^- \gg k_i$ and $k_i^- \gg k_{i+1}$ then the relaxation of the initial cycle can be approximated by relaxation of the auxiliary acyclic system.

Asymptotic equivalence (for $k_i^- \gg k_i,\ k_{i+1}$) of the reaction network $A_i \leftrightarrow A_{i+1} \to A_{i+2}$ with rate constants k_i, k_i^- and k_{i+1} to the reaction network $A_{i+1} \to A_i \to A_{i+2}$ with rate constants k_i^- (for the reaction $A_{i+1} \to A_i$) and $k_{i+1}k_i/k_i^-$ (for the reaction $A_i \to A_{i+2}$) is simple, but slightly surprising fact. The kinetic matrix for the first network in coordinates c_i, c_{i+1} and c_{i+2} is

$$K = \begin{bmatrix} -k_i & k_i^- & 0 \\ k_i & -(k_i^- + k_{i+1}) & 0 \\ 0 & k_{i+1} & 0 \end{bmatrix}$$

The eigenvalues are 0 and

$$\lambda_{1,2} = \frac{1}{2}\left[-(k_i + k_i^- + k_{i+1}) \pm \sqrt{(k_i + k_i^- + k_{i+1})^2 - 4k_ik_{i+1}}\right]$$

$\lambda_1 = -k_{i+1}k_i(1 + o(1))/k_i^-$, $\lambda_2 = -k_i^-(1 + o(1))$, where $o(1) \ll 1$. Right eigenvector r^0 for zero eigenvalue is (0, 0, 1) (we write vector columns in rows). For λ_1 the eigenvector is $r^1 = (1, 0, -1) + o(1)$, and for λ_2 it is $r^2 = (1, -1, 0) + o(1)$. For the linear chain of reactions, $A_{i+1} \to A_i \to A_{i+2}$, with rate constants k_i^- and $k_{i+1}k_i/k_i^-$ eigenvalues are $-k_i^-$ and $-k_{i+1}k_i/k_i^-$. These values approximate eigenvalues of the initial system with small relative error. The linear chain has the same zero-one asymptotic of the correspondent eigenvectors.

This construction, a small cycle inside a big system, a quasi-steady state in the small cycle, and deletion of the limiting step with reattaching of reactions (see Figure 1 below) appears in this chapter many times in much general settings. The uniform estimates that we need for approximation of eigenvalues and eigenvectors by these procedures are proven in Appendices.

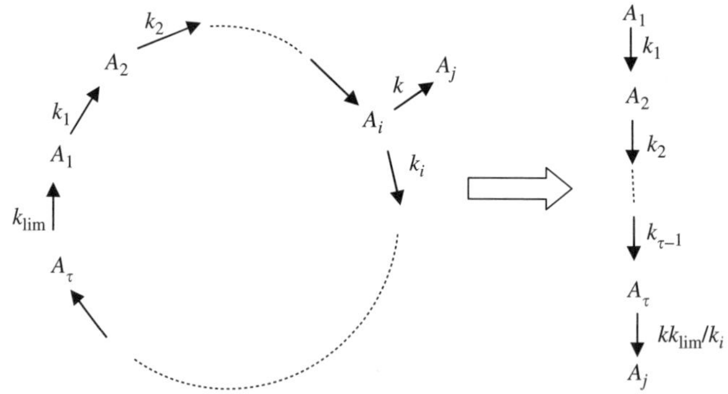

Figure 1 The main operation of the cycle surgery: on a step back we get a cycle $A_1 \to \ldots \to A_\tau \to A_1$ with the limiting step $A_\tau \to A_1$ and one outgoing reaction $A_i \to A_j$. We should delete the limiting step, reattach ("recharge") the outgoing reaction $A_i \to A_j$ from A_i to A_τ and change its rate constant k to the rate constant $kk_{\lim}/k_i$. The new value of reaction rate constant is always smaller than the initial one: $kk_{\lim}/k_i < k$ if $k_{\lim} \neq k_i$. For this operation only one condition $k \ll k_i$ is necessary (k should be small with respect to reaction $A_i \to A_{i+1}$ rate constant, and can exceed any other reaction rate constant).

3. MULTISCALE ENSEMBLES AND FINITE-ADDITIVE DISTRIBUTIONS

3.1 Ensembles with well-separated constants, formal approach

In previous section, ensembles with well-separated constants appear. We represented them by a log-uniform distribution in a sufficiently big interval log $k \in [\alpha, \beta]$, but we were not interested in most of probability distribution properties, and did not use them. The only property we really used is: if $k_i > k_j$, then $k_i/k_j \gg 1$ (with probability close to one). It means that we can assume that $k_i/k_j \gg a$ for any preassigned value of a that does not depend on k values. One can interpret this property as an asymptotic one for $\alpha \to -\infty$, $\beta \to \infty$.

That property allows us to simplify algebraic formulas. For example, $k_i + k_j$ could be substituted by $\max\{k_i, k_j\}$ (with small relative error), or

$$\frac{ak_i + bk_j}{ck_i + dk_j} \approx \begin{cases} a/c, & \text{if } k_i \gg k_j; \\ b/d, & \text{if } k_i \ll k_j \end{cases}$$

for nonzero a, b, c, d (see, e.g. Equation (4)).

Of course, some ambiguity can be introduced, for example, what is it, $(k_1 + k_2) - k_1$, if $k_1 \gg k_2$? If we first simplify the expression in brackets, it is zero, but if we open brackets without simplification, it is k_2. This is a standard difficulty in use of relative errors for round-off. If we estimate the error in the final answer, and then simplify, we shall avoid this difficulty. Use of o and $\mathcal{O}$ symbols also helps to control the error qualitatively: if $k_1 \gg k_2$, then we can write $(k_1 + k_2) = k_1(1 + o(1))$ and $k_1(1 + o(1)) - k_1 = k_1 o(1)$. The last expression is neither zero nor absolutely small — it is just relatively small with respect to k_1.

The formal approach is: for any ordering of rate constants, we use relations $\gg$ and $\ll$, and assume that $k_i/k_j \gg a$ for any preassigned value of a that does not depend on k values. This approach allows us to perform asymptotic analysis of reaction networks. A special version of this approach consists of group ordering: constants are separated on several groups, inside groups they are comparable, and between groups there are relations $\gg$ or $\ll$. An example of such group ordering was discussed at the end of previous section (several limiting constants in a cycle).

3.2 Probability approach: finite additive measures

The asymptotic analysis of multiscale systems for log-uniform distribution of independent constants on an interval log $k \in [\alpha, \beta]$ $(-\alpha, \beta \to \infty)$ is possible, but parameters α, β do not present in any answer, they just should be sufficiently big. A natural question arises, what is the limit? It is a log-uniform distribution on a line, or, for n independent identically distributed constants, a log-uniform distribution on $\mathbb{R}^n$).

It is well known that the uniform distribution on $\mathbb{R}^n$ is impossible: if a cube has positive probability $\varepsilon > 0$ (i.e. the distribution has positive density) then the union of $N > 1/\varepsilon$ such disjoint cubes has probability bigger than 1 (here we use the finite-additivity of probability). This is impossible. But if that cube has probability zero, then the whole space has also zero probability, because it can be covered by countable family of the cube translation. Hence, translation invariance and σ-additivity (countable additivity) are in contradiction (if we have no doubt about probability normalization).

Nevertheless, there exists finite-additive probability which is invariant with respect to Euclidean group $E(n)$ (generated by rotations and translations). Its values are densities of sets.

Let λ be the Lebesgue measure and $D \subset \mathbb{R}^n$ be a Lebesgue measurable subset. Density of D is the limit (if it exists):

$$\rho(D) = \lim_{r\to\infty} \frac{\lambda(D \cap \mathbb{B}_r^n)}{\lambda(\mathbb{B}_r^n)} \tag{24}$$

where $\mathbb{B}_r^n$ is a ball with radius r and center at origin. Density of $\mathbb{R}^n$ is 1, density of every half-space is ½, density of bounded set is zero, density of a cone is its solid angle (measured as a sphere surface fractional area). Density (24) and translation and rotational invariant. It is finite-additive: if densities $\rho(D)$ and $\rho(H)$ (24) exist and $D \cap H = \emptyset$ then $\rho(D \cup H)$ exists and $\rho(D \cup H) = \rho(D) + \rho(H)$.

Every polyhedron has a density. A polyhedron could be defined as the union of a finite number of convex polyhedra. A convex polyhedron is the intersection of a finite number of half-spaces. It may be bounded or unbounded. The family of polyhedra is closed with respect to union, intersection and subtraction of sets. For our goals, polyhedra form sufficiently rich class. It is important that in definition of polyhedron *finite* intersections and unions are used. If one uses countable unions, he gets too many sets including all open sets, because open convex polyhedra (or just cubes with rational vertices) form a basis of standard topology.

Of course, not every measurable set has density. If it is necessary, we can use the Hahn–Banach theorem (Rudin, 1991) and study extensions ρ_{Ex} of ρ with the following property:

$$\underline{\rho}(D) \leq \rho_{\mathrm{Ex}}(D) \leq \bar{\rho}(D)$$

where

$$\underline{\rho}(D) = \lim_{r\to\infty} \inf \frac{\lambda(D \cap \mathbb{B}_r^n)}{\lambda(\mathbb{B}_r^n)},$$
$$\bar{\rho}(D) = \lim_{r\to\infty} \sup \frac{\lambda(D \cap \mathbb{B}_r^n)}{\lambda(\mathbb{B}_r^n)}$$

Functionals $\underline{\rho}(D)$ and $\bar{\rho}(D)$ are defined for all measurable D. We should stress that such extensions are not unique. Extension of density (24) using the Hahn–Banach theorem for picking up a random integer was used in a very recent work by Adamaszek (2006).

One of the most important concepts of any probability theory is the conditional probability. In the density-based approach we can introduce the conditional density. If densities $\rho(D)$ and $\rho(H)$ (24) exist, $\rho(H) \neq 0$ and the following limit $\rho(D \mid H)$ exists, then we call it conditional density:

$$\rho(D|H) = \lim_{r\to\infty} \frac{\lambda(D \cap H \cap \mathbb{B}_r^n)}{\lambda(H \cap \mathbb{B}_r^n)} \tag{25}$$

For polyhedra the situation is similar to usual probability theory: densities $\rho(D)$ and $\rho(H)$ always exist and if $\rho(H) \neq 0$ then conditional density exists too. For general measurable sets the situation is not so simple, and existence of $\rho(D)$ and $\rho(H) \neq 0$ does not guarantee existence of $\rho(D \mid H)$.

On a line, convex polyhedra are just intervals, finite or infinite. The probability defined on polyhedra is: for finite intervals and their finite unions it is zero, for half-lines $x > \alpha$ or $x < \alpha$ it is ½, and for the whole line $\mathbb{R}$ the probability is 1. If one takes a set of positive probability and adds or subtracts a zero-probability set, the probability does not change.

If independent random variables x and y are uniformly distributed on a line, then their linear combination $z = \alpha x + \beta y$ is also uniformly distributed on a line. (Indeed, vector (x, y) is uniformly distributed on a plane (by definition), a set $z > \gamma$ is a half-plane, the correspondent probability is ½.) This is a simple, but useful stability property. We shall use this result in the following form. If independent random variables $k_1, \ldots, k_n$ are log-uniformly distributed on a line, then the monomial $\prod_{i=1}^{n} k_i^{\alpha_i}$ for real α_i is also log-uniformly distributed on a line, if some of $\alpha_i \neq 0$.

3.3 Carroll's obtuse problem and paradoxes of conditioning

Lewis Carroll's Pillow Problem #58 (Carroll, 1958): "Three points are taken at random on an infinite plane. Find the chance of their being the vertices of an obtuse-angled triangle".

A random triangle on an infinite plane is presented by a point equidistributed in $\mathbb{R}^6$. Owing to the density — based definition, we should take and calculate the density of the set of obtuse-angled triangles in $\mathbb{R}^6$. This is equivalent to the problem: find a fraction of the sphere $\mathbb{S}^5 \subset \mathbb{R}^6$ that corresponds to obtuse-angled triangles. Just integrate .... But there remains a problem. Vertices of triangle are independent. Let us use the standard logic for discussion of independent trials: we take the first point A at random, then the second point B and then the third point C. Let us draw the first side AB. Immediately we find that for almost all positions of the the third point C the triangle is obtuse-angled (Guy, 1993). Carroll proposed to take another condition: let AB be the longest side and let C be uniformly distributed in the allowed area. The answer then is easy — just a ratio of areas of two simple figures. But there are absolutely no reasons for uniformity of C distribution. And it is more important that the absolutely standard reasoning for independently chosen points gives another answer than could be found on the base of joint distribution. Why these approaches are in disagreement now? Because there is no classical Fubini theorem for our finite-additive probabilities, and we cannot easily transfer from a multiple integral to a repeated one.

There exists a much simpler example. Let x and y be independent positive real number. This means that vector (x, y) is uniformly distributed in the first quadrant. What is probability that $x \geqslant y$? Following the definition of probability based on the density of sets, we take the correspondent angle and find immediately that this probability is ½. This meets our intuition well. But let us take the first number x and look for possible values of y. The result: for given x the second number y is uniformly distributed on $[0, \infty)$, and only a finite interval $[0, x]$ corresponds to $x \geqslant y$. For the infinite rest we have $x<y$. Hence, $x<y$ with probability 1. This is nonsense because of symmetry. So, for our finite-additive measure we cannot use repeated integrals (or, may be, should use them in a very peculiar manner).

3.4 Law of total probability and orderings

For polyhedra, there appear no conditioning problems. The law of total probabilities holds: if $\mathbb{R}^n = \cup_{i=1}^m H_i$, H_i are polyhedra, $\rho(H_i)>0$, $\rho(H_i \cap H_j) = 0$ for $i \neq j$, and $D \subset \mathbb{R}^n$ is a polyhedron, then

$$\rho(D) = \sum_{i=1}^{m} \rho(D \cap H_i) = \sum_{i=1}^{m} \rho(D|H_i)\rho(H_i) \tag{26}$$

Our basic example of multiscale ensemble is log-uniform distribution of reaction constants in $\mathbb{R}_+^n$ ($\log k_i$ are independent and uniformly distributed on the line). For every ordering $k_{j_1} > k_{j_2} > \cdots > k_{j_n}$ a polyhedral cone $H_{j_1, j_2, \ldots, j_n}$ in $\mathbb{R}^n$ is defined. These cones have equal probabilities $\rho(H_{j_1, j_2, \ldots, j_n}) = 1/n!$ and probability of intersection of cones for different orderings is zero. Hence, we can apply the law of total probability (26). This means that we can study every event D conditionally, for different orderings, and then combine the results of these studies in the final answer (26).

For example, if we study a simple cycle then formula (13) for steady state is valid with any given accuracy with unite probability for any ordering with the given minimal element k_n.

For cycle with given ordering of constants we can find zero-one approximation of left and right eigenvectors (5). This approximation is valid with any given accuracy for this ordering with probability one.

If we consider sufficiently wide log-uniform distribution of constants on a bounded interval instead of the infinite axis then these statements are true with probability close to 1.

For general system that we study below the situation is slightly more complicated: new terms, auxiliary reactions with monomial rate constants $k_\varsigma = \prod_i k_i^{\varsigma_i}$ could appear with integer (but not necessary positive) ς_i, and we should include these k_ς in ordering. It follows from stability property that these monomials are log-uniform distributed on infinite interval, if k_i are. Therefore the situation seems to be similar to ordering of constants, but there is a significant difference: monomials are not independent, they depend on k_i with $\varsigma_i \neq 0$.

Happily, in the forthcoming analysis when we include auxiliary reactions with constant k_ς, we always exclude at least one of the reactions with rate constant k_i and $\varsigma_i \neq 0$. Hence, for we always can use the following statement (for the new list of constants, or for the old one): if $k_{j_1} > k_{j_2} > \cdots > k_{j_n}$ then $k_{j_1} \gg k_{j_2} \gg \cdots \gg k_{j_n}$, where $a \gg b$ for positive a, b means: for any given $\varepsilon > 0$ the inequality $\varepsilon a > b$ holds with probability one.

If we use sufficiently wide but finite log-uniform distribution then ε could not be arbitrarily small (this depends on the interval with), and probability is not unite but close to one. For given $\varepsilon>0$ probability tends to one when the interval width goes to infinity. It is important that we use only finite number of auxiliary reactions with monomial constants, and this number is bounded from above for given number of elementary reactions. For completeness, we should mention here general algebraic theory of orderings that is necessary in more sophisticated cases (Greuel and Pfister, 2002; Robbiano, 1985).

4. RELAXATION OF MULTISCALE NETWORKS AND HIERARCHY OF AUXILIARY DISCRETE DYNAMICAL SYSTEMS

4.1 Definitions, notations and auxiliary results

4.1.1 Notations

In this section, we consider a general network of linear (monomolecular) reactions. This network is represented as a directed graph (digraph): vertices correspond to components A_i, edges correspond to reactions $A_i \to A_j$ with kinetic constants $k_{ji}>0$. For each vertex, A_i, a positive real variable c_i (concentration) is defined. A basis vector e^i corresponds to A_i with components $e^i_j = \delta_{ij}$, where δ_{ij} is the Kronecker delta. The kinetic equation for the system is

$$\frac{\mathrm{d}c_i}{\mathrm{d}t} = \sum_j (k_{ij}c_j - k_{ji}c_i) \tag{27}$$

or in vector form: $\dot{c} = Kc$.

To write another form of Equation (27) we use stoichiometric vectors: for a reaction $A_i \to A_j$ the stoichiometric vector γ_{ji} is a vector in concentration space with ith coordinate -1, jth coordinate 1 and other coordinates 0. The reaction rate $w_{ji} = k_{ji}c_i$. The kinetic equation has the form

$$\frac{\mathrm{d}c}{\mathrm{d}t} = \sum_{i,j} w_{ji}\gamma_{ji} \tag{28}$$

where c is the concentration vector. One more form of Equation (27) describes directly dynamics of reaction rates:

$$\frac{\mathrm{d}w_{ji}}{\mathrm{d}t}\left(= k_{ji}\frac{\mathrm{d}c_i}{\mathrm{d}t}\right) = k_{ji}\sum_l (w_{il} - w_{li}) \tag{29}$$

It is necessary to mention that, in general, system (29) is not equivalent to system (28), because there are additional connections between variables w_{ji}. If there exists at least one A_i with two different outgoing reactions, $A_i \to A_j$ and $A_i \to A_l$ $(j \neq l)$, then $w_{ji}/w_{li} \equiv k_{ji}/k_{li}$. If the reaction network generates a discrete dynamical system $A_i \to A_j$ on the set of A_i (see below), then the variables w_{ji} are independent, and Equation (29) gives equivalent representation of kinetics.

For analysis of kinetic systems, linear conservation laws and positively invariant polyhedra are important. A linear conservation law is a linear function defined on the concentrations $b(c) = \sum_i b_i c_i$, whose value is preserved by the dynamics (27). The conservation laws coefficient vectors b_i are left eigenvectors of the matrix K corresponding to the zero eigenvalue. The set of all the conservation laws forms the left kernel of the matrix K. Equation (27) always has a linear conservation law: $b^0(c) = \sum_i c_i = \text{constant}$. If there is no other independent linear conservation law, then the system is *weakly ergodic*.

A set E is *positively invariant* with respect to kinetic equations (27), if any solution $c(t)$ that starts in E at time $t_0 (c(t_0) \in E)$ belongs to E for $t > t_0$ $(c(t) \in E$ if $t > t_0)$. It is straightforward to check that the standard simplex $\Sigma = \{c | c_i \geq 0, \sum_i c_i = 1\}$ is positively invariant set for kinetic equation (27): just to check that if $c_i = 0$ for some i, and all $c_j \geq 0$ then $\dot{c}_i \geq 0$. This simple fact immediately implies the following properties of K:

- All eigenvalues λ of K have non-positive real parts, $Re\lambda \leqslant 0$, because solutions cannot leave Σ in positive time.
- If $Re\lambda = 0$ then $\lambda = 0$, because intersection of Σ with any plain is a polygon, and a polygon cannot be invariant with respect of rotations to sufficiently small angles.
- The Jordan cell of K that corresponds to zero eigenvalue is diagonal — because all solutions should be bounded in Σ for positive time.
- The shift in time operator $\exp(Kt)$ is a contraction in the l_1 norm for $t > 0$: for positive t and any two solutions of Equation (27) $c(t)$, $c'(t) \in \Sigma$

$$\sum_i |c_i(t) - c_i'(t)| \leq \sum_i |c_i(0) - c_i'(0)|$$

Two vertices are called adjacent if they share a common edge. A path is a sequence of adjacent vertices. A graph is connected if any two of its vertices are linked by a path. A maximal connected subgraph of graph G is called a connected component of G. Every graph can be decomposed into connected components.

A directed path is a sequence of adjacent edges where each step goes in direction of an edge. A vertex A is *reachable* by a vertex B, if there exists an oriented path from B to A.

A nonempty set V of graph vertexes forms a *sink*, if there are no oriented edges from $A_i \in V$ to any $A_j \notin V$. For example, in the reaction graph $A_1 \leftarrow A_2 \rightarrow A_3$ the one-vertex sets $\{A_1\}$ and $\{A_3\}$ are sinks. A sink is minimal if it does not contain a strictly smaller sink. In the previous example, $\{A_1\}$ and $\{A_3\}$ are minimal sinks. Minimal sinks are also called ergodic components.

A digraph is strongly connected, if every vertex A is reachable by any other vertex B. Ergodic components are maximal strongly connected subgraphs of the graph, but inverse is not true: there may exist maximal strongly connected subgraphs that have outgoing edges and, therefore, are not sinks.

We study ensembles of systems with a given graph and independent and well-separated kinetic constants k_{ij}. This means that we study asymptotic behavior of ensembles with independent identically distributed constants, log-uniform distributed in sufficiently big interval $\log k \in [\alpha, \beta]$, for $\alpha \rightarrow -\infty$, $\beta \rightarrow \infty$, or just a log-uniform distribution on infinite axis, $\log k \in \mathbb{R}$.

4.1.2 Sinks and ergodicity

If there is no other independent linear conservation law, then the system is weakly ergodic. The weak ergodicity of the network follows from its topological properties.

The following properties are equivalent and each one of them can be used as an alternative definition of weak ergodicity:

(i) There exist the only independent linear conservation law for kinetic equations (27) (this is $b^0(\mathrm{c}) = \sum_i c_i = \text{constant}$).
(ii) For any normalized initial state $c(0)$ ($b^0(c) = 1$) there exists a limit state

$$c^* = \lim_{t\to\infty} \exp(Kt)c(0)$$

that is the same for all normalized initial conditions: For all c,

$$\lim_{t\to\infty} \exp(Kt)c = b^0(c)c^*$$

(iii) For each two vertices A_i and $A_j (i \neq j)$ we can find such a vertex A_k that is reachable both by A_i and by A_j. This means that the following structure exists:

$$A_i \to \ldots \to A_k \leftarrow \ldots \leftarrow A_j$$

One of the paths can be degenerated: it may be $i = k$ or $j = k$.
(iv) The network has only one minimal sink (one ergodic component).

For every monomolecular kinetic system, the Jordan cell for zero eigenvalue of matrix K is diagonal and the maximal number of independent linear conservation laws (i.e. the geometric multiplicity of the zero eigenvalue of the matrix K) is equal to the maximal number of disjoint ergodic components (minimal sinks).

Let $G = \{A_{i_1}, \ldots, A_{i_l}\}$ be an ergodic component. Then there exists a unique vector (normalized invariant distribution) c^G with the following properties: $c_i^G = 0$ for $i \notin \{i_1, \ldots, i_l\}$, $c_i^G > 0$ for all $i \in \{i_1, \ldots, i_l\}$, $b^0(c^G) = 1$, $Kc^G = 0$.

If $G_1, \ldots, G_m$ are all ergodic components of the system, then there exist m independent positive linear functionals $b^1(c), \ldots, b^m(c)$, such that $\sum_{i=1}^m b^i = b^0$ and for each c

$$\lim_{t\to\infty} \exp(Kt)c = \sum_{i=1}^{m} b^i(c)c^{G_i} \tag{30}$$

So, for any solution of kinetic equations (27), $c(t)$, the limit at $t \to \infty$ is a linear combination of normalized invariant distributions c^{G_i} with coefficients $b^i(c(0))$. In the simplest example, $A_1 \leftarrow A_2 \to A_3$, $G_1 = \{A_1\}$, $G_2 = \{A_3\}$, components of vectors c^{G_1}, c^{G_2} are (1, 0, 0) and (0, 0, 1), correspondingly. For functionals $b^{1,2}$ we get:

$$b^1(c) = c_1 + \frac{k_1}{k_1 + k_2} c_2; \quad b^2(c) = \frac{k_2}{k_1 + k_2} c_2 + c_3 \tag{31}$$

where k_1 and k_2 are rate constants for reaction $A_2 \to A_1$ and $A_2 \to A_3$, correspondingly. We can mention that for well-separated constants either $k_1 \gg k_2$

or $k_1 \ll k_2$. Hence, one of the coefficients $k_1/(k_1+k_2)$ and $k_2/(k_1+k_2)$ is close to 0, another is close to 1. This is an example of the general zero-one law for multiscale systems: for any l, i, the value of functional b^l (30) on basis vector e^i, $b^l(e^i)$, is either close to 1 or close to 0 (with probability close to 1).

We can understand better this asymptotics by using the Markov chain language. For nonseparated constants a particle in A_2 has nonzero probability to reach A_1 and nonzero probability to reach A_3. The zero-one law in this simplest case means that the dynamics of the particle becomes deterministic: with probability one it chooses to go to one of vertices A_2, A_3 and to avoid another. Instead of branching, $A_2 \to A_1$ and $A_2 \to A_3$, we select only one way: either $A_2 \to A_1$ or $A_2 \to A_3$. Graphs without branching represent discrete dynamical systems.

4.1.3 Decomposition of discrete dynamical systems

Discrete dynamical system on a finite set $V = \{A_1, A_2, \ldots, A_n\}$ is a semigroup $1, \phi, \phi^2, \ldots$, where ϕ is a map $\phi: V \to V$. $A_i \in V$ is a periodic point, if $\phi^l(A_i) = A_i$ for some $l > 0$; else A_i is a transient point. A cycle of period l is a sequence of l distinct periodic points $A, \phi(A), \phi^2(A), \ldots, \phi^{l-1}(A)$ with $\phi^l(A) = A$. A cycle of period one consists of one fixed point, $\phi(A) = A$. Two cycles, C and C' either coincide or have empty intersection.

The set of periodic points, V^{p}, is always nonempty. It is a union of cycles: $V^{\mathrm{p}} = \cup_j C_j$. For each point $A \in V$ there exist such a positive integer $\tau(A)$ and a cycle $C(A) = C_j$ that $\phi^q(A) \in C_j$ for $q \geqslant \tau(A)$. In that case we say that A belongs to basin of attraction of cycle C_j and use notation $\mathrm{Att}(C_j) = \{A | C(A) = C_j\}$. Of course, $C_j \subset \mathrm{Att}(C_j)$. For different cycles, $\mathrm{Att}(C_j) \cap \mathrm{Att}(C_l) = \varnothing$. If A is periodic point then $\tau(A) = 0$. For transient points $\tau(A) > 0$.

So, the phase space V is divided onto subsets $\mathrm{Att}(C_j)$. Each of these subsets includes one cycle (or a fixed point, that is a cycle of length 1). Sets $\mathrm{Att}(C_j)$ are ϕ-invariant: $\phi(\mathrm{Att}(C_j)) \subset \mathrm{Att}(C_j)$. The set $\mathrm{Att}(C_j) \backslash C_j$ consist of transient points and there exists such positive integer τ that $\phi^q(\mathrm{Att}(C_j)) = C_j$ if $q \geqslant \tau$.

4.2 Auxiliary discrete dynamical systems and relaxation analysis

4.2.1 Auxiliary discrete dynamical system

For each A_i, we define κ_i as the maximal kinetic constant for reactions $A_i \to A_j : \kappa_i = \max_j\{k_{ji}\}$. For correspondent j we use notation $\phi(i) : \phi(i) = \arg\max_j\{k_{ji}\}$. The function $\phi(i)$ is defined under condition that for A_i outgoing reactions $A_i \to A_j$ exist. Let us extend the definition: $\phi(i) = i$ if there exist no such outgoing reactions.

The map ϕ determines discrete dynamical system on a set of components $V = \{A_i\}$. We call it the auxiliary discrete dynamical system for a given network of monomolecular reactions. Let us decompose this system and find the cycles C_j and their basins of attraction, $\mathrm{Att}(C_j)$.

Notice that for the graph that represents a discrete dynamic system, attractors are ergodic components, whereas basins are connected components.

An auxiliary reaction network is associated with the auxiliary discrete dynamical system. This is the set of reactions $A_i \rightarrow A_{\phi(i)}$ with kinetic constants κ_i. The correspondent kinetic equation is

$$\dot{c}_i = -\kappa_i c_i + \sum_{\phi(j)=i} \kappa_j c_j \tag{32}$$

or in vector notations (28)

$$\frac{\mathrm{d}c}{\mathrm{d}t} = \tilde{K}c = \sum_i \kappa_i c_i \gamma_{\phi(i)i}; \ \tilde{K}_{ij} = -\kappa_j \delta_{ij} + \kappa_j \delta_{i\phi(j)} \tag{33}$$

For deriving of the auxiliary discrete dynamical system we do not need the values of rate constants. Only the ordering is important. Below we consider multiscale ensembles of kinetic systems with given ordering and with well-separated kinetic constants ($k_{\sigma(1)} \gg k_{\sigma(2)} \gg \cdots$ for some permutation σ).

In the following, we analyze first the situation when the system is connected and has only one attractor. This can be a point or a cycle. Then, we discuss the general situation with any number of attractors.

4.2.2 Eigenvectors for acyclic auxiliary kinetics

Let us study kinetics (32) for acyclic discrete dynamical system (each vertex has one or zero outgoing reactions, and there are no cycles). Such acyclic reaction networks have many simple properties. For example, the nonzero eigenvalues are exactly minus reaction rate constants, and it is easy to find all left and right eigenvectors in explicit form. Let us find left and right eigenvectors of matrix $\tilde{K}$ of auxiliary kinetic system (32) for acyclic auxiliary dynamics. In this case, for any vertex A_i there exists an eigenvector. If A_i is a fixed point of the discrete dynamical system (i.e. $\phi(i) = i$) then this eigenvalue is zero. If A_i is not a fixed point (i.e. $\phi(i) \neq i$ and reaction $A_i \rightarrow A_{\phi(i)}$ has nonzero rate constant κ_i) then this eigenvector corresponds to eigenvalue $-\kappa_i$. For left and right eigenvectors of $\tilde{K}$ that correspond to A_i we use notations l^i (vector-raw) and r^i (vector-column), correspondingly, and apply normalization condition $r_i^i = l_i^i = 1$.

First, let us find the eigenvectors for zero eigenvalue. Dimension of zero eigenspace is equal to the number of fixed points in the discrete dynamical system. If A_i is a fixed point then the correspondent eigenvalue is zero, and the right eigenvector r^i has only one nonzero coordinate, concentration of $A_i : r_j^i = \delta_{ij}$.

To construct the correspondent left eigenvectors l^i for zero eigenvalue (for fixed point A_i), let us mention that l_j^i could have nonzero value only if there exists such $q \geqslant 0$ that $\phi^q(j) = i$ (this q is unique because absence of cycles). In that case (for $q > 0$),

$$(l^i \tilde{K})_j = -\kappa_j l_j^i + \kappa_j l_{\phi(j)}^i = 0$$

Hence, $l_j^i = l_{\phi(j)}^i$ and $l_j^i = 1$ if $\phi^q(j) = i$ for some $q > 0$.

For nonzero eigenvalues, right eigenvectors will be constructed by recurrence starting from the vertex A_i and moving in the direction of the flow. The construction is in opposite direction for left eigenvectors. Nonzero eigenvalues of $\tilde{K}$ (32) are $-\kappa_i$.

For given i, τ_i is the minimal integer such that $\phi^{\tau_i}(i) = \phi^{\tau_i+1}(i)$ (this is a *relaxation time* i.e. the number of steps to reach a fixed point). All indices $\{\phi^k(i)|k = 0, 1, \ldots, \tau_i\}$ are different. For right eigenvector r^i only coordinates $r^i_{\phi^k(i)} (k = 0, 1, \ldots, \tau_i)$ could have nonzero values, and

$$\begin{aligned}(\tilde{K}r^i)_{\phi^{\kappa+1}(i)} &= -\kappa_{\phi^{\kappa+1}(i)} r^i_{\phi^{\kappa+1}(i)} + \kappa_{\phi^{\kappa}(i)} r^i_{\phi^{\kappa}(i)} \\ &= -\kappa_i r^i_{\phi^{\kappa+1}(i)}\end{aligned}$$

Hence,

$$\begin{aligned}r^i_{\phi^{k+1}(i)} &= \frac{\kappa_{\phi^k(i)}}{\kappa_{\phi^{k+1}(i)} - \kappa_i} r^i_{\phi^k(i)} = \prod_{j=0}^{k} \frac{\kappa_{\phi^j(i)}}{\kappa_{\phi^{j+1}(i)} - \kappa_i} \\ &= \frac{\kappa_i}{\kappa_{\phi^{k+1}(i)} - \kappa_i} \prod_{j=0}^{k-1} \frac{\kappa_{\phi^{j+1}(i)}}{\kappa_{\phi^{j+1}(i)} - \kappa_i}\end{aligned} \tag{34}$$

The last transformation is convenient for estimation of the product for well-separated constants (compare to Equation (4)):

$$\begin{aligned}\frac{\kappa_{\phi^{j+1}(i)}}{\kappa_{\phi^{j+1}(i)} - \kappa_i} &\approx \begin{cases} 1, & \text{if } \kappa_{\phi^{j+1}(i)} \gg \kappa_i, \\ 0, & \text{if } \kappa_{\phi^{j+1}(i)} \ll \kappa_i; \end{cases} \\ \frac{\kappa_i}{\kappa_{\phi^{k+1}(i)} - \kappa_i} &\approx \begin{cases} -1, & \text{if } \kappa_i \gg \kappa_{\phi^{k+1}(i)}, \\ 0, & \text{if } \kappa_i \ll \kappa_{\phi^{k+1}(i)} \end{cases}\end{aligned} \tag{35}$$

For left eigenvector l^i coordinate l^i_j could have nonzero value only if there exists such $q \geqslant 0$ that $\phi^q(j) = i$ (this q is unique because the auxiliary dynamical system has no cycles). In that case (for $q > 0$),

$$(l^i\tilde{K})_j = -\kappa_j l^i_j + \kappa_j l^i_{\phi(j)} = -\kappa_i l^i_j$$

Hence,

$$l^i_j = \frac{\kappa_j}{\kappa_j - \kappa_i} l^i_{\phi(j)} = \prod_{k=0}^{q-1} \frac{\kappa_{\phi^k(j)}}{\kappa_{\phi^k(j)} - \kappa_i} \tag{36}$$

For every fraction in Equation (36) the following estimate holds:

$$\frac{\kappa_{\phi^k(j)}}{\kappa_{\phi^k(j)} - \kappa_i} \approx \begin{cases} 1, & \text{if } \kappa_{\phi^k(j)} \gg \kappa_i, \\ 0, & \text{if } \kappa_{\phi^k(j)} \ll \kappa_i \end{cases} \tag{37}$$

As we can see, every coordinate of left and right eigenvectors of $\tilde{K}$ (34), (36) is either 0 or ± 1, or close to 0 or to ± 1 (with probability close to 1). We can write this asymptotic representation explicitly (analogously to Equation (5)). For left eigenvectors, $l^i_i = 1$ and $l^i_j = 1$ (for $i \neq j$) if there exists such q that $\phi^q(j) = i$, and $\kappa_{\phi^d(j)} > \kappa_i$ for all $d = 0, \ldots, q-1$, else $l^i_j = 0$. For right eigenvectors, $r^i_i = 1$ and

$r^i_{\phi^k(i)} = -1$ if $\kappa_{\phi^k(i)} < \kappa_i$ and for all positive $m < k$ inequality $\kappa_{\phi^m(i)} > \kappa_i$ holds, i.e. k is first such positive integer that $\kappa_{\phi^k(i)} < \kappa_i$ (for fixed point A_p we use $\kappa_p = 0$). Vector r^i has not more than two nonzero coordinates. It is straightforward to check that in this asymptotic $l^i r^j = \delta_{ij}$.

In general, coordinates of eigenvectors l^i_j and r^i_j are simultaneously nonzero only for one value $j = i$ because the auxiliary system is acyclic. However, $l^i r^j = 0$ if $i \neq j$, just because that are eigenvectors for different eigenvalues, κ_i and κ_j. Hence, $l^i r^j = \delta_{ij}$.

For example, let us find the asymptotic of left and right eigenvectors for a branched acyclic system of reactions:

$$A_1 \xrightarrow{7} A_2 \xrightarrow{5} A_3 \xrightarrow{6} A_4 \xrightarrow{2} A_5 \xrightarrow{4} A_8, \quad A_6 \xrightarrow{1} A_7 \xrightarrow{3} A_4$$

where the upper index marks the order of rate constants: $\kappa_6 > \kappa_4 > \kappa_7 > \kappa_5 > \kappa_2 > \kappa_3 > \kappa_1$ (κ_i is the rate constant of reaction $A_i \to \ldots$).

For zero eigenvalue, the left and right eigenvectors are

$$l^8 = (1,1,1,1,1,1,1,1,1), \quad r^8 = (0,0,0,0,0,0,0,0,1)$$

For left eigenvectors, rows l^i, that correspond to nonzero eigenvalues we have the following asymptotics:

$$\begin{aligned} &l^1 \approx (1,0,0,0,0,0,0,0), \quad l^2 \approx (0,1,0,0,0,0,0,0), \\ &l^3 \approx (0,1,1,0,0,0,0,0), \quad l^4 \approx (0,0,0,1,0,0,0,0), \\ &l^5 \approx (0,0,0,1,1,1,1,0), \quad l^6 \approx (0,0,0,0,0,1,0,0), \\ &l^7 \approx (0,0,0,0,0,1,1,0) \end{aligned} \tag{38}$$

For the correspondent right eigenvectors, columns r^i, we have the following asymptotics (we write vector-columns in rows):

$$\begin{aligned} &r^1 \approx (1,0,0,0,0,0,0,-1), \quad r^2 \approx (0,1,-1,0,0,0,0,0), \\ &r^3 \approx (0,0,1,0,0,0,0,-1), \quad r^4 \approx (0,0,0,1,-1,0,0,0), \\ &r^5 \approx (0,0,0,0,1,0,0,-1), \quad r^6 \approx (0,0,0,0,0,1,-1,0), \\ &r^7 \approx (0,0,0,0,-1,0,1,0) \end{aligned} \tag{39}$$

4.2.3 The first case: auxiliary dynamical system is acyclic and has one attractor

In the simplest case, the auxiliary discrete dynamical system for the reaction network $\mathcal{W}$ is acyclic and has only one attractor, a fixed point. Let this point be A_n (n is the number of vertices). The correspondent eigenvectors for zero eigenvalue are $r^n_j = \delta_{nj}$ and $l^n_j = 1$. For such a system, it is easy to find explicit analytic solution of kinetic equation (32).

Acyclic auxiliary dynamical system with one attractor have a characteristic property among all auxiliary dynamical systems: the stoichiometric vectors of reactions $A_i \to A_{\phi(i)}$ form a basis in the subspace of concentration space with $\sum_i c_i = 0$. Indeed, for such a system there exist $n-1$ reactions, and their stoichiometric vectors are independent. However, existence of cycles implies linear connections between stoichiometric vectors, and existence of two attractors in acyclic system implies that the number of reactions is less $n-1$, and their stoichiometric vectors could not form a basis in $n-1$-dimensional space.

Let us assume that the auxiliary dynamical system is acyclic and has only one attractor, a fixed point. This means that stoichiometric vectors $\gamma_{\phi(i)i}$ form a basis in a subspace of concentration space with $\sum_i c_i = 0$. For every reaction $A_i \to A_l$ the following linear operators Q_{il} can be defined:

$$Q_{il}(\gamma_{\phi(i)i}) = \gamma_{li}, \quad Q_{il}(\gamma_{\phi(p)p}) = 0 \text{ for } p \neq i \tag{40}$$

The kinetic equation for the whole reaction network (28) could be transformed in the form

$$\begin{aligned} \frac{\mathrm{d}c}{\mathrm{d}t} &= \sum_i \left(1 + \sum_{l, l \neq \phi(i)} \frac{k_{li}}{\kappa_i} Q_{il}\right) \gamma_{\phi(i)\ i} \kappa_i c_i \\ &= \left(1 + \sum_{j,l(l \neq \phi(j))} \frac{k_{lj}}{\kappa_j} Q_{jl}\right) \sum_i \gamma_{\phi(i)i} \kappa_i c_i \\ &= \left(1 + \sum_{j,l(l \neq \phi(j))} \frac{k_{lj}}{\kappa_j} Q_{jl}\right) \tilde{K} c \end{aligned} \tag{41}$$

where $\tilde{K}$ is kinetic matrix of auxiliary kinetic equation (33). By construction of auxiliary dynamical system, $k_{li} \ll \kappa_i$ if $l \neq \phi(i)$. Notice also that $|Q_{jl}|$ does not depend on rate constants.

Let us represent system (41) in eigenbasis of $\tilde{K}$ obtained in previous subsection. Any matrix B in this eigenbasis has the form $B = (\tilde{b}_{ij})$, $\tilde{b}_{ij} = l^i B r^j = \sum_{qs} l^i_q b_{qs} r^j_s$, where (b_{qs}) is matrix B in the initial basis, l^i and r^j are left and right eigenvectors of $\tilde{K}$ (34), (36). In eigenbasis of $\tilde{K}$ the Gershgorin estimates of eigenvalues and estimates of eigenvectors are much more efficient than in original coordinates: the system is stronger diagonally dominant. Transformation to this basis is an effective preconditioning for perturbation theory that uses auxiliary kinetics as a first approximation to the kinetics of the whole system.

First of all, we can exclude the conservation law. Any solution of (41) has the form $c(t) = br^n + \tilde{c}(t)$, where $b = l^n c(t) = l^n c(0)$ and $\sum_i \tilde{c}_i(t) = 0$. On the subspace of concentration space with $\sum_i c_i = 0$ we get

$$\frac{\mathrm{d}c}{\mathrm{d}t} = (1 + \mathscr{E})\mathrm{diag}\{-\kappa_1, \ldots, -\kappa_{n-1}\}c \tag{42}$$

where $\mathscr{E} = (\varepsilon_{ij})$, $|\varepsilon_{ij}| \ll 1$ and $\mathrm{diag}\{-\kappa_1, \ldots, -\kappa_{n-1}\}$ is diagonal matrix with $-\kappa_1, \ldots, -\kappa_{n-1}$ on the main diagonal. If $|\varepsilon_{ij}| \ll 1$ then we can use the Gershgorin theorem and state that eigenvalues of matrix $(1 + \mathscr{E})\mathrm{diag}\{-\kappa_1, \ldots, -\kappa_{n-1}\}$ are real and have the form $\lambda_i = -\kappa_i + \theta_i$ with $|\theta_i| \ll \kappa_i$.

To prove inequality $|\varepsilon_{ij}| \ll 1$ (for ensembles with well-separated constants, with probability close to 1) we use that the left and right eigenvectors of $\tilde{K}$ (34), (36) are uniformly bounded under some non-degeneracy conditions and those conditions are true for well-separated constants. For ensembles with well-separated constants, for any given positive $g<1$ and all i, j ($i \neq j$) the following inequality is true with probability close to 1: $|\kappa_i - \kappa_j| > g\kappa_i$. Let us select a value of g and assume that this *diagonal gap condition* is always true. In this case, for every fraction in (34), (36) we have estimate

$$\frac{\kappa_i}{|\kappa_j - \kappa_i|} < \frac{1}{g}$$

Therefore, for coordinates of right and left eigenvectors of $\tilde{K}$ (34), (36) we get

$$|r^i_{\phi^{\kappa+1}(i)}| < \frac{1}{g^k} < \frac{1}{g^n}, \quad |l^i_j| < \frac{1}{g^q} < \frac{1}{g^n} \tag{43}$$

We can estimate $|\varepsilon_{ij}|$ and $|\theta_i|/\kappa_i$ from above as constant $\times$ $\max_{l \neq \phi(s)}\{k_{ls}/\kappa_s\}$. So, the eigenvalues for kinetic matrix of the whole system (41) are real and close to eigenvalues of auxiliary kinetic matrix $\tilde{K}$ (33). For eigenvectors, the Gershgorin theorem gives no result, and additionally to diagonal dominance we must assume the diagonal gap condition. Based on this assumption, we proved the Gershgorin type estimate of eigenvectors in Appendix 1. In particular, according to this estimate, eigenvectors for the whole reaction network are arbitrarily close to eigenvectors of $\tilde{K}$ (with probability close to 1).

So, if the auxiliary discrete dynamical system is acyclic and has only one attractor (a fixed point), then the relaxation of the whole reaction network could be approximated by the auxiliary kinetics (32):

$$c(t) = (l^n c(0))r^n + \sum_{i=1}^{n-1}(l^i c(0))r^i \exp(-\kappa_i t) \tag{44}$$

For l^i and r^i one can use exact formulas (34) and (36) or zero-one asymptotic representations based on Equations (37) and (35) for multiscale systems. This approximation (44) could be improved by iterative methods, if necessary.

4.2.4 The second case: auxiliary system has one cyclic attractor

The second simple particular case on the way to general case is a reaction network with components $A_1, \ldots, A_n$ whose auxiliary discrete dynamical system has one attractor, a cycle with period $\tau>1$: $A_{n-\tau+1} \to A_{n-\tau+2} \to \ldots A_n \to A_{n-\tau+1}$ (after some change of enumeration). We assume that the limiting step in this cycle (reaction with minimal constant) is $A_n \to A_{n-\tau+1}$. If auxiliary discrete dynamical system has only one attractor then the whole network is weakly ergodic. But the attractor of the auxiliary system may not coincide with a sink of the reaction network.

There are two possibilities:

(i) In the whole network, all the outgoing reactions from the cycle have the form $A_{n-\tau+i} \to A_{n-\tau+j}$ (i, $j>0$). This means that the cycle vertices $A_{n-\tau+1}$, $A_{n-\tau+2}, \ldots, A_n$ form a sink for the whole network.

(ii) There exists a reaction from a cycle vertex $A_{n-\tau+i}$ to A_m, $m \leqslant n-\tau$. This means that the set $\{A_{n-\tau+1}, A_{n-\tau+2}, \ldots, A_n\}$ is not a sink for the whole network.

In the first case, the limit (for $t \to \infty$) distribution for the auxiliary kinetics is the well-studied stationary distribution of the cycle $A_{n-\tau+1}, A_{n-\tau+2}, \ldots, A_n$ described in Section 2 (11)–(13), (15). The set $\{A_{n-\tau+1}, A_{n-\tau+2}, \ldots, A_n\}$ is the only ergodic component for the whole network too, and the limit distribution for that system is nonzero on vertices only. The stationary distribution for the cycle $A_{n-\tau+1} \to A_{n-\tau+2} \to \ldots A_n \to A_{n-\tau+1}$ approximates the stationary distribution for the whole system. To approximate the relaxation process, let us delete the limiting step $A_n \to A_{n-\tau+1}$ from this cycle. By this deletion we produce an acyclic system with one fixed point, A_n, and auxiliary kinetic equation (33) transforms into

$$\frac{\mathrm{d}c}{\mathrm{d}t} = \tilde{K}_0 c = \sum_{i=1}^{n-1} \kappa_i c_i \gamma_{\phi(i)i} \tag{45}$$

As it is demonstrated, dynamics of this system approximates relaxation of the whole network in subspace $\sum_i c_i = 0$. Eigenvalues for Equation (45) are $-\kappa_i$ ($i<n$), the corresponded eigenvectors are represented by Equations (34), (36) and zero-one multiscale asymptotic representation is based on Equations (37) and (35).

In the second case, the set

$$\{A_{n-\tau+1}, A_{n-\tau+2}, \ldots, A_n\}$$

is not a sink for the whole network. This means that there exist outgoing reactions from the cycle, $A_{n-\tau+i} \to A_j$ with $A_j \notin \{A_{n-\tau+1}, A_{n-\tau+2}, \ldots, A_n\}$. For every cycle vertex $A_{n-\tau+i}$ the rate constant $\kappa_{n-\tau+i}$ that corresponds to the cycle reaction $A_{n-\tau+i} \to A_{n-\tau+i+1}$ is much bigger than any other constant $k_{j,n-\tau+i}$ that corresponds to a "side" reaction $A_{n-\tau+i} \to A_j$ ($j \neq n-\tau+i+1$): $\kappa_{n-\tau+i} \gg k_{j,n-\tau+i}$. This is because definition of auxiliary discrete dynamical system and assumption of ensemble with well-separated constants (multiscale asymptotics). This inequality allows us to separate motion and to use for computation of the rates of outgoing reaction $A_{n-\tau+i} \to A_j$ the quasi-steady-state distribution in the cycle. This means that we can glue the cycle into one vertex $A^1_{n-\tau+1}$ with the correspondent concentration $c^1_{n-\tau+1} = \sum_{1 \leq i \leq \tau} c_{n-\tau+i}$ and substitute the reaction $A_{n-\tau+i} \to A_j$ by $A^1_{n-\tau+1} \to A_j$ with the rate constant renormalization: $k^1_{j,n-\tau+1} = k_{j,n-\tau+i} c^{QS}_{n-\tau+i} / c^1_{n-\tau+1}$. By the superscript QS we mark here the quasi-stationary concentrations for given total cycle concentration $c^1_{n-\tau+1}$. Another possibility is to recharge the link $A_{n-\tau+i} \to A_j$ to another vertex of the cycle (usually to A_n): we can substitute the reaction $A_{n-\tau+i} \to A_j$ by the reaction $A_{n-\tau+q} \to A_j$ with the rate constant renormalization:

$$k_{j,n-\tau+q} = k_{j,n-\tau+i} c^{QS}_{n-\tau+i} / c^{QS}_{n-\tau+q} \tag{46}$$

The new rate constant is smaller than the initial one: $k_{j,n-\tau+q} \leq k_{j,n-\tau+i}$, because $c^{QS}_{n-\tau+i} \leq c^{QS}_{n-\tau+q}$ due to definition.

We apply this approach now and demonstrate its applicability in more details later in this section. For the quasi-stationary distribution on the cycle we get $c_{n-\tau+i} = c_n\kappa_n/\kappa_{n-\tau+i}$ $(1 \leqslant i < \tau)$. The original reaction network is transformed by gluing the cycle $\{A_{n-\tau+1}, A_{n-\tau+2}, \ldots, A_n\}$, into a point $A^1_{n-\tau+1}$. We say that components $A_{n-\tau+1}, A_{n-\tau+2}, \ldots, A_n$ of the original system belong to the component $A^1_{n-\tau+1}$ of the new system. All the reactions $A_i \to A_j$ with $i,j \leqslant n-\tau$ remain the same with rate constant k_{ji}. Reactions of the form $A_i \to A_j$ with $i \leqslant n-\tau$, $j > n-\tau$ (incoming reactions of the cycle $\{A_{n-\tau+1}, A_{n-\tau+2}, \ldots, A_n\}$) transform into $A_i \to A^1_{n-\tau+1}$ with the same rate constant k_{ji}. Reactions of the form $A_i \to A_j$ with $i > n-\tau$, $j \leqslant n-\tau$ (outgoing reactions of the cycle $\{A_{n-\tau+1}, A_{n-\tau+2}, \ldots, A_n\}$) transform into reactions $A^1_{n-\tau+1} \to A_j$ with the "quasi-stationary" rate constant $k^{\rm QS}_{ji} = k_{ji}\kappa_n/\kappa_{n-\tau+i}$. After that, we select the maximal $k^{\rm QS}_{ji}$ for given j: $k^{(1)}_{j,n-\tau+1} = \max_{i>n-\tau} k^{QS}_{ji}$. This $k^{(1)}_{j,n-\tau+1}$ is the rate constant for reaction $A^1_{n-\tau+1} \to A_j$ in the new system. Reactions $A_i \to A_j$ with $i,j > n-\tau$ (internal reactions of the site) vanish.

Among rate constants for reactions of the form $A_{n-\tau+i} \to A_m$ $(m \geqslant n-\tau)$ we find

$$\kappa^{(1)}_{n-\tau+i} = \max_{i,m}\{k_{m,n-\tau+i}\kappa_n/\kappa_{n-\tau+i}\} \tag{47}$$

Let the correspondent i, m be i_1, m_1.

After that, we create a new auxiliary discrete dynamical system for the new reaction network on the set $\{A_1, \ldots, A_{n-\tau}, A^1_{n-\tau+1}\}$. We can describe this new auxiliary system as a result of transformation of the first auxiliary discrete dynamical system of initial reaction network. All the reactions from this first auxiliary system of the form $A_i \to A_j$ with $i,j \leqslant n-\tau$ remain the same with rate constant κ_i. Reactions of the form $A_i \to A_j$ with $i \leqslant n-\tau$, $j > n-\tau$ transform into $A_i \to A^1_{n-\tau+1}$ with the same rate constant κ_i. One more reaction is to be added: $A^1_{n-\tau+1} \to A_{m_1}$ with rate constant $\kappa^{(1)}_{n-\tau+i}$. We "glued" the cycle into one vertex, $A^1_{n-\tau+1}$, and added new reaction from this vertex to A_{m_1} with maximal possible constant (47). Without this reaction the new auxiliary dynamical system has only one attractor, the fixed point $A^1_{n-\tau+1}$. With this additional reaction that point is not fixed, and a new cycle appears: $A_{m_1} \to \ldots A^1_{n-\tau+1} \to A_{m_1}$.

Again we should analyze, whether this new cycle is a sink in the new reaction network, etc. Finally, after a chain of transformations, we should come to an auxiliary discrete dynamical system with one attractor, a cycle, that is the sink of the transformed whole reaction network. After that, we can find stationary distribution by restoring of glued cycles in auxiliary kinetic system and applying formulas (11)–(13) and (15) from Section 2. First, we find the stationary state of the cycle constructed on the last iteration, after that for each vertex A^k_j that is a glued cycle we know its concentration (the sum of all concentrations) and can find the stationary distribution, then if there remain some vertices that are glued cycles we find distribution of concentrations in these cycles, etc. At the end of this process we find all stationary concentrations with high accuracy, with probability close to one.

As a simple example we use the following system, a chain supplemented by three reactions:

$$\begin{gathered}A_1 \xrightarrow{1} A_2 \xrightarrow{2} A_3 \xrightarrow{3} A_4 \xrightarrow{4} A_5 \xrightarrow{5} A_6,\\ A_6 \xrightarrow{6} A_4,\ \ A_5 \xrightarrow{7} A_2 \text{ and } A_3 \xrightarrow{8} A_1\end{gathered} \tag{48}$$

where the upper index marks the order of rate constants.

Auxiliary discrete dynamical system for the network (48) includes the chain and one reaction:

$$A_1 \xrightarrow{1} A_2 \xrightarrow{2} A_3 \xrightarrow{3} A_4 \xrightarrow{4} A_5 \xrightarrow{5} A_6 \xrightarrow{6} A_4$$

It has one attractor, the cycle $A_4 \xrightarrow{4} A_5 \xrightarrow{5} A_6 \xrightarrow{6} A_4$. This cycle is not a sink for the whole system, because there exists an outgoing reaction $A_5 \xrightarrow{7} A_2$.

By gluing the cycle $A_4 \xrightarrow{4} A_5 \xrightarrow{5} A_6 \xrightarrow{6} A_4$ into a vertex A_4^1 we get new network with a chain supplemented by two reactions:

$$A_1 \xrightarrow{1} A_2 \xrightarrow{2} A_3 \xrightarrow{3} A_4^1,\ \ A_4^1 \xrightarrow{?} A_2 \text{ and } A_3 \xrightarrow{?} A_1 \tag{49}$$

Here the new rate constant is $k_{24}^{(1)} = k_{25}\kappa_6/\kappa_5$ ($\kappa_6 = k_{46}$ is the limiting step of the cycle $A_4 \xrightarrow{4} A_5 \xrightarrow{5} A_6 \xrightarrow{6} A_4$, $\kappa_5 = k_{65}$).

Here we can make a simple but important observation: the new constant $k_{24}^1 = k_{25}\kappa_6/\kappa_5$ has the same log-uniform distribution on the whole axis as constants k_{25}, κ_6 and κ_5 have. The new constant k_{24}^1 depends on k_{25} and the internal cycle constants κ_6 and κ_5, and is independent from other constants.

Of course, $k_{24}^{(1)} < \kappa_5$, but relations between $k_{24}^{(1)}$ and k_{13} are a priori unknown. Both orderings, $k_{24}^{(1)} > k_{13}$ and $k_{24}^{(1)} < k_{13}$, are possible, and should be considered separately, if necessary. But for both orderings the auxiliary dynamical system for network (49) is

$$A_1 \xrightarrow{1} A_2 \xrightarrow{2} A_3 \xrightarrow{3} A_4^1 \xrightarrow{?} A_2$$

(of course, $\kappa_4^{(1)} < \kappa_3 < \kappa_2 < \kappa_1$). It has one attractor, the cycle $A_2 \xrightarrow{2} A_3 \xrightarrow{3} A_4^1 \xrightarrow{?} A_2$. This cycle is not a sink for the whole system, because there exists an outgoing reaction $A_3 \xrightarrow{?} A_1$. The limiting constant for this cycle is $\kappa_4^{(1)} = k_{24}^{(1)} = k_{25}k_{46}/k_{65}$. We glue this cycle into one point, A_2^2. The new transformed system is very simple, it is just a two-step cycle: $A_1 \xrightarrow{1} A_2^2 \xrightarrow{?} A_1$. The new reaction constant is $k_{12}^{(2)} = k_{13}\kappa_4^{(1)}/\kappa_3 = k_{13}k_{25}k_{46}/(k_{65}k_{43})$. The auxiliary discrete dynamical system is the same graph $A_1 \xrightarrow{1} A_2^2 \xrightarrow{?} A_1$, this is a cycle, and we do not need further transformations.

Let us find the steady state on the way back, from this final auxiliary system to the original one. For steady state of each cycle we use formula (13).

The steady state for the final system is $c_1 = bk_{12}^{(2)}/k_{21}$ and $c_2^2 = b(1 - k_{12}^{(2)}/k_{21})$. The component A_2^2 includes the cycle $A_2 \xrightarrow{2} A_3 \xrightarrow{3} A_4^1 \xrightarrow{?} A_2$. The steady state of this cycle is $c_2 = c_2^{(2)}k_{24}^{(1)}/k_{32}$, $c_3 = c_2^{(2)}k_{24}^{(1)}/k_{43}$ and $c_4^{(1)} = c_2^{(2)}(1 - k_{24}^{(1)}/k_{32} - k_{24}^{(1)}/k_{43})$. The component A_4^1 includes the cycle $A_4 \xrightarrow{4} A_5 \xrightarrow{5} A_6 \xrightarrow{6} A_4$.

The steady state of this cycle is $c_4 = c_4^{(1)} k_{46}/k_{54}$, $c_5 = c_4^{(1)} k_{46}/k_{65}$ and $c_6 = c_4^{(1)}(1 - k_{46}/k_{54} - k_{46}/k_{65})$.

For one catalytic cycle, relaxation in the subspace $\sum_i c_i = 0$ is approximated by relaxation of a chain that is produced from the cycle by cutting the limiting step (Section 2). For reaction networks under consideration (with one cyclic attractor in auxiliary discrete dynamical system) the direct generalization works: for approximation of relaxation in the subspace $\sum_i c_i = 0$ it is sufficient to perform the following procedures:

- to glue iteratively attractors (cycles) of the auxiliary system that are not sinks of the whole system;
- to restore these cycles from the end of the first procedure to its beginning. For each of cycles (including the last one that is a sink) the limited step should be deleted, and the outgoing reaction should be reattached to the head of the limiting steps (with the proper normalization), if it was not deleted before as a limiting step of one of the cycles.

The heads of outgoing reactions of that cycles should be reattached to the heads of the limiting steps. Let for a cycle this limiting step be $A_m \to A_q$. If for a glued cycle A^k there exists an outgoing reaction $A^k \to A_j$ with the constant κ (47), then after restoration we add the outgoing reaction $A_m \to A_j$ with the rate constant κ. Kinetic of the resulting acyclic system approximates relaxation of the initial networks (under assumption of well-separated constants, for given ordering, with probability close to 1).

Let us construct this acyclic network for the same example (48). The final cycle is $A_1 \xrightarrow{1} A_2^2 \xrightarrow{?} A_1$. The limiting step in this cycle is $A_2^2 \xrightarrow{?} A_1$. After cutting we get $A_1 \xrightarrow{1} A_2^2$. The component A_2^2 is glued cycle $A_2 \xrightarrow{2} A_3 \xrightarrow{3} A_4^1 \xrightarrow{?} A_2$. The reaction $A_1 \xrightarrow{1} A_2^2$ corresponds to the reaction $A_1 \xrightarrow{1} A_2$ (in this case, this is the only reaction from A_1 to cycle; in other case one should take the reaction from A_1 to cycle with maximal constant). The limiting step in the cycle is $A_4^1 \xrightarrow{?} A_2$. After cutting, we get a system $A_1 \xrightarrow{1} A_2 \xrightarrow{2} A_3 \xrightarrow{3} A_4^1$. The component A_4^1 is the glued cycle $A_4 \xrightarrow{4} A_5 \xrightarrow{5} A_6 \xrightarrow{6} A_4$ from the previous step. The limiting step in this cycle is $A_6 \xrightarrow{6} A_4$. After restoring this cycle and cutting the limiting step, we get an acyclic system $A_1 \xrightarrow{1} A_2 \xrightarrow{2} A_3 \xrightarrow{3} A_4 \xrightarrow{4} A_5 \xrightarrow{5} A_6$ (as one can guess from the beginning: this coincidence is provided by the simple constant ordering selected in Equation (48)). Relaxation of this system approximates relaxation of the whole initial network.

To demonstrate possible branching of described algorithm for cycles surgery (gluing, restoring and cutting) with necessity of additional orderings, let us consider the following system:

$$A_1 \xrightarrow{1} A_2 \xrightarrow{6} A_3 \xrightarrow{2} A_4 \xrightarrow{3} A_5 \xrightarrow{4} A_3, \quad A_4 \xrightarrow{5} A_2 \tag{50}$$

The auxiliary discrete dynamical system for reaction network (50) is

$$A_1 \xrightarrow{1} A_2 \xrightarrow{6} A_3 \xrightarrow{2} A_4 \xrightarrow{3} A_5 \xrightarrow{4} A_3$$

It has only one attractor, a cycle $A_3 \xrightarrow{2} A_4 \xrightarrow{3} A_5 \xrightarrow{4} A_3$. This cycle is not a sink for the whole network (50) because reaction $A_4 \xrightarrow{5} A_2$ leads from that cycle. After gluing the cycle into a vertex A_3^1 we get the new network $A_1 \xrightarrow{1} A_2 \xrightarrow{6} A_3^1 \xrightarrow{?} A_2$. The rate constant for the reaction $A_3^1 \to A_2$ is $k_{23}^1 = k_{24}k_{35}/k_{54}$, where k_{ij} is the rate constant for the reaction $A_j \to A_i$ in the initial network (k_{35} is the cycle limiting reaction). The new network coincides with its auxiliary system and has one cycle, $A_2 \xrightarrow{6} A_3^1 \xrightarrow{?} A_2$. This cycle is a sink, hence, we can start the back process of cycles restoring and cutting. One question arises immediately: which constant is smaller, k_{32} or k_{23}^1. The smallest of them is the limiting constant, and the answer depends on this choice. Let us consider two possibilities separately: (1) $k_{32} > k_{23}^1$ and (2) $k_{32} < k_{23}^1$. Of course, for any choice the stationary concentration of the source component A_1 vanishes: $c_1 = 0$.

(1) Let us assume that $k_{32} > k_{23}^1$. In this case, the steady state of the cycle $A_2 \xrightarrow{6} A_3^1 \xrightarrow{?} A_2$ is (according to Equation (13)) $c_2 = bk_{23}^1/k_{32}$ and $c_3^1 = b(1 - k_{23}^1/k_{32})$, where $b = \sum c_i$. The component A_3^1 is a glued cycle $A_3 \xrightarrow{2} A_4 \xrightarrow{3} A_5 \xrightarrow{4} A_3$. Its steady state is $c_3 = c_3^1 k_{35}/k_{43}$, $c_4 = c_3^1 k_{35}/k_{54}$ and $c_5 = c_3^1(1 - k_{35}/k_{43} - k_{35}/k_{54})$.

Let us construct an acyclic system that approximates relaxation of Equation (50) under the same assumption (1) $k_{32} > k_{23}^1$. The final auxiliary system after gluing cycles is $A_1 \xrightarrow{1} A_2 \xrightarrow{6} A_3^1 \xrightarrow{?} A_2$. Let us delete the limiting reaction $A_3^1 \xrightarrow{?} A_2$ from the cycle. We get an acyclic system $A_1 \xrightarrow{1} A_2 \xrightarrow{6} A_3^1$. The component A_3^1 is the glued cycle $A_3 \xrightarrow{2} A_4 \xrightarrow{3} A_5 \xrightarrow{4} A_3$. Let us restore this cycle and delete the limiting reaction $A_5 \xrightarrow{4} A_3$. We get an acyclic system $A_1 \xrightarrow{1} A_2 \xrightarrow{6} A_3 \xrightarrow{2} A_4 \xrightarrow{3} A_5$. Relaxation of this system approximates relaxation of the initial network (50) under additional condition $k_{32} > k_{23}^1$.

(2) Let as assume now that $k_{32} < k_{23}^1$. In this case, the steady state of the cycle $A_2 \xrightarrow{6} A_3^1 \xrightarrow{?} A_2$ is (according to Equation (13)) $c_2 = b(1 - k_{32}/k_{23}^1)$ and $c_3^1 = bk_{32}/k_{23}^1$. The further analysis is the same as it was above: $c_3 = c_3^1 k_{35}/k_{43}$, $c_4 = c_3^1 k_{35}/k_{54}$ and $c_5 = c_3^1(1 - k_{35}/k_{43} - k_{35}/k_{54})$ (with another c_3^1).

Let us construct an acyclic system that approximates relaxation of Equation (50) under assumption (2) $k_{32} < k_{23}^1$. The final auxiliary system after gluing cycles is the same, $A_1 \xrightarrow{1} A_2 \xrightarrow{6} A_3^1 \xrightarrow{?} A_2$, but the limiting step in the cycle is different, $A_2 \xrightarrow{6} A_3^1$. After cutting this step, we get acyclic system $A_1 \xrightarrow{1} A_2 \xleftarrow{?} A_3^1$, where the last reaction has rate constant k_{23}^1.

The component A_3^1 is the glued cycle

$$A_3 \xrightarrow{2} A_4 \xrightarrow{3} A_5 \xrightarrow{4} A_3$$

Let us restore this cycle and delete the limiting reaction $A_5 \xrightarrow{4} A_3$. The connection from glued cycle $A_3^1 \xrightarrow{?} A_2$ with constant k_{23}^1 transforms into connection $A_5 \xrightarrow{?} A_2$ with the same constant k_{23}^1.

We get the acyclic system:

$$A_1 \xrightarrow{1} A_2, A_3 \xrightarrow{2} A_4 \xrightarrow{3} A_5 \xrightarrow{?} A_2$$

The order of constants is now known: $k_{21} > k_{43} > k_{54} > k_{23}^1$, and we can substitute the sign "?" by "4": $A_3 \xrightarrow{2} A_4 \xrightarrow{3} A_5 \xrightarrow{4} A_2$.

For both cases, $k_{32} > k_{23}^1 (k_{23}^1 = k_{24}k_{35}/k_{54})$ and $k_{32} < k_{23}^1$ it is easy to find the eigenvectors explicitly and to write the solution to the kinetic equations in explicit form.

4.3 The general case: cycles surgery for auxiliary discrete dynamical system with arbitrary family of attractors

In this subsection, we summarize results of relaxation analysis and describe the algorithm of approximation of steady state and relaxation process for arbitrary reaction network with well-separated constants.

4.3.1 Hierarchy of cycles gluing

Let us consider a reaction network $\mathcal{W}$ with a given structure and fixed ordering of constants. The set of vertices of $\mathcal{W}$ is $\mathcal{A}$ and the set of elementary reactions is $\mathcal{R}$. Each reaction from $\mathcal{R}$ has the form $A_i \to A_j$, $A_i, A_j \in \mathcal{A}$. The correspondent constant is k_{ji}. For each $A_i \in \mathcal{A}$ we define $\kappa_i = \max_j\{k_{ji}\}$ and $\phi(i) = \arg\max_j\{k_{ji}\}$. In addition, $\phi(i) = i$ if $k_{ji} = 0$ for all j.

The auxiliary discrete dynamical system for the reaction network $\mathcal{W}$ is the dynamical system $\Phi = \Phi_{\mathcal{W}}$ defined by the map ϕ on the set $\mathcal{A}$. Auxiliary reaction network $\mathcal{V} = \mathcal{V}_{\mathcal{W}}$ has the same set of vertices $\mathcal{A}$ and the set of reactions $A_i \to A_{\phi(i)}$ with reaction constants κ_i. Auxiliary kinetics is described by $\dot{c} = \tilde{K}c$, where $\tilde{K}_{ij} = -\kappa_j\delta_{ij} + \kappa_j\delta_{i\phi(j)}$.

Every fixed point of $\Phi_{\mathcal{W}}$ is also a sink for the reaction network $\mathcal{W}$. If all attractors of the system $\Phi_{\mathcal{W}}$ are fixed points $A_{f1}, A_{f2}, \ldots \in \mathcal{A}$ then the set of stationary distributions for the initial kinetics as well as for the auxiliary kinetics is the set of distributions concentrated the set of fixed points $\{A_{f1}, A_{f2}, \ldots\}$. In this case, the auxiliary reaction network is acyclic, and the auxiliary kinetics approximates relaxation of the whole network $\mathcal{W}$.

In general case, let the system $\Phi_{\mathcal{W}}$ have several attractors that are not fixed points, but cycles $C_1, C_2, \ldots$ with periods $\tau_1, \tau_2, \ldots > 1$. By gluing these cycles in points, we transform the reaction network $\mathcal{W}$ into $\mathcal{W}^1$. The dynamical system $\Phi_{\mathcal{W}}$ is transformed into Φ^1. For these new system and network, the connection $\Phi^1 = \Phi_{\mathcal{W}^1}$ persists: Φ^1 is the auxiliary discrete dynamical system for $\mathcal{W}^1$.

For each cycle, C_i, we introduce a new vertex A^i. The new set of vertices, $\mathcal{A}^1 = \mathcal{A} \cup \{A^1, A^2, ...\} \setminus (\cup_i C_i)$ (we delete cycles C_i and add vertices A^i).

All the reaction between $A \to B (A, B \in \mathcal{A})$ can be separated into 5 groups:

(i) both $A, B \notin \cup_i C_i$;
(ii) $A \notin \cup_i C_i$, but $B \in C_i$;
(iii) $A \in C_i$, but $B \notin \cup_i C_i$;
(iv) $A \in C_i$, $B \in C_j$, $i \neq j$;
(v) $A, B \in C_i$.

Reactions from the first group do not change. Reaction from the second group transforms into $A \to A^i$ (to the whole glued cycle) with the same constant. Reaction of the third type changes into $A^i \to B$ with the rate constant renormalization (46): let the cycle C^i be the following sequence of reactions $A_1 \to A_2 \to \dots A_{\tau_i} \to A_1$, and the reaction rate constant for $A_i \to A_{i+1}$ is k_i (k_{τ_i} for $A_{\tau_i} \to A_1$). For the limiting reaction of the cycle C_i we use notation $k_{\lim\, i}$. If $A = A_j$ and k is the rate reaction for $A \to B$, then the new reaction $A^i \to B$ has the rate constant $kk_{\lim\, i}/k_j$. This corresponds to a quasi-stationary distribution on the cycle (13). It is obvious that the new rate constant is smaller than the initial one: $kk_{\lim\ i}/k_j < k$, because $k_{\lim\ i} < k_j$ due to definition of limiting constant. The same constant renormalization is necessary for reactions of the fourth type. These reactions transform into $A^i \to A^j$. Finally, reactions of the fifth type vanish.

After we glue all the cycles of auxiliary dynamical system in the reaction network $\mathcal{W}$, we get $\mathcal{W}^1$. Strictly speaking, the whole network $\mathcal{W}^1$ is not necessary, and in efficient realization of the algorithm for large networks the computation could be significantly reduced. What we need, is the correspondent auxiliary dynamical system $\Phi^1 = \Phi_{\mathcal{W}^1}$ with auxiliary kinetics.

To find the auxiliary kinetic system, we should glue all cycles in the first auxiliary system, and then add several reactions: for each A^i it is necessary to find in $\mathcal{W}^1$ the reaction of the form $A^i \to B$ with maximal constant and add this reaction to the auxiliary network. If there is no reaction of the form $A^i \to B$ for given i then the point A^i is the fixed point for $\mathcal{W}^1$ and vertices of the cycle C_i form a sink for the initial network.

After that, we decompose the new auxiliary dynamical system, find cycles and repeat gluing. Terminate when all attractors of the auxiliary dynamical system Φ^m become fixed points.

4.3.2 Reconstruction of steady states

After this termination, we can find all steady-state distributions by restoring cycles in the auxiliary reaction network $\mathcal{V}^m$. Let $A^m_{f1}, A^m_{f2}, \dots$ be fixed points of Φ^m. The set of steady states for $\mathcal{V}^m$ is the set of all distributions on the set of fixed points $\{A^m_{f1}, A^m_{f2}, \dots\}$. Let us take one of these distributions, $c = (c^m_{f1}, c^m_{f2}, \dots)$ (we mark the concentrations by the same indexes as the vertex has; other $c_i = 0$).

To make a step of cycle restoration we select those vertexes A^m_{fi} that are glued cycles and substitute them in the list $A^m_{f1}, A^m_{f2}, \dots$ by all the vertices of these cycles. For each of those cycles we find the limiting rate constant and redistribute the concentration c^m_{fi} between the vertices of the correspondent cycle by the rule (13) (with $b = c^m_{fi}$). As a result, we get a set of vertices and a distribution on this set of vertices. If among these vertices there are glued cycles, then we repeat the procedure of cycle restoration. Terminate when there is no glued cycles in the support of the distribution. The resulting distribution is the approximation to a steady state of $\mathcal{W}$, and all steady states for $\mathcal{W}$ can be approximated by this method.

To construct the approximation to the basis of stationary distributions of $\mathcal{W}$, it is sufficient to apply the described algorithm to distributions concentrated on a single fixed point A^m_{fi}, $c^m_{fj} = \delta_{ij}$, for every i.

The steady-state approximation on the base of the rule (13) is a linear function of the restored-and-cut cycles rate-limiting constants. It is the first-order approximation.

The zero-order approximation also makes sense. For one cycle gives Equation (14): all the concentration is collected at the start of the limiting step. The algorithm for the zero-order approximation is even simpler than for the first order. Let us start from the distributions concentrated on a single fixed point A^m_{fi}, $c^m_{fj} = \delta_{ij}$ for some i. If this point is a glued cycle then restore that cycle, and find the limiting step. The new distribution is concentrated at the starting vertex of that step. If this vertex is a glued cycle, then repeat. If it is not then terminate. As a result we get a distribution concentrated in one vertex of $\mathcal{A}$.

4.3.3 Dominant kinetic system for approximation of relaxation

To construct an approximation to the relaxation process in the reaction network $\mathcal{W}$, we also need to restore cycles, but for this purpose we should start from the whole glued network $\mathcal{V}^m$ on $\mathcal{A}^m$ (not only from fixed points as we did for the steady-state approximation). On a step back, from the set $\mathcal{A}^m$ to $\mathcal{A}^{m-1}$ and so on some of glued cycles should be restored and cut. On each step we build an acyclic reaction network, the final network is defined on the initial vertex set and approximates relaxation of $\mathcal{W}$.

To make one step back from $\mathcal{V}^m$ let us select the vertices of $\mathcal{A}^m$ that are glued cycles from $\mathcal{V}^{m-1}$. Let these vertices be $A^m_1, A^m_2, \ldots$ Each A^m_i corresponds to a glued cycle from $\mathcal{V}^{m-1}$, $A^{m-1}_{i1} \to A^{m-1}_{i2} \to \ldots A^{m-1}_{i\tau_i} \to A^{m-1}_{i1}$, of the length τ_i. We assume that the limiting steps in these cycles are $A^{m-1}_{i\tau_i} \to A^{m-1}_{i1}$. Let us substitute each vertex A^m_i in $\mathcal{V}^m$ by τ_i vertices $A^{m-1}_{i1}, A^{m-1}_{i2}, \ldots, A^{m-1}_{i\tau_i}$ and add to $\mathcal{V}^m$ reactions $A^{m-1}_{i1} \to A^{m-1}_{i2} \to \ldots A^{m-1}_{i\tau_i}$ (that are the cycle reactions without the limiting step) with correspondent constants from $\mathcal{V}^{m-1}$.

If there exists an outgoing reaction $A^m_i \to B$ in $\mathcal{V}^m$ then we substitute it by the reaction $A^{m-1}_{i\tau_i} \to B$ with the same constant, i.e. outgoing reactions $A^m_i \to \ldots$ are reattached to the heads of the limiting steps. Let us rearrange reactions from $\mathcal{V}^m$ of the form $B \to A^m_i$. These reactions have prototypes in $\mathcal{V}^{m-1}$ (before the last gluing). We simply restore these reactions. If there exists a reaction $A^m_i \to A^m_j$ then we find the prototype in $\mathcal{V}^{m-1}$, $A \to B$ and substitute the reaction by $A^{m-1}_{i\tau_i} \to B$ with the same constant, as for $A^m_i \to A^m_j$.

After that step is performed, the vertices set is $\mathcal{A}^{m-1}$, but the reaction set differs from the reactions of the network $\mathcal{V}^{m-1}$: the limiting steps of cycles are excluded and the outgoing reactions of glued cycles are included (reattached to the heads of the limiting steps). To make the next step, we select vertices of $\mathcal{A}^{m-1}$ that are glued cycles from $\mathcal{V}^{m-2}$, substitute these vertices by vertices of cycles, delete the limiting steps, attach outgoing reactions to the heads of the limiting steps, and for incoming reactions restore their prototypes from $\mathcal{V}^{m-2}$ and so on.

After all, we restore all the glued cycles, and construct an acyclic reaction network on the set $\mathcal{A}$. This acyclic network approximates relaxation of the network $\mathcal{W}$. We call this system the dominant system of $\mathcal{W}$ and use notation dom mod ($\mathcal{W}$).

4.4 Example: a prism of reactions

Let us demonstrate work of the algorithm on a typical example, a prism of reaction that consists of two connected cycles (Figures 2 and 3). Such systems appear in many areas of biophysics and biochemistry (see, e.g. the paper of Kurzynski, 1998).

For the first example we use the reaction rate constants ordering presented in Figure 2a. For this ordering, the auxiliary dynamical system consists of two cycles (Figure 2b) with the limiting constants k_{54} and k_{32}, correspondingly. These cycles are connected by four reactions (Figure 2c). We glue the cycles into new components A_1^1 and A_2^1 (Figure 2d), and the reaction network is transformed into $A_1^1 \leftrightarrow A_2^1$. Following the general rule ($k^1 = kk_{\lim}/k_j$), we determine the rate constants: for reaction $A_1^1 \to A_2^1$

$$k_{21}^1 = \max\{k_{41}k_{32}/k_{21},\ k_{52},\ k_{63}k_{32}/k_{13}\}$$

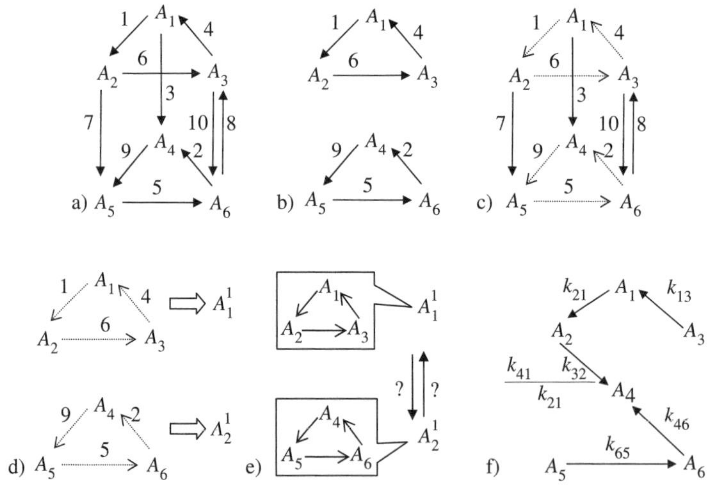

Figure 2 Gluing of cycles for the prism of reactions with a given ordering of rate constants in the case of two attractors in the auxiliary dynamical system: (a) initial reaction network, (b) auxiliary dynamical system that consists of two cycles, (c) connection between cycles, (d) gluing cycles into new components, (e) network $\mathcal{W}^1$ with glued vertices and (f) an example of dominant system in the case when $k_{21}^1 = k_{41}k_{32}/k_{21}$ and $k_{21}^1 > k_{12}^1$ (by definition, $k_{21}^1 = \max\{k_{41}k_{32}/k_{21}, k_{52}, k_{63}k_{32}/k_{13}\}$ and $k_{12}^1 = k_{36}k_{54}/k_{46}$), the order of constants in the dominant system is: $k_{21} > k_{46} > k_{13} > k_{65} > k_{41}k_{32}/k_{21}$.

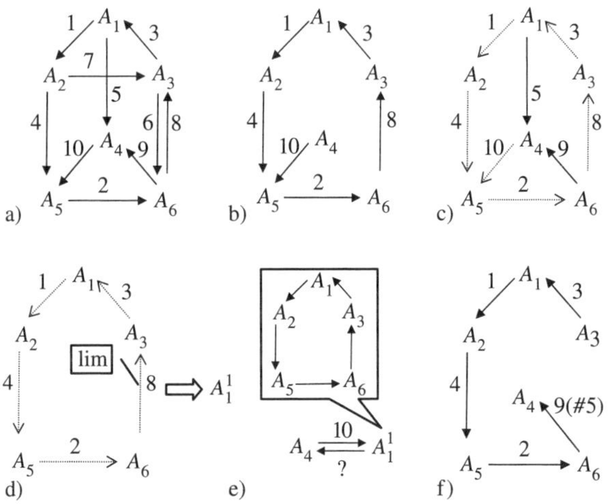

Figure 3 Gluing of a cycle for the prism of reactions with a given ordering of rate constants in the case of one attractors in the auxiliary dynamical system: (a) initial reaction network, (b) auxiliary dynamical system that has one attractor, (c) outgoing reactions from a cycle, (d) gluing of a cycle into new component, (e) network $\mathcal{W}^1$ with glued vertices and (f) an example of dominant system in the case when $k^1 = k_{46}$, and, therefore $k^1 > k_{54}$ (by definition, $k^1 = \max\{k_{41}k_{36}/k_{21}, k_{46}\}$); this dominant system is a linear chain that consists of some reactions from the initial system (no nontrivial monomials among constants). Only one reaction rate constant has in the dominant system new number (number 5 instead of 9).

and for reaction $A_2^1 \to A_1^1$

$$k_{12}^1 = k_{36}k_{54}/k_{46}$$

There are six possible orderings of the constant combinations: three possibilities for the choice of k_{21}^1 and for each such a choice there exist two possibilities: $k_{21}^1 > k_{12}^1$ or $k_{21}^1 < k_{12}^1$.

The zero-order approximation of the steady state depends only on the sign of inequality between k_{21}^1 and k_{12}^1. If $k_{21}^1 \gg k_{12}^1$ then almost all concentration in the steady state is accumulated inside A_2^1. After restoring the cycle $A_4 \to A_5 \to A_6 \to A_4$ we find that in the steady state almost all concentration is accumulated in A_4 (the component at the beginning of the limiting step of this cycle, $A_4 \to A_5$). Finally, the eigenvector for zero eigenvalue is estimated as the vector column with coordinates (0,0,0,1,0,0).

If, inverse, $k_{21}^1 \ll k_{12}^1$ then almost all concentration in the steady state is accumulated inside A_1^1. After restoring the cycle $A_1 \to A_2 \to A_3 \to A_1$ we find that in the steady state almost all concentration is accumulated in A_2 (the component at the beginning of the limiting step of this cycle, $A_2 \to A_3$). Finally, the eigenvector for zero eigenvalue is estimated as the vector column with coordinates (0,1,0,0,0,0).

Let us find the first-order (in rate limiting constants) approximation to the steady states. If $k_{21}^1 \gg k_{12}^1$ then k_{12}^1 is the rate-limiting constant for the cycle $A_1^1 \leftrightarrow A_2^1$ and almost all concentration in the steady state is accumulated inside $A_2^1 : c_2^1 \approx 1 - k_{12}^1/k_{21}^1$ and $c_1^1 \approx k_{12}^1/k_{21}^1$. Let us restore the glued cycles (Figure 2). In the upper cycle the rate-limiting constant is k_{32}, hence, in steady state almost all concentration of the upper cycle, c_1^1, is accumulated in $A_2 : c_2 \approx c_1^1(1 - k_{32}/k_{13} - k_{32}/k_{21})$, $c_3 \approx c_1^1 k_{32}/k_{13}$ and $c_1 \approx c_1^1 k_{32}/k_{21}$. In the bottom cycle the rate-limiting constant is k_{54}, hence, $c_4 \approx c_2^1(1 - k_{54}/k_{65} - k_{54}/k_{46})$, $c_5 \approx c_2^1 k_{54}/k_{65}$ and $c_6 \approx c_2^1 k_{54}/k_{46}$.

If, inverse, $k_{21}^1 \ll k_{12}^1$ then k_{21}^1 is the rate-limiting constant for the cycle $A_1^1 \leftrightarrow A_2^1$ and almost all concentration in the steady state is accumulated inside $A_1^1 : c_1^1 \approx 1 - k_{21}^1/k_{12}^1$ and $c_2^1 \approx k_{21}^1/k_{12}^1$. For distributions of concentrations in the upper and lower cycles only the prefactors c_1^1 and c_2^1 change their values.

For analysis of relaxation, let us analyze one of the six particular cases separately.

1. $k_{21}^1 = k_{41}k_{32}/k_{21}$ and $k_{21}^1 > k_{12}^1$

In this case, the finite acyclic auxiliary dynamical system, $\Phi^m = \Phi^1$, is $A_1^1 \to A_2^1$ with reaction rate constant $k_{21}^1 = k_{41}k_{32}/k_{21}$, and $\mathcal{W}^1$ is $A_1^1 \leftrightarrow A_2^1$. We restore both cycles and delete the limiting reactions $A_2 \to A_3$ and $A_4 \to A_5$. This is the common step for all cases. Following the general procedure, we substitute the reaction $A_1^1 \to A_2^1$ by $A_2 \to A_4$ with the rate constant $k_{21}^1 = k_{41}k_{32}/k_{21}$ (because A_2 is the head of the limiting step for the cycle $A_1 \to A_2 \to A_3 \to A_1$, and the prototype of the reaction $A_1^1 \to A_2^1$ is in that case $A_1 \to A_4$.

We find the dominant system for relaxation description: reactions $A_3 \to A_1 \to A_2$ and $A_5 \to A_6 \to A_4$ with original constants, and reaction $A_2 \to A_4$ with the rate constant $k_{21}^1 = k_{41}k_{32}/k_{21}$.

This dominant system graph is acyclic and, moreover, represents a discrete dynamical system, as it should be (not more than one outgoing reaction for any component). Therefore, we can estimate the eigenvalues and eigenvectors on the base of formulas (35) and (37). It is easy to determine the order of constants because $k_{21}^1 = k_{41}k_{32}/k_{21}$: this constant is the smallest nonzero constant in the obtained acyclic system. Finally, we have the following ordering of constants: $A_3 \xrightarrow{3} A_1 \xrightarrow{1} A_2 \xrightarrow{5} A_4$ and $A_5 \xrightarrow{4} A_6 \xrightarrow{2} A_4$.

So, the eigenvalues of the prism of reaction for the given ordering are (with high accuracy, with probability close to one) $-k_{21} < -k_{46} < -k_{13} < -k_{65} < -k_{41}k_{32}/k_{21}$. The relaxation time is $\tau \approx k_{21}/(k_{41}k_{32})$.

We use the same notations as in previous sections: eigenvectors l^i and r^i correspond to the eigenvalue $-\kappa_i$, where κ_i is the reaction rate constant for the reaction $A_i \to \ldots$ The left eigenvectors l^i are:

$$\begin{aligned} &l^1 \approx (1,0,0,0,0,0), \ l^2 \approx (1,1,1,0,0,0), \\ &l^3 \approx (0,0,1,0,0,0), \ l^4 \approx (1,1,1,1,1,1), \\ &l^5 \approx (0,0,0,0,1,0), \ l^6 \approx (0,0,0,0,0,1) \end{aligned} \tag{51}$$

The right eigenvectors r^i are (we represent vector columns as rows):

$$\begin{aligned} &r^1 \approx (1,-1,0,0,0,0), \;\; r^2 \approx (0,1,0,-1,0,0), \\ &r^3 \approx (0,-1,1,0,0,0), \;\; r^4 \approx (0,0,0,1,0,0), \\ &r^5 \approx (0,0,0,-1,1,0), \;\; r^6 \approx (0,0,0,-1,0,1) \end{aligned} \tag{52}$$

The vertex A_4 is the fixed point for the discrete dynamical system. There is no reaction $A_4 \to \ldots$ For convenience, we include the eigenvectors l^4 and r^4 for zero eigenvalue, $\kappa_4 = 0$. These vectors correspond to the steady state: r^4 is the steady-state vector, and the functional l^4 is the conservation law.

The correspondent approximation to the general solution of the kinetic equation for the prism of reaction (Figure 2a) is:

$$c(t) = \sum_{i=1}^{6} r^i(l^i, c(0)) \exp(-\kappa_i t) \tag{53}$$

Analysis of other five particular cases is similar. Of course, some of the eigenvectors and eigenvalues can differ.

Of course, different ordering can lead to very different approximations. For example, let us consider the same prism of reactions, but with the ordering of constants presented in Figure 3a. The auxiliary dynamical system has one cycle (Figure 3b) with the limiting constant k_{36}. This cycle is not a sink to the initial network, there are outgoing reactions from its vertices (Figure 3c). After gluing, this cycles transforms into a vertex A_1^1 (Figure 3d). The glued network, $\mathcal{W}^1$ (Figure 3e), has two vertices, A_4 and A_1^1 the rate constant for the reaction $A_4 \to A_1^1$ is k_{54}, and the rate constant for the reaction $A_1^1 \to A_4$ is $k^1 = \max\{k_{41}k_{36}/k_{21}, k_{46}\}$. Hence, there are not more than four possible versions: two possibilities for the choice of k^1 and for each such a choice there exist two possibilities: $k^1 > k_{54}$ or $k^1 < k_{54}$ (one of these four possibilities cannot be realized, because $k_{46} > k_{54}$).

Exactly as it was in the previous example, the zero-order approximation of the steady state depends only on the sign of inequality between k^1 and k_{54}. If $k^1 \ll k_{54}$ then almost all concentration in the steady state is accumulated inside A^1. After restoring the cycle $A_3 \to A_1 \to A_2 \to A_5 \to A_6 \to A_3$ we find that in the steady state almost all concentration is accumulated in A_6 (the component at the beginning of the limiting step of this cycle, $A_6 \to A_3$). The eigenvector for zero eigenvalue is estimated as the vector column with coordinates (0,0,0,0,0,1).

If $k^1 \gg k_{54}$ then almost all concentration in the steady state is accumulated inside A^4. This vertex is not a glued cycle, and immediately we find the approximate eigenvector for zero eigenvalue, the vector column with coordinates (0,0,0,1,0,0).

Let us find the first-order (in rate-limiting constants) approximation to the steady states. If $k^1 \ll k_{54}$ then k^1 is the rate-limiting constant for the cycle $A_1^1 \leftrightarrow A_4$ and almost all concentration in the steady state is accumulated inside A_1^1: $c_1^1 \approx 1 - k^1/k_{54}$ and $c_4 \approx k^1/k_{54}$. Let us restore the glued cycle (Figure 3). The limiting constant for that cycle is k_{36}, $c_6 \approx c_1^1(1 - k_{36}/k_{13} - k_{36}/k_{21} - k_{36}/k_{52} - k_{36}/k_{65})$, $c_3 \approx c_1^1 k_{36}/k_{13}$, $c_1 \approx c_1^1 k_{36}/k_{21}$, $c_2 \approx c_1^1 k_{36}/k_{52}$ and $c_5 \approx c_1^1 k_{36}/k_{65}$.

If $k^1 \gg k_{54}$ then k_{54} is the rate-limiting constant for the cycle $A_1^1 \leftrightarrow A_4$ and almost all concentration in the steady state is accumulated inside A_4: $c_4 \approx 1 - k_{54}/k^1$ and $c_1^1 \approx k_{54}/k^1$. In distribution of concentration inside the cycle only the prefactor c_1^1 changes.

Let us analyze the relaxation process for one of the possibilities: $k^1 = k_{46}$, and, therefore $k^1 > k_{54}$. We restore the cycle, delete the limiting step, transform the reaction $A_1^1 \to A_4$ into reaction $A_6 \to A_4$ with the same constant $k^1 = k_{46}$ and get the chain with ordered constants: $A_3 \xrightarrow{3} A_1 \xrightarrow{1} A_2 \xrightarrow{4} A_5 \xrightarrow{2} A_6 \xrightarrow{5} A_4$. Here the nonzero rate constants k_{ij} have the same value as for the initial system (Figure 3a). The relaxation time is $\tau \approx 1/k_{46}$. Left eigenvectors are (including l^4 for the zero eigenvalue):

$$\begin{aligned} &l^1 \approx (1,0,0,0,0,0), \;\; l^2 \approx (1,1,1,0,0,0),\\ &l^3 \approx (0,0,1,0,0,0), \;\; l^4 \approx (1,1,1,1,1,1),\\ &l^5 \approx (0,0,0,0,1,0) \;\; l^6 \approx (1,1,1,0,1,1) \end{aligned} \tag{54}$$

Right eigenvectors are (including r^4 for the zero eigenvalue):

$$\begin{aligned} &r^1 \approx (1,-1,0,0,0,0), \;\; r^2 \approx (0,1,0,0,0,-1),\\ &r^3 \approx (0,-1,1,0,0,0), \;\; r^4 \approx (0,0,0,1,0,0),\\ &r^5 \approx (0,0,0,0,1,-1), \;\; r^6 \approx (0,0,0,-1,0,1) \end{aligned} \tag{55}$$

Here we represent vector columns as rows.

For the approximation of relaxation in that order we can use Equation (53).

5. THE REVERSIBLE TRIANGLE OF REACTIONS: THE SIMPLE EXAMPLE CASE STUDY

In this section, we illustrate the analysis of dominant systems on a simple example, the reversible triangle of reactions.

$$A_1 \leftrightarrow A_2 \leftrightarrow A_3 \leftrightarrow A_1 \tag{56}$$

This triangle appeared in many works as an ideal object for a case study. Our favorite example is the work of Wei and Prater (1962). Now in our study the triangle (56) is not obligatory a closed system. We can assume that it is a subsystem of a larger system, and any reaction $A_i \to A_j$ represents a reaction of the form $\cdots + A_i \to A_j + \cdots$, where unknown but slow components are substituted by dots. This means that there are no obligatory relations between reaction rate constants, first of all, no detailed balance relations, and six reaction rate constants are arbitrary nonnegative numbers.

There exist $6! = 720$ orderings of six reaction rate constants for this triangle, but, of course, it is not necessary to consider all these orderings. First of all, because of the permutation symmetry, we can select an arbitrary reaction as the fastest one. Let the reaction rate constant k_{21} for the reaction $A_1 \to A_2$ is the largest. (If it is not, we just have to change the enumeration of reagents.)

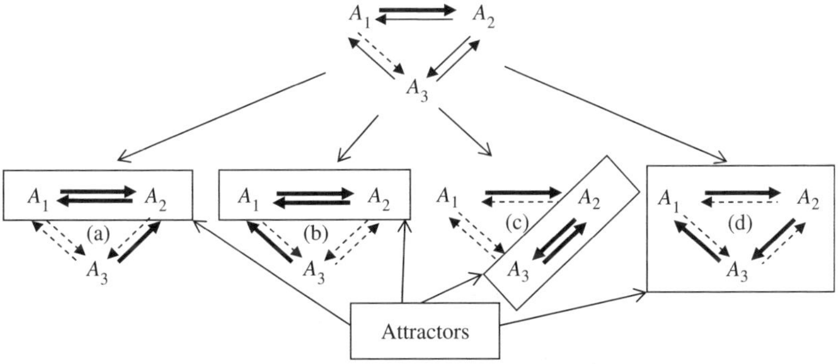

Figure 4 Four possible auxiliary dynamical systems for the reversible triangle of reactions with $k_{21} > k_{ij}$ for $(i,j) \neq (2,1)$: (a) $k_{12} > k_{32}$, $k_{23} > k_{13}$; (b) $k_{12} > k_{32}$, $k_{13} > k_{23}$; (c) $k_{32} > k_{12}$, $k_{23} > k_{13}$ and (d) $k_{32} > k_{12}$, $k_{13} > k_{23}$. For each vertex the outgoing reaction with the largest rate constant is represented by the solid bold arrow, and other reactions are represented by the dashed arrows. The digraphs formed by solid bold arrows are the auxiliary discrete dynamical systems. Attractors of these systems are isolated in frames.

First of all, let us describe all possible auxiliary dynamical systems for the triangle (56). For each vertex, we have to select the fastest outgoing reaction. For A_1, it is always $A_1 \to A_2$, because of our choice of enumeration (the higher scheme in Figure 4). There exist two choices of the fastest outgoing reaction for two other vertices and, therefore, only four versions of auxiliary dynamical systems for Equation (56) (Figure 4).

Because of the choice of enumeration, the vectors of logarithms of reaction rate constants form a convex cone in R^6 which is described by the system of inequalities $\ln k_{21} > \ln k_{ij}$, $(i,j) \neq (2,1)$. For each of the possible auxiliary systems (Figure 4) additional inequalities between constants should be valid, and we get four correspondent cones in R^6. These cones form a partitions of the initial one (we neglect intersections of faces which have zero measure). Let us discuss the typical behavior of systems from these cones separately. (Let us remind that if in a cone for some values of coefficients θ_{ij}, $\zeta_{ij} \sum_{ij} \theta_{ij} \ln k_{ij} < \sum_{ij} \zeta_{ij} \ln k_{ij}$, then, typically in this cone $\sum_{ij} \theta_{ij} \ln k_{ij} < K + \sum_{ij} \zeta_{ij} \ln k_{ij}$ for any positive K. This means that typically $\prod_{ij} k_{ij}^{\theta_{ij}} \ll \prod_{ij} k_{ij}^{\zeta_{ij}}$.)

5.1 Auxiliary system (a): $A_1 \leftrightarrow A_2 \leftarrow A_3$; $k_{12} > k_{32}$, $k_{23} > k_{13}$

5.1.1 Gluing cycles

The attractor is a cycle (with only two vertices) $A_1 \leftrightarrow A_2$. This is not a sink, because two outgoing reactions exist: $A_1 \to A_3$ and $A_2 \to A_3$. They are relatively slow: $k_{31} \ll k_{21}$ and $k_{32} \ll k_{12}$. The limiting step in this cycle is $A_2 \to A_1$ with the rate constant k_{12}. We have to glue the cycle $A_1 \leftrightarrow A_2$ into one new component A_1^1 and to add a new reaction $A_1^1 \to A_3$ with the rate constant

$$k_{31}^1 = \max\{k_{32},\ k_{31}k_{12}/k_{21}\} \tag{57}$$

This is a particular case of Equations (46) and (47).

As a result, we get a new system, $A_1^1 \leftrightarrow A_3$ with reaction rate constants k_{31}^1 (for $A_1^1 \to A_3$) and initial k_{23} (for $A_1^1 \leftarrow A_3$). This cycle is a sink, because it has no outgoing reactions (the whole system is a trivial example of a sink).

5.1.2 Steady states

To find the steady state, we have to compute the stationary concentrations for the cycle $A_1^1 \leftrightarrow A_3$, c_1^1 and c_3. We use the standard normalization condition $c_1^1 + c_3 = 1$. On the base of the general formula for a simple cycle (11) we obtain:

$$w = \frac{1}{(1/k_{31}^1) + (1/k_{23})}, \quad c_1^1 = \frac{w}{k_{31}^1}, \quad c_3 = \frac{w}{k_{23}} \tag{58}$$

After that, we can calculate the concentrations of A_1 and A_2 with normalization $c_1 + c_2 = c_1^1$. Formula (11) gives:

$$w' = \frac{c_1^1}{(1/k_{21}) + (1/k_{12})}, \quad c_1 = \frac{w'}{k_{21}}, \quad c_2 = \frac{w'}{k_{12}} \tag{59}$$

We can simplify the answer using inequalities between constants, as it was done in formulas (12) and (13). For example, $(1/k_{21}) + (1/k_{21}) \approx (1/k_{21})$, because $k_{21} \gg k_{12}$. It is necessary to stress that we have used the inequalities between constants $k_{21} > k_{ij}$ for $(i,j) \neq (2,1)$, $k_{12} > k_{32}$ and $k_{23} > k_{13}$ to obtain the simple answer (58), (59), hence if we even do not use these inequalities for the further simplification, this does not guarantee the higher accuracy of formulas.

5.1.3 Eigenvalues and eigenvectors

At the next step, we have to restore and cut the cycles. First cycle to cut is the result of cycle gluing, $A_1^1 \leftrightarrow A_3$. It is necessary to delete the limiting step, i.e. the reaction with the smallest rate constant. If $k_{31}^1 > k_{23}$, then we get $A_1^1 \to A_3$. If, inverse, $k_{23} > k_{31}^1$, then we obtain $A_1^1 \leftarrow A_3$.

After that, we have to restore and cut the cycle which was glued into the vertex A_1^1. This is the two-vertices cycle $A_1 \leftrightarrow A_2$. The limiting step for this cycle is $A_1 \leftarrow A_2$, because $k_{21} \gg k_{12}$. If $k_{31}^1 > k_{23}$, then following the rule visualized by Figure 1, we get the dominant system $A_1 \to A_2 \to A_3$ with reaction rate constants k_{21} for $A_1 \to A_2$ and k_{31}^1 for $A_2 \to A_3$. If $k_{23} > k_{31}^1$ then we obtain $A_1 \to A_2 \leftarrow A_3$ with reaction rate constants k_{21} for $A_1 \to A_2$ and k_{23} for $A_2 \leftarrow A_3$. All the procedure is illustrated by Figure 5.

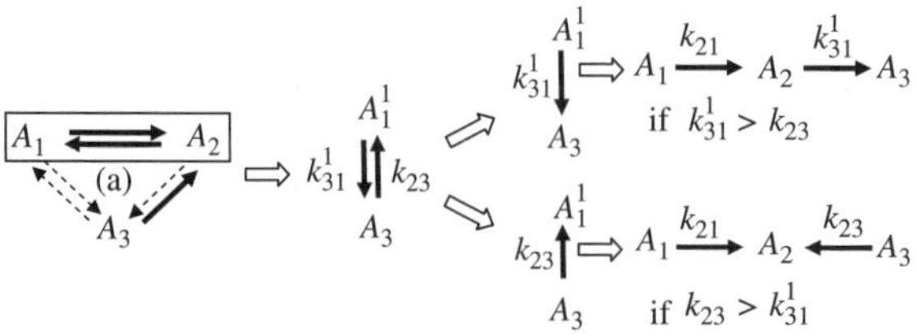

Figure 5 Dominant systems for case (a) (defined in Figure 4).

The eigenvalues and the correspondent eigenvectors for dominant systems in case (a) are represented below in zero-one asymptotic.

1. $k_{31}^1 > k_{23}$, the dominant system $A_1 \to A_2 \to A_3$,

$$\begin{aligned} \lambda_0 &= 0, & r^0 &\approx (0,0,1), & l^0 &= (1,1,1); \\ \lambda_1 &\approx -k_{21}, & r^1 &\approx (1,-1,0), & l^1 &\approx (1,0,0); \\ \lambda_2 &\approx -k_{31}^1, & r^2 &\approx (0,1,-1), & l^2 &\approx (1,1,0) \end{aligned} \tag{60}$$

2. $k_{23} > k_{31}^1$, the dominant system $A_1 \to A_2 \leftarrow A_3$,

$$\begin{aligned} \lambda_0 &= 0, & r^0 &\approx (0,1,0), & l^0 &= (1,1,1); \\ \lambda_1 &\approx -k_{21}, & r^1 &\approx (1,-1,0), & l^1 &\approx (1,0,0); \\ \lambda_2 &\approx -k_{23}, & r^2 &\approx (0,-1,1), & l^2 &\approx (0,0,1) \end{aligned} \tag{61}$$

Here, the value of k_{31}^1 is given by formula (57).

With higher accuracy, in case (a)

$$r^0 \approx \left(\frac{w'}{k_{21}}, \frac{w'}{k_{12}}, \frac{w}{k_{23}}\right) \tag{62}$$

where

$$w = \frac{1}{(1/k_{31}^1) + (1/k_{23}^1)}, \quad w' = \frac{c_1^1}{(1/k_{21}^1) + (1/k_{12}^1)}, \quad c_1^1 = \frac{w}{k_{31}^1}$$

in according to Equations (58), (59).

5.2 Auxiliary system (b): $A_3 \to A_1 \leftrightarrow A_2$; $k_{12} > k_{32}$, $k_{13} > k_{23}$

5.2.1 Gluing cycles

The attractor is a cycle $A_1 \leftrightarrow A_2$ again, and this is not a sink. We have to glue the cycle $A_1 \leftrightarrow A_2$ into one new component A_1^1 and to add a new reaction $A_1^1 \to A_3$ with the rate constant k_{31}^1 given by formula (57). As a result, we get a new system, $A_1^1 \leftrightarrow A_3$ with reaction rate constants k_{31}^1 (for $A_1^1 \to A_3$) and initial k_{13} (for $A_1^1 \leftarrow A_3$). At this stage, the only difference from the case (a) is the reaction $A_1^1 \leftarrow A_3$ rate constant k_{13} instead of k_{23}.

5.2.2 Steady states

For the steady states we have to repeat formulas (58) and (59) with minor changes (just use k_{13} instead of k_{23}):

$$\begin{aligned} w &= \frac{1}{(1/k_{31}^1) + (1/k_{13}^1)}, \quad c_1^1 = \frac{w}{k_{31}^1}, \quad c_3 = \frac{w}{k_{13}}; \\ w' &= \frac{c_1^1}{(1/k_{21}^1) + (1/k_{12}^1)}, \quad c_1 = \frac{w'}{k_{21}}, \quad c_2 = \frac{w'}{k_{12}} \end{aligned} \tag{63}$$

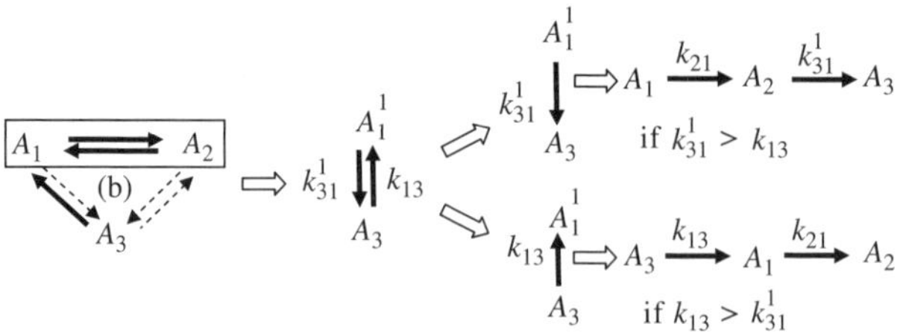

Figure 6 Dominant systems for case (b) (defined in Figure 4).

5.2.3 Eigenvalues and eigenvectors

The structure of the dominant system depends on the limiting step of the cycle $A_1^1 \leftrightarrow A_3$ (Figure 6). If $k_{31}^1 > k_{13}$, then in the dominant system remains the reaction $A_1^1 \to A_3$ from this cycle. After restoring the glued cycle $A_1 \leftrightarrow A_2$ it is necessary to delete the slowest reaction from this cycle too. This is always $A_1 \leftarrow A_2$, because $A_1 \to A_2$ is the fastest reaction. The reaction $A_1^1 \to A_3$ transforms into $A_2 \to A_3$, because A_2 is the head of the limiting step $A_1 \leftarrow A_2$ (see Figure 1). Finally, we get $A_1 \to A_2 \to A_3$.

If $k_{13} > k_{31}^1$, then in the dominant system remains the reaction $A_3 \to A_1$, and the dominant system is $A_3 \to A_1 \to A_2$ (Figure 6).

The eigenvalues and the correspondent eigenvectors for dominant systems in case (b) are represented below in zero-one asymptotic.

(i) $k_{31}^1 > k_{13}$, the dominant system $A_1 \to A_2 \to A_3$,

$$\begin{aligned} &\lambda_0 = 0, && r^0 \approx (0,0,1), && l^0 = (1,1,1); \\ &\lambda_1 \approx -k_{21}, && r^1 \approx (1,-1,0), && l^1 \approx (1,0,0); \\ &\lambda_2 \approx -k_{31}^1, && r^2 \approx (0,1,-1), && l^2 \approx (1,1,0) \end{aligned} \tag{64}$$

(ii) $k_{13} > k_{31}^1$, the dominant system $A_3 \to A_1 \to A_2$,

$$\begin{aligned} &\lambda_0 = 0, && r^0 \approx (0,1,0), && l^0 = (1,1,1); \\ &\lambda_1 \approx -k_{21}, && r^1 \approx (1,-1,0), && l^1 \approx (1,0,0); \\ &\lambda_2 \approx -k_{13}, && r^2 \approx (0,-1,1), && l^2 \approx (0,0,1) \end{aligned} \tag{65}$$

Here, the value of k_{31}^1 is given by formula (57). The only difference from case (a) is the rate constant k_{23} instead of k_{13}.

With higher accuracy, in case (b)

$$r^0 \approx \left(\frac{w'}{k_{21}}, \frac{w'}{k_{12}}, \frac{w}{k_{13}}\right) \tag{66}$$

where w and w' are given by formula (63).

5.3 Auxiliary system (c): $A_1 \to A_2 \leftrightarrow A_3$; $k_{32} > k_{12}$, $k_{23} > k_{13}$

5.3.1 Gluing cycles

The attractor is a cycle $A_2 \leftrightarrow A_3$. This is not a sink, because two outgoing reactions exist: $A_2 \to A_1$ and $A_3 \to A_1$. We have to glue the cycle $A_2 \leftrightarrow A_3$ into one new component A_2^1 and to add a new reaction $A_2^1 \to A_1$ with the rate constant k_{12}^1. The definition of this new constant depends on the normalized steady-state distribution in this cycle. If c_2^*, c_3^* are the steady-state concentrations (with normalization $c_2^* + c_3^* = 1$), then

$$k_{12}^1 \approx \max\{k_{12}c_2^*, k_{13}c_3^*\}$$

If we use limitation in the glued cycle explicitly, then we get the direct analog of Equation (57) in two versions: one for $k_{32} > k_{23}$, another for $k_{23} > k_{32}$. But we can skip this simplification and write

$$k_{12}^1 \approx \max\{k_{12}w^*/k_{32}, k_{13}w^*/k_{23}\} \tag{67}$$

where

$$w^* = \frac{1}{(1/k_{32}) + (1/k_{23})}$$

5.3.2 Steady states

Exactly as in the cases (a) and (b) we can find approximation of steady state using steady states in cycles $A_1 \leftrightarrow A_2^1$ and $A_2 \leftrightarrow A_3$:

$$\begin{aligned} w &= \frac{1}{(1/k_{12}^1) + (1/k_{21})}, \quad c_2^1 = \frac{w}{k_{12}^1}, \quad c_1 = \frac{w}{k_{21}}; \\ w' &= \frac{c_2^1}{(1/k_{32}) + (1/k_{23})}, \quad c_2 = \frac{w'}{k_{32}}, \quad c_3 = \frac{w'}{k_{23}} \end{aligned} \tag{68}$$

5.3.3 Eigenvalues and eigenvectors

The limiting step in the cycle $A_1 \leftrightarrow A_2^1$ in known, this is $A_1 \leftarrow A_2^1$. There are two possibilities for the choice on limiting step in the cycle $A_2 \leftrightarrow A_3$. If $k_{32} > k_{23}$, then this limiting step is $A_2 \leftarrow A_3$, and the dominant system is $A_1 \to A_2 \to A_3$. If $k_{23} > k_{32}$, then the dominant system is $A_1 \to A_2 \leftarrow A_3$ (Figure 7).

The eigenvalues and the correspondent eigenvectors for dominant systems in case (b) are represented below in zero-one asymptotic.

(i) $k_{32} > k_{23}$, the dominant system $A_1 \to A_2 \to A_3$,

$$\begin{aligned} &\lambda_0 = 0, && r^0 \approx (0,0,1), && l^0 = (1,1,1); \\ &\lambda_1 \approx -k_{21}, && r^1 \approx (1,-1,0), && l^1 \approx (1,0,0); \\ &\lambda_2 \approx -k_{32}, && r^2 \approx (0,1,-1), && l^2 \approx (1,1,0) \end{aligned} \tag{69}$$

(ii) $k_{23} > k_{32}$, the dominant system $A_1 \to A_2 \leftarrow A_3$,

$$\begin{aligned} &\lambda_0 = 0, && r^0 \approx (0,1,0), && l^0 = (1,1,1); \\ &\lambda_1 \approx -k_{21}, && r^1 \approx (1,-1,0), && l^1 \approx (1,0,0); \\ &\lambda_2 \approx -k_{23}, && r^2 \approx (0,-1,1), && l^2 \approx (0,0,1) \end{aligned} \tag{70}$$

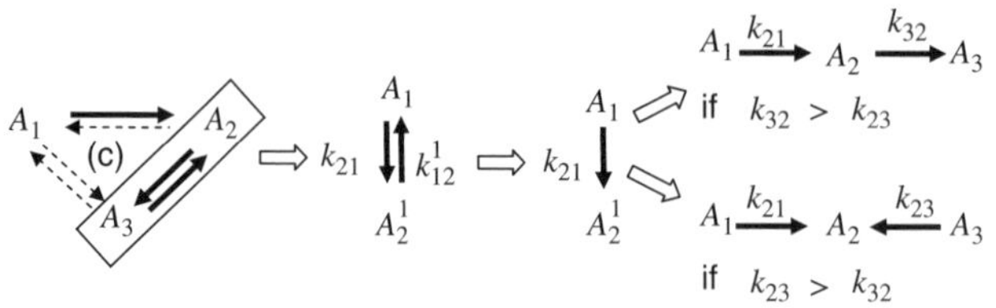

Figure 7 Dominant systems for case (c) (defined in Figure 4).

With higher accuracy the value of r^0 is given by formula of the steady-state concentrations (68).

5.4 Auxiliary system (d): $A_1 \to A_2 \to A_3 \to A_1$; $k_{32} > k_{12}$, $k_{13} > k_{23}$

This is a simple cycle. We discussed this case in details several times. To get the dominant system it is sufficient just to delete the limiting step. Everything is determined by the choice of the minimal constant in the couple $\{k_{32}, k_{13}\}$. Formulas for steady state are well known too: Equations (11)–(13).

This is not necessary to discuss all orderings of constants, because some of them are irrelevant to the final answer. For example, in this case (d) interrelations between constants k_{31}, k_{23} and k_{12} are not important.

5.5 Resume: zero-one multiscale asymptotic for the reversible reaction triangle

We found only three topologically different version of dominant systems for the reversible reaction triangle: (i) $A_1 \to A_2 \to A_3$, (ii) $A_1 \to A_2 \leftarrow A_3$ and (iii) $A_3 \to A_1 \to A_2$. Moreover, there exist only two versions of zero-one asymptotic for eigenvectors: the fastest eigenvalue is always $-k_{21}$ (because our choice of enumeration), the correspondent right and left eigenvectors (fast mode) are: $r^1 \approx (1, -1, 0)$ and $l^1 = (1, 0, 0)$. (The difference between systems (ii) and (iii) appears in the first order of the slow/fast constants ratio.)

If in the steady state (almost) all mass is concentrated in A_2 (this means that $r^0 \approx (0, 1, 0)$, dominant systems (ii) or (iii)), then $r^2 \approx (0, -1, 1)$ and $l^2 \approx (0, 0, 1)$. If in the steady state (almost) all mass is concentrated in A_3 (this means that $r^0 \approx (0, 0, 1)$, dominant system (i)), then $r^2 \approx (0, 1, -1)$ and $l^2 \approx (0, 1, 0)$. We can see that the dominant systems of the forms (ii) and (iii) produce the same zero-one asymptotic of eigenvectors. Moreover, the right eigenvectors $r^2 \approx (0, 1, -1)$ coincide for all cases (there is no difference between r^2 and $-r^2$), and the difference appears in the left eigenvector l^2. Of course, this peculiarity (everything is regulated by the steady-state asymptotic) results from the simplicity of this example.

In the zero-one asymptotic, the reversible reaction triangle is represented by one of the reaction mechanisms, (i) or (iii). The rate constant of the first reaction $A_1 \to A_2$ is always k_{12}. The direction of the second reaction is determined by a system of linear uniform inequalities between *logarithms* of rate constants. The logarithm of effective constant of this reaction is the piecewise linear function of the logarithms of reaction rate constants, and the switching between different pieces is regulated by linear inequalities. These inequalities are described in this section, and most of them are represented in Figures 4–7. One can obtain the first-order approximation of eigenvectors in the slow/fast constants ratio from the Appendix 1 formulas.

6. THREE ZERO-ONE LAWS AND NONEQUILIBRIUM PHASE TRANSITIONS IN MULTISCALE SYSTEMS

6.1 Zero-one law for steady states of weakly ergodic reaction networks

Let us take a weakly ergodic network $\mathcal{W}$ and apply the algorithms of auxiliary systems construction and cycles gluing. As a result we obtain an auxiliary dynamic system with one fixed point (there may be only one minimal sink). In the algorithm of steady-state reconstruction (Section 4.3) we always operate with one cycle (and with small auxiliary cycles inside that one, as in a simple example in Section 2.9). In a cycle with limitation almost all concentration is accumulated at the start of the limiting step (13), (14). Hence, in the whole network almost all concentration will be accumulated in one component. The dominant system for a weekly ergodic network is an acyclic network with minimal element. The minimal element is such a component $A_{\min}$ that there exists an oriented path in the dominant system from any element to $A_{\min}$. Almost all concentration in the steady state of the network $\mathcal{W}$ will be concentrated in the component $A_{\min}$.

6.2 Zero-one law for nonergodic multiscale networks

The simplest example of nonergodic but connected reaction network is $A_1 \leftarrow A_2 \rightarrow A_3$ with reaction rate constants k_1 and k_2. For this network, in addition to $b^0(c) = c_1 + c_2 + c_3$ a kinetic conservation law exist, $b^k(c) = (c_1/k_1) - (c_3/k_2)$. The result of time evolution, $\lim_{t\to\infty} \exp(Kt)c$ (30), is described by simple formula (31):

$$\lim_{t\to\infty} \exp(Kt)c = b^1(c)(1,0,0) + b^2(c)(0,1,1)$$

where $b^1(c) + b^2(c) = b^0(c)$ and $((k_1 + k_2)/k_1)b^1(c) - ((k_1 + k_2)/k_2)b^2(c) = b^k(c)$. If $k_1 \gg k_2$ then $b^1(c) \approx c_1 + c_2$ and $b^2(c) \approx c_3$. If $k_1 \ll k_2$ then $b^1(c) \approx c_1$ and $b^2(c) \approx c_2 + c_3$. This simple zero-one law (either almost all amount of A_2 transforms into A_1, or almost all amount of A_2 transforms into A_3) can be generalized onto all nonergodic multiscale systems.

Let us take a multiscale network and perform the iterative process of auxiliary dynamic systems construction and cycle gluing, as it is prescribed in Section 4.3. After the final step the algorithm gives the discrete dynamical system Φ^m with fixed points A^m_{fi}.

The fixed points A^m_{fi} of the discrete dynamical system Φ^m are the glued ergodic components $G_i \subset \mathcal{A}$ of the initial network $\mathcal{W}$. At the same time, these points are attractors of Φ^m. Let us consider the correspondent decomposition of this system with partition $\mathcal{A}^m = \cup_i Att(A^m_{fi})$. In the cycle restoration during construction of dominant system dom mod($\mathcal{W}$) this partition transforms into partition of $\mathcal{A} : \mathcal{A} = \cup_i U_i$, Att($A^m_{fi}$) transforms into U_i and $G_i \subset U_i$ (and U_i transforms into Att(A^m_{fi}) in hierarchical gluing of cycles).

It is straightforward to see that during construction of dominant systems for $\mathcal{W}$ from the network $\mathcal{V}^m$ no connection between U_i are created. Therefore, the reaction network dom mod($\mathcal{W}$) is a union of networks on sets U_i without any link between sets.

If $G_1, \ldots, G_m$ are all ergodic components of the system, then there exist m independent positive linear functionals $b^1(c), \ldots, b^m(c)$, that describe asymptotical behavior of kinetic system when $t \to \infty$ (30). For dom mod($\mathcal{W}$) these functionals are: $b^l(c) = \sum_{A \in U_l} c_A$ where c_A is concentration of A. Hence, for the initial reaction network $\mathcal{W}$ with well-separated constants

$$b^l(c) \approx \sum_{A \in U_l} c_A \tag{71}$$

This is the zero-one law for multiscale networks: for any l,i, the value of functional b^l (30) on basis vector e^i, $b^l(e^i)$, is either close to one or close to zero (with probability close to 1). We already mentioned this law in discussion of a simple example (31). The approximate equality (71) means that for each reagent $A \in \mathcal{A}$ there exists such an ergodic component G of $\mathcal{W}$ that A transforms when $t \to \infty$ preferably into elements of G even if there exist paths from A to other ergodic components of $\mathcal{W}$.

6.3 Dynamic limitation and ergodicity boundary

Dominant systems are acyclic. All the stationary rates in the first order are limited by limiting steps of some cycles. Those cycles are glued in the hierarchical cycle gluing procedure, and their limiting steps are deleted in the cycles surgery procedures (see Section 4.3 and Figure 1).

Relaxation to steady state of the network is multiexponential, and now we are interested in estimate of the longest relaxation time τ:

$$\tau = 1/\min\{-Re\lambda_i | \lambda_i \neq 0\} \tag{72}$$

Is there a constant that limits the relaxation time? The general answer for multiscale system is: $1/\tau$ is equal to the minimal reaction rate constant of the dominant system. It is impossible to guess a priori, before construction of the dominant system, which constant it is. Moreover, this may be not a rate constant for a reaction from the initial network, but a monomial of such constants.

Nevertheless, sometimes it is possible to point the reaction rate constant that is limiting for the relaxation in the following sense. For known topology of reaction network and given ordering of reaction rate constants we find such a constant (ergodicity boundary) k_τ that

$$\tau \approx \frac{1}{ak_\tau} \tag{73}$$

with $a \lesssim 1$ is a function of constants $k_j > k_\tau$. This means that $1/k_\tau$ gives the lower estimate of the relaxation time, but τ could be larger. In addition, we show that there is a zero-one alternative too: if the constants are well separated then either $a \approx 1$ or $a \ll 1$.

We study a multiscale system with a given reaction rate constants ordering, $k_{j_1} > k_{j_2} > \cdots > k_{j_n}$. Let us suppose that the network is weakly ergodic (when there are several ergodic components, each one has its longest relaxation time that can be found independently). We say that k_{j_r}, $1 \leqslant r \leqslant n$ is the *ergodicity boundary* k_τ if the network of reactions with parameters $k_{j_1}, k_{j_2}, \ldots, k_{j_r}$ (when $k_{j_{r+1}} = \ldots k_{j_n} = 0$) is weakly ergodic, but the network with parameters $k_{j_1}, k_{j_2}, \ldots, k_{j_r-1}$ (when $k_{j_r} = k_{j_{r+1}} = \ldots k_{j_n} = 0$) it is not. In other words, when eliminating reactions in decreasing order of their characteristic times, starting with the slowest one, the ergodicity boundary is the constant of the first reaction whose elimination breaks the ergodicity of the reaction digraph. This reaction we also call the "ergodicity boundary".

Let us describe the possible location of the ergodicity boundary in the general multiscale reaction network ($\mathcal{W}$). After deletion of reactions with constants $k_{j_r}, k_{j_{r+1}}, \ldots, k_{j_n}$ from the network two ergodic components (minimal sinks) appear, G_1 and G_2. The ergodicity boundary starts in one of the ergodic components, say G_1, and ends at the such a reagent B that another ergodic component, G_2, is reachable by B (there exists an oriented path from B to some element of G_2).

An estimate of the longest relaxation time can be obtained by applying the perturbation theory for linear operators to the degenerated case of the zero eigenvalue of the matrix K. We have $K = K_{<r}(k_{j_1}, k_{j_2}, \ldots, k_{j_r-1}) + k_{j_r} Q + o(k_r)$, where $K_{<r}$ is obtained from K by letting $k_r = k_{r+1} = \ldots k_n = 0$, Q is a constant matrix of rank 1, and $o(k_r)$ includes terms that are negligible relative to k_r. The zero eigenvalue is twice degenerate in $K_{<r}$ and not degenerate in $K_{<r} + k_r Q$. One gets the following estimate:

$$\bar{a}\frac{1}{k_\tau} \geq \tau \geq \underline{a}\frac{1}{k_\tau} \tag{74}$$

where $\bar{a}$ and $\underline{a} > 0$ are some positive functions of $k_1, k_2, \ldots, k_{r-1}$ (and of the reaction graph topology).

Two simplest examples demonstrate two types of dependencies of τ on k_τ:

(i) For the reaction mechanism Figure 8a

$$\min_{\lambda \neq 0}\{-Re\lambda\} = \varepsilon$$

if $\varepsilon < k_1 + k_2$.

(ii) For the reaction mechanism Figure 8b

$$\min_{\lambda \neq 0}\{-Re\lambda\} = \varepsilon k_2/(k_1 + k_2) + o(\varepsilon)$$

if $\varepsilon < k_1 + k_2$. For well-separated parameters there exists as a zero-one (trigger) alternative: if $k_1 \ll k_2$ then $\min_{\lambda \neq 0}\{-Re\lambda\} \approx \varepsilon$; if, inverse, $k_1 \gg k_2$ then $\min_{\lambda \neq 0}\{-Re\lambda\} = o(\varepsilon)$.

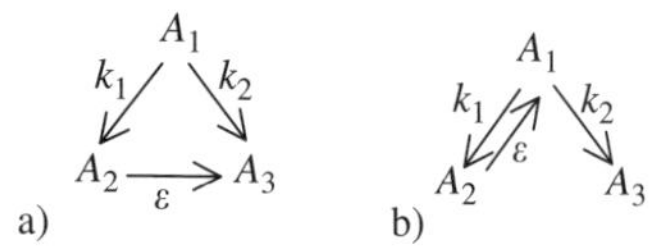

Figure 8 Two basic examples of ergodicity boundary reaction: (a) Connection between ergodic components and (b) Connection from one ergodic component to element that is connected to the both ergodic components by oriented paths. In both cases, for $\varepsilon = 0$, the ergodic components are $\{A_2\}$ and $\{A_3\}$.

In general multiscale network, two type of obstacles can violate approximate equality $\tau \approx 1/k_\tau$. Following the zero-one law for nonergodic multiscale networks (previous subsection) we can split the set of all vertices into two subsets, U_1 and U_2. The dominant reaction network dom mod($\mathcal{W}$) is a union of networks on sets $U_{1,2}$ without any link between sets.

If the ergodicity boundary reaction starts in the ergodic component G_1 and ends at B which does not belong to the "opposite" basin of attraction U_2, then $\tau \gg 1/k_\tau$. This is the first possible obstacle.

Let the ergodicity boundary reaction start at $A \in G_1$ and end at $B \in U_2$. We define the maximal linear chain of reactions in dominant system with start at B: $B \to \ldots$ This chain belongs to U_2. Let us extend this chain from the left by the ergodicity boundary: $A \to B \to \ldots$ Relaxation time for the network of r reactions (with the kinetic matrix $K_{\leq r} = K_{<r}(k_{j_1}, k_{j_2}, \ldots, k_{j_r-1}) + k_{j_r}Q$) is, approximately, the relaxation time of this chain, i.e. $1/k$, where k is the minimal constant in the chain. There may appear a monomial constant $k \ll k_\tau$. In that case, $\tau \gg 1/k_\tau$, and relaxation is limited by this minimal k or by some of constants k_{j_p}, $p > r$ or by some of their combinations. This existence of a monomial constant $k \ll k_\tau$ in the maximal chain $A \to B \to \ldots$ from the dominant system is the second possible obstacle for approximate equality $\tau \approx 1/k_\tau$.

If there is neither the first obstacle, nor the second one, then $\tau \approx 1/k_\tau$. The possibility of these obstacles depends on the definition of multiscale ensembles we use. For example for the log-uniform distribution of rate constants in the ordering cone $k_{j_1} > k_{j_2} > \cdots > k_{j_n}$ (Section 3.3) the both obstacles have nonzero probability, if they are topologically possible. However, if we study asymptotic of relaxation time at $\varepsilon \to 0$ for $k_{i_r} = \varepsilon k_{j_r-1}$ for given values of $k_{j_1}, k_{j_2}, \ldots, k_{j_r-1}$, then for sufficiently small $\varepsilon > 0$ the second obstacle is impossible.

Thus, the well-known concept of stationary reaction rates *limitation* by "narrow places" or "limiting steps" (slowest reaction) should be complemented by the *ergodicity boundary* limitation of relaxation time. It should be stressed that the relaxation process is limited not by the classical limiting steps (narrow places), but by reactions that may be absolutely different. The simplest example of this kind is an irreversible catalytic cycle: the stationary rate is limited by the slowest reaction (the smallest constant), but the relaxation time is limited by the reaction constant with the second lowest value (in order to break the weak ergodicity of a cycle two reactions must be eliminated).

6.4 Zero-one law for relaxation modes (eigenvectors) and lumping analysis

For kinetic systems with well-separated constants the left and right eigenvectors can be explicitly estimated. Their coordinates are close to ± 1 or 0. We analyzed these estimates first for linear chains and cycles (5) and then for general acyclic auxiliary dynamical systems (34), (36) (35), (37). The distribution of zeros and ± 1 in the eigenvectors components depends on the rate constant ordering and may be rather surprising. Perhaps, the simplest example gives the asymptotic equivalence (for $k_i^- \gg k_i, k_{i+1}$) of the reaction network $A_i \leftrightarrow A_{i+1} \to A_{i+2}$ with rate constants k_i, k_i^- and k_{i+1} to the reaction network $A_{i+1} \to A_i \to A_{i+2}$ with rate constants k_i^- (for the reaction $A_{i+1} \to A_i$) and $k_{i+1}k_i/k_i^-$ (for the reaction $A_i \to A_{i+2}$) presented in Section 2.9.

For reaction networks with well-separated constants coordinates of left eigenvectors l^i are close to 0 or 1. We can use the left eigenvectors for coordinate change. For the new coordinates $z_i = l^i c$ (eigenmodes) the simplest equations hold: $\dot{z}_i = \lambda_i z_i$. The zero-one law for left eigenvectors means that the eigenmodes are (almost) sums of some components: $z_i = \sum_{i \in V_i} c_i$ for some sets of numbers V_i. Many examples, Equations (6), (38), (51), (54), demonstrate that some of z_i can include the same concentrations: it may be that $V_i \cap V_j \neq \emptyset$ for some $i \neq j$. Aggregation of some components (possibly with some coefficients) into new group components for simplification of kinetics is the major task of lumping analysis.

Wei and Kuo studied conditions for exact (Wei and Kuo, 1969) and approximate (Kuo and Wei, 1969) linear lumping. More recently, sensitivity analysis and Lie group approach were applied to lumping analysis (Li and Rabitz, 1989; Toth et al., 1997), and more general nonlinear forms of lumped concentrations are used (e.g. z_i could be rational function of c). The power of lumping using a timescale-based approach was demonstrated by Whitehouse et al. (2004) and by Liao and Lightfoot (1988). This computationally cheap approach combines ideas of sensitivity analysis with simple and useful grouping of species with similar lifetimes and similar topological properties caused by connections of the species in the reaction networks. The lumped concentrations in this approach are simply sums of concentrations in groups.

Kinetics of multiscale systems studied in this chapter and developed theory of dynamic limitation demonstrates that in multiscale limit lumping analysis can work (almost) exactly. Lumped concentrations are sums in groups, but these groups can intersect and usually there exist several intersections.

6.5 Nonequilibrium phase transitions in multiscale systems

For each zero-one law specific sharp transitions exist: if two systems in a one-parametric family have different zero-one steady states or relaxation modes, then somewhere between a point of jump exists. Of course, for given finite values of parameters this will be not a point of discontinuity, but rather a thin zone of fast change. At such a point the dominant system changes. We can call this change a *nonequilibrium phase transition*. Here we identify a "multiscale nonequilibrium phase" with a dominant system.

A point of phase transition can be a point where the order of parameters changes. But not every change of order causes the change of dominant systems. However, change of order of some monomials can change the dominant system even if the order of parameters persists (examples are presented in previous section). Evolution of a parameter-dependent multiscale reaction network can be represented as a sequence of sharp change of dominant system. Between such sharp changes there are periods of evolution of dominant system parameters without qualitative changes.

7. LIMITATION IN MODULAR STRUCTURE AND SOLVABLE MODULES

7.1 Modular limitation

The simplest one-constant limitation concept cannot be applied to all systems. There is another very simple case based on exclusion of "fast equilibria" $A_i \rightleftharpoons A_j$. In this limit, the ratio of reaction constants $K_{ij} = k_{ij}/k_{ji}$ is bounded, $0<a<K_{ij}<b<\infty$, but for different pairs (i,j), (l,s) one of the inequalities $k_{ij} \ll k_{ls}$ or $k_{ij} \gg k_{ls}$ holds. (One usually calls these K "equilibrium constant", even if there is no relevant thermodynamics.) Ray (1983) discussed that case systematically for some real examples. Of course, it is possible to create the theory for that case very similarly to the theory presented above. This should be done, but it is worth to mention now that the limitation concept can be applied to *any* modular structure of reaction network. Let for the reaction network $\mathcal{W}$ the set of elementary reactions $\mathcal{R}$ is partitioned on some modules: $\mathcal{R} = \cup_i \mathcal{R}_i$. We can consider the related multiscale ensemble of reaction constants: let the ratio of any two-rate constants inside each module be bounded (and separated from zero, of course), but the ratios between modules form a well-separated ensemble. This can be formalized by multiplication of rate constants of each module $\mathcal{R}_i$ on a timescale coefficient k_i. If we assume that $\ln k_i$ are uniformly and independently distributed on a real line (or k_i are independently and log-uniformly distributed on a sufficiently large interval) then we come to the problem of modular limitation. The problem is quite general: describe the typical behavior of multiscale ensembles for systems with given modular structure: each module has its own timescale and these time scales are well separated.

Development of such a general theory is outside the scope of our chapter, and here we just find building blocks for the future theory, *solvable reaction modules*. There may be many various criteria of selection of the reaction modules. Here are several possible choices: individual reactions (we developed the theory of multiscale ensembles of individual reactions in this chapter), couples of mutually inverse reactions, as we mentioned earlier, acyclic reaction networks, ...

Among the possible reasons for selection the class of reaction mechanisms for this purpose, there is one formal, but important: the possibility to solve the kinetic equation for every module in explicit analytical (algebraic) form with quadratures. We call these systems "solvable".

7.2 Solvable reaction mechanisms

Let us describe all solvable reaction systems (with mass action law), linear and nonlinear.

Formally, we call the set of reaction solvable, if there exists a linear transformation of coordinates $c \mapsto a$ such that kinetic equation in new coordinates for all values of reaction constants has the triangle form:

$$\frac{\mathrm{d}a_i}{\mathrm{d}t} = f_i(a_1, a_2, \ldots, a_i) \tag{75}$$

This system has the lower triangle Jacobian matrix $\partial \dot{a}_i / \partial a_j$.

To construct the general mass action law system we need: the list of components, $\mathscr{A} = \{A_1, \ldots, A_n\}$ and the list of reactions (the reaction mechanism):

$$\sum_i \alpha_{ri} A_i \rightarrow \sum_k \beta_{rk} A_k \tag{76}$$

where r is the reaction number, α_{ri} and β_{rk} nonnegative integers (stoichiometric coefficients). Formally, it is possible that all $\beta_k = 0$ or all $\alpha_i = 0$. We allow such reactions. They can appear in reduced models or in auxiliary systems.

A real variable c_i is assigned to every component A_i, c_i is the concentration of A_i and c the concentration vector with coordinates c_i. The reaction kinetic equations are

$$\frac{\mathrm{d}c}{\mathrm{d}t} = \sum_r \gamma_r w_r(c) \tag{77}$$

where γ_r is the reaction stoichiometric vector with coordinates $\gamma_{ri} = \beta_{ri} - \alpha_{ri}$, $w_r(c)$ is the reaction rate. For mass action law,

$$w_r(c) = k_r \prod_i c_i^{\alpha_{ri}} \tag{78}$$

where k_r is the reaction constant.

Physically, equations (77) correspond to reactions in fixed volume, and in more general case a multiplier V (volume) is necessary:

$$\frac{\mathrm{d}(Vc)}{\mathrm{d}t} = V \sum_r \gamma_r w_r(c)$$

Here we study the systems (77) and postpone any further generalization.

The first example of solvable systems give the sets of reactions of the form

$$\alpha_{ri} A_i \rightarrow \sum_{k, k>i} \beta_{rk} A_k \tag{79}$$

(components A_k on the right-hand side have higher numbers k than the component A_i on the left-hand side, $i<k$). For these systems, kinetic equations (77) have the triangle form from the very beginning.

The second standard example gives the couple of mutually inverse reactions:

$$\sum_i \alpha_i A_i \rightleftharpoons \sum_k \beta_k A_k \tag{80}$$

these reactions have stoichiometric vectors $\pm\gamma$, $\gamma_i = \beta_i - \alpha_i$. The kinetic equation $\dot{c} = (w^+ - w^-)\gamma$ has the triangle form Equation (75) in any orthogonal coordinate system with the last coordinate $a_n = (\gamma, c) = \sum_i \gamma_i c_i$. Of course, if there are several reactions with proportional stoichiometric vectors, the kinetic equations have the triangle form in the same coordinate systems.

The general case of solvable systems is essentially a combination of that two Equations (79) and (80), with some generalization. Here we follow the book by Gorban et al. (1986) and present an algorithm for analysis of reaction network solvability. First, we introduce a relation between reactions "rth reaction directly affects the rate of sth reaction" with notation $r \to s$: $r \to s$ if there exists such A_i that $\gamma_{ri}\alpha_{si} \neq 0$. This means that concentration of A_i changes in the rth reaction ($\gamma_{ri} \neq 0$) and the rate of the sth reaction depends on A_i concentration ($\alpha_{si} \neq 0$). For that relation we use $r \to s$. For transitive closure of this relation we use notation $r \succcurlyeq s$ ("rth reaction affects the rate of sth reaction"): $r \succcurlyeq s$ if there exists such a sequence $s_1, s_2, \ldots, s_q$ that $r \to s_1 \to s_2 \to \ldots s_q \to s$.

The *hanging component* of the reaction network $\mathcal{W}$ is such $A_i \in \mathcal{A}$ that for all reactions $\alpha_{ri} = 0$. This means that all reaction rates do not depend on concentration of A_i. The *hanging reaction* is such element of $\mathcal{R}$ with number r that $r \succcurlyeq s$ only if $\gamma_s = \lambda\gamma_r$ for some number λ. An example of hanging components gives the last component A_n for the triangle network (79). An example of hanging reactions gives a couple of reactions (80) if they do not affect any other reaction.

To check solvability of the reaction network $\mathcal{W}$ we should find all hanging components and reactions and delete them from $\mathcal{A}$ and $\mathcal{R}$, correspondingly. After that, we get a new system, $\mathcal{W}_1$ with the component set $\mathcal{A}_1$ and the reaction set $\mathcal{R}_1$. Next, we should find all hanging components and reactions for $\mathcal{W}_1$ and delete them from $\mathcal{A}_1$ and $\mathcal{R}_1$. Iterate until no hanging components or hanging reactions could be found. If the final set of components is empty, then the reaction network $\mathcal{W}$ is solvable. If it is not empty, then $\mathcal{W}$ is not solvable.

For example, let us consider the reaction mechanism with $\mathcal{A} = \{A_1, A_2, A_3, A_4\}$ and reactions $A_1 + A_2 \to 2A_3$, $A_1 + A_2 \to A_3 + A_4$, $A_3 \to A_4$ and $A_4 \to A_3$. There are no hanging components, but two hanging reactions, $A_3 \to A_4$ and $A_4 \to A_3$. After deletion of these two reactions, two hanging components appear, A_3 and A_4. After deletion these two components, we get two hanging reactions, $A_1 + A_2 \to 0$ and $A_1 + A_2 \to 0$ (they coincide). We delete these reactions and get two components A_1 and A_2 without reactions. After deletion these hanging components we obtain the empty system. The reaction network is solvable.

An oriented cycle of the length more than two is not solvable. For each number of vertices one can calculate the set of all maximal solvable mechanisms. For example, for five components there are two maximal solvable mechanisms of monomolecular reactions:

(i) $A_1 \to A_2 \to A_4$, $A_1 \to A_4$, $A_2 \to A_3$, $A_1 \to A_3 \to A_5$, $A_1 \to A_5$, $A_4 \leftrightarrow A_5$ and
(ii) $A_1 \to A_2$, $A_1 \to A_3$, $A_1 \to A_4$, $A_1 \to A_5$, $A_2 \leftrightarrow A_3$, $A_4 \leftrightarrow A_5$.

It is straightforward to check solvability of these mechanisms. The first mechanism has a couple of hanging reactions, $A_4 \leftrightarrow A_5$. After deletion of these reactions, the system becomes acyclic, of the form Equation (79). The second mechanism has two couples of hanging reactions, $A_2 \leftrightarrow A_3$ and $A_4 \leftrightarrow A_5$. After deletion of these reactions, the system also transforms into the triangle form Equation (79). It is impossible to add any new monomolecular reactions between $\{A_1, A_2, A_3, A_4, A_5\}$ to these mechanisms with preservation of solvability, and any solvable monomolecular reaction network with five reagents is a subset of one of these mechanisms.

Finally, we should mention connections between solvable reaction networks and solvable Lie algebras (de Graaf, 2000; Jacobson, 1979). Let us remind that matrices $M_1, \ldots, M_q$ generate a solvable Lie algebra if and only if they could be transformed simultaneously into a triangle form by a change of basis.

The Jacobian matrix for the mass action law kinetic equation (77) is:

$$J = \left(\frac{\partial c_i}{\partial c_j}\right) = \sum_r w_r J_r = \sum_{rj} \frac{w_r}{c_j} M_{rj} \tag{81}$$

where

$$J_r = \gamma_r \alpha_r^\top \mathrm{diag}\left\{\frac{1}{c_1}, \frac{1}{c_2}, \ldots, \frac{1}{c_n}\right\} = \sum_j \frac{1}{c_j} M_{rj},$$
$$M_{rj} = \alpha_{rj} \gamma_r e^{j\top} \tag{82}$$

$\alpha_r^\top$ is the vector row $(\alpha_{r1}, \ldots, \alpha_{rn})$, $e^{j\top}$ the jth basis vector row with coordinates $e_k^{j\top} = \delta_{jk}$.

The Jacobian matrix (81) should have the lower triangle form in coordinates α_i (75) for all nonnegative values of rate constants and concentrations. This is equivalent to the lower triangle form of all matrices M_{rj} in these coordinates. Because usually there are many zero matrices among M_{rj}, it is convenient to describe the set of nonzero matrices.

For the rth reaction $I_r = \{i | \alpha_{ri} \neq 0\}$. The reaction rate w_r depends on c_i if and only if $i \in I_r$. For each $i = 1, \ldots, n$ we define a matrix

$$m_{ri} = \left[0, 0, \ldots, \underbrace{\gamma_r}_{i}, \ldots, 0\right]$$

The ith column of this matrix coincides with the vector column γ_r. Other columns are equal to zero. For each r we define a set of matrices $\mathcal{M}_r = \{m_{ri} | i \in I_r\}$ and $\mathcal{M} = \cup_r \mathcal{M}_r$. The reaction network $\mathcal{W}$ is solvable if and only if the finite set of matrices $\mathcal{M}$ generates a solvable Lie algebra.

Classification of finite dimensional solvable Lie algebras remains a difficult problem (de Graaf, 2000, 2005). It seems plausible that the classification of solvable algebras associated with reaction networks can bring new ideas into this field of algebra.

8. CONCLUSION: CONCEPT OF LIMIT SIMPLIFICATION IN MULTISCALE SYSTEMS

In this chapter, we study networks of linear reactions. For any ordering of reaction rate constants we look for the dominant kinetic system. The dominant system is, by definition, the system that gives us the main asymptotic terms of the stationary state and relaxation in the limit for well-separated rate constants. In this limit any two constants are connected by the relation $\gg$ or $\ll$.

The topology of dominant systems is rather simple; they are those networks which are graphs of discrete dynamical systems on the set of vertices. In such graphs each vertex has no more than one outgoing reaction. This allows us to construct the explicit asymptotics of eigenvectors and eigenvalues. In the limit of well-separated constants, the coordinates of eigenvectors for dominant systems can take only three values: ± 1 or 0. All algorithms are represented topologically by transformation of the graph of reaction (labeled by reaction rate constants). We call these transformations "cycles surgery", because the main operations are gluing cycles and cutting cycles in graphs of auxiliary discrete dynamical systems.

In the simplest case, the dominant system is determined by the ordering of constants. But for sufficiently complex systems we need to introduce auxiliary elementary reactions. They appear after cycle gluing and have monomial rate constants of the form $k_{\varsigma} = \prod_i k_i^{\varsigma_i}$. The dominant system depends on the place of these monomial values among the ordered constants.

Construction of the dominant system clarifies the notion of limiting steps for relaxation. There is an exponential relaxation process that lasts much longer than the others in Equations (44) and (53). This is the slowest relaxation and it is controlled by one reaction in the dominant system, the limiting step. The limiting step for relaxation is not the slowest reaction, or the second slowest reaction of the whole network, but the slowest reaction of the dominant system. That limiting step constant is not necessarily a reaction rate constant for the initial system, but can be represented by a monomial of such constants as well.

The idea of dominant subsystems in asymptotic analysis was proposed by Newton and developed by Kruskal (1963). A modern introduction with some historical review is presented by White. In our analysis we do not use the powers of small parameters (as it was done by Akian et al., 2004; Kruskal, 1963; Lidskii, 1965; Vishik and Ljusternik, 1960; White, 2006), but operate directly with the rate constants ordering.

To develop the idea of systems with well-separated constants to the state of a mathematical notion, we introduce multiscale ensembles of constant tuples. This notion allows us to discuss rigorously uniform distributions on infinite space and gives the answers to a question: what does it mean "to pick a multiscale system at random".

Some of results obtained are rather surprising and unexpected. First of all is the zero-one asymptotic of eigenvectors. Then, the good approximation to eigenvectors does not give approximate eigenvectors (the inverse situation is more common: an approximate eigenvector could be far from the eigenvector). The almost exact lumping analysis provided by the zero-one approximation of eigenvectors has an unexpected property: the lumped groups for different eigenvalues can intersect. Rather unexpected seems the change of reaction sequence when we construct the dominant systems. For example, asymptotic equivalence (for $k_i^- \gg k_i, \ k_{i+1}$) of the reaction network $A_i \leftrightarrow A_{i+1} \rightarrow A_{i+2}$ with rate constants k_i, k_i^- and k_{i+1} to the reaction network $A_{i+1} \rightarrow A_i \rightarrow A_{i+2}$ with rate constants k_i^- (for the reaction $A_{i+1} \rightarrow A_i$) and $k_{i+1}k_i/k_i^-$ (for the reaction $A_i \rightarrow A_{i+2}$) is simple, but surprising (Section 2.9). And, of course, it was surprising to observe how the dynamics of linear multiscale networks transforms into the dynamics on finite sets of reagent names.

Now we have the complete theory and the exhaustive construction of algorithms for linear reaction networks with well-separated rate constants. There are several ways of using the developed theory and algorithms:

(i) For direct computation of steady states and relaxation dynamics; this may be useful for complex systems because of the simplicity of the algorithm and resulting formulas and because often we do not know the rate constants for complex networks, and kinetics that is ruled by orderings rather than by exact values of rate constants may be very useful.
(ii) For planning of experiments and mining the experimental data — the observable kinetics is more sensitive to reactions from the dominant network, and much less sensitive to other reactions, the relaxation spectrum of the dominant network is explicitly connected with the correspondent reaction rate constants, and the eigenvectors ("modes") are sensitive to the constant ordering, but not to exact values.
(iii) The steady states and dynamics of the dominant system could serve as a robust first approximation in perturbation theory or as a preconditioning in numerical methods.

The developed methods are computationally cheap, for example, the algorithm for construction of dominant system has linear complexity ($\sim$ number of reactions). From a practical point of view, it is attractive to use exact rational expressions for the dominant system modes (3), (34) and (36) instead of the zero-one approximation. Also, we can use exact formula (11) for irreversible cycle steady state instead of linear approximation (13). These improvements are computationally cheap and may enhance accuracy of computations.

From a theoretical point of view the outlook is more important. Let us answer the question: what has to be done, but is not done yet? Three directions for further development are clear now:

(i) Construction of dominant systems for the reaction network that has a group of constants with comparable values (without relations $\gg$ between them). We considered cycles with several comparable constants in Section 2.2, but the general theory still has to be developed.
(ii) Construction of dominant systems for reaction networks with modular structure. We can assume that the ratio of any two-rate constants inside each module be bounded and separated from zero, but the ratios between

modules form a well-separated ensemble. A reaction network that has a group of constants with comparable values gives us an example of the simplest modular structure: one module includes several reactions and other modules arise from one reaction. In Section 7.7 we describe all solvable modules such that it is possible to solve the kinetic equation for every module in explicit analytical (algebraic) form with quadratures (even for nonconstant in time reaction rate constants).

(iii) Construction of dominant systems for nonlinear reaction networks. The first idea here is the representation of a nonlinear reaction as a pseudomonomolecular reaction: if for reaction $A+B \to \ldots$ concentrations c_A and c_B are well separated, say, $c_A \gg c_B$, then we can consider this reaction as $B \to \ldots$ with rate constant dependent on c_A. The relative change of c_A is slow, and we can consider this reaction as pseudomonomolecular until the relation $c_A \gg c_B$ changes to $c_A \sim c_B$. We can assume that in the general case only for small fraction of nonlinear reactions the pseudomonomolecular approach is not applicable, and this set of genuinely nonlinear reactions changes in time, but remains small. For nonlinear systems, even the realization of the limiting step idea for steady states of a one-route mechanism of a catalytic reaction is nontrivial and was developed through the concept of kinetic polynomial (Lazman and Yablonskii, 1988).

Finally, the concept of "limit simplification" will be developed. For multiscale nonlinear reaction networks the expected dynamical behavior is to be approximated by the system of dominant networks. These networks may change in time but remain small enough.

This hypothetical picture should give an answer to a very practical question: how to describe kinetics beyond the standard quasi-steady-state and quasi-equilibrium approximations (Schnell and Maini, 2002). We guess that the answer has the following form: during almost all time almost everything could be simplified and the whole system behaves as a small one. But this picture is also nonstationary: this small system change in time. Almost always "something is very small and something is very big", but due to nonlinearity this ordering can change in time. The whole system walks along small subsystems, and constants of these small subsystems change in time under control of the whole system state. The dynamics of this walk supplements the dynamics of individual small subsystems.

The corresponding structure of fast–slow time separation in phase space is not necessarily a smooth slow invariant manifold, but may be similar to a "crazy quilt" and may consist of fragments of various dimensions that do not join smoothly or even continuously.

ACKNOWLEDGEMENT

This work was supported by British Council Alliance Franco-British Research Partnership Programme.

REFERENCES

Adamaszek, M. *Newsl. Eur. Math. Soc.* **62**(December), 21–23 (2006).
Akian, M., Bapat, R., and Gaubert, S. (2004). Min-plus methods in eigenvalue perturbation theory and generalised Lidskii–Vishik–Ljusternik theorem, arXiv e-print math.SP/0402090.
Albeverio, S., Fenstad, J., Hoegh-Krohn, R., and Lindstrom, T., "Nonstandard Methods in Stochastic Analysis and Mathematical Physics". Academic Press, Orlando etc. (1986).
Birkhoff, G. *Composito Math.* **3**, 427–430 (1936).
Boyd, R. K. *J. Chem. Educ.* **55**, 84–89 (1978).
Brown, G. C., and Cooper, C. E. *Biochem. J.* **294**, 87–94 (1993).
Carnap, R., "Logical Foundations of Probability". University of Chicago Press, Chicago (1950).
Carroll, L. (Dodgson C. L.)., Mathematical Recreations of Lewis Carroll: Pillow Problems and a Tangled Tale. Dover (1958).
Cheresiz, V. M., and Yablonskii, G. S. *React. Kinet. Catal. Lett.* **22**, 69–73 (1983).
Cornish-Bowden, A., and Cardenas, M. L., "Control on Metabolic Processes". Plenum Press, New York (1990).
de Graaf, W. A., "Lie Algebras: Theory and Algorithms, North-Holland Mathematical Library, 36". Elsevier, Amsterdam (2000).
de Graaf, W. A. *Exp. Math.* **14**, 15–25 (2005).
Eigen, M., Immeasurably fast reactions, Nobel Lecture, December 11, 1967, *in* Nobel Lectures, Chemistry 1963–1970," pp. 170–203. Amsterdam, Elsevier (1972).
Eisenberg, B., and Sullivan, R. *Am. Math. Mon.* **103**, 308–318 (1996).
Falk, R., and Samuel-Cahn, E. *Teaching Stat.* **23**, 72–75 (2001).
Feng, X-j., Hooshangi, S., Chen, D., Li, G., Weiss, R., and Rabitz, H. *Biophys. J.* **87**, 2195–2202 (2004).
Gorban, A. *Physica A* **374**, 85–102 (2006).
Gorban, A. N., Bykov, V. I., and Yablonskii, G. S., "Essays on Chemical Relaxation". Nauka, Novosibirsk (1986).
Gorban, A. N., and Karlin, I. V. *Chem. Eng. Sci.* **58**, 4751–4768 (2003).
Gorban, A. N., and Karlin, I. V., "Invariant Manifolds for Physical and Chemical Kinetics". Springer, Berlin, Volume 660 of Lect. Notes Phys. (2005).
Gorban, A. N., and Radulescu, O. *IET Syst. Biol.* **1**, 238–246 (2007).
Greuel, G.-M., and Pfister, G., "A Singular Introduction to Commutative Algebra". Springer, Berlin (2002).
Gromov, M., "Metric Structures for Riemannian and Non-Riemannian Spaces. Progress in Mathematics, 152". Birkhauser, Boston (1999).
Guy, R. K. *Math. Mag.* **66**, 175–179 (1993).
Hewitt, E., and Ross, A., "Abstract Harmonic Analysis". Vol. 1, Springer, Berlin (1963).
Jacobson, N., "Lie Algebras". Dover, New York (1979).
Johnston, H. S., "Gas Phase Reaction Theory". Roland Press, New York (1966).
Kholodenko, B. N., Westerhoff, H. V., and Brown, G. C. *FEBS Lett.* **349**, 131–134 (1994).
Khrennikov, A. Yu. *Theory Probab. Appl.* **46**, 256–273 (2002).
Kruskal, M. D., Asymptotology, *in* "Mathematical Models in Physical Sciences" (S. Dobrot Ed.), pp. 17–48. Prentice-Hall, Englewood Cliffs, NJ (1963).
Kuo, J. C., and Wei, J. *Ind. Eng. Chem. Fundam.* **8**, 124–133 (1969).
Kurzynski, M. *Prog. Biophys. Mol. Biol.* **69**, 23–82 (1998).
Lazman, M. Z., and Yablonskii, G. S. *React. Kinet. Catal. Lett.* **37**, 379–384 (1988).
Lazman, M. Z., and Yablonskii, G. S., Kinetic polynomial: A new concept of chemical kinetics, *in* "Patterns and Dynamics in Reactive Media, The IMA Volumes in Mathematics and its Applications", pp. 117–150. Springer, Berlin (1991).
Li, G., and Rabitz, H. *Chem. Eng. Sci.* **44**, 1413–1430 (1989).
Li, G., Rosenthal, C., and Rabitz, H. *J. Phys. Chem. A.* **105**, 7765–7777 (2001).
Li, G., Wang, S.-W., Rabitz, H., Wang, S., and Jaffe, P. *Chem. Eng. Sci.* **57**, 4445–4460 (2002).
Liao, J. C., and Lightfoot, E. N. Jr. *Biotechnol. Bioeng.* **31**, 869–879 (1988).
Lidskii, V. *USSR Comput. Math. Math. Phys.* **6**, 73–85 (1965).
Litvinov, G. L., and Maslov, V. P. (Eds.), "Idempotent Mathematics and Mathematical Physics, Contemporary Mathematics". AMS, Providence (2005).

Loeb, P. A. *Trans. Am. Math. Soc.* **211**, 113–122 (1975).
Marcus, M., and Minc, H., "A Survey of Matrix Theory and Matrix Inequalities". Dover, New York (1992).
Neumann, W. D., Hilbert's 3rd problem and invariants of 3-manifolds, *in* "Geometry & Topology Monographs, Vol. 1: The Epstein Birthday Schrift" pp. 383–411. University of Warwick, Coventry, UK. (1998).
Northrop, D. B. *Biochemistry* **20**, 4056–4061 (1981).
Northrop, D. B. *Methods* **24**, 117–124 (2001).
Portnoy, S. *Stat. Sci.* **9**, 279–284 (1994).
Rate-controlling step (2007). In: IUPAC Compendium of Chemical Terminology, Electronic version, http://goldbook.iupac.org/R05139.html
Rate-determining step (rate-limiting step) (2007). In: IUPAC Compendium of Chemical Terminology, Electronic version, http://goldbook.iupac.org/R05140.html
Ray, W. J. Jr. *Biochemistry* **22**, 4625–4637 (1983).
Robbiano, L., Term orderings on the polynomial ring, *in* "Proceedings of the EUROCAL 85" (B. F. Caviness Ed.), *Lec. Notes in Computer Sciences 204,* Vol. 2, pp. 513–518. Springer, Berlin (1985).
Rudin, W., "Functional Analysis". McGraw-Hill, New York (1991).
Schnell, S., and Maini, P. K. *Math. Comput. Model.* **35**, 137–144 (2002).
Toth, J., Li, G., Rabitz, H., and Tomlin, A. S. *SIAM J. Appl. Math.* **57**, 1531–1556 (1997).
Varga, R. S., "Gerschgorin and His Circles, Springer Series in Computational Mathematics, 36". Springer, Berlin (2004).
Vishik, M. I., and Ljusternik, L. A. *Russ. Math. Surv.* **15**, 1–73 (1960).
von Mises, R., "The Mathematical Theory of Probability and Statistics". Academic Press, London (1964).
Wei, J., and Kuo, J. C. *Ind. Eng. Chem. Fundam.* **8**, 114–123 (1969).
Wei, J., and Prater, C. *Adv. Catal.* **13**, 203–393 (1962).
White, R. B., "Asymptotic Analysis of Differential Equations". Imperial College Press & World Scientific, London (2006).
Whitehouse, L. E., Tomlin, A. S., and Pilling, M. J. *Atmos. Chem. Phys.* **4**, 2057–2081 (2004).
Yablonskii, G. S., Bykov, V. I., Gorban, A. N., and Elokhin, V. I., *in* "Kinetic Models of Catalytic Reactions. Comprehensive Chemical Kinetics" (R. G. Compton Ed.), Vol. 32, Elsevier, Amsterdam (1991).
Yablonskii, G. S., and Cheresiz, V. M. *React. Kinet. Catal. Lett.* **24**, 49–53 (1984).
Yablonskii, G. S., Lazman, M. Z., and Bykov, V. I. *React. Kinet. Catal. Lett.* **20**, 73–77 (1982).

APPENDIX 1. ESTIMATES OF EIGENVECTORS FOR DIAGONALLY DOMINANT MATRICES WITH DIAGONAL GAP CONDITION

The famous Gershgorin theorem gives estimates of eigenvalues. The estimates of correspondent eigenvectors are not so well-known. In the chapter we use some estimates of eigenvectors of kinetic matrices. Here we formulate and prove these estimates for general matrices. Below $A = (a_{ij})$ is a complex $n \times n$ matrix, $P_i = \sum_{j,j \neq i} |a_{ij}|$ (sums of nondiagonal elements in rows), $Q_i = \sum_{j,j \neq i} |a_{ji}|$ (sums of nondiagonal elements in columns).

Gershgorin theorem (Marcus and Minc, 1992, p. 146): The characteristic roots of A lie in the closed region G^P of the z-plane

$$G^P = \bigcup_i G_i^P \left(G_i^P = \{z \,|\, |z - a_{ii}| \leq P_i\} \right) \tag{83}$$

Analogously, the characteristic roots of A lie in the closed region G^Q of the z-plane

$$G^Q = \cup_i G_i^Q \left(G_i^Q \{z||z - a_{ii}| \le Q_i\} \right) \quad (84)$$

Areas G_i^P and G_i^Q are the Gershgorin discs.

Gershgorin discs G_i^P $(i = 1,\dots,n)$ are isolated, if $G_i^P \cap G_j^P = \varnothing$ for $i \neq j$. If discs G_i^P $(i = 1,\dots,n)$ are isolated, then the spectrum of A is simple, and each Gershgorin disc G_i^P contains one and only one eigenvalue of A (Marcus and Minc, 1992, p. 147). The same is true for discs G_i^Q.

Below we assume that Gershgorin discs G_i^Q $(i = 1,\dots,n)$ are isolated, this means that for all i,j

$$|a_{ii} - a_{jj}| > Q_i + Q_j \quad (85)$$

Let us introduce the following notations:

$$\frac{Q_i}{|a_{ii}|} = \varepsilon_i, \quad \frac{|a_{ij}|}{|a_{jj}|} = \chi_{ij} \quad \left(\varepsilon_i = \sum_l \delta_{li} \right), \qquad \min_j \frac{|a_{ii} - a_{jj}|}{|a_{ii}|} = g_i \quad (86)$$

Usually, we consider ε_i and χ_{ij} as sufficiently small numbers. In contrary, g_i should not be small, (this is the *gap condition*). For example, if for any two diagonal elements a_{ii} and a_{jj} either $a_{ii} \gg a_{jj}$ or $a_{ii} \ll a_{jj}$, then $g_i \gtrsim 1$ for all i.

Let $\lambda_1 \in G_1^Q$ be the eigenvalue of A $(|\lambda_1 - a_{11}| < Q_1)$. Let us estimate the correspondent right eigenvector $x^{(1)} = (x_i)$: $Ax^{(1)} = \lambda_1 x^{(1)}$. We take $x_1 = 1$ and write equations for x_i $(i \neq 1)$:

$$(a_{ii} - a_{11} - \theta_1)x_i + \sum_{j, j \neq 1, i} a_{ij} x_j = -a_{i1} \quad (87)$$

where $\theta_1 = \lambda_1 - a_{11}$, $|\theta_1| < Q_1$.

Let us introduce new variables

$$\tilde{x} = (\tilde{x}_i), \quad \tilde{x}_i = x_i(a_{ii} - a_{11}) \ (i = 2, \dots, n)$$

In these variables,

$$\left(1 - \frac{\theta_1}{a_{ii} - a_{11}}\right)\tilde{x}_i + \sum_{j, j \neq 1, i} \frac{a_{ij}}{a_{jj} - a_{11}} \tilde{x}_j = -a_{i1} \quad (88)$$

or in matrix notations: $(1 - B)\tilde{x} = -\tilde{a}_1$, where $\tilde{a}_1$ is a vector column with coordinates a_{i1}. Because of gap condition and smallness of ε_i and χ_{ij} we can consider matrix B as a small matrix, assume that $\|B\| < 1$ and $(1 - B)$ is reversible (for detailed estimate of $\|B\|$ see below).

For $\tilde{x}$ we obtain:

$$\tilde{x} = -\tilde{a}_1 - B(1 - B)^{-1}\tilde{a}_1 \quad (89)$$

and for residual estimate

$$\|B(1-B)^{-1}\tilde{a}_1\| \le \frac{\|B\|}{1-\|B\|}\|\tilde{a}_1\| \tag{90}$$

For eigenvector coordinates we get from Equation (89):

$$x_i = -\frac{a_{i1}}{a_{ii}-a_{11}} - \frac{(B(1-B)^{-1}\tilde{a}_1)_i}{a_{ii}-a_{11}} \tag{91}$$

and for residual estimate

$$\frac{|(B(1-B)^{-1}\tilde{a}_1)_i|}{|a_{ii}-a_{11}|} \le \frac{\|B\|}{1-\|B\|}\frac{\|\tilde{a}_1\|}{|a_{ii}-a_{11}|} \tag{92}$$

Let us give more detailed estimate of residual. For vectors we use l_1 norm: $\|x\| = \sum |x_i|$. The correspondent operator norm of matrix B is

$$\|B\| = \max_{|x|=1}\|Bx\| \le \sum_i \max_j |b_{ij}|$$

With the last estimate for matrix B (88) we find:

$$\begin{aligned} |b_{ii}| &\le \frac{Q_1}{|a_{ii}-a_{11}|} \le \frac{\varepsilon_1}{g_1} \le \frac{\varepsilon}{g}, \\ |b_{ij}| &= \frac{|a_{ij}|}{|a_{jj}-a_{11}|} \le \frac{\chi_{ij}}{g_j} \le \frac{\chi}{g}\ (i\neq j) \end{aligned} \tag{93}$$

where $\varepsilon = \max_i \varepsilon_i$, $\chi = \max_{i,j}\chi_{ij}$ and $g = \min_i g_i$. By definition, $\varepsilon \geqslant \chi$, and for all i,j the simple estimate holds: $|b_{ij}| \le \varepsilon/g$. Therefore, $\|Bx\| \leqslant n\varepsilon/g$ and $\|B\|/(1-\|B\|) \leqslant n\varepsilon/(g\text{-}n\varepsilon)$ (under condition $g > n\varepsilon$). Finally, $\|\tilde{a}_1\| = Q_1$ and for residual estimate we get:

$$\left|x_i + \frac{a_{i1}}{a_{ii}-a_{11}}\right| \le \frac{n\varepsilon^2}{g(g-n\varepsilon)}(i\neq 1) \tag{94}$$

More accurate estimate can be produced from inequalities (93), if it is necessary. For our goals it is sufficient to use the following consequence of Equation (94):

$$|x_i| \le \frac{\chi}{g} + \frac{n\varepsilon^2}{g(g-n\varepsilon)}(i\neq 1) \tag{95}$$

With this accuracy, eigenvectors of A coincide with standard basis vectors, i.e. with eigenvectors of diagonal part of A, $\mathrm{diag}\{a_{11},\ldots,a_{nn}\}$.

APPENDIX 2. TIME SEPARATION AND AVERAGING IN CYCLES

In Section 2, we analyzed relaxation of a simple cycle with limitation as a perturbation of the linear chain relaxation by one more step that closes the chain into the cycle. The reaction rate constant for this perturbation is the smallest one. For this analysis we used explicit estimates (5) of the chain eginvectors for reactions with well-separated constants.

Of course, one can use estimates (34)–(37) to obtain a similar perturbation analysis for more general acyclic systems (instead of a linear chain). If we add a reaction to an acyclic system (after that a cycle may appear) and assume that the reaction rate constant for additional reaction is smaller than all other reaction constants, then the generalization is easy.

This smallness with respect to all constants is required only in a very special case when the additional reaction has a form $A_i \to A_j$ (with the rate constant k_{ji}) and there is no reaction of the form $A_i \to \ldots$ in the nonperturbed system. In Section 7 and Appendix 1 we demonstrated that if in a nonperturbed acyclic system there exists another reaction of the form $A_i \to \ldots$ with rate constant κ_i, then we need inequality $k_{ji} \ll \kappa_i$ only. This inequality allows us to get the uniform estimates of eigenvectors for all possible values of other rate constants (under the diagonally gap condition in the nonperturbed system).

For substantiation of cycle's surgery we need additional perturbation analysis for zero eigenvalues. Let us consider a simple cycle $A_1 \to A_2 \to \ldots \to A_n \to A_1$ with reaction $A_i \to \ldots$ rate constants κ_i. We add a perturbation $A_1 \to 0$ (from A_1 to nothing) with rate constant $\varepsilon\kappa_1$. Our goal is to demonstrate that the zero eigenvalue moves under this perturbation to $\lambda_0 = -\varepsilon w^*(1+\chi_w)$, the correspondent left and right eigenvectors r^0 and l^0 are $r_i^0 = c_i^*(1+\chi_{ri})$ and $l_i^0 = 1+\chi_{li}$, and χ_w, χ_{ri} and χ_{li} are uniformly small for a given sufficiently small ε under all variations of rate constants. Here, w^* is the stationary cycle reaction rate and c_i^* are stationary concentrations for a cycle (11) normalized by condition $\sum_i c_i^* = 1$. The estimate εw^* for $-\lambda_0$ is ε-small with respect to any reaction of the cycle: $w^* = \kappa_i c_i^* < \kappa_i$ for all i (because $c_i^* < 1$) and $\varepsilon w^* \ll \kappa_i$ for all i.

The kinetic equation for the perturbed system is:

$$\begin{aligned} \dot{c}_1 &= -(1+\varepsilon)\kappa_1 c_1 + \kappa_n c_n, \\ \dot{c}_i &= -\kappa_i c_i + \kappa_{i-1} c_{i-1} \text{ (for } i \neq 1) \end{aligned} \tag{96}$$

In the matrix form we can write

$$\dot{c} = Kc = (K_0 - \varepsilon k_1 e^1 e^{1\top})c \tag{97}$$

where K_0 is the kinetic matrix for nonperturbed cycle. To estimate the right perturbed eigenvector r^0 and eigenvalue λ_0 we are looking for transformation of matrix K into the form $K = K_r - \theta r e^{1\top}$, where K is a kinetic matrix for extended reaction system with components $A_1, \ldots, A_n$, $K_r r = 0$ and $\sum_i r_i = 1$. In that case, r is the eigenvector, and $\lambda = -\theta r_1$ is the correspondent eigenvalue.

To find vector r, we add to the cycle new reactions $A_1 \to A_i$ with rate constants $\varepsilon\kappa_1 r_i$ and subtract the correspondent kinetic terms from the perturbation term $\varepsilon e^1 e^{1\top} c$. After that, we get $K = K_r - \theta r e^{1\top}$ with $\theta = \varepsilon k_1$ and

$$\begin{aligned} (K_r c)_1 &= -k_1 c_1 - \varepsilon k_1 (1 - r_1) c_1 + k_n c_n, \\ (K_r c)_i &= -k_i c_i + \varepsilon k_1 r_i c_1 + k_{i-1} c_{i-1} \text{ for } i > 1 \end{aligned} \tag{98}$$

We have to find a positive normalized solution $r_i>0$, $\sum_i r_i = 1$ to equation $K_r r = 0$. This is the fixed-point equation: for every positive normalized r there exists unique positive normalized steady state $c^*(r)$: $K_r c^*(r) = 0$, $c_i^*>0$ and $\sum_i c_i^*(r) = 1$. We have to solve the equation $r = c^*(r)$. The solution exists because the Brauer fixed point theorem.

If $r = c^*(r)$ then $k_i r_i - \varepsilon k_1 r_i r_1 = k_{i-1} r_{i-1}$. We use notation $w_i^*(r)$ for the correspondent stationary reaction rate along the "nonperturbed route": $w_i^*(r) = k_i r_i$. In this notation, $w_i^*(r) - \varepsilon r_i w_1^*(r) = w_{i-1}^*(r)$. Hence, $|w_i^*(r) - w_1^*(r)| < \varepsilon w_1^*(r)$ (or $|k_i r_i - k_1 r_1| < \varepsilon k_1 r_1$). Assume $\varepsilon < 1/4$ (to provide $1-2\varepsilon < 1/(1 \pm \varepsilon) < 1+2\varepsilon$). Finally,

$$r_i = \frac{1}{k_i} \frac{1+\chi_i}{\sum\limits_j (1/k_j)} = (1+\chi_i) c_i^* \tag{99}$$

where the relative errors $|\chi_i| < 3\varepsilon$ and $c_i^* = c_i^*(0)$ is the normalized steady state for the nonperturbed system. For cycles with limitation, $r_i \approx (1+\chi_i) k_{\lim}/k_i$ with $|\chi_i| < 3\varepsilon$. For the eigenvalue we obtain

$$\begin{aligned} \lambda_0 &= -\varepsilon w_1^*(r) = -\varepsilon w_i^*(r)(1+\varsigma_i) \\ &= -\varepsilon w^*(1+\chi) = -\varepsilon k_i c_i^*(0)(1+\chi) \end{aligned} \tag{100}$$

for all i, with $|\varsigma_i| < \varepsilon$ and $|\chi| < 3\varepsilon$. $|\chi| < 3\varepsilon$. Therefore, λ_0 is ε-small rate constant k_i of the nonperturbed cycle. This implies that λ_0 is ε-small with respect to the real part of every nonzero eigenvalue of the nonperturbed kinetic matrix K_0 (for given number of components n). For the cycles from multiscale ensembles these eigenvalues are typically real and close to $-k_i$ for nonlimiting rate constants, hence we proved for λ_0 even more than we need.

Let us estimate the correspondent left eigenvector l^0 (a vector row). The eigenvalue is known, hence it is easy to do just by solution of linear equations. This system of $n-1$ equations is:

$$\begin{aligned} -l_1(1+\varepsilon)k_1 + l_2 k_1 &= \lambda_0 l_1 \\ -l_i k_i + l_{i+1} k_i &= \lambda_0 l_i,\ i = 2, \ldots, n-1 \end{aligned} \tag{101}$$

For normalization, we take $l_1 = 1$ and find:

$$l_2 = \left(\frac{\lambda_0}{k_1} + 1 + \varepsilon\right) l_1, \quad l_{i+1} = \left(\frac{\lambda_0}{k_i} + 1\right) l_i \quad i>2 \tag{102}$$

Formulas (99), (100) and (102) give the backgrounds for surgery of cycles with outgoing reactions. The left eigenvector gives the slow variable: if there are some incomes to the cycle, then

$$\begin{aligned} \dot{c}_1 &= -(1+\varepsilon)\kappa_1 c_1 + \kappa_n c_n + \phi_1(t), \\ \dot{c}_i &= -\kappa_i c_i + \kappa_{i-1} c_{i-1} + \phi_i(t) \text{ (for } i \neq 1) \end{aligned} \tag{103}$$

and for slow variable $\tilde{c} = \sum l_i c_i$ we get

$$\frac{d\tilde{c}}{dt} = \lambda_0 \tilde{c} + \sum_i l_i \phi_i(t) \tag{104}$$

This is the kinetic equation for a glued cycle. In the leading term, all the outgoing reactions $A_i \to 0$ with rate constants $k = \varepsilon k_i$ give the same eigenvalue $-\varepsilon w^*$ (100).

Of course, similar results for perturbations of zero eigenvalue are valid for more general ergodic chemical reaction network with positive steady state, and not only for simple cycles, but for cycles we get simple explicit estimates, and this is enough for our goals.

CHAPTER 4

Multiscale Theorems

Liqiu Wang[1,*], **Mingtian Xu**[2] and **Xiaohao Wei**[1]

Contents

[1] Department of Mechanical Engineering, The University of Hong Kong, Pokfulam Road, Hong Kong, P. R. China

[2] Institute of Thermal Science and Technology, Shandong University, Jinan, P. R. China

*Corresponding author.
E-mail address: lqwang@hku.hk

Advances in Chemical Engineering, Volume 34
ISSN 0065-2377, DOI 10.1016/S0065-2377(08)00004-5

Abstract

We present 71 multiscale theorems that transform various derivatives of a function from one scale to another and contain all 128 such theorems in the literature. These theorems are grouped into integration theorems and averaging theorems. The former refers to those which change any or all spatial scales of a derivative from the microscale (or any continuum scale) to the megascale by integration. The integration domain is allowed to translate and deform with time, and can be a volume, a surface or a curve. The latter is those which change any or all spatial scales of a derivative from microscale to macroscale by averaging. An averaging volume may be located in space and integration is performed over volumes, surfaces, or curves contained within the averaging volume. These theorems not only provide simple tools to model multiscale phenomena at various scales and to interchange among those scales, but they also form generalized and universal integration theorems. As with all mathematical theorems, the tools used to prove each theorem and the logic behind the proof itself also hold significant value. Furthermore, these theorems are endowed with important information regarding geometrical and topological structures of and interactions among various entities such as phases, interfaces, common curves, and common points. Therefore, they also provide critical information for resolving the closure problems that are routinely encountered in multiscale science.

In Section 1, we present some examples of multiscale issues in materials science, physics, hydrology, chemical engineering, transport phenomena in biological systems, and the universe. We then discuss scales and scaling and introduce the multiscale theorems as powerful tools for model-scaling. Section 2 covers the derivatives of a function, including various gradients, divergences, curls, and partial time derivatives. In particular, we denote all derivatives using three orthonormal vectors and develop relations among them. In Section 3, we extend the familiar Heaviside step function to form indicator functions for identifying portions of a curve, a surface, or a space. We then develop some useful identities involving indicator functions and their derivatives. Finally, we present two major applications of indicator functions in developing the multiscale theorems: the selection and interchange of integration domains. Section 4 gives the detailed proofs of the 38 integration theorems that convert integrals of the gradient, divergence, curl, and partial time derivatives of a microscale function to some combination of derivatives of integrals of the function and integrals over domains of reduced dimensionality. These theorems are powerful tools for scaling between the microscale (or any continuum scale) and the megascale. In Sectiion 5, we develop 33 averaging theorems that relate the integral of a derivative of a microscale function to the derivative of that integral. These theorems are powerful tools for scaling among microscales, macroscales, and megascales. Section 6 covers applications of the multiscale theorems for modeling single-phase turbulent flow, heat conduction in two-phase systems, transport in porous and multiphase systems and for developing the thermodynamically constrained averaging theory (TCAT) approach for modeling flow and transport phenomena in multiscale porous-medium systems. These applications lead to transport-phenomena models, which contain important information for developing closure, and some new results such as the scale-by-scale energy budget equations in real space for turbulent flows, the energy flux rates among various groups of eddies and the intrinsic equivalence between dual-phase-lagging heat conduction and Fourier heat conduction in two-phase systems that are subject to a lack of local thermal equilibrium. In the last section, we conclude with some remarks regarding advances in multiscale science and multiscale theorems (their features; tools and logic behind their proof; their applications).

1. INTRODUCTION

We first present some multiscale phenomena arising from various fields followed by scales and scaling, the key issues in multiscale science. We then discuss model-scaling, which is an important subgroup of scaling. Finally, we define some basic concepts of systems and scales formally and introduce multiscale theorems, powerful tools for model-scaling.

1.1 Multiscale phenomena

Multiscale science is the study of multiscale phenomena which couple distinct length or time scales (Abdallah et al., 2004; Al-Ghoul et al., 2004; Attinger and Koumoutsakos 2004; Barth et al., 2002; Brueckner et al., 2005; Chai, 2005; Charpentier and McKenna, 2004; Costanza, 2003; Coveney, 2003b; Gerde and Marder, 2001; Glimm and Sharp, 1997; Kretzschmar and Consolini, 2006; Krumhansl, 2000; Li and Kwauk, 2003, 2004; Li et al., 2004b; Mack, 2001; Marin, 2005; Meakin, 1998; Myers et al., 2004, 2005; Novak, 2004; Robinson and Brink, 2005; Skar, 2003; Skar and Coveney, 2003a, b; Tretter and Jones, 2003; Vlachos, 2005; Waldrop, 1992). It cuts across virtually all fields of sciences and technologies. In this section, we briefly list and describe some multiscale issues in order to demonstrate the scope of multiscale science. The readers are referred to Al-Ghoul et al. (2004), Attinger and Koumoutsakos (2004), Charpentier and McKenna (2004), Coveney (2003b), Gerde and Marder (2001), Kretzschmar and Consolini (2006), Mack (2001), Marin (2005), Robinson and Brink (2005), Abdullaev and Kraenkel (2000), Adams et al. (2003), Ahn et al. (2004), Alarcon et al. (2004), Alber et al. (2006), Aleman-Flores and Alvarez-Leon (2003), Alkire and Braatz (2004), Allen (2005), Antonic et al. (2002), Bar-Joseph et al. (2001), Belashchenko and Antropov (2002b), Benassi et al. (2002), Bent et al. (2003), Billock et al. (2001), Bonilla et al. (2000), Briere (2006), Brun et al. (2004), Byrne et al. (2006), Cavallotti et al. (2005), Cheng and Wang (1996), Ciofalo (1994), Collis et al. (2004), Cushman (1997), Diego et al. (2004), Dokholyan (2006), Dollet et al. (2004), Drolon et al. (2000), Drolon et al. (2003), Fauchais and Vardelle (2000), Federov (2005), Feng et al. (1997), Flad et al. (2005), Fox-Rabinovitz et al. (2006), Frantziskonis (2002b), Fredberg and Kamm (2006), Freeden and Michel (2005), Galves and Jonalasinio (1991), Gatto et al. (2002), Geers et al. (2005), Giardino et al. (2004), Gusev (2005), Gutkowski and Kowalewski (2005), Hermes and Buhmann (2003), Hidy (2002), Hubert et al. (2002), Humby et al. (2002), Jaubert and Stein (2003), Kafer et al. (2006), Kaminski (2005), Karsch et al. (1997), Klein (2005), Kok et al. (2003), Konstandopoulos et al. (2006), Lam and Vlachos (2001), Lanfredi et al. (2003), Lartigue-Korinek et al. (2002), Lechelle et al. (2004), Lpinoux et al. (2000), Levitas (2004), Liao et al. (2007), Lipowsky and Klumpp (2005), Lui (2002), Madec et al. (2003), Masi et al. (2003), Masi (2001), Mezzacappa (2005), Moseler et al. (2005), Narayanan et al. (2004), Nash and Ragsdale (2001), Neale et al. (2003), Nicot (2003), Nieminen (1999), Patil et al. (2004), Paulson (2003), Paulson et al. (2003), Perotto (2006), Pingault et al. (2003), Pont et al. (2003), Rafii-Tabar et al. (2006), Ren et al. (2003), Ren and Otsuka (2002), Robinson et al. (2005), Rouch and Ladeveze (2003), Sagar and Rao (2003), Salis et al. (2006), Siegert et al. (2001), Sornette and Zhou (2006), Sudderth et al. (2004), Truskey et al. (2004), Tung et al. (2004), Tworek (2003), Ukhorskiy et al. (2003), Van Dommelen et al. (2004), Vignon and Taylor (2004), Vinogradov and Hashimoto (2001), Wang (1999), Wang and Cheng (1995, 1996, 1997), Wang and Yang (2004), Watkins et al. (2001), Yang and Wang (2002, 2003), Yoon and MacGregor (2004), Zhang et al. (2006), and Zhu et al. (2005) for more examples of multiscale phenomena.

1.1.1 Materials science and technology

Multiscale issues are central to materials science. This can be shown by solid-state metals which are commonly and most usefully not pure crystals. Metals have important structures on length scales above that of their constituent atoms: ranging from voids and fractures at the large scale to granular structures, dislocations, and point defects such as inclusions, substitutions and vacancies on the atomic scale. Their macroscopic properties such as conductivity, corrosion resistance, strength and toughness depend strongly on these structures at various length scales. For example, the strength of a pure crystal typically has an order of magnitude higher than that of a polycrystal made from the same material, while the fracture toughness and ductility are much lower. Plastic deformation of metals has its origin in the flow of dislocations. However, dislocation theory by itself cannot predict plastic flow rates and yield strength, as the interaction of dislocations with grain boundaries, point defects and atomic vibrations is responsible for the phenomena such as strain hardening and thermal softening. Culminating with the program of materials by design, it is a grand dream for material science to predict macroscopic properties of solids on the basis of their microstructural state combined with their atomic level composition.

Multiscale issues are also often encountered in materials technology. A typical multiscale problem occurs in chemical vapor deposition, physical vapor deposition, plasma-enhanced chemical vapor deposition, and electroplating for the deposition of many layers of materials on a silicon wafer in fabricating integrated circuits. Each such deposition process is performed in a special chamber usually on the order of 0.1–1 m in size where one or more wafers are accommodated. On the other end of the scale, each wafer contain *millions* of microelectronic components that contain critical features on the order of 0.1 μm or less. The challenge is to fabricate microscopic structures uniform over the entire wafer of 0.2–0.3 m in diameter using only the macroscopic controls available with each chamber such as pressure, temperature, gas composition, and flow rate. Since each wafer may be worth several thousand dollars and each chamber may cost several million dollars to purchase and install, there is considerable incentive to optimize the operation on the macroscale to obtain the greatest possible yield of components and circuits on the microscopic level.

1.1.2 Physics

Multiscale problems abound in physics. A typical example is turbulent flow and transport. In the simplest statistically homogeneous and isotropic turbulent flow, the ratio of the characteristic length of the most energetic scale L over that of the smallest dynamically active scale η is evaluated by the relation (Ciofalo, 1994),

$$\frac{L}{\eta} = O(Re^{3/4})$$

in which Re is the Reynolds number and O the order symbol. Applications in the aeronautical field deal with Reynolds numbers of order as high as 10^8. The aeronautical turbulent flow thus contains a spatial scale range across six orders of magnitude. At the microscale, smaller than the size of turbulence eddies, the flow could be either viscosity-dominated inside the eddies or inertia-dominated at the interface between eddies. Without the dominance of viscosity inside the eddies, the eddies could not move as a whole; however, if viscosity played a dominant role at the interface between eddies, the eddies could not move independently. Without paying attention to its scale-dependent nature, it is hardly possible to properly understand turbulent flow.

The physics of fluid mixtures in porous media is incredibly complex, largely due to multiscale heterogeneity in the underlying porous matrix. This is relevant to all of science and engineering, since everything is porous given the right scale of observation. This is especially true in the hydrologic sciences wherein all geologic formations are porous with heterogeneity evolving over many orders of magnitude in space and time. Heterogeneity (non-uniformity) gives rise to velocity fluctuations and pressure excitations over a multitude of scales. The manifestations of these fluctuations and excitations are anomalous flow and transport of fluid mixtures, and anomalous deformation of the underlying matrix itself. One of the primary goals of theoreticians involved with natural geological and biological media, as well as engineering porous media, is to develop accurate predictive capabilities on a variety of natural scales, and/or between natural scales.

Climatology and meteorology are dominated by subgrid modeling issues. Typical computational grid sizes for general circulation models are on the order of 100 km. Cloud formation and snow cover play a large role in reflectivity, hence in the energy balance and the entire dynamics of the atmosphere. But local precipitation, moisture content and some features of storm systems are typically subgrid phenomena, which must be modeled at a subgrid level. These subjects have thus important multiscale aspects. Subgrid modeling, whenever it occurs, is indicative of multiscale issues, and is an effort to deal with them. The modeling of petroleum reservoirs and chemically reactive fluids (such as occur in turbulent combustion) uses subgrid modeling extensively. So do plate tectonics and astrophysics simulations. Even perfectly mixed (spatially homogeneous) chemical reactions are temporally multiscale, a fact which leads to stiff ordinary differential equation (ODE) solvers and adaptive time stepping for their solution.

High energy physics has long recognized that each theory fails to predict its own parameters such as particle masses and coupling constants, which must thus come from a still more fundamental theory. The program of computational lattice quantum chromodynamics (QCD) is just such a multiscale program, with the goal of predicting the mass of the proton, for example, considered as a bound state in a theory whose *elementary* constituents are quarks and gluons.

1.1.3 Hydrology

Hydrological processes occur at a wide range of scales, from unsaturated flow in a 1-m soil profile to floods in river systems of a million square kilometers; from flashfloods of several minutes duration to flow in aquifers over hundreds of years. Hydrological processes span, actually, about eight orders of magnitude in space and time. A good discussion of various scale issues in hydrology is available in Blöschl (2001, 2004), Blöschl and Sivapalan (1995), and Blöschl et al. (1997). Here we just briefly list some of them.

Figure 1 is based on both data and heuristic considerations and attempts a classification of hydrological processes according to typical length and time scales. Precipitation is one of the forcings driving the hydrological cycle. Precipitation phenomena range from cell (associated with cumulus convection) at scales of 1 km and several minutes, to synoptic areas (frontal systems) at scales of 1,000 km and more than a day. Many hydrological processes operate at similar length scales, in response to precipitation, but the timescales are delayed.

As in the case of atmospheric processes, different hydrological processes occur at different length scales (Figure 1). Runoff generation associated with rainfall intensities exceeding infiltration capacities (which produces infiltration excess/Horton overland flow) is a *point phenomena* and can, as such, be defined at

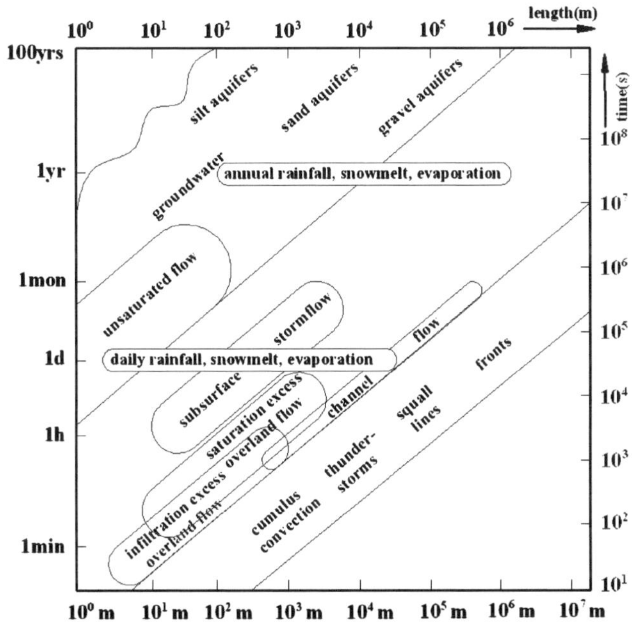

Figure 1 Spatial and temporal scales of hydrological processes (after Blöschl and Sivapalan, 1995).

a very small length scale. Saturation excess runoff (i.e. saturation overland flow) is an integrating process and needs a certain minimum catchment area to be operative. This is because, typically, the main mechanism for raising the groundwater table (which in turn produces saturation overland flow) is lateral percolation above an impeding horizon. Also, subsurface stormflow needs a certain minimum catchment area to be operative. Channel flow typically occurs at larger scales above a channel initiation area up to the length scales of the largest river basins.

1.1.4 Chemical engineering

Heterogeneous catalysis reactions occur in a typical multiscale system. Several different physical and chemical phenomena can be distinguished on various characteristic dimensional scales during the reaction. The chemical events such as adsorption, desorption, and surface reaction occur within the microscale level (nanoscale and subnanoscale). At the catalyst particle scale, mass transfer on the surface or inside the catalyst particle is significant. At a reactor scale, the hydrodynamics and external mass transfer play important roles.

Multiscale structures also appear in gas–solid two-phase flows that are often encountered in chemical processes such as coal combustion. When observation or measurement is limited to a single particle-size scale either in the particle-rich dense phase or in the gas-rich dilute phase, what is observed is exclusively the interaction between gas and individual particles. When observation or measurement is extended to the cluster-size scale, we see the existence of particle clusters

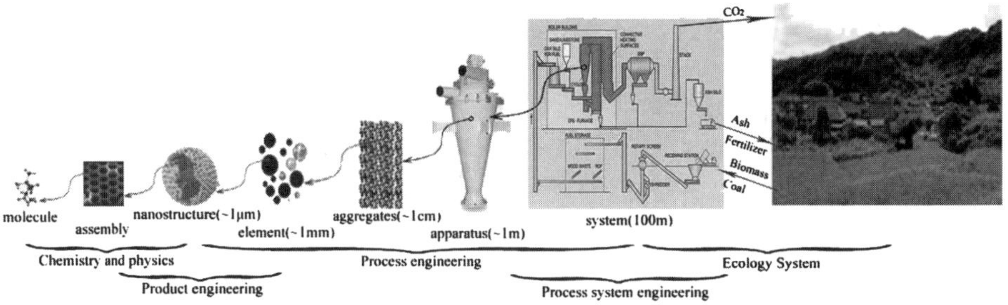

Figure 2 Multiscale chemical processes (after Li et al., 2004b) (see Plate 2 in Color Plate Section at the end of this book).

and their interaction with the surrounding diluted broth in addition to what we observed at the particle scale. Furthermore, when the observation or measurement reaches the vessel scale, the vessel wall effect is also involved.

The multiscale nature of chemical processes often requires multiscale control. To have a low-emission combustion, for example, temperature must be controlled below 900°C at the particle-size scale to satisfy the deSOx and deNOx requirements. At the cluster-size scale, the alternative change of the solid-rich dense phase and the gas-rich dilute phase is favorable to reduce NOx. At the unit scale, staged air supply could further reduce the NOx emission. The system-scale control is also involved if recirculating flue gas is applied.

The multiscale nature of chemical engineering can be further demonstrated by a typical chemical system illustrated in Figure 2 ranging from the molecular scale through factory scales to the whole ecological system scale. A molecular system studied by chemists and physicists is within a reactor system studied by engineers that is, in turn, within an ecological system studied by ecologists. Each of these three systems is also a multiscale system. Bridging these three systems, product engineering correlates the molecular behavior with the processes occurring in reactors. Process system engineering, however, integrates different chemical reactors, processes and the environment to minimize energy consumption, maximize efficiency and profits and protect our environment. Clearly, all three levels of the chemical system are also characterized by their own multiscale structures: the molecular system consisting of atoms, molecules, and assemblies; the reactor system involving the multiscale nature of a particle/droplet/bubble scale, the aggregate scale and apparatus scale; and ecological systems containing the process apparatus, factories, and the environment.

1.1.5 Bio-transport

Length scales in biological systems range over eight orders of magnitude (Table 1). Clearly, no single transport process can function efficiently over these length scales. At short distances, diffusion can be quite rapid. As distance increases, the diffusion time increases as the square of this distance; thus, diffusion becomes increasingly less efficient. In biological systems, convection typically transports molecules over distances for which diffusion is too slow. The cardiovascular system uses, for example, convection to optimize oxygen delivery

Table 1 Relevant length scales in biological systems

Quantity	Length scale (m)
Proteins and nucleic acids	10^{-8}
Organelles	10^{-7}
Cells	10^{-5} to 10^{-6}
Capillary spacing	10^{-4}
Organs	10^{-1}
Whole body	10^{0}

to the various organs. The dynamic response of other molecules (e.g. hormones) that are transported through the blood is limited by the oxygen-delivery requirements. The elapsed time for the delivery of these molecules is on the order of a few minutes, sufficient to meet normal demands. Body movement, however, requires a much faster response time than can be accomplished by convective transport through blood and diffusion in tissues. Therefore, the nervous system uses electrical conduction of signals through transmembrane ion movement and the release of neurotransmitters, at speeds as high as 500 m/s.

In general, drug and gene delivery to tumor cells must be studied at four different length scales (Figure 3; Ukhorskiy et al., 2003): (a) the body level ($\sim$1 m), (b) the tissue level ($\sim$1 cm), (c) the microvessel level ($\sim$0.01 cm), and (d) the cell level ($\sim$10 μm). The whole-body distribution of drugs and genes must be analyzed in order to understand which organs are targeted by these agents. Determining local distributions of drugs and genes within a tumor or around a microvessel is important in understanding whether the drug concentration around all tumor cells is higher than a threshold level that is required to kill tumor cells. If targets of drugs and genes are inside tumor cells or inside the nucleus, then transport in tumor cells becomes important. Barriers to intracellular transport include the plasma membrane, the membrane of vesicles in cells, the cytoskeleton, and the nuclear envelope. It is a challenge to facilitate drug and gene transport across these barriers.

1.1.6 Universe

Broadly speaking, the hierarchical multiscale structures are an inherent nature of the universe (Figure 4; Li et al., 2004b). Elementary particles were organized into more than 100 kinds of atoms as listed in the element periodic table. Starting with these atoms, biotic and abiotic worlds were formed and evolved, each with bifurcations during evolution such as animal and plant for the biotic world, and land, ocean and atmosphere for the abiotic world in nature. Further bifurcation led to the biodiversity of life and to different landscapes in nature. The activities of human beings have created various industries, agricultures, and buildings, which also show bifurcations in different engineering fields. Therefore, the biotic, abiotic, and artificial worlds are all characterized by their hierarchical multiscale nature, starting with chemical elements and finally emerging into the whole ecological system and the universe. The defining challenge for science and

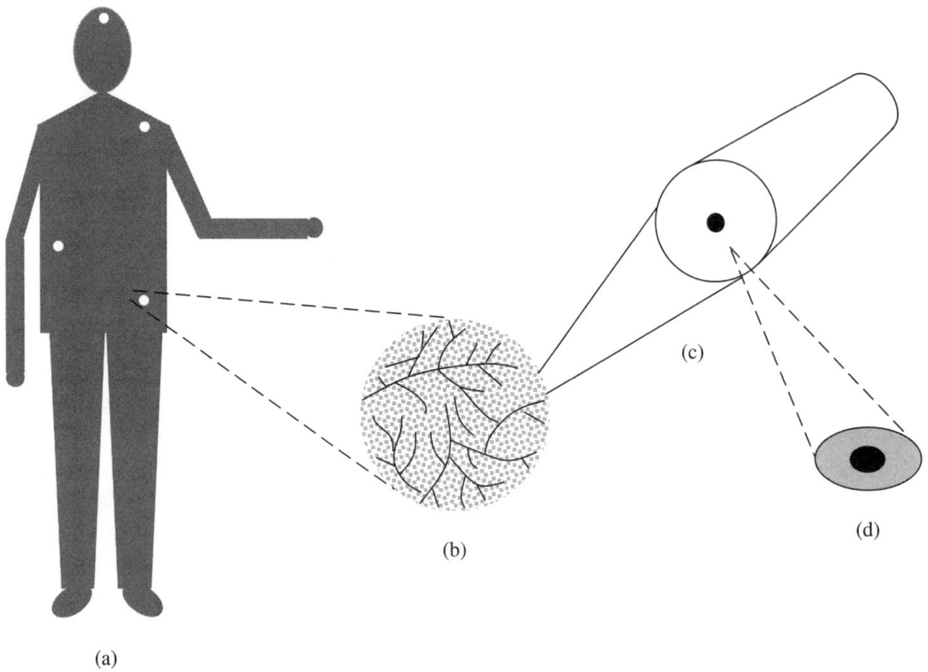

Figure 3 Drug delivery to tumor cells at four different levels (open circles in (a) represent solid tumors; solid curves in (b) represent tumor vessels; inner cylinder in (c) represents a microvessel; outer cylinder in (c) represents the region in tumor tissues that receives nutrients from this microvessel; part (d) represents a tumor cell with a nucleus; after Truskey et al., 2004).

engineering in the 21st century is to unify understandings of these bifurcations of hierarchical multiscale phenomena and to correlate them to microscopic elementary particles and to the megascopic universe.

1.2 Scales and scaling

Predicting and analyzing multiscale phenomena are research challenges of rare potential but daunting difficulty. The potential comes from both practical and scientific opportunities as shown by examples in Section 1.1. The difficulty reflects the issues related to scale and scaling. In this section, we briefly discuss scales and scaling, which are the keys to multiscale science.

1.2.1 Scales

A *scale* may be defined as a characteristic dimension (or size), in either space or time or both, of an observation, a process, or a model of a process (Blöschl and Sivapalan, 1995; Jewitt and Grgens, 2000). Intuitively, a scale is an indication of an order of magnitude rather than a specific value. We need to discuss types of scales and their characteristics as well as the interactions between the types of scales. Here we largely follow Blöschl and Sivapalan (1995), and Jewitt and

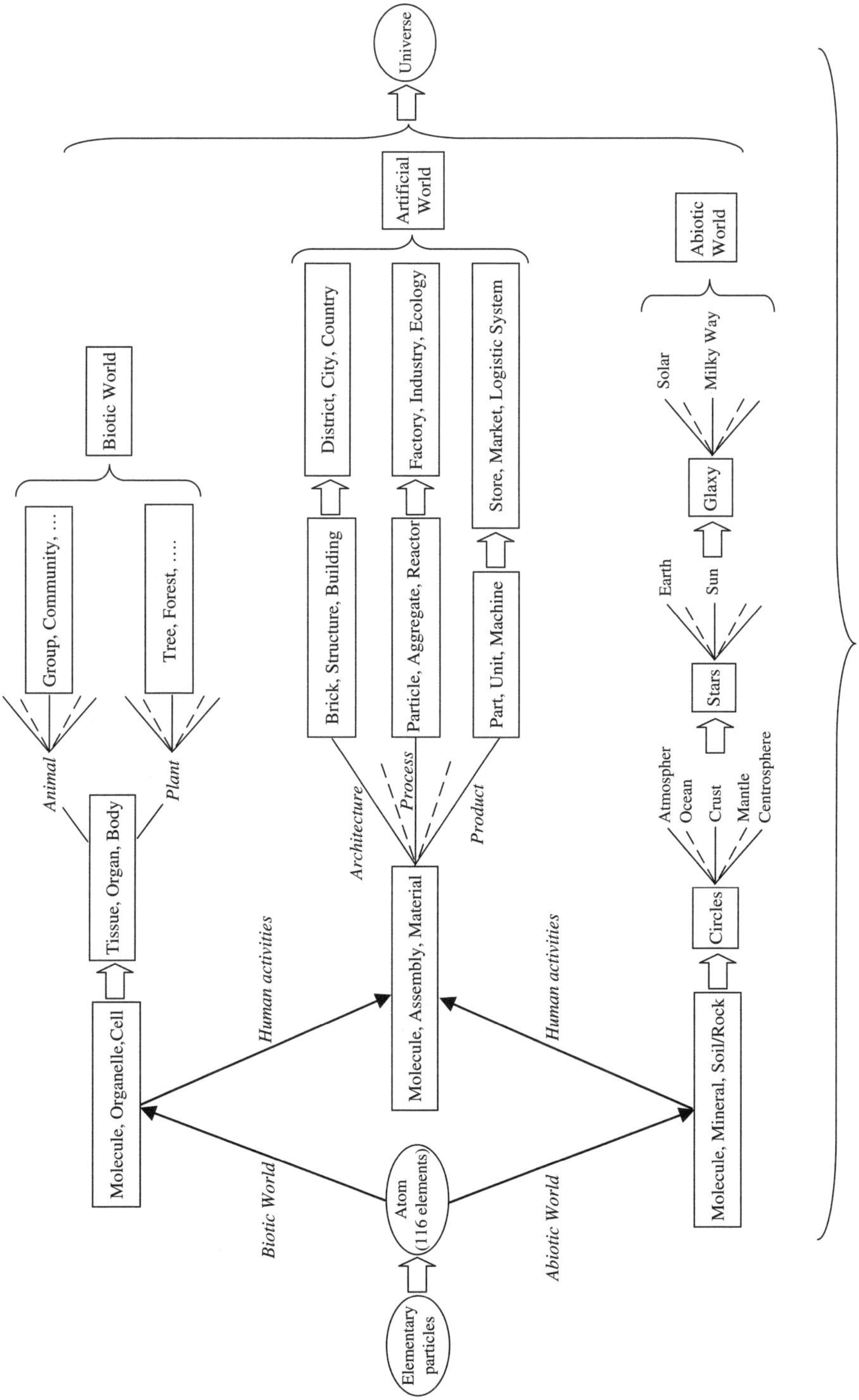

Figure 4 Multiscale nature of complex systems (after Li et al., 2004b).

Grgens (2000) in briefly describing concepts regarding scales. The readers are referred to Al-Ghoul et al. (2004), Attinger and Koumoutsakos (2004), Brueckner et al. (2005), Coveney (2003b), Marin (2005), Meakin (1998), Novak (2004), Robinson and Brink (2005), Skar (2003), Skar and Coveney (2003a, b), Antonic et al. (2002), Feng et al. (1997), Frantziskonis (2002b), Blöschl (2001, 2004), Blöschl and Sivapalan (1995), Blöschl et al. (1997), Jewitt and Grgens (2000), Becker et al. (1999), Blöschl (1999), Borri-Brunetto et al. (2004), Borucki (2003), Chave and Levin (2003), Costanza (2003), Crawley and Harral (2001), Donoho and Huo (2001), Ehleringer and Field (1992), Farina (1998), Field and Ehleringer (1993), Gallegher and Appenzeller (1999), Gibson et al. (2000), Golledge and Stimson (1997), Granbakken et al. (1991), Grayson and Blöschl (2000), Grayson et al. (1997), Haarberg et al. (1990), Kalma and Sivapalan (1995), Ladevze (2005), Levin (1992, 1993, 2000, 2002), Levin et al. (1997), Lindeberg (1994a, b, 1998, 1999), Lindeberg and Romeny (1994), Lovell et al. (2002), Lu and Fu (2001), Perry and Ommer (2003), Schulze (2000), Skoien et al. (2003), Sposito (1998), Van Gardingen et al. (1997), Vidakovic et al. (2000), Wang (1998, 2000, 2002), Western and Blöschl (1999), and Western et al. (2002) for more comprehensive discussions.

Process scales are defined as the scales at which natural phenomena occur. They are intrinsic to the system and the process and are beyond our control. These scales are not fixed, but vary with the process. The intrinsic relationships between time and space scales are well illustrated by the so-called scope diagrams of which Figure 1 is an example for processes relevant in hydrology. Process scales have the following important features:

1. Processes and their scales overlap; process scales are, therefore, relative rather than absolute in time and space.
2. Events of a small spatial scale are associated with short temporal scales.
3. Small-scale events display more variability than large-scale ones.

Temporally, we have *intermittent processes*, which have a certain lifetime; *periodic processes*, which have a cycle, or period, and *stochastic processes*, i.e. probabilistic processes that have a correlation length which is often expressed by the recurrence interval.

Spatially, however, process scales exhibit *spatial extent* (e.g. the area over which the thunderstorm rainfall occurred), *space period* (e.g. the area over which a certain seasonality of rainfall occurs), and *correlation space* (e.g. the area over which the 1:10 year drought left its mark).

Some processes exhibit one or more preferred scales, i.e. certain length (or time) scales are more likely to occur than others. These preferred scales are also called *natural scales* (Blöschl and Sivapalan, 1995). In a power spectrum representation preferred scales appear as peaks of spectral variance (Padmanabhan and Rao, 1988). Scales that are less likely to occur than others are often related to as a spectral gap. This name comes from the power spectrum representation in which the spectral gap appears as a minimum in spectral variances (Stull, 1988). The existence of a spectral gap is tantamount to the existence of a *separation of scales* (Gelhar, 1986; Wang, 2000). Separations of scales refer to a process consisting of a small-scale (or fast) component superimposed on a much larger (or slow) component with a minimum in spectral variance in between.

In the time domain, many hydrological processes (such as snow melt) exhibit preferred timescales of one day and one year with a spectral gap in between. Clearly, this relates to the periodicity of solar radiation. In the space domain, there is no clear evidence for the existence of preferred scales. Precipitation, generally, does not seem to exhibit preferred scales and/or spectral gaps (Gupta and Waymire, 1993). As suggested in Wood et al. (1988), catchment runoff may show a spectral gap at a 1-km^2 catchment area. The work in Blöschl et al. (1995), however, indicates that both the existence and size of a spectral gap is highly dependent on specific catchment properties and climatic conditions.

Observation scales are those at which humans choose to collect samples of observations and to study phenomena (Blöschl and Sivapalan, 1995). We are free to choose the observation scale within the constraints of technical and logistical types. Observation scales are normally determined by logistical constraints (e.g. access to places of observation), technology (e.g. the cost of state-of-the-art instrumentation), and perceptions (i.e. what is perceived to be important for a study at a particular time). The observational scale is usually quite inflexible in a given circumstance. It can be characterized by its *extent*, *resolution*, and *grain* (Blöschl and Sivapalan, 1995).

There is often a mismatch between the observation and process scales. Ideally, observations should be made at the scale at which the processes are taking place. However, this is seldom possible since processes operate over a range of scales, as well as having logistical, technological and perceptual constraints. Problems arise when observations are made at incorrect scales, so that we then have an idea of what is happening, but not why or where within the study area.

Furthermore, processes should be observed over a large extent, with high resolution and a fine grain in order to allow any signal within the process to be observed at the appropriate timescale. However, this is rare and thus begs the question of what the appropriate (or commensurate) scale is.

Yet another scale is the *modeling scale* at which the process and the phenomena are modeled. Models play a fundamental role because they allow the integration of processes operating at diverse scales. However, the art of developing models to act across widely different scales of space, time, or organizational complexity involves more than just the inclusion of every possible detail at every possible scale. In the study of general circulation models, for example, building models with finer and finer spatial resolution are not only limited by computational complexity, but also introduce detail that can confound interpretation. The same applies to the analysis of any non-linear system, in which the challenge in moving across scales is not to include as much detail as possible, but to find ways to suppress irrelevant detail and simplify system description (Levin, 1991; Levin and Pacala, 1997; Ludwig and Walters, 1985). No sensible scientist would try to build a model of the behavior of an individual organism by accounting for the processes within every cell, tied together in a network of interaction of numbing complexity. Similarly, no sensible scientist should try to build models of ecological systems in which one reproduces the behavior of every organism, or even every species. The goal rather should be to identify relevant details and to describe the dynamics of whole systems in terms of the statistical properties of their units.

Finding the relevant scale for modeling is an important step in the analysis of a multiscale system. Usually, only a few scales are relevant, each corresponding with a process that drives the dynamics of the system. Many efforts have been made for detecting relevant scales for modeling (Al-Ghoul et al., 2004; Antonic et al., 2002; Attinger and Koumoutsakos, 2004; Becker et al., 1999; Blöschl, 1999, 2001, 2004; Blöschl and Sivapalan, 1995; Blöschl et al., 1995, 1997; Borri-Brunetto et al., 2004; Borucki, 2003; Brueckner et al., 2005; Chave and Levin, 2003; Costanza, 2003; Coveney (2003b); Crawley and Harral, 2001; Donoho and Huo, 2001; Ehleringer and Field, 1992; Farina, 1998; Feng et al., 1997; Field and Ehleringer, 1993; Frantziskonis, 2002b; Gallegher and Appenzeller, 1999; Gelhar, 1986; Gibson et al., 2000; Golledge and Stimson, 1997; Granbakken et al., 1991; Grayson and Blöschl, 2000; Grayson et al., 1997; Gupta and Waymire, 1993; Haarberg et al., 1990; Jewitt and Grgens, 2000; Kalma and Sivapalan, 1995; Ladevze, 2005; Levin, 1991, 1992, 1993, 2000, 2002; Levin and Pacala, 1997; Levin et al., 1997; Lindeberg, 1994a, b, 1998, 1999; Lindeberg and Romeny, 1994; Lovell et al., 2002; Lu and Fu, 2001; Ludwig and Walters, 1985; Marin, 2005; Meakin, 1998; Novak, 2004; Padmanabhan and Rao, 1988; Perry and Ommer, 2003; Robinson and Brink, 2005; Schulze, 2000; Skar, 2003; Skar and Coveney, 2003a, b; Stull, 1988; Skoien et al., 2003; Sposito, 1998; Van Gardingen et al., 1997; Vidakovic et al., 2000; Wang, 1998, 2000, 2002; Western and Blöschl, 1999; Western et al., 2002; Wood et al., 1988). In many cases, the investigators must perform a rather sophisticated analysis to detect them, usually involving a careful variability study of variables across scales, an understanding of the mechanisms determining and governing systems and processes, a careful investigation of the interplay among processes operating at diverse scales and the way the information transfers across the scales.

Therefore, it normally takes the efforts of several generations to arrive at some modeling scales generally agreed upon within the scientific community. Clearly, such agreed modeling scales change as our fundamental understanding of multiscale systems and processes deepens and technology progresses. For example, four modeling scales that are normally used in modeling landscape ecology are (Farina, 1998): (1) *Microscale*: This scale considers a period from 1 to 500 years and a space from $1–10^6\,m^2$. Scientists working at this scale are geomorphologists, plant succession and animal ecologists and planners. At this scale, we can model disturbances (fires, windthrow, and clearcutting), biological processes (the cycles of animal populations, gap-phase replacement in forests and succession after abandonment) and geomorphic processes (soil creep, movement of sand dunes, debris avalanches, slumps, fluvial transport and exposition, and cryoturbation); (2) *Mesoscale*: This scale extends from 500 to 10,000 years and in space from 10^6 to $10^{10}\,m^2$. This period is comprised of events such as the last interglacial interval. The cultural evolution of humans also occurs at this scale; (3) *Macroscale*: This scale extends from 10,000 to 1,000,000 years, with spatial extension from 10^{10} to $10^{12}\,m^2$. The glacial-interglacial cycles occur, and speciation and extinction operate at this scale; and (4) *Megascale*: This scale extends from 10^6 to 4.6 billion years, with an extension $>10^{12}\,m^2$, covering the entire American continent and interacting geological events such as plate tectonic movements.

In modeling transport processes by continuous representations of mass, momentum, energy, and entropy balances, we normally work with three spatial modeling scales that are of importance in describing various dynamic mechanisms: microscale, macrosacle, and megascale (Gray et al., 1993; Wang, 2000). Briefly, the *microscale* refers to the smallest scale at which a system can be viewed as a continuum. Below this scale, the materials under consideration are viewed as a discrete collection of molecules or particles such that modeling approaches that do not explicitly involve spatial gradients would be required. The next scale above microscale, the *macroscale,* is larger than the microscale but smaller than the scale of the system under study. It often represents a scale at which a multiphase system can be idealized as sets of overlapping continua. Finally, the length scale of the *megascale* is on the order of the system dimensions. If a system is modeled as being completely megascopic, spatial variations are not explicitly considered as the system is characterized only in terms of average values. Although these three scales can be conceptually described, the transition from one scale to another is not clearly defined and the scales have not been identified with precision. Thus the modeling scale is often problem dependent with conservation laws for a particular problem being required and implemented at the appropriate scale.

Models developed on a modeling scale may be either *predictive* (to obtain a specific answer to a specific problem) or *investigative* (to further our understanding of processes) (Grayson et al., 1992; O'Connell, 1991). Typically, investigative models need more data, are more sophisticated in structure and estimates are less robust, but allow more insight into the system behavior. The development of both types of model has traditionally followed a set pattern involving the following steps (O'Connell, 1991): (1) collecting and analyzing data; (2) developing a conceptual model (in the researcher's mind) which describes the important characteristics of systems; (3) translating the conceptual model into a mathematical model; (4) calibrating the mathematical model to fit a part of the available data by adjusting various coefficients; and (5) validating the model against the remaining data set. If the validation is not satisfying, one or more of the previous steps needs to be repeated. If, however, the results are sufficiently close to the observations, the model is considered to be ready for use in a predictive model.

1.2.2 Scaling

Scaling represents the transcending concepts that link processes at different scales in time and space (Al-Ghoul et al., 2004; Attinger and Koumoutsakos, 2004; Blöschl, 1999, 2001, 2004; Blöschl and Sivapalan, 1995; Blöschl et al., 1997; Becker et al., 1999; Borri-Brunetto et al., 2004; Brueckner et al., 2005; Chave and Levin, 2003; Costanza, 2003; Coveney, 2003a, b; Crawley and Harral, 2001; DeCoursey, 1996; Donoho and Huo, 2001; Ehleringer and Field, 1992; Farina, 1998; Field and Ehleringer, 1993; Freeden and Michel, 2004; Gallegher and Appenzeller, 1999;

Gibson et al., 2000; Granbakken et al., 1991; Grayson and Blöschl, 2000; Grayson et al., 1997; Gupta et al., 1986; Haarberg et al., 1990; Kalma and Sivapalan, 1995; Levin, 1993; Lindeberg, 1994a, b, 1998, 1999; Lindeberg and Romeny, 1994; Lovell et al., 2002; Lu and Fu, 2001; Marin, 2005; Meakin, 1998; Novak, 2004; Perry and Ommer, 2003; Robinson and Brink, 2005; Schneider, 2002; Schulze, 2000; Sih, 2001; Skar, 2003; Skar and Coveney, 2003a, b; Skoien et al., 2003; Sposito, 1998; Tian et al., 2004; Van Gardingen et al., 1997; Vidakovic et al., 2000; Wang, 1998a, 2000b, 2002, 2007; Western and Blöschl, 1999; Western et al., 2002; Yang and Wang, 2000). We need to distinguish between *upscaling* and *downscaling*. The former refers to transferring information from a given scale to a large scale, whereas the latter refers to transferring information to a smaller scale (Gupta et al., 1986). For example, measuring hydraulic conductivity in a borehole and assuming it applies to the surrounding area involves upscaling. Also, estimating a 100-year flood from a 10-year record involves upscaling. Conversely, using runoff coefficients derived from a large catchment for culvert design on a small catchment involves downscaling. One of the factors that make scaling difficult is the heterogeneity of material systems and the variability of processes.

To develop a good scaling approach, we must successfully address the following questions (Jewitt and Grgens, 2000; Schulze, 2000): (1) When is simple aggregation (i.e. linear addition of elements) sufficiently accurate for upscaling? (2) Are processes which are observed or models which are formulated, at points or small spatial scales transferable to larger scales? (3) Where such scaling is possible, how should it be done? (4) How do these means change with scale? (5) How does the variability change with scale? (6) How does the sensitivity change with scale? (7) Under what circumstances would non-linear responses be either amplified or dampened as scales change? (8) How does heterogeneity change with scale? (9) How does predictability change with space and time scales? (10) What types of conceptual errors are involved, wittingly or unwittingly, by scientists in their up- or downscaling assumptions? (11) How can observations made at two scales be reconciled? Various scaling errors may occur in up- and downscaling. They can be classified as the errors of omission and the errors of commission (Schulze, 2000). The former comes mainly from: (1) failing to ask some relevant questions, which are not obvious at the scale being considered, in the model; (2) failing by focusing on a single scale to include some important processes that only become obvious at either finer or broader scales; (3) failing to ask questions about cumulative impacts on a broader scale than that being studied; and (4) failing to examine large-scale impacts on a smaller scale. The latter may arise by asking questions at a scale at which meaningful answers cannot be provided or by assuming that the appropriate scale for a given process is the same for all component processes of the system.

Much effort has been made on the development of up- and downscaling technologies in various fields. Because of the complexities of the problem, there is no universal systematic approach available. The available approaches are more or less haphazard and ad hoc with applications usually limited to some particular multiscale problems. A review of these approaches is out of the scope of the present monography. The readers are referred to Al-Ghoul et al. (2004),

Attinger and Koumoutsakos (2004), Marin (2005), Antonic et al. (2002), Cushman (1997), Fredberg and Kamm (2006), Karsch et al. (1997), Van Gardingen et al. (1997), Wang (1997, 2000b, 2002, 2007); Freeden and Michel (2004), Schneider (2002), Sih (2001), Yang and Wang (2000), Aanonsen and Eydinov (2006), Aarnes et al. (2005), Aarnes et al. (2006), Anderson et al. (2003, 2004a, b), Arbogast (2004), Arbogast and Boyd (2006), Artus and Noetinger (2004), Artus et al. (2004), Ates et al. (2005), Babadagli (2006), Baraka-Lokmane and Liedl, 2006; Barthel et al. (2005), Basquet et al. (2004), Berentsen et al. (2005), Berkowitz et al. (2006), Berryman (2005), Bierkens et al. (2000), Blöschl (2005), Blum et al. (2005), Borges (2005), Bourgeat et al. (2004), Bramble et al. (2003), Brandt (2005), Braun et al. (2005), Busch et al. (2004), Byun et al. (2005), Cardells-Tormo and Arnabat-Benedicto (2006), Cassiraga et al. (2005), Chalon et al. (2004), Chastanet et al. (2004), Chawathe and Taggart (2004), Chen and Durlofsky (2006a, b), Chen (2006), Chen et al., 2005a, b, Chen and Yue (2003), Coveney and Wattis (2006), Corwin et al. (2006), Curtarolo and Ceder (2002), Crow et al. (2005), Cushman et al. (2002), Droujinine (2006), Daly et al. (2004a, b), Darrah et al. (2006), DasGupta et al. (2006), Das and Hassanizadeh (2005), Das et al. (2004), Davis et al. (2006), Dean and Russell (2004), Ding (2004), Dittmann et al. (2005), Drew and Passman (1999), Dunbabin et al. (2006), Eberhard (2004, 2005a, b), Eberhard et al. (2004, 2005), Efendiev et al. (2005), Efendiev and Pankov (2004a, b), Egermann and Lenormand (2005), Flodin et al. (2004), Frantziskonis (2002a, b), Gasda and Celia (2005), Geindreau et al. (2004), Geraerts et al. (2005), Gerritsen and Durlofsky (2005), Gill et al. (2006), Givon et al. (2004), Groenenberg et al. (2004), Gueguen et al. (2006), Hall et al. (2004), Held et al. (2005), Hilfer and Helmig (2004), Hontans and Terpolilli (2005), Hou (2005), Hui et al. (2005), Idris et al. (2004), Illman (2004), Imkeller and Von Storch (2001), Javaux and Vanclooster (2006), Jenny et al. (2004), Karssenberg (2006), Kfoury et al. (2004, 2006), Kim et al. (2005, 2006), Knudby et al. (2006), Korostyshevskaya and Minkoff (2006), Larachi and Desvigne (2006), Lewandowska et al. (2005), Li et al. (2004a, 2006), Liu et al. (2005), Lock et al. (2004), MacLachlan and Moulton (2006), Magesa et al. (2005), Mander et al. (2005), Ma et al. (2006), McLeary et al. (2006), Mezghani and Roggero (2004), Mezzenga et al. (2004), Miehle et al. (2006), Mueller and Jochen (2006), Nardin and Schrefler (2005), Neuweiler and Cirpka (2005), Niedda (2004), Noetinger et al. (2004, 2005), Noetinger and Gallouet (2004), Noetinger and Zargar (2004), Nordbotten et al. (2006), Novikov (2004), Ogunlana and Mohanty (2005), Oja et al. (2005b), Orgeas et al. (2006), Painter and Cvetkovic (2005), Park and Parker (2005), Park and Cushman (2006), Pathak et al. (2005), Pennock et al. (2005), Perfect et al. (2006), Petts et al. (2006), Pickup et al. (2005), Prevost et al. (2005), Pruhs et al. (2006), Qi and Hesketh (2004a, b, 2005a, b); Quintard et al. (2006), Robinson et al. (2004), Rodgers et al. (2005), Ronayne and Gorelick (2006), Rundle et al. (2003), Russo (2003), Samantray et al. (2006), Seeboonruang and Ginn (2006a, b), Schlecht and Hiernaux (2004), Sitnov et al. (2002), Sun et al. (2004), Temizel and Vlachos (2005), Toniolo et al. (2006), Toomanian et al. (2004), Tureyen and Caers (2005), Valluzzi et al. (2003), Vdovina et al. (2005), Verburg et al. (2006a), Vitale et al. (2005), Walker et al. (2005), Wang (1997), Ward et al. (2006), Wett (2006), Whitaker (1999), Wong and Asseng (2006), Xiao et al. (2006),

Xie et al. (2004), Yong (2004), Yu and Christakos (2006), Zbib and de la Rubia (2002), Zbib et al. (2002), Zhang et al. (2004a, b), Zhu et al. (2004a, b), Zhu and Mohanty (2004), and Zlokarnik (2006) for upscaling techniques and Al-Ghoul et al. (2004), Attinger and Koumoutsakos (2004), Antonic et al. (2002), Cushman (1997), Fredberg and Kamm (2006), Karsch et al. (1997), Van Gardingen et al. (1997), Freeden and Michel (2004), Schneider (2002), Sih (2001), Anderson et al. (2003), Bierkens et al. (2000), Blöschl (2005), Bramble et al. (2003), Valluzzi et al. (2003), Zbib and de la Rubia (2002), Zbib et al. (2002), Aanonsen and Eydinov (2006), Abildtrup et al. (2006), Anderson et al. (2004), Antic et al. (2004, 2006), Araujo et al. (2005), Artus et al. (2004), Auclair et al. (2006), Bacchi and Ranzi (2003), Badaroglu et al. (2006), Badas et al. (2006), Bardossy et al. (2005), Bergant and Kajfez-Bogataj (2005), Bruinink et al. (2006), Burger and Chen (2005), Busuioc et al. (2006), Cardells-Tormo and Arnabat-Benedicto (2006), Charlton et al. (2006), Chen et al. (2006), Corwin et al. (2006), Coulibaly et al. (2005), Coulibaly (2004), Deidda et al. (2006), Deidda et al. (2004), De Rooy and Kok (2004), Diaz-Nieto and Wilby (2005), Dibike and Coulibaly (2006), Dibike and Coulibaly (2005), Diez et al. (2005), Eidsvik (2005), Enke et al. (2005a, b), Feddersen and Andersen (2005), Fischer et al. (2004), Fung and Siu (2006), Gaffin et al. (2004), Gangopadhyay et al. (2005), Gaslikova and Weisse (2006), Ghosh and Mujumdar (2006), Ghan and Shippert (2006), Ghan et al. (2006), Gibson et al. (2005), Gutierrez et al. (2005), Hanssen-Bauer et al. (2005), Harpham and Wilby (2005), Haylock et al. (2006), Hewitson and Crane (2006), Hong and Root (2006), Hong et al. (2006), Hoofman et al. (2005), Huth (2004, 2005), Jungen et al. (2006), Kaipio (2005), Kettle and Thompson (2004), Khan et al. (2006a, b), Knight et al. (2004), Kunstmann et al. (2004), Lambert et al. (2004), Liang et al. (2006), Ludwig et al. (2003), Mander et al. (2005), Matulla (2005), Meier (2006), Meier et al. (2006), Merlin et al. (2006), Mika et al. (2005), Miksovsky and Raidl (2005), Moriondo and Bindi (2006), Mounier et al. (2005), Muller et al. (2006), Mateos et al. (2005), Medjdoub et al. (2005), Mehrotra and Sharma (2005, 2006a–c), Mehrotra et al. (2004), Nuttinck (2006), Nuttinck et al. (2006), Oja et al. (2005a), Olsson et al. (2004), Paeth et al. (2005), Pai et al. (2004), Pardo-Iguzquiza et al. (2006), Pasini et al. (2006), Pavan et al. (2005), Pettorelli et al. (2005), Pradhan et al. (2006), Pryor et al. (2005a, b), Raderschall (2004), Rebora et al. (2006a, b), Rengel et al. (2004), Riitters (2005), Salathe (2005), Schmidli et al. (2006), Shin et al. (2006), Sivapalan et al. (2003), Solecki and Oliveri (2004), Sun et al. (2006, 2004), Takeuchi (2004), Timbal (2004), Tononi (2005), Valenza et al. (2004), Veneziano et al. (2006), Verburg et al. (2006b), Wang et al. (2004), Wilby (2005), Wood et al. (2004), Woth et al. (2006), Xoplaki et al. (2004), Xu et al. (2005), Yang et al. (2005), Zagar et al. (2006), Zhang (2005), and Zhu et al. (2004a, b) for downscaling techniques.

1.3 Model-scaling

The demand for multiscale theorems in Sections 4 and 5 comes originally from the model-upscaling and model-downscaling. As their names indicate, they refer to finding a model at a large (small) scale based on a model at a smaller (larger) scale, respectively. They are a subgroup of scaling. In this section, we briefly discuss them to demonstrate values of multiscale theorems.

1.3.1 Model-upscaling

The following is the basic set-up of model-upscaling. We are given a system whose small-scale behaviors and processes, with state/process variable or parameter v, are described by a given small-scale model. This small-scale model is too inefficient to be used in full detail for various reasons. However, we are only interested in the large-scale behavior of the system, described by the state/process variable or parameter V. V and v are linked together by compression and reconstruction operators Q and R, with the property $QR = \mathbf{I}$ (the identity operator),

$$V = Qv, \qquad v = RV$$

Here the compression operators are, in general, local/ensemble averages, projection to low order moments, or slow manifolds. The reconstruction operator does the opposite, and is not unique in general. *Model-upscaling* refers to finding the large-scale model for V based on a small-scale model for v.

A related issue, *multiscale modeling*, arises when the large-scale model for V is either not explicitly available or invalid in some parts of the domain of interest. The basic strategy in multiscale modeling is to use the small-scale model as a supplement to provide the necessary information for extracting the large-scale behavior of the system, with the hope that the combined larger small-scale modeling will be much more efficient than solving the full small-scale model in detail.

Model-upscaling can completely transform a small-scale model to the large-scale in all spatial dimensions and temporal dimension. It can also partially transform the small-scale model to the large-scale in some of the spatial dimensions or temporal dimension. The large-scale model in the latter is actually a multiscale model involving a large scale in some dimensions and a small-scale in the others.

Both model-upscaling and multiscale modeling are important and relevant for multiscale science. Over the last few decades a number of powerful approaches have been developed for model-upscaling (Aanonsen and Eydinov, 2006; Aarnes et al., 2005, 2006; Abgrall and Perrier, 2006; Alarcon et al., 2004; Alber et al., 2006; Aldama, 1990; Al-Ghoul et al., 2004; Alkire, 2003; Anderson et al., 2003, 2004a, b; Antonic et al., 2002; Arbogast, 2004; Arbogast and Boyd, 2006; Artus and Noetinger, 2004; Artus et al., 2004, 2005; Attinger and Koumoutsakos, 2004; Babadagli, 2006; Bacon and Osetsky, 2004; Baraka-Lokmane and Liedl, 2006; Barth et al., 2002; Barthel et al., 2005; Basquet et al., 2004; Bassi et al., 2004; Belashchenko and Antropov, 2002a; Berentsen et al., 2005; Berkowitz et al., 2006; Berryman, 2005; Bierkens et al., 2000; Blöschl, 2005; Blum et al., 2005; Borges, 2005; Bourgeat et al., 2004; Bramble et al., 2003; Brandt, 2005; Braun et al., 2005; Busch et al., 2004; Byun et al., 2005; Cardells-Tormo and Arnabat-Benedicto, 2006; Cassiraga et al., 2005; Cavallotti et al., 2005; Chalon et al., 2004; Chastanet et al., 2004; Chawathe and Taggart, 2004; Chen, 2006; Chen and

Durlofsky, 2006a, b; Chen and Yue, 2003; Chen et al., 2005a, b; Chiaravalloti et al., 2006; Corwin et al., 2006; Coveney and Wattis, 2006; Crow et al., 2005; Curtarolo and Ceder, 2002; Cushman, 1990, 1997; Cushman et al., 2002; Daly et al., 2004a, b; Darrah et al., 2006; Das et al., 2004; DasGupta et al., 2006; Das and Hassanizadeh, 2005; Davis et al., 2006; Dean and Russell, 2004; Defranoux et al., 2005; Ding, 2004; Dittmann et al., 2005; Dobrovitski et al., 2000, 2003; Dormieux and Ulm, 2005; Dowell and Tang, 2003; Drew and Passman, 1999; Droujinine, 2006; Dunbabin et al., 2006; Eberhard, 2005a, b; Eberhard et al., 2004, 2005; Efendiev and Pankov, 2004a, b; Efendiev et al., 2005; Egermann and Lenormand, 2005; Engquist and Runborg, 2005; Flodin et al., 2004; Fredberg and Kamm, 2006; Gasda and Celia, 2005; Geindreau et al., 2004; Geraerts et al., 2005; Gerritsen and Durlofsky, 2005; Gill et al., 2006; Givon et al., 2004; Groenenberg et al., 2004; Gray and Miller, 2005; Grigoriev and Dargush, 2003; Grohens et al., 2001; Gueguen et al., 2006; Guo and Tang, 2006; Hall et al., 2004; Harari, 2006; Hauke, 2002; Hauke and Doweidar, 2005a, b, 2006; Hauke and Garcia-Olivares, 2001; Held et al., 2005; Hilfer and Helmig, 2004; Hontans and Terpolilli, 2005; Hou, 2005; Hui et al., 2005; Idris et al., 2004; Illman, 2004; Imkeller and Von Storch, 2001; Javaux and Vanclooster, 2006; Jenny et al., 2004; Karsch et al., 1997; Karssenberg, 2006; Katsoulakis et al., 2005; Kfoury et al., 2004, 2006; Kim et al., 2005, 2006; Knudby et al., 2006; Kolaczyk and Huang, 2001; Korostyshevskaya and Minkoff, 2006; Koumoutsakos, 2005; Kruzik and Prohl, 2006; Larachi, 2005; Larachi and Desvigne, 2006; Lemaire et al., 2006; Leveque et al., 2005; Lewandowska et al., 2005; Li et al., 2004, 2005, 2006; Likos and Lu, 2006; Liu et al., 2005; Lock et al., 2004; Ma et al., 2006; MacLachlan and Moulton, 2006; Magesa et al., 2005; Mander et al., 2005; Marin, 2005; Maroudas and Gungor, 2002; Maroudas et al., 2002; McLeary et al., 2006; Mezghani and Roggero, 2004; Mezzacappa, 2005; Mezzenga et al., 2004; Miehle et al., 2006; Moin and Apte, 2006; Mueller and Jochen, 2006; Murad and Cushman, 2000; Nardin and Schrefler, 2005; Neuweiler and Cirpka, 2005; Nicot, 2004; Niedda, 2004; Noetinger and Gallouet, 2004; Noetinger and Zargar, 2004; Noetinger et al., 2004, 2005; Nordbotten et al., 2006; Novikov, 2004; Ogunlana and Mohanty, 2005; Oja et al., 2005b; Orgeas et al., 2006; Osetsky and Bacon, 2003; Painter and Cvetkovic, 2005; Park and Cushman, 2006; Park and Parker, 2005; Pathak et al., 2005; Pennock et al., 2005; Perfect et al., 2006; Petts et al., 2006; Pickup et al., 2005; Ponziani et al., 2003; Pozdnyakova et al., 2005; Prevost et al., 2005; Pruhs et al., 2006; Qi and Hesketh, 2004a, b, 2005a, b; Quintard et al., 2006; Robinson et al., 2004; Rodgers et al., 2005; Ronayne and Gorelick, 2006; Rundle et al., 2003; Russo, 2003; Saedi, 2006; Sagaut, 2006b; Samantray et al., 2006; Schiehlen and Seifried, 2004; Schlecht and Hiernaux, 2004; Seeboonruang and Ginn, 2006a, b; Shehadeh et al., 2005; Sinha and Goodson, 2005; Sitnov et al., 2002; Soulard et al., 2004; Sun et al., 2004; Temizel and Vlachos, 2005; Teppola and Minkkinen, 2000; Tewfik et al., 1993; Tidriri, 2001; Toniolo et al., 2006; Toomanian et al., 2004; Tureyen and Caers, 2005; Uva and Salerno, 2006; Van Gardingen et al., 1997; Vdovina et al., 2005; Verburg et al., 2006a; Vitale et al., 2005; Voyiadjis et al., 2003; Walker et al., 2005; Wang, 1997, 2000b, 2002, 2007; Ward et al., 2006; Wett, 2006; Whitaker, 1999; Wong and Asseng, 2006; Xie et al., 2004, 2006; Yong, 2004; Yu and Christakos, 2006; Zhang et al., 2004a, b; Zhu and Mohanty, 2004; Zhu et al., 2004a, b;

Zlokarnik, 2006) and multiscale modeling (Abdallah et al., 2004; Abgrall and Perrier, 2006; Acharya, 2005; Aldama, 1990; Alkire, 2003; Al-Ghoul et al., 2004; Attinger and Koumoutsakos, 2004; Alarcon et al., 2004; Alber et al., 2006; Antonic et al., 2002; Antropov and Belashchenko, 2003; Bacon and Osetsky, 2004; Banks and Pinter, 2005; Barth et al., 2002; Bassi et al., 2004; Belashchenko and Antropov, 2002a; Bezzo et al., 2004; Boghosian et al., 2001, 2003; Brandt, 2002; Bramble et al., 2003; Breakspear and Stam, 2005; Buiron et al., 1999; Buldum et al., 2005a, b; Byrne et al., 2006; Cavallotti et al., 2005; Chatterjee et al., 2004; Chaturvedi et al., 2005; Chen et al., 1996; Coveney, 2003a; Coveney and Fowler, 2005; Cushman, 1997; Dadvar and Sahimi, 2003; Dahmen et al., 1997; Daniel et al., 2004; De Fabritiis et al., 2002; Defranoux et al., 2005; De la Rubia et al., 2000; Diego et al., 2004; Dowell and Tang, 2003; Drews et al., 2004; Dumoulin et al., 2003; Engquist and Runborg, 2005; Fabry et al., 2001; Filippova et al., 2001; Fish and Yu, 2001; Flekkoy et al., 2000; Fouque et al., 2003; Frantziskonis, 2002a; Fredberg and Kamm, 2006; Ghoniem et al., 2003; Gray and Miller, 2005; Grigoriev and Dargush, 2003; Grohens et al., 2001; Guo and Tang, 2006; Harari, 2006; Hauke, 2002; Hauke and Doweidar, 2005a, b, 2006; Hauke and Garcia-Olivares, 2001; Hayes et al., 2005; Hoffman and Coveney, 2001; Hou, 2005; Hughes et al., 2005; Imkeller and Von Storch, 2001; Ingram et al., 2004; Israeli and Goldenfeld, 2004; Juanes and Patzek, 2004; Kadowaki and Liu, 2005; Kang et al., 2002; Karniadakis et al., 2005; Karsch et al., 1997; Katsoulakis et al., 2005; Khurram and Masud, 2006; Kolaczyk and Huang, 2001; Koumoutsakos, 2005; Krause and Rank, 2003; Kruzik and Prohl, 2006; Larachi, 2005; Lassila et al., 2003; Lee et al., 2004; Liou and Fang, 2006; Love et al., 2003; Lemaire et al., 2006; Leveque et al., 2005; Marin, 2005; Maroudas and Gungor, 2002; Maroudas et al., 2002; Mezzacappa, 2005; Miller et al., 2004; Mizuseki et al., 2002; Moore, 2001; Mosler, 2005; Nicot, 2004; Niu and Lin, 2006; Noetinger and Zargar, 2004; Osetsky and Bacon, 2003; Ponziani et al., 2003; Pozdnyakova et al., 2005; Qin et al., 2006; Quarteroni and Veneziani, 2003; Raimondeau and Vlachos, 2002; Ren and Malik, 2002; Ren and Weinan, 2005; Robinson et al., 2005; Rusli et al., 2004; Saedi, 2006; Sakiyama et al., 2004, 2006; Salis et al., 2006; Sanyal et al., 2006; Schiehlen and Seifried, 2004; Segall et al., 2006; Shehadeh et al., 2005; Shilkrot et al., 2004; Sinha and Goodson, 2005; Succi et al., 2001; Tewfik et al., 1993; Tidriri, 2001; Uva and Salerno, 2006; Varshney and Armaou, 2005; Voigt, 2005; Vvedensky, 2004; Voyiadjis et al., 2003; Wang, 2002; Weinan et al., 2003; Xiao and Yang, 2006; Xu and Wang, 2005; Xu and Subramaniam, 2006). Examples of the former are mathematical homogenization, mixture and hybrid mixture theory, spatial averaging, filtering techniques, moment methods, central limit or Martingale methods, Stochastic-convective approaches, projection operators, renormalization group techniques, variational approaches, space transformation methods, continuous time random walks, and Eulerian and Lagrangian perturbation schemes. The latter includes heterogeneous multiscale method, serial method, onion-type hybrid method, coarse-graining method, multigrid-type hybrid method, parallel approach, dynamic method, and concurrent method. Here we briefly outline three model-upscaling methods used in transport phenomena in porous media: mixture theory, local volume-averaging and hybrid mixture theory (Wang, 2000).

The microscale model for transport in porous media is well-known. It consists of field equations and constitutive equations. Field equations are the fundamental conservation equations which govern the motion of all materials and include the conservation of mass, or continuity equation; the balance of linear and angular momentum equations; the conservation of energy; and in charged or magnetic bodies, Maxwells equations of electrodynamics. When formulated appropriately, the second law of thermodynamics can also be considered a field equation as it can be written as a balance of entropy with the entropy generation being taken into account. Constitutive equations are specific for the material under considerations and the exact form of these equations is generally determined experimentally. Examples of constitutive equations include Fourier's law of heat conduction, Newton's law of viscosity, and Fick's law of mass diffusion (Wang, 1999). Historically, constitutive equations have been empirically derived, but when multiple processes involved for a complicated material, as is the case for many problems involving porous media, it is necessary to return to first principles (the principle of material frame indifference and the second law of thermodynamics in particular) to obtain them (Wang, 1999).

For transport in porous media, the macroscale (so-called Darcy Scale in the literature) is a phenomenological scale that is much larger than the microscale of pores and grains and much smaller than the system length scale. The interest in the macroscale rather than the microscale comes from the fact that a prediction at the microscale is complicated because of complex microscale geometry of porous media, and that we are usually more interested in large scales of transport for practical applications. Existence of such a macroscale description equivalent to the microscale behavior requires a good separation of length scale and has been well discussed in Auriault (1991).

To develop a macroscale model of transport in porous media, the mixture theory views fluid and porous media as a mixture of continuum deformable media, and introduces conservation equations directly at the macroscale with additional terms accounting for the interaction among phases. Required constitutive equations are supplied initially by direct postulation of desirable relations and later by the Coleman and Noll method to exploit consequences of entropy inequality (Coleman and Noll, 1963) or of constitutive-modeling constraints such as the material frame indifference, the principle of phase separation, immiscibility, the volume fraction constraint, and the viscous dissipation inequality (Rajagopal and Tao, 1995). Readers are referred to (Bedford and Drumheller, 1983; Bowen, 1982; Dobran, 1984; Marle, 1982; Prvost, 1980; Rajagopal and Tao, 1995; Sampaio and Williams, 1979; Thigpen and Berryman, 1985) for details of this approach. Therefore, the mixture theory is an *implicit* approach of model-upscaling, in which scale-dependent features are taken into account while developing model equations, so as to formulate the model according to requirement of its particular site.

Development by the local volume-averaging method starts, on the other hand, with a microscale description. Both conservation and constitutive equations are introduced at the microscale. Resulting microscale field equations are then averaged over a representative elementary volume (REV), the smallest differential volume resulting in statistically meaningful local average properties, to obtain macroscale field equations Whitaker (1999), Goyeau et al. (1997), Haro et al. (1996), Lasseux et al. (1996), Quintard and Whitaker (1993), Quintard and Whitaker (1994a), Quintard and Whitaker (1996), Whitaker (1986), and Whitaker (1996). Both weighted and unweighted averages have been used in the literature (Goyeau et al., 1997; Whitaker, 1986; Whitaker, 1996). Alternatively, the microscale field equations are first formed into two-scale expansions by double scale asymptotic method to render the procedure of averaging route, and then averaged over a cell (Auriault, 1980, 1991; Auriault et al., 1985; Mei and Auriault, 1989, 1991). Here the cell is either the period when the medium is periodic or the REV when it is random. In the process of averaging, *averaging theorems* are used to convert integrals of gradient, divergence, curl, and partial time derivatives of a function to some combination of gradient, divergence, curl, and partial time derivatives of integrals of the function and integrals over the boundary of the REV (Gray et al., 1993; Whitaker, 1969). Often in order to make averaging procedure tractable and to obtain desirable results some assumption are made before, during, and after averaging. These assumptions typically relate to the spatial and/or temporal distribution of properties, expected order of magnitude of various terms, and existence of certain relations among various properties, all based on intuitive and somewhat heuristic arguments.

The macroscale field equations obtained are not a closed system for determination of velocity and pressure because of unclosed terms reflecting the microscale effect. To form a closed system, the approach used for Reynolds-stress closures in turbulence is usually employed to develop governing differential equations and boundary conditions for spatial derivations of pressure and velocity, the difference between microscale and macroscale values (Goyeau et al., 1997; Haro et al., 1996; Quintard and Whitaker, 1994b; Wang, 1997, 2002, 2007; Whitaker, 1986, 1996). Resulting closure model is a set of differential equations defined on the microscale, which is difficult to solve due to complex microsacle geometry. The closure problem is usually solved over a unit cell for a spatially periodic model of a porous medium. This leads to a local closure problem in terms of closure variables and a method of predicting the permeability tensor. Physically, the closure issue for porous media flows is actually much simpler than the closures in turbulence. This enables the application of some simpler tools to calculate the closure relations for a number of upscaling problems in porous media (Goyeau et al., 1997; Haro et al., 1996; Quintard and Whitaker, 1994b; Whitaker, 1986, 1996).

The hybrid mixture theory is a combination of classic mixture theory and the local volume-averaging method. Conservation equations are introduced at the microscale, but without microscale constitutive relations. These microscale equations are then averaged over the REV. The required constitutive equations are introduced at macroscale initially by direct postulation of desirable relations and later by the Coleman and Noll method. The constitutive relations are usually linearized to yield a set of equations solvable upon experimental specification of material coefficients. Readers are referred to Marle (1982), Achanta and Cushman (1994), Achanta et al. (1994), Bennethum and Cushman (1996a, b), Drew (1971), Gray and Hassanizadeh (1989, 1991a, b), Hassanizadeh and Gray (1979a, b, 1980,

1990), Ishii (1975), Murad et al. (1995), Murad and Cushman (1996), and Nigmatulin (1979) for details of this approach. The advantage and disadvantage of these three approaches have also been well discussed in Hassanizadeh and Gray (1990) and Kaviany (1995).

To show the values of multiscale theorems explicitly, consider the volume-averaging method to obtain the macroscale balance equation for a porous medium consisting of a solid phase and a number of fluid phases. The phases are supposed to be immiscible and to have distinct thermodynamics properties. The phases are separated by very thin transition regions which are usually modeled as two-dimensional interfacial surfaces. Each interface has its own thermodynamics properties distinct from those of phases and other interface types. When a system consists of three or more phases, common lines may also exist. These are regions of transition where three interfaces come together. Common lines are one-dimensional regions with thermodynamic properties of their own. The space is occupied by phases, interfaces, and common lines which exist in mutually exclusive domains. This conceptualization of a multiphase system is referred to as the microscopic picture.

At each and every point within any particular phase, the general microscale conservation equation for the single fluid phase property ψ may be written as Cushman et al. (2002),

$$\frac{\partial(\rho\psi)}{\partial t} + \nabla \cdot (\rho \mathbf{v}\psi) - \nabla \cdot \mathbf{i} - \rho f = \rho G \tag{1}$$

where ψ is a phase property defined per unit mass of the phase at the microscale, ρ the density, t the time, $\mathbf{v}$ the velocity vector, f the external supply of ψ, and G the net production ψ of within the volume. For each of the respective field equations, the quantities given in Table 2 are used. Other conservation equations are needed for interfaces and common lines to describe thermodynamic processes at the microscopic scale. However, microscopic details in a complex system are often not needed and are almost impossible to model. Therefore, the microscopic picture needs to be replaced with an averaged description of process.

Averaged, or macroscale, properties are commonly defined by the integral of microscale properties over a representative element volume (REV). The concepts

Table 2 Quantities for Equation (1)

Quantity	ψ	$\mathbf{i}$	f	G
Mass	1	0	0	0
Linear momentum	v	$\mathbf{t}$	g	0
Angular momentum	$\mathbf{r} \times v$	$\mathbf{r} \times \mathbf{t}$	$\mathbf{r} \times g$	0
Energy	e+$1/2v^2$	$\mathbf{t} \cdot \mathbf{v}$+$\mathbf{q}$	$g \cdot \mathbf{v}$+h	0
Entropy	η	$\boldsymbol{\phi}$	b	Λ

Notes: The quantities are defined as: e, internal energy density (J/kg); g, external supply of momentum (gravity) (m/s^2); h, external supply of energy (J/(kg s)); $\mathbf{q}$, heat flux vector (J/(m^{2}s)); $\mathbf{r}$, microscale spatial moment arm (m); $\mathbf{t}$, stress tensor (N/m^2); $\mathbf{v}$, velocity (m/s); η, entropy (J/(kg K)); $\boldsymbol{\phi}$, entropy flux vector (J/(m^2 s K)); ρ, mass density (kg/m^3); b, external entropy source (J/(kg s K)); Λ, net entropy production (J/(kg s K)).

behind the REV and its properties are discussed by many authors (Auriault, 1991; Gray and Miller, 2005; Gray et al., 1993; Whitaker, 1999). Of particular importance are the features that the size and shape of the REV do not vary with space or time. Also its length scale l_{ma} must be much greater than the pore (microscale) scale l_{mi}, but much less than the scale l_{me} of the full system under study. If the REV is designated as V, the portion of this volume that is occupied by the α-phase is indicated as V_α. The union of interfacial regions within the REV between the α- and the β-phases is designated as $S_{\alpha\beta}$ and the unit vector normal to this surface oriented outward from the α-phase is designated as $\mathbf{n}^\alpha$. Within this framework, the macroscale-governing equation can be developed by averaging microscopic balance laws, such as Equation (1), over the REV.

To average Equation (1), two theorems are commonly used that transform the average of a derivative to the derivative of average. The first of these theorems is applied to a time derivative and may be stated as follows.

Time average theorem.

$$\int_{V_\alpha} \frac{\partial F}{\partial t} \mathrm{d}V = \frac{\partial}{\partial t} \int_{V_\alpha} F \mathrm{d}V - \underbrace{\sum}_{\beta \neq \alpha} \int_{S_{\alpha\beta}} \mathbf{n}^\alpha \cdot \mathbf{w}_{\text{b}} F|_\alpha \mathrm{d}S \tag{2}$$

The second theorem is applied to the divergence operator and is given as follows.

Divergence average theorem.

$$\int_{V_\alpha} \nabla \cdot \mathbf{B} \mathrm{d}V = \nabla \cdot \int_{V_\alpha} \mathbf{B} \mathrm{d}V + \underbrace{\sum}_{\beta \neq \alpha} \int_{S_{\alpha\beta}} \mathbf{n}^\alpha \cdot \mathbf{B}|_\alpha \mathrm{d}S \tag{3}$$

In Equations (2) and (3) $F|_\alpha$ and $\mathbf{B}|_\alpha$ indicate that the quantities F and $\mathbf{B}$ are microscale properties of the α-phase (i.e. properties defined using length scale d) that are being integrated over the α–β interface, $\sum_{\beta \neq \alpha}$ denotes a summation over all phases except the α-phase, and $\mathbf{w}_{\text{b}}$ is the boundary velocity of $S_{\alpha\beta}$.

Integration of Equation (1) over V_α and application of Equations (2) and (3) to the appropriate derivatives yields

$$\frac{\partial}{\partial t} \int_{V_\alpha} \rho\psi \mathrm{d}V + \nabla \cdot \int_{V_\alpha} [\rho \mathbf{v} \psi - \mathbf{i}] \mathrm{d}v + \underbrace{\sum}_{\beta \neq \alpha} \int_{S_{\alpha\beta}} \mathbf{n}^\alpha \cdot [\rho(\mathbf{v} - \mathbf{w})\psi - \mathbf{i}]|_\alpha \mathrm{d}S$$
$$- \int_{V_\alpha} \rho f \mathrm{d}V = \int_{V_\alpha} G \mathrm{d}V \tag{4}$$

This is the macroscale balance equation and may be written in terms of average quantities after dividing by the constant macroscale volume $V = V_\alpha / \varepsilon^\alpha$, where ε^α is the α-phase volume fraction:

$$\frac{\partial(\varepsilon^\alpha \langle \rho\psi \rangle^\alpha)}{\partial t} + \nabla \cdot (\varepsilon^\alpha \langle \rho \mathbf{v} \psi \rangle^\alpha) - \nabla \cdot (\varepsilon^\alpha \langle \mathbf{i} \rangle^\alpha)$$
$$+ \frac{\varepsilon^\alpha}{V_\alpha} \underbrace{\sum}_{\beta \neq \alpha} \int_{S_{\alpha\beta}} \mathbf{n}^\alpha \cdot [\rho(\mathbf{v} - \mathbf{w})\psi - \mathbf{i}]|_\alpha \mathrm{d}S - \varepsilon^\alpha \langle \rho f \rangle^\alpha = \varepsilon^\alpha \langle G \rangle^\alpha \tag{5}$$

where the notation to indicate an average over a volume has been introduced such that

$$\langle F\rangle^{\alpha} = \frac{1}{V_{\alpha}} \int_{V_{\alpha}} F \mathrm{d}V \tag{6}$$

It is also useful to introduce the mass weighted average given by

$$\bar{F}^{\alpha} = \frac{1}{\langle\rho\rangle^{\alpha} V_{\alpha}} \int_{V_{\alpha}} \rho F \mathrm{d}V = \frac{\langle\rho F\rangle^{\alpha}}{\langle\rho\rangle^{\alpha}} \tag{7}$$

and to define the deviation of a microscopic quantity from the macroscopic mean value as

$$\hat{F}^{\alpha} = F - \bar{F}^{\alpha} \tag{8}$$

where this relation applies only for a microscale point in the α-phase. Introduction of the notation in Equations (7) and (8) into Equation (5) yields the macroscale balance equation in the following form.

1.3.1.1 Macroscale balance equation in terms of averaged quantities

$$\begin{aligned}
&\frac{\partial(\varepsilon^{\alpha}\langle\rho\rangle^{\alpha}\bar{\psi}^{\alpha})}{\partial t} + \nabla\cdot(\varepsilon^{\alpha}\langle\rho\rangle^{\alpha}\bar{\mathbf{v}}^{\alpha}\bar{\psi}^{\alpha}) \\
&\quad - \nabla\cdot\left\{\varepsilon^{\alpha}\left[\langle\mathbf{i}\rangle^{\alpha} - \langle\rho\rangle^{\alpha}\left(\overline{\bar{\mathbf{v}}^{\alpha}\hat{\psi}}^{\alpha} + \overline{\hat{\mathbf{v}}^{\alpha}\bar{\psi}}^{\alpha} + \overline{\hat{\mathbf{v}}^{\alpha}\hat{\psi}}^{\alpha}\right)\right]\right\} \\
&\quad - \varepsilon^{\alpha}\langle\rho\rangle^{\alpha}\bar{f}^{\alpha} = \varepsilon^{\alpha}\langle G\rangle^{\alpha} + \underbrace{\sum}_{\beta\neq\alpha}(\hat{e}^{\alpha}_{\alpha\beta}\bar{\psi}^{\alpha} + \hat{I}^{\alpha}_{\alpha\beta})
\end{aligned} \tag{9}$$

where

$$\hat{e}^{\mathrm{i}}_{\alpha\beta} = \frac{1}{V}\int_{S_{\alpha\beta}} \mathbf{n}^{\mathrm{i}}\cdot[\rho(\mathbf{w}-\mathbf{v})]|_i \mathrm{d}S$$

and

$$\hat{I}^{\mathrm{i}}_{\alpha\beta} = \frac{1}{V}\int_{S_{\alpha\beta}} \mathbf{n}^{\mathrm{i}}\cdot\left[\mathbf{i} - \rho(\mathbf{v}-\mathbf{w})\hat{\psi}^{\mathrm{i}}\right]|_i \mathrm{d}S$$

The quantity $\hat{e}^{\mathrm{i}}_{\alpha\beta}$ accounts for mass exchange between the i-phase ($\mathrm{i} = \alpha, \beta$) and the α–β interface. The term $\hat{I}^{\mathrm{i}}_{\alpha\beta}$ deals with non-convective interaction of the i-phase ($\mathrm{i} = \alpha, \beta$) with the interface.

1.3.2 Model-downscaling

The basic set-up of model-downscaling is as follows. We are given a system whose large-scale behaviors and processes, with state/process variable or parameter V, are described by a given large-scale model. We are interested in the small-scale behavior of the system, described by the state/process variables or parameters v. v and V are linked together by compression and reconstruction operators Q and R with the property $QR = \mathbf{I}$ (the identity operator),

$$V = Qv, \qquad v = RV$$

Model-downscaling refers to finding the small-scale model for v based on the large-scale model for V. *Multiscale modeling* also arises when the small-scale model for v is either not explicitly available or invalid in some parts of the domain of interest. The basic strategy in multiscale modeling is to use the large-scale model as a supplement to provide the necessary information for extracting the small-scale behavior of the system, with the hope that the combined large-small scale modeling will be much more efficient.

Similar to model-upscaling, we can also have complete model-downscaling in all spatial dimensions and temporal dimension or partial model-downscaling in some spatial or temporal dimensions. The small-scale model in the latter is also a multiscale one.

Available techniques for model-downscaling can be found in Al-Ghoul et al. (2004), Attinger and Koumoutsakos (2004), Barth et al. (2002), Alarcon et al. (2004), Alber et al. (2006), Antonic et al. (2002), Cavallotti et al. (2005), Cushman (1997), Fredberg and Kamm (2006), Karsch et al. (1997), Mezzacappa (2005), Wang (2002), Anderson et al. (2003), Bierkens et al. (2000), Blöschl (2005), Bramble et al. (2003), Aanonsen and Eydinov (2006), Abildtrup et al. (2006), Anderson et al. (2004), Antic et al. (2006), Antic et al. (2004), Araujo et al. (2005), Artus et al. (2004), Auclair et al. (2006), Bacchi and Ranzi (2003), Badaroglu et al. (2006), Badas et al. (2006), Bardossy et al. (2005), Bergant and Kajfez-Bogataj (2005), Bruinink et al. (2006), Burger and Chen (2005), Busuioc et al. (2006), Cardells-Tormo and Arnabat-Benedicto (2006), Charlton et al. (2006), Chen et al. (2006), Corwin et al. (2006), Coulibaly et al. (2005), Coulibaly (2004), Deidda et al. (2006), Deidda et al. (2004), De Rooy and Kok (2004), Diaz-Nieto and Wilby (2005), Dibike and Coulibaly (2005, 2006), Diez et al. (2005), Eidsvik (2005), Enke et al. (2005a, b), Feddersen and Andersen (2005), Fischer et al. (2004), Fung and Siu (2006), Gaffin et al. (2004), Gangopadhyay et al. (2005), Gaslikova and Weisse (2006), Ghosh and Mujumdar (2006), Ghan and Shippert (2006), Ghan et al. (2006), Gibson et al. (2005), Gutierrez et al. (2005), Hanssen-Bauer et al. (2005), Harpham and Wilby (2005), Haylock et al. (2006), Hewitson and Crane (2006), Hong and Root (2006), Hong et al. (2006), Hoofman et al. (2005), Huth (2004, 2005), Jungen et al. (2006), Kaipio (2005), Kettle and Thompson (2004), Khan et al. (2006a, b), Knight et al. (2004), Kunstmann et al. (2004), Lambert et al. (2004), Liang et al. (2006), Ludwig et al. (2003), Mander et al. (2005), Matulla (2005), Meier (2006), Meier et al. (2006), Merlin et al. (2006), Mika et al. (2005), Miksovsky and Raidl (2005), Moriondo and Bindi (2006), Mounier et al. (2005), Muller et al. (2006), Mateos et al. (2005), Medjdoub et al. (2005), Mehrotra and Sharma (2006a–c, 2005), Mehrotra et al. (2004), Nuttinck (2006), Nuttinck et al. (2006), Oja et al. (2005a), Olsson et al. (2004), Paeth et al. (2005), Pasini et al. (2006), Pavan et al. (2005), Pettorelli et al. (2005), Pradhan et al. (2006), Pryor et al. (2005a, b), Raderschall (2004), Rebora et al. (2006a, b), Rengel et al. (2004), Riitters (2005), Salathe (2005), Schmidli et al. (2006), Shin et al. (2006), Sivapalan et al. (2003), Solecki and Oliveri (2004), Sun et al. (2004, 2006), Takeuchi (2004), Timbal (2004), Tononi (2005), Valenza et al. (2004), Veneziano et al. (2006), Verburg et al. (2006b), Wang et al. (2004), Wilby (2005), Wood et al. (2004), Woth et al. (2006), Xoplaki et al. (2004), Xu et al. (2005), Yang et al. (2005), Zagar et al. (2006), Zhang (2005),

Zhu et al. (2004a), Alkire (2003), Belashchenko and Antropov (2002a), Defranoux et al. (2005), Engquist and Runborg (2005), Grigoriev and Dargush (2003), Grohens et al. (2001), Guo and Tang (2006), Harari (2006), Hauke and Doweidar, (2005a, b, 2006), Katsoulakis et al. (2005), Koumoutsakos (2005), Kruzik and Prohl (2006), Larachi (2005), Lemaire et al. (2006), Maroudas and Gungor (2002), Maroudas et al. (2002), Nicot (2004), Ponziani et al. (2003), Pozdnyakova et al. (2005), Saedi (2006), Schiehlen and Seifried (2004), Shehadeh et al. (2005), Sinha and Goodson (2005), Teppola and Minkkinen (2000), Tewfik et al. (1993), Tidriri (2001), Uva and Salerno (2006), Voyiadjis et al. (2003), Fabry et al. (2001), Hayes et al. (2005), Kadowaki and Liu (2005), Lee et al. (2004), and Bramble et al. (2003). Here we develop Equation (1) from its large-scale counterpart by model-downscaling using a localization approach to illustrate applications of multiscale theorems (Gray and Hassanizadeh, 1998).

Consider a single-phase continuum occupying a non-material global volume V at a given time instant t. The boundary of V, denoted by S, may have a velocity $\mathbf{w}_b$. Consider a phase property ψ defined per unit mass of the phase at the microscale. The general conservation equation for the single fluid phase property, ψ, may be stated as (Batchelor, 1967; Eringen, 1980)

$$\frac{\mathrm{d}}{\mathrm{d}t}\int_V \rho\psi \mathrm{d}V + \int_S \mathbf{n}^* \cdot [\rho(\mathbf{v} - \mathbf{w}_b)\psi - \mathbf{i}]\mathrm{d}S - \int_V \rho f \mathrm{d}V = \int_V \rho G \mathrm{d}V \tag{10}$$

where ρ is the fluid density, $\mathbf{v}$ the fluid velocity, $\mathbf{i}$ the diffusive flux of ψ across the boundary, $\mathbf{n}^*$ the unit vector normal to S and pointing outward from V, f the external supply of ψ, and G the term accounting for production of ψ within the volume. This equation is a mathematical statement of the physical principle that the rate of change of some property in a volume is equal to the net flux of that property across the boundary of the volume plus the external supply plus production of the property.

To transform the global large-scale balance Equation (10) to a point form, two mathematical theorems are typically applied to the integrals in this equation, the transport theorem and divergence theorem, which are available in virtually all undergraduate mathematical and fluid text books. The former, written for some function F and global volume V, is as follows.

Transport theorem.

$$\int_V \frac{\partial F}{\partial t}\mathrm{d}V = \frac{\mathrm{d}}{\mathrm{d}t}\int_V F\mathrm{d}V - \int_S \mathbf{n}^* \cdot \mathbf{w}_b F \mathrm{d}S \tag{11}$$

The latter for a vector function B is as follows.

Divergence theorem.

$$\int_V \nabla \cdot \mathbf{B}\mathrm{d}V = \int_S \mathbf{n}^* \cdot \mathbf{B}\mathrm{d}S \tag{12}$$

Application of these theorems to Equation (10) brings the time derivative inside the integral and converts the surface integrals to volume integrals so that

$$\int_V \left[\frac{\partial(\rho\psi)}{\partial t} + \nabla \cdot (\rho\mathbf{v}\psi) - \nabla \cdot \mathbf{i} - \rho f\right]\mathrm{d}V = \int_V \rho G \mathrm{d}V \tag{13}$$

Because the size of the volume is arbitrary, by the localization theorem (Wang, 1997; Wang and Zhou, 2000), the integrands themselves in Equation (13) must be equal so that the point equation can be obtained as follows.

Point microscale balance equation

$$\frac{\partial(\rho\psi)}{\partial t} + \nabla \cdot (\rho \mathbf{v}\psi) - \nabla \cdot \mathbf{i} - \rho f = \rho G \tag{14}$$

which is the same as Equation (1).

1.4 Multiscale theorems

Multiscale theorems are powerful mathematical tools for facilitating scaling among various modeling scales. We start this section with a formal definition of fundamental concepts regarding physical systems and scales and end with an introduction of multiscale theorems.

1.4.1 Systems and scales

Figure 5 illustrates the physical components of a microscale system composed of a solid phase s, a wetting fluid phase w, and a non-wetting fluid phase n. The figure provides a telescoped view such that the microscale components are seen to be elements within the macroscale averaging volume, which in turn exists in an aquifer. The solid is connected and fills a portion of the domain Ω, and a connected pore space within the solid fills the remaining portion of Ω. The solid phase may be rigid or deform as a function of an applied stress.

In this example, the pore space contains the w and n fluid phases, each of which may be incompressible or compressible. Two-dimensional interfaces within Ω form the boundaries between pairs of immiscible phases and are denoted as wn, ws, and ns interfaces with the designation referring to the phases involved. One-dimensional common curves may be identified as the location where three different interfaces or three different phases meet, here denoted as the wns common curve. Zero-dimensional common points could exist where four common curves, six interfaces, or four phases meet. Since Figure 5 depicts a system with only three phases, no common points exist. Interfaces, common curves, and common points are idealizations that account for regions of transition in physicochemical characteristics between or among distinct phases. Transfer of mass, momentum, energy, and entropy between phases occurs at these locations. We refer to the collection of phase volumes, interfaces, common curves, and common points as entities.

We follow Miller and Gray (2005) in defining these physical components formally.

Definition 1 (Domain). The domain of interest is $\Omega \subset \Re^3$ with boundary Γ, and the external closure of the domain is $\bar{\Omega} = \Omega \cup \Gamma$. The extent of Ω has a measure of volume denoted as V.

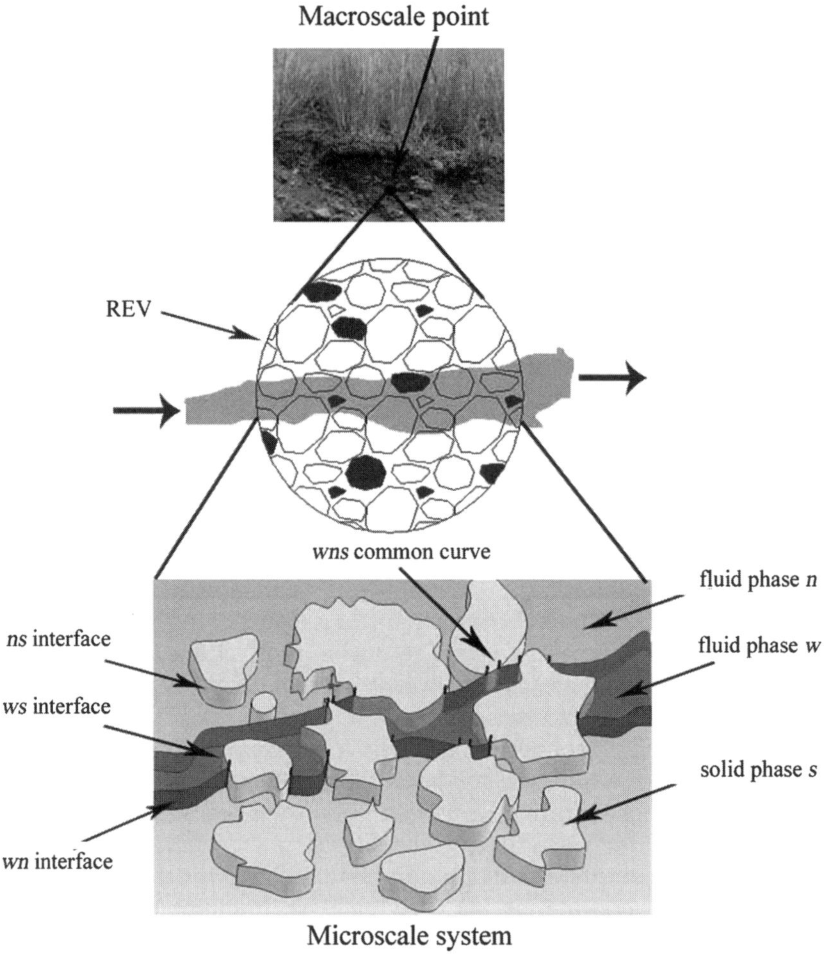

Figure 5 Three-phase microscale system (s, solid phase; w, wetting fluid phase; n, non-wetting fluid phase; after Gray, 1999) (see Plate 3 in Color Plate Section at the end of this book).

Definition 2 (Phase volume). Phase volumes are regions occupied by distinct material (either fluid or solid) denoted as $\Omega_\alpha \subset \Omega \subset \Re^3$ for a material α. The set of all types of phase volumes is denoted as $\mathcal{E}_P = \{\Omega_i | i \in \mathcal{I}_P\}$, and $\mathcal{I}_P$ is the index set of all types of phase volumes with individual entries consisting of an index corresponding to a phase volume of the form α and having n_P members. Here n_P is the number of distinct materials or phase volumes. The closure of Ω_α is denoted as $\bar{\Omega}_\alpha = \Omega_\alpha \cup \Gamma_{\alpha \mathrm{e}} \cup \Gamma_{\alpha \mathrm{i}}$, where the external boundary $\Gamma_{\alpha \mathrm{e}} = \bar{\Omega}_\alpha \cap \Gamma$, and the internal boundary $\Gamma_{\alpha \mathrm{i}} = \cup_{\beta \neq \alpha} \bar{\Omega}_\alpha \cap \bar{\Omega}_\beta$. The extent of Ω_α has a measure of volume denoted as V_α.

Definition 3 (Interface). Interfaces are regions within Ω formed by the intersection of two distinct phase volumes and denoted as $\Omega_{\alpha\beta} = \Omega_\alpha \cap \Omega_\beta \subset \Re^2$.

The set of all types of interfaces is denoted as $\mathcal{E}_I = \{\Omega_i | i \in \mathcal{I}_I\}$ and $\mathcal{I}_I$ is the index set of all types of interfaces with individual entries consisting of an index corresponding to a pair of phase volumes of the form $\alpha\beta$ and having $n_I \leq \binom{n_P}{2}$ members, where the order of the phase volume qualifiers comprising an interface qualifier is irrelevant. The closure of $\Omega_{\alpha\beta}$ is denoted as $\bar{\Omega}_{\alpha\beta} = \Omega_{\alpha\beta} \cup \Omega_{\alpha\beta e} \cup \Omega_{\alpha\beta i}$, where the external boundary $\Omega_{\alpha\beta e} = \bar{\Omega}_\alpha \cap \bar{\Omega}_\beta \cap \Gamma$, and the internal boundary $\Gamma_{\alpha\beta i} = \cup_{\varepsilon \neq \beta,\alpha} \bar{\Omega}_\alpha \cap \bar{\Omega}_\beta \cap \bar{\Omega}_\varepsilon$. The extent of $\Omega_{\alpha\beta}$ has a measure of area denoted as $S_{\alpha\beta}$.

Definition 4 (Common curve). Common curves are regions within Ω formed by the intersection of three distinct phase volumes and denoted as $\Omega_{\alpha\beta\varepsilon} = \Omega_\alpha \cap \Omega_\beta \cap \Omega_\varepsilon - \subset \Re^1$. The set of all types of common curves is denoted as $\mathcal{E}_C = \{\Omega_i | i \in \mathcal{I}_C\}$, and $\mathcal{I}_C$ is the index set of all types of common curves with individual entries consisting of an index corresponding to a group of three phases of the form $\alpha\beta\varepsilon$ and having $n_C \leq \binom{n_P}{3}$ members, where the order of the phase volume qualifiers comprising a common curve qualifier is irrelevant. The closure of $\Omega_{\alpha\beta\varepsilon}$ is denoted as $\bar{\Omega}_{\alpha\beta\varepsilon} = \Omega_{\alpha\beta\varepsilon} \cup \Omega_{\alpha\beta\varepsilon e} \cup \Omega_{\alpha\beta\varepsilon i}$, where the external boundary $\Omega_{\alpha\beta\varepsilon e} = \bar{\Omega}_\alpha \cap \bar{\Omega}_\beta \cap \bar{\Omega}_\varepsilon \cap \Gamma$, and the internal boundary $\Gamma_{\alpha\beta\varepsilon i} = \cup_{\delta \neq \beta,\alpha,\varepsilon} \bar{\Omega}_\alpha \cap \bar{\Omega}_\beta \cap \bar{\Omega}_\varepsilon \cap \bar{\Omega}_\delta$. The extent of $\Omega_{\alpha\beta\varepsilon}$ has a measure of length denoted as $C_{\alpha\beta\varepsilon}$.

Definition 5 (Common points). Common points are regions within Ω formed by the intersection of four distinct phase volumes and denoted as $\Omega_{\alpha\beta\varepsilon\delta} = \Omega_\alpha \cap \Omega_\beta \cap \Omega_\varepsilon \cap \Omega_\delta \subset \Re^0$. The set of all types of common points is denoted as $\mathcal{E}_{Pt} = \{\Omega_i | i \in \mathcal{I}_{Pt}\}$, and $\mathcal{I}_{Pt}$ is the index set of all types of common points with individual entries consisting of an index corresponding to a group of four phases of the form $\alpha\beta\varepsilon\delta$ and having $n_{Pt} \leq \binom{n_P}{4}$ members, where the order of the phase volume qualifiers comprising a common point qualifier is irrelevant. The extent of $\Omega_{\alpha\beta\varepsilon\delta}$ has a measure of number denoted as $N_{\alpha\beta\varepsilon\delta}$.

Definition 6 (Entities). The set of all types of entities is denoted as $\mathcal{E} = \mathcal{E}_P \cup \mathcal{E}_I \cup \mathcal{E}_C \cup \mathcal{E}_{Pt}$, and the index set of all types of entities is $\mathcal{I} = \mathcal{I}_P \cup \mathcal{I}_I \cup \mathcal{I}_C \cup \mathcal{I}_{Pt}$, giving $\mathcal{E} = \{\mathcal{E}_i | i \in \mathcal{I}\} = \{\Omega_i | i \in \mathcal{I}\}$.

The multiscale theorems in Sections 4 and 5 are concerned primary with three distinct length scales: the microscale l_{mi}, macroscale, l_{ma}, and the megascale l_{me}, all of which are much longer than the molecular scale l_{mo}, characterized by the mean free path between molecular collisions. These scales have the following formal definitions.

Definition 7 (Molecular scale). The molecular length scale l_{mo} is defined as the length scale for molecular collisions for a phase in a system of concern.

Definition 8 (Microscale). The microscale l_{mi} is the smallest length scale at which laws of continuum mechanics can be developed with $\|[\mathcal{P}_i(l_{mi} + \delta l_{mi}) - \mathcal{P}_i(l_{mi})]\| \leq \varepsilon_{mi}, \forall i$, where $\mathcal{P}_i(l)$ is a microscale property estimated by a well-defined average over length scale l, δl_{mi} a change in the length scale, and ε_{mi} a specified precision of the estimate of $\mathcal{P}_i$.

Definition 9 (Macroscale). The macroscale, l_{ma}, is the length scale at which the set of averaged properties of concern for the system can be rigorously defined and $\|[\mathcal{P}_i(l_{ma} + \delta l_{ma}) - \mathcal{P}_i(l_{ma})]\| \leq \varepsilon_{ma}, \forall i$, where $\mathcal{P}_i(l)$ is a macroscale property estimated by a well-defined average over length scale l, δl_{ma} a change in the length scale, and ε_{ma} a specified precision of the estimate of $\mathcal{P}_i$.

Definition 10 (Megascale). The megascale l_{me} is the length scale corresponding to the domain of interest Ω.

We also adopt the following axiom regarding hierarchical spatial scales.

Axiom 1 (hierarchical spatial scales). A clear hierarchy of separate length scales exists and is of the form $l_{mo} \ll l_{mi} \ll l_{ma} \ll l_{me}$ where the four scales are, respectively, the molecular scale, the microscale, the macroscale, and the megascale.

Although a clear discrete set of separated length scales has been stipulated, we note that most natural systems consist of a hierarchy of many different length scales that may not have a clear separation (Cushman et al., 2002). Whereas such systems occur routinely and are important, these systems are outside the scope of our current focus. However, we believe that the approach used in developing the multiscale theorems can be employed for the study of such systems.

It is interesting to note that length scales relate to the physical properties of the system. Selection of various length scales can result in the system being modeled as heterogeneous, for example, with flow in pores within a solid or in a gas as homogeneous. Here the system is being viewed with all entities existing at a point but with different densities. The boundaries between entities must be accounted for. Conversely, the time domain is continuous without heterogeneities; there are no boundaries in the time domain. Therefore, averaging over time as well as space does not alter the form of the equations in comparison to averaging only over space, although it does alter the meaning of the terms that appear in the equation. For this reason, time averaging is not explicitly applied in developing multiscale theorems.

1.4.2 Integration and averaging theorems

Modeling at various scales, multiscale modeling and model-scaling of multiscale systems and processes can be significantly facilitated by a set of mathematical theorems that support change of scale operations. Typical examples of such theorems are the time averaging theorem (Equations (2) and (11)) and the divergence averaging theorem (Equations (3) and (12)) used in the last section. The fundamental function of such theorems is to transform the integrals of space and time derivatives to derivatives of integral quantities. Other functions include interconversion of integrals over curves, surfaces, and volumes (integration dimension ascent and descent) and interchange of the order of differentiation and integration with a variable integration domain. Their applications not only facilitate mathematical manipulation in the modeling and scaling of multiscale systems and processes but also offer valuable insight into the physics of a problem.

We term these theorems as multiscale theorems. Note that time is continuous and the entities each occupy all of time within an increment studied. Integration and averaging over a small region of time only alters the interpretation of large-scale quantities but not the form of the theorems. Therefore, we can focus exclusively on spatial integration and averaging in developing these theorems. We also limit our discussion to those involving three spatial scales: microscale, macroscale, and megascale. This comes from the facts that most applications only involve these three scales and that it is straightforward to extend to the case involving more spatial scales.

The multiscale theorems can be grouped into *integration theorems* and *averaging theorems*. The former refers to those which change any or all spatial scales of a differential operator from the microscale (or any continuum scale) to the megascale by integration. The integration domain is allowed to translate and deform with time, and can be a volume, a surface (a plane as its special case) and a curve (a straight line as its special case). The latter is those which change any or all spatial scales of a differential operator from the microscale to the macroscale by averaging. An averaging volume, also called a REV, may be located in space; integration is performed over volumes, surfaces, or curves contained within the averaging volume. Because the REV is located at every macroscale position in space, the averages obtained have a function dependence on macroscale coordinates. If the REV is megascale in one or two spatial directions, the averaging also changes the scale of the differential operator to the megascale along those directions. Therefore, the averaging theorems can generally involve all three scales while the macroscale does not appear in the integration theorems. The microscale differential operator in the integrand for both types of multiscale theorems can be gradient, divergence, curl, and the time derivative. Each of these four operators can in turn be either spatial, surficial, or curvilinear.

The REV is one of the most useful conceptualizations that has permeated the development of a macroscale view of multiscale systems with a complex microscale structure such as porous medium systems. This concept is very similar to the microscale approach whereby a system is viewed as a continuum with each *point* deriving its properties from the large collection of molecules associated with that point. In fact, the point is a very small region in time and space. The number of molecules within the region is considered to be so large that variations in properties such as density are negligible for small changes in the size and duration of the measurement of the sampling device. In essence, the objective of making this definition is to identify measurement regions such that the value of a quantity measured can be meaningfully reported without having to stipulate the size of the sample examined.

The REV as applied to a macroscopic formulation makes use of a representative region, which is large enough to include all phases present. The region is assumed to be of sufficient size that the values of the averages that characterize a phase are independent of that size. Further, the size of the REV is considered to have a characteristic length scale that is much smaller than the system length scale such that the gradients of macroscale quantities within the system are meaningful. However, since this REV includes multiple phases, interfaces, common curves, and common points, geometric densities must also be defined. These should stipulate, for example, the fraction of the REV occupied by a particular phase or the amount of interfacial area between two phases that exists within the volume. Practically, the transition from a microscale to a macroscale point eliminates distributions of function values within a phase in favor of average values of those distributions over a portion of that phase. For example, the velocity distribution of a phase within the pores is replaced in favor of an average volumetric flow over a cross-sectional area. Certainly, this macroscale velocity is conceptually easier to compute, especially in light of the fact that the actual distribution of the pore space is typically unknown. However, in practice this simplicity is countered by the need to deal with the geometric densities and the need to obtain closure relations in terms of the macroscale variables.

The REV approach is sometimes criticized because it is invoked without proving the existence of the REV for the system under study and without determining that the REV employed is appropriate for all quantities being studied. In some instances when a porous medium exhibits fractal behavior, or when larger scale heterogeneities preclude the identification of an REV, approaches that are dependent upon a set of discrete and separable length scales fail (Cushman et al., 2002). Thus the common assumption of the existence of discrete and separable length scales is a key premise which affects the applicability of models developed. However, even in case of when an REV does not exist, an explicit definition of macroscale variables in terms of microscale quantities can be retained, allowing for the development of larger scale variables whose dependence on the size of the averaging volume used can be investigated. We must be aware of these limitations when applying the multiscale theorems, although we can ignore them when developing them.

Prototypical classical examples of multiscale theorems are the Gauss divergence theorem, the Reynolds transport theorem, and the Leibnitz rule. While there is extremely high demand for them, the development of new theorems has been very limited. Only a few additional theorems such as surface transport theorems have been added to the list over the years (Aris, 1962; Stone, 1990). The development of the volume-averaging method starting in the 1960s has furnished a few more, a spatial-averaging theorem (a three-dimensional analog to the Leibnitz rule) and particularly a less restrictive transport theorem (Anderson and Jackson, 1967; Bachmat, 1972; Gray and Lee, 1977; Slattery, 1967b; Whitaker, 1967). The reason behind this dissatisfactory status is the lack of a proper systematic approach for deriving such theorems. The classical approach is based on geometric arguments. For example, the spatial divergence theorem is traditionally derived by identifying a volume, integrating each term of the divergence of a vector over that volume, and then converting the volume integral to a surface integral using the projection of differential areas. Extension of this approach to more complex geometries encountered in multiphase systems is difficult and has been inhibited by the arduous task of keeping track of deforming regions and interfaces between phases.

A novel approach has been developed in Gray and Hassanizadeh (1989), Hassanizadeh and Gray (1979a), Hassanizadeh and Gray (1990), and Gray and Lee (1977). It uses indicator functions (generalized functions) to identify a region of interest by taking on the value of one in the interior and zero in the exterior of the region. By using the indicator functions, line, surface and volume integrals can be transformed to integrals over all space with these functions and/or their derivatives appearing in the integrand. The spatial and temporal dependence of limits of integration is thus moved into the integrand. All subsequent manipulations involving the integrand necessitate minimal concern for the limits of integration. Also the differential and integral properties of indicator functions provide a simplified route to changes in scale. The indicator functions are mathematical catalysts: they facilitate the derivations but do not appear in the end product. As in chemistry wherein a reaction would not take place without the presence of a catalyst, the derivations here rely on the catalytic activity of the indicator functions.

Fifty-six integration theorems and 72 averaging theorems have been listed in Gray et al. (1993) with no proof available for most of them. Whereas they form an extremely powerful tool for changing scale in the analysis of physical systems, they have not received as much attention and applications as they should. This is mainly due to the lack of proofs for them. Without detailed proofs, users have little structured access to the physics, application conditions and the ways of improving the theorems. In an attempt to improve this unsatisfying situation, we have developed 252 multiscale theorems including some averaging theorems applicable to variable REV. To keep the volume length within a reasonable level, we present here only 71 of them (38 integration theorems and 33 averaging theorems) with the proof details. As with all mathematical theorems, the proof of each theorem itself offers significant insight into the physics of a problem and thus holds considerable value. These 71 theorems are carefully selected to cover all 128 theorems in Gray et al. (1993) and are likely to be adequate for many applications. The other 181 multiscale theorems can be developed either by following the same approach as that used in developing the 71 theorems or by combining some of them and making use of some standard relations from calculus.

2. DERIVATIVES

Multiscale theorems transform derivatives from one scale to another. The derivatives can be spatial or temporal. The former appears normally in the form of gradient, divergence, or curl. Each of these three spatial operators can be spatial, surficial, or curvilineal. The latter is the partial derivative with respect to time of a function defined in space, on a surface or along a curve. The objective of this section is twofold: expressing these derivatives by three orthonormal vectors and developing relations among various derivatives.

2.1 Orthonormal vectors

To describe arbitrary volumes, surfaces and curves in space, we must define a set of orthogonal unit vectors. Such a set of vectors also forms the basis of expressing vector functions and various derivatives. We first define three orthonormal vectors and develop their fundamental properties in this section. We then use them to express vector functions.

2.1.1 Orthonormal vectors and their properties

Consider a right-handed triple of orthogonal unit vectors, $\boldsymbol{\lambda}$, $\mathbf{n}$, and $\boldsymbol{\nu}$ in Figure 6. Here, $\boldsymbol{\lambda}$ indicates the direction along a curve. The normal direction to a surface or volume is usually given by $\mathbf{n}$, and $\boldsymbol{\nu}$ can be used to represent a coordinate direction within a surface. We limit ourselves to orientable surfaces and simple curves only in this chapter (Gray et al., 1993). Since they are orthogonal and have unit magnitude, we have the following group of properties:

$$\boldsymbol{\lambda}\cdot\boldsymbol{\lambda} = \mathbf{n}\cdot\mathbf{n} = \boldsymbol{\nu}\cdot\boldsymbol{\nu} = 1 \tag{15}$$

$$\boldsymbol{\lambda}\cdot\mathbf{n} = \boldsymbol{\lambda}\cdot\boldsymbol{\nu} = \mathbf{n}\cdot\boldsymbol{\nu} = 0 \tag{16}$$

$$\boldsymbol{\lambda}\times\mathbf{n} = \boldsymbol{\nu} \tag{17}$$

$$\mathbf{n}\times\boldsymbol{\nu} = \boldsymbol{\lambda} \tag{18}$$

$$\boldsymbol{\nu}\times\boldsymbol{\lambda} = \mathbf{n} \tag{19}$$

$$\boldsymbol{\lambda}\times\boldsymbol{\lambda} = \mathbf{n}\times\mathbf{n} = \boldsymbol{\nu}\times\boldsymbol{\nu} = 0 \tag{20}$$

Their second group of properties involve the spatial del operator and a temporal derivative and read

$$\nabla\boldsymbol{\lambda}\cdot\boldsymbol{\lambda} = \nabla\mathbf{n}\cdot\mathbf{n} = \nabla\boldsymbol{\nu}\cdot\boldsymbol{\nu} = 0 \tag{21}$$

$$\frac{\partial\boldsymbol{\lambda}}{\partial t}\cdot\boldsymbol{\lambda} = \frac{\partial\mathbf{n}}{\partial t}\cdot\mathbf{n} = \frac{\partial\boldsymbol{\nu}}{\partial t}\cdot\boldsymbol{\nu} = 0 \tag{22}$$

$$\nabla\mathbf{n}\cdot\boldsymbol{\nu} = -\nabla\boldsymbol{\nu}\cdot\mathbf{n} \tag{23}$$

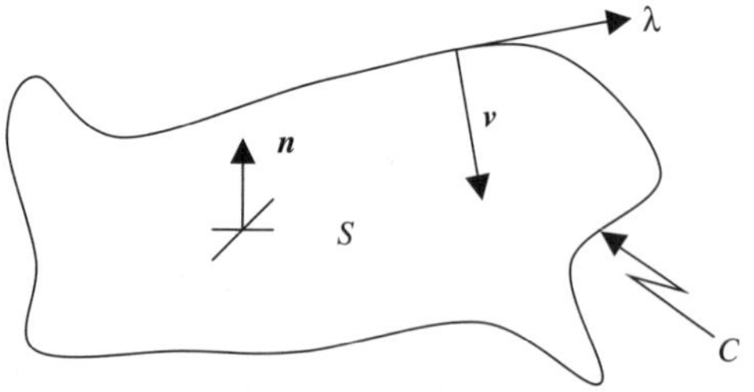

Figure 6 Orthonormal set of vectors $\boldsymbol{\lambda}$, $\mathbf{n}$, and $\boldsymbol{\nu}$ ($\boldsymbol{\lambda}$ is tangent to C; $\mathbf{n}$ normal to S; $\boldsymbol{\nu}$ not only tangent to S but also normal to C).

$$\nabla \boldsymbol{\nu} \cdot \boldsymbol{\lambda} = -\nabla \boldsymbol{\lambda} \cdot \boldsymbol{\nu} \tag{24}$$

$$\nabla \boldsymbol{\lambda} \cdot \mathbf{n} = -\nabla \mathbf{n} \cdot \boldsymbol{\lambda} \tag{25}$$

A proof of these properties is straightforward. We demonstrate this by proving (21) as an example.

By Equation (15),

$$\boldsymbol{\lambda} \cdot \boldsymbol{\lambda} = 1$$

A gradient operation on it leads to

$$\nabla \boldsymbol{\lambda} \cdot \boldsymbol{\lambda} = 0$$

where the chain rule has been used. Similarly, we have

$$\nabla \mathbf{n} \cdot \mathbf{n} = \nabla \boldsymbol{\nu} \cdot \boldsymbol{\nu} = 0$$

The third group of properties involve curl operation and are:

$$\nabla \times \boldsymbol{\lambda} = (\boldsymbol{\lambda}\boldsymbol{\lambda} \cdot \nabla) \times \boldsymbol{\lambda} = -\boldsymbol{\lambda} \cdot \nabla \boldsymbol{\lambda} \times \boldsymbol{\lambda} \tag{26}$$

$$\nabla \times \mathbf{n} = (\mathbf{nn} \cdot \nabla) \times \mathbf{n} = -\mathbf{n} \cdot \nabla \mathbf{n} \times \mathbf{n} \tag{27}$$

$$\nabla \times \boldsymbol{\nu} = (\boldsymbol{\nu}\boldsymbol{\nu} \cdot \nabla) \times \boldsymbol{\nu} = -\boldsymbol{\nu} \cdot \nabla \boldsymbol{\nu} \times \boldsymbol{\nu} \tag{28}$$

Proof. Since $\nabla = \boldsymbol{\lambda}\boldsymbol{\lambda} \cdot \nabla + \boldsymbol{\nu}\boldsymbol{\nu} \cdot \nabla + \mathbf{nn} \cdot \nabla$ (see next section), we have

$$\nabla \times \boldsymbol{\lambda} = (\boldsymbol{\lambda}\boldsymbol{\lambda} \cdot \nabla) \times \boldsymbol{\lambda} + (\boldsymbol{\nu}\boldsymbol{\nu} \cdot \nabla) \times \boldsymbol{\lambda} + (\mathbf{nn} \cdot \nabla) \times \boldsymbol{\lambda} \tag{29}$$

From Equation (17), it follows that

$$(\boldsymbol{\nu}\boldsymbol{\nu} \cdot \nabla) \times \boldsymbol{\lambda} = \boldsymbol{\nu} \times (\boldsymbol{\nu} \cdot \nabla \boldsymbol{\lambda}) = \boldsymbol{\lambda} \times \mathbf{n} \times (\boldsymbol{\nu} \cdot \nabla \boldsymbol{\lambda}) \tag{30}$$

Note that, for any three vectors **a**, **b**, and **c**,

$$\mathbf{a} \times (\mathbf{b} \times \mathbf{c}) = (\mathbf{a} \cdot \mathbf{c})\mathbf{b} - (\mathbf{a} \cdot \mathbf{b})\mathbf{c}$$

Equation (30) becomes

$$(\boldsymbol{\nu}\boldsymbol{\nu} \cdot \nabla) \times \boldsymbol{\lambda} = \boldsymbol{\lambda} \cdot (\boldsymbol{\nu} \cdot \nabla \boldsymbol{\lambda})\mathbf{n} - [\mathbf{n} \cdot (\boldsymbol{\nu} \cdot \nabla \boldsymbol{\lambda})]\boldsymbol{\lambda}$$

Therefore, we have, by applying Equation (21),

$$(\boldsymbol{\nu}\boldsymbol{\nu} \cdot \nabla) \times \boldsymbol{\lambda} = -(\mathbf{n} \cdot (\boldsymbol{\nu} \cdot \nabla \boldsymbol{\lambda}))\boldsymbol{\lambda} \tag{31}$$

Similarly, we have

$$(\mathbf{nn} \cdot \nabla) \times \boldsymbol{\lambda} = (\boldsymbol{\nu} \cdot (\mathbf{n} \cdot \nabla \boldsymbol{\lambda}))\boldsymbol{\lambda} \tag{32}$$

Substituting Equations (31) and (32) into Equation (29) yields

$$\nabla \times \boldsymbol{\lambda} = (\boldsymbol{\lambda}\boldsymbol{\lambda} \cdot \nabla) \times \boldsymbol{\lambda}$$

Also,

$$(\boldsymbol{\lambda}\boldsymbol{\lambda} \cdot \nabla) \times \boldsymbol{\lambda} = \boldsymbol{\lambda} \times (\boldsymbol{\lambda} \cdot \nabla \boldsymbol{\lambda}) = -\boldsymbol{\lambda} \cdot \nabla \boldsymbol{\lambda} \times \boldsymbol{\lambda}$$

Equation (26) thus holds. By following a similar approach, we can also prove Equations (27) and (28).

In terms of the orthonormal vectors, the identity (unit) tensor reads,

$$\boldsymbol{\lambda\lambda} + \boldsymbol{\nu\nu} + \mathbf{nn} = \mathbf{I} \tag{33}$$

2.1.2 Vector functions

By the orthonormality of $\boldsymbol{\lambda}$, $\mathbf{n}$, and $\boldsymbol{\nu}$, we can write any three-dimensional vector function $\mathbf{f}$ in Euclidean space in its component form as:

$$\mathbf{f} = \boldsymbol{\lambda\lambda}\cdot\mathbf{f} + \mathbf{nn}\cdot\mathbf{f} + \boldsymbol{\nu\nu}\cdot\mathbf{f} \tag{34}$$

The surface vector function,

$$\mathbf{f}_{\mathbf{n}}^{\mathrm{S}} = \mathbf{f} - \mathbf{nn}\cdot\mathbf{f} \tag{35}$$

is the vector components of $\mathbf{f}$ on a surface S with the unit normal $\mathbf{n}$. Similarly, the vector component of $\mathbf{f}$ along a curve C with tangent vector $\boldsymbol{\lambda}$ is defined by:

$$\mathbf{f}_{\lambda}^{\mathrm{C}} = \boldsymbol{\lambda\lambda}\cdot\mathbf{f} \tag{36}$$

We also have

$$\mathbf{f} = \mathbf{f}_{\mathbf{n}}^{\mathrm{S}} + \mathbf{f}_{\mathbf{n}}^{\mathrm{C}} \tag{37}$$

2.2 Gradient

In this section, we first express spatial, surface, and curvilineal gradients in terms of their components. We then develop a relation among these three types of gradients.For a scalar function f, consider its gradient

$$\mathbf{h} = \nabla f \tag{38}$$

By Equation (34), it can be written as,

$$\mathbf{h} = \boldsymbol{\lambda\lambda}\cdot\mathbf{h} + \mathbf{nn}\cdot\mathbf{h} + \boldsymbol{\nu\nu}\cdot\mathbf{h} \tag{39}$$

Applying Equation (38) in Equation (39) yields

$$\begin{aligned}\mathbf{h} &= \boldsymbol{\lambda\lambda}\cdot\nabla f + \mathbf{nn}\cdot\nabla f + \boldsymbol{\nu\nu}\cdot\nabla f\\ &= (\boldsymbol{\lambda\lambda}\cdot\nabla + \mathbf{nn}\cdot\nabla + \boldsymbol{\nu\nu}\cdot\nabla)f\end{aligned} \tag{40}$$

By comparing Equation (40) with Equation (38),

$$\nabla = \boldsymbol{\lambda\lambda}\cdot\nabla + \mathbf{nn}\cdot\nabla + \boldsymbol{\nu\nu}\cdot\nabla \tag{41}$$

Similar to the definition of $\mathbf{f}^{\mathrm{s}}$ and $\mathbf{f}^{\mathrm{c}}$, we define $\nabla_{\mathbf{n}}^{\mathrm{S}}$, the surficial gradient operator on a surface S with the unit normal vector $\mathbf{n}$, and $\nabla_{\lambda}^{\mathrm{c}}$, the curvilineal gradient operator along a curve C with tangent vector $\boldsymbol{\lambda}$, as

$$\nabla_{\mathbf{n}}^{\mathrm{S}} = \nabla - \mathbf{nn}\cdot\nabla \tag{42}$$

$$\nabla_{\lambda}^{\mathrm{C}} = \boldsymbol{\lambda\lambda}\cdot\nabla \tag{43}$$

Therefore, we have the spatial gradient and the surficial and curvilineal gradients of a scalar function f as

$$\nabla f = \boldsymbol{\lambda\lambda}\cdot\nabla f + \mathbf{nn}\cdot\nabla f + \boldsymbol{\nu\nu}\cdot\nabla f \tag{44}$$

Table 3 Gradient geometric terms in standard coordinate systems

Coordinate system	λ	**n**	ν	$\lambda \cdot \nabla_{\lambda}^{c} f$	$\mathbf{n} \cdot \nabla_{\mathbf{n}}^{c} f$	$\nu \cdot \nabla_{\nu}^{c} f$
Cartesian	$\mathbf{i}$	$\mathbf{j}$	$\mathbf{k}$	$\partial f/\partial x$	$\partial f/\partial y$	$\partial f/\partial z$
Cylindrical	$\mathbf{e}_r$	$\mathbf{e}_\theta$	$\mathbf{e}_z$	$\partial f/\partial r$	$\dfrac{1}{r}\dfrac{\partial f}{\partial \theta}$	$\partial f/\partial z$
Spherical	$\mathbf{e}_r$	$\mathbf{e}_\theta$	$\mathbf{e}_\phi$	$\partial f/\partial r$	$\dfrac{1}{r}\dfrac{\partial f}{\partial \theta}$	$\dfrac{1}{r\sin\theta}\dfrac{\partial f}{\partial \phi}$

$$\nabla_{\mathbf{n}}^{S} f = \lambda\lambda \cdot \nabla f + \nu\nu \cdot \nabla f \tag{45}$$

$$\nabla_{\lambda}^{c} f = \lambda\lambda \cdot \nabla f \tag{46}$$

The relation among them is

$$\nabla f = \nabla^{S} f + \nabla^{c} f \tag{47}$$

Table 3 lists the particular forms of evaluating the gradient in three common systems: Cartesian, cylindrical, and spherical coordinates.

2.3 Divergence

We may encounter three types of divergences in developing multiscale theorems: spatial divergence of a spatial vector function, surface divergence of a surface vector function and curvilineal divergence of a curvilineal vector function. We first express these divergences in vector operator notation and then develop various relations among them. The readers are referred to Table 4 for the details of various terms in divergences for Cartesian, cylindrical and spherical systems.

2.3.1 Spatial divergence

By Equations (34) and (41), the divergence of a spatial vector function **f** reads

$$\begin{aligned} \nabla \cdot \mathbf{f} &= (\lambda\lambda \cdot \nabla + \mathbf{nn} \cdot \nabla + \nu\nu \cdot \nabla) \cdot (\lambda\lambda \cdot \mathbf{f} + \mathbf{nn} \cdot \mathbf{f} + \nu\nu \cdot \mathbf{f}) \\ &= (\lambda\lambda \cdot \nabla) \cdot (\lambda\lambda \cdot \mathbf{f}) + (\lambda\lambda \cdot \nabla) \cdot (\mathbf{nn} \cdot \mathbf{f}) + (\lambda\lambda \cdot \nabla) \cdot (\nu\nu \cdot \mathbf{f}) \\ &\quad + (nn \cdot \nabla) \cdot (\lambda\lambda \cdot \mathbf{f}) + (nn \cdot \nabla) \cdot (\mathbf{nn} \cdot \mathbf{f}) + (nn \cdot \nabla) \cdot (\nu\nu \cdot \mathbf{f}) \\ &\quad + (\nu\nu \cdot \nabla) \cdot (\lambda\lambda \cdot \mathbf{f}) + (\nu\nu \cdot \nabla) \cdot (\mathbf{nn} \cdot \mathbf{f}) + (\nu\nu \cdot \nabla) \cdot (\nu\nu \cdot \mathbf{f}) \end{aligned} \tag{48}$$

Applying the chain rule yields

$$\begin{aligned} \nabla \cdot \mathbf{f} &= \lambda \cdot (\lambda \cdot \nabla\lambda)\lambda \cdot \mathbf{f} + \lambda \cdot (\lambda(\lambda \cdot \nabla(\lambda \cdot \mathbf{f}))) + \lambda \cdot (\lambda \cdot \nabla nn \cdot \mathbf{f}) \\ &\quad + \lambda \cdot (n(\lambda \cdot \nabla(n \cdot \mathbf{f}))) + \lambda \cdot (\lambda \cdot \nabla\nu)\nu \cdot \mathbf{f} + \lambda \cdot (\nu\lambda \cdot \nabla(\nu \cdot \mathbf{f})) \\ &\quad + n \cdot (n \cdot \nabla\lambda)\lambda \cdot \mathbf{f} + n \cdot (\lambda(\mathbf{n} \cdot \nabla(\lambda \cdot \mathbf{f}))) + n \cdot (n \cdot \nabla n)n \cdot \mathbf{f} \\ &\quad + n \cdot (nn \cdot \nabla(n \cdot \mathbf{f})) + n \cdot (n \cdot \nabla\nu)\nu \cdot \mathbf{f} + n \cdot (\nu n \cdot \nabla(\nu \cdot \mathbf{f})) \\ &\quad + \nu \cdot (\nu \cdot \nabla\lambda)\lambda \cdot \mathbf{f} + \nu \cdot (\lambda\nu \cdot \nabla(\lambda \cdot \mathbf{f})) + \nu \cdot (\nu \cdot \nabla n)n \cdot \mathbf{f} \\ &\quad + \nu \cdot (n\nu \cdot \nabla(n \cdot \mathbf{f})) + \nu \cdot (\nu \cdot \nabla\nu)\nu \cdot \mathbf{f} + \nu \cdot (\nu\nu \cdot \nabla(\nu \cdot \mathbf{f})) \end{aligned} \tag{49}$$

Table 4 Divergence geometric terms in standard coordinate systems

Coordinate System	$\boldsymbol{\lambda}$	$\mathbf{n}$	$\boldsymbol{\nu}$	$\nabla^c_{\lambda} \cdot \mathbf{f}^c_{\lambda}$	$\nabla^c_{\mathrm{n}} \cdot \mathbf{f}^c_{\mathrm{n}}$	$\nabla^c_{\nu} \cdot \mathbf{f}^c_{\nu}$	$\nabla \cdot \boldsymbol{\lambda}$	$\nabla \cdot \mathbf{n}$	$\nabla \cdot \boldsymbol{\nu}$	$\boldsymbol{\lambda} \cdot \nabla\boldsymbol{\lambda}$	$\mathbf{n} \cdot \nabla\mathbf{n}$	$\boldsymbol{\nu} \cdot \nabla\boldsymbol{\nu}$
Cartesian	$\mathbf{i}$	$\mathbf{j}$	$\mathbf{k}$	$\partial f_x/\partial x$	$\partial f_y/\partial y$	$\partial f_z/\partial z$	0	0	0	0	0	0
Cylindrical	$\mathbf{e}_r$	$\mathbf{e}_\theta$	$\mathbf{e}_z$	$\partial f_r/\partial r$	$\frac{1}{r}\frac{\partial f_\theta}{\partial \theta}$	$\partial f_z/\partial z$	$1/r$	0	0	0	$-(e_r/r)$	0
Spherical	$\mathbf{e}_r$	$\mathbf{e}_\theta$	e_ϕ	$\partial f_r/\partial r$	$\frac{1}{r}\frac{\partial f_\theta}{\partial \theta}$	$\frac{1}{r \sin\theta}\frac{\partial f_\phi}{\partial \phi}$	$2/r$	$\frac{1}{r \tan\theta}$	0	0	$-(e_r/r)$	$-\frac{e_r}{r} - \frac{e_\theta}{r \tan\theta}$

It becomes, by Equations (15), (16), and (21),

$$\begin{aligned}\nabla\cdot f &= \boldsymbol{\lambda}\cdot\nabla(\boldsymbol{\lambda}\cdot f)+\boldsymbol{\lambda}\cdot\nabla \boldsymbol{n}\cdot\boldsymbol{\lambda}\boldsymbol{n}\cdot f+\boldsymbol{\lambda}\cdot\nabla \boldsymbol{v}\cdot\boldsymbol{\lambda}\boldsymbol{v}\cdot f\\ &\quad+\boldsymbol{n}\cdot\nabla\boldsymbol{\lambda}\cdot\boldsymbol{n}\boldsymbol{\lambda}\cdot f+\boldsymbol{n}\cdot\nabla(\boldsymbol{n}\cdot f)+\boldsymbol{n}\cdot\nabla\boldsymbol{v}\cdot\boldsymbol{n}\boldsymbol{v}\cdot f\\ &\quad+\boldsymbol{v}\cdot\nabla\boldsymbol{\lambda}\cdot\boldsymbol{v}\boldsymbol{\lambda}\cdot f+\boldsymbol{v}\cdot\nabla\boldsymbol{n}\cdot\boldsymbol{v}\boldsymbol{n}\cdot f+\boldsymbol{v}\cdot\nabla(\boldsymbol{v}\cdot f)\end{aligned}\tag{50}$$

By Equations (23), (24), and (25), we have

$$\begin{aligned}\boldsymbol{\lambda}\cdot\nabla\boldsymbol{n}\cdot\boldsymbol{\lambda}\boldsymbol{n}\cdot f &= -\boldsymbol{\lambda}\cdot\nabla\boldsymbol{\lambda}\cdot\boldsymbol{n}\boldsymbol{n}\cdot f\\ \boldsymbol{\lambda}\cdot\nabla\boldsymbol{v}\cdot\boldsymbol{\lambda}\boldsymbol{v}\cdot f &= -\boldsymbol{\lambda}\cdot\nabla\boldsymbol{\lambda}\cdot\boldsymbol{v}\boldsymbol{v}\cdot f\\ \boldsymbol{n}\cdot\nabla\boldsymbol{\lambda}\cdot\boldsymbol{n}\boldsymbol{\lambda}\cdot f &= -\boldsymbol{n}\cdot\nabla\boldsymbol{n}\cdot\boldsymbol{\lambda}\boldsymbol{\lambda}\cdot f\\ \boldsymbol{n}\cdot\nabla\boldsymbol{v}\cdot\boldsymbol{n}\boldsymbol{v}\cdot f &= -\boldsymbol{n}\cdot\nabla\boldsymbol{n}\cdot\boldsymbol{v}\boldsymbol{v}\cdot f\\ \boldsymbol{v}\cdot\nabla\boldsymbol{\lambda}\cdot\boldsymbol{v}\boldsymbol{\lambda}\cdot f &= -\boldsymbol{v}\cdot\nabla\boldsymbol{v}\cdot\boldsymbol{\lambda}\boldsymbol{\lambda}\cdot f\\ \boldsymbol{v}\cdot\nabla\boldsymbol{n}\cdot\boldsymbol{v}\boldsymbol{n}\cdot f &= -\boldsymbol{v}\cdot\nabla\boldsymbol{v}\cdot\boldsymbol{n}\boldsymbol{n}\cdot f\end{aligned}$$

Substituting these into Equation (50) yields

$$\begin{aligned}\nabla\cdot\mathbf{f} &= \boldsymbol{\lambda}\cdot\nabla(\boldsymbol{\lambda}\cdot\mathbf{f})-\boldsymbol{\lambda}\cdot\nabla\boldsymbol{\lambda}\cdot(\boldsymbol{n}\boldsymbol{n}\cdot f+\boldsymbol{v}\boldsymbol{v}\cdot f)\\ &\quad+\boldsymbol{n}\cdot\nabla(\boldsymbol{n}\cdot\mathbf{f})-\boldsymbol{n}\cdot\nabla\boldsymbol{n}\cdot(\boldsymbol{v}\boldsymbol{v}\cdot f+\boldsymbol{\lambda}\boldsymbol{\lambda}\cdot f)\\ &\quad+\boldsymbol{v}\cdot\nabla(\boldsymbol{v}\cdot\mathbf{f})-\boldsymbol{v}\cdot\nabla\boldsymbol{v}\cdot(\boldsymbol{\lambda}\boldsymbol{\lambda}\cdot f+\boldsymbol{n}\boldsymbol{n}\cdot f)\end{aligned}\tag{51}$$

which can be rewritten, by Equation (34), as

$$\begin{aligned}\nabla\cdot\mathbf{f} &= \boldsymbol{\lambda}\cdot\nabla(\boldsymbol{\lambda}\cdot\mathbf{f})-\boldsymbol{\lambda}\cdot\nabla\boldsymbol{\lambda}\cdot(f-\boldsymbol{\lambda}\boldsymbol{\lambda}\cdot f)\\ &\quad+\boldsymbol{n}\cdot\nabla(\boldsymbol{n}\cdot\mathbf{f})-\boldsymbol{n}\cdot\nabla\boldsymbol{n}\cdot(f-\boldsymbol{n}\boldsymbol{n}\cdot f)\\ &\quad+\boldsymbol{v}\cdot\nabla(\boldsymbol{v}\cdot\mathbf{f})-\boldsymbol{v}\cdot\nabla\boldsymbol{v}\cdot(f-\boldsymbol{v}\boldsymbol{v}\cdot f)\end{aligned}\tag{52}$$

Since

$$\begin{aligned}\boldsymbol{\lambda}\cdot\nabla(\boldsymbol{\lambda}\cdot\mathbf{f}) &= \boldsymbol{\lambda}\cdot\nabla\boldsymbol{\lambda}\cdot f+\boldsymbol{\lambda}\cdot(\nabla f\cdot\boldsymbol{\lambda})\\ \boldsymbol{n}\cdot\nabla(\boldsymbol{n}\cdot\mathbf{f}) &= \boldsymbol{n}\cdot\nabla\boldsymbol{n}\cdot f+\boldsymbol{n}\cdot(\nabla f\cdot\boldsymbol{n})\\ \boldsymbol{v}\cdot\nabla(\boldsymbol{v}\cdot\mathbf{f}) &= \boldsymbol{v}\cdot\nabla\boldsymbol{v}\cdot f+\boldsymbol{v}\cdot(\nabla f\cdot\boldsymbol{v})\end{aligned}$$

Equation (52) becomes

$$\begin{aligned}\nabla\cdot f &= \boldsymbol{\lambda}\cdot\nabla f\cdot\boldsymbol{\lambda}+\boldsymbol{\lambda}\cdot\nabla\boldsymbol{\lambda}\cdot\boldsymbol{\lambda}\boldsymbol{\lambda}\cdot f\\ &\quad+\boldsymbol{n}\cdot\nabla f\cdot\boldsymbol{n}+\boldsymbol{n}\cdot\nabla\boldsymbol{n}\cdot\boldsymbol{n}\boldsymbol{n}\cdot f\\ &\quad+\boldsymbol{v}\cdot\nabla f\cdot\boldsymbol{v}+\boldsymbol{v}\cdot\nabla\boldsymbol{v}\cdot\boldsymbol{v}\cdot f\end{aligned}$$

Finally, an application of Equation (21) yields

$$\nabla\cdot\mathbf{f}=\boldsymbol{\lambda}\cdot\nabla f\cdot\boldsymbol{\lambda}+\boldsymbol{n}\cdot\nabla f\cdot\boldsymbol{n}+\boldsymbol{v}\cdot\nabla f\cdot\boldsymbol{v}\tag{53}$$

2.3.2 Surficial divergence

For a surficial vector $\mathbf{f}_\mathbf{n}^s$ on a surface S with $\mathbf{n}$ as the unit normal, its surficial divergence is

$$\begin{aligned}\nabla_\mathbf{n}^S\cdot\mathbf{f}_\mathbf{n}^s &= (\nabla-\mathbf{nn}\cdot\nabla)\cdot(\mathbf{f}-\mathbf{nn}\cdot\mathbf{f})\\ &= \nabla\cdot\mathbf{f}-\nabla\cdot(\mathbf{nn}\cdot\mathbf{f})-(\mathbf{nn}\cdot\nabla)\cdot\mathbf{f}+(\mathbf{nn}\cdot\nabla)\cdot(\mathbf{nn})\cdot\mathbf{f}\end{aligned}\tag{54}$$

It reduces to, by Equation (21),

$$\begin{aligned}\nabla^S_{\mathbf{n}} \cdot \mathbf{f}^s_{\mathbf{n}} &= \nabla \cdot \mathbf{f} - \mathbf{n} \cdot \nabla \mathbf{f} \cdot \mathbf{n} - \mathbf{n} \cdot \mathbf{f}\, \nabla \cdot \mathbf{n} - \mathbf{n} \cdot \nabla(\mathbf{n} \cdot \mathbf{f}) + \mathbf{n} \cdot \nabla(\mathbf{n} \cdot \mathbf{f}) \\ &= \nabla \cdot \mathbf{f} - \mathbf{n} \cdot \nabla \mathbf{f} \cdot \mathbf{n} - \mathbf{n} \cdot \mathbf{f}\, \nabla \cdot \mathbf{n}\end{aligned} \tag{55}$$

which by applying the chain rule to the last two terms, can be rewritten as,

$$\nabla^S_{\mathbf{n}} \cdot \mathbf{f}^s_{\mathbf{n}} = \nabla \cdot \mathbf{f} - \nabla \cdot (\mathbf{nf}) \cdot \mathbf{n} \tag{56}$$

2.3.3 Curvilineal divergence

For a curvilineal vector $\mathbf{f}^c_{\boldsymbol{\lambda}}$ along a curve C with the unit tangent vector $\boldsymbol{\lambda}$, its curvilineal divergence is

$$\begin{aligned}\nabla^c_{\boldsymbol{\lambda}} \cdot \mathbf{f}^c_{\boldsymbol{\lambda}} &= (\boldsymbol{\lambda\lambda} \cdot \nabla) \cdot (\boldsymbol{\lambda\lambda} \cdot \mathbf{f}) \\ &= \boldsymbol{\lambda} \cdot \nabla \boldsymbol{\lambda} \cdot \boldsymbol{\lambda\lambda} \cdot \mathbf{f} + \boldsymbol{\lambda} \cdot \boldsymbol{\lambda\lambda} \cdot \nabla(\boldsymbol{\lambda} \cdot \mathbf{f})\end{aligned} \tag{57}$$

Equations (15) and (21) reduce it into

$$\nabla^c_{\boldsymbol{\lambda}} \cdot \mathbf{f}^c_{\boldsymbol{\lambda}} = \boldsymbol{\lambda} \cdot \nabla(\boldsymbol{\lambda} \cdot \mathbf{f}) \tag{58}$$

2.3.4 Spatial vs curvilineal divergences

By Equation (58), we have

$$\nabla^c_{\boldsymbol{\lambda}} \cdot \mathbf{f}^c_{\boldsymbol{\lambda}} = \boldsymbol{\lambda} \cdot \nabla(\boldsymbol{\lambda} \cdot \mathbf{f}) \tag{59}$$

$$\nabla^c_{\mathbf{n}} \cdot \mathbf{f}^c_{\mathbf{n}} = \mathbf{n} \cdot \nabla(\mathbf{n} \cdot \mathbf{f}) \tag{60}$$

$$\nabla^c_{\boldsymbol{v}} \cdot \mathbf{f}^c_{\boldsymbol{v}} = \boldsymbol{v} \cdot \nabla(\boldsymbol{v} \cdot \mathbf{f}) \tag{61}$$

which becomes, by the chain rule,

$$\nabla^c_{\boldsymbol{\lambda}} \cdot \mathbf{f}^c_{\boldsymbol{\lambda}} = \nabla \cdot (\boldsymbol{\lambda\lambda} \cdot \mathbf{f}) - \boldsymbol{\lambda} \cdot \mathbf{f} \nabla \cdot \boldsymbol{\lambda} \tag{62}$$

$$\nabla^c_{\mathbf{n}} \cdot \mathbf{f}^c_{\mathbf{n}} = \nabla \cdot (\mathbf{nn} \cdot \mathbf{f}) - \mathbf{n} \cdot \mathbf{f} \nabla \cdot \mathbf{n} \tag{63}$$

$$\nabla^c_{\boldsymbol{v}} \cdot \mathbf{f}^c_{\boldsymbol{v}} = \nabla \cdot (\boldsymbol{vv} \cdot \mathbf{f}) - \boldsymbol{v} \cdot \mathbf{f} \nabla \cdot \boldsymbol{v} \tag{64}$$

Therefore,

$$\nabla^c_{\boldsymbol{\lambda}} \cdot \mathbf{f}^c_{\boldsymbol{\lambda}} + \nabla^c_{\mathbf{n}} \cdot \mathbf{f}^c_{\mathbf{n}} + \nabla^c_{\boldsymbol{v}} \cdot \mathbf{f}^c_{\boldsymbol{v}} = \nabla \cdot \mathbf{f} - \boldsymbol{\lambda} \cdot \mathbf{f} \nabla \cdot \boldsymbol{\lambda} - \mathbf{n} \cdot \mathbf{f} \nabla \cdot \mathbf{n} - \boldsymbol{v} \cdot \mathbf{f} \nabla \cdot \boldsymbol{v} \tag{65}$$

in which Equation (34) has been used. Equation (65) can also be rearranged as:

$$\nabla \cdot \mathbf{f} = \nabla^c_{\boldsymbol{\lambda}} \cdot \mathbf{f}^c_{\boldsymbol{\lambda}} + \nabla^c_{\mathbf{n}} \cdot \mathbf{f}^c_{\mathbf{n}} + \nabla^c_{\boldsymbol{v}} \cdot \mathbf{f}^c_{\boldsymbol{v}} + \nabla \cdot \boldsymbol{\lambda\lambda} \cdot \mathbf{f} + \nabla \cdot \boldsymbol{nn} \cdot \mathbf{f} + \nabla \cdot \boldsymbol{vv} \cdot \mathbf{f} \tag{66}$$

Note that, by the chain rule,

$$\boldsymbol{\lambda} \nabla \cdot \boldsymbol{\lambda} = \nabla \cdot (\boldsymbol{\lambda\lambda}) - \boldsymbol{\lambda} \cdot \nabla \boldsymbol{\lambda} \tag{67}$$

$$\mathbf{n} \nabla \cdot \mathbf{n} = \nabla \cdot (\mathbf{nn}) - \mathbf{n} \cdot \nabla \mathbf{n} \tag{68}$$

$$\boldsymbol{v} \nabla \cdot \boldsymbol{v} = \nabla \cdot (\boldsymbol{vv}) - \boldsymbol{v} \cdot \nabla \boldsymbol{v} \tag{69}$$

Substituting Equations (67)–(69) into Equation (66) yields

$$\nabla \cdot \mathbf{f} = \nabla_{\lambda}^{c} \cdot \mathbf{f}_{\lambda}^{c} + \nabla_{\mathbf{n}}^{c} \cdot \mathbf{f}_{\mathbf{n}}^{c} + \nabla_{\nu}^{c} \cdot \mathbf{f}_{\nu}^{c} + \nabla \cdot (\boldsymbol{\lambda}\boldsymbol{\lambda}) \cdot \mathbf{f} - \boldsymbol{\lambda} \cdot \nabla\boldsymbol{\lambda} \cdot \mathbf{f}$$
$$+ \nabla \cdot (\mathbf{nn}) \cdot \mathbf{f} - \mathbf{n} \cdot \nabla\mathbf{n} \cdot \mathbf{f} + \nabla \cdot (\boldsymbol{\nu}\boldsymbol{\nu}) \cdot \mathbf{f} - \boldsymbol{\nu} \cdot \nabla\boldsymbol{\nu} \cdot \mathbf{f}$$

Also,

$$\nabla \cdot (\boldsymbol{\lambda}\boldsymbol{\lambda}) + \nabla \cdot (\mathbf{nn}) + \nabla \cdot (\boldsymbol{\nu}\boldsymbol{\nu}) = \nabla \cdot \mathbf{I} = 0$$

Finally,

$$\nabla \cdot \mathbf{f} = \nabla_{\lambda}^{c} \cdot \mathbf{f}_{\lambda}^{c} + \nabla_{\mathbf{n}}^{c} \cdot \mathbf{f}_{\mathbf{n}}^{c} + \nabla_{\nu}^{c} \cdot \mathbf{f}_{\nu}^{c} - \boldsymbol{\lambda} \cdot \nabla\boldsymbol{\lambda} \cdot \mathbf{f} - \mathbf{n} \cdot \nabla\mathbf{n} \cdot \mathbf{f} - \boldsymbol{\nu} \cdot \nabla\boldsymbol{\nu} \cdot \mathbf{f} \tag{70}$$

2.3.5 Spatial vs surficial and curvilineal divergences

Equations (56) and (60) give

$$\nabla_{\mathbf{n}}^{S} \cdot \mathbf{f}_{\mathbf{n}}^{s} + \nabla_{\mathbf{n}}^{c} \cdot \mathbf{f}_{\mathbf{n}}^{c} = \nabla \cdot \mathbf{f} - \nabla \cdot (\mathbf{nf}) \cdot \mathbf{n} + \mathbf{n} \cdot \nabla(\mathbf{n} \cdot \mathbf{f}) \tag{71}$$

which can be rearranged as

$$\nabla \cdot \mathbf{f} = \nabla_{\mathbf{n}}^{S} \cdot \mathbf{f}_{\mathbf{n}}^{s} + \nabla_{\mathbf{n}}^{c} \cdot \mathbf{f}_{\mathbf{n}}^{c} + \nabla \cdot (\mathbf{nf}) \cdot \mathbf{n} - \mathbf{n} \cdot \nabla(\mathbf{n} \cdot \mathbf{f})$$

An application of the chain rule to the last two terms yields

$$\nabla \cdot \mathbf{f} = \nabla_{\mathbf{n}}^{S} \cdot \mathbf{f}_{\mathbf{n}}^{s} + \nabla_{\mathbf{n}}^{c} \cdot \mathbf{f}_{\mathbf{n}}^{c} + \nabla \cdot \mathbf{nf} \cdot \mathbf{n} + \mathbf{n} \cdot \nabla\mathbf{f} \cdot \mathbf{n} - \mathbf{n} \cdot \nabla\mathbf{n} \cdot \mathbf{f} - \mathbf{n} \cdot (\mathbf{n} \cdot \nabla\mathbf{f})$$
$$= \nabla_{\mathbf{n}}^{S} \cdot \mathbf{f}_{\mathbf{n}}^{s} + \nabla_{\mathbf{n}}^{c} \cdot \mathbf{f}_{\mathbf{n}}^{c} + \nabla \cdot \mathbf{nf} \cdot \mathbf{n} - \mathbf{n} \cdot \nabla\mathbf{n} \cdot \mathbf{f} \tag{72}$$

The last two terms in this expression come from the curvature of the coordinate normal to the surface in the direction of the curve. For a Cartsian coordinate system, these terms vanish.

2.3.6 Curvilineal vs surficial divergences

Let $\boldsymbol{\lambda}$ and $\boldsymbol{\nu}$ be the coordinates on the surface S shown in Figure 7. By Equations (35) and (45), we have

$$\nabla_{\lambda}^{c} \cdot \mathbf{f}_{\lambda}^{c} = (\nabla_{\mathbf{n}}^{S} - \boldsymbol{\nu}\boldsymbol{\nu} \cdot \nabla_{\mathbf{n}}^{S}) \cdot (\mathbf{f}_{\mathbf{n}}^{s} - \boldsymbol{\nu}\boldsymbol{\nu} \cdot \mathbf{f}_{\mathbf{n}}^{s})$$
$$= \nabla_{\mathbf{n}}^{S} \cdot \mathbf{f}_{\mathbf{n}}^{s} - \nabla_{\mathbf{n}}^{S} \cdot (\boldsymbol{\nu}\boldsymbol{\nu} \cdot \mathbf{f}_{\mathbf{n}}^{s}) - (\boldsymbol{\nu}\boldsymbol{\nu} \cdot \nabla_{\mathbf{n}}^{S}) \cdot \mathbf{f}_{\mathbf{n}}^{s} + (\boldsymbol{\nu}\boldsymbol{\nu} \cdot \nabla_{\mathbf{n}}^{S}) \cdot (\boldsymbol{\nu}\boldsymbol{\nu} \cdot \mathbf{f}_{\mathbf{n}}^{s})$$

An application of the chain rule yields

$$\nabla_{\lambda}^{c} \cdot \mathbf{f}_{\lambda}^{c} = \nabla_{\mathbf{n}}^{S} \cdot \mathbf{f}_{\mathbf{n}}^{s} - \nabla_{\mathbf{n}}^{S} \cdot \boldsymbol{\nu}\boldsymbol{\nu} \cdot \mathbf{f}_{\mathbf{n}}^{s} - \boldsymbol{\nu} \cdot \nabla_{\mathbf{n}}^{S}(\boldsymbol{\nu} \cdot \mathbf{f}_{\mathbf{n}}^{s}) - \boldsymbol{\nu} \cdot \nabla_{\mathbf{n}}^{S}\mathbf{f}_{\mathbf{n}}^{s} \cdot \boldsymbol{\nu}$$
$$+ \boldsymbol{\nu} \cdot \nabla_{\mathbf{n}}^{S}\boldsymbol{\nu} \cdot \boldsymbol{\nu}\boldsymbol{\nu} \cdot \mathbf{f}_{\mathbf{n}}^{s} + \boldsymbol{\nu} \cdot \{\boldsymbol{\nu} \cdot [\boldsymbol{\nu} \cdot \nabla_{\mathbf{n}}^{S}(\boldsymbol{\nu} \cdot \mathbf{f}_{\mathbf{n}}^{s})]\} \tag{73}$$

By Equations (16), (21) and (42), we obtain

$$\boldsymbol{\nu} \cdot \nabla_{\mathbf{n}}^{S}\boldsymbol{\nu} \cdot \boldsymbol{\nu} = \boldsymbol{\nu} \cdot (\nabla - \mathbf{nn} \cdot \nabla)\boldsymbol{\nu} \cdot \boldsymbol{\nu} = \boldsymbol{\nu} \cdot \nabla\boldsymbol{\nu} \cdot \boldsymbol{\nu} - \boldsymbol{\nu} \cdot \mathbf{nn} \cdot \nabla\boldsymbol{\nu} \cdot \boldsymbol{\nu} = 0$$

Also

$$\boldsymbol{\nu} \cdot \{\boldsymbol{\nu} \cdot [\boldsymbol{\nu} \cdot \nabla_{\mathbf{n}}^{S}(\boldsymbol{\nu} \cdot \mathbf{f}_{\mathbf{n}}^{s})]\} = \boldsymbol{\nu} \cdot \nabla_{\mathbf{n}}^{S}(\boldsymbol{\nu} \cdot \mathbf{f}_{\mathbf{n}}^{s})$$

These two relations allow us to rewrite Equation (73) as

$$\nabla_{\lambda}^{c} \cdot \mathbf{f}_{\lambda}^{c} = \nabla_{\mathbf{n}}^{S} \cdot \mathbf{f}_{\mathbf{n}}^{s} - \nabla_{\mathbf{n}}^{S} \cdot (\boldsymbol{\nu}\mathbf{f}_{\mathbf{n}}^{s}) \cdot \boldsymbol{\nu} \tag{74}$$

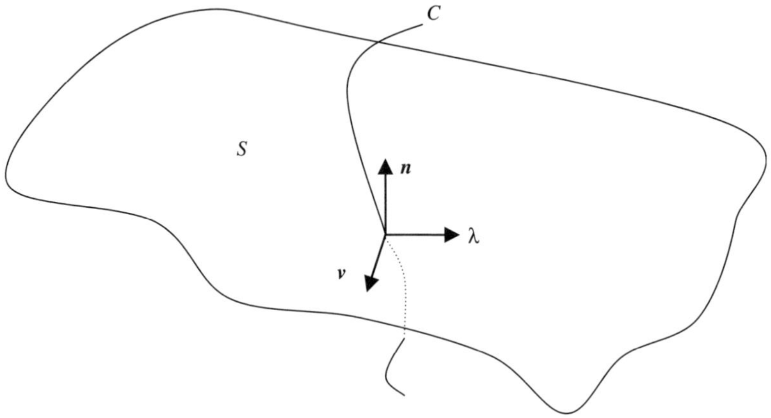

Figure 7 A curve C orthogonally intersecting surface S (**n** is unit normal vector of S and tangent to C; orthonormal coordinate directions $\boldsymbol{\lambda}$ and $\boldsymbol{\nu}$ are tangent to the surface S).

Similarly,

$$\nabla^c_{\mathbf{v}} \cdot \mathbf{f}^c_{\mathbf{v}} = \nabla^S_{\mathbf{n}} \cdot \mathbf{f}^s_{\mathbf{n}} - \nabla^S_{\mathbf{n}} \cdot (\boldsymbol{\lambda}\mathbf{f}^s_{\mathbf{n}}) \cdot \boldsymbol{\lambda} \tag{75}$$

2.3.7 Surficial vs curvilineal divergences

By Equation (65), we have

$$\nabla^c_{\boldsymbol{\lambda}} \cdot \mathbf{f}^c_{\boldsymbol{\lambda}} + \nabla^c_{\mathbf{n}} \cdot \mathbf{f}^c_{\mathbf{n}} + \nabla^c_{\mathbf{v}} \cdot \mathbf{f}^c_{\mathbf{v}} = \nabla \cdot \mathbf{f} - \boldsymbol{\lambda} \cdot \mathbf{f}\nabla \cdot \boldsymbol{\lambda} - \mathbf{n} \cdot \mathbf{f}\nabla \cdot \mathbf{n} - \boldsymbol{\nu} \cdot \mathbf{f}\nabla \cdot \boldsymbol{\nu} \tag{76}$$

which can be rearranged as

$$\nabla \cdot \mathbf{f} = \nabla^c_{\boldsymbol{\lambda}} \cdot \mathbf{f}^c_{\boldsymbol{\lambda}} + \nabla^c_{\mathbf{n}} \cdot \mathbf{f}^c_{\mathbf{n}} + \nabla^c_{\mathbf{v}} \cdot \mathbf{f}^c_{\mathbf{v}} + \boldsymbol{\lambda} \cdot \mathbf{f}\nabla \cdot \boldsymbol{\lambda} + \mathbf{n} \cdot \mathbf{f}\nabla \cdot \mathbf{n} + \boldsymbol{\nu} \cdot \mathbf{f}\nabla \cdot \boldsymbol{\nu}$$

An application of Equation (56) yields

$$\nabla \cdot \mathbf{f} = \nabla^S_{\mathbf{n}} \cdot \mathbf{f}^s_{\mathbf{n}} + \nabla \cdot \mathbf{nf} \cdot \mathbf{n} + \mathbf{n} \cdot \nabla\mathbf{f} \cdot \mathbf{n} \tag{77}$$

Substituting Equation (77) into Equation (76) leads to

$$\begin{aligned} &\nabla^c_{\boldsymbol{\lambda}} \cdot \mathbf{f}^c_{\boldsymbol{\lambda}} + \nabla^c_{\mathbf{n}} \cdot \mathbf{f}^c_{\mathbf{n}} + \nabla^c_{\mathbf{v}} \cdot \mathbf{f}^c_{\mathbf{v}} \\ &= \nabla^S_{\mathbf{n}} \cdot \mathbf{f}^s_{\mathbf{n}} + \mathbf{n} \cdot \nabla\mathbf{f} \cdot \mathbf{n} - \boldsymbol{\lambda} \cdot \mathbf{f}\nabla \cdot \boldsymbol{\lambda} - \boldsymbol{\nu} \cdot \mathbf{f}\nabla \cdot \boldsymbol{\nu} \end{aligned} \tag{78}$$

which can be rearranged into

$$\nabla^S_{\mathbf{n}} \cdot \mathbf{f}^s_{\mathbf{n}} = \nabla^c_{\boldsymbol{\lambda}} \cdot \mathbf{f}^c_{\boldsymbol{\lambda}} + \nabla^c_{\mathbf{n}} \cdot \mathbf{f}^c_{\mathbf{n}} + \nabla^c_{\mathbf{v}} \cdot \mathbf{f}^c_{\mathbf{v}} - \mathbf{n} \cdot \nabla\mathbf{f} \cdot \mathbf{n} + \boldsymbol{\lambda} \cdot \mathbf{f}\nabla \cdot \boldsymbol{\lambda} + \boldsymbol{\nu} \cdot \mathbf{f}\nabla \cdot \boldsymbol{\nu} \tag{79}$$

Note that

$$\begin{aligned} \nabla^c_{\mathbf{n}} \cdot \mathbf{f}^c_{\mathbf{n}} &= (\mathbf{nn} \cdot \nabla) \cdot (\mathbf{nn} \cdot \mathbf{f}) \\ &= \mathbf{n} \cdot \nabla\mathbf{n} \cdot \mathbf{nn} \cdot \mathbf{f} + \mathbf{n} \cdot \mathbf{n}(\mathbf{n} \cdot \nabla(\mathbf{n} \cdot \mathbf{f})) \end{aligned}$$

which becomes, by using Equations (15) and (21),

$$\begin{aligned} \nabla^c_{\mathbf{n}} \cdot \mathbf{f}^c_{\mathbf{n}} &= \mathbf{n} \cdot \nabla(\mathbf{n} \cdot \mathbf{f}) \\ &= \mathbf{n} \cdot \nabla\mathbf{n} \cdot \mathbf{f} + \mathbf{n} \cdot \nabla\mathbf{f} \cdot \mathbf{n} \end{aligned}$$

Therefore, Equation (79) becomes

$$\begin{aligned}\nabla_{\mathbf{n}}^{S}\cdot\mathbf{f}_{\mathbf{n}}^{s} &= \nabla_{\lambda}^{c}\cdot\mathbf{f}_{\lambda}^{c}+\nabla_{\nu}^{c}\cdot\mathbf{f}_{\nu}^{c}-\mathbf{n}\cdot\nabla\mathbf{f}\cdot\mathbf{n}+\boldsymbol{\lambda}\cdot\mathbf{f}\nabla\cdot\boldsymbol{\lambda}+\boldsymbol{\nu}\cdot\mathbf{f}\nabla\cdot\boldsymbol{\nu}+\mathbf{n}\cdot\nabla\mathbf{n}\cdot\mathbf{f}+\mathbf{n}\cdot\nabla\mathbf{f}\cdot\mathbf{n}\\ &= \nabla_{\lambda}^{c}\cdot\mathbf{f}_{\lambda}^{c}+\nabla_{\nu}^{c}\cdot\mathbf{f}_{\nu}^{c}+\boldsymbol{\lambda}\cdot\mathbf{f}\nabla\cdot\boldsymbol{\lambda}+\boldsymbol{\nu}\cdot\mathbf{f}\nabla\cdot\boldsymbol{\nu}+\mathbf{n}\cdot\nabla\mathbf{n}\cdot\mathbf{f}\end{aligned}\tag{80}$$

By Equations (16) and (21) and

$$\mathbf{f}=\mathbf{f}_{\mathbf{n}}^{s}+\mathbf{nn}\cdot\mathbf{f}$$

we have

$$\boldsymbol{\lambda}\cdot\mathbf{f}\,\nabla\cdot\boldsymbol{\lambda}=\boldsymbol{\lambda}\cdot(\mathbf{f}_{\mathbf{n}}^{s}+\mathbf{nn}\cdot\mathbf{f})\nabla\cdot\boldsymbol{\lambda}=\boldsymbol{\lambda}\cdot\mathbf{f}_{\mathbf{n}}^{s}\nabla\cdot\boldsymbol{\lambda}\tag{81}$$

$$\boldsymbol{\nu}\cdot\mathbf{f}\nabla\cdot\boldsymbol{\nu}=\boldsymbol{\nu}\cdot(\mathbf{f}_{\mathbf{n}}^{s}+\mathbf{nn}\cdot\mathbf{f})\nabla\cdot\boldsymbol{\nu}=\boldsymbol{\nu}\cdot\mathbf{f}_{\mathbf{n}}^{s}\nabla\cdot\boldsymbol{\nu}\tag{82}$$

$$\mathbf{n}\cdot\nabla\mathbf{n}\cdot\mathbf{f}=\mathbf{n}\cdot\nabla\mathbf{n}\cdot(\mathbf{f}_{\mathbf{n}}^{s}+\mathbf{nn}\cdot\mathbf{f})=\mathbf{n}\cdot\nabla\mathbf{n}\cdot\mathbf{f}_{\mathbf{n}}^{s}\tag{83}$$

Finally, substituting Equations (81)–(83) into Equation (80) yields

$$\nabla_{\mathbf{n}}^{S}\cdot\mathbf{f}_{\mathbf{n}}^{s}=\nabla_{\lambda}^{c}\cdot\mathbf{f}_{\lambda}^{c}+\nabla_{\nu}^{c}\cdot\mathbf{f}_{\nu}^{c}+\boldsymbol{\lambda}\cdot\mathbf{f}_{\mathbf{n}}^{s}\,\nabla\cdot\boldsymbol{\lambda}+\boldsymbol{\nu}\cdot\mathbf{f}_{\mathbf{n}}^{s}\,\nabla\cdot\boldsymbol{\nu}+\boldsymbol{n}\cdot\nabla\boldsymbol{n}\cdot\mathbf{f}_{\mathbf{n}}^{s}\tag{84}$$

2.4 Curl

We first discuss three types of curls: spatial curl of a spatial vector function, surface curl of a surface vector function, and curvilineal curl of a curvilineal vector function. These curls are often encountered in developing multiscale theorems. We then develop their relations. The details of various terms in curls are available in Table 5 for Cartesian, cylindrical, and spherical coordinates.

2.4.1 Spatial curl

By Equations (34) and (41), the curl of a vector function $\mathbf{f}$ can be written as,

$$\begin{aligned}\nabla\times\mathbf{f} &= (\boldsymbol{\lambda\lambda}\cdot\nabla+\mathbf{nn}\cdot\nabla+\boldsymbol{\nu\nu}\cdot\nabla)\times(\boldsymbol{\lambda\lambda}\cdot\mathbf{f}+\mathbf{nn}\cdot\mathbf{f}+\boldsymbol{\nu\nu}\cdot\mathbf{f})\\ &= (\boldsymbol{\lambda\lambda}\cdot\nabla)\times(\boldsymbol{\lambda\lambda}\cdot\mathbf{f})+(\boldsymbol{\lambda\lambda}\cdot\nabla)\times(\mathbf{nn}\cdot\mathbf{f})+(\boldsymbol{\lambda\lambda}\cdot\nabla)\times(\boldsymbol{\nu\nu}\cdot\mathbf{f})\\ &\quad+(\boldsymbol{nn}\cdot\nabla)\times(\boldsymbol{\lambda\lambda}\cdot\mathbf{f})+(\boldsymbol{nn}\cdot\nabla)\times(\mathbf{nn}\cdot\mathbf{f})+(\boldsymbol{nn}\cdot\nabla)\times(\boldsymbol{\nu\nu}\cdot\mathbf{f})\\ &\quad+(\boldsymbol{\nu\nu}\cdot\nabla)\times(\boldsymbol{\lambda\lambda}\cdot\mathbf{f})+(\boldsymbol{\nu\nu}\cdot\nabla)\times(\mathbf{nn}\cdot\mathbf{f})+(\boldsymbol{\nu\nu}\cdot\nabla)\times(\boldsymbol{\nu\nu}\cdot\mathbf{f})\end{aligned}\tag{85}$$

which contains nine cross product terms. The first term reads, by the chain rule and Equation (20),

$$\begin{aligned}(\boldsymbol{\lambda\lambda}\cdot\nabla)\times(\boldsymbol{\lambda\lambda}\cdot\mathbf{f}) &= \boldsymbol{\lambda}\times(\boldsymbol{\lambda}\cdot\nabla\boldsymbol{\lambda})\boldsymbol{\lambda}\cdot\mathbf{f}+\boldsymbol{\lambda}\times\boldsymbol{\lambda}(\boldsymbol{\lambda}\cdot\nabla(\boldsymbol{\lambda}\cdot\mathbf{f}))\\ &= \boldsymbol{\lambda}\times(\boldsymbol{\lambda}\cdot\nabla\boldsymbol{\lambda})\boldsymbol{\lambda}\cdot\mathbf{f}\end{aligned}\tag{86}$$

Note that

$$\boldsymbol{\lambda}\times(\boldsymbol{\lambda}\cdot\nabla\boldsymbol{\lambda})=\boldsymbol{\lambda}\times(\boldsymbol{\lambda}\cdot\nabla\boldsymbol{\lambda})\cdot\boldsymbol{\lambda\lambda}+\boldsymbol{\lambda}\times(\boldsymbol{\lambda}\cdot\nabla\boldsymbol{\lambda})\cdot\boldsymbol{\nu\nu}+\boldsymbol{\lambda}\times(\boldsymbol{\lambda}\cdot\nabla\boldsymbol{\lambda})\cdot\mathbf{nn}\tag{87}$$

Equation (86) thus reduces to

$$\begin{aligned}(\boldsymbol{\lambda\lambda}\cdot\nabla)\times(\boldsymbol{\lambda\lambda}\cdot\mathbf{f}) &= \boldsymbol{\lambda}\times(\boldsymbol{\lambda}\cdot\nabla\boldsymbol{\lambda})\cdot\boldsymbol{\lambda\lambda\lambda}\cdot\mathbf{f}+\boldsymbol{\lambda}\times(\boldsymbol{\lambda}\cdot\nabla\boldsymbol{\lambda})\cdot\boldsymbol{\nu\nu\lambda}\cdot\mathbf{f}\\ &\quad+\boldsymbol{\lambda}\times(\boldsymbol{\lambda}\cdot\nabla\boldsymbol{\lambda})\cdot\mathbf{nn}\boldsymbol{\lambda}\cdot\mathbf{f}\end{aligned}\tag{88}$$

Table 5 Curl geometric terms in standard coordinate systems

Coordinated System	$\boldsymbol{\lambda}$	$\mathbf{n}$	$\boldsymbol{\nu}$	$\boldsymbol{\lambda}\cdot\nabla$	$\mathbf{n}\cdot\nabla$	$\boldsymbol{\nu}\cdot\nabla$	$\mathbf{n}\cdot\nabla\boldsymbol{\lambda}$	$\boldsymbol{\lambda}\cdot\nabla\boldsymbol{\lambda}$	$\boldsymbol{\nu}\cdot\nabla\mathbf{n}$	$\boldsymbol{\lambda}\cdot\nabla\boldsymbol{\nu}$	$\mathrm{n}\cdot\nabla\boldsymbol{\nu}$	
Cartesian	$\mathbf{i}$	$\mathbf{j}$	$\mathbf{k}$	∂/∂_x	∂/∂_y	∂/∂_z	0	0	0	0	0	0
Cylindrical	$\mathbf{e}_r$	$\mathbf{e}_\theta$	$\mathbf{e}_z$	∂/∂_r	$\frac{1}{r}\frac{\partial}{\partial\theta}$	∂/∂_z	$e_{\theta/r}$	0	0	0	0	0
Spherical	$\mathbf{e}_r$	$\mathbf{e}_\theta$	$\mathbf{e}_\o$	∂/∂_r	$\frac{1}{r}\frac{\partial}{\partial\theta}$	$\frac{1}{r\sin\theta}\frac{\partial}{\partial\phi}$	$e_{\theta/r}$	$e_{\phi/r}$	0	$e_{\phi/r\ tan\theta}$	0	0

Also, for any vectors **a**, **b** and **c**

$$\mathbf{a} \times \mathbf{b} \cdot \mathbf{c} = -\mathbf{b} \cdot \mathbf{a} \times \mathbf{c} \tag{89}$$

Therefore,

$$\boldsymbol{\lambda} \times (\boldsymbol{\lambda} \cdot \nabla\boldsymbol{\lambda}) \cdot \boldsymbol{\lambda} = -\boldsymbol{\lambda} \cdot \nabla\boldsymbol{\lambda} \cdot (\boldsymbol{\lambda} \times \boldsymbol{\lambda}) \tag{90}$$

$$\boldsymbol{\lambda} \times (\boldsymbol{\lambda} \cdot \nabla\boldsymbol{\lambda}) \cdot \boldsymbol{\nu} = -\boldsymbol{\lambda} \cdot \nabla\boldsymbol{\lambda} \cdot (\boldsymbol{\lambda} \times \boldsymbol{\nu}) \tag{91}$$

$$\boldsymbol{\lambda} \times (\boldsymbol{\lambda} \cdot \nabla\boldsymbol{\lambda}) \cdot \mathbf{n} = -\boldsymbol{\lambda} \cdot \nabla\boldsymbol{\lambda} \cdot (\boldsymbol{\lambda} \times \mathbf{n}) \tag{92}$$

Substituting these into Equation (88) leads to

$$\begin{aligned}(\boldsymbol{\lambda}\boldsymbol{\lambda} \cdot \nabla) \times (\boldsymbol{\lambda}\boldsymbol{\lambda} \cdot \mathbf{f}) = & -\boldsymbol{\lambda} \cdot \nabla\boldsymbol{\lambda} \cdot (\boldsymbol{\lambda} \times \boldsymbol{\lambda})\boldsymbol{\lambda}\boldsymbol{\lambda} \cdot \mathbf{f} \\ & -\boldsymbol{\lambda} \cdot \nabla\boldsymbol{\lambda} \cdot (\boldsymbol{\lambda} \times \boldsymbol{\nu})\boldsymbol{\nu}\boldsymbol{\lambda} \cdot \mathbf{f} - \boldsymbol{\lambda} \cdot \nabla\boldsymbol{\lambda} \cdot (\boldsymbol{\lambda} \times \mathbf{n})\mathbf{n}\boldsymbol{\lambda} \cdot \mathbf{f}\end{aligned}$$

By Equations (17), (19), (20) and (21), we have

$$(\boldsymbol{\lambda}\boldsymbol{\lambda} \cdot \nabla) \times (\boldsymbol{\lambda}\boldsymbol{\lambda} \cdot \mathbf{f}) = \boldsymbol{\lambda} \cdot \nabla\boldsymbol{\lambda} \cdot \mathbf{n}\boldsymbol{\nu}\boldsymbol{\lambda} \cdot \mathbf{f} - \boldsymbol{\lambda} \cdot \nabla\boldsymbol{\lambda} \cdot \boldsymbol{\nu}\mathbf{n}\boldsymbol{\lambda} \cdot \mathbf{f}$$

Thus, by Equations (23) and (25),

$$(\boldsymbol{\lambda}\boldsymbol{\lambda} \cdot \nabla) \times (\boldsymbol{\lambda}\boldsymbol{\lambda} \cdot \mathbf{f}) = \boldsymbol{\lambda} \cdot \nabla\boldsymbol{\nu} \cdot \boldsymbol{\lambda}\boldsymbol{\lambda} \cdot \mathbf{f}\mathbf{n} - \boldsymbol{\lambda} \cdot \nabla\mathbf{n} \cdot \boldsymbol{\lambda}\boldsymbol{\lambda} \cdot \mathbf{f}\boldsymbol{\nu} \tag{93}$$

Similarly,

$$(\boldsymbol{\nu}\boldsymbol{\nu} \cdot \nabla) \times (\boldsymbol{\nu}\boldsymbol{\nu} \cdot \mathbf{f}) = -\boldsymbol{\nu} \cdot \nabla\boldsymbol{\lambda} \cdot \boldsymbol{\nu}\boldsymbol{\nu} \cdot \mathbf{f}\mathbf{n} + \boldsymbol{\nu} \cdot \nabla\mathbf{n} \cdot \boldsymbol{\nu}\boldsymbol{\nu} \cdot \mathbf{f}\boldsymbol{\lambda} \tag{94}$$

$$(\mathbf{n}\mathbf{n} \cdot \nabla) \times (\mathbf{n}\mathbf{n} \cdot \mathbf{f}) = -\mathbf{n} \cdot \nabla\boldsymbol{\nu} \cdot \mathbf{n}\mathbf{n} \cdot \mathbf{f}\boldsymbol{\lambda} + \mathbf{n} \cdot \nabla\boldsymbol{\lambda} \cdot \mathbf{n}\mathbf{n} \cdot \mathbf{f}\boldsymbol{\nu} \tag{95}$$

Also,

$$(\boldsymbol{\lambda}\boldsymbol{\lambda} \cdot \nabla) \times (\mathbf{n}\mathbf{n} \cdot \mathbf{f}) = -(\boldsymbol{\lambda} \cdot \nabla\mathbf{n}) \times \boldsymbol{\lambda}\mathbf{n} \cdot \mathbf{f} + \boldsymbol{\nu}(\boldsymbol{\lambda} \cdot \nabla(\mathbf{n} \cdot \mathbf{f})) \tag{96}$$

Note that

$$(\boldsymbol{\lambda} \cdot \nabla\mathbf{n}) \times \boldsymbol{\lambda} = (\boldsymbol{\lambda} \cdot \nabla\mathbf{n}) \times \boldsymbol{\lambda} \cdot \boldsymbol{\lambda}\boldsymbol{\lambda} + (\boldsymbol{\lambda} \cdot \nabla\mathbf{n}) \times \boldsymbol{\lambda} \cdot \boldsymbol{\nu}\boldsymbol{\nu} + (\boldsymbol{\lambda} \cdot \nabla\mathbf{n}) \times \boldsymbol{\lambda} \cdot \mathbf{n}\mathbf{n} \tag{97}$$

A substitution of Equation (97) into Equation (96) gives

$$\begin{aligned}(\boldsymbol{\lambda}\boldsymbol{\lambda} \cdot \nabla) \times (\mathbf{n}\mathbf{n} \cdot \mathbf{f}) = & -(\boldsymbol{\lambda} \cdot \nabla\mathbf{n}) \times \boldsymbol{\lambda} \cdot \boldsymbol{\lambda}\boldsymbol{\lambda}\mathbf{n} \cdot \mathbf{f} - (\boldsymbol{\lambda} \cdot \nabla\mathbf{n}) \times \boldsymbol{\lambda} \cdot \boldsymbol{\nu}\boldsymbol{\nu}\mathbf{n} \cdot \mathbf{f} \\ & -(\boldsymbol{\lambda} \cdot \nabla\mathbf{n}) \times \boldsymbol{\lambda} \cdot \mathbf{n}\mathbf{n}n \cdot \mathbf{f} + \boldsymbol{\nu}(\boldsymbol{\lambda} \cdot \nabla(\mathbf{n} \cdot \mathbf{f}))\end{aligned}$$

which can be rewritten as, by Equation (89),

$$\begin{aligned}(\boldsymbol{\lambda}\boldsymbol{\lambda} \cdot \nabla) \times (\mathbf{n}\mathbf{n} \cdot \mathbf{f}) = & -(\boldsymbol{\lambda} \cdot \nabla\mathbf{n}) \cdot \boldsymbol{\lambda} \times \boldsymbol{\lambda}\boldsymbol{\lambda}\mathbf{n} \cdot \mathbf{f} - (\boldsymbol{\lambda} \cdot \nabla\mathbf{n}) \cdot \boldsymbol{\lambda} \times \boldsymbol{\nu}\boldsymbol{\nu}\mathbf{n} \cdot \mathbf{f} \\ & -(\boldsymbol{\lambda} \cdot \nabla\mathbf{n}) \cdot \boldsymbol{\lambda} \times \mathbf{n}\mathbf{n}\mathbf{n} \cdot \mathbf{f} + \boldsymbol{\nu}(\boldsymbol{\lambda} \cdot \nabla(\mathbf{n} \cdot \mathbf{f}))\end{aligned} \tag{98}$$

Equation (98) becomes, by Equations (17), (19) and (20),

$$(\boldsymbol{\lambda}\boldsymbol{\lambda} \cdot \nabla) \times (\mathbf{n}\mathbf{n} \cdot \mathbf{f}) = (\boldsymbol{\lambda} \cdot \nabla\mathbf{n}) \cdot \mathbf{n}\boldsymbol{\nu}\mathbf{n} \cdot \mathbf{f} - (\boldsymbol{\lambda} \cdot \nabla\mathbf{n}) \cdot \boldsymbol{\nu}\mathbf{n} \cdot \mathbf{f}\mathbf{n} + \boldsymbol{\nu}(\boldsymbol{\lambda} \cdot \nabla(\mathbf{n} \cdot \mathbf{f})) \tag{99}$$

which becomes, by Equation (21),

$$(\boldsymbol{\lambda}\boldsymbol{\lambda} \cdot \nabla) \times (\mathbf{n}\mathbf{n} \cdot \mathbf{f}) = -\boldsymbol{\lambda} \cdot \nabla\mathbf{n} \cdot \boldsymbol{\nu}\mathbf{n} \cdot \mathbf{f}\mathbf{n} + \boldsymbol{\nu}(\boldsymbol{\lambda} \cdot \nabla(\mathbf{n} \cdot \mathbf{f})) \tag{100}$$

Similarly,

$$(\boldsymbol{\lambda}\boldsymbol{\lambda} \cdot \nabla) \times (\boldsymbol{\nu}\boldsymbol{\nu} \cdot \mathbf{f}) = (\boldsymbol{\lambda} \cdot \nabla\boldsymbol{\nu}) \cdot \mathbf{n}(\boldsymbol{\nu} \cdot \mathbf{f})\boldsymbol{\nu} - \mathbf{n}\boldsymbol{\lambda} \cdot \nabla(\boldsymbol{\nu} \cdot \mathbf{f}) \tag{101}$$

$$(\mathbf{n}\mathbf{n} \cdot \nabla) \times (\boldsymbol{\lambda}\boldsymbol{\lambda} \cdot \mathbf{f}) = \mathbf{n} \cdot \nabla\boldsymbol{\lambda} \cdot \boldsymbol{\nu}\boldsymbol{\lambda} \cdot \mathbf{f}\boldsymbol{\lambda} - \boldsymbol{\nu}(\mathbf{n} \cdot \nabla(\boldsymbol{\lambda} \cdot \mathbf{f})) \tag{102}$$

$$(\mathbf{nn}\cdot\nabla)\times(\boldsymbol{\nu}\boldsymbol{\nu}\cdot\mathbf{f}) = -\mathbf{n}\cdot\nabla\boldsymbol{\nu}\cdot\boldsymbol{\lambda}\boldsymbol{\nu}\cdot\mathbf{f}\boldsymbol{\nu} + \boldsymbol{\lambda}(\mathbf{n}\cdot\nabla(\boldsymbol{\nu}\cdot\mathbf{f})) \tag{103}$$

$$(\boldsymbol{\nu}\boldsymbol{\nu}\cdot\nabla)\times(\boldsymbol{\lambda}\boldsymbol{\lambda}\cdot\mathbf{f}) = -(\boldsymbol{\nu}\cdot\nabla\boldsymbol{\lambda})\cdot\mathbf{n}\boldsymbol{\lambda}\cdot\mathbf{f}\boldsymbol{\lambda} + \mathbf{n}(\boldsymbol{\nu}\cdot\nabla(\boldsymbol{\lambda}\cdot\mathbf{f})) \tag{104}$$

$$(\boldsymbol{\nu}\boldsymbol{\nu}\cdot\nabla)\times(\mathbf{nn}\cdot\mathbf{f}) = \boldsymbol{\nu}\cdot\nabla\mathbf{n}\cdot\boldsymbol{\lambda}\mathbf{n}\cdot\mathbf{fn} - \boldsymbol{\lambda}(\boldsymbol{\nu}\cdot\nabla(\mathbf{n}\cdot\mathbf{f})) \tag{105}$$

Substitution of Equations (93)–(95), (100)–(105) into Equation (85) leads to

$$\begin{aligned}\nabla\times\mathbf{f} = {} & \boldsymbol{\lambda}\cdot\nabla\boldsymbol{\nu}\cdot\boldsymbol{\lambda}\boldsymbol{\lambda}\cdot\mathbf{fn} - \boldsymbol{\lambda}\cdot\nabla\mathbf{n}\cdot\boldsymbol{\lambda}\boldsymbol{\lambda}\cdot\mathbf{f}\boldsymbol{\nu} - \mathbf{n}\cdot\nabla\boldsymbol{\nu}\cdot\mathbf{nn}\cdot\mathbf{f}\boldsymbol{\lambda}\\ & + \mathbf{n}\cdot\nabla\boldsymbol{\lambda}\cdot\mathbf{nn}\cdot\mathbf{f}\boldsymbol{\nu} - \boldsymbol{\nu}\cdot\nabla\boldsymbol{\lambda}\cdot\boldsymbol{\nu}\boldsymbol{\nu}\cdot\mathbf{fn} + \boldsymbol{\nu}\cdot\nabla\mathbf{n}\cdot\boldsymbol{\nu}\boldsymbol{\nu}\cdot\mathbf{f}\boldsymbol{\lambda}\\ & - (\boldsymbol{\lambda}\cdot\nabla\mathbf{n})\cdot\boldsymbol{\nu}\mathbf{n}\cdot\mathbf{fn} + \boldsymbol{\nu}(\boldsymbol{\lambda}\cdot\nabla(\mathbf{n}\cdot\mathbf{f})) + (\boldsymbol{\lambda}\cdot\nabla\boldsymbol{\nu})\cdot\mathbf{n}(\boldsymbol{\nu}\cdot\mathbf{f})\boldsymbol{\nu}\\ & - \mathbf{n}(\boldsymbol{\lambda}\cdot\nabla(\boldsymbol{\nu}\cdot\mathbf{f})) + \mathbf{n}\cdot\nabla\boldsymbol{\lambda}\cdot\boldsymbol{\nu}\boldsymbol{\lambda}\cdot\mathbf{f}\boldsymbol{\lambda} - \boldsymbol{\nu}(\mathbf{n}\cdot\nabla(\boldsymbol{\lambda}\cdot\mathbf{f}))\\ & - \mathbf{n}\cdot\nabla\boldsymbol{\nu}\cdot\boldsymbol{\lambda}\boldsymbol{\nu}\cdot\mathbf{f}\boldsymbol{\nu} + \boldsymbol{\lambda}(\mathbf{n}\cdot\nabla(\boldsymbol{\nu}\cdot\mathbf{f})) - \boldsymbol{\nu}\cdot\nabla\boldsymbol{\lambda}\cdot\mathbf{n}\boldsymbol{\lambda}\cdot\mathbf{f}\boldsymbol{\lambda}\\ & + \mathbf{n}(\boldsymbol{\nu}\cdot\nabla(\boldsymbol{\lambda}\cdot\mathbf{f})) + \boldsymbol{\nu}\cdot\nabla\mathbf{n}\cdot\boldsymbol{\lambda}\mathbf{n}\cdot\mathbf{fn} - \boldsymbol{\lambda}(\boldsymbol{\nu}\cdot\nabla(\mathbf{n}\cdot\mathbf{f}))\end{aligned}$$

Grouping terms containing $\boldsymbol{\lambda}$, $\mathbf{n}$ and $\boldsymbol{\nu}$, respectively, yields

$$\begin{aligned}\nabla\times\mathbf{f} = {} & \boldsymbol{\lambda}[-\mathbf{n}\cdot\nabla\boldsymbol{\nu}\cdot\mathbf{nn}\cdot\mathbf{f} + \boldsymbol{\nu}\cdot\nabla\mathbf{n}\cdot\boldsymbol{\nu}\boldsymbol{\nu}\cdot\mathbf{f} + \mathbf{n}\cdot\nabla(\boldsymbol{\nu}\cdot\mathbf{f}) - \boldsymbol{\nu}\cdot\nabla(\mathbf{n}\cdot\mathbf{f})]\\ & + \mathbf{n}[\boldsymbol{\lambda}\cdot\nabla\boldsymbol{\nu}\cdot\boldsymbol{\lambda}\boldsymbol{\lambda}\cdot\mathbf{f} - \boldsymbol{\nu}\cdot\nabla\boldsymbol{\lambda}\cdot\boldsymbol{\nu}\boldsymbol{\nu}\cdot\mathbf{f} + \boldsymbol{\nu}\cdot\nabla(\boldsymbol{\lambda}\cdot\mathbf{f}) - \boldsymbol{\lambda}\cdot\nabla(\boldsymbol{\nu}\cdot\mathbf{f})]\\ & + \boldsymbol{\nu}[-\boldsymbol{\lambda}\cdot\nabla\mathbf{n}\cdot\boldsymbol{\lambda}\boldsymbol{\lambda}\cdot\mathbf{f} + \mathbf{n}\cdot\nabla\boldsymbol{\lambda}\cdot\mathbf{nn}\cdot\mathbf{f} + \boldsymbol{\lambda}\cdot\nabla(\mathbf{n}\cdot\mathbf{f}) - \mathbf{n}\cdot\nabla(\boldsymbol{\lambda}\cdot\mathbf{f})]\end{aligned} \tag{106}$$

2.4.2 Surficial curl

For a surface vector $\mathbf{f}_{\mathbf{n}}^{s} = \mathbf{f} - \mathbf{nn}\cdot\mathbf{f}$, its surficial curl is

$$\begin{aligned}\nabla_{\mathbf{n}}^{S}\times\mathbf{f}_{\mathbf{n}}^{s} &= (\nabla - \mathbf{nn}\cdot\nabla)\times(\mathbf{f} - \mathbf{nn}\cdot\mathbf{f})\\ &= \nabla\times\mathbf{f} + \mathbf{n}\cdot\nabla\mathbf{f}\times\mathbf{n} - \nabla\times\mathbf{nn}\cdot\mathbf{f} + \mathbf{n}\times\nabla(\mathbf{n}\cdot\mathbf{f})\\ &\quad - \mathbf{n}\cdot\nabla\mathbf{n}\times\mathbf{nn}\cdot\mathbf{f} - \nabla(\mathbf{n}\cdot\mathbf{f})\times\mathbf{n}\end{aligned} \tag{107}$$

in which Equation (16) has been used. Also, the third term cancels the fifth term by Equation (27). Therefore, we have

$$\nabla_{\mathbf{n}}^{S}\times\mathbf{f}_{\mathbf{n}}^{s} = \nabla\times\mathbf{f} + \mathbf{n}\cdot\nabla\mathbf{f}\times\mathbf{n} + \mathbf{n}\times\nabla(\mathbf{n}\cdot\mathbf{f}) \tag{108}$$

2.4.3 Curvilineal curl

For a curvilineal vector $\mathbf{f}_{\lambda}^{c} = \boldsymbol{\lambda}\boldsymbol{\lambda}\cdot\mathbf{f}$, its curvilineal curl is

$$\begin{aligned}\nabla_{\lambda}^{c}\times\mathbf{f}_{\lambda}^{c} &= (\boldsymbol{\lambda}\boldsymbol{\lambda}\cdot\nabla)\times(\boldsymbol{\lambda}\boldsymbol{\lambda}\cdot\mathbf{f})\\ &= \boldsymbol{\lambda}\times(\boldsymbol{\lambda}\cdot\nabla(\boldsymbol{\lambda}\boldsymbol{\lambda}\cdot\mathbf{f}))\\ &= \boldsymbol{\lambda}\times(\boldsymbol{\lambda}\cdot\nabla\boldsymbol{\lambda})\boldsymbol{\lambda}\cdot\mathbf{f} + \boldsymbol{\lambda}\times\boldsymbol{\lambda}(\boldsymbol{\lambda}\cdot\nabla(\boldsymbol{\lambda}\cdot\mathbf{f}))\end{aligned} \tag{109}$$

It becomes, by Equation (20),

$$\nabla_{\lambda}^{c}\times\mathbf{f}_{\lambda}^{c} = -(\boldsymbol{\lambda}\cdot\nabla\boldsymbol{\lambda})\times\boldsymbol{\lambda}\boldsymbol{\lambda}\cdot\mathbf{f} \tag{110}$$

If the curve is a straight line, $\nabla_{\lambda}^{c}\times\mathbf{f}_{\lambda}^{c}$ will vanish because $\boldsymbol{\lambda}\cdot\nabla\boldsymbol{\lambda} = 0$.

2.4.4 Spatial vs curvilineal curls

Note that

$$\begin{aligned}\nabla\times\mathbf{f} &= (\nabla^c_{\lambda}+\nabla^c_{\mathbf{n}}+\nabla^c_{\nu})\times(\mathbf{f}^c_{\lambda}+\mathbf{f}^c_{\mathbf{n}}+\mathbf{f}^c_{\nu})\\ &= \nabla^c_{\lambda}\times\mathbf{f}^c_{\lambda}+\nabla^c_{\lambda}\times\mathbf{f}^c_{\mathbf{n}}+\nabla^c_{\lambda}\times\mathbf{f}^c_{\nu}+\nabla^c_{\mathbf{n}}\times\mathbf{f}^c_{\lambda}+\nabla^c_{\mathbf{n}}\times\mathbf{f}^c_{\mathbf{n}}+\nabla^c_{\mathbf{n}}\times\mathbf{f}^c_{\nu}\\ &\quad+\nabla^c_{\nu}\times\mathbf{f}^c_{\lambda}+\nabla^c_{\nu}\times\mathbf{f}^c_{\mathbf{n}}+\nabla^c_{\nu}\times\mathbf{f}^c_{\nu}\end{aligned} \tag{111}$$

Since, by Equation (110),

$$\nabla^c_{\lambda}\times\mathbf{f}^c_{\lambda} = -(\boldsymbol{\lambda}\cdot\nabla\boldsymbol{\lambda})\times\boldsymbol{\lambda}\boldsymbol{\lambda}\cdot\mathbf{f} = -(\boldsymbol{\lambda}\cdot\nabla\boldsymbol{\lambda})\times\mathbf{f}^c_{\lambda} \tag{112}$$

$$\nabla^c_{\mathbf{n}}\times\mathbf{f}^c_{\mathbf{n}} = -(\mathbf{n}\cdot\nabla\mathbf{n})\times\mathbf{n}n\cdot\mathbf{f} = -(\mathbf{n}\cdot\nabla\mathbf{n})\times\mathbf{f}^c_{\mathbf{n}} \tag{113}$$

$$\nabla^c_{\nu}\times\mathbf{f}^c_{\nu} = -(\boldsymbol{\nu}\cdot\nabla\boldsymbol{\nu})\times\boldsymbol{\nu}\boldsymbol{\nu}\cdot\mathbf{f} = -(\boldsymbol{\nu}\cdot\nabla\boldsymbol{\nu})\times\mathbf{f}^c_{\nu} \tag{114}$$

we have

$$\begin{aligned}\nabla\times\mathbf{f} = {} & \left[(\nabla^c_{\lambda}\times\mathbf{f}^c_{\mathbf{n}})+(\nabla^c_{\mathbf{n}}\times\mathbf{f}^c_{\lambda})\right]+\left[(\nabla^c_{\mathbf{n}}\times\mathbf{f}^c_{\nu})+(\nabla^c_{\nu}\times\mathbf{f}^c_{\mathbf{n}})\right]\\ &+\left[(\nabla^c_{\nu}\times\mathbf{f}^c_{\lambda})+(\nabla^c_{\lambda}\times\mathbf{f}^c_{\nu})\right]-\boldsymbol{\lambda}\cdot\nabla\boldsymbol{\lambda}\times\mathbf{f}^c_{\lambda}\\ &-\mathbf{n}\cdot\nabla\mathbf{n}\times\mathbf{f}^c_{\mathbf{n}}-\boldsymbol{\nu}\cdot\nabla\boldsymbol{\nu}\times\mathbf{f}^c_{\nu}\end{aligned} \tag{115}$$

2.4.5 A formula of surficial curl

Note that

$$\begin{aligned}\nabla^S_{\mathbf{n}} &= \nabla^c_{\nu}+\nabla^c_{\lambda}\\ \mathbf{f}^s_{\mathbf{n}} &= \mathbf{f}^c_{\nu}+\mathbf{f}^c_{\lambda}\end{aligned}$$

Thus,

$$\begin{aligned}\nabla^S_{\mathbf{n}}\times\mathbf{f}^s_{\mathbf{n}} &= (\nabla^c_{\nu}+\nabla^c_{\lambda})\times(\mathbf{f}^c_{\nu}+\mathbf{f}^c_{\lambda})\\ &= \nabla^c_{\nu}\times\mathbf{f}^c_{\nu}+\nabla^c_{\nu}\times\mathbf{f}^c_{\lambda}+\nabla^c_{\lambda}\times\mathbf{f}^c_{\nu}+\nabla^c_{\lambda}\times\mathbf{f}^c_{\lambda}\end{aligned} \tag{116}$$

Also

$$\nabla^c_{\nu}\times\mathbf{f}^c_{\nu} = (\boldsymbol{\nu}\boldsymbol{\nu}\cdot\nabla)\times(\boldsymbol{\nu}\boldsymbol{\nu}\cdot\mathbf{f})$$

which becomes, by the chain rule,

$$\nabla^c_{\nu}\times\mathbf{f}^c_{\nu} = \boldsymbol{\nu}\times(\boldsymbol{\nu}\cdot\nabla\boldsymbol{\nu})\boldsymbol{\nu}\cdot\mathbf{f}+\boldsymbol{\nu}\times\boldsymbol{\nu}(\boldsymbol{\nu}\cdot\nabla(\boldsymbol{\nu}\cdot\mathbf{f}))$$

By Equation (20), we have

$$\nabla^c_{\nu}\times\mathbf{f}^c_{\nu} = -(\boldsymbol{\nu}\cdot\nabla\boldsymbol{\nu})\times\boldsymbol{\nu}\boldsymbol{\nu}\cdot\mathbf{f}$$

It becomes, by $\boldsymbol{\nu}\boldsymbol{\nu}+\boldsymbol{n}\boldsymbol{n}+\boldsymbol{\lambda}\boldsymbol{\lambda}=\mathbf{I}$,

$$\nabla^c_{\nu}\times\mathbf{f}^c_{\nu} = -[(\boldsymbol{\nu}\cdot\nabla\boldsymbol{\nu})\times\boldsymbol{\nu}\cdot\boldsymbol{\nu}\boldsymbol{\nu}+(\boldsymbol{\nu}\cdot\nabla\boldsymbol{\nu})\times\boldsymbol{\nu}\cdot\boldsymbol{\lambda}\boldsymbol{\lambda}+(\boldsymbol{\nu}\cdot\nabla\boldsymbol{\nu})\times\boldsymbol{\nu}\cdot\mathbf{nn}]\boldsymbol{\nu}\cdot\mathbf{f}$$

Therefore, we have by Equations (18), (19) and (89)

$$\nabla^c_{\nu}\times\mathbf{f}^c_{\nu} = -(\boldsymbol{\nu}\cdot\nabla\boldsymbol{\nu})\cdot\mathbf{n}\boldsymbol{\nu}\cdot\mathbf{f}\boldsymbol{\lambda}+(\boldsymbol{\nu}\cdot\nabla\boldsymbol{\nu})\cdot\boldsymbol{\lambda}\boldsymbol{\nu}\cdot\mathbf{fn} \tag{117}$$

Similarly

$$\nabla^c_{\nu}\times\mathbf{f}^c_{\nu} = -(\boldsymbol{\nu}\cdot\nabla\boldsymbol{\lambda})\cdot\mathbf{n}\boldsymbol{\lambda}\cdot\mathbf{f}\boldsymbol{\lambda}+\mathbf{n}\boldsymbol{\nu}\cdot\nabla(\boldsymbol{\lambda}\cdot\mathbf{f}) \tag{118}$$

$$\nabla^{c}_{\lambda} \times \mathbf{f}^{c}_{\nu} = \boldsymbol{\lambda} \cdot \nabla \boldsymbol{\nu} \cdot \mathbf{n}\boldsymbol{\nu} \cdot \mathbf{f}\boldsymbol{\nu} - \mathbf{n}(\boldsymbol{\lambda} \cdot \nabla(\boldsymbol{\nu} \cdot \mathbf{f})) \tag{119}$$

$$\nabla^{c}_{\lambda} \times \mathbf{f}^{c}_{\lambda} = \boldsymbol{\lambda} \cdot \nabla \boldsymbol{\lambda} \cdot \mathbf{n}\boldsymbol{\nu}\boldsymbol{\lambda} \cdot \mathbf{f} - \boldsymbol{\lambda} \cdot \nabla \boldsymbol{\lambda} \cdot \boldsymbol{\nu}\mathbf{n}\boldsymbol{\lambda} \cdot \mathbf{f} \tag{120}$$

A substitution of Equations (117)–(120) into Equation (116) yields

$$\begin{aligned} \nabla^{S}_{\mathbf{n}} \times \mathbf{f}^{s}_{\mathbf{n}} = & \boldsymbol{\lambda}[-\boldsymbol{\nu} \cdot \nabla \boldsymbol{\nu} \cdot \mathbf{n}\boldsymbol{\nu} \cdot \mathbf{f} - \boldsymbol{\nu} \cdot \nabla \boldsymbol{\lambda} \cdot \mathbf{n}\boldsymbol{\lambda} \cdot \mathbf{f}] \\ & + [\boldsymbol{\lambda} \cdot \nabla \boldsymbol{\nu} \cdot \mathbf{n}\boldsymbol{\nu} \cdot \mathbf{f} + \boldsymbol{\lambda} \cdot \nabla \boldsymbol{\lambda} \cdot \mathbf{n}\boldsymbol{\lambda} \cdot \mathbf{f}]\boldsymbol{\nu} \\ & + [\boldsymbol{\nu} \cdot \nabla \boldsymbol{\nu} \cdot \boldsymbol{\lambda}\boldsymbol{\nu} \cdot \mathbf{f} - \boldsymbol{\lambda} \cdot \nabla \boldsymbol{\lambda} \cdot \boldsymbol{\nu}\boldsymbol{\lambda} \cdot \mathbf{f} \\ & + \boldsymbol{\nu} \cdot \nabla(\boldsymbol{\lambda} \cdot \mathbf{f}) - \boldsymbol{\lambda} \cdot \nabla(\boldsymbol{\nu} \cdot \mathbf{f})]\mathbf{n} \end{aligned} \tag{121}$$

By the definition of $\nabla^{S}_{\mathbf{n}}$, we have

$$\begin{aligned} & [(\boldsymbol{\lambda}\boldsymbol{\nu} - \boldsymbol{\nu}\boldsymbol{\lambda}) \cdot \nabla^{S}_{\mathbf{n}}\mathbf{n} \cdot \mathbf{f}^{s}_{\mathbf{n}}] + [\mathbf{n}(\boldsymbol{\nu} \cdot \nabla^{S}_{\mathbf{n}}\mathbf{f}^{s}_{\mathbf{n}} \cdot \boldsymbol{\lambda} - \boldsymbol{\lambda} \cdot \nabla^{S}_{\mathbf{n}}\mathbf{f}^{s}_{\mathbf{n}} \cdot \boldsymbol{\nu})] \\ & \quad = (\boldsymbol{\lambda}\boldsymbol{\nu} - \boldsymbol{\nu}\boldsymbol{\lambda}) \cdot (\boldsymbol{\lambda}\boldsymbol{\lambda} \cdot \nabla + \boldsymbol{\nu}\boldsymbol{\nu} \cdot \nabla)\mathbf{n} \cdot \mathbf{f}^{s}_{\mathbf{n}} + \mathbf{n}\,(\boldsymbol{\nu} \cdot (\boldsymbol{\nu}\boldsymbol{\nu} \cdot \nabla + \boldsymbol{\lambda}\boldsymbol{\lambda} \cdot \nabla)\mathbf{f}^{s}_{\mathbf{n}} \cdot \boldsymbol{\lambda} \\ & \qquad - \boldsymbol{\lambda} \cdot (\boldsymbol{\nu}\boldsymbol{\nu} \cdot \nabla + \boldsymbol{\lambda}\boldsymbol{\lambda} \cdot \nabla)\mathbf{f}^{s}_{\mathbf{n}} \cdot \boldsymbol{\nu}) \end{aligned}$$

which becomes, by Equation (15),

$$\begin{aligned} & [(\boldsymbol{\lambda}\boldsymbol{\nu} - \boldsymbol{\nu}\boldsymbol{\lambda}) \cdot \nabla^{S}_{\mathbf{n}}\mathbf{n} \cdot \mathbf{f}^{s}_{\mathbf{n}}] + [\mathbf{n}(\boldsymbol{\nu} \cdot \nabla^{S}_{\mathbf{n}}\mathbf{f}^{s}_{\mathbf{n}} \cdot \boldsymbol{\lambda} - \boldsymbol{\lambda} \cdot \nabla^{S}_{\mathbf{n}}\mathbf{f}^{s}_{\mathbf{n}} \cdot \boldsymbol{\nu})] \\ & \quad = \boldsymbol{\lambda}\boldsymbol{\nu} \cdot \nabla \mathbf{n} \cdot \mathbf{f}^{s}_{\mathbf{n}} - \boldsymbol{\nu}\boldsymbol{\lambda} \cdot \nabla \mathbf{n} \cdot \mathbf{f}^{s}_{\mathbf{n}} + \mathbf{n}[\boldsymbol{\nu} \cdot \nabla \boldsymbol{\nu} \cdot \boldsymbol{\lambda}\boldsymbol{\nu}\mathbf{f} + \boldsymbol{\nu} \cdot \nabla(\boldsymbol{\nu} \cdot \mathbf{f})\,\boldsymbol{\nu} \cdot \boldsymbol{\lambda} \\ & \qquad + \boldsymbol{\nu} \cdot \nabla \boldsymbol{\lambda} \cdot \boldsymbol{\lambda}\boldsymbol{\lambda} \cdot \mathbf{f} + \boldsymbol{\nu} \cdot \nabla(\mathbf{f} \cdot \boldsymbol{\lambda})\boldsymbol{\lambda} \cdot \boldsymbol{\lambda} - \boldsymbol{\lambda} \cdot \nabla \boldsymbol{\nu} \cdot \boldsymbol{\nu}\boldsymbol{\nu} \cdot \mathbf{f} \\ & \qquad - \boldsymbol{\lambda} \cdot \nabla(\boldsymbol{\nu} \cdot \mathbf{f})\boldsymbol{\nu} \cdot \boldsymbol{\nu} - \boldsymbol{\lambda} \cdot \nabla \boldsymbol{\lambda} \cdot \boldsymbol{\nu}\boldsymbol{\lambda} \cdot \mathbf{f} - \boldsymbol{\lambda} \cdot \nabla(\boldsymbol{\lambda} \cdot \mathbf{f})\boldsymbol{\lambda} \cdot \boldsymbol{\nu}] \\ & \quad = \boldsymbol{\lambda}[\boldsymbol{\nu} \cdot \nabla \mathbf{n} \cdot (\boldsymbol{\lambda}\boldsymbol{\lambda} \cdot \mathbf{f} + \boldsymbol{\nu}\boldsymbol{\nu} \cdot \mathbf{f})] - \boldsymbol{\nu}[\boldsymbol{\lambda} \cdot \nabla \mathbf{n} \cdot (\boldsymbol{\lambda}\boldsymbol{\lambda} \cdot \mathbf{f} + \boldsymbol{\nu}\boldsymbol{\nu} \cdot \mathbf{f})] \\ & \qquad + \mathbf{n}[\boldsymbol{\nu} \cdot \nabla \boldsymbol{\nu} \cdot \boldsymbol{\lambda}\boldsymbol{\nu} \cdot \mathbf{f} + \boldsymbol{\nu} \cdot \nabla \boldsymbol{\lambda} \cdot \boldsymbol{\lambda}\boldsymbol{\lambda} \cdot \mathbf{f} + \boldsymbol{\nu} \cdot \nabla(\boldsymbol{\lambda} \cdot \mathbf{f}) - \boldsymbol{\lambda} \cdot \nabla(\boldsymbol{\nu} \cdot \mathbf{f}) - \boldsymbol{\lambda} \cdot \nabla \boldsymbol{\lambda} \cdot \boldsymbol{\nu}\boldsymbol{\lambda} \cdot \mathbf{f}] \\ & \quad = \boldsymbol{\lambda}[\boldsymbol{\nu} \cdot \nabla \mathbf{n} \cdot \boldsymbol{\lambda}\boldsymbol{\lambda} \cdot \mathbf{f} + \boldsymbol{\nu} \cdot \nabla \mathbf{n} \cdot \boldsymbol{\nu}\boldsymbol{\nu} \cdot \mathbf{f}] - \boldsymbol{\nu}[\boldsymbol{\lambda} \cdot \nabla \mathbf{n} \cdot \boldsymbol{\lambda}\boldsymbol{\lambda} \cdot \mathbf{f} + \boldsymbol{\lambda}\nabla \mathbf{n} \cdot \boldsymbol{\nu}\boldsymbol{\nu} \cdot \mathbf{f}] \\ & \qquad + \mathbf{n}[\boldsymbol{\nu} \cdot \nabla \boldsymbol{\nu} \cdot \boldsymbol{\lambda}\boldsymbol{\nu} \cdot \mathbf{f} + \boldsymbol{\nu} \cdot \nabla(\boldsymbol{\lambda} \cdot \mathbf{f}) - \boldsymbol{\lambda} \cdot \nabla(\boldsymbol{\nu} \cdot \mathbf{f}) - \boldsymbol{\lambda} \cdot \nabla \boldsymbol{\lambda} \cdot \boldsymbol{\nu}\boldsymbol{\lambda} \cdot \mathbf{f}] \end{aligned}$$

Rearranging this expression yields

$$\begin{aligned} & [(\boldsymbol{\lambda}\boldsymbol{\nu} - \boldsymbol{\nu}\boldsymbol{\lambda}) \cdot \nabla^{S}_{\mathbf{n}}\mathbf{n} \cdot \mathbf{f}^{s}_{\mathbf{n}}] + [\mathbf{n}(\boldsymbol{\nu} \cdot \nabla^{S}_{\mathbf{n}}\mathbf{f}^{s}_{\mathbf{n}} \cdot \boldsymbol{\lambda} - \boldsymbol{\lambda} \cdot \nabla^{S}_{\mathbf{n}}\mathbf{f}^{s}_{\mathbf{n}} \cdot \boldsymbol{\nu})] \\ & \quad = \boldsymbol{\lambda}[-\boldsymbol{\nu} \cdot \nabla \boldsymbol{\nu} \cdot \mathbf{n}\boldsymbol{\nu} \cdot \mathbf{f} - \boldsymbol{\nu} \cdot \nabla \boldsymbol{\lambda} \cdot \mathbf{n}\boldsymbol{\lambda} \cdot \mathbf{f}] \\ & \qquad + [\boldsymbol{\lambda} \cdot \nabla \boldsymbol{\nu} \cdot \mathbf{n}\boldsymbol{\nu} \cdot \mathbf{f} + \boldsymbol{\lambda} \cdot \nabla \boldsymbol{\lambda} \cdot \mathbf{n}\boldsymbol{\lambda} \cdot \mathbf{f}]\boldsymbol{\nu} \\ & \qquad + [\boldsymbol{\nu} \cdot \nabla \boldsymbol{\nu} \cdot \boldsymbol{\lambda}\boldsymbol{\nu} \cdot \mathbf{f} - \boldsymbol{\lambda} \cdot \nabla \boldsymbol{\lambda} \cdot \boldsymbol{\nu}\boldsymbol{\lambda} \cdot \mathbf{f} + \boldsymbol{\nu} \cdot \nabla(\boldsymbol{\lambda} \cdot \mathbf{f}) - \boldsymbol{\lambda} \cdot \nabla(\boldsymbol{\nu} \cdot \mathbf{f})]\mathbf{n} \end{aligned} \tag{122}$$

A comparison of Equations (121) with (122) leads to

$$\nabla^{S}_{\mathbf{n}} \times \mathbf{f}^{s}_{\mathbf{n}} = [(\boldsymbol{\lambda}\boldsymbol{\nu} - \boldsymbol{\nu}\boldsymbol{\lambda}) \cdot \nabla^{S}_{\mathbf{n}}\mathbf{n} \cdot \mathbf{f}^{s}_{\mathbf{n}}] + [\mathbf{n}(\boldsymbol{\nu} \cdot \nabla^{S}_{\mathbf{n}}\mathbf{f}^{s}_{\mathbf{n}} \cdot \boldsymbol{\lambda} - \boldsymbol{\lambda} \cdot \nabla^{S}_{\mathbf{n}}\mathbf{f}^{s}_{\mathbf{n}} \cdot \boldsymbol{\nu})] \tag{123}$$

2.5 Time derivatives

Partial time derivatives of a function can be taken as fixed to a position in space, on a surface, or along a curve. We discuss these three temporal derivatives and develop relationships among them via the total time derivative of the function in this section.

2.5.1 Partial time derivative in a space

The partial derivative with respect to time of a function $f(\mathbf{x}, t)$ is written as,

$$\frac{\partial f(\mathbf{x},t)}{\partial t} = \left.\frac{\partial f}{\partial t}\right|_{\mathbf{x}} \tag{124}$$

where the vertical bar with subscript $\mathbf{x}$ indicates that the spatial coordinates are held constant. It measures the *local rate* of change in a function with time at a point fixed in space.

The total time derivative measures the *total* rate of change in a function with time due to both the local rate of change and the change of the location at which the function is evaluated. The relation between the partial and the total time derivatives is

$$\frac{\mathrm{d}f}{\mathrm{d}t} = \left.\frac{\partial f}{\partial t}\right|_{\mathbf{x}} + \mathbf{w} \cdot \nabla f \tag{125}$$

where

$$\frac{\mathrm{d}x}{\mathrm{d}t} = \mathbf{w}$$

2.5.2 Partial time derivative on a surface

Consider a scalar function f of time t and points on a surface S specified by their surficial coordinates $\mathbf{u}$. The partial time derivative of this function at a position fixed on the surface is

$$\frac{\partial f(\mathbf{u},t)}{\partial t} = \left.\frac{\partial f}{\partial t}\right|_{\mathbf{u}} \tag{126}$$

where the vertical bar with subscript $\mathbf{u}$ indicates that the surficial coordinates are held constant. Note that the partial derivative in Equation (126) mandates that although the surface coordinates are being held constant, movement in the direction normal to the surface at the velocity of the surface would be necessary in order to remain on the surface and measure the indicated partial derivative.

If the time derivative of $f(\mathbf{u}, t)$ is to be calculated while moving on the surface at some velocity $\mathrm{d}\mathbf{u}/\mathrm{d}t = \mathbf{w}_{\mathbf{n}}^{\mathrm{s}}$ (subscript $\mathbf{n}$ indicates the unit normal of the surface S), the rate of change of the surficial coordinate position must be accounted for. This time derivative of f is also a total time derivative and may be expressed as:

$$\frac{\mathrm{d}f}{\mathrm{d}t} = \left.\frac{\partial f}{\partial t}\right|_{\mathbf{u}} + \mathbf{w}_{\mathbf{n}}^{\mathrm{s}} \cdot \nabla_{\mathbf{n}}^{\mathrm{S}} f \tag{127}$$

where $\nabla_{\mathbf{n}}^{\mathrm{S}}$ is the surficial gradient operator given by Equation (42).

Note that the surficial coordinate may be expressed in terms of spatial coordinates such that $\mathbf{u} = \mathbf{u}(\mathbf{x})$. A point on a surface can also be viewed as a point specified by $\mathbf{x}$. Therefore, Equations (125) and (127) lead to

$$\left.\frac{\partial f}{\partial t}\right|_{\mathbf{x}} + \mathbf{w} \cdot \nabla f = \left.\frac{\partial f}{\partial t}\right|_{\mathbf{u}} + \mathbf{w}_{\mathbf{n}}^{\mathrm{s}} \cdot \nabla_{\mathbf{n}}^{\mathrm{S}} f \tag{128}$$

Because $\mathbf{w}_{\mathbf{n}}^{s}$ has no component in the $\mathbf{n}$ direction, we have

$$\mathbf{w}_{\mathbf{n}}^{s} \cdot \nabla_{\mathbf{n}}^{S} = \mathbf{w}_{\mathbf{n}}^{s} \cdot (\nabla - \nabla_{\mathbf{n}}^{c}) = \mathbf{w}_{\mathbf{n}}^{s} \cdot \nabla = \mathbf{w} \cdot \nabla - \mathbf{w} \cdot \mathbf{nn} \cdot \nabla \tag{129}$$

and Equation (128) becomes:

$$\left.\frac{\partial f}{\partial t}\right|_{\mathbf{x}} = \left.\frac{\partial f}{\partial t}\right|_{\mathbf{u}} - \mathbf{w} \cdot \mathbf{nn} \cdot \nabla f \tag{130}$$

This relates the partial time derivative of a function fixed in space to one fixed to a point on a surface. It plays an important role in developing multiscale theorems. Note that for the case where the surface normal velocity $\mathbf{w} \cdot \mathbf{n}$ is zero, the two partial time derivatives are identical.

2.5.3 Partial time derivative along a curve

Consider a scalar function f of time t and points on a curve C specified by their curvilineal coordinates $\mathbf{l}$. The partial time derivative of f at a point on the curve is

$$\frac{\partial f(\mathbf{l},t)}{\partial t} = \left.\frac{\partial f}{\partial t}\right|_{\mathbf{l}} \tag{131}$$

For this time derivative, although the coordinate along the curve is held constant, the evaluation takes place while moving with the curve as it translates in directions normal to $\mathbf{l}$.

The total time derivative is the rate of change of f calculated while moving at some velocity $\mathrm{d}\mathbf{l}/\mathrm{d}t = \mathbf{w}_{\lambda}^{c}$ along the curve. Note that $\mathbf{w}_{\lambda}^{c}$ is tangent to the curve and may also be written as $\mathbf{w}_{\lambda}^{c} = \boldsymbol{\lambda\lambda} \cdot \mathbf{w}$. Thus the total time derivative is:

$$\frac{\mathrm{d}f}{\mathrm{d}t} = \left.\frac{\partial f}{\partial t}\right|_{\mathbf{l}} + \mathbf{w}_{\lambda}^{c} \cdot \nabla_{\lambda}^{c} f \tag{132}$$

where ∇_{λ}^{c} is the gradient operator along the curve given by Equation (43). If the velocity at which the curve moves through space in a direction normal to the curve is indicated as $\mathbf{w} - \mathbf{w}_{\lambda}^{c}$ the total derivatives of f given by Equations (125) and (132) lead to

$$\left.\frac{\partial f}{\partial t}\right|_{\mathbf{x}} + \mathbf{w} \cdot \nabla f = \left.\frac{\partial f}{\partial t}\right|_{\mathbf{l}} + \mathbf{w}_{\lambda}^{c} \cdot \nabla_{\lambda}^{c} f \tag{133}$$

Note that

$$\mathbf{w}_{\lambda}^{c} \cdot \nabla_{\lambda}^{c} = \mathbf{w}_{\lambda}^{c} \cdot \nabla = \mathbf{w} \cdot \boldsymbol{\lambda\lambda} \cdot \nabla$$

Equation (133) becomes

$$\left.\frac{\partial f}{\partial t}\right|_{\mathbf{x}} = \left.\frac{\partial f}{\partial t}\right|_{\mathbf{l}} - (\mathbf{w} - \mathbf{w} \cdot \boldsymbol{\lambda\lambda}) \cdot \nabla f \tag{134}$$

This equation expresses the relationship between a time derivative of f at a point fixed in space to the one at a point fixed on a curve that moves through space with velocity $\mathbf{w} - \mathbf{w} \cdot \boldsymbol{\lambda\lambda}$.

By equating the total time derivatives in Equations (127) and (133) and employing similar manipulations to those above, we can also obtain the following relationships between the time derivative of a function at a point on a curve and the one at a point on the surface containing the curve:

$$\left.\frac{\partial f}{\partial t}\right|_{\mathbf{u}} = \left.\frac{\partial f}{\partial t}\right|_{1} - (\mathbf{w}_{\mathbf{n}}^{\mathrm{s}} - \mathbf{w}_{\mathbf{n}}^{\mathrm{s}} \cdot \boldsymbol{\lambda}\boldsymbol{\lambda}) \cdot \nabla_{\mathbf{n}}^{\mathrm{S}} f \tag{135}$$

$$\left.\frac{\partial f}{\partial t}\right|_{\mathbf{u}} = \left.\frac{\partial f}{\partial t}\right|_{1} - (\mathbf{w}_{\mathbf{n}}^{\mathrm{s}} - \mathbf{w}_{\mathbf{n}}^{\mathrm{s}} \cdot \boldsymbol{\lambda}\boldsymbol{\lambda}) \cdot \nabla f \tag{136}$$

$$\left.\frac{\partial f}{\partial t}\right|_{\mathbf{u}} = \left.\frac{\partial f}{\partial t}\right|_{1} - (\mathbf{w} - \mathbf{w} \cdot \boldsymbol{\lambda}\boldsymbol{\lambda}) \cdot \nabla_{\mathbf{n}}^{\mathrm{S}} f \tag{137}$$

where $\mathbf{w}_{\mathbf{n}}^{\mathrm{s}}$ is the velocity of the surface and $\nabla_{\mathbf{n}}^{\mathrm{S}}$ is the surface gradient operator.

2.6 Derivatives of unit vectors

We present three types of identities involving spatial and temporal derivatives of a unit vector in terms of the unit vector **n.** All these identities are useful in developing multiscale theorems and apply to any of the unit vectors with suitable permutation of the coordinated directions.

Identity 1.

$$\nabla \cdot \mathbf{n} = \nabla_{\mathbf{n}}^{\mathrm{S}} \cdot \mathbf{n} \tag{138}$$

Proof. By Equation (42), we have

$$\begin{aligned} \nabla \cdot \mathbf{n} &= \nabla_{\mathbf{n}}^{\mathrm{S}} \cdot \mathbf{n} + (\mathbf{n}\mathbf{n} \cdot \nabla) \cdot \mathbf{n} \\ &= \nabla_{\mathbf{n}}^{\mathrm{S}} \cdot \mathbf{n} + \mathbf{n} \cdot (\mathbf{n} \cdot \nabla \mathbf{n}) = \nabla_{\mathbf{n}}^{\mathrm{S}} \cdot \mathbf{n} + \mathbf{n} \cdot \nabla \mathbf{n} \cdot \mathbf{n} \end{aligned} \tag{139}$$

Since $\nabla \mathbf{n} \cdot \mathbf{n} = \frac{1}{2} \nabla(\mathbf{n} \cdot \mathbf{n}) = 0$, the last term in Equation (139) is zero so that Identity 1 is proven.

Identity 2.

$$\nabla_{\mathbf{n}}^{\mathrm{S}} \times \mathbf{n} = 0 \tag{140}$$

Proof. By the definition of operator $\nabla_{\mathbf{n}}^{\mathrm{S}}$

$$\nabla^{\mathrm{S}} = \boldsymbol{\lambda}\boldsymbol{\lambda} \cdot \nabla + \boldsymbol{\nu}\boldsymbol{\nu} \cdot \nabla$$

Thus,

$$\nabla_{\mathbf{n}}^{\mathrm{S}} \times \mathbf{n} = (\boldsymbol{\lambda}\boldsymbol{\lambda} \cdot \nabla + \boldsymbol{\nu}\boldsymbol{\nu} \cdot \nabla) \times \mathbf{n} \tag{141}$$

Distribution of the terms of the right-side yields,

$$\nabla_{\mathbf{n}}^{\mathrm{S}} \times \mathbf{n} = -(\boldsymbol{\lambda} \cdot \nabla \mathbf{n}) \times \boldsymbol{\lambda} - (\boldsymbol{\nu} \cdot \nabla \mathbf{n}) \times \boldsymbol{\nu} \tag{142}$$

It becomes, by Equations (17) and (18),

$$\nabla_{\mathbf{n}}^{\mathrm{S}} \times \mathbf{n} = -(\boldsymbol{\lambda} \cdot \nabla \mathbf{n}) \times (\mathbf{n} \times \boldsymbol{\nu}) - (\boldsymbol{\nu} \cdot \nabla \mathbf{n}) \times (\boldsymbol{\lambda} \times \mathbf{n}) \tag{143}$$

Applying, for any vectors **a**, **b**, and **c**,

$$\mathbf{a} \times (\mathbf{b} \times \mathbf{c}) = (\mathbf{a} \cdot \mathbf{c})\mathbf{b} - (\mathbf{a} \cdot \mathbf{b})\mathbf{c} \tag{144}$$

we obtain,

$$\nabla_{\mathbf{n}}^{\mathrm{S}} \times \mathbf{n} = -(\boldsymbol{\lambda} \cdot \nabla \mathbf{n} \cdot \boldsymbol{\nu})\mathbf{n} + (\boldsymbol{\lambda} \cdot \nabla \mathbf{n} \cdot \mathbf{n})\boldsymbol{\nu} - (\boldsymbol{\nu} \cdot \nabla \mathbf{n} \cdot \mathbf{n})\boldsymbol{\lambda} + (\boldsymbol{\nu} \cdot \nabla \mathbf{n} \cdot \boldsymbol{\lambda})\mathbf{n} \tag{145}$$

Since $\nabla \mathbf{n} \cdot \mathbf{n} = 0$, this equation reduces to:

$$\nabla_{\mathbf{n}}^{S} \times \mathbf{n} = -(\boldsymbol{\lambda} \cdot \nabla \mathbf{n} \cdot \boldsymbol{\nu} - \boldsymbol{\nu} \cdot \nabla \mathbf{n} \cdot \boldsymbol{\lambda})\mathbf{n} \tag{146}$$

However, $\nabla_{\mathbf{n}}^{S} \times \mathbf{n}$ must be orthogonal to $\mathbf{n}$ so that the term in parentheses in Equation (146) must be zero and the desired identity is obtained.

Identity 3a.

$$\left.\frac{\partial \mathbf{n}}{\partial t}\right|_{\mathbf{x}} = -\nabla_{\mathbf{n}}^{S}\mathbf{w} \cdot \mathbf{n} - \mathbf{w} \cdot \nabla \mathbf{n} \tag{147}$$

Identity 3b.

$$\frac{\mathrm{d}\mathbf{n}}{\mathrm{d}t} = -\nabla_{\mathbf{n}}^{S}\mathbf{w} \cdot \mathbf{n} \tag{148}$$

where the total derivative is defined by,

$$\frac{\mathrm{d}\mathbf{n}}{\mathrm{d}t} = \left.\frac{\partial \mathbf{n}}{\partial t}\right|_{\mathbf{x}} + \mathbf{w} \cdot \nabla \mathbf{n} \tag{149}$$

The proof of Identity 3a by a classical approach is more difficult than Identities 1 and 2 and is available in Anderson and Jackson (1967), Ferreira (1997), and Hadsell (1999). An application of indicator functions in Section 3 will, however, simplify the proof significantly. We therefore defer its proof to Section 3.

3. INDICATOR FUNCTIONS

Indicator functions come from extensions of Heaviside step function. They play a critical role in developing multiscale theorems. We first define indicator functions for identifying portions of a curve, a surface, or a space. We then develop a number of useful identities involving indicator functions and their derivatives. Finally, we present a brief outline of applications of indicator functions in developing multiscale theorems.

3.1 Definitions and derivatives

In this section we define indicator functions for specifying portions of a straight line, a space curve, a surface, and a volume. We also develop relations between the temporal and spatial derivatives of these functions and integral properties of their spatial derivatives. We employ a heuristic approach for a clearer understanding of important relations while minimizing mathematical complexities.

3.1.1 Straight line

In terms of the Heaviside function H(x), we define indicator function $\gamma^{L}(x,t)$ to identify a portion of a straight line in the domain a $a<x<b$ as

$$\gamma^{L}(x,t) = H[x - a(t)] - H[x - b(t)] \tag{150}$$

where the dependence of γ^L on time t would occur if either a or b were a function of time t. From the definition, it follows that γ^L is equal to 1 in $a<x<b$ and 0 outside of this domain on the line.

By the generalized function γ^L, the behavior of a function $f(x, t)$ where x is the coordinate of a straight line and t the time in the interval $a<x<b$ can be isolated by forming the product $\gamma^L f(x, t)$ which satisfies:

$$\gamma^L f(x,t) = \begin{cases} 0 & x<a \\ f(x,t) & a<x<b \\ 0 & x>b \end{cases} \tag{151}$$

Inclusion of the endpoints is irrelevant because all manipulations of $\gamma^L(x, t)$ in developing multiscale theorems will be performed under integration and the value of an integral is not affected by a finite value of the integrand at a point.

Consider the total time derivative of $\gamma^L(x, t)$,

$$\frac{d\gamma^L}{dt} = \left.\frac{\partial \gamma^L}{\partial t}\right|_x + \frac{\partial \gamma^L}{\partial x}\frac{dx}{dt} \tag{152}$$

where the first term on the right-hand side represents the rate of change of γ^L at a fixed point on the line, while the second term accounts for changes in γ^L as the point of evaluation moves along the line. Thus the quantity $\mathrm{d}\gamma^L/\mathrm{d}t$ is the total change in γ^L that an observer moving with velocity $\mathrm{d}x/\mathrm{d}t$ would see. Note that γ^L is used to identify the region of interest $a<x<b$ where the boundary points of this region, a and b, may be functions of time. Further, note that $\mathrm{d}x/\mathrm{d}t$ is the velocity of an observer at any point in the domain $-\infty<x<\infty$ and may exhibit spatial and/or temporal variation. Generally, $\mathrm{d}\gamma^L/\mathrm{d}t$ will be non-zero. However, $\mathrm{d}\gamma^L/\mathrm{d}t$ will be zero if the velocity is constrained such that $\mathrm{d}x/\mathrm{d}t = w_x$ where w_x is defined as follows: (1) $w_x|_a = \mathrm{d}a/\mathrm{d}t$ and $w_x|_b = \mathrm{d}b/\mathrm{d}t$ (i.e. an observer at the boundary of the region of interest moves with the boundary as the region deforms) and (2) w_x is chosen such that an observer not on the boundary never crosses the boundary (e.g. an observer at x moves with the same velocity w_x as the point x). Such an observer will detect no change in the value of γ^L with time so that

$$\left.\frac{\partial \gamma^L}{\partial t}\right|_x = -\frac{\partial \gamma^L}{\partial x} w_x \tag{153}$$

Note that (Vladimirov, 1971; Vladimirov, 2002; Wang and Zhou, 2000),

$$\frac{\partial \gamma^L(x,t)}{\partial x} = \delta(x-a) - \delta(x-b) \tag{154}$$

where $\delta(x - x_0)$ is the Dirac delta function defined as

$$\delta(x - x_0) = \begin{cases} 0 & x \neq x_0 \\ \infty & x = x_0 \end{cases}$$

Also (Vladimirov, 1971; Vladimirov, 2002; Wang and Zhou, 2000),

$$\int_{-\infty}^{+\infty} f(x)\delta(x - x_0)\mathrm{d}x = f(x_0)$$

Thus

$$-\int_{-\infty}^{+\infty} f(x,t)\frac{\partial \gamma^{\mathrm{L}}}{\partial x}\mathrm{d}x = f(b,t) - f(a,t) \tag{155}$$

The readers are referred to Wang and Zhou (2000), Vladimirov (1971, 2002), Antoniou and Lumer (1999), Demidov (2001), Farassat (1977), Ferreira (1997), Hadsell (1999), Kanwal (2004), Kinnmark and Gray (1984), and Shilov (1968) for a systematic discussion of the Heaviside function $H(x)$ and delta function $\delta(x)$.

3.1.2 Space curve

To identify a portion of a space curve, define an indicator function $\gamma^{\mathrm{c}}(l,t)$ as

$$\gamma^{\mathrm{c}}(l,t) = H[g^{\mathrm{c}}(l;a)] - H[g^{\mathrm{c}}(l;b)] \tag{156}$$

such that

$$\gamma^{\mathrm{c}}(l,t) = \begin{cases} 0 & l<a(t) \\ 1 & a(t)<l<b(t) \\ 0 & l>b(t) \end{cases} \tag{157}$$

where the superscript c is used to indicate that the indicator function is defined along a simple space curve C, l denotes the coordinate along the curve $g^{\mathrm{c}}(l;a) = l - a(t)$, $g^{\mathrm{c}}(l;b) = l - b(t)$ (conceptual extensions of the functions x–a and x–b used in Equation (150)). As such, it will be referred to as the position function. Note that g^{c} depends on the independent variable as well as a parameter defining a point on the curve. H is a conceptual extension of the Heaviside function undergoing a step change in value at $g^{\mathrm{c}} = 0$. Time dependence of $\gamma^{\mathrm{c}}(l,t)$ occurs through the dependence of a and b on time. Figure 8 illustrates use of γ^{c} to specify a portion of a space curve.

The total time derivative of γ^{c} along the curve is

$$\frac{\mathrm{d}\gamma^{\mathrm{c}}}{\mathrm{d}t} = \left.\frac{\partial \gamma^{\mathrm{c}}}{\partial t}\right|_{l} + \frac{\mathrm{d}l}{\mathrm{d}t}\boldsymbol{\lambda}\cdot\nabla^{\mathrm{c}}\gamma^{\mathrm{c}} \tag{158}$$

Similarly to the straight line case, this total derivative will be zero if $\mathrm{d}l/\mathrm{d}t$ is set equal to an observation velocity along the curve, $\mathbf{w}^{\mathrm{c}}$, under the following constraints: (1) $\mathbf{w}^{\mathrm{c}}$ is tangent to the curve and at the boundary of the portion of the curve of interest it is equal to the velocity of that boundary; and (2) at other points on the curve, $\mathbf{w}^{\mathrm{c}}$ is such that the points in the region specified by γ^{c} do not cross the boundary of the portion of the curve of interest, forever. Under these constraints:

$$\left.\frac{\partial \gamma^{\mathrm{c}}}{\partial t}\right|_{l} = -\mathbf{w}^{\mathrm{c}}\cdot\nabla^{\mathrm{c}}_{\lambda}\gamma^{\mathrm{c}} \tag{159}$$

For a space curve, the orientation is not necessarily constant, so that the curvature must be accounted for in evaluating the spatial derivative of γ^{c}. If $\boldsymbol{\lambda}$ is a unit vector tangent to a curve C, the change of γ^{c} with position along the curve is $\boldsymbol{\lambda}\cdot\nabla^{\mathrm{c}}_{\lambda}\gamma^{\mathrm{c}}$. In terms of γ^{c} defined in Equation (156)

$$\boldsymbol{\lambda}\cdot\nabla^{\mathrm{c}}_{\lambda}\gamma^{\mathrm{c}} = \delta[g^{\mathrm{c}}(l;a)] - \delta[g^{\mathrm{c}}(l;b)] \tag{160}$$

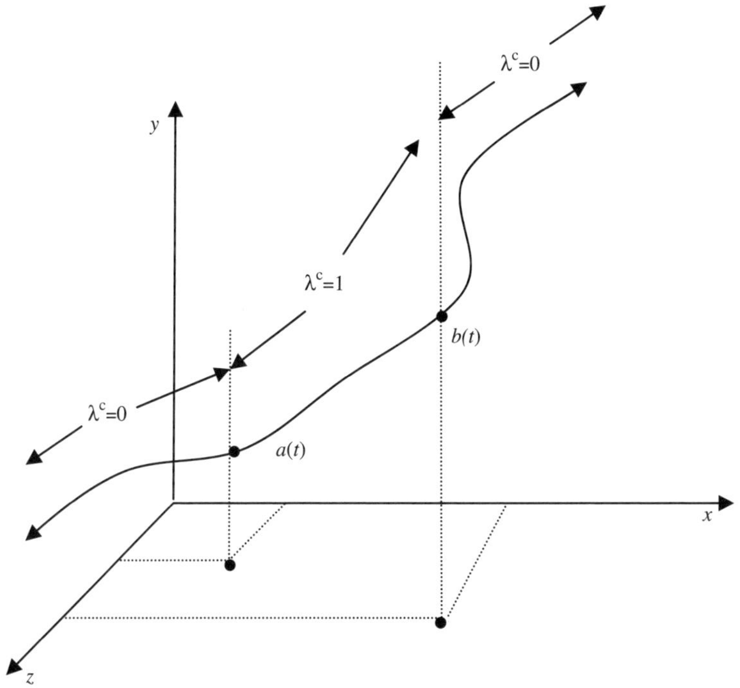

Figure 8 γ^c for identifying a portion of a space curve.

Let **e** be a unit vector which is tangent to **λ** but pointing outward from the region of interest $a<l<b$ such that $\mathbf{e} = -\boldsymbol{\lambda}$ at $l = a$ and $\mathbf{e} = \boldsymbol{\lambda}$ at $l = b$ then

$$\nabla^c_{\lambda}\gamma^c = -\mathbf{e}|_{l=a}\delta[g^c(l;a)] - \mathbf{e}|_{l=b}\delta[g^c(l;b)] \tag{161}$$

In some cases, a generalized function γ^c will be used to indicate a number of segments along a curve. If the N points across which the function undergoes a change in value from 1 to 0 (or vice versa) are indicated as a_i, then

$$\nabla^c_{\lambda}\gamma^c = -\sum_{i=1}^{N}\mathbf{e}|_{l=a_i}\,\delta[g^c(l;a_i)] \tag{162}$$

For the case where the curve of interest is a closed curve such that $\gamma^c = 1$ along the entire curve (i.e., $N = 0$ in Equation (162)), $\nabla^c_{\lambda}\gamma^c = 0$ along the entire curve.

For a continuous function f in $a<l<b$, a direct extension of Equation (155) yields

$$-\int_{-\infty}^{\infty} f\boldsymbol{\lambda}\cdot\nabla^c\gamma^c \mathrm{d}c = f|_{l=b} - f|_{l=a} \tag{163}$$

For the more general case where γ^c is used to identify more than one segment of a curve C such that Equation (162) applies, it becomes

$$-\int_{-\infty}^{\infty} f\boldsymbol{\lambda}\cdot\nabla^c_{\lambda}\gamma^c \mathrm{d}c = \sum_{i=1}^{N}(\boldsymbol{\lambda}\cdot\mathbf{e}f)|_{l=a_i} \tag{164}$$

or

$$-\int_{-\infty}^{\infty} f\nabla_{\lambda}^{c}\gamma^{c}\mathrm{d}c = \sum_{i=1}^{N}(\mathbf{e}f)\Big|_{l=a_i} \tag{165}$$

Similarly, we have

$$-\int_{-\infty}^{\infty} f\mathbf{w}\cdot\nabla_{\lambda}^{c}\gamma^{c}\mathrm{d}c = \sum_{i=1}^{N}(\mathbf{w}\cdot\mathbf{e}f)\Big|_{l=a_i} \tag{166}$$

$$-\int_{-\infty}^{\infty} f\mathbf{w}\times\nabla_{\lambda}^{c}\gamma^{c}\mathrm{d}c = \sum_{i=1}^{N}(\mathbf{w}\times\mathbf{e}f)\Big|_{l=a_i} \tag{167}$$

where $\mathbf{w}$ is an arbitrary vector function. Note also that $\mathbf{e}$ points outward from the region identified by $\gamma^c = 1$ at each of the points a_i such that $\mathbf{e}$ will equal either $\boldsymbol{\lambda}$ or $-\boldsymbol{\lambda}$.

3.1.3 Surface

Consider an orientable surface in space. Any position on the surface can be described in terms of two orthogonal surface coordinates which are denoted as $\mathbf{u} = (u_1, u_2)$. In turn, $\mathbf{u}$ will depend on the spatial coordinates $\mathbf{x} = (x, y, z)$. A generalized function γ^s can be defined in terms of $\mathbf{u}$ and time to specify any region on the surface as

$$\gamma^s(\mathbf{u},t) = \begin{cases} 1 & \text{on the portion of the surface which is of interest} \\ 0 & \text{on the remainder of the surface} \end{cases} \tag{168}$$

Here, the *surface of interest* is restricted to a finite number of non-overlapping simple regions (Figure 9). The dependence of γ^s on time accounts for the expansion or contraction of the region of interest on the entire surface.

The total time derivative of γ^s in the surface is

$$\frac{\mathrm{d}\gamma^s}{\mathrm{d}t} = \frac{\partial\gamma^s}{\partial t}\bigg|_{\mathbf{u}} + \frac{\mathrm{d}\mathbf{u}}{\mathrm{d}t}\cdot\nabla_{\mathbf{n}}^{S}\gamma^s \tag{169}$$

This total derivative will be zero if $\mathrm{d}\mathbf{u}/\mathrm{d}t$ is specified to the observation velocity in the surface, $\mathbf{w}^s$, constrained such that: (1) on the boundary between the portion of the surface of interest and the extended surface (i.e., where $g^s(\mathbf{u}; \mathbf{u}^c) = 0$), the observation velocity $\mathbf{w}^s$ equals the velocity of the boundary; and (2) the velocity of the observer at other locations on the surface is such that the boundary curve of the portion of the surface of interest is never crossed. Under these constraints, no change in the value of γ^s will be observed so that

$$\frac{\partial\gamma^s}{\partial t}\bigg|_{\mathbf{u}} = -\mathbf{w}^s\cdot\nabla_{\mathbf{n}}^{S}\gamma^s \tag{170}$$

By the definition of a surface gradient, $\nabla_{\mathbf{n}}^{S}\gamma^s$ is a vector pointing in the direction of largest change in γ^s so that it is normal to the region defined by $\gamma^s = 1$. For a surface where position is specified by two orthogonal surface coordinates, the surficial gradient of γ^s is zero everywhere on the surface except along the

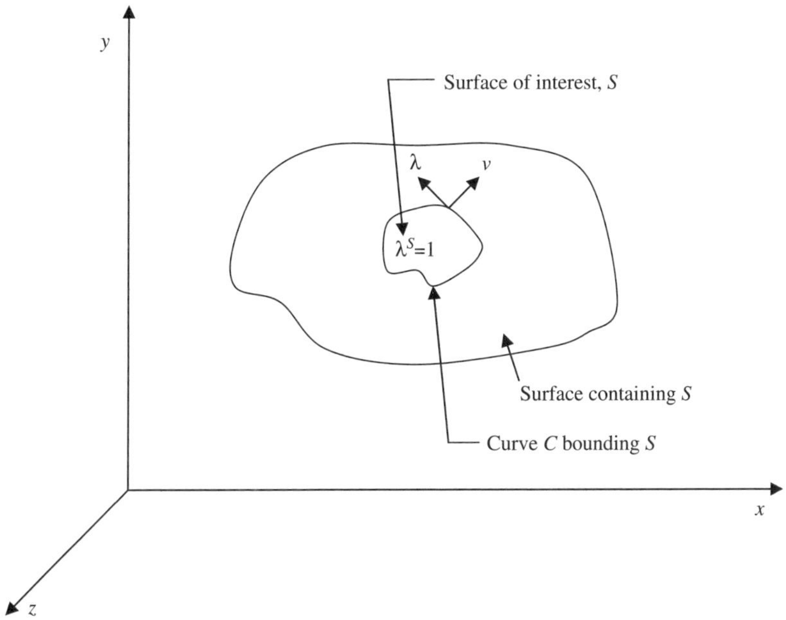

Figure 9 γ^s for identifying a portion of a surface.

curve C dividing the portion of the surface of interest from the extension of that surface. Accordingly, the surficial gradient of γ^s is

$$\nabla_{\mathbf{n}}^{S}\gamma^{s} = -\boldsymbol{\nu}\delta[g^{s}(\mathbf{u};\mathbf{u}^{c})] \tag{171}$$

where $\boldsymbol{\nu}$ is a unit vector tangent to the surface and outwardly normal to the region where $\gamma^s = 1$ (see Figure 9). The points $\mathbf{u}^c$ lie on the curve C; and the position function $g^c(\mathbf{u};\mathbf{u}^c) = 0$ defines curve C in terms of the surface coordinate $\mathbf{u}$. If γ^s is used to identify N separate portions of a surface then Equation (171) can be extended to

$$\nabla_{\mathbf{n}}^{S}\gamma^{s} = -\sum_{i=1}^{N}\boldsymbol{\nu}\delta[g_{i}^{s}(\mathbf{u};\mathbf{u}_{i}^{c})] \tag{172}$$

where the subscript i is used to denote each of the surfaces of interest. Note also that when the surface of interest is a closed surface such that $\gamma^s = 1$ along the entire surface (i.e., $N = 0$ in Equation (172) because no boundary curve exists), $\nabla_{\mathbf{n}}^{S}\gamma^{s} = 0$ on the entire surface.

If γ^s is used to identify one section of a very large surface, S_∞, and the boundary curve of this section is C, then

$$\begin{aligned} -\int_{S_\infty} f\boldsymbol{\nu}\cdot\nabla_{\mathbf{n}}^{S}\gamma^{s}\mathrm{d}s &= -\int_{S_\infty} f\boldsymbol{\nu}\cdot(-\boldsymbol{\nu}\delta[g^{s}(\mathbf{u};\mathbf{u}^{c})])\mathrm{d}S \\ &= \int_{S_\infty} f\delta[g^{s}(\mathbf{u};\mathbf{u}^{c})]\mathrm{d}S = \int_{C} f\mathrm{d}C \end{aligned} \tag{173}$$

If γ^s is used to identify N separate regions of the surface S_∞ where region i is bounded by curve C_i, Equation (173) becomes

$$-\int_{S_\infty} f\mathbf{v}\cdot\nabla_{\mathbf{n}}^{S}\gamma^s \mathrm{d}S = \sum_{i=1}^{N}\int_{C_i} f\mathrm{d}C \tag{174}$$

or

$$-\int_{S_\infty} f\nabla_{\mathbf{n}}^{S}\gamma^s \mathrm{d}s = \sum_{i=1}^{N}\int_{C_i} f\mathbf{v}\mathrm{d}C \tag{175}$$

Similarly, we have

$$-\int_{S_\infty} f\mathbf{w}\cdot\nabla_{\mathbf{n}}^{S}\gamma^s \mathrm{d}s = \sum_{i=1}^{N}\int_{C_i} f\mathbf{w}\cdot\mathbf{v}\mathrm{d}C \tag{176}$$

$$-\int_{S_\infty} f\mathbf{w}\times\nabla_{\mathbf{n}}^{S}\gamma^s \mathrm{d}s = \sum_{i=1}^{N}\int_{C_i} f\mathbf{w}\times\mathbf{v}\mathrm{d}C \tag{177}$$

where $\mathbf{w}$ is an arbitrary vector function.

3.1.4 Volume

In three-dimensional space, one way in which any point may be located is by specifying its cartesian coordinates $\mathbf{x} = (x, y, z)$. A spatial indicator function γ is defined by

$$\gamma(\mathbf{x},t) = \begin{cases} 1 & \text{in the spatial region of interest} \\ 0 & \text{in the rest of space} \end{cases} \tag{178}$$

The region of interest may consist of one or more non-overlapping finite volumes, each bounded by an orientable surface. The dependence of γ on time accounts for the expansion, contraction, and deformation of the spatial region of interest. Figure 10 illustrates use of γ to identify a region in space.

The total time derivative of γ is

$$\frac{\mathrm{d}\gamma}{\mathrm{d}t} = \left.\frac{\partial\gamma}{\partial t}\right|_{\mathbf{x}} + \frac{\mathrm{d}\mathbf{x}}{\mathrm{d}t}\cdot\nabla\gamma \tag{179}$$

This total derivative will be zero if the observation velocity, $\mathrm{d}\mathbf{x}/\mathrm{d}t$, is set equal to $\mathbf{w}$ such that: (1) on the boundary S, the velocity of the observer $\mathbf{w}$ equals the velocity of the boundary (i.e., an observer located at $g(\mathbf{x};\mathbf{x}^s) = 0$ remains at a point where this equation is satisfied); and (2) the velocity of an observer not on the boundary S is such that the boundary is never crossed resulting in the observer seeing no change in γ. With these constraints, $\mathrm{d}\gamma/\mathrm{d}t$ will be zero so that

$$\left.\frac{\partial\gamma}{\partial t}\right|_{\mathbf{x}} = -\mathbf{w}\cdot\nabla\gamma \tag{180}$$

which relates time derivatives to spatial derivatives of γ.

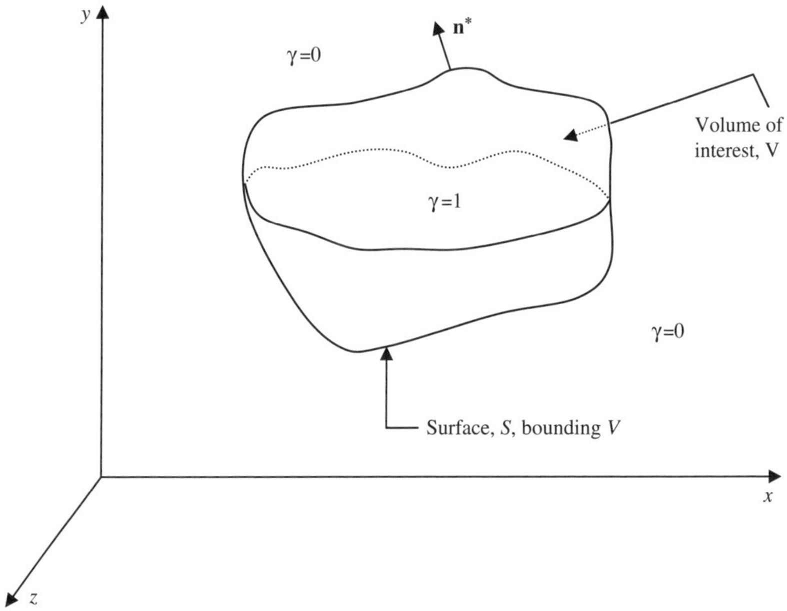

Figure 10 γ for identifying a portion of space.

The spatial gradient of γ will be a vector pointing in the direction of largest change in γ. Within and exterior to the region of interest, γ is constant so that $\nabla\gamma$ will be zero. At the boundary of the region of interest, the direction of largest change in γ is normal to the boundary and the magnitude of the change is characterized by Dirac delta function. Therefore, for a simple volume with boundary S, the gradient of γ is

$$\nabla\gamma = -\mathbf{n}^*\delta[g(\mathbf{x};\mathbf{x}^s)] \tag{181}$$

where $\mathbf{n}^*$ is the outwardly directed unit vector normal to the region where $\gamma = 1$ (Figure 10); coordinates $\mathbf{x}^s$ are on the surface S; and $g(\mathbf{x};\mathbf{x}^s) = 0$ defines surface S in terms of the spatial coordinates. If the spatial generalized function is used to identify N distinct regions in space, Equation (181) can then be extended to

$$\nabla\gamma = -\sum_{i=1}^{N}\mathbf{n}_i^*\delta[g_i(\mathbf{x};\mathbf{x}_i^s)] \tag{182}$$

where i is used to indicate each of the regions of interest and the unit normal vector $\mathbf{n}^*$ varies with position on the surface.

If γ is used to identify one volume, V, which is constrained in V_∞ and has a bounding surface S, we have

$$-\int_{V_\infty} f\mathbf{n}^*\cdot\nabla\gamma \mathrm{d}V = \int_S f\mathrm{d}S \tag{183}$$

If γ is used to identify N separate regions of volume V_∞, as in Equation (182), where S_i is the boundary of volume V_i, Equation (183) becomes

$$-\int_{V_\infty} f\mathbf{n}^* \cdot \nabla\gamma \mathrm{d}V = \sum_{i=1}^{N} \int_{S_i} f\mathrm{d}S \tag{184}$$

or

$$-\int_{V_\infty} f\nabla\gamma \mathrm{d}V = \sum_{i=1}^{N} \int_{S_i} f\mathbf{n}^*\mathrm{d}S \tag{185}$$

Similarly, we have

$$\int_{V_\infty} f\nabla\gamma \cdot \mathbf{w}\mathrm{d}V = -\sum_{i=1}^{N} \int_{S_i} f\mathbf{n}^* \cdot \mathbf{w}\mathrm{d}S \tag{186}$$

and

$$\int_{V_\infty} f\nabla\gamma \times \mathbf{w}\mathrm{d}V = -\sum_{i=1}^{N} \int_{S_i} f\mathbf{n}^* \times \mathbf{w}\mathrm{d}S \tag{187}$$

where $\mathbf{w}$ is an arbitrary vector function. Note also that f in Equations (185)–(187) can be one component of a vector or a tensor.

3.1.5 Proof of Identities 2.3a and 2.3b

Let $\mathbf{n}$ be a unit normal vector to some surface, S, in space. The integral of the partial time derivative in space of $\mathbf{n}$ over the surface may be transformed to an integral over an infinite volume by using the surficial indicator function, γ^{s}, and the spatial indicator function, γ to obtain

$$\int_S \left.\frac{\partial \mathbf{n}}{\partial t}\right|_{\mathbf{x}} \mathrm{d}S = -\int_{V_\infty} \left.\frac{\partial \mathbf{n}}{\partial t}\right|_{\mathbf{x}} \gamma^{\mathrm{s}}\mathbf{n} \cdot \nabla\gamma \mathrm{d}V \tag{188}$$

Applying the chain rule to the right side of this expression yields

$$\int_S \left.\frac{\partial \mathbf{n}}{\partial t}\right|_{\mathbf{x}} \mathrm{d}S = -\int_{V_\infty} \left.\frac{\partial(\mathbf{nn}\cdot\nabla\gamma)}{\partial t}\right|_{\mathbf{x}} \gamma^{\mathrm{s}}\mathrm{d}v + \int_{V_\infty} \mathbf{n}\left.\frac{\partial(\mathbf{n}\cdot\nabla\gamma)}{\partial t}\right|_{\mathbf{x}} \gamma^{\mathrm{s}}\mathrm{d}V \tag{189}$$

Since $\nabla\gamma$ is a vector in the $\mathbf{n}$ direction,

$$\mathbf{nn}\cdot\nabla\gamma = \nabla\gamma$$

By Equation (194), we have

$$\nabla\gamma \cdot \frac{\partial \mathbf{n}}{\partial t} = 0$$

Therefore, Equation (189) becomes

$$\int_S \left.\frac{\partial \mathbf{n}}{\partial t}\right|_{\mathbf{x}} \mathrm{d}S = -\int_{V_\infty} \nabla_{\mathbf{n}}^{\mathrm{S}}\left(\left.\frac{\partial \gamma}{\partial t}\right|_{\mathbf{x}}\right)\gamma^{\mathrm{s}}\mathrm{d}V \tag{190}$$

By Equation (180),

$$\frac{\partial \gamma}{\partial t} = -\mathbf{w}\cdot\nabla\gamma$$

where $\mathbf{w}$ is the velocity of the surface. Substitution of this expression into Equation (190) and application of the surface gradient operator yield

$$\int_S \left.\frac{\partial \mathbf{n}}{\partial t}\right|_{\mathbf{x}} ds = \int_{V_\infty} \nabla_{\mathbf{n}}^{S}\mathbf{w} \cdot \nabla\gamma\gamma^{s} dv + \int_{V_\infty} \nabla_{\mathbf{n}}^{S}\nabla\gamma \cdot \mathbf{w}\gamma^{s} dV \tag{191}$$

Substituting Equation (256) into the second integral on the right side of this equation yields:

$$\int_S \left.\frac{\partial \mathbf{n}}{\partial t}\right|_{\mathbf{x}} \mathrm{d}S = \int_{V_\infty} \nabla^{S}\mathbf{w} \cdot \nabla\gamma\gamma^{s} \mathrm{d}V + \int_{V_\infty} \mathbf{w} \cdot \nabla\mathbf{n}(\mathbf{n} \cdot \nabla\gamma)\gamma^{s} \mathrm{d}V \tag{192}$$

Now the volume integrals may be converted back to surface integrals so that

$$\int_S \left.\frac{\partial \mathbf{n}}{\partial t}\right|_{\mathbf{x}} \mathrm{d}S = -\int_S \nabla^{S}\mathbf{w} \cdot \mathbf{n} \mathrm{d}S - \int_S \mathbf{w} \cdot \nabla\mathbf{n} \mathrm{d}S \tag{193}$$

As the surface is arbitrary, the localization theorem thus leads to (Wang, 1997; Wang and Zhou, 2000)

$$\left.\frac{\partial \mathbf{n}}{\partial t}\right|_{\mathbf{x}} = -\nabla_{\mathbf{n}}^{S}\mathbf{w} \cdot \mathbf{n} - \mathbf{w} \cdot \nabla\mathbf{n}$$

or

$$\frac{\mathrm{d}\mathbf{n}}{\mathrm{d}t} = -\nabla_{\mathbf{n}}^{S}\mathbf{w} \cdot \mathbf{n}$$

where the total derivative is defined by

$$\frac{\mathrm{d}\mathbf{n}}{\mathrm{d}t} = \left.\frac{\partial \mathbf{n}}{\partial t}\right|_{\mathbf{x}} + \mathbf{w} \cdot \nabla\mathbf{n}$$

3.2 Identities involving indicator functions

In this section we develop identities involving indicator functions and their derivatives. They are very useful for derivation of multiscale theorems.

3.2.1 Orthogonality relations

Identity 1.

$$\frac{\partial \mathbf{n}}{\partial t} \cdot \nabla\gamma = 0 \tag{194}$$

Identity 2.

$$\nabla\mathbf{n} \cdot \nabla\gamma = 0 \tag{195}$$

where γ is the indicator function used to identify a volume in space, $\mathbf{n}$ the unit vector normal to the surface bounding the volume, and ∇ the spatial gradient operator.

Proof. By Equation (181), we have

$$\nabla\gamma = -\mathbf{n}\delta[g(\mathbf{x};\mathbf{x}^s)] \tag{196}$$

Therefore,

$$\frac{\partial \mathbf{n}}{\partial t}\cdot\nabla\gamma = -\frac{\partial \mathbf{n}}{\partial t}\cdot\mathbf{n}\delta[g(\mathbf{x};\mathbf{x}^s)] \tag{197}$$

$$\nabla\mathbf{n}\cdot\nabla\gamma = -\nabla\mathbf{n}\cdot\mathbf{n}\delta[g(\mathbf{x};\mathbf{x}^s)] \tag{198}$$

However,

$$\mathbf{n}\cdot\mathbf{n} = 1$$

Therefore,

$$\frac{\partial \mathbf{n}}{\partial t}\cdot\mathbf{n} = 0$$

and

$$\nabla\mathbf{n}\cdot\mathbf{n} = 0$$

Therefore,

$$\frac{\partial \mathbf{n}}{\partial t}\cdot\nabla\gamma = 0$$

$$\nabla\mathbf{n}\cdot\nabla\gamma = 0$$

so that Identities 1 and 2 are proven.

Identity 3.

$$\frac{\partial \boldsymbol{\nu}}{\partial t}\cdot\nabla^S_{\mathbf{n}}\gamma^s = \frac{\partial \boldsymbol{\nu}}{\partial t}\cdot\nabla\gamma^s = 0 \tag{199}$$

Identity 4.

$$\nabla\boldsymbol{\nu}\cdot\nabla^S_{\mathbf{n}}\gamma^s = \nabla\boldsymbol{\nu}\cdot\nabla\gamma^s = 0 \tag{200}$$

where γ^s is the indicator function used to identify a simple surface that is a portion of a much larger surface, $\mathbf{n}$ a unit vector normal to the simple surface, $\boldsymbol{\nu}$ a unit vector tangent to the simple surface such that $\boldsymbol{\nu}\cdot\mathbf{n} = 0$ and also normal to the curve that describes the boundary of this surface, ∇ the spatial gradient operator, and $\nabla^S_{\mathbf{n}}$ is the surface gradient operator, $\nabla^S_{\mathbf{n}} = \nabla - \mathbf{nn}\cdot\nabla$.

Proof. By Equations (22) and (171), we have

$$\frac{\partial \boldsymbol{\nu}}{\partial t}\cdot\nabla^S_{\mathbf{n}}\gamma^s = -\frac{\partial \boldsymbol{\nu}}{\partial t}\cdot\boldsymbol{\nu}[g^s(\mathbf{u};\mathbf{u}^c)] = 0 \tag{201}$$

To prove the second part of Equation (199), an application of the chain rule yields

$$\frac{\partial \boldsymbol{\nu}}{\partial t}\cdot\nabla\gamma^s = \frac{\partial}{\partial t}(\boldsymbol{\nu}\cdot\nabla\gamma^s) - \boldsymbol{\nu}\cdot\frac{\partial}{\partial t}\nabla\gamma^s \tag{202}$$

Applying

$$\nabla = \nabla^S + \mathbf{nn}\cdot\nabla$$

and

$$\boldsymbol{v} \cdot \mathbf{n} = 0$$

Equation (202) becomes

$$\begin{aligned}
\frac{\partial \boldsymbol{v}}{\partial t} \cdot \nabla \gamma^{\mathrm{s}} &= \frac{\partial}{\partial t}(\boldsymbol{v} \cdot \nabla^{\mathrm{S}} \gamma^{\mathrm{s}}) - \boldsymbol{v} \cdot \frac{\partial}{\partial t} \nabla \gamma^{\mathrm{s}} \\
&= \frac{\partial \boldsymbol{v}}{\partial t} \cdot \nabla^{\mathrm{S}} \gamma^{\mathrm{s}} + \boldsymbol{v} \cdot \frac{\partial}{\partial t} \nabla^{\mathrm{S}} \gamma^{\mathrm{s}} - \boldsymbol{v} \cdot \frac{\partial}{\partial t} \nabla \gamma^{\mathrm{s}} \\
&= \boldsymbol{v} \cdot \frac{\partial}{\partial t} (\nabla^{\mathrm{S}} - \nabla) \gamma^{\mathrm{s}} \\
&= -\boldsymbol{v} \cdot \frac{\partial}{\partial t} (\mathbf{nn} \cdot \nabla \gamma^{\mathrm{s}}) = -\boldsymbol{v} \cdot \mathbf{nn} \cdot \frac{\partial}{\partial t} \nabla \gamma^{\mathrm{s}} = 0
\end{aligned}$$

This completes the proof of Identity 3.

The more difficult part of the proof of Equation (200) is $\nabla \boldsymbol{v} \cdot \nabla \gamma^{\mathrm{s}} = 0$. Applying the chain rule yields

$$\nabla \boldsymbol{v} \cdot \nabla \gamma^{\mathrm{s}} = \nabla(\boldsymbol{v} \cdot \nabla \gamma^{\mathrm{s}}) - \boldsymbol{v} \cdot \nabla \nabla \gamma^{\mathrm{s}} \tag{203}$$

Since

$$\nabla^{\mathrm{S}}_{\mathbf{n}} = \nabla - \mathbf{nn} \cdot \nabla$$

the first term on the right side of Equation (203) becomes

$$\nabla(\boldsymbol{v} \cdot \nabla \gamma^{\mathrm{s}}) = \nabla(\boldsymbol{v} \cdot \nabla^{\mathrm{S}} \gamma^{\mathrm{s}} + \boldsymbol{v} \cdot \mathbf{nn} \cdot \nabla \gamma^{\mathrm{s}}) \tag{204}$$

or, since $\boldsymbol{v}$ and $\mathbf{n}$ are orthogonal:

$$\nabla(\boldsymbol{v} \cdot \nabla \gamma^{\mathrm{s}}) = \nabla(\boldsymbol{v} \cdot \nabla^{\mathrm{S}} \gamma^{\mathrm{s}}) \tag{205}$$

Substitution of Equation (205) into Equation (203), differentiation of the resulting expression, and rearrangement lead to

$$\begin{aligned}
\nabla \boldsymbol{v} \cdot \nabla \gamma^{\mathrm{s}} &= \nabla \boldsymbol{v} \cdot \nabla^{\mathrm{S}}_{\mathbf{n}} \gamma^{\mathrm{s}} + \boldsymbol{v} \cdot \nabla \nabla^{\mathrm{S}}_{\mathbf{n}} \gamma^{\mathrm{s}} - \boldsymbol{v} \cdot \nabla \nabla \gamma^{\mathrm{s}} \\
&= \boldsymbol{v} \cdot \nabla^{\mathrm{S}}_{\mathbf{n}} \nabla \gamma^{\mathrm{s}} - \boldsymbol{v} \cdot \nabla \nabla \gamma^{\mathrm{s}} + \nabla \boldsymbol{v} \cdot \nabla^{\mathrm{S}}_{\mathbf{n}} \gamma^{\mathrm{s}} \\
&= \nabla \boldsymbol{v} \cdot \nabla^{\mathrm{S}}_{\mathbf{n}} \gamma^{\mathrm{s}} + \boldsymbol{v} \cdot (\nabla^{\mathrm{S}}_{\mathbf{n}} - \nabla) \nabla \gamma^{\mathrm{s}}
\end{aligned} \tag{206}$$

By Equation (42), we have

$$\begin{aligned}
\nabla \boldsymbol{v} \cdot \nabla \gamma^{\mathrm{s}} &= \nabla \boldsymbol{v} \cdot \nabla^{\mathrm{S}}_{\mathbf{n}} \gamma^{\mathrm{s}} - \boldsymbol{v} \cdot \mathbf{nn} \cdot \nabla \nabla \gamma^{\mathrm{s}} \\
&= \nabla \boldsymbol{v} \cdot \nabla^{\mathrm{S}}_{\mathbf{n}} \gamma^{\mathrm{s}}
\end{aligned}$$

Then application of Equations (21) and (171) yields the second part of Identity 4.

Identity 5.

$$\frac{\partial \boldsymbol{\lambda}}{\partial t} \cdot \nabla^{\mathrm{c}}_{\lambda} \gamma^{\mathrm{c}} = \frac{\partial \boldsymbol{\lambda}}{\partial t} \cdot \nabla \gamma^{\mathrm{c}} = 0 \tag{207}$$

Identity 6.

$$\nabla \boldsymbol{\lambda} \cdot \nabla^{\mathrm{c}}_{\lambda} \gamma^{\mathrm{c}} = \nabla \boldsymbol{\lambda} \cdot \nabla \gamma^{\mathrm{c}} = 0 \tag{208}$$

where γ^c is the indicator function used to identify a simple curve that is a segment of a longer curve, $\boldsymbol{\lambda}$ the unit vector tangent to the simple curve, ∇ is the spatial gradient operator, and ∇_λ^c is the curvilineal gradient operator $\nabla_\lambda^c = \boldsymbol{\lambda\lambda}\cdot\nabla$.

Proof. By Equation (162), we have

$$\nabla_\lambda^c \gamma^c = -\sum_{i=1}^{N} \mathbf{e}|_{l=a_i}\delta[g^c(l;a_i)]$$

where $\mathbf{e}$ is a unit vector and equal to $\boldsymbol{\lambda}$ or $-\boldsymbol{\lambda}$. Since $(\partial\boldsymbol{\lambda}/\partial t)\cdot\boldsymbol{\lambda} = 0$, we obtain

$$\frac{\partial\boldsymbol{\lambda}}{\partial t}\cdot\mathbf{e} = 0$$

Therefore,

$$\frac{\partial\boldsymbol{\lambda}}{\partial t}\cdot\nabla_\lambda^c\gamma^c = 0 \tag{209}$$

To prove the second part of Identity 5, applying the chain rule yields

$$\frac{\partial\boldsymbol{\lambda}}{\partial t}\cdot\nabla\gamma^c = \frac{\partial}{\partial t}(\boldsymbol{\lambda}\cdot\nabla\gamma^c) - \boldsymbol{\lambda}\cdot\frac{\partial}{\partial t}\nabla\gamma^c \tag{210}$$

Applying

$$\nabla = \nabla_\lambda^c + \nabla_\lambda^S$$

and $\boldsymbol{\lambda}\cdot\mathbf{n} = 0$, $\boldsymbol{\lambda}\cdot\boldsymbol{\nu} = 0$, Equation (210) becomes

$$\begin{aligned}
\frac{\partial\boldsymbol{\lambda}}{\partial t}\cdot\nabla\gamma^c &= \frac{\partial\boldsymbol{\lambda}}{\partial t}\cdot(\nabla_\lambda^c + \nabla_\lambda^S)\gamma^c = \frac{\partial\boldsymbol{\lambda}}{\partial t}\cdot\nabla_\lambda^S\gamma^c \\
&= \frac{\partial\boldsymbol{\lambda}}{\partial t}\cdot(\mathbf{nn}\cdot\nabla\gamma^c) + \frac{\partial\boldsymbol{\lambda}}{\partial t}\cdot(\boldsymbol{\nu\nu}\cdot\nabla\gamma^c) \\
&= \frac{\partial}{\partial t}\cdot(\boldsymbol{\lambda}\cdot\mathbf{nn}\cdot\nabla\gamma^c) - \boldsymbol{\lambda}\cdot\frac{\partial}{\partial t}\cdot(\mathbf{nn}\cdot\nabla\gamma^c) \\
&\quad + \frac{\partial}{\partial t}\cdot(\boldsymbol{\lambda}\cdot\boldsymbol{\nu\nu}\cdot\nabla\gamma^c) - \boldsymbol{\lambda}\cdot\frac{\partial}{\partial t}\cdot(\boldsymbol{\nu\nu}\cdot\nabla\gamma^c) \\
&= -\boldsymbol{\lambda}\cdot\mathbf{nn}\cdot\nabla\frac{\partial\gamma^c}{\partial t} - \boldsymbol{\lambda}\cdot\boldsymbol{\nu\nu}\cdot\nabla\frac{\partial\gamma^c}{\partial t} = 0
\end{aligned} \tag{211}$$

This completes the proof of Identity 5. Similarly, we can prove Identity 6.

3.2.2 Gradients of indicator functions

Identity 1.

$$\mathbf{n}\cdot\nabla\gamma^s = 0 \tag{212}$$

where γ^s is a generalized function for a surface and $\mathbf{n}$ the normal to the surface under consideration.

Proof. By Equation (199), we have

$$\frac{\partial\boldsymbol{\nu}}{\partial t}\cdot(\nabla - \nabla_\mathbf{n}^S)\gamma^s = 0 \tag{213}$$

A substitution of Equation (42) into this expression yields:

$$\frac{\partial \boldsymbol{\nu}}{\partial t} \cdot \mathbf{nn} \cdot \nabla \gamma^{s} = 0 \tag{214}$$

Note that Equation (214) is valid for any simple surface, S, and $(\partial \boldsymbol{\nu}/\partial t) \cdot \mathbf{n}$ is in general non-zero. Therefore,

$$\mathbf{n} \cdot \nabla \gamma^{s} = 0$$

Identity 2.

$$\boldsymbol{\nu} \cdot \nabla \gamma^{c} = \mathbf{n} \cdot \nabla \gamma^{c} = 0 \tag{215}$$

where γ^{c} is a generalized function for a curve, $\mathbf{n}$ a unit vector normal to the curve, and $\boldsymbol{\nu}$ another unit vector normal to the curve but also normal to $\mathbf{n}$ such that $\mathbf{n} \cdot \boldsymbol{\nu} = 0$.

Proof. By $\nabla_{\mathbf{n}}^{S} = \nabla_{\lambda}^{c} + \boldsymbol{\nu\nu} \cdot \nabla$, we have

$$\begin{aligned}
\frac{\partial \boldsymbol{\lambda}}{\partial t} \cdot \nabla_{\mathbf{n}}^{S} \gamma^{c} &= \frac{\partial \boldsymbol{\lambda}}{\partial t} \cdot (\nabla_{\lambda}^{c} + \mathbf{vv} \cdot \nabla) \gamma^{c} \\
&= \frac{\partial \boldsymbol{\lambda}}{\partial t} \cdot \mathbf{vv} \cdot \nabla \gamma^{c} \\
&= \frac{\partial}{\partial t}(\boldsymbol{\lambda} \cdot \mathbf{vv} \cdot \nabla \gamma^{c}) - \boldsymbol{\lambda} \cdot \frac{\partial}{\partial t}(\mathbf{vv} \cdot \nabla \gamma^{c}) \\
&= -\boldsymbol{\lambda} \cdot \mathbf{vv} \cdot \nabla \frac{\partial}{\partial t} \gamma^{c} \\
&= 0
\end{aligned}$$

Therefore,

$$\frac{\partial \boldsymbol{\lambda}}{\partial t} \cdot \boldsymbol{\nu\nu} \cdot \nabla \gamma^{c} = 0$$

Since $(\partial \boldsymbol{\lambda}/\partial t) \cdot \boldsymbol{\nu}$ is in general non-zero, we obtain

$$\boldsymbol{\nu} \cdot \nabla \gamma^{c} = 0$$

Similarly,

$$\mathbf{n} \cdot \nabla \gamma^{c} = 0$$

Identity 1.

$$\nabla \gamma^{s} = \nabla_{\mathbf{n}}^{S} \gamma^{s} \tag{216}$$

Proof. Note that

$$\begin{aligned}
\nabla \gamma^{s} &= (\nabla_{\mathbf{n}}^{S} + \mathbf{nn} \cdot \nabla) \gamma^{s} \\
&= \nabla^{S} \gamma^{s} + \mathbf{nn} \cdot \nabla \gamma^{s}
\end{aligned}$$

By Equation (212), this equation becomes

$$\nabla \gamma^{s} = \nabla^{S} \gamma^{s}$$

Identity 2.

$$\nabla\gamma^{c} = \nabla^{S}_{\mathbf{n}}\gamma^{c} = \nabla^{c}_{\boldsymbol{\lambda}}\gamma^{c} \tag{217}$$

Proof. Note that

$$\begin{aligned}\nabla\gamma^{c} &= (\nabla^{S}_{\mathbf{n}} + \mathbf{nn}\cdot\nabla)\gamma^{c}\\ &= \nabla^{S}_{\mathbf{n}}\gamma^{c} + \mathbf{nn}\cdot\nabla\gamma^{c}\end{aligned}$$

Applying $\mathbf{n}\cdot\nabla\gamma^{c} = 0$, we thus obtain the first part of Identity 2.

$$\nabla\gamma^{c} = \nabla^{S}\gamma^{c}$$

To prove the second part, apply

$$\begin{aligned}\nabla^{S}_{\mathbf{n}}\gamma^{c} &= (\nabla^{c}_{\boldsymbol{\lambda}} + \boldsymbol{\nu\nu}\cdot\nabla)\gamma^{c}\\ &= \nabla^{c}_{\boldsymbol{\lambda}}\gamma^{c} + \boldsymbol{\nu\nu}\cdot\nabla\gamma^{c}\end{aligned}$$

Note also that

$$\boldsymbol{\nu}\cdot\nabla\gamma^{c} = 0$$

Therefore,

$$\nabla^{S}\gamma^{c} = \nabla^{c}\gamma^{c}$$

This completes the proof of Identity 2.

Identity 1.

$$\mathbf{n}\nabla\gamma = \nabla\gamma\mathbf{n} \tag{218}$$

Proof. By Equation (181), we have

$$\nabla\gamma = -\mathbf{n}\delta[g(\mathbf{x};\mathbf{x}^{s})]$$

Thus,

$$\begin{aligned}\mathbf{n}\nabla\gamma &= -\,\mathbf{nn}\delta[g(\mathbf{x};\mathbf{x}^{s})]\\ &= -\,\mathbf{n}\delta[g(\mathbf{x};\mathbf{x}^{s})]\mathbf{n}\\ &= \nabla\gamma\mathbf{n}\end{aligned}$$

This completes the proof of Identity 1.

The following identities can also be proven by a similar manner:

Identity 2.

$$\boldsymbol{\nu}\nabla^{S}_{\mathbf{n}}\gamma^{s} = (\nabla^{S}\gamma^{s})\boldsymbol{\nu} \tag{219}$$

Identity 3.

$$\boldsymbol{\nu}\nabla\gamma^{s} = (\nabla\gamma^{s})\boldsymbol{\nu} \tag{220}$$

Identity 4.

$$\boldsymbol{\lambda}\nabla^{c}_{\boldsymbol{\lambda}}\gamma^{c} = (\nabla^{c}\gamma^{c})\boldsymbol{\lambda} \tag{221}$$

Identity 5.

$$\boldsymbol{\lambda}\nabla^{S}_{\mathbf{n}}\gamma^{c} = (\nabla^{S}\gamma^{c})\boldsymbol{\lambda} \tag{222}$$

Identity 6.

$$\boldsymbol{\lambda}\nabla\gamma^{c} = (\nabla\gamma^{c})\boldsymbol{\lambda} \tag{223}$$

3.2.3 Time derivatives of indicator functions

Identity 1.

$$\left.\frac{\partial\gamma^{s}}{\partial t}\right|_{\mathbf{x}} = \left.\frac{\partial\gamma^{s}}{\partial t}\right|_{\mathbf{u}} = -\mathbf{w}\cdot\nabla\gamma^{s} \tag{224}$$

Proof. First, by Equation (170) we have

$$\left.\frac{\partial\gamma^{s}}{\partial t}\right|_{\mathbf{u}} = -\mathbf{w}^{s}\cdot\nabla_{\mathbf{n}}^{S}\gamma^{s} \tag{225}$$

where $\mathbf{w}^{s}$ is the velocity of the surface in the directions tangent to the surface and $\nabla_{\mathbf{n}}^{S} = \nabla - \mathbf{nn}\cdot\nabla$ the surface gradient operator.

Applying Equation (216) (Identity 1) yields

$$\left.\frac{\partial\gamma^{s}}{\partial t}\right|_{\mathbf{u}} = -\mathbf{w}^{s}\cdot\nabla\gamma^{s} \tag{226}$$

Note that $\mathbf{w}^{s} = \mathbf{w} - \mathbf{nn}\cdot\mathbf{w}$ and $\mathbf{n}\cdot\nabla\gamma^{s} = 0$. Therefore,

$$\left.\frac{\partial\gamma^{s}}{\partial t}\right|_{\mathbf{u}} = -\mathbf{w}\cdot\nabla\gamma^{s} \tag{227}$$

By Equation (130), we have

$$\left.\frac{\partial\gamma^{s}}{\partial t}\right|_{\mathbf{x}} = \left.\frac{\partial\gamma^{s}}{\partial t}\right|_{\mathbf{u}} - \mathbf{w}\cdot\mathbf{nn}\cdot\nabla\gamma^{s} \tag{228}$$

However, by Equation (212) (Identity 1), the last term in this expression vanishes. Thus combination of Equations (227) and (228) yields

$$\left.\frac{\partial\gamma^{s}}{\partial t}\right|_{\mathbf{x}} = \left.\frac{\partial\gamma^{s}}{\partial t}\right|_{\mathbf{u}} = -\mathbf{w}\cdot\nabla\gamma^{s}$$

This completes the proof of Identity 1.

Similarly, we can obtain the following identity for the time derivatives of the curvilineal indicator function γ^{c}

Identity 2.

$$\left.\frac{\partial\gamma^{c}}{\partial t}\right|_{\mathbf{x}} = \left.\frac{\partial\gamma^{c}}{\partial t}\right|_{\mathbf{u}} = \left.\frac{\partial\gamma^{c}}{\partial t}\right|_{\mathbf{l}} = -\mathbf{w}\cdot\nabla\gamma^{c} \tag{229}$$

3.2.4 Integrands containing the Del operator

In developing some multiscale theorems, we require two coordinate systems: macroscale (global) system $\mathbf{X}$ to locate REVs in space and microscale (local) system $\boldsymbol{\xi}$ to account for variation within an REV (Figure 11). The latter can be characterized by its cartesian coordinates (ξ, η, ζ) (Figure 11).

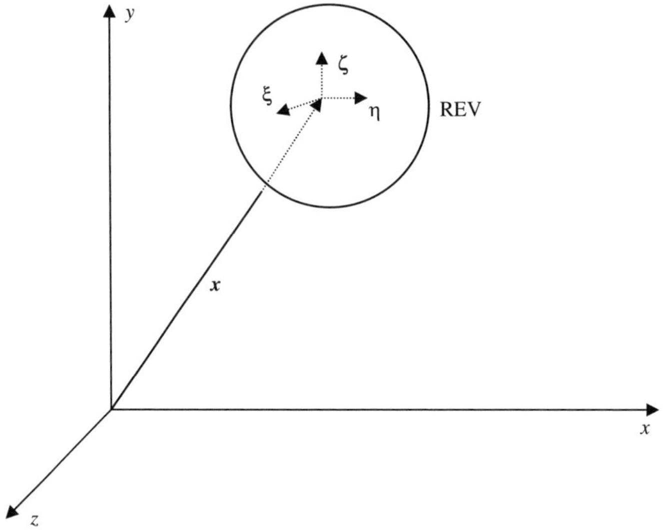

Figure 11 Global coordinate system, *x*, used to locate REV in space and local coordinate system, ξ, within each REV.

Identity 1.

$$\int_{V_\infty} \nabla_\xi(\gamma f) \mathrm{d}V_\xi = 0 \tag{230}$$

where V_∞ stands for an infinite volume.

Proof. Note that

$$\begin{aligned}\int_{V_\infty} \frac{\partial(\gamma f)}{\partial \xi} d\eta d\xi d\zeta &= \int_{-\infty}^{+\infty}\int_{-\infty}^{+\infty}\int_{-\infty}^{+\infty} \frac{\partial(\gamma f)}{\partial \xi} \mathrm{d}\eta \mathrm{d}\xi \mathrm{d}\zeta \\ &= \int_{-\infty}^{+\infty}\int_{-\infty}^{+\infty} (\gamma f)|_{-\infty}^{+\infty} \mathrm{d}\eta \mathrm{d}\zeta = 0\end{aligned}$$

in which the definition of γ is used.

Similarly,

$$\int_{V_\infty} \frac{\partial(\gamma f)}{\partial \eta} \mathrm{d}\eta \mathrm{d}\xi \mathrm{d}\zeta = 0$$

$$\int_{V_\infty} \frac{\partial(\gamma f)}{\partial \zeta} \mathrm{d}\xi \mathrm{d}\eta \mathrm{d}\zeta = 0$$

Thus,

$$\int_{V_\infty} \nabla_\xi(\gamma f) \mathrm{d}V_\xi = 0$$

This completes the proof of Identity 1.

Similarly, we can prove the following two identities for any vector or tensor **f**

Identity 2.

$$\int_{V_\infty} \nabla_\xi \cdot (\gamma \mathbf{f}) \mathrm{d}V_\xi = 0 \tag{231}$$

Identity 3.

$$\int_{V_\infty} \nabla_\xi \times (\gamma \mathbf{f}) \mathrm{d}V_\xi = 0 \tag{232}$$

Identity 4.

$$\int_{S_\infty} \nabla_{\mathbf{u}}^{\mathrm{S}} (\gamma^{\mathrm{s}} \mathbf{f}^{\mathrm{s}}) \mathrm{d}S_{\mathbf{u}} = 0 \tag{233}$$

where S_∞ stands for an infinite area and the subscript **u** is explicitly used to denote the fact that differentiation and integration are performed in the same coordinate system.

Proof. By Equation (183), we have

$$\int_{S_\infty} \nabla_{\mathbf{u}}^{\mathrm{S}} (\gamma^{\mathrm{s}} \mathbf{f}^{\mathrm{s}}) \mathrm{d}S_{\mathbf{u}} = -\int_{V_\infty} \nabla_{\mathbf{u}}^{\mathrm{S}} (\gamma^{\mathrm{s}} \mathbf{f}^{\mathrm{s}}) \mathbf{n} \cdot \nabla \gamma \mathrm{d}V \tag{234}$$

Note that

$$\nabla_{\mathbf{u}}^{\mathrm{S}} = \nabla - \mathbf{n}\mathbf{n} \cdot \nabla$$

Therefore,

$$\int_{S_\infty} \nabla_{\mathbf{u}}^{\mathrm{S}} (\gamma^{\mathrm{s}} \mathbf{f}^{\mathrm{s}}) \mathrm{d}S_{\mathbf{u}} = -\int_{V_\infty} \nabla (\gamma^{\mathrm{s}} \mathbf{f}^{\mathrm{s}}) \mathbf{n} \cdot \nabla \gamma \mathrm{d}V + \int_{V_\infty} \mathbf{n}\mathbf{n} \cdot \nabla (\gamma^{\mathrm{s}} \mathbf{f}^{\mathrm{s}}) \mathbf{n} \cdot \nabla \gamma \mathrm{d}V \tag{235}$$

By Equation (181), Equation (235) becomes

$$\int_{S_\infty} \nabla_{\mathbf{u}}^{\mathrm{S}} (\gamma^{\mathrm{s}} \mathbf{f}^{\mathrm{s}}) \mathrm{d}S_{\mathbf{u}} = -\int_{V_\infty} \nabla (\gamma^{\mathrm{s}} \mathbf{f}^{\mathrm{s}}) \mathbf{n} \cdot \nabla \gamma \mathrm{d}V + \int_{V_\infty} \nabla \gamma \mathbf{n} \cdot \nabla (\gamma^{\mathrm{s}} \mathbf{f}^{\mathrm{s}}) \mathrm{d}V \tag{236}$$

Since $\mathbf{n} \cdot \nabla (\gamma^{\mathrm{s}} \mathbf{f}^{\mathrm{s}}) = 0$, we have

$$\int_{S_\infty} \nabla_{\mathbf{u}}^{\mathrm{S}} (\gamma^{\mathrm{s}} \mathbf{f}^{\mathrm{s}}) \mathrm{d}S_{\mathbf{u}} = -\int_{V_\infty} \nabla (\gamma^{\mathrm{s}} \mathbf{f}^{\mathrm{s}}) \mathbf{n} \cdot \nabla \gamma \mathrm{d}V \tag{237}$$

An application of the chain rule yields

$$\int_{S_\infty} \nabla_{\mathbf{u}}^{\mathrm{S}} (\gamma^{\mathrm{s}} \mathbf{f}^{\mathrm{s}}) \mathrm{d}S_{\mathbf{u}} = -\int_{V_\infty} \nabla (\gamma^{\mathrm{s}} \mathbf{f}^{\mathrm{s}} \mathbf{n} \cdot \nabla \gamma) \mathrm{d}V + \int_{V_\infty} \gamma^{\mathrm{s}} \mathbf{f}^{\mathrm{s}} \nabla (\mathbf{n} \cdot \nabla \gamma) \mathrm{d}V \tag{238}$$

By Equation (230), Equation (238) becomes

$$\begin{aligned}\int_{S_\infty} \nabla_{\mathbf{u}}^{\mathrm{S}} (\gamma^{\mathrm{s}} \mathbf{f}^{\mathrm{s}}) \mathrm{d}S_{\mathbf{u}} &= -\int_{V_\infty} \gamma^{\mathrm{s}} \mathbf{f}^{\mathrm{s}} \nabla (\mathbf{n} \cdot \nabla \gamma) \mathrm{d}V \\ &= -\int_{V_\infty} \gamma^{\mathrm{s}} \mathbf{f}^{\mathrm{s}} \nabla \mathbf{n} \cdot \nabla \gamma \mathrm{d}V - \int_{V_\infty} \gamma^{\mathrm{s}} \mathbf{f}^{\mathrm{s}} \nabla \nabla \gamma \cdot \mathbf{n} \mathrm{d}V\end{aligned} \tag{239}$$

Note also that $\nabla \mathbf{n} \cdot \nabla \gamma = 0$, and $\nabla \nabla \gamma \cdot \mathbf{n} = 0$. Therefore,

$$\int_{S_\infty} \nabla_{\mathbf{u}}^{S} (\gamma^s \mathbf{f}^s) \mathrm{d}S_{\mathbf{u}} = 0$$

This completes the proof of Identity 4.

Identity 5.

$$\int_{S_\infty} \nabla_{\mathbf{u}}^{S} \cdot (\gamma^s \mathbf{f}^s) \mathrm{d}S_{\mathbf{u}} = 0 \tag{240}$$

Proof. Consider S_∞ to be a large closed surface for which $\gamma^s = 1$ on some portion, and $\gamma^s = 0$ on the rest. The unit normal vector pointing outward from S_∞ is $\mathbf{n}$. Now consider an indicator function in space, γ, defined such that $\gamma = 1$ within S_∞ and $\gamma = 0$ in the rest of space. By Equation (183), we have

$$\int_{S_\infty} \nabla_{\mathbf{u}}^{S} \cdot (\gamma^s \mathbf{f}^s) \mathrm{d}S_{\mathbf{u}} = -\int_{V_\infty} \nabla_{\mathbf{u}}^{S} \cdot (\gamma^s \mathbf{f}^s) \mathbf{n} \cdot \nabla \gamma \mathrm{d}V \tag{241}$$

Converting the surface divergence of a surface vector to a spatial divergence of a space vector and applying Equation (38) yield

$$\int_{S_\infty} \nabla_{\mathbf{u}}^{S} \cdot (\gamma^s \mathbf{f}^s) \mathrm{d}S_{\mathbf{u}} = -\int_{V_\infty} \nabla \cdot (\gamma^s \mathbf{f}) \mathbf{n} \cdot \nabla \gamma \mathrm{d}V + \int_{V_\infty} \nabla \cdot (\gamma^s \mathbf{n}\mathbf{f}) \cdot \mathbf{n}\mathbf{n} \cdot \nabla \gamma \mathrm{d}V \tag{242}$$

An application of the chain rule leads to

$$\begin{aligned}
\int_{S_\infty} \nabla_{\mathbf{u}}^{S} \cdot (\gamma^s \mathbf{f}^s) \mathrm{d}S_{\mathbf{u}} = & -\int_{V_\infty} \nabla \cdot (\gamma^s \mathbf{f}\mathbf{n} \cdot \nabla \gamma) \mathrm{d}v + \int_{V_\infty} \gamma^s \mathbf{f} \cdot \nabla (\mathbf{n} \cdot \nabla \gamma) \mathrm{d}v \\
& + \int_{V_\infty} \nabla \cdot (\gamma^s \mathbf{n}\mathbf{f} \cdot \mathbf{n}\mathbf{n} \cdot \nabla \gamma) \mathrm{d}v \\
& - \int_{V_\infty} \gamma^s \mathbf{n} \cdot \nabla (\mathbf{n}\mathbf{n} \cdot \nabla \gamma) \cdot \mathbf{f} \mathrm{d}V
\end{aligned} \tag{243}$$

By applying $\mathbf{n}\mathbf{n} \cdot \nabla \gamma = -\mathbf{n}\mathbf{n} \cdot \mathbf{n}\delta[g(\mathbf{x}; \mathbf{x}^s)] = -\mathbf{n}\delta[g(\mathbf{x}; \mathbf{x}^s)] = \nabla \gamma$ and Equation (231), we have

$$\begin{aligned}
\int_{S_\infty} \nabla_{\mathbf{u}}^{S} \cdot (\gamma^s \mathbf{f}^s) \mathrm{d}S_{\mathbf{u}} = & \int_{V_\infty} \gamma^s \mathbf{f} \cdot \nabla \mathbf{n} \cdot \nabla \gamma \mathrm{d}V + \int_{V_\infty} \gamma^s \mathbf{f} \cdot (\nabla \nabla \gamma) \cdot \mathbf{n} \mathrm{d}V \\
& - \int_{V_\infty} \gamma^s \mathbf{n} \cdot (\nabla \nabla \gamma) \cdot \mathbf{f} \mathrm{d}V = \int_{V_\infty} \gamma^s \mathbf{f} \cdot \nabla \mathbf{n} \cdot \nabla \gamma \mathrm{d}V
\end{aligned} \tag{244}$$

Note that $\nabla \mathbf{n} \cdot \nabla \gamma = 0$. Therefore,

$$\int_{V_\infty} \gamma^s \mathbf{f} \cdot \nabla \mathbf{n} \cdot \nabla \gamma \mathrm{d}V = 0$$

This completes the proof of Identity 5.

Similarly, we can prove the following identities:

Identity 6.

$$\int_{S_\infty} \nabla_{\mathbf{u}}^{S} \times (\gamma^s \mathbf{f}^s) \mathrm{d}S_{\mathbf{u}} = 0 \tag{245}$$

Identity 7.

$$\int_{C_\infty} \nabla_1^c(\gamma^c \mathbf{f}^c)dC_1 = 0 \tag{246}$$

Identity 8.

$$\int_{C_\infty} \nabla_1^c \cdot (\gamma^c \mathbf{f}^c)dC_1 = 0 \tag{247}$$

Identity 9.

$$\int_{C_\infty} \nabla_1^c \times (\gamma^c \mathbf{f}^c)dC_1 = 0 \tag{248}$$

where **u** and **l** are explicitly used to denote the fact that differentiation and integration are carried out in the same coordinate system.

3.2.5 Dyadic Del of indicator functions

Identity 1.

$$(\nabla\nabla\gamma) \cdot \boldsymbol{v} = -\nabla\boldsymbol{v} \cdot \nabla\gamma \tag{249}$$

Proof. Applying the chain rule leads to

$$(\nabla\nabla\gamma) \cdot \boldsymbol{v} = \nabla(\boldsymbol{v} \cdot \nabla\gamma) - \nabla\boldsymbol{v} \cdot \nabla\gamma$$

Since $\nabla\gamma$ is a vector tangent to **n**, $\boldsymbol{v} \cdot \nabla\gamma = 0$. Therefore,

$$(\nabla\nabla\gamma) \cdot \boldsymbol{v} = -\nabla\boldsymbol{v} \cdot \nabla\gamma$$

Identity 2.

$$(\nabla\nabla\gamma) \cdot \boldsymbol{\lambda} = -\nabla\boldsymbol{\lambda} \cdot \nabla\gamma \tag{250}$$

Proof. Applying the chain rule yields

$$(\nabla\nabla\gamma) \cdot \boldsymbol{\lambda} = \nabla(\boldsymbol{\lambda} \cdot \nabla\gamma) - \nabla\boldsymbol{\lambda} \cdot \nabla\gamma \tag{251}$$

Because $\nabla\gamma$ is a vector tangent to **n**, $\boldsymbol{\lambda} \cdot \nabla\gamma = 0$. Therefore

$$(\nabla\nabla\gamma) \cdot \boldsymbol{\lambda} = -\nabla\boldsymbol{\lambda} \cdot \nabla\gamma$$

Identity 3.

$$\nabla\nabla\gamma^s \cdot \mathbf{n} = -\nabla\mathbf{n} \cdot \nabla\gamma^s \tag{252}$$

Proof. Applying the chain rule yields

$$\nabla\nabla\gamma^s \cdot \mathbf{n} = \nabla(\mathbf{n} \cdot \gamma^s) - \nabla\mathbf{n} \cdot \nabla\gamma^s$$

Since $\mathbf{n} \cdot \nabla\gamma^s = 0$, we have

$$\nabla\nabla\gamma^s \cdot \mathbf{n} = -\nabla\mathbf{n} \cdot \nabla\gamma^s$$

Similarly, we can prove Identities 31–33:

Identity 4.

$$\nabla\nabla\gamma^s \cdot \boldsymbol{\lambda} = -\nabla\boldsymbol{\lambda} \cdot \nabla\gamma^s \tag{253}$$

Identity 5.

$$\nabla\nabla\gamma^{c}\cdot\mathbf{n} = -\nabla\mathbf{n}\cdot\nabla\gamma^{c} \tag{254}$$

Identity 6.

$$\nabla\nabla\gamma^{c}\cdot\boldsymbol{v} = -\nabla\boldsymbol{v}\cdot\nabla\gamma^{c} \tag{255}$$

Identity 1.

$$\nabla_{\mathbf{n}}^{S}\nabla\gamma = \mathbf{n}\cdot\nabla\gamma(\nabla\mathbf{n})^{T} \tag{256}$$

Proof. By Equation (42), we have

$$\begin{aligned}\nabla^{S}\nabla\gamma &= (\nabla - \mathbf{nn}\cdot\nabla)\nabla\gamma\\ &= \nabla\nabla\gamma - \mathbf{nn}\cdot\nabla\nabla\gamma\\ &= \nabla\nabla\gamma - (\nabla(\mathbf{n}\cdot\nabla\gamma\mathbf{n}))^{T} + (\nabla\mathbf{n})^{T}\mathbf{n}\cdot\nabla\gamma + (\nabla\mathbf{n}\cdot\nabla\gamma\mathbf{n})^{T}\end{aligned} \tag{257}$$

By $\nabla\mathbf{n}\cdot\nabla\gamma = 0$, and, $\mathbf{n}\cdot\mathbf{n} = 1$, Equation (257) becomes

$$\begin{aligned}\nabla^{S}\nabla\gamma &= \nabla\nabla\gamma - (\nabla\nabla\gamma)^{T} + (\nabla\mathbf{n})^{T}\mathbf{n}\cdot\nabla\gamma\\ &= \mathbf{n}\cdot\nabla\gamma(\nabla\mathbf{n})^{T}\end{aligned}$$

This completes the proof of Identity 1.

Identity 2.

$$\nabla_{\lambda}^{c}\nabla\gamma = -\boldsymbol{\lambda}\nabla\boldsymbol{\lambda}\cdot\nabla\gamma \tag{258}$$

Proof. Since $\nabla_{\lambda}^{c} = \boldsymbol{\lambda\lambda}\cdot\nabla$, we have

$$\nabla_{\lambda}^{c}\nabla\gamma = -\boldsymbol{\lambda\lambda}\cdot\nabla\nabla\gamma \tag{259}$$

Applying the chain rule yields

$$\nabla_{\lambda}^{c}\nabla\gamma = \boldsymbol{\lambda}\nabla(\boldsymbol{\lambda}\cdot\nabla\gamma) - \boldsymbol{\lambda}\nabla\boldsymbol{\lambda}\cdot\nabla\gamma$$

Applying $\nabla\gamma\cdot\boldsymbol{\lambda} = 0$, we have

$$\nabla_{\lambda}^{c}\nabla\gamma = -\boldsymbol{\lambda}\nabla\boldsymbol{\lambda}\cdot\nabla\gamma$$

This completes the proof of Identity 2.

Identity 3.

$$\nabla_{\mathbf{n}}^{S}\nabla\gamma^{s} = \nabla\nabla\gamma^{s} + \mathbf{n}\nabla\mathbf{n}\cdot\nabla\gamma^{s} \tag{260}$$

Proof. By Equation (42), we have

$$\begin{aligned}\nabla_{\mathbf{n}}^{S}\nabla\gamma^{s} &= (\nabla - \mathbf{nn}\cdot\nabla)\nabla\gamma^{s}\\ &= \nabla\nabla\gamma^{s} - \mathbf{nn}\cdot\nabla\nabla\gamma^{s}\end{aligned}$$

Applying the chain rule yields

$$\nabla_{\mathbf{n}}^{S}\nabla\gamma^{s} = \nabla\nabla\gamma^{s} - \mathbf{n}\nabla(\mathbf{n}\cdot\nabla\gamma^{s}) + \mathbf{n}\nabla\mathbf{n}\cdot\nabla\gamma^{s}$$

Note that

$$\mathbf{n}\cdot\nabla\gamma^{s} = 0$$

Therefore,

$$\nabla^{S}_{\mathbf{n}}\nabla\gamma^{s} = \nabla\nabla\gamma^{s} + \mathbf{n}\nabla\mathbf{n}\cdot\nabla\gamma^{s}$$

This completes the proof of Identity 3.

Identity 4.

$$\nabla^{c}_{\lambda}\nabla\gamma^{s} = -\boldsymbol{\lambda}\nabla\boldsymbol{\lambda}\cdot\nabla\gamma^{s} \tag{261}$$

Proof. Since $\nabla^{c}_{\lambda} = \boldsymbol{\lambda}\boldsymbol{\lambda}\cdot\nabla$, we have

$$\nabla^{c}_{\lambda}\nabla\gamma^{s} = \boldsymbol{\lambda}\boldsymbol{\lambda}\cdot\nabla\nabla\gamma^{s}$$

Applying the chain rule leads to

$$\nabla^{c}_{\lambda}\nabla\gamma^{s} = \boldsymbol{\lambda}\nabla(\boldsymbol{\lambda}\cdot\nabla\gamma^{s}) - \boldsymbol{\lambda}\nabla\boldsymbol{\lambda}\cdot\nabla\gamma^{s}$$

Note that

$$\boldsymbol{\lambda}\cdot\nabla\gamma^{s} = \boldsymbol{\lambda}\cdot\nabla^{S}\gamma^{s} = 0$$

Therefore,

$$\nabla^{c}\nabla\gamma^{s} = -\boldsymbol{\lambda}\nabla\boldsymbol{\lambda}\cdot\nabla\gamma^{s}$$

This completes the proof of Identity 4.

Identity 5.

$$\nabla^{S}_{\mathbf{n}}\nabla\gamma^{c} = \nabla\nabla\gamma^{c} + \mathbf{n}\nabla\mathbf{n}\cdot\nabla\gamma^{c} \tag{262}$$

Proof. Since $\nabla^{S} = \nabla - \mathbf{nn}\cdot\nabla$, we have

$$\begin{aligned}\nabla^{S}_{\mathbf{n}}\nabla\gamma^{c} &= (\nabla - \mathbf{nn}\cdot\nabla)\nabla\gamma^{c}\\ &= \nabla\nabla\gamma^{c} - \mathbf{nn}\cdot\nabla\nabla\gamma^{c}\end{aligned}$$

Applying the chain rule yields

$$\begin{aligned}\nabla^{S}_{\mathbf{n}}\nabla\gamma^{c} &= \nabla\nabla\gamma^{c} - \mathbf{nn}\cdot\nabla\gamma^{c}\\ &= \nabla\nabla\gamma^{c} - \mathbf{n}\nabla(\mathbf{n}\cdot\nabla\gamma^{c}) + \mathbf{n}\nabla\mathbf{n}\cdot\nabla\gamma^{c}\end{aligned}$$

Note that $\mathbf{n}\cdot\nabla\gamma^{c} = \mathbf{n}\cdot\nabla^{c}\gamma^{c} = 0$. Therefore,

$$\nabla^{S}\nabla\gamma^{c} = \nabla\nabla\gamma^{c} + \mathbf{n}\nabla\mathbf{n}\cdot\nabla\gamma^{c}$$

This completes the proof of Identity 5.

Identity 6.

$$\nabla^{c}_{\lambda}\nabla\gamma^{c} = \nabla\nabla\gamma^{c} - \boldsymbol{\lambda}\cdot\nabla\gamma^{c}(\nabla\boldsymbol{\lambda})^{T} \tag{263}$$

Proof. As $\nabla^{c}_{\lambda} = \boldsymbol{\lambda}\boldsymbol{\lambda}\cdot\nabla$, we have

$$\nabla^{c}_{\lambda}\nabla\gamma^{c} = \boldsymbol{\lambda}\boldsymbol{\lambda}\cdot\nabla\nabla\gamma^{c}$$

Applying the chain rule yields

$$\nabla^{c}_{\lambda}\nabla\gamma^{c} = (\nabla(\boldsymbol{\lambda}\boldsymbol{\lambda}\cdot\nabla\gamma^{c}))^{T} - (\nabla\boldsymbol{\lambda})^{T}\boldsymbol{\lambda}\cdot\nabla\gamma^{c} - (\boldsymbol{\lambda}\nabla\boldsymbol{\lambda}\cdot\nabla\gamma^{c})^{T}$$

Note that $\nabla^c \gamma^c = \nabla \gamma^c$, $\nabla \boldsymbol{\lambda} \cdot \nabla \gamma^c = \nabla \boldsymbol{\lambda} \cdot \nabla^c \gamma^c = 0$. Therefore,

$$\nabla_{\lambda}^{c} \nabla \gamma^c = \nabla \nabla \gamma^c - (\nabla \boldsymbol{\lambda})^T \boldsymbol{\lambda} \cdot \nabla \gamma^c$$

This completes the proof of Identity 6.

3.3 Examples of indicator functions

The fundamental importance of indicator functions in developing multiscale theorems lies in their ability to identify regions of interest. In this section we present some indicator functions of specifying spherical REVs, phases, interfaces, and common curves in multiphase systems.

3.3.1 Spherical REVs

To identify a spherical volume of radius ρ which is independent of time and its location in space, we can define an indicator function (Figure 11)

$$\gamma(\xi) = \begin{cases} 1 & \text{when} \quad |\xi| < \rho \\ 0 & \text{when} \quad |\xi| > \rho \end{cases} \tag{264}$$

If the size of the sphere depends on time but not the position of the sphere, we may define an indicator function such that

$$\gamma(\xi, t) = \begin{cases} 1 & \text{when} \quad |\xi| < \rho(t) \\ 0 & \text{when} \quad |\xi| > \rho(t) \end{cases} \tag{265}$$

When the size of the sphere is a function of position but not time, we can define an indicator function as

$$\gamma(\xi, \mathbf{x}) = \begin{cases} 1 & \text{when} \quad |\xi| < \rho(\mathbf{x}) \\ 0 & \text{when} \quad |\xi| > \rho(\mathbf{x}) \end{cases} \tag{266}$$

If the size of the sphere depends on both time and position (Figure 12), its indicator function is

$$\gamma(\xi, \mathbf{x}, t) = \begin{cases} 1 & \text{when} \quad |\xi| < \rho(\mathbf{x}, t) \\ 0 & \text{when} \quad |\xi| > \rho(\mathbf{x}, t) \end{cases} \tag{267}$$

The spheres identified by these indicator functions are particularly useful as REVs for macroscopic averaging.

3.3.2 Phases

Suppose that the region occupied by one phase, referred to as the α-phase in a multiphase system, is to be identified. For this case, the value of the indicator function depends on position in the domain, such that if a point is in the α-phase, $\gamma = 1$; otherwise $\gamma = 0$. Note that the position of the α-phase in general changes with time, and can be identified by $\mathbf{x}+\xi$ uniquely, therefore, $\gamma^\alpha = \gamma^\alpha(\mathbf{x} + \xi, t)$ can be used to specify the α-phase, where superscript α indicates that this indicator function identifies the α-phase in

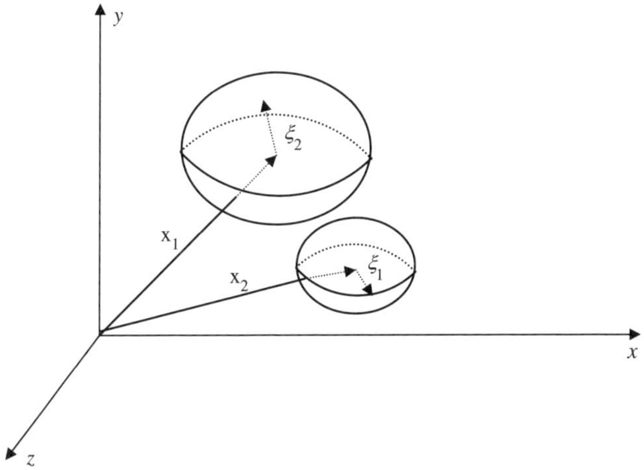

Figure 12 Spatially dependent REV for which γ is a function of **x** and ξ rather than **x**+ξ as with a constant REV.

entire space:

$$\gamma^\alpha(\mathbf{x} + \xi,t) = \begin{cases} 1 & \text{in } \alpha\text{-phase} \\ 0 & \text{elsewhere} \end{cases} \tag{268}$$

All of the α-phase inside an REV can be identified by the product of two indicator functions. We may use $\gamma(\xi)\gamma^\alpha(\mathbf{x} + \xi, t)$, for example, to identify the α-phase inside a constant spherical REV. Here $\gamma(\xi)$ and $\gamma^\alpha(\mathbf{x} + \xi, t)$ are defined by Equations (264) and (268), respectively.

3.3.3 Interfaces

We can also use an indicator function to specify an interface between two phases in multiphase systems. Let $S_{\alpha\beta}$ be the interphase between the α-phase and the β-phase (Figure 13). To identify this surface as opposed to all interfaces between phases, the following indicator function can be defined

$$\gamma^s[\mathbf{u}(\mathbf{x} + \xi,t)] = \begin{cases} 1 & \text{when} \quad (\mathbf{u} \in S_{\alpha\beta}) \\ 0 & \text{when} \quad (\mathbf{u} \notin S_{\alpha\beta}) \end{cases} \tag{269}$$

Note that $S_{\alpha\beta}$ could be a finite number of non-overlapping simple regions.

A product of two indicator functions can be used to restrict $S_{\alpha\beta}$ to an REV. For example consider $\gamma\gamma^s$ where γ and γ^s are defined by Equations (264) and (269), respectively. This product identifies all $\alpha\beta$-interfaces inside a constant spherical REV.

3.3.4 Common phase curve

To identify a common curve $C_{\alpha\beta\varepsilon}$ among three distinct phases α, β, and ε (Figure 13), we can define an indicator function such that

$$\gamma^c[\mathbf{l}(\mathbf{x} + \xi),t] = \begin{cases} 1 & \text{when} \quad \mathbf{l} \in C_{\alpha\beta\varepsilon} \\ 0 & \text{when} \quad \mathbf{l} \notin C_{\alpha\beta\varepsilon} \end{cases} \tag{270}$$

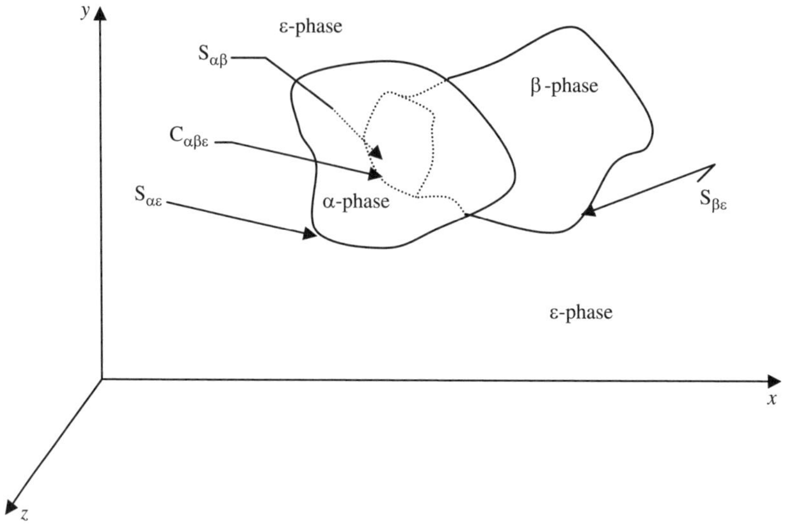

Figure 13 Interface $S_{\alpha\beta}$ between adjacent α- and β-phases.

where $C_{\alpha\beta\varepsilon}$ is the curve formed by the intersection of the $\alpha\beta$-, $\beta\varepsilon$-, and $\alpha\varepsilon$-interfaces. Note that $C_{\alpha\beta\varepsilon}$ could represent a finite number of non-intersecting simple curves.

A product of two indicator functions can be used to restrict $C_{\alpha\beta\varepsilon}$ to certain regions in space. For example, $\gamma\gamma^{c}$, where γ and γ^{c} are defined by Equations (264) and (270), respectively, identifies all $\alpha\beta\varepsilon$-curves inside a constant-sized spherical REV.

3.4 Interchange of integration dimensions

When a gradient of an indicator function appears in the integrand, a volume integral becomes a surface integral; a surface integral becomes an integral along a curve; a curvilineal integral becomes point evaluations at points where the derivative of the indicator function is non-zero. We can use this property to reduce or increase the integration dimension. In this section we discuss this interchange of integration dimensions which is of considerable importance in developing multiscale theorems.

3.4.1 Surficial to spatial integral transformation

A surface integration may be transformed to a volume integration using integral properties in Equations (183)–(187). For example, consider an orientable surface S viewed as a portion of a closed simple surface S_{∞} (Figure 14). By the surficial indicator function γ^{s}, we may convert an integral over S to an integral over S_{∞} as

$$\int_{S} f \mathrm{d}S = \int_{S_{\infty}} f\gamma^{s} \mathrm{d}S \tag{271}$$

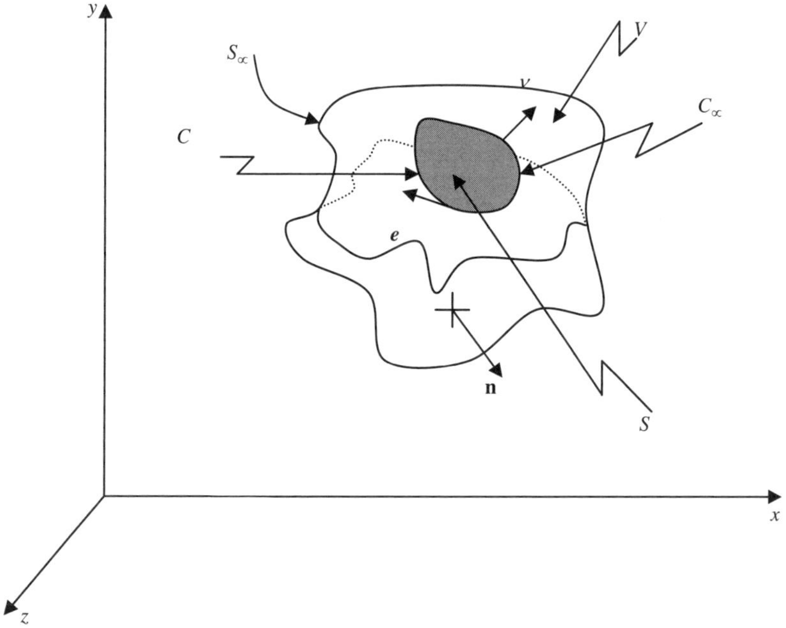

Figure 14 Surface S as a portion of a simple closed surface S_∞ that bounds volume V (C_∞ is a closed curve and the boundary of S; C a portion of C_∞; $\mathbf{n}$ the unit normal of S; $\boldsymbol{\nu}$ the unit normal of C_∞ and tangent to S_∞; and $\mathbf{e}$ the unit vector tangent to both C and S_∞).

Because S_∞ is a closed surface, it may be viewed as the boundary of a volume V. A spatial generalized function, γ, can be used to identify this volume. Then by Equation (183), we have

$$\int_S f \mathrm{d}S = -\int_{V_\infty} f\gamma^{\mathrm{s}} \mathbf{n} \cdot \nabla\gamma \mathrm{d}V \tag{272}$$

where $\mathbf{n}$ is the outwardly directed unit vector normal to the volume.

3.4.2 Curvilineal to surficial integral transformation

Integral properties in Equations (173)–(177) can be used to change an integration over spatial curves to a surface integration. For example, consider an integration of some function f over some finite simple curve C. This curve may be extended to form a simple closed loop C_∞. An integral over C can be written as an integral over C_∞ by making use of the curvilineal indicator function γ^{c} defined to be equal to 1 over C and 0 over the rest of C_∞ such that

$$\int_C f \mathrm{d}C = \int_{C_\infty} f\gamma^{\mathrm{c}} \mathrm{d}C \tag{273}$$

Now the closed loop C_∞ can be viewed as the bounding curve of a simple orientable surface S. The unit normal vector to C_∞ which is tangent to and pointing outward from S is $\boldsymbol{\nu}$. Also let S_∞ be a closed surface which contains S.

If γ^s is defined to equal 1 on S but zero on the rest of S_∞, Equation (173) can be used to transform the right side of Equation (273) to a surface integral such that

$$\int_C f\mathrm{d}C = -\int_{S_\infty} f\gamma^c(\mathbf{v}\cdot\nabla^S\gamma^s)\mathrm{d}S \tag{274}$$

Next the closed surface S_∞ can be considered to be the boundary surface of a volume V. This volume is a subregion of all space V_∞. Let γ be the indicator function identifying volume V such that γ equals 1 in V but is 0 in the rest of V_∞. Equation (183) can thus be applied to change the right side of Equation (274) to an integral over V_∞ so that

$$\int_C f\mathrm{d}C = \int_{V_\infty} f\gamma^c(\mathbf{v}\cdot\nabla^S\gamma^s)(\mathbf{n}\cdot\nabla\gamma)\mathrm{d}V \tag{275}$$

where $\mathbf{n}$ is the outwardly directed unit normal to V.

Equations (273)–(275) provide the successive steps for changing an integral over a curve to an integral over a surface, and the integral over that surface to an integral over a volume. It is important to have an understanding of these manipulations and to be able to relate them to a physical system under study. For example, the integration of a function $f(\mathbf{x}+\xi, t)$ over the interface $S_{\alpha\beta}$ between the α- and β-phases within an REV in a multiphase system can be written as, in terms of an integration over V_∞,

$$\int_{S_{\alpha\beta}} f\mathrm{d}S = -\int_{V_\infty} f\gamma^s\gamma(\mathbf{n}\cdot\nabla\gamma^\alpha)\mathrm{d}V$$

where $\gamma(\xi)$ is a spatial indicator function for an REV defined by Equation (264), $\gamma^\alpha(\mathbf{x}+\xi,t)$ the spatial indicator function that locates the α-phase in space defined by Equation (268), and $\gamma^s[\mathbf{u}(\mathbf{x}+\xi),t]$ the surficial indicator function that identifies $S_{\alpha\beta}$ in Equation (269).

One other important relation between integrals over a surface and a curve is obtained by considering the integral of the surface gradient of a function f over a surface S. If integration is to be extended from being over S to integration over a closed surface, S_∞, that contains S, an indicator function γ^s must be introduced such that

$$\int_S \nabla_{\mathbf{n}}^S f\mathrm{d}S = \int_{S_\infty} (\nabla_{\mathbf{n}}^S f)\gamma^s\mathrm{d}S \tag{276}$$

An alternative approach to facilitating the transformation from integration over a finite surface to integration over S_∞ is to take the surface of integration as being one that intersects a volume of interest as in Figure 15. Then the generalized function for the volume may be used to identify the portion of S_∞ intersecting that volume to form the surface of interest. Thus an alternative expression to Equation (276) is

$$\int_S \nabla_{\mathbf{n}}^S f\mathrm{d}S = \int_{S_\infty} (\nabla_{\mathbf{n}}^S f)\gamma\mathrm{d}S \tag{277}$$

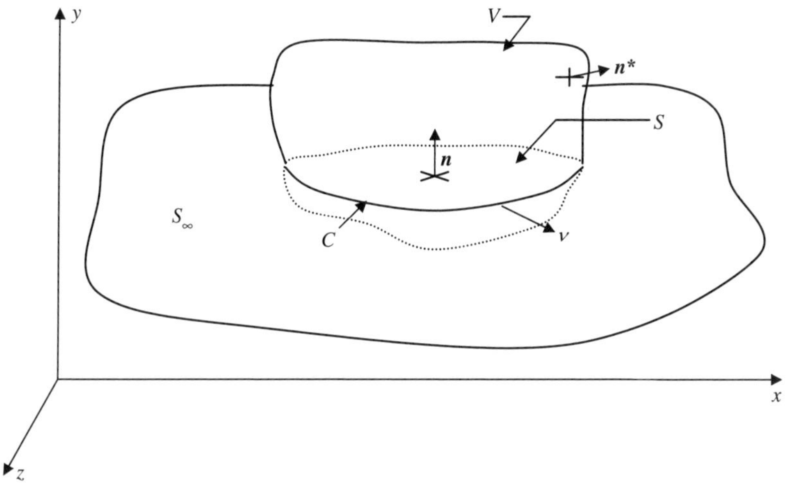

Figure 15 Surface S as a portion of surface S_∞ formed as the intersection of S_∞ with a volume V (Curve C is the intersection of the boundary of V with S_∞; $\mathbf{n}$ the unit normal of S, ν the unit normal of C and tangent to S_∞; and $\mathbf{n}^*$ the unit normal of the surface of V).

Although the difference between these expressions may appear to be minor, in fact some important information may be obtained. The normal to the surface will be denoted as $\mathbf{n}$ while the normal to the edge of the surface that is also tangent to the surface will be denoted as $\boldsymbol{\nu}$. These two vectors are orthogonal at the edge of the surface. The normal to the volume at the edge of the surface is denoted as $\mathbf{n}^*$ and is equal to $\boldsymbol{\nu}$ only if the surface of interest is orthogonal to the boundary of the volume. The right sides of Equations (276) and (277) are identical. Thus integration by parts is applied while preserving the equality to obtain

$$\int_{S_\infty} \nabla_{\mathbf{n}}^{\mathrm{S}}(f\gamma^{\mathrm{s}})\mathrm{d}S - \int_{S_\infty} f\nabla_{\mathbf{n}}^{\mathrm{S}}\gamma^{\mathrm{s}}\mathrm{d}S = \int_{S_\infty} \nabla_{\mathbf{n}}^{\mathrm{S}}(f\gamma)\mathrm{d}S - \int_{S_\infty} f\nabla_{\mathbf{n}}^{\mathrm{S}}\gamma\mathrm{d}S \tag{278}$$

The first terms on both sides of this equation are zero (Identity 4). Note also that

$$\nabla_{\mathbf{n}}^{\mathrm{S}} = \nabla - \mathbf{nn}\cdot\nabla$$

Thus Equation (278) becomes

$$\int_{S_\infty} f\nabla_{\mathbf{n}}^{\mathrm{S}}\gamma^{\mathrm{s}}\mathrm{d}S = \int_{S_\infty} f\nabla\gamma\mathrm{d}S - \int_{S_\infty} f\mathbf{nn}\cdot\nabla\gamma\mathrm{d}S \tag{279}$$

By Equation (175), the left side of this equation is readily converted to an integral over the closed curve C bounding S. Note that $\nabla\gamma$, which is non-zero only on the boundary of the volume, will be non-zero only where S_∞ and the boundary of the volume intersect, i.e., only on C. However, because $\nabla\gamma$ is the spatial derivative of a spatial indicator function, the magnitude of its integral, though appropriate for conversion of a volume integral to a surface integral, is not known when converting a surface integral to an integral over a curve. For convenience, denote this magnitude as μ such that the integrals in Equation (279) convert to integrals over a curve as follows:

$$-\int_C f\boldsymbol{\nu}\mathrm{d}C = -\int_C f\mu\mathbf{n}^*\mathrm{d}C + \int_C f\mu\mathbf{n}\mathbf{n}\cdot\mathbf{n}^*\mathrm{d}C \tag{280}$$

Because the integration here is independent on any particular curve geometry, the localization theorem yields (Wang, 1997; Wang and Zhou, 2000)

$$\boldsymbol{\nu} = \mu\mathbf{n}^* - \mu\mathbf{n}\mathbf{n}\cdot\mathbf{n}^* \tag{281}$$

Because $\boldsymbol{\nu}$ is orthogonal to $\mathbf{n}$, the dot product of $\boldsymbol{\nu}$ with this equation yields

$$\mu = \frac{1}{\boldsymbol{\nu}\cdot\mathbf{n}^*} \tag{282}$$

From this expression, and equating the corresponding first terms on the right sides of Equations (279) and (280), we obtain

$$\int_{S_\infty} f\nabla\gamma\mathrm{d}S = -\int_C \frac{f\mathbf{n}^*}{\boldsymbol{\nu}\cdot\mathbf{n}^*}\mathrm{d}C \tag{283}$$

Similarly, we have

$$\int_{S_\infty} f\mathbf{a}\cdot\nabla\gamma\mathrm{d}S = -\int_C \frac{f\mathbf{a}\cdot\mathbf{n}^*}{\boldsymbol{\nu}\cdot\mathbf{n}^*}\mathrm{d}C \tag{284}$$

$$\int_{S_\infty} f\mathbf{a}\times\nabla\gamma\mathrm{d}S = -\int_C \frac{f\mathbf{a}\times\mathbf{n}^*}{\boldsymbol{\nu}\cdot\mathbf{n}^*}\mathrm{d}C \tag{285}$$

where $\mathbf{a}$ is an arbitrary vector function.

These equations will play an important role in the derivation of integration theorems when integration is over a planar surface.

3.4.3 Curvilineal integration to point-value evaluation

This kind of transformation is needed when integrating a curvilineal gradient of a function over a simple curve. For a curve C (the part of a closed curve C_∞; Figure 14) identified by γ^c, for example, we have

$$\int_C \nabla_\lambda^c f\mathrm{d}C = \int_{C_\infty} (\nabla_\lambda^c f)\gamma^c\mathrm{d}C \tag{286}$$

Applying the chain rule to the right side of this equation and noting that integration of $\nabla_\lambda^c(f\gamma^c)$ along a closed curve will be zero, we have

$$\int_C \nabla_\lambda^c f\mathrm{d}C = -\int_{C_\infty} f\nabla_\lambda^c\gamma^c\mathrm{d}C \tag{287}$$

Then by Equation (165), the right side of this equation can be converted to an evaluation of the integrand at the ends of C such that

$$\int_C \nabla_\lambda^c f\mathrm{d}C = (\mathbf{e}f)\big|_{\text{ends}} \tag{288}$$

where **e** is a vector tangent to the curve and pointing outward from C at each end. If C is a closed curve such that it has no 'ends', the right side of Equation (288) will be zero.

When integration over a curve C is considered, the indicator function for a surface S may be used to identify a portion of the curve C as lying in S (Figure 16). If γ^c is defined as $\gamma^c = 1$ over C but $\gamma^c = 0$ over the rest of C, we have

$$\int_C \nabla_\lambda^c f \mathrm{d}C = \int_{C_\infty} \nabla_\lambda^c f \gamma^c \mathrm{d}C \tag{289}$$

Another alternative form is

$$\int_C \nabla_\lambda^c f \mathrm{d}C = \int_{C_\infty} \nabla_\lambda^c f \gamma^s \mathrm{d}C \tag{290}$$

in which we view C as a part of the surface S identified by the indicator function γ^s and isolated from the curve C_∞. **e** is tangent to C pointing outward from S. The right side of Equations (289) and (290) must be identical. Thus integration by parts leads to

$$\int_{C_\infty} \nabla_\lambda^c (f\gamma^c) \mathrm{d}C - \int_{C_\infty} f \nabla_\lambda^c \gamma^c \mathrm{d}C = \int_{C_\infty} \nabla_\lambda^c (f\gamma^s) \mathrm{d}c - \int_{C_\infty} f \nabla_\lambda^c \gamma^s \mathrm{d}C \tag{291}$$

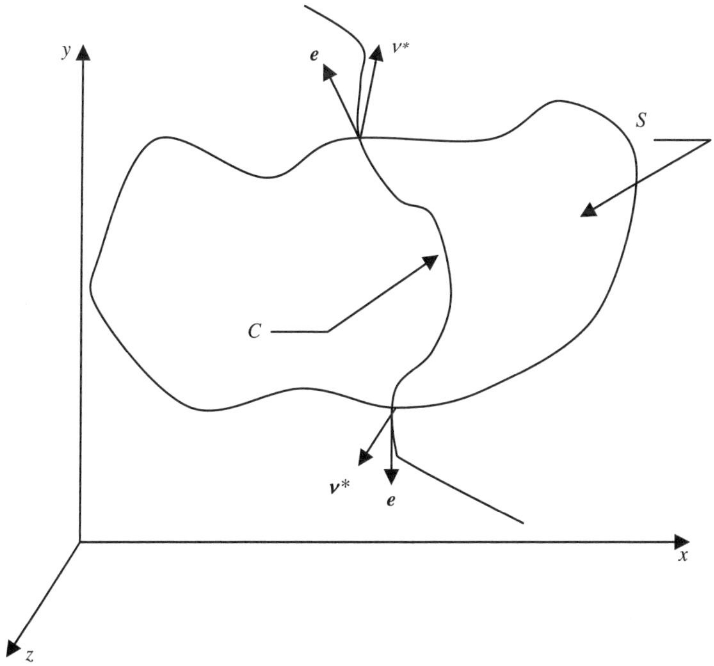

Figure 16 Curve C as a portion of a curve C_∞ lying in a surface S (the endpoints of curve C are the intersection of the boundary of S with C_∞; **e** the tangent to C pointing outward from S; and $\boldsymbol{\nu}^*$ the normal to the boundary of S and tangent to S).

The first terms on both sides of Equation (291) are zero. Note also that

$$\nabla^{c}_{\lambda} = \nabla^{S}_{\mathbf{n}} - \boldsymbol{\nu}\boldsymbol{\nu} \cdot \nabla^{S} \tag{292}$$

where $\boldsymbol{\nu}$ is a unit vector tangent to the surface S and orthogonal to $\mathbf{e}$. Therefore, Equation (291) becomes

$$\int_{C_\infty} f\nabla^{c}_{\lambda}\gamma^{c}\mathrm{d}C = \int_{C_\infty} f\nabla^{S}_{\mathbf{n}}\gamma^{s}\mathrm{d}C - \int_{C_\infty} f\boldsymbol{\nu\nu} \cdot \nabla^{S}_{\mathbf{n}}\gamma^{s}\mathrm{d}C \tag{293}$$

Note that $\nabla^{S}_{\mathbf{n}}\gamma^{s}$ is non-zero only on the boundary of the surface S. Therefore, it is non-zero only where C and the boundary of S intersect. However, the magnitude of $\nabla^{S}_{\mathbf{n}}\gamma^{s}$ at the ends of C where the boundary of S and C intersect is unknown. For convenience, denote this magnitude as μ such that

$$-(\mathbf{e}f)|_{\mathrm{ends}} = -(f\mu\boldsymbol{\nu}^{*})|_{\mathrm{ends}} + (f\boldsymbol{\nu\nu} \cdot \boldsymbol{\nu}^{*}\mu)|_{\mathrm{ends}} \tag{294}$$

where $\boldsymbol{\nu}^{*}$ is the outwardly directed normal to the boundary of the surface and tangent to the surface at the intersection points of C and the boundary of the surface (Figure 16).

Because Equation (294) is valid for all ends, the localization theorem thus concludes (Wang, 1997b; Wang and Zhou, 2000)

$$-\mathbf{e}f = -f\mu\boldsymbol{\nu}^{*} + f\boldsymbol{\nu\nu} \cdot \boldsymbol{\nu}^{*}\mu \tag{295}$$

Since $\mathbf{e}$ is orthogonal to $\boldsymbol{\nu}$, the dot product of $\mathbf{e}$ with this equation yields

$$1 = \mu\boldsymbol{\nu}^{*} \cdot \mathbf{e} \tag{296}$$

or

$$\mu = \frac{1}{\boldsymbol{\nu}^{*} \cdot \mathbf{e}} \tag{297}$$

From this equation, and equating the corresponding first terms on the right sides of Equations (293) and (294), we have

$$\int_{C_\infty} f\nabla^{S}_{\mathbf{n}}\gamma^{s}\mathrm{d}C = \left.\frac{-f\boldsymbol{\nu}^{*}}{\mathbf{e} \cdot \boldsymbol{\nu}^{*}}\right|_{\mathrm{ends}} \tag{298}$$

Similarly, we obtain

$$\int_{C_\infty} f\mathbf{a} \cdot \nabla^{S}_{\mathbf{n}}\gamma^{s}\mathrm{d}C = \left.\frac{-f\mathbf{a} \cdot \boldsymbol{\nu}^{*}}{\mathbf{e} \cdot \boldsymbol{\nu}^{*}}\right|_{\mathrm{ends}} \tag{299}$$

$$\int_{C_\infty} f\mathbf{a} \times \nabla^{S}_{\mathbf{n}}\gamma^{s}\mathrm{d}C = \left.\frac{-f\mathbf{a} \times \boldsymbol{\nu}^{*}}{\mathbf{e} \cdot \boldsymbol{\nu}^{*}}\right|_{\mathrm{ends}} \tag{300}$$

where $\mathbf{a}$ is an arbitrary vector function.

Let $\gamma^{c} = 1$ along C but $\gamma^{c} = 0$ on the rest of C_∞ (Figure 17); thus we have

$$\int_{C} \nabla^{c}_{\lambda}f\mathrm{d}C = \int_{C_\infty} \nabla^{c}_{\lambda}f\gamma^{c}\mathrm{d}C \tag{301}$$

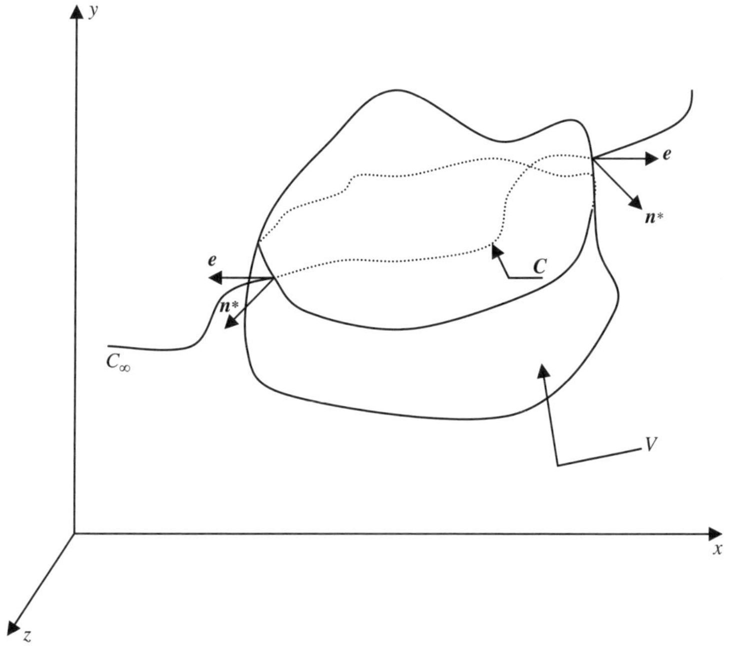

Figure 17 Curve C as a portion of a curve C_∞ intersecting a volume V (the endpoints of curve C are the intersection of the boundary of V with C_∞; **e** the unit tangent of C pointing outward from V; and **n*** the unit normal to the boundary of V.

Note also that C intersects a volume V identified by an indicator function γ (Figure 17). Therefore,

$$\int_C \nabla_\lambda^c f \mathrm{d}C = \int_{C_\infty} \nabla_\lambda^c f \gamma \mathrm{d}C \tag{302}$$

In deriving Equation (302), C is viewed as a part of the space V so that it can be isolated by the indicator function γ from the rest of curve C_∞. Unit vectors are defined such that **e** is tangent to C pointing outward from C and **n*** is normal to the boundary surface of V. The right side of Equations (301) and (302) must be identical. Thus, after applying integration by parts,

$$\int_{C_\infty} \nabla_\lambda^c (f\gamma^c) \mathrm{d}C - \int_{C_\infty} f \nabla_\lambda^c \gamma^c \mathrm{d}C = \int_{C_\infty} \nabla_\lambda^c (f\gamma) \mathrm{d}C - \int_{C_\infty} f \nabla_\lambda^c \gamma \mathrm{d}C \tag{303}$$

The first terms on both sides of Equation (303) are zero. Note also that

$$\nabla_\lambda^c = \boldsymbol{\lambda\lambda} \cdot \nabla$$

where $\boldsymbol{\lambda}$ is a unit vector tangent to the curve C. Therefore, Equation (303) becomes

$$\int_{C_\infty} f \nabla_\lambda^c \gamma^c \mathrm{d}C = \int_{C_\infty} f\, \boldsymbol{\lambda\lambda} \cdot \nabla \gamma \mathrm{d}C \tag{304}$$

Note that $\nabla\gamma$ is non-zero only on the boundary of the volume V. Therefore, it is non-zero only where C and the boundary of V intersect. Let μ denote the magnitude of $\nabla\gamma$ at the ends of C where the boundary of V and C intersect. Thus,

$$\mathbf{e}f\big|_{\text{ends}} = f\mu\boldsymbol{\lambda}\boldsymbol{\lambda}\cdot\mathbf{n}^*\big|_{\text{ends}} \tag{305}$$

Since this is valid for all ends, the localization theorem yields (Wang, 1997b; Wang and Zhou, 2000)

$$\mathbf{e}f = f\mu\boldsymbol{\lambda}\boldsymbol{\lambda}\cdot\mathbf{n}^* \tag{306}$$

The dot product of $\mathbf{e}$ with this equation leads to

$$\mu = \frac{1}{\mathbf{e}\cdot\mathbf{n}^*} \tag{307}$$

Therefore,

$$\int_{C_\infty} f\nabla\gamma \mathrm{d}C = \frac{-f\mathbf{n}^*}{\mathbf{e}\cdot\mathbf{n}^*}\bigg|_{\text{ends}} \tag{308}$$

where $\mathbf{n}^*$ is the outwardly directed unit normal to the surface of the volume at the points where the curve intersects the surface.

Similarly, we have

$$\int_{C_\infty} f\mathbf{a}\cdot\nabla\gamma \mathrm{d}C = \frac{-f\mathbf{a}\cdot\mathbf{n}^*}{\mathbf{e}\cdot\mathbf{n}^*}\bigg|_{\text{ends}} \tag{309}$$

$$\int_{C_\infty} f\mathbf{a}\times\nabla\gamma \mathrm{d}C = \frac{-f\mathbf{a}\times\mathbf{n}^*}{\mathbf{e}\cdot\mathbf{n}^*}\bigg|_{\text{ends}} \tag{310}$$

where $\mathbf{a}$ is an arbitrary vector function.

4. INTEGRATION THEOREMS

We develop 38 integration theorems that convert integrals of gradient, divergence, curl, and partial time derivatives of a microscale function to some combination of derivatives of integrals of the function and integrals over domains of reduced dimensionality. These theorems are powerful tools for the scaling between microscale (or any continuum scale) and megascale.

In the derivation, integration volumes are taken to be bounded by closed regular surfaces; integration surfaces are orientable; and integration curves are simple curves. Functions being integrated are assumed to be continuous and have continuous first partial derivatives. For a concise derivation, we also introduce three differential operators: spatial operator $\nabla\circ F$, surficial operator $\nabla^S\circ F$, and curvilineal operator $\nabla^c\circ F$. In these operators, F is either a scalar function f or a vector function $\mathbf{f}$. Each of these three operators is a unified form of gradient, divergence, and curl such that $\nabla\circ F$ contains ∇f, $\nabla\cdot\mathbf{f}$, and $\nabla\times\mathbf{f}$; $\nabla^S\circ F$ covers $\nabla^S f$, $\nabla^S\cdot\mathbf{f}$, and $\nabla^S\times\mathbf{f}$; and $\nabla^c\circ F$ stands for $\nabla^c f$, $\nabla^c\cdot\mathbf{f}$, or $\nabla^c\times\mathbf{f}$.

4.1 Theorems for integration over a volume

In this section we develop six integration theorems regarding integration over a volume V of spatial, surficial, or curvilineal differential operators of a microscale function f or $\mathbf{f}$. Here $f(\mathbf{x}, t)$ and $\mathbf{f}(\mathbf{x}, t)$ are microscale scalar and vector functions of position vector $\mathbf{x}$ and time t, respectively.

4.1.1 Spatial integration of spatial operators

This group of integration theorems involve integrations of $(\partial f/\partial t)|_{\mathbf{x}}$, ∇f, $\nabla \cdot \mathbf{f}$ and $\nabla \times \mathbf{f}$ over a volume V (Figure 18).

Theorem 1.

$$\int_V \left.\frac{\partial f}{\partial t}\right|_{\mathbf{x}} \mathrm{d}v = \frac{\mathrm{d}}{\mathrm{d}t}\int_V f\mathrm{d}v - \int_S f\mathbf{n}^* \cdot \mathbf{w}\mathrm{d}s \tag{311}$$

where V is the integration volume, S the boundary surface of V, $\mathbf{n}^*$ the unit normal vector outward from V on the surface S, and $\mathbf{w}$ the velocity on the surface S.

Proof. Define an indicator function $\gamma(\mathbf{x})$ such that $\gamma = 1$ within V and $\gamma = 0$ outside V. Then we have

$$\int_V \left.\frac{\partial f}{\partial t}\right|_{\mathbf{x}} \mathrm{d}v = \int_{V_\infty} \left.\frac{\partial f}{\partial t}\right|_{\mathbf{x}} \gamma \mathrm{d}v \tag{312}$$

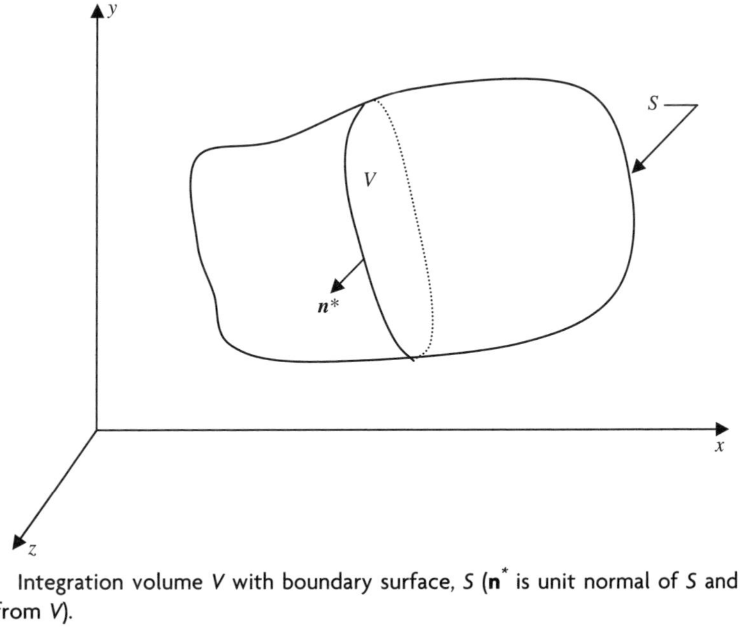

Figure 18 Integration volume V with boundary surface, S ($\mathbf{n}^*$ is unit normal of S and directed outward from V).

where V_∞ represents the entire three-dimensional space. An application of the chain rule to the right side of Equation (312) yields

$$\int_V \left.\frac{\partial f}{\partial t}\right|_{\mathbf{x}} dv = \int_{V_\infty} \left.\frac{\partial (f\gamma)}{\partial t}\right|_{\mathbf{x}} dv - \int_{V_\infty} f\left.\frac{\partial \gamma}{\partial t}\right|_{\mathbf{x}} dv \tag{313}$$

Since V_∞ is independent on the time variable t and by Equation (180), we obtain

$$\int_V \left.\frac{\partial f}{\partial t}\right|_{\mathbf{x}} dv = \frac{\partial}{\partial t}\int_{V_\infty} f\gamma dv + \int_{V_\infty} f\mathbf{w}\cdot\nabla\gamma dv \tag{314}$$

By Equation (186), Equation (314) becomes

$$\int_V \left.\frac{\partial f}{\partial t}\right|_{\mathbf{x}} dv = \frac{\partial}{\partial t}\int_V f dv - \int_S f\mathbf{n}^*\cdot\mathbf{w}ds$$

Theorem 2.

$$\int_V \nabla\circ F dv = \int_S \mathbf{n}^*\circ F ds \tag{315}$$

where $\nabla\circ F$ can be any of the three operators: ∇f, $\nabla\cdot\mathbf{f}$ and $\nabla\times\mathbf{f}$.

Proof. By defining an indicator function $\gamma(\mathbf{x})$ such that $\gamma = 1$ inside V and $\gamma = 0$ in the rest of the space, the left side of Equation (315) may be rewritten as

$$\int_V \nabla\circ F dv = \int_{V_\infty} \nabla\circ F\gamma dv \tag{316}$$

An application of the chain rule on the right side of Equation (316) yields

$$\int_{V_\infty} \nabla\circ F\gamma dv = \int_{V_\infty} \nabla\circ(F\gamma)dv - \int_{V_\infty} \nabla\gamma\circ F dv \tag{317}$$

where we should mention that according to the definition of the symbol $\nabla\circ F$, Equation (317) actually imply the following three formulas:

$$\int_{V_\infty} \nabla f\gamma dv = \int_{V_\infty} \nabla(f\gamma)dv - \int_{V_\infty} (\nabla\gamma) f dv$$

$$\int_{V_\infty} \nabla\cdot\mathbf{f}\gamma dv = \int_{V_\infty} \nabla\cdot(\mathbf{f}\gamma)dv - \int_{V_\infty} \nabla\gamma\cdot\mathbf{f}dv$$

$$\int_{V_\infty} \nabla\times\mathbf{f}\gamma dv = \int_{V_\infty} \nabla\times(\mathbf{f}\gamma)dv - \int_{V_\infty} \nabla\gamma\times\mathbf{f}dv$$

The validity of these formulas is obvious. Proving the multiscale theorems regarding gradient, divergence, and curl frequently appears similar, so we introduce $\nabla\circ F$, $\nabla^S\circ F$, and $\nabla^c\circ F$ for a concise presentation.

From Equations (230), (231), and (232), it follows that

$$\int_{V_\infty} \nabla\circ(F\gamma)dv = 0$$

Thus Equation (317) becomes

$$\int_{V_\infty} \nabla \circ F\gamma \mathrm{d}v = -\int_{V_\infty} \nabla\gamma \circ F\mathrm{d}v \tag{318}$$

By Equations (185)–(187), we conclude that

$$\int_{V} \nabla \circ F\mathrm{d}v = \int_{V_\infty} \nabla \circ F\gamma \mathrm{d}v = \int_{S} \mathbf{n}^* \circ F\mathrm{d}s$$

Remark. Equation (315) contains the following three formulas:
Gradient

$$\int_{V} \nabla f\mathrm{d}v = \int_{S} \mathbf{n}^* f\mathrm{d}s \tag{319}$$

Divergence

$$\int_{V} \nabla \cdot \mathbf{f}\mathrm{d}v = \int_{S} \mathbf{n}^* \cdot \mathbf{f}\mathrm{d}s \tag{320}$$

Curl

$$\int_{V} \nabla \times \mathbf{f}\mathrm{d}v = \int_{S} \mathbf{n}^* \times \mathbf{f}\mathrm{d}s \tag{321}$$

4.1.2 Spatial integration of surficial operators

This group of integration theorems is regarding integrations of $(\partial f/\partial t)|_{\mathbf{u}}$, $\nabla^S f$, $\nabla^S \cdot \mathbf{f}$, and $\nabla^S \times \mathbf{f}$ over a volume V (Figure 18).

Theorem 3.

$$\int_{V} \left.\frac{\partial f}{\partial t}\right|_{\mathbf{u}} \mathrm{d}v = \frac{\partial}{\partial t}\int_{V} f\mathrm{d}v - \int_{V} f\nabla \cdot \mathbf{w}^c \mathrm{d}v + \int_{S} \mathbf{n}^* \cdot \mathbf{w}^c f\mathrm{d}s - \int_{S} \mathbf{n}^* \cdot \mathbf{w} f\mathrm{d}s \tag{322}$$

where V is the integration volume, S the boundary surface of V, and $\mathbf{n}$ a unit vector in one of the general orthogonal coordinate directions (so that Equation (322) is also valid after replacing $\mathbf{n}$ by $\boldsymbol{\lambda}$ or $\boldsymbol{\nu}$). Also, $\mathbf{n}^*$ is the unit normal vector outward from V on the surface S, $\mathbf{w}$ the velocity of surface S, and $\mathbf{w}^c$ the velocity of the coordinates orthogonal to $\mathbf{n}$ such that $\mathbf{w}^c \times \mathbf{n} = 0$.

Proof. By Equation (130), we have

$$\int_{V} \left.\frac{\partial f}{\partial t}\right|_{\mathbf{u}} \mathrm{d}v = \int_{V} \left.\frac{\partial f}{\partial t}\right|_{\mathbf{x}} \mathrm{d}v + \int_{V} \mathbf{w} \cdot \mathbf{nn} \cdot \nabla f\mathrm{d}v \tag{323}$$

By defining an indicator function $\gamma(\mathbf{x})$ such that $\gamma = 1$ within V, $\gamma = 0$ outside of V, we obtain

$$\int_{V} \left.\frac{\partial f}{\partial t}\right|_{\mathbf{u}} \mathrm{d}v = \int_{V_\infty} \left.\frac{\partial f}{\partial t}\right|_{\mathbf{x}} \gamma\mathrm{d}v + \int_{V_\infty} \mathbf{w} \cdot \mathbf{nn} \cdot \nabla f\gamma\mathrm{d}v \tag{324}$$

For the first term on the right side of Equation (324), an application of the chain rule yields

$$\int_{V_\infty} \left.\frac{\partial f}{\partial t}\right|_{\mathbf{x}} \gamma dv = \int_{V_\infty} \left.\frac{\partial (f\gamma)}{\partial t}\right|_{\mathbf{x}} dv - \int_{V_\infty} f \left.\frac{\partial \gamma}{\partial t}\right|_{\mathbf{x}} dv \tag{325}$$

By Equation (180), Equation (325) becomes

$$\int_{V_\infty} \left.\frac{\partial f}{\partial t}\right|_{\mathbf{x}} \gamma dv = \frac{\partial}{\partial t}\int_{V_\infty} f\gamma dv + \int_{V_\infty} f\mathbf{w}\cdot\nabla\gamma dv \tag{326}$$

By the definition of γ and Equation (186), we have

$$\int_{V_\infty} \left.\frac{\partial f}{\partial t}\right|_{\mathbf{x}} \gamma dv = \frac{\partial}{\partial t}\int_{V} f dv - \int_{S} f\mathbf{w}\cdot\mathbf{n}^* ds \tag{327}$$

For the second term on the right side of Equation (324), an application of the chain rule yields

$$\int_{V_\infty} \mathbf{w}\cdot\mathbf{nn}\cdot\nabla f\gamma dv = \int_{V_\infty} \nabla\cdot(\mathbf{w}\cdot\mathbf{nn}f\gamma)dv - \int_{V_\infty} \nabla\cdot(\mathbf{w}\cdot\mathbf{nn})f\gamma dv - \int_{V_\infty} \mathbf{w}\cdot\mathbf{nn}\cdot\nabla\gamma f dv$$

By Equations (231) and (186), we obtain

$$\int_{V_\infty} \mathbf{w}\cdot\mathbf{nn}\cdot\nabla f\gamma dv = -\int_{V_\infty} \nabla\cdot(\mathbf{w}\cdot\mathbf{nn})f\gamma dv + \int_{S} \mathbf{w}\cdot\mathbf{nn}\cdot\mathbf{n}^* f ds \tag{328}$$

By the definitions of the function γ and $\mathbf{w}^c(\mathbf{w}^c \equiv \mathbf{w}\cdot\mathbf{nn})$, Equation (328) becomes

$$\int_{V_\infty} \mathbf{w}\cdot\mathbf{nn}\cdot\nabla f\gamma dv = -\int_{V} \nabla\cdot\mathbf{w}^c f dv + \int_{S} \mathbf{w}^c\cdot\mathbf{n}^* f ds \tag{329}$$

Substituting Equations (327) and (329) into Equation (324) leads to

$$\int_V \left.\frac{\partial f}{\partial t}\right|_{\mathbf{u}} dv = \frac{\partial}{\partial t}\int_V f dv - \int_V f\nabla\cdot\mathbf{w}^c dv + \int_S \mathbf{n}^*\cdot\mathbf{w}^c f ds - \int_S \mathbf{n}^*\cdot\mathbf{w} f ds$$

Theorem 4.

$$\int_V \nabla^S \circ F dv = \int_V (\nabla^S\cdot\mathbf{n})\mathbf{n}\circ F dv + \int_V (\mathbf{n}\cdot\nabla^c\mathbf{n})\circ F dv - \int_S (\mathbf{n}^*\cdot\mathbf{n})\mathbf{n}\circ F ds + \int_S \mathbf{n}^*\circ F ds \tag{330}$$

where $\nabla^S \circ F$ can be the gradient, divergence, or curl. ∇^c is the curvilineal operator such that $\nabla^c = \mathbf{nn}\cdot\nabla\cdot\mathbf{n}$ is a unit vector in one of the general orthogonal coordinate directions so that Equation (330) is also valid after replacing $\mathbf{n}$ by $\boldsymbol{\lambda}$ or $\boldsymbol{\nu}$. ∇^S is the surficial operator such that $\nabla^S = \nabla - \nabla^c$.

Proof. We define an indicator function $\gamma(\mathbf{x})$ such that $\gamma = 1$ within V, $\gamma = 0$ elsewhere. As such, the left side of Equation (330) can be rewritten as

$$\int_V \nabla^S \circ F dv = \int_{V_\infty} \nabla^S \circ F\gamma dv \tag{331}$$

By the definition of the operator ∇^S, this equation becomes

$$\begin{aligned}\int_V \nabla^S \circ F dv &= \int_{V_\infty} (\nabla - \mathbf{nn} \cdot \nabla) \circ F\gamma dv \\ &= \int_{V_\infty} \nabla \circ F\gamma dv - \int_{V_\infty} (\mathbf{nn} \cdot \nabla) \circ F\gamma dv \end{aligned} \tag{332}$$

For the first term on the right side of Equation (332), applying the chain rule yields

$$\int_{V_\infty} \nabla \circ F\gamma dv = \int_{V_\infty} \nabla \circ (F\gamma) dv - \int_{V_\infty} \nabla\gamma \circ F dv \tag{333}$$

By Equations (185)–(187) and Equations (230)–(232), Equation (333) leads to

$$\int_{V_\infty} \nabla \circ F\gamma dv = \int_S \mathbf{n}^* \circ F ds \tag{334}$$

For the second term on the right side of Equation (332), an application of the chain rule yields

$$\begin{aligned}\int_{V_\infty} (\mathbf{nn} \cdot \nabla) \circ F\gamma dv &= \int_{V_\infty} \nabla \cdot (\mathbf{nn} \circ F\gamma) dv - \int_{V_\infty} \nabla \cdot \mathbf{nn} \circ F\gamma dv \\ &\quad - \int_{V_\infty} (\mathbf{n} \cdot \nabla\mathbf{n}) \circ F\gamma dv - \int_{V_\infty} \mathbf{n} \cdot \nabla\gamma \mathbf{n} \circ F dv \end{aligned} \tag{335}$$

By Equations (231) and (138) as the relation, $\nabla = \nabla^c + \nabla^S$, we obtain

$$\begin{aligned}\int_{V_\infty} (\mathbf{nn} \cdot \nabla) \circ F\gamma dv &= -\int_{V_\infty} \nabla^S \cdot \mathbf{nn} \circ F\gamma dv - \int_{V_\infty} \mathbf{n} \cdot \nabla^c \mathbf{n} \circ F\gamma dv \\ &\quad - \int_{V_\infty} \mathbf{n} \cdot \nabla\gamma \mathbf{n} \circ F dv \end{aligned} \tag{336}$$

Substituting Equations (334) and (336) into Equation (332) and using Equation (186), we conclude

$$\begin{aligned}\int_V \nabla^S \circ F dv &= \int_V \nabla^S \cdot \mathbf{nn} \circ F dv + \int_V (\mathbf{n} \cdot \nabla^c \mathbf{n}) \circ F dv \\ &\quad - \int_S (\mathbf{n} \cdot \mathbf{n}^*) \mathbf{n} \circ F ds + \int_S \mathbf{n}^* \circ F ds \end{aligned} \tag{337}$$

Remark. Equation (330) contains the following three formulas:

Gradient

$$\int_V \nabla^S f dv = \int_V (\nabla^S \cdot \mathbf{n}) \mathbf{n} f dv + \int_V (\mathbf{n} \cdot \nabla^c \mathbf{n}) f dv - \int_S (\mathbf{n}^* \cdot \mathbf{n}) \mathbf{n} f ds + \int_S \mathbf{n}^* f ds \tag{338}$$

Divergence

$$\int_V \nabla^S \cdot \mathbf{f}dv = \int_V (\nabla^S \cdot \mathbf{n})\mathbf{n} \cdot \mathbf{f}dv + \int_V (\mathbf{n} \cdot \nabla^c \mathbf{n}) \cdot \mathbf{f}dv - \int_S (\mathbf{n}^* \cdot \mathbf{n})\mathbf{n} \cdot \mathbf{f}ds + \int_S \mathbf{n}^* \cdot \mathbf{f}ds \tag{339}$$

Curl

$$\begin{aligned}\int_V \nabla^S \times \mathbf{f}dv = \int_V (\nabla^S \cdot \mathbf{n})\mathbf{n} \times \mathbf{f}dv + \int_V (\mathbf{n} \cdot \nabla^c \mathbf{n}) \times \mathbf{f}dv \\ - \int_S (\mathbf{n}^* \cdot \mathbf{n})\mathbf{n} \times \mathbf{f}ds + \int_S \mathbf{n}^* \times \mathbf{f}ds\end{aligned} \tag{340}$$

4.1.3 Spatial integration of curvilineal operators

This group of integration theorems are regarding integrations of $(\partial f/\partial t)|_1, \nabla^c f,\ \nabla^c \cdot \mathbf{f}$ and $\nabla^c \times$ f over a volume V (Figure 18).

Theorem 5.

$$\int_V \frac{\partial f}{\partial t}\bigg|_1 dv = \frac{\partial}{\partial t}\int_V f dv - \int_V f\nabla \cdot \mathbf{w}^s dv + \int_S \mathbf{n}^* \cdot \mathbf{w}^s f ds - \int_S \mathbf{n}^* \cdot \mathbf{w} f ds \tag{341}$$

where V is the integration volume, S the boundary surface of V, $\mathbf{w}$ the velocity of surface S, and $\mathbf{w}^s$ the velocity of the surface S in the direction normal to $\boldsymbol{\lambda}$ such that $\mathbf{w}^s \cdot \boldsymbol{\lambda} = 0$.

Proof. Defining an indicator function $\gamma(\mathbf{x})$ such that $\gamma = 1$ within V and $\gamma = 0$ outside of V, we have

$$\int_V \frac{\partial f}{\partial t}\bigg|_1 dv = \int_{V_\infty} \frac{\partial f}{\partial t}\bigg|_1 \gamma dv \tag{342}$$

By Equation (134), we obtain

$$\int_V \frac{\partial f}{\partial t}\bigg|_1 dv = \int_{V_\infty} \left(\frac{\partial f}{\partial t}\bigg|_{\mathbf{x}} + (\mathbf{w} - \mathbf{w} \cdot \boldsymbol{\lambda\lambda}) \cdot \nabla f\right)\gamma dv \tag{343}$$

An application of the chain rule to the right side of Equation (343) yields

$$\begin{aligned}\int_V \frac{\partial f}{\partial t}\bigg|_1 dv = \int_{V_\infty} \frac{\partial (f\gamma)}{\partial t}\bigg|_{\mathbf{x}} dv - \int_{V_\infty} f\frac{\partial \gamma}{\partial t}\bigg|_{\mathbf{x}} dv + \int_{V_\infty} \nabla \cdot (\mathbf{w}f\gamma)dv \\ - \int_{V_\infty} \nabla \cdot \mathbf{w}f\gamma dv - \int_{V_\infty} f\nabla\gamma \cdot \mathbf{w}dv - \int_{V_\infty} \nabla \cdot (\boldsymbol{\lambda}\mathbf{w} \cdot \boldsymbol{\lambda}f\gamma)dv \\ + \int_{V_\infty} \nabla \cdot (\mathbf{w} \cdot \boldsymbol{\lambda\lambda})f\gamma dv + \int_{V_\infty} \mathbf{w} \cdot \boldsymbol{\lambda}f\boldsymbol{\lambda} \cdot \nabla\gamma dv\end{aligned} \tag{344}$$

By Equations (180) and (231) and noting that V_∞ is independent of t, Equation (344) becomes

$$\begin{aligned}\int_V \frac{\partial f}{\partial t}\bigg|_1 dv = \frac{\partial}{\partial t}\int_{V_\infty} f\gamma dv + \int_{V_\infty} f\mathbf{w} \cdot \nabla\gamma dv - \int_{V_\infty} \nabla \cdot \mathbf{w}f\gamma dv - \int_{V_\infty} f\nabla\gamma \cdot \mathbf{w}dv \\ + \int_{V_\infty} \nabla \cdot (\boldsymbol{\lambda}\mathbf{w} \cdot \boldsymbol{\lambda})f\gamma dv + \int_{V_\infty} \mathbf{w} \cdot \boldsymbol{\lambda}f\boldsymbol{\lambda} \cdot \nabla\gamma dv\end{aligned} \tag{345}$$

Rearranging the right side of Equation (345) yields

$$\int_V \frac{\partial f}{\partial t}\bigg|_1 dv = \frac{\partial}{\partial t}\int_{V_\infty} f\gamma dv - \int_{V_\infty} \nabla\cdot(\mathbf{w}-\mathbf{w}\cdot\boldsymbol{\lambda\lambda})f\gamma dv$$
$$-\int_{V_\infty} f\nabla\gamma\cdot(\mathbf{w}-\mathbf{w}\cdot\boldsymbol{\lambda\lambda})dv + \int_{V_\infty} f\mathbf{w}\cdot\nabla\gamma dv$$

By Equation (186), the definition of the indicator function γ and $\mathbf{w}^s = \mathbf{w}-\mathbf{w}\cdot\boldsymbol{\lambda\lambda}$, we finally obtain

$$\int_V \frac{\partial f}{\partial t}\bigg|_1 dv = \frac{\partial}{\partial t}\int_V f dv - \int_V f\nabla\cdot\mathbf{w}^s dv + \int_S \mathbf{n}^*\cdot\mathbf{w}^s f ds - \int_S \mathbf{n}^*\cdot\mathbf{w} f ds$$

Theorem 6.

$$\int_V \nabla^c\circ F dv = -\int_V(\nabla^S\cdot\boldsymbol{\lambda})\boldsymbol{\lambda}\circ F dv - \int_V(\boldsymbol{\lambda}\cdot\nabla^c\boldsymbol{\lambda})\circ F dv + \int_S(\mathbf{n}^*\cdot\boldsymbol{\lambda})\boldsymbol{\lambda}\circ F ds \quad (346)$$

where $\nabla^c\circ F$ stands for $\nabla^c f$, $\nabla^c\cdot\mathbf{f}$, or $\nabla^c\times\mathbf{f}$, $\boldsymbol{\lambda}$ is a unit vector tangent to one of the general orthogonal coordinate directions; $\mathbf{n}^*$ the unit normal vector outward from V on surface S; ∇^c the curvilineal operator such that $\nabla^c = \boldsymbol{\lambda\lambda}\cdot\nabla$; and ∇^S the surficial operator such that $\nabla^S = \nabla - \nabla^c$.

Proof. Defining an indicator function V such that $\gamma(\mathbf{x}) = 1$ within V and $\gamma = 0$ elsewhere, we have

$$\int_V \nabla^c\circ F dv = \int_{V_\infty}\nabla^c\circ F\gamma dv \quad (347)$$

Since $\nabla^c = \boldsymbol{\lambda\lambda}\cdot\nabla$, Equation (347) becomes

$$\int_V \nabla^c\circ F dv = \int_{V_\infty}(\boldsymbol{\lambda\lambda}\cdot\nabla)\circ F\gamma dv \quad (348)$$

An application of the chain rule to the right side of Equation (348) yields

$$\int_V \nabla^c\circ F dv = \int_{V_\infty}\nabla\cdot(\boldsymbol{\lambda\lambda}\circ F\gamma)dv - \int_{V_\infty}\nabla\cdot\boldsymbol{\lambda\lambda}\circ F\gamma dv$$
$$-\int_{V_\infty}(\boldsymbol{\lambda}\cdot\nabla\boldsymbol{\lambda})\circ F\gamma dv - \int_{V_\infty}\boldsymbol{\lambda}\cdot\nabla\gamma\boldsymbol{\lambda}\circ F dv \quad (349)$$

By the definition of ∇^c, Equations (230)–(232) and Equation (138) with $\mathbf{n}$ replaced by $\boldsymbol{\lambda}$, we obtain

$$\int_V \nabla^c\circ F dv = -\int_{V_\infty}\nabla^S\cdot\boldsymbol{\lambda\lambda}\circ F\gamma dv - \int_{V_\infty}\boldsymbol{\lambda}\cdot\nabla^c\boldsymbol{\lambda}\circ F dv - \int_{V_\infty}\boldsymbol{\lambda}\cdot\nabla\gamma\boldsymbol{\lambda}\circ F dv \quad (350)$$

By Equation (186) and the definition of the indicator function γ, Equation (350) leads to

$$\int_V \nabla^c\circ F dv = -\int_V(\nabla^S\cdot\boldsymbol{\lambda})\boldsymbol{\lambda}\circ F dv - \int_V(\boldsymbol{\lambda}\cdot\nabla^c\boldsymbol{\lambda})\circ F dv + \int_S(\mathbf{n}^*\cdot\boldsymbol{\lambda})\boldsymbol{\lambda}\circ F ds$$

Remark. Equation (346) covers the following three formulas:

Gradient

$$\int_V \nabla^c f dv = -\int_V (\nabla^S \cdot \boldsymbol{\lambda})\boldsymbol{\lambda} f dv - \int_V (\boldsymbol{\lambda} \cdot \nabla^c \boldsymbol{\lambda}) f dv + \int_S (\mathbf{n}^* \cdot \boldsymbol{\lambda})\boldsymbol{\lambda} f ds \tag{351}$$

Divergence

$$\int_V \nabla^c \cdot \mathbf{f} dv = -\int_V (\nabla^S \cdot \boldsymbol{\lambda})\boldsymbol{\lambda} \cdot \mathbf{f} dv - \int_V (\boldsymbol{\lambda} \cdot \nabla^c \boldsymbol{\lambda}) \cdot \mathbf{f} dv + \int_S (\mathbf{n}^* \cdot \boldsymbol{\lambda})\boldsymbol{\lambda} \cdot \mathbf{f} ds \tag{352}$$

Curl

$$\int_V \nabla^c \times \mathbf{f} dv = -\int_V (\nabla^S \cdot \boldsymbol{\lambda})\boldsymbol{\lambda} \times \mathbf{f} dv - \int_V (\boldsymbol{\lambda} \cdot \nabla^c \boldsymbol{\lambda}) \times \mathbf{f} dv + \int_S (\mathbf{n}^* \cdot \boldsymbol{\lambda})\boldsymbol{\lambda} \times \mathbf{f} ds \tag{353}$$

4.2 Theorems for integration over a surface

In this section we develop 14 integration theorems regarding integration over a surface S or a plane A of spatial, surficial, or curvilineal differential operators of a microscale function f or $\mathbf{f}$.

4.2.1 Integration of spatial operators over a plane fixed in space

This group of integration theorems involves integrations of $(\partial f/\partial t)|_{\mathbf{x}}, \nabla f, \nabla \cdot \mathbf{f}$, and $\nabla \times \mathbf{f}$ over a plane fixed in space (Figure 19). Here $f(\mathbf{x}, t)$ and $\mathbf{f}(\mathbf{x}, t)$ are microscale scalar and vector functions of position vector $\mathbf{x}$ and time t, respectively.

In these integration theorems A is a planar integration surface fixed in space, and C is not only the curve bounding A but also the curve of intersection of an extension of plane A with a volume (Figure 19). $\mathbf{N}$ is the unit vector normal to planar surface A, $\boldsymbol{\nu}^*$ the unit vector normal to curve C positive outward from A such that $\mathbf{N} \cdot \boldsymbol{\nu}^* = 0$, and $\mathbf{n}^*$ the unit vector normal to curve C that is also normal to the boundary of a volume intersected by an extension of plane A to form curve C (Figure 19). ∇^c is the curvilineal operator such that $\nabla^c = \mathbf{NN} \cdot \nabla$. $\mathbf{w}$ is the velocity of C such that $\mathbf{w} = \boldsymbol{\nu}^*\boldsymbol{\nu}^* \cdot \mathbf{w}$, $\mathbf{f}^c$ is a vector normal to A such that $\mathbf{f}^c = \mathbf{NN} \cdot \mathbf{f}$, and f^s is a vector tangent to A such that $\mathbf{f}^s = \mathbf{f} - \mathbf{f}^c$.

Theorem 7.

$$\int_A \frac{\partial f}{\partial t}\bigg|_{\mathbf{x}} dA = \frac{\partial}{\partial t}\int_A f dA - \int_C \frac{\mathbf{n}^* \cdot \mathbf{w} f}{\boldsymbol{\nu}^* \cdot \mathbf{n}^*} dc \tag{354}$$

Proof. Defining an indicator function $\gamma(\mathbf{x})$ such that $\gamma = 1$ within V and $\gamma = 0$ elsewhere, we obtain

$$\int_A \frac{\partial f}{\partial t}\bigg|_{\mathbf{x}} dA = \int_{A_\infty} \frac{\partial f}{\partial t}\bigg|_{\mathbf{x}} \gamma dA \tag{355}$$

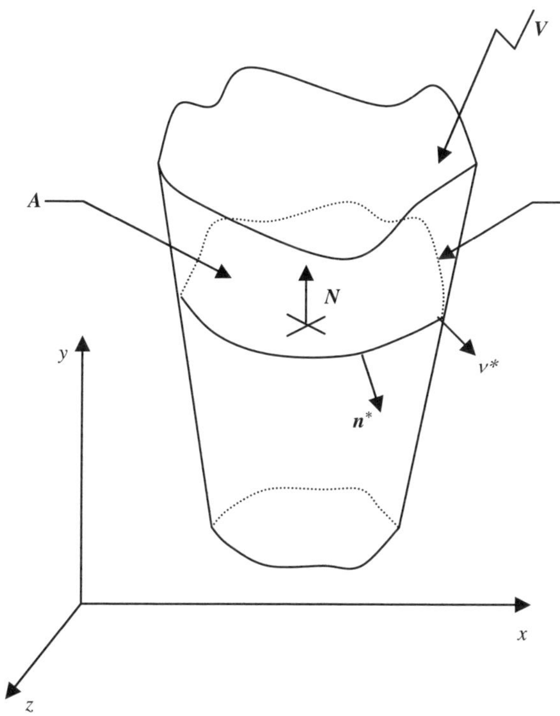

Figure 19 Integration plane *A* with boundary curve *C*, that is the intersection of *A* with the boundary of volume *V* ($\mathbf{n}^*$ is unit vector normal to *C* and to the boundary of *V* and **N** the unit normal of *A*).

An application of the chain rule to the right side of Equation (355) yields

$$\begin{aligned}\int_A \frac{\partial f}{\partial t}\bigg|_{\mathbf{x}} \mathrm{d}A &= \int_{A_\infty} \frac{\partial(f\gamma)}{\partial t}\bigg|_{\mathbf{x}} \mathrm{d}A - \int_{A_\infty} f\frac{\partial \gamma}{\partial t}\bigg|_{\mathbf{x}} \mathrm{d}A \\ &= \int_{A_\infty} \frac{\partial(f\gamma)}{\partial t}\bigg|_{\mathbf{x}} \mathrm{d}A + \int_{A_\infty} f\mathbf{w}\cdot\nabla\gamma \mathrm{d}A \\ &= \int_{A_\infty} \left(\frac{\partial(f\gamma)}{\partial t}\bigg|_{\mathbf{u}} - \mathbf{w}\cdot\mathbf{NN}\cdot\nabla(f\gamma)\right)\mathrm{d}A + \int_{A_\infty} f\mathbf{w}\cdot\nabla\gamma \mathrm{d}A \end{aligned} \quad (356)$$

In deriving Equation (356), Equations (130) and (180) have been used. By using $\mathbf{w} = \nu^*\nu^*\cdot\mathbf{w}$, $\mathbf{N}\cdot\nu^* = 0$ and noting that A_∞ is independent of t, we have

$$\int_A \frac{\partial f}{\partial t}\bigg|_{\mathbf{x}} \mathrm{d}A = \frac{\partial}{\partial t}\int_{A_\infty} f\gamma \mathrm{d}A + \int_{A_\infty} f\mathbf{w}\cdot\nabla\gamma \mathrm{d}A \quad (357)$$

From Equation (284), it follows that

$$\int_A \frac{\partial f}{\partial t}\bigg|_{\mathbf{x}} \mathrm{d}A = \frac{\partial}{\partial t}\int_A f\mathrm{d}A - \int_C \frac{\mathbf{n}^*\cdot\mathbf{w}f}{\nu^*\cdot\mathbf{n}^*}\mathrm{d}c$$

Theorem 8.

$$\int_A \nabla f \mathrm{d}A = \nabla^{\mathrm{c}} \int_A f \mathrm{d}A + \int_C \frac{\mathbf{n}^* f}{\mathbf{v}^* \cdot \mathbf{n}^*} \mathrm{d}c \tag{358}$$

Proof. Defining an indicator function $\gamma(\mathbf{x})$ such that $\gamma = 1$ within V and $\gamma = 0$ outside of V, we have

$$\int_A \nabla f \mathrm{d}A = \int_{A_\infty} \nabla f \gamma \mathrm{d}A \tag{359}$$

An application of the chain rule on the right side of Equation (359) yields

$$\begin{aligned}\int_A \nabla f \mathrm{d}A &= \int_{A_\infty} \nabla(f\gamma) \mathrm{d}A - \int_{A_\infty} f \nabla \gamma \mathrm{d}A = \int_{A_\infty} (\nabla^{\mathrm{S}} + \nabla^{\mathrm{c}})(f\gamma) \mathrm{d}A - \int_{A_\infty} f \nabla \gamma \mathrm{d}A \\ &= \int_{A_\infty} \nabla^{\mathrm{S}}(f\gamma) \mathrm{d}A + \int_{A_\infty} \nabla^{\mathrm{c}}(f\gamma) \mathrm{d}A - \int_{A_\infty} f \nabla \gamma \mathrm{d}A \end{aligned} \tag{360}$$

Since ∇^{c} is the directional derivative along $\mathbf{N}$ which is normal to the planar surface A, the second term of the third line of Equation (360) becomes

$$\int_{A_\infty} \nabla^{\mathrm{c}}(f\gamma) \mathrm{d}A = \nabla^{\mathrm{c}} \int_{A_\infty} f\gamma \mathrm{d}A = \nabla^{\mathrm{c}} \int_A f \mathrm{d}A \tag{361}$$

Here the role of γ is the same as that of γ^{s}. Therefore Equation (233) allows us to conclude that the first term of the third line of Equation (360) is equal to zero. Regarding the third term, by Equation (283) we have

$$\int_{A_\infty} f \nabla \gamma \mathrm{d}A = -\int_C \frac{f \mathbf{n}^*}{\mathbf{v}^* \cdot \mathbf{n}^*} \mathrm{d}c \tag{362}$$

Therefore, Equation (360) yields

$$\int_A \nabla f \mathrm{d}A = \nabla^{\mathrm{c}} \int_A f \mathrm{d}A + \int_C \frac{\mathbf{n}^* f}{\mathbf{v}^* \cdot \mathbf{n}^*} \mathrm{d}c$$

Theorem 9.

$$\int_A \nabla \cdot \mathbf{f} \mathrm{d}A = \nabla^{\mathrm{c}} \cdot \int_A \mathbf{f}^{\mathrm{c}} \mathrm{d}A - \int_A (\mathbf{N} \cdot \nabla^{\mathrm{c}} \mathbf{N}) \cdot \mathbf{f}^{\mathrm{s}} \mathrm{d}A + \int_C \frac{\mathbf{n}^* . \mathbf{f}}{\mathbf{v}^* \cdot \mathbf{n}^*} \mathrm{d}c \tag{363}$$

Proof. Defining an indicator function $\gamma(\mathbf{x})$ such that $\gamma = 1$ within V and $\gamma = 0$ outside of V, we have

$$\int_A \nabla \cdot \mathbf{f} \mathrm{d}A = \int_{A_\infty} \nabla \cdot \mathbf{f} \gamma \mathrm{d}A \tag{364}$$

Replacing $\mathbf{n}$ by $\mathbf{N}$, by Equation (72) we have

$$\begin{aligned}\int_{A_\infty} \nabla \cdot \mathbf{f} \gamma \mathrm{d}A &= \int_{A_\infty} (\nabla^{\mathrm{S}} \cdot \mathbf{f}^{\mathrm{s}} + \nabla^{\mathrm{c}} \cdot \mathbf{f}^{\mathrm{c}} + \nabla \cdot \mathbf{N} \mathbf{f} \cdot \mathbf{N} - \mathbf{N} \cdot \nabla \mathbf{N} \cdot \mathbf{f}) \gamma \mathrm{d}A \\ &= \int_{A_\infty} \nabla^{\mathrm{S}} \cdot \mathbf{f}^{\mathrm{s}} \gamma \mathrm{d}A + \int_{A_\infty} \nabla^{\mathrm{c}} \cdot \mathbf{f}^{\mathrm{c}} \gamma \mathrm{d}A \\ &\quad + \int_{A_\infty} \nabla \cdot \mathbf{N} \mathbf{f} \cdot \mathbf{N} \gamma \mathrm{d}A - \int_{A_\infty} \mathbf{N} \cdot \nabla \mathbf{N} \cdot \mathbf{f} \gamma \mathrm{d}A \end{aligned} \tag{365}$$

Applying the chain rule, the first term on the right side of Equation (365) becomes

$$\int_{A_\infty} \nabla^S \cdot \mathbf{f}^s \gamma \mathrm{d}A = \int_{A_\infty} \nabla^S \cdot (\mathbf{f}^s \gamma)\mathrm{d}A - \int_{A_\infty} \mathbf{f}^s \cdot \nabla^S \gamma \mathrm{d}A \tag{366}$$

By Equation (240) and noting that in this case the role of γ is the same as that of γ^s, we conclude

$$\int_{A_\infty} \nabla^S \cdot \mathbf{f}^s \gamma \mathrm{d}A = -\int_{A_\infty} \mathbf{f}^s \cdot \nabla^S \gamma \mathrm{d}A = -\int_{A_\infty} \mathbf{f}^s \cdot \nabla \gamma \mathrm{d}A \tag{367}$$

In deriving Equation (367), we have used $\mathbf{f}^s \cdot \nabla^c = 0$. An application of the chain rule leads the second term on the right side of Equation (365) to

$$\int_{A_\infty} \nabla^c \cdot \mathbf{f}^c \gamma \mathrm{d}A = \int_{A_\infty} \nabla^c \cdot (\mathbf{f}^c \gamma)\mathrm{d}A - \int_{A_\infty} \nabla^c \gamma \cdot \mathbf{f}^c \mathrm{d}A \tag{368}$$

Note that $\nabla^c = \mathbf{NN} \cdot \nabla$ is the gradient along $\mathbf{N}$ which is perpendicular to the plane A. Therefore,

$$\int_{A_\infty} \nabla^c \cdot (\mathbf{f}^c \gamma)\mathrm{d}A = \nabla^c \cdot \int_{A_\infty} \mathbf{f}^c \gamma \mathrm{d}A = \nabla^c \cdot \int_{A} \mathbf{f}^c \mathrm{d}A \tag{369}$$

In deriving this equation, the definition of γ has been used. Thus Equation (368) becomes

$$\begin{aligned} \int_{A_\infty} \nabla^c \cdot \mathbf{f}^c \gamma \mathrm{d}A &= \nabla^c \cdot \int_{A} \mathbf{f}^c \mathrm{d}A - \int_{A_\infty} \mathbf{f}^c \cdot \nabla^c \gamma \mathrm{d}A \\ &= \nabla^c \cdot \int_{A} \mathbf{f}^c \mathrm{d}A - \int_{A_\infty} \mathbf{f}^c \cdot \nabla \gamma \mathrm{d}A \end{aligned} \tag{370}$$

In deriving Equation (370), we have used $\mathbf{f}^c \cdot \nabla^S = 0$.

By Equation (138), we have

$$\nabla \cdot \mathbf{N} = \nabla^S \cdot \mathbf{N} \tag{371}$$

Note that, on the plane A, $\mathbf{N}$ is a constant unit vector such that $\nabla^S \cdot \mathbf{N} = 0$. Therefore, we have the third term on the right side of Equation (365)

$$\int_{A_\infty} \nabla \cdot \mathbf{Nf} \cdot \mathbf{N} \gamma \mathrm{d}A = 0 \tag{372}$$

By the relation

$$\mathbf{N} \cdot \nabla \mathbf{N} \cdot \mathbf{f} = \mathbf{N} \cdot \nabla^c \mathbf{N} \cdot \mathbf{f}^s \tag{373}$$

the last term on the right side of Equation (365) becomes

$$-\int_{A_\infty} \mathbf{N} \cdot \nabla \mathbf{N} \cdot \mathbf{f} \gamma \mathrm{d}A = -\int_{A} \mathbf{N} \cdot \nabla^c \mathbf{N} \cdot \mathbf{f}^s \mathrm{d}A \tag{374}$$

Substituting Equations (367), (370), (372), and (374) into Equation (365) yields

$$\int_{A} \nabla \cdot \mathbf{f} \mathrm{d}A = -\int_{A_\infty} \mathbf{f}^s \cdot \nabla \gamma \mathrm{d}A - \int_{A_\infty} \mathbf{f}^c \cdot \nabla \gamma \mathrm{d}A + \nabla^c \cdot \int_{A} \mathbf{f}^c \mathrm{d}A - \int_{A} \mathbf{N} \cdot \nabla^c \mathbf{N} \cdot \mathbf{f}^s \mathrm{d}A \tag{375}$$

By Equation (284) and $\mathbf{f} = \mathbf{f}^s + \mathbf{f}^c$, we finally obtain

$$\int_A \nabla \cdot \mathbf{f} dA = \nabla^c \cdot \int_A \mathbf{f}^c dA - \int_A (\mathbf{N} \cdot \nabla^c \mathbf{N}) \cdot \mathbf{f}^s dA + \int_c \frac{\mathbf{n}^* \cdot \mathbf{f}}{\boldsymbol{\nu}^* \cdot \mathbf{n}^*} dc$$

Theorem 10.

$$\int_A \nabla \times \mathbf{f} dA = \nabla^c \times \int_A \mathbf{f}^s dA - \int_A (\mathbf{N} \cdot \nabla^c \mathbf{N}) \times \mathbf{f}^c dA + \int_C \frac{\mathbf{n}^* \times \mathbf{f}}{\boldsymbol{\nu}^* \cdot \mathbf{n}^*} dc \tag{376}$$

Proof. Defining an indicator function $\gamma(\mathbf{x})$ such that $\gamma = 1$ within V and $\gamma = 0$ outside of V, we have

$$\int_A \nabla \times \mathbf{f} dA = \int_{A_\infty} \nabla \times \mathbf{f}\gamma dA \tag{377}$$

Replacing $\mathbf{n}$ by $\mathbf{N}$, by Equation (108) we have

$$\int_A \nabla \times \mathbf{f}\gamma dA = \int_{A_\infty} \nabla^S \times \mathbf{f}^s \gamma dA - \int_{A_\infty} \mathbf{N} \cdot \nabla \mathbf{f} \times \mathbf{N}\gamma dA - \int_{A_\infty} \mathbf{N} \times \nabla(\mathbf{N} \cdot \mathbf{f})\gamma dA \tag{378}$$

An application of the chain rule leads the first term on the right side of Equation (378) to

$$\int_{A_\infty} \nabla^S \times \mathbf{f}^s \gamma dA = \int_{A_\infty} \nabla^S \times (\mathbf{f}^s \gamma) dA - \int_{A_\infty} \nabla^S \gamma \times \mathbf{f}^s dA \tag{379}$$

Using Equation (245) and noting that here γ plays the same role as γ^s, we obtain

$$\int_{A_\infty} \nabla^S \times \mathbf{f}^s \gamma dA = -\int_{A_\infty} \nabla^S \gamma \times \mathbf{f}^s dA \tag{380}$$

The second term on the right side of Equation (378) is

$$-\int_{A_\infty} \mathbf{N} \cdot \nabla \mathbf{f} \times \mathbf{N}\gamma dA = \int_{A_\infty} \mathbf{N} \times (\mathbf{N} \cdot \nabla \mathbf{f})\gamma dA = \int_{A_\infty} (\mathbf{NN} \cdot \nabla) \times \mathbf{f}\gamma dA = \int_{A_\infty} \nabla^c \times \mathbf{f}\gamma dA$$

An application of the chain rule yields

$$-\int_{A_\infty} \mathbf{N} \cdot \nabla \mathbf{f} \times \mathbf{N}\gamma dA = \int_{A_\infty} \nabla^c \times (\mathbf{f}\gamma) dA - \int_{A_\infty} \nabla^c \gamma \times \mathbf{f} dA$$

By $\mathbf{f} = \mathbf{f}^s + \mathbf{f}^c$ and the definitions of ∇^c and $\mathbf{f}^c$, we have

$$\begin{aligned}
-\int_{A_\infty} & \mathbf{N} \cdot \nabla \mathbf{f} \times \mathbf{N}\gamma dA \\
&= \int_{A_\infty} \nabla^c \times (\mathbf{f}^s \gamma) dA + \int_{A_\infty} \nabla^c \times (\mathbf{f}^c \gamma) dA - \int_{A_\infty} \nabla^c \gamma \times \mathbf{f} dA \\
&= \nabla^c \times \int_{A_\infty} \mathbf{f}^s \gamma dA + \int_{A_\infty} (\mathbf{NN} \cdot \nabla) \times (\mathbf{Nf} \cdot \mathbf{N}\gamma) dA - \int_{A_\infty} \nabla^c \gamma \times \mathbf{f} dA \\
&= \nabla^c \times \int_{A_\infty} \mathbf{f}^s \gamma dA + \int_{A_\infty} \mathbf{N} \times (\mathbf{N} \cdot \nabla \mathbf{N})\mathbf{f} \cdot \mathbf{N}\gamma dA - \int_{A_\infty} \nabla^c \gamma \times \mathbf{f} dA \\
&= \nabla^c \times \int_A \mathbf{f}^s dA + \int_{A_\infty} \mathbf{f}^c \times (\mathbf{N} \cdot \nabla^c \mathbf{N})\gamma dA - \int_{A_\infty} \nabla^c \gamma \times \mathbf{f} dA \\
&= \nabla^c \times \int_A \mathbf{f}^s dA - \int_A (\mathbf{N} \cdot \nabla^c \mathbf{N}) \times \mathbf{f}^c dA - \int_{A_\infty} \nabla^c \gamma \times \mathbf{f} dA
\end{aligned} \tag{381}$$

The third term on the right side of Equation (378) is

$$\begin{aligned}
-\int_{A_\infty} \mathbf{N}\times\nabla(\mathbf{N}\cdot\mathbf{f})\gamma \mathrm{d}A &= \int_{A_\infty} \nabla(\mathbf{N}\cdot\mathbf{f})\times\mathbf{N}\gamma \mathrm{d}A \\
&= \int_{A_\infty} \nabla^{\mathrm{S}}\times(\mathbf{NN}\cdot\mathbf{f}\gamma)\mathrm{d}A - \int_{A_\infty} \nabla^{\mathrm{S}}\times\mathbf{NN}\cdot\mathbf{f}\gamma \mathrm{d}A \\
&\quad - \int_{A_\infty} \mathbf{N}\cdot\mathbf{f}\nabla^{\mathrm{S}}\gamma\times\mathbf{N}\mathrm{d}A \\
&= \int_{A_\infty} \nabla^{\mathrm{S}}\times(\mathbf{NN}\cdot\mathbf{f}\gamma)\mathrm{d}A - \int_{A_\infty} \nabla^{\mathrm{S}}\times\mathbf{NN}\cdot\mathbf{f}\gamma \mathrm{d}A \\
&\quad - \int_{A_\infty} \nabla^{\mathrm{S}}\gamma\times\mathbf{f}^{\mathrm{c}}\mathrm{d}A
\end{aligned} \tag{382}$$

In deriving this equation, we have used the definition of ∇^{S} and the chain rule. Equations (245) and (140) enable us to further obtain

$$-\int_{A_\infty} \mathbf{N}\times\nabla(\mathbf{N}\cdot\mathbf{f})\gamma \mathrm{d}A = -\int_{A_\infty} \nabla^{\mathrm{S}}\gamma\times\mathbf{f}^{\mathrm{c}}\mathrm{d}A \tag{383}$$

Substituting Equations (380), (381), and (383) into Equation (378) yields

$$\begin{aligned}
\int_A \nabla\times\mathbf{f}\mathrm{d}A &= -\int_{A_\infty} \nabla^{\mathrm{S}}\gamma\times\mathbf{f}^{\mathrm{s}}\mathrm{d}A - \int_{A_\infty} \nabla^{\mathrm{S}}\gamma\times\mathbf{f}^{\mathrm{c}}\mathrm{d}A - \int_{A_\infty} \nabla^{\mathrm{c}}\gamma\times\mathbf{f}\mathrm{d}A \\
&\quad + \nabla^{\mathrm{c}}\times\int_A \mathbf{f}^{\mathrm{s}}\mathrm{d}A - \int_A (\mathbf{N}\cdot\nabla^{\mathrm{c}}\mathbf{N})\times\mathbf{f}^{\mathrm{c}}\mathrm{d}A \\
&= -\int_{A_\infty} \nabla\gamma\times\mathbf{f}\mathrm{d}A + \nabla^{\mathrm{c}}\times\int_A \mathbf{f}^{\mathrm{s}}\mathrm{d}A - \int_A (\mathbf{N}\cdot\nabla^{\mathrm{c}}\mathbf{N})\times\mathbf{f}^{\mathrm{c}}\mathrm{d}A
\end{aligned} \tag{384}$$

By Equation (285), we thus obtain

$$\int_A \nabla\times\mathbf{f}\mathrm{d}A = \nabla^{\mathrm{c}}\times\int_A \mathbf{f}^{\mathrm{s}}\mathrm{d}A - \int_A (\mathbf{N}\cdot\nabla^{\mathrm{c}}\mathbf{N})\times\mathbf{f}^{\mathrm{c}}\mathrm{d}A + \int_C \frac{\mathbf{n}^*\times\mathbf{f}}{\boldsymbol{\nu}^*\cdot\mathbf{n}^*}\mathrm{d}c$$

4.2.2 Surficial integration of surficial operators

This group of integration theorems involves integrations of $(\partial f/\partial t)|_{\mathbf{u}}$, $\nabla^{\mathrm{S}} f$, $\nabla^{\mathrm{S}}\cdot\mathbf{f}$, and $\nabla^{\mathrm{S}}\times\mathbf{f}$ over a surface S (Figure 20). Here $f(\mathbf{u}, t)$ and $\mathbf{f}(\mathbf{u}, t)$ are microscale scalar and vector functions of surficial position vector $\mathbf{u}$ and time t, respectively.

Theorem 11.

$$\int_S \left.\frac{\partial f}{\partial t}\right|_{\mathbf{u}} \mathrm{d}s = \frac{\mathrm{d}}{\mathrm{d}t}\int_S f\mathrm{d}s - \int_S (\nabla^{\mathrm{S}}\cdot\mathbf{n})\mathbf{n}\cdot\mathbf{w}f\mathrm{d}s - \int_C \boldsymbol{\nu}^*\cdot\mathbf{w}f\mathrm{d}c \tag{385}$$

where S is the integration surface, C the curve bounding S, $\mathbf{n}$ the unit vector normal to surface S, and $\boldsymbol{\nu}^*$ the unit vector normal to curve C which is positive in the outward direction from S such that $\mathbf{n}\cdot\boldsymbol{\nu}^* = 0$. ∇^{S} is the surficial operator such that $\nabla^{\mathrm{S}} = \nabla - \mathbf{nn}\cdot\nabla$, and $\mathbf{w}$ is the velocity of S.

Proof. Define an indicator function $\gamma^{\mathrm{s}}(\mathbf{u})$ such that $\gamma^{\mathrm{s}} = 1$ on the surface S, otherwise $\gamma^{\mathrm{s}} = 0$, and another indicator function $\gamma(\mathbf{x})$ such that $\gamma = 1$ in the

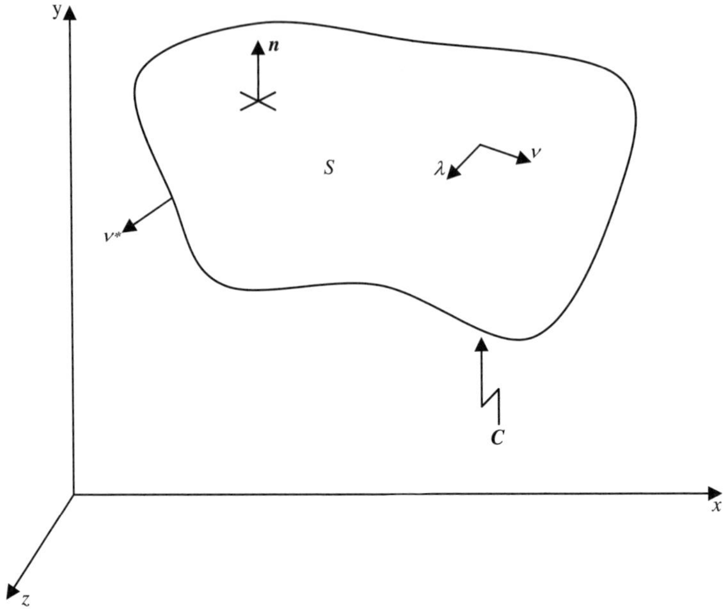

Figure 20 Integration surface S with boundary curve C (**n** is unit normal of S; $\boldsymbol{\nu}^*$ the unit normal of C, tangent to S, and directed outward from S).

volume V enclosed by the surface S, and $\gamma = 0$ outside V. Therefore,

$$\int_S \frac{\partial f}{\partial t}\bigg|_{\mathbf{u}} ds = -\int_{V_\infty} \frac{\partial f}{\partial t}\bigg|_{\mathbf{u}} \gamma^s \mathbf{n} \cdot \nabla \gamma dv \tag{386}$$

By Equation (130), we have

$$\begin{aligned}\int_S \frac{\partial f}{\partial t}\bigg|_{\mathbf{u}} \mathrm{d}s = &-\int_{V_\infty} \frac{\partial f}{\partial t}\bigg|_{\mathbf{x}} \gamma^s \mathbf{n} \cdot \nabla \gamma \mathrm{d}v \\ &-\int_{V_\infty} \mathbf{w} \cdot \mathbf{nn} \cdot \nabla f \gamma^s \mathbf{n} \cdot \nabla \gamma dv\end{aligned} \tag{387}$$

An application of the chain rule to the both terms on the right side of Equation (387) yields

$$\begin{aligned}\int_S \frac{\partial f}{\partial t}\bigg|_{\mathbf{u}} \mathrm{d}s = &-\int_{V_\infty} \frac{\partial}{\partial t}(f\gamma^s \mathbf{n} \cdot \nabla \gamma)\big|_{\mathbf{x}} \mathrm{d}v + \int_{V_\infty} f \frac{\partial \gamma^s}{\partial t}\bigg|_{\mathbf{x}} \mathbf{n} \cdot \nabla \gamma \mathrm{d}v \\ &+ \int_{V_\infty} f\gamma^s \frac{\partial}{\partial t}(\mathbf{n} \cdot \nabla \gamma)\big|_{\mathbf{x}} \mathrm{d}v - \int_{V_\infty} \nabla \cdot (\mathbf{n} f \mathbf{w} \cdot \mathbf{n} \gamma^s \mathbf{n} \cdot \nabla \gamma) \mathrm{d}v \\ &+ \int_{V_\infty} \nabla \cdot \mathbf{n} f \mathbf{w} \cdot \mathbf{n} \gamma^s \mathbf{n} \cdot \nabla \gamma \mathrm{d}v + \int_{V_\infty} f \mathbf{n} \cdot \nabla (\mathbf{w} \cdot \mathbf{n}) \gamma^s \mathbf{n} \cdot \nabla \gamma \mathrm{d}v \\ &+ \int_{V_\infty} f \mathbf{w} \cdot \mathbf{nn} \cdot \nabla \gamma^s \mathbf{n} \cdot \nabla \gamma \mathrm{d}v \\ &+ \int_{V_\infty} f \mathbf{w} \cdot \mathbf{n} \gamma^s \mathbf{n} \cdot \nabla (\mathbf{n} \cdot \nabla \gamma) \mathrm{d}v\end{aligned} \tag{388}$$

By the definition of ∇^S, Equations (231) and (212), and $\nabla(\mathbf{n}\cdot\nabla\gamma) = \nabla\mathbf{n}\cdot\nabla\gamma + \nabla\nabla\gamma\cdot\nabla\mathbf{n}$, we may obtain

$$\begin{aligned}\int_S \frac{\partial f}{\partial t}\bigg|_{\mathbf{u}} &= -\frac{\partial}{\partial t}\int_{V_\infty} f\gamma^s \mathbf{n}\cdot\nabla\gamma dv + \int_{V_\infty} f\frac{\partial\gamma^s}{\partial t}\bigg|_{\mathbf{x}} \mathbf{n}\cdot\nabla\gamma dv \\ &+ \int_{V_\infty} f\gamma^s\left(\frac{\partial\mathbf{n}}{\partial t}\cdot\nabla\gamma + \mathbf{n}\cdot\nabla\frac{\partial\gamma}{\partial t}\right)dv + \int_{V_\infty}\nabla\cdot\mathbf{n}f\mathbf{w}\cdot\mathbf{n}\gamma^s\mathbf{n}\cdot\nabla\gamma dv \\ &+ \int_{V_\infty} f\mathbf{n}\cdot\nabla\mathbf{w}\cdot\mathbf{n}\gamma^s\mathbf{n}\cdot\nabla\gamma dv + \int_{V_\infty} f\mathbf{n}\cdot\nabla\mathbf{n}\cdot\mathbf{w}\gamma^s\mathbf{n}\cdot\nabla\gamma dv \\ &+ \int_{V_\infty} f\mathbf{w}\cdot\mathbf{n}\gamma^s\mathbf{n}\cdot\nabla\mathbf{n}\cdot\nabla\gamma dv + \int_{V_\infty} f\mathbf{w}\cdot\mathbf{n}\gamma^s\mathbf{n}\cdot\nabla\nabla\gamma\cdot\mathbf{n}dv \end{aligned} \tag{389}$$

By Equations (138), (180), (194), and (224), we have

$$\begin{aligned}\int_S \frac{\partial f}{\partial t}\bigg|_{\mathbf{u}} &= -\frac{\partial}{\partial t}\int_{V_\infty} f\gamma^s \mathbf{n}\cdot\nabla\gamma dv + \int_{V_\infty}\nabla^S\cdot\mathbf{n}f\mathbf{w}\cdot\mathbf{n}\gamma^s\mathbf{n}\cdot\nabla\gamma dv \\ &- \int_{V_\infty} f\mathbf{w}\cdot\nabla\gamma^s\mathbf{n}\cdot\nabla\gamma dv - \int_{V_\infty} f\gamma^s\mathbf{n}\cdot\nabla(\mathbf{w}\cdot\nabla\gamma)dv \\ &+ \int_{V_\infty} f\mathbf{n}\cdot\nabla\mathbf{w}\cdot\mathbf{n}\gamma^s\mathbf{n}\cdot\nabla\gamma dv + \int_{V_\infty} f\mathbf{n}\cdot\nabla\mathbf{n}\cdot\mathbf{w}\gamma^s\mathbf{n}\cdot\nabla\gamma dv \\ &+ \int_{V_\infty} f\mathbf{w}\cdot\mathbf{n}\gamma^s\mathbf{n}\cdot\nabla\mathbf{n}\cdot\nabla\gamma dv + \int_{V_\infty} f\mathbf{w}\cdot\mathbf{n}\gamma^s\mathbf{n}\cdot\nabla\nabla\gamma\cdot\mathbf{n}dv \end{aligned} \tag{390}$$

In deriving Equation (390), we have used

$$\nabla\mathbf{n}\cdot\nabla\gamma = 0$$

and

$$\nabla(\mathbf{w}\cdot\nabla\gamma) = \nabla\mathbf{w}\cdot\nabla\gamma + \nabla\nabla\gamma\cdot\mathbf{w}$$

By applying $\nabla^S = \nabla - \mathbf{nn}\cdot\nabla$ and $(\nabla\nabla\gamma)^T - \nabla\nabla\gamma$, we have

$$\begin{aligned}\int_S \frac{\partial f}{\partial t}\bigg|_{\mathbf{u}} ds &= -\frac{\partial}{\partial t}\int_{V_\infty} f\gamma^s \mathbf{n}\cdot\nabla\gamma dv + \int_{V_\infty}\nabla^S\cdot\mathbf{n}f\mathbf{w}\cdot\mathbf{n}\gamma^s\mathbf{n}\cdot\nabla\gamma dv \\ &- \int_{V_\infty} f\mathbf{w}\cdot\nabla\gamma^s\mathbf{n}\cdot\nabla\gamma dv - \int_{V_\infty} f\gamma^s\mathbf{n}\cdot\nabla\mathbf{w}\cdot\nabla\gamma dv \\ &+ \int_{V_\infty} f\mathbf{n}\cdot\nabla\mathbf{w}\cdot\mathbf{n}\gamma^s\mathbf{n}\cdot\nabla\gamma dv + \int_{V_\infty} f\mathbf{n}\cdot\nabla\mathbf{n}\cdot\mathbf{w}\gamma^s\mathbf{n}\cdot\nabla\gamma dv \\ &+ \int_{V_\infty} f\mathbf{w}\cdot(\mathbf{nn}\cdot\nabla - \nabla)\nabla\gamma\cdot\mathbf{n}\gamma^s dv \\ &= -\frac{\partial}{\partial t}\int_{V_\infty} f\gamma^s \mathbf{n}\cdot\nabla\gamma dv + \int_{V_\infty}\nabla^S\cdot\mathbf{n}f\mathbf{w}\cdot\mathbf{n}\gamma^s\mathbf{n}\cdot\nabla\gamma dv \\ &- \int_{V_\infty} f\mathbf{w}\cdot\nabla\gamma^s\mathbf{n}\cdot\nabla\gamma dv - \int_{V_\infty} f\gamma^s\mathbf{n}\cdot\nabla\mathbf{w}\cdot\nabla\gamma dv \\ &+ \int_{V_\infty} f\mathbf{n}\cdot\nabla\mathbf{w}\cdot\mathbf{n}\gamma^s\mathbf{n}\cdot\nabla\gamma dv + \int_{V_\infty} f\mathbf{n}\cdot\nabla\mathbf{n}\cdot\mathbf{w}\gamma^s\mathbf{n}\cdot\nabla\gamma dv \\ &- \int_{V_\infty} f\mathbf{w}\cdot\nabla^S\nabla\gamma\cdot\mathbf{n}\gamma^s dv \end{aligned} \tag{391}$$

By Equation (256), we have

$$\begin{aligned}
\int_S \frac{\partial f}{\partial t}\bigg|_{\mathbf{u}} ds = & -\frac{\partial}{\partial t}\int_{V_\infty} f\gamma^s \mathbf{n}\cdot\nabla\gamma dv + \int_{V_\infty} \nabla^S\cdot\mathbf{n} f\mathbf{w}\cdot\mathbf{n}\gamma^s\mathbf{n}\cdot\nabla\gamma dv \\
& - \int_{V_\infty} f\mathbf{w}\cdot\nabla\gamma^s\mathbf{n}\cdot\nabla\gamma dv - \int_{V_\infty} f\gamma^s\mathbf{n}\cdot\nabla\mathbf{w}\cdot\nabla\gamma dv \\
& + \int_{V_\infty} f\mathbf{n}\cdot\nabla\mathbf{w}\cdot\mathbf{n}\gamma^s\mathbf{n}\cdot\nabla\gamma dv + \int_{V_\infty} f\mathbf{n}\cdot\nabla\mathbf{n}\cdot\mathbf{w}\gamma^s\mathbf{n}\cdot\nabla\gamma dv \\
& - \int_{V_\infty} f\mathbf{w}\cdot\left(\mathbf{n}\cdot\nabla\gamma(\nabla\mathbf{n})^T\right)\cdot\mathbf{n}\gamma^s dv \\
= & -\frac{\partial}{\partial t}\int_{V_\infty} f\gamma^s \mathbf{n}\cdot\nabla\gamma dv + \int_{V_\infty} \nabla^S\cdot\mathbf{n} f\mathbf{w}\cdot\mathbf{n}\gamma^s\mathbf{n}\cdot\nabla\gamma dv \\
& - \int_{V_\infty} f\mathbf{w}\cdot\nabla\gamma^s\mathbf{n}\cdot\nabla\gamma dv - \int_{V_\infty} f\gamma^s\mathbf{n}\cdot\nabla\mathbf{w}\cdot\nabla\gamma dv \\
& + \int_{V_\infty} f\mathbf{n}\cdot\nabla\mathbf{w}\cdot\mathbf{n}\gamma^s\mathbf{n}\cdot\nabla\gamma dv + \int_{V_\infty} f\mathbf{n}\cdot\nabla\mathbf{n}\cdot\mathbf{w}\gamma^s\mathbf{n}\cdot\nabla\gamma dv \\
& - \int_{V_\infty} f\mathbf{n}\cdot\nabla\mathbf{n}\cdot\mathbf{w}\gamma^s\mathbf{n}\cdot\nabla\gamma dv \\
= & -\frac{\partial}{\partial t}\int_{V_\infty} f\gamma^s \mathbf{n}\cdot\nabla\gamma dv + \int_{V_\infty} \nabla^S\cdot\mathbf{n} f\mathbf{w}\cdot\mathbf{n}\gamma^s\mathbf{n}\cdot\nabla\gamma dv \\
& - \int_{V_\infty} f\mathbf{w}\cdot\nabla\gamma^s\mathbf{n}\cdot\nabla\gamma dv - \int_{V_\infty} f\gamma^s\mathbf{n}\cdot\nabla\mathbf{w}\cdot\nabla\gamma dv \\
& + \int_{V_\infty} f\mathbf{n}\cdot\nabla\mathbf{w}\cdot\mathbf{n}\gamma^s\mathbf{n}\cdot\nabla\gamma dv
\end{aligned} \tag{392}$$

By Equations (272) and (275), we finally obtain

$$\begin{aligned}
\int_S \frac{\partial f}{\partial t}\bigg|_{\mathbf{u}} ds = & \frac{\partial}{\partial t}\int_S f ds - \int_S \nabla^S\cdot\mathbf{n}\mathbf{n}\cdot\mathbf{w} f ds \\
& - \int_C \mathbf{v}^*\cdot\mathbf{w} f dc + \int_S f\mathbf{n}\cdot\nabla\mathbf{w}\cdot\mathbf{n} dv - \int_S f\mathbf{n}\cdot\nabla\mathbf{w}\cdot\mathbf{n} dv \\
= & \frac{\partial}{\partial t}\int_S f ds - \int_S (\nabla^S\cdot\mathbf{n})\mathbf{n}\cdot\mathbf{w} f ds - \int_C \mathbf{v}^*\mathbf{w} f dc
\end{aligned}$$

Theorem 12.

$$\int_S \nabla^S\circ F ds = \int_S (\nabla^S\cdot\mathbf{n})\mathbf{n}\circ F ds + \int_C \mathbf{v}^*\circ F dc \tag{393}$$

where $\nabla^S\circ F$ can be $\nabla^S f$, $\nabla^S\cdot\mathbf{f}$, or $\nabla^S\times\mathbf{f}$.

Proof. Define an indicator function $\gamma^s(\mathbf{u})$ such that $\gamma^s = 1$ on S, $\gamma^s = 0$ outside S and another indicator function $\gamma(\mathbf{x})$ such that $\gamma = 1$ in V enclosed by S_∞ and $\gamma = 0$ outside V. Thus,

$$\int_S \nabla^S\circ F ds = -\int_{V_\infty} \nabla^S\circ F\gamma^s\mathbf{n}\cdot\nabla\gamma dv \tag{394}$$

Since $\nabla^S = \nabla - \mathbf{nn} \cdot \nabla$, Equation (394) becomes

$$\int_S \nabla^S \circ F ds = -\int_{V_\infty} (\nabla - \mathbf{nn} \cdot \nabla) \circ F\gamma^s \mathbf{n} \cdot \nabla\gamma dv$$
$$= -\int_{V_\infty} \nabla \circ F\gamma^s \mathbf{n} \cdot \nabla\gamma dv + \int_{V_\infty} (\mathbf{nn} \cdot \nabla) \circ F\gamma^s \mathbf{n} \cdot \nabla\gamma dv \qquad (395)$$

An application of the chain rule to both the terms on the right side of Equation (395) yields

$$\begin{aligned}\int_S \nabla \circ F ds = &-\int_{V_\infty} \nabla \circ (F\gamma^s \mathbf{n} \cdot \nabla\gamma) dv + \int_{V_\infty} \nabla\gamma^s \circ F\mathbf{n} \cdot \nabla\gamma dv \\ &+ \int_{V_\infty} \gamma^s \nabla(\mathbf{n} \cdot \nabla\gamma) \circ F dv + \int_{V_\infty} \nabla \cdot (\mathbf{nn} \circ F\gamma^s \mathbf{n} \cdot \nabla\gamma) dv \\ &- \int_{V_\infty} \nabla \cdot \mathbf{nn} \circ F\gamma^s \mathbf{n} \cdot \nabla\gamma dv - \int_{V_\infty} \mathbf{n} \cdot \nabla\mathbf{n} \circ F\gamma^s \mathbf{n} \cdot \nabla\gamma dv \\ &- \int_{V_\infty} \mathbf{n} \circ F\mathbf{n} \cdot \nabla\gamma^s \mathbf{n} \cdot \nabla\gamma dv - \int_{V_\infty} \mathbf{n} \circ F\gamma^s \mathbf{n} \cdot \nabla(\mathbf{n} \cdot \nabla\gamma) dv \end{aligned} \qquad (396)$$

By Equations (230)–(232) and $\mathbf{n} \cdot \nabla\gamma^s = \mathbf{n} \cdot \nabla^S \gamma^s$, we obtain

$$\begin{aligned}\int_S \nabla^S \circ F ds = &\int_{V_\infty} \nabla\gamma^s \circ F\mathbf{n} \cdot \nabla\gamma dv + \int_{V_\infty} \gamma^s \nabla(\mathbf{n} \cdot \nabla\gamma) \circ F dv \\ &- \int_{V_\infty} \nabla \cdot \mathbf{nn} \circ F\gamma^s \mathbf{n} \cdot \nabla\gamma dv - \int_{V_\infty} \mathbf{n} \cdot \nabla\mathbf{n} \circ F\gamma^s \mathbf{n} \cdot \nabla\gamma dv \\ &- \int_{V_\infty} \mathbf{n} \circ F\gamma^s \mathbf{n} \cdot \nabla(\mathbf{n} \cdot \nabla\gamma) dv \end{aligned} \qquad (397)$$

By Equation (138) and $\nabla(\mathbf{n} \cdot \nabla\gamma) = \nabla\mathbf{n} \cdot \nabla\gamma + \nabla\nabla\gamma \cdot \mathbf{n}$, we have

$$\begin{aligned}\int_S \nabla^S \circ F ds = &\int_{V_\infty} \nabla\gamma^s \circ F\mathbf{n} \cdot \nabla\gamma dv - \int_{V_\infty} \nabla^S \cdot \mathbf{nn} \circ F\gamma^s \mathbf{n} \cdot \nabla\gamma dv \\ &+ \int_{V_\infty} \gamma^s (\nabla\mathbf{n} \cdot \nabla\gamma) \circ F dv + \int_{V_\infty} \gamma^s (\nabla\nabla\gamma \cdot \mathbf{n}) \circ F dv \\ &- \int_{V_\infty} \mathbf{n} \cdot \nabla\mathbf{n} \circ F\gamma^s \mathbf{n} \cdot \nabla\gamma dv \\ &- \int_{V_\infty} \mathbf{n} \circ F\gamma^s \mathbf{n} \cdot \nabla\mathbf{n} \cdot \nabla\gamma dv \\ &- \int_{V_\infty} \mathbf{n} \circ F\gamma^s \mathbf{n} \cdot \nabla\nabla\gamma \cdot \mathbf{n} dv \end{aligned} \qquad (398)$$

By Equation (195), Equation (398) becomes

$$\begin{aligned}\int_S \nabla^S \circ F ds = &\int_{v_\infty} \nabla\gamma^s \circ F\mathbf{n} \cdot \nabla\gamma dv - \int_{V_\infty} \nabla^S \cdot \mathbf{nn} \circ F\gamma^s \mathbf{n} \cdot \nabla\gamma dv \\ &+ \int_{V_\infty} \gamma^s (\nabla - \mathbf{nn} \cdot \nabla)\nabla\gamma \cdot \mathbf{n} \circ F dv \\ &- \int_{V_\infty} \mathbf{n} \cdot \nabla\mathbf{n} \circ F\gamma^s \mathbf{n} \cdot \nabla\gamma dv \end{aligned} \qquad (399)$$

Note that $\nabla - \mathbf{nn}\cdot\nabla = \nabla^S$. Equation (399) thus becomes

$$\int_S \nabla^S \circ F ds = \int_{V_\infty} \nabla\gamma^s \circ F\mathbf{n}\cdot\nabla\gamma dv - \int_{V_\infty} \nabla^S\cdot\mathbf{nn}\circ F\gamma^s\mathbf{n}\cdot\nabla\gamma dv + \int_{V_\infty} \gamma^s\nabla^S\nabla\gamma\cdot\mathbf{n}\circ F dv - \int_{V_\infty} \mathbf{n}\cdot\nabla\mathbf{n}\circ F\gamma^s\mathbf{n}\cdot\nabla\gamma dv \quad (400)$$

By Equation (256), Equation (400) becomes

$$\begin{aligned}\int_S \nabla^S \circ F ds &= \int_{V_\infty} \nabla\gamma^s \circ F\mathbf{n}\cdot\nabla\gamma dv - \int_{V_\infty} \nabla^S\cdot\mathbf{nn}\circ F\gamma^s\mathbf{n}\cdot\nabla\gamma dv \\ &+ \int_{V_\infty} \gamma^s\left(\mathbf{n}\cdot\nabla\gamma(\nabla\mathbf{n}^T)\right)\cdot\mathbf{n}\circ F dv - \int_{V_\infty} \mathbf{n}\cdot\nabla\mathbf{n}\circ F\gamma^s\mathbf{n}\cdot\nabla\gamma dv \\ &= \int_{V_\infty} \nabla\gamma^s \circ F\mathbf{n}\cdot\nabla\gamma dv - \int_{V_\infty} \nabla^S\cdot\mathbf{nn}\circ F\gamma^s\mathbf{n}\cdot\nabla\gamma dv \\ &+ \int_{V_\infty} \gamma^s(\mathbf{n}\cdot\nabla\mathbf{n})\circ F\mathbf{n}\cdot\nabla\gamma dv - \int_{V_\infty} \mathbf{n}\cdot\nabla\mathbf{n}\circ F\gamma^s\mathbf{n}\cdot\nabla\gamma dv\end{aligned} \quad (401)$$

By applying Equations (272) and (275) to the right side of Equation (401), we obtain

$$\int_S \nabla^S \circ F ds = \int_S (\nabla^S\cdot\mathbf{n})\mathbf{n}\circ F ds + \int_C \mathbf{v}^*\circ F dc$$

Remark. Equation (393) includes the following three formulas:
Gradient

$$\int_S \nabla^S f ds = \int_S (\nabla^S\cdot\mathbf{n})\mathbf{n} f ds + \int_C \mathbf{v}^* f dc \quad (402)$$

Divergence

$$\int_S \nabla^S\cdot\mathbf{f} ds = \int_S (\nabla^S\cdot\mathbf{n})\mathbf{n}\cdot\mathbf{f} ds + \int_C \mathbf{v}^*\cdot\mathbf{f} dc \quad (403)$$

Curl

$$\int_S \nabla^S\times\mathbf{f} ds = \int_S (\nabla^S\cdot\mathbf{n})\mathbf{n}\times\mathbf{f} ds + \int_C \mathbf{v}^*\times\mathbf{f} dc \quad (404)$$

4.2.3 Integration of surficial operators over a plane fixed in space

This group of integration theorems involves integrations of $(\partial f/\partial t)|_{\mathbf{u}}$, $\nabla^S f$, $\nabla^S\cdot\mathbf{f}$ and $\nabla^S\times\mathbf{f}$ over a plane fixed in space S (Figure 19). Here $f(\mathbf{u},t)$ and $\mathbf{f}(\mathbf{u},t)$ are microscale scalar and vector functions of surficial position vector $\mathbf{u}$ and time t, respectively. They are the special cases of Theorems 11 and 12. For a plane fixed in space, where $\mathbf{N}$ is a unit vector normal to planar surface A, $\mathbf{N}$ is constant such that $\nabla^S\cdot\mathbf{N} = 0$. Applying this to Theorems 11 and 12 yields:

Theorem 13.

$$\int_A \left.\frac{\partial f}{\partial t}\right|_{\mathbf{u}} dA = \frac{\partial}{\partial t}\int_A f dA - \int_C \mathbf{v}^*\cdot\mathbf{w} f dc \quad (405)$$

where A is the planar surface of integration fixed in space. C is the curve bounding A and is also the curve of intersection of an extension of plane A with a volume. $\boldsymbol{\nu}^*$ is the unit vector normal to curve C with positive outward direction from A such that $\mathbf{N} \cdot \boldsymbol{\nu}^* = 0$. $\mathbf{w}$ is the velocity of C such that $\mathbf{w} = \boldsymbol{\nu}^* \boldsymbol{\nu}^* \cdot \mathbf{w}$.

Theorem 14.

$$\int_A \nabla^S \circ F \mathrm{d}A = \int_C \boldsymbol{\nu}^* \circ F \mathrm{d}c \tag{406}$$

where $\nabla^S \circ F$ can be $\nabla^S f$, $\nabla^S \cdot \mathbf{f}$ or $\nabla^S \times \mathbf{f}$. ∇^S is the surficial operator such that $\nabla^S = \nabla - \mathbf{NN} \cdot \nabla$. $\mathbf{f}$ is a surficial vector $\mathbf{f}(\mathbf{u},t)$, $\mathbf{f}^c$ a vector normal to A such that $\mathbf{f}^c = \mathbf{NN} \cdot \mathbf{f}$, and $\mathbf{f}^s$ a vector tangent to A such that $\mathbf{f}^s = \mathbf{f} - \mathbf{f}^c$.

Remark. Equation (406) includes the following three formulas:

Gradient

$$\int_A \nabla^S f \mathrm{d}A = \int_C \boldsymbol{\nu}^* f \mathrm{d}c \tag{407}$$

Divergence

$$\int_A \nabla^S \cdot \mathbf{f} \mathrm{d}A = \int_C \boldsymbol{\nu}^* \cdot \mathbf{f}^s \mathrm{d}c \tag{408}$$

Curl

$$\int_A \nabla^S \times \mathbf{f} \mathrm{d}A = \int_C \boldsymbol{\nu}^* \times \mathbf{f} \mathrm{d}c \tag{409}$$

4.2.4 Surficial integration of spatial operators

This group of integration theorems involves integrations of $(\partial f/\partial t)|_{\mathbf{x}}$, ∇f, $\nabla \cdot \mathbf{f}$ and $\nabla \times f$ over a surface S (Figure 20). Here $f(\mathbf{x},t)$ and $\mathbf{f}(\mathbf{x},t)$ are microscale scalar and vector functions of position vector $\mathbf{x}$ and time t.

In this group of integration theorems, S is the integration surface, C the curve bounding S, and $\mathbf{n}$ the unit vector normal to surface S. $\boldsymbol{\nu}^*$ is the unit vector normal to cure C which is positive in the outward direction from S such that $\mathbf{n} \cdot \boldsymbol{\nu}^* = 0$. ∇^c is the curvilineal operator such that $\nabla^c = \mathbf{nn} \cdot \nabla$ and ∇^S the surficial operator such that $\nabla^S = \nabla - \nabla^c$. $\mathbf{w}$ is the velocity of S and $\mathbf{w}^c$ the normal component of $\mathbf{w}$ such that $\mathbf{w}^c = \mathbf{nn} \cdot \mathbf{w}$. $\mathbf{f}^c$ is a vector normal to S such that $\mathbf{f}^c = \mathbf{nn} \cdot \mathbf{f}$ and $\mathbf{f}^s$ a vector tangent to S such that $\mathbf{f}^s = \mathbf{f} - \mathbf{f}^c$.

Theorem 15.

$$\begin{aligned}\int_S \left.\frac{\partial f}{\partial t}\right|_{\mathbf{x}} \mathrm{d}s = \frac{\partial}{\partial t}\int_S f \mathrm{d}s - \int_S (\nabla^S \cdot \mathbf{n})\mathbf{w}^c \cdot \mathbf{n} f \mathrm{d}s \\ - \int_S \mathbf{w}^c \cdot \nabla^c f \mathrm{d}s - \int_C \boldsymbol{\nu}^* \cdot \mathbf{w} f \mathrm{d}c\end{aligned} \tag{410}$$

Proof. By Equation (130), we have

$$\int_S \frac{\partial f}{\partial t}\bigg|_{\mathbf{x}} ds = \int_S \frac{\partial f}{\partial t}\bigg|_{\mathbf{u}} ds - \int_S \mathbf{w}\cdot\mathbf{nn}\cdot\nabla f ds$$
$$= \int_S \frac{\partial f}{\partial t}\bigg|_{\mathbf{u}} ds - \int_S \mathbf{w}\cdot\nabla^c f ds$$
$$= \int_S \frac{\partial f}{\partial t}\bigg|_{\mathbf{u}} ds - \int_S \mathbf{w}^c\cdot\nabla^c f ds \tag{411}$$

In deriving Equation (411), we used $\mathbf{w}\cdot\nabla^c f = (\mathbf{w}^s + \mathbf{w}^c)\cdot\nabla^c f = \mathbf{w}^c\cdot\nabla^c f$. By Equation (385),

$$\int_S \frac{\partial f}{\partial t}\bigg|_{\mathbf{x}} ds = \frac{\partial}{\partial t}\int_S f ds - \int_S (\nabla^S\cdot\mathbf{n})\mathbf{w}^c\cdot\mathbf{n} f ds - \int_S \mathbf{w}^c\cdot\nabla^c f ds - \int_C \boldsymbol{v}^*\cdot\mathbf{w} f dc$$

Theorem 16.

$$\int_S \nabla f ds = \int_S \nabla^c f ds + \int_S (\nabla^S\cdot\mathbf{n})\mathbf{n} f ds + \int_C \boldsymbol{v}^* f dc \tag{412}$$

Proof. Note that $\nabla = \nabla^c + \nabla^S$. Therefore

$$\int_S \nabla f ds = \int_S (\nabla^c + \nabla^S) f ds = \int_S \nabla^c f ds + \int_S \nabla^S f ds$$

By Equation (402), we obtain

$$\int_S \nabla f ds = \int_S \nabla^c f ds + \int_S (\nabla^S\cdot\mathbf{n})\mathbf{n} f ds + \int_C \boldsymbol{v}^* f dc$$

Theorem 17.

$$\int_S \nabla\cdot\mathbf{f} ds = \int_S \nabla^c\cdot\mathbf{f}^c ds - \int_S \mathbf{n}\cdot\nabla^c\mathbf{n}\cdot\mathbf{f}^s ds + \int_S \nabla^S\cdot\mathbf{nn}\cdot\mathbf{f}^c ds + \int_C \boldsymbol{v}^*\cdot\mathbf{f} dc \tag{413}$$

Proof. Since $\nabla = \nabla^c + \nabla^S$, we have

$$\int_S \nabla\cdot\mathbf{f} ds = \int_S (\nabla^c + \nabla^S)\cdot\mathbf{f} ds = \int_S \nabla^c\cdot\mathbf{f} ds + \int_S \nabla^S\cdot\mathbf{f} ds \tag{414}$$

By $\mathbf{f} = \mathbf{f}^c + \mathbf{f}^s$,

$$\int_S \nabla\cdot\mathbf{f} ds = \int_S \nabla^c\cdot\mathbf{f}^c ds + \int_S \nabla^c\cdot\mathbf{f}^s ds + \int_S \nabla^S\cdot\mathbf{f} ds \tag{415}$$

Define an indicator function $\gamma^s(\mathbf{u})$, such that $\gamma^s = 1$ on S and $\gamma^s = 0$ outside of S, and another indicator function $\gamma(x)$ such that $\gamma = 1$ in V enclosed by S_∞ and $\gamma = 0$ outside of V. Thus the second term on the right side of Equation (415) is equivalent to

$$\int_S \nabla^c\cdot\mathbf{f}^s ds = -\int_{V_\infty} (\mathbf{nn}\cdot\nabla)\cdot\mathbf{f}^s\gamma^s\mathbf{n}\cdot\nabla\gamma dv \tag{416}$$

In deriving Equation (416), we applied Equation (272) and $\nabla^c = \mathbf{nn}\cdot\nabla$. An application of the chain rule to the right side of Equation (416) yields

$$\begin{aligned}\int_S \nabla^c\cdot\mathbf{f}^s ds = &-\int_{V_\infty}\nabla\cdot(\mathbf{nn}\cdot\mathbf{f}^s\gamma^s\mathbf{n}\cdot\nabla\gamma)dv + \int_{V_\infty}\nabla\cdot\mathbf{nn}\cdot\mathbf{f}^s\gamma^s\mathbf{n}\cdot\nabla\gamma dv\\ &+\int_{V_\infty}\mathbf{n}\cdot\nabla\mathbf{n}\cdot\mathbf{f}^s\gamma^s\mathbf{n}\cdot\nabla\gamma dv + \int_{V_\infty}\mathbf{n}\cdot\mathbf{f}^s\mathbf{n}\cdot\nabla\gamma^s\mathbf{n}\cdot\nabla\gamma dv\\ &+\int_{V_\infty}\mathbf{n}\cdot\mathbf{f}^s\gamma^s\mathbf{n}\cdot\nabla(\mathbf{n}\cdot\nabla\gamma)dv \end{aligned}\tag{417}$$

By Equations (212), (231) and $\mathbf{n}\cdot\mathbf{f}^s = 0$, we have

$$\begin{aligned}\int_S \nabla^c\cdot\mathbf{f}^s ds &= \int_{V_\infty}\mathbf{n}\cdot\nabla\mathbf{n}\cdot\mathbf{f}^s\gamma^s\mathbf{n}\cdot\nabla\gamma dv\\ &= \int_{V_\infty}\mathbf{n}\cdot\nabla^c\mathbf{n}\cdot\mathbf{f}^s\gamma^s\mathbf{n}\cdot\nabla\gamma dv\\ &= -\int_S \mathbf{n}\cdot\nabla^c\mathbf{n}\cdot\mathbf{f}^s ds\end{aligned}\tag{418}$$

In deriving Equation (418), we used Equation (272) and $\mathbf{n}\cdot\nabla\mathbf{n} = \mathbf{n}\cdot(\nabla^c + \nabla^S)\mathbf{n} = \mathbf{n}\cdot\nabla^c\mathbf{n}$.

Substituting Equation (418) into Equation (415) and using Equation (403) lead to

$$\int_S \nabla\cdot\mathbf{f}ds = \int_S \nabla^c\cdot\mathbf{f}^c ds - \int_S \mathbf{n}\cdot\nabla^c\mathbf{n}\cdot\mathbf{f}^s ds + \int_S \nabla^S\cdot\mathbf{nn}\cdot\mathbf{f}^c ds + \int_C \mathbf{v}^*\cdot\mathbf{f}dc$$

Theorem 18.

$$\int_S \nabla\times\mathbf{f}ds = \int_S \nabla^c\times\mathbf{f}^s ds - \int_S (\mathbf{n}\cdot\nabla^c\mathbf{n})\times\mathbf{f}^c ds + \int_S (\nabla^S\cdot\mathbf{n})\mathbf{n}\times\mathbf{f}^s ds + \int_C \mathbf{v}^*\times\mathbf{f}dc \tag{419}$$

Proof. Since $\nabla = \nabla^c + \nabla^S$,

$$\begin{aligned}\int_S \nabla\times\mathbf{f}ds &= \int_S(\nabla^c + \nabla^S)\times\mathbf{f}ds\\ &= \int_S \nabla^c\times\mathbf{f}ds + \int_S \nabla^S\times\mathbf{f}ds\\ &= \int_S \nabla^c\times\mathbf{f}^c ds + \int_S \nabla^c\times\mathbf{f}^s + \int_S \nabla^S\times\mathbf{f}ds\end{aligned}\tag{420}$$

Define an indicator function $\gamma^s(\mathbf{u})$ such that $\gamma^s = 1$ on the surface S and $\gamma^s = 0$ elsewhere, and another indicator function $\gamma(\mathbf{x})$ such that $\gamma = 1$ within the volume V enclosed by S_∞ and $\gamma = 0$ outside of V. The first term on the right side of Equation (420) is equivalent to

$$\int_S \nabla^c\times\mathbf{f}^c ds = -\int_{V_\infty}(\mathbf{nn}\cdot\nabla)\times\mathbf{f}^c\gamma^s\mathbf{n}\cdot\nabla\gamma dv \tag{421}$$

An application of the chain rule to the right side of Equation (421) yields

$$\begin{aligned}\int_S \nabla^c \times \mathbf{f}^c ds = & -\int_{V_\infty} \nabla \cdot (\mathbf{nn} \times \mathbf{f}^c \gamma^s \mathbf{n} \cdot \nabla\gamma) dv + \int_{V_\infty} \nabla \cdot \mathbf{nn} \times \mathbf{f}^c \gamma^s \mathbf{n} \cdot \nabla\gamma dv \\ & + \int_{V_\infty} \mathbf{n} \cdot \nabla \mathbf{n} \times \mathbf{f}^c \gamma^s \mathbf{n} \cdot \nabla\gamma dv + \int_{V_\infty} \mathbf{n} \times \mathbf{f}^c \mathbf{n} \cdot \nabla\gamma^s \mathbf{n} \cdot \nabla\gamma dv \\ & + \int_{V_\infty} \mathbf{n} \times \mathbf{f}^c \gamma^s \mathbf{n} \cdot \nabla(\mathbf{n} \cdot \nabla\gamma) dv \end{aligned} \tag{422}$$

By Equation (231) and $\mathbf{n} \times \mathbf{f}^c = 0$, we obtain

$$\int_S \nabla^c \times \mathbf{f}^c ds = \int_{V_\infty} \mathbf{n} \cdot \nabla \mathbf{n} \times \mathbf{f}^c \gamma^s \mathbf{n} \cdot \nabla\gamma dv \tag{423}$$

Note that $\mathbf{n} \cdot \nabla \mathbf{n} = \mathbf{n} \cdot (\nabla^c + \nabla^S)\mathbf{n} = \mathbf{n} \cdot \nabla^c \mathbf{n}$. Thus

$$\int_S \nabla^c \times \mathbf{f}^c ds = -\int_S \mathbf{n} \cdot \nabla^c \mathbf{n} \times \mathbf{f}^c ds \tag{424}$$

In deriving Equation (424), we used Equation (272).

Substituting Equation (424) into Equation (420) and applying Equation (404) finally yield

$$\int_S \nabla \times \mathbf{f} = \int_S \nabla^c \times \mathbf{f}^s ds - \int_S (\mathbf{n} \cdot \nabla^c \mathbf{n}) \times \mathbf{f}^c ds + \int_S (\nabla^S \cdot \mathbf{n})\mathbf{n} \times \mathbf{f}^s ds + \int_C \boldsymbol{\nu}^* \times \mathbf{f} dc$$

4.2.5 Surficial integration of curvilineal operators

This group of integration theorems involves integrations of $(\partial f/\partial t)|_1$, $\nabla^c f$, $\nabla^c \cdot \mathbf{f}$ and $\nabla^c \times \mathbf{f}$ over a surface S (Figure 20). Here $f(\mathbf{u},t)$ and $\mathbf{f}(\mathbf{u},t)$ are microscale scalar and vector functions of surficial position vector $\mathbf{u}$ and time t.

Theorem 19.

$$\int_S \left.\frac{\partial f}{\partial t}\right|_1 ds = \frac{\partial}{\partial t}\int_S f ds - \int_S (\nabla^S \cdot \mathbf{n})\mathbf{n} \cdot \mathbf{w} ds - \int_S f(\nabla^S \cdot \mathbf{w}^\nu) ds - \int_C (\boldsymbol{\nu}^* \cdot \boldsymbol{\lambda})\boldsymbol{\lambda} \cdot \mathbf{w} f ds \tag{425}$$

where S is the integration surface, C the curve bounding S, $\boldsymbol{\lambda}$ the unit vector tangent to one of the surficial coordinates on S, and $\mathbf{n}$ the unit vector normal to S. $\boldsymbol{\nu}^*$ is the unit vector normal to curve C which is positive in the outward direction from S such that $\mathbf{n} \cdot \boldsymbol{\nu}^* = 0$. ∇^S is the surficial operator such that $\nabla^S = \nabla - \mathbf{nn} \cdot \nabla$. $\mathbf{w}$ is the velocity of C and $\mathbf{w}^\nu$ the velocity of the $\boldsymbol{\nu}$-coordinate in the direction normal to $\boldsymbol{\lambda}$ such that $\mathbf{w}^\nu = \boldsymbol{\nu}\boldsymbol{\nu} \cdot \mathbf{w}$.

Proof. Define an indicator function $\gamma^s = (\mathbf{u})$ such that $\gamma^s = 1$ on the surface S and $\gamma^s = 0$ elsewhere. Thus

$$\int_S \left.\frac{\partial f}{\partial t}\right|_1 ds = \int_{S_\infty} \left.\frac{\partial f}{\partial t}\right|_1 \gamma^s ds \tag{426}$$

By Equation (137), we have

$$\int_S \left.\frac{\partial f}{\partial t}\right|_1 ds = \int_{S_\infty} \left.\frac{\partial f}{\partial t}\right|_1 \gamma^s\, ds + \int_{S_\infty} \mathbf{w}\cdot\nabla^S f\gamma^s ds - \int_{S_\infty} (\mathbf{w}\cdot\boldsymbol{\lambda\lambda})\cdot\nabla^S f ds \tag{427}$$

By Equation (227) and applying the chain rule to the first term on the right side of Equation (427), we have

$$\int_{S_\infty} \left.\frac{\partial f}{\partial t}\right|_{\mathbf{u}} \gamma^s ds = \int_{S_\infty} \left.\frac{\partial (f\gamma^s)}{\partial t}\right|_{\mathbf{u}} ds - \int_{S_\infty} f\left.\frac{\partial \gamma^s}{\partial t}\right|_{\mathbf{u}} ds = \frac{\partial}{\partial t}\int_S f ds + \int_{S_\infty} f\mathbf{w}\cdot\nabla\gamma^s ds \tag{428}$$

In deriving this equation, we used the fact that S_∞ is independent of t. An application of the chain rule to the second term of Equation (428) yields

$$\begin{aligned}
\int_{S_\infty} f\mathbf{w}\cdot\nabla\gamma^s ds &= \int_{S_\infty} \nabla^S\cdot(\mathbf{w}f\gamma^s) ds - \int_{S_\infty} \nabla^S\cdot\mathbf{w}f\gamma^s ds - \int_{S_\infty} f\mathbf{w}\cdot\nabla^S\gamma^s ds \\
&= \int_{S_\infty} \nabla^S\cdot\mathbf{w}^{\nu}f\gamma^s ds - \int_{S_\infty} \nabla^S\cdot\mathbf{w}^{n}f\gamma^s ds \\
&\quad - \int_{S_\infty} \nabla^S\cdot\mathbf{w}^{\lambda}f\gamma^s ds - \int_{S_\infty} f\mathbf{w}\cdot\nabla^S\gamma^s ds \\
&= \int_S \nabla^S\cdot\mathbf{w}^{\nu}f ds - \int_{S_\infty} \nabla^S\cdot(\mathbf{nn}\cdot\mathbf{w})f\gamma^s ds \\
&\quad - \int_{S_\infty} \nabla^S\cdot(\boldsymbol{\lambda\lambda}\cdot\mathbf{w})f\gamma^s ds - \int_{S_\infty} f\mathbf{w}\cdot\nabla^S\gamma^s ds \\
&= \int_S \nabla^S\cdot\mathbf{w}^{\nu}f ds - \int_{S_\infty} \nabla^S\cdot\mathbf{nn}\cdot\mathbf{w}f\gamma^s ds \\
&\quad - \int_{S_\infty} \mathbf{n}\cdot\nabla^S(\mathbf{n}\cdot\mathbf{w})f\gamma^s ds \\
&\quad - \int_{S_\infty} \nabla^S\cdot(\boldsymbol{\lambda\lambda}\cdot\mathbf{w})f\gamma^s ds - \int_{S_\infty} f\mathbf{w}\cdot\nabla^S\gamma^s ds \\
&= \int_S \nabla^S\cdot\mathbf{w}^{\nu}f ds - \int_{S_\infty} \nabla^S\cdot\mathbf{nn}\cdot\mathbf{w}f\gamma^s ds \\
&\quad - \int_{S_\infty} \nabla^S\cdot(\boldsymbol{\lambda\lambda}\cdot\mathbf{w})f\gamma^s ds - \int_{S_\infty} f\mathbf{w}\cdot\nabla^S\gamma^s ds
\end{aligned} \tag{429}$$

in which we have used the definition of ∇^S and $\mathbf{w} = \mathbf{w}^{\nu} + \mathbf{w}^{n} + \mathbf{w}^{\lambda}$.

An application of the chain rule to the third term on the right side of Equation (427) yields

$$\begin{aligned}
-\int_{S_\infty} (\mathbf{w}\cdot\boldsymbol{\lambda\lambda})\cdot\nabla^S f\gamma^s ds &= -\int_{S_\infty} \nabla^S\cdot(\boldsymbol{\lambda}\mathbf{w}\cdot\boldsymbol{\lambda}f\gamma^s) ds \\
&\quad + \int_{S_\infty} \nabla^S\cdot(\boldsymbol{\lambda\lambda}\cdot\mathbf{w})f\gamma^s ds + \int_{S_\infty} f\mathbf{w}\cdot\boldsymbol{\lambda\lambda}\cdot\nabla^S\gamma^s ds \\
&= \int_{S_\infty} \nabla^S\cdot(\boldsymbol{\lambda\lambda}\cdot\mathbf{w})f\gamma^s ds + \int_{S_\infty} f\mathbf{w}\cdot\boldsymbol{\lambda\lambda}\cdot\nabla^S\gamma^s ds
\end{aligned} \tag{430}$$

In deriving Equation (430), we have used Equation (431). By Equations (176) and (427)–(430), we have

$$\begin{aligned}\int_S \frac{\partial f}{\partial t}\bigg|_1 ds = &\frac{\partial}{\partial t}\int_S f ds + \int_{S_\infty} f\mathbf{w}\cdot\nabla\gamma^s ds - \int_S \nabla^S\cdot\mathbf{w}^v f ds \\ &- \int_S \nabla^S\cdot\mathbf{nn}\cdot\mathbf{w}f ds - \int_{S_\infty} f\mathbf{w}\cdot\nabla^S\gamma^s ds \\ &+ \int_{S_\infty} f\mathbf{w}\cdot\boldsymbol{\lambda\lambda}\cdot\nabla^S\gamma^s ds \\ = &\frac{\partial}{\partial t}\int_S f ds - \int_S \nabla^S\cdot\mathbf{w}^v f ds - \int_S \nabla^S\cdot\mathbf{nn}\cdot\mathbf{w}f ds \\ &- \int_C \boldsymbol{\lambda}\cdot\boldsymbol{v}^* f\mathbf{w}\cdot\boldsymbol{\lambda} dc \end{aligned} \tag{431}$$

Equation (216) enables us rewrite Equation (431) as

$$\int_S \frac{\partial f}{\partial t}\bigg|_1 ds = \frac{\partial}{\partial t}\int_S f ds - \int_S (\nabla^S\cdot\mathbf{n})\mathbf{n}\cdot\mathbf{w}ds - \int_S f(\nabla^S\cdot\mathbf{w}^v)ds - \int_C (\boldsymbol{v}^*\cdot\boldsymbol{\lambda})\boldsymbol{\lambda}\cdot\mathbf{w}f dc$$

Theorem 20.

$$\int_S \nabla^c\circ F ds = -\int_S (\nabla^S\cdot\boldsymbol{\lambda})\boldsymbol{\lambda}\circ F ds - \int_S (\boldsymbol{\lambda}\cdot\nabla^c\boldsymbol{\lambda})\circ F ds + \int_C \boldsymbol{v}^*\cdot\boldsymbol{\lambda\lambda}\circ F dc \tag{432}$$

where $\nabla^c\circ F$ can be $\nabla^c f$, $\nabla^c\cdot\mathbf{f}$ or $\nabla^c\times\mathbf{f}$. ∇^c is the curvilineal operator such that $\nabla^c = \boldsymbol{\lambda\lambda}\cdot\nabla$ and ∇^S the surficial operator such that $\nabla^S = \nabla - \mathbf{nn}\cdot\nabla$.

Proof. Define two indicator functions $\gamma^s(\mathbf{u})$ and $\gamma(\mathbf{x})$, such that $\gamma^s = 1$ on the surface S and $\gamma^s = 0$ elsewhere; and $\gamma = 1$ within V enclosed by the S_∞ and $\gamma = 0$ elsewhere. Thus

$$\int_S \nabla^c\circ F ds = -\int_{V_\infty} \gamma^s\nabla^c\circ F\mathbf{n}\cdot\nabla\gamma dv \tag{433}$$

Applying $\nabla^c = \boldsymbol{\lambda\lambda}\cdot\nabla$ and the chain rule yields

$$\begin{aligned}\int_S \nabla^c\circ F ds = &-\int_{V_\infty} \gamma^s(\boldsymbol{\lambda\lambda}\cdot\nabla)\circ F\mathbf{n}\cdot\nabla\gamma dv \\ = &-\int_{V_\infty} \nabla\cdot(\boldsymbol{\lambda\lambda}\circ F\gamma^s\mathbf{n}\cdot\nabla\gamma)dv + \int_{V_\infty} \nabla\cdot\boldsymbol{\lambda\lambda}\circ F\gamma^s\mathbf{n}\cdot\nabla\gamma dv \\ &+ \int_{V_\infty} \boldsymbol{\lambda}\cdot\nabla\boldsymbol{\lambda}\circ F\gamma^s\mathbf{n}\cdot\nabla\gamma dv + \int_{V_\infty} \boldsymbol{\lambda}\cdot\nabla\gamma^s\boldsymbol{\lambda}\circ F\mathbf{n}\cdot\gamma dv \\ &+ \int_{V_\infty} \boldsymbol{\lambda}\circ F\gamma^s\boldsymbol{\lambda}\cdot\nabla(\mathbf{n}\cdot\nabla\gamma)dv \end{aligned} \tag{434}$$

Note that

$$\begin{aligned}\nabla\cdot\boldsymbol{\lambda} = \nabla^S\cdot\boldsymbol{\lambda} + \mathbf{n}\cdot\nabla\boldsymbol{\lambda}\cdot\mathbf{n},\ \boldsymbol{\lambda}\cdot\nabla\boldsymbol{\lambda} = \boldsymbol{\lambda}\cdot\nabla^c\boldsymbol{\lambda},\ \nabla(\mathbf{n}\cdot\nabla\gamma) \\ = \nabla\mathbf{n}\cdot\nabla\gamma + \nabla\nabla\gamma\cdot\mathbf{n} = \nabla\nabla\gamma\cdot\mathbf{n}\end{aligned}$$

Also by Equation (230), (231) and (232), we obtain

$$\begin{aligned}\int_S \nabla^c \circ F ds = & \int_{V_\infty} \nabla^S \cdot \boldsymbol{\lambda\lambda} \circ F\gamma^s \mathbf{n} \cdot \nabla\gamma dv + \int_{V_\infty} \boldsymbol{\lambda} \cdot \nabla^c \boldsymbol{\lambda} \circ F\gamma^s \mathbf{n} \cdot \nabla\gamma dv \\ & + \int_{V_\infty} \boldsymbol{\lambda} \cdot \nabla\gamma^s \boldsymbol{\lambda} \circ F\mathbf{n} \cdot \nabla\gamma dv + \int_{V_\infty} (\mathbf{nn} \cdot \nabla) \cdot \boldsymbol{\lambda\lambda} \circ F\gamma^s \mathbf{n} \cdot \nabla\gamma dv \\ & + \int_{V_\infty} \boldsymbol{\lambda} \circ F\gamma^s \boldsymbol{\lambda} \cdot \nabla\nabla\gamma \cdot \mathbf{n} dv \end{aligned} \tag{435}$$

By Equation (250), we have

$$\begin{aligned}\int_S \nabla^c \circ F ds = & \int_{V_\infty} \nabla^S \cdot \boldsymbol{\lambda\lambda} \circ F\gamma^s \mathbf{n} \cdot \nabla\gamma dv + \int_{V_\infty} \boldsymbol{\lambda} \cdot \nabla^c \boldsymbol{\lambda} \circ F\gamma^s \mathbf{n} \cdot \nabla\gamma dv \\ & + \int_{V_\infty} \boldsymbol{\lambda} \cdot \nabla\gamma^s \boldsymbol{\lambda} \circ F\mathbf{n} \cdot \nabla\gamma dv + \int_{V_\infty} \mathbf{n} \cdot \nabla\boldsymbol{\lambda} \cdot \mathbf{n}\boldsymbol{\lambda} \circ F\gamma^s \mathbf{n} \cdot \nabla\gamma dv \\ & - \int_{V_\infty} \boldsymbol{\lambda} \circ F(\nabla\boldsymbol{\lambda} \cdot \nabla\gamma) \cdot \mathbf{n}\gamma^s dv \\ = & \int_{V_\infty} \nabla^S \cdot \boldsymbol{\lambda\lambda} \circ F\gamma^s \mathbf{n} \cdot \nabla\gamma dv + \int_{V_\infty} \boldsymbol{\lambda} \cdot \nabla^c \boldsymbol{\lambda} \circ F\gamma^s \mathbf{n} \cdot \nabla\gamma dv \\ & + \int_{V_\infty} \boldsymbol{\lambda} \cdot \nabla\gamma^s \boldsymbol{\lambda} \circ F\mathbf{n} \cdot \nabla\gamma dv + \int_{V_\infty} \mathbf{n} \cdot \nabla\boldsymbol{\lambda} \cdot \mathbf{n}\boldsymbol{\lambda} \circ F\gamma^s \mathbf{n} \cdot \nabla\gamma dv \\ & - \int_{V_\infty} F\boldsymbol{\lambda} \circ F\mathbf{n} \cdot \nabla\boldsymbol{\lambda} \cdot \nabla\gamma\gamma^s dv \end{aligned} \tag{436}$$

By Equations (272) and (275), we finally obtain

$$\begin{aligned}\int_S \nabla^c \circ F ds = & -\int_S \nabla^S \cdot \boldsymbol{\lambda\lambda} \circ F ds - \int_S \boldsymbol{\lambda} \cdot \nabla^c \boldsymbol{\lambda} \circ F ds + \int_C \boldsymbol{\lambda} \cdot \boldsymbol{v}^* \boldsymbol{\lambda} \circ F dc \\ & - \int_S \mathbf{n} \cdot \nabla\boldsymbol{\lambda} \cdot \mathbf{n}\boldsymbol{\lambda} \circ F ds + \int_S \boldsymbol{\lambda} \circ F\mathbf{n} \cdot \nabla\boldsymbol{\lambda} \cdot \mathbf{n} ds \\ = & -\int_S \nabla^S \cdot \boldsymbol{\lambda\lambda} \circ F ds - \int_S (\boldsymbol{\lambda} \cdot \nabla^c \boldsymbol{\lambda}) \circ F ds + \int_S \boldsymbol{v}^* \cdot \boldsymbol{\lambda\lambda} \circ F dc \end{aligned}$$

Remark. Equation (432) contains the following three formulas:

Gradient

$$\int_S \nabla^c f ds = -\int_S (\nabla^S \cdot \boldsymbol{\lambda})\boldsymbol{\lambda} f ds - \int_S (\boldsymbol{\lambda} \cdot \nabla^c \boldsymbol{\lambda}) f ds + \int_C (\boldsymbol{v}^* \cdot \boldsymbol{\lambda})\boldsymbol{\lambda} f dc \tag{437}$$

Divergence

$$\int_S \nabla^c \cdot \mathbf{f} ds = -\int_S (\nabla^S \cdot \boldsymbol{\lambda})\boldsymbol{\lambda} \cdot \mathbf{f} ds - \int_S (\boldsymbol{\lambda} \cdot \nabla^c \boldsymbol{\lambda}) \cdot \mathbf{f} ds + \int_C (\boldsymbol{v}^* \cdot \boldsymbol{\lambda})\boldsymbol{\lambda} \cdot \mathbf{f} dc \tag{438}$$

Curl

$$\int_S \nabla^c \times \mathbf{f} ds = -\int_S (\nabla^S \cdot \boldsymbol{\lambda})\boldsymbol{\lambda} \times \mathbf{f} ds - \int_S (\boldsymbol{\lambda} \cdot \nabla^c \boldsymbol{\lambda}) \times \mathbf{f} ds + \int_C (\boldsymbol{v}^* \cdot \boldsymbol{\lambda})\boldsymbol{\lambda} \times \mathbf{f} dc \tag{439}$$

4.3 Theorems for integration over a curve

In this section we develop 18 integration theorems regarding integration over a curve C (or a straight line segment L) of spatial, surficial, or curvilineal differential operators of a microscale function f or $\mathbf{f}$.

4.3.1 Curvilineal integration of spatial operators

This group of integration theorems involves integrations of $(\partial f/\partial t)|_{\mathbf{x}}$, ∇f, $\nabla \cdot \mathbf{f}$, and $\nabla \times \mathbf{f}$ over a curve C (Figure 21). Here $f(\mathbf{x}, t)$ and $\mathbf{f}(\mathbf{x}, t)$ are microscale scalar and vector functions of position vector $\mathbf{x}$ and time t. Generally, the curve C can deform and translate with time.

Theorem 21.

$$\int_C \left.\frac{\partial f}{\partial t}\right|_{\mathbf{x}} \mathrm{d}c = \frac{\partial}{\partial t}\int_C f \mathrm{d}c - \int_C \mathbf{w}^{\mathrm{s}} \cdot \nabla^{\mathrm{S}} f \mathrm{d}c + \int_C (\boldsymbol{\lambda} \cdot \nabla^{\mathrm{c}} \boldsymbol{\lambda}) \cdot \mathbf{w}^{\mathrm{s}} f \mathrm{d}c - (\mathbf{e} \cdot \mathbf{w}^{\mathrm{c}} f)|_{\mathrm{ends}} \quad (440)$$

where C is the integration curve, $\boldsymbol{\lambda}$ the unit vector tangent to curve C, and $\mathbf{e}$ the particular unit vector tangent to curve C at the endpoints which is positive in the outward direction from the curve. ∇^{c} is the curvilineal operator such that $\nabla^{\mathrm{c}} = \boldsymbol{\lambda}\boldsymbol{\lambda} \cdot \nabla$, ∇^{S} is the surficial operator such that $\nabla^{\mathrm{S}} = \nabla - \nabla^{\mathrm{c}}$, $\mathbf{w}$ the velocity of C, $\mathbf{w}^{\mathrm{c}}$ the velocity tangent to C such that $\mathbf{w}^{\mathrm{c}} = \boldsymbol{\lambda}\boldsymbol{\lambda} \cdot \mathbf{w}$ and $\mathbf{w}^{\mathrm{s}}$ the velocity normal to C such that $\mathbf{w}^{\mathrm{s}} = \mathbf{w} - \mathbf{w}^{\mathrm{c}}$.

Proof. Extend the curve C to form a closed curve C_∞. Denote the surface enclosed by C_∞ as S. Extend the surface S to form a closed surface S_∞. Let V be the volume enclosed by S_∞.

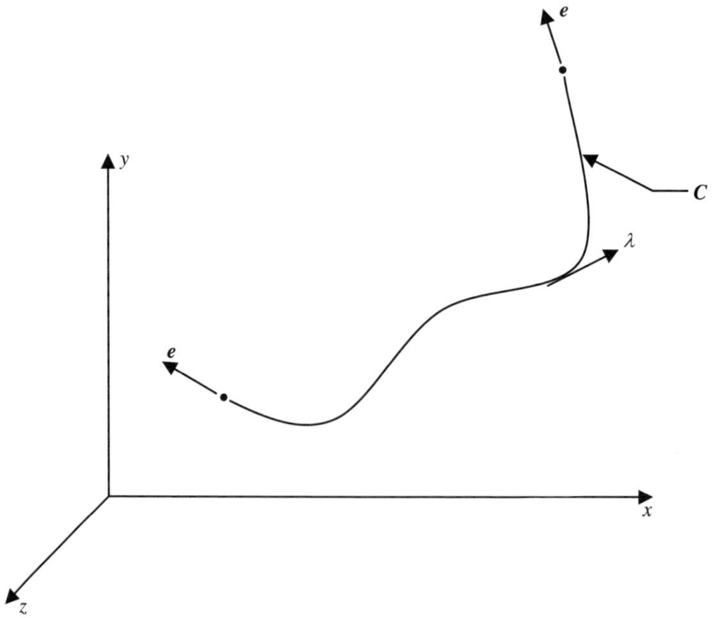

Figure 21 Simple integration curve C (λ is unit tangent of C; $\mathbf{e}$ unit vector tangent of C, and pointing outward from C at the endpoints).

Define three indicator functions $\gamma^{c}(\mathbf{l})$, $\gamma^{s}(\mathbf{u})$, and $\gamma(\mathbf{x})$ such that $\gamma^{c} = 1$ on the curve C and $\gamma^{c} = 0$ elsewhere, $\gamma^{s} = 1$ on the surface S and $\gamma^{s} = 0$ elsewhere and $\gamma = 1$ within the volume V with $\gamma = 0$ outside of V. Therefore

$$\int_{C} \left.\frac{\partial f}{\partial t}\right|_{\mathbf{x}} dc = \int_{V_\infty} \left.\frac{\partial f}{\partial t}\right|_{\mathbf{x}} \gamma^{c}\boldsymbol{v} \cdot \nabla\gamma^{s}\mathbf{n} \cdot \nabla\gamma dv \tag{441}$$

An application of the chain rule to the right-hand side of Equation (441) yields

$$\begin{aligned}\int_{C} \left.\frac{\partial f}{\partial t}\right|_{\mathbf{x}} dc = &\int_{V_\infty} \left.\frac{\partial(f\gamma^{c}\boldsymbol{v} \cdot \nabla\gamma^{s}\mathbf{n} \cdot \nabla\gamma)}{\partial t}\right|_{\mathbf{x}} dv - \int_{V_\infty} f\left.\frac{\partial\gamma^{c}}{\partial t}\right|_{\mathbf{x}} \boldsymbol{v} \cdot \nabla\gamma^{s}\mathbf{n} \cdot \nabla\gamma dv \\ &- \int_{V_\infty} f\gamma^{c}\left(\frac{\partial\boldsymbol{v}}{\partial t} \cdot \nabla\gamma^{s} + \boldsymbol{v} \cdot \nabla\frac{\partial\gamma^{s}}{\partial t}\right)\mathbf{n} \cdot \nabla\gamma dv \\ &- \int_{V_\infty} f\gamma^{c}\boldsymbol{v} \cdot \nabla\gamma^{s}\left(\frac{\partial\mathbf{n}}{\partial t} \cdot \nabla\gamma + \mathbf{n} \cdot \nabla\frac{\partial\gamma}{\partial t}\right) dv\end{aligned} \tag{442}$$

Note that V_∞ is independent of t so that the first term on the right side of Equation (442) becomes

$$\int_{V_\infty} \left.\frac{\partial(f\gamma^{c}v \cdot \nabla\gamma^{s}\mathbf{n} \cdot \nabla\gamma)}{\partial t}\right|_{\mathbf{x}} dv = \frac{\partial}{\partial t}\int_{C} f dc \tag{443}$$

By Equations (166), (229) and (275), the second term is

$$\int_{V_\infty} f\left.\frac{\partial\gamma^{c}}{\partial t}\right|_{\mathbf{x}} \boldsymbol{v} \cdot \nabla\gamma^{s}\mathbf{n} \cdot \nabla\gamma dv = -\int_{V_\infty} f\mathbf{w} \cdot \nabla\gamma^{c}\boldsymbol{v} \cdot \nabla\gamma^{s}\mathbf{n} \cdot \nabla\gamma dv = (\mathbf{e} \cdot \mathbf{w}f)|_{\text{ends}}$$

Note that

$$\mathbf{e} \cdot \mathbf{w}^{s} = 0$$

Therefore

$$\int_{V_\infty} f\left.\frac{\partial\gamma^{s}}{\partial t}\right|_{\mathbf{x}} \boldsymbol{v} \cdot \nabla\gamma^{s}\mathbf{n} \cdot \nabla\gamma dv = (\mathbf{e} \cdot \mathbf{w}^{c}f)|_{\text{ends}} \tag{444}$$

By Equations (199) and (224), the third term becomes

$$\int_{V_\infty} f\gamma^{c}\left(\frac{\partial\boldsymbol{v}}{\partial t} \cdot \nabla\gamma^{s} + \boldsymbol{v} \cdot \nabla\frac{\partial\gamma^{s}}{\partial t}\right)\mathbf{n} \cdot \nabla\gamma dv = -\int_{V_\infty} f\gamma^{c}\boldsymbol{v} \cdot \nabla(\mathbf{w} \cdot \nabla\gamma^{s})\mathbf{n} \cdot \nabla\gamma dv \tag{445}$$

An application of the chain rule to the right-hand side of Equation (445) yields

$$\begin{aligned}&\int_{V_\infty} f\gamma^{c}\left(\frac{\partial\boldsymbol{v}}{\partial t} \cdot \nabla\gamma^{s} + \boldsymbol{v} \cdot \nabla\frac{\partial\gamma^{s}}{\partial t}\right)\mathbf{n} \cdot \nabla\gamma dv \\ &= -\int_{V_\infty} \nabla \cdot (\boldsymbol{v}f\gamma^{c}(\mathbf{w} \cdot \nabla\gamma^{s})\mathbf{n} \cdot \nabla\gamma) dv + \int_{V_\infty} \nabla \cdot \boldsymbol{v}f\gamma^{c}(\mathbf{w} \cdot \nabla\gamma^{s})\mathbf{n} \cdot \nabla\gamma dv \\ &\quad + \int_{V_\infty} \boldsymbol{v} \cdot \nabla f\gamma^{c}(\mathbf{w} \cdot \nabla\gamma^{s})\mathbf{n} \cdot \nabla\gamma dv + \int_{V_\infty} f\boldsymbol{v} \cdot \nabla\gamma^{c}(\mathbf{w} \cdot \nabla\gamma^{s})\mathbf{n} \cdot \nabla\gamma dv \\ &\quad + \int_{V_\infty} f\gamma^{c}(\mathbf{w} \cdot \nabla\gamma^{s})\boldsymbol{v} \cdot (\nabla\mathbf{n} \cdot \nabla\gamma + \nabla\nabla\gamma \cdot \mathbf{n}) dv\end{aligned} \tag{446}$$

By Equation (231), we have

$$\int_{V_\infty} \nabla \cdot (\boldsymbol{v}f\gamma^{c}(\mathbf{w} \cdot \nabla\gamma^{s})\mathbf{n} \cdot \nabla\gamma) dv = 0 \tag{447}$$

By Equation (275), we also have

$$\begin{aligned}&\int_{V_\infty} \nabla \cdot \boldsymbol{\nu} f \gamma^{c} (\mathbf{w} \cdot \nabla \gamma^{s}) \mathbf{n} \cdot \nabla \gamma dv + \int_{V_\infty} \boldsymbol{\nu} \cdot \nabla f \gamma^{c} (\mathbf{w} \cdot \nabla \gamma^{s}) \mathbf{n} \cdot \nabla \gamma dv \\ &\quad = \int_C \nabla \cdot \boldsymbol{\nu} f \mathbf{w} \cdot \boldsymbol{\nu} dc + \int_C \boldsymbol{\nu} \cdot \nabla f \mathbf{w} \cdot \boldsymbol{\nu} dc\end{aligned} \tag{448}$$

Applying Equation (215) yields

$$\int_{V_\infty} \boldsymbol{\nu} \cdot \nabla \gamma^{c} f (\mathbf{w} \cdot \nabla \gamma^{s}) \mathbf{n} \cdot \nabla \gamma dv = 0 \tag{449}$$

By Equations (195), we have

$$\begin{aligned}&\int_{V_\infty} f \gamma^{c} (\mathbf{w} \cdot \nabla \gamma^{s}) \boldsymbol{\nu} \cdot (\nabla \mathbf{n} \cdot \nabla \gamma + \nabla \nabla \gamma \cdot \mathbf{n}) dv \\ &\quad = -\int_{V_\infty} f \gamma^{c} (\mathbf{w} \cdot \nabla \gamma^{s}) \boldsymbol{\nu} \cdot \nabla \nabla \gamma \cdot \mathbf{n} dv\end{aligned}$$

Also, by Equation (259),

$$\int_{V_\infty} \gamma^{c} \mathbf{w} \cdot \nabla \gamma^{s} \boldsymbol{\nu} \cdot (\nabla \mathbf{n} \cdot \nabla \gamma + \nabla \nabla \gamma \cdot \mathbf{n}) dv = -\int_C f \mathbf{w} \cdot \boldsymbol{\nu} \mathbf{n} \cdot \nabla \boldsymbol{\nu} \cdot \mathbf{n} dc \tag{450}$$

Substituting Equations (447)–(450) into Equation (446) thus yields

$$\begin{aligned}&\int_{V_\infty} f \gamma^{c} \left(\frac{\partial \boldsymbol{\nu}}{\partial t} \cdot \nabla \gamma^{s} + \boldsymbol{\nu} \cdot \nabla \frac{\partial \gamma^{s}}{\partial t} \right) \mathbf{n} \cdot \nabla \gamma dv = \int_C \nabla \cdot \boldsymbol{\nu} f \mathbf{w} \cdot \boldsymbol{\nu} dc \\ &\quad + \int_C \boldsymbol{\nu} \cdot \nabla f \mathbf{w} \cdot \boldsymbol{\nu} dc - \int_C f \mathbf{w} \cdot \boldsymbol{\nu} \mathbf{n} \cdot \nabla \boldsymbol{\nu} \cdot \mathbf{n} dc\end{aligned} \tag{451}$$

Similarly,

$$\begin{aligned}&\int_{V_\infty} f \gamma^{c} \boldsymbol{\nu} \cdot \nabla \gamma^{s} \left(\frac{\partial \boldsymbol{n}}{\partial t} \cdot \nabla \gamma + \boldsymbol{\nu} \cdot \nabla \frac{\partial \gamma^{s}}{\partial t} \right) \mathbf{n} \cdot \nabla \gamma dv = \int_C \nabla \cdot \mathbf{n} f \mathbf{w} \cdot \mathbf{n} dc \\ &\quad + \int_C \mathbf{n} \cdot \nabla f \mathbf{w} \cdot \mathbf{n} dc - \int_C f \mathbf{w} \cdot \mathbf{n} \boldsymbol{\nu} \cdot \nabla \mathbf{n} \cdot \boldsymbol{\nu} dc\end{aligned} \tag{452}$$

Substituting Equations (443), (444), (451) and (452) into Equation (442) yields

$$\begin{aligned}\int_C \left. \frac{\partial f}{\partial t} \right|_{\mathbf{x}} dc = &\frac{\partial}{\partial t} \int_C f dc - (\mathbf{e} \cdot \mathbf{w}^{c} f)|_{\text{ends}} - \int_C (\nabla \cdot \boldsymbol{\nu} - \mathbf{n} \cdot \nabla \boldsymbol{\nu} \cdot \mathbf{n}) f \mathbf{w} \cdot \boldsymbol{\nu} dc \\ &- \int_C \boldsymbol{\nu} \cdot \nabla f \mathbf{w} \cdot \boldsymbol{\nu} dc - \int_C (\nabla \cdot \mathbf{n} - \boldsymbol{\nu} \cdot \nabla \mathbf{n} \cdot \boldsymbol{\nu}) f \mathbf{w} \cdot \mathbf{n} dc \\ &- \int_C \mathbf{n} \cdot \nabla f \mathbf{w} \cdot \mathbf{n} dc\end{aligned} \tag{453}$$

By Equation (53), we have

$$\nabla \cdot \boldsymbol{\nu} - \mathbf{n} \cdot \nabla \boldsymbol{\nu} \cdot \mathbf{n} = \boldsymbol{\lambda} \cdot \nabla \boldsymbol{\nu} \cdot \boldsymbol{\lambda} \tag{454}$$

$$\nabla \cdot \mathbf{n} - \boldsymbol{\nu} \cdot \nabla \mathbf{n} \cdot \boldsymbol{\nu} = \boldsymbol{\lambda} \cdot \nabla \mathbf{n} \cdot \boldsymbol{\lambda} \tag{455}$$

Note also that

$$\mathbf{w} \cdot \boldsymbol{\nu} \boldsymbol{\nu} + \mathbf{w} \cdot \mathbf{n} \mathbf{n} = \mathbf{w}^{s}$$

Therefore

$$\int_C \boldsymbol{\nu} \cdot \nabla f \mathbf{w} \cdot \boldsymbol{\nu} dc + \int_C \mathbf{n} \cdot \nabla f \mathbf{w} \cdot \mathbf{n} dc = \int_C \mathbf{w}^s \cdot \nabla f dc = \int_C \mathbf{w}^s \cdot \nabla^S f dc \tag{456}$$

Substituting Equations (454)–(456) into Equation (453) yields

$$\begin{aligned}\int_C \left.\frac{\partial f}{\partial t}\right|_{\mathbf{x}} dc = \frac{\partial}{\partial t}\int_C f dc - (\mathbf{e}\cdot\mathbf{w}^c f)|_{\text{ends}} - \int_C \boldsymbol{\lambda}\cdot\nabla\boldsymbol{\nu}\cdot\boldsymbol{\lambda} f\mathbf{w}\cdot\boldsymbol{\nu} dc \\ - \int_C \boldsymbol{\lambda}\cdot\nabla\mathbf{n}\cdot\boldsymbol{\lambda} f\mathbf{w}\cdot\mathbf{n} dc - \int_C \mathbf{w}^s\cdot\nabla^S f dc\end{aligned} \tag{457}$$

By Equations (24) and (25), we have

$$\begin{aligned}&\textstyle\int_C \boldsymbol{\lambda}\cdot\nabla\boldsymbol{\nu}\cdot\boldsymbol{\lambda} f\mathbf{w}\cdot\boldsymbol{\nu} dc + \int_C \boldsymbol{\lambda}\cdot\nabla\mathbf{n}\cdot\boldsymbol{\lambda} f\mathbf{w}\cdot\mathbf{n} dc \\ &\quad = -\textstyle\int_C \boldsymbol{\lambda}\cdot\nabla\boldsymbol{\lambda}\cdot\boldsymbol{\nu}\mathbf{w}\cdot\boldsymbol{\nu} f dc - \int_C \boldsymbol{\lambda}\cdot\nabla\boldsymbol{\lambda}\cdot\mathbf{n} f\mathbf{w}\cdot\mathbf{n} dc \\ &\quad = -\textstyle\int_C \boldsymbol{\lambda}\cdot\nabla\boldsymbol{\lambda}\cdot\mathbf{w}^s f dc = -\int_C \boldsymbol{\lambda}\cdot\nabla^c\boldsymbol{\lambda}\cdot\mathbf{w}^s f dc\end{aligned} \tag{458}$$

By substituting Equation (458) into Equation (457), we finally obtain

$$\int_C \left.\frac{\partial f}{\partial t}\right|_{\mathbf{x}} dc = \frac{\partial}{\partial t}\int_C f dc - \int_C \mathbf{w}^s\cdot\nabla^S f dc + \int_C (\boldsymbol{\lambda}\cdot\nabla^c\boldsymbol{\lambda})\cdot\mathbf{w}^s f dc - (\mathbf{e}\cdot\mathbf{w}^c f)|_{\text{ends}}$$

Theorem 22.

$$\int_C \nabla \circ F dc = \int_C \nabla^S \circ F dc - \int_C (\boldsymbol{\lambda}\cdot\nabla^c\boldsymbol{\lambda}) \circ F dc + (\mathbf{e}\circ F)\big|_{\text{ends}} \tag{459}$$

where the symbol $\nabla \circ F$ can be ∇f, $\nabla \cdot \mathbf{f}$ or $\nabla \times \mathbf{f}$.

Proof. Since $\nabla = \nabla^S + \nabla^c$, we have

$$\int_C \nabla \circ F dc = \int_C \nabla^S \circ F dc + \int_C \nabla^c \circ F dc \tag{460}$$

Extend C to form a closed curve C_∞. Denote the surface enclosed by C_∞ as S. Extend the surface S to form a closed surface S_∞. Let V be the volume enclosed by S_∞.

Define three indicator functions $\gamma^c(1)$, $\gamma^s(\mathbf{u})$ and $\gamma(\mathbf{x})$ such that $\gamma^c = 1$ on C and $\gamma^c = 0$ elsewhere, $\gamma^s = 1$ on the surface S and $\gamma^s = 0$ elsewhere and $\gamma = 1$ within V with $\gamma = 0$ outside of V. Thus

$$\int_C \nabla^c \circ F dc = \int_{V_\infty} \nabla^c \circ F \gamma^c \boldsymbol{\nu} \cdot \nabla^S \gamma^s \mathbf{n} \cdot \nabla \gamma dv \tag{461}$$

The definition of the operator ∇^c enables us to rewrite Equation (461) as

$$\begin{aligned}\int_C \nabla^c \circ F dc &= \int_{V_\infty} (\boldsymbol{\lambda}\boldsymbol{\lambda}\cdot\nabla) \circ F \gamma^c \boldsymbol{\nu} \cdot \nabla^S \gamma^s \mathbf{n} \cdot \nabla \gamma dv \\ &= \int_{V_\infty} \boldsymbol{\lambda} \circ (\boldsymbol{\lambda}\cdot\nabla F) \gamma^c \boldsymbol{\nu} \cdot \nabla^S \gamma^s \mathbf{n} \cdot \nabla \gamma dv\end{aligned} \tag{462}$$

An application of the chain rule to the right side of Equation (462) yields

$$\begin{aligned}\int_C \nabla^c \circ F dc = &\int_{V_\infty} \nabla \cdot \big(\boldsymbol{\lambda\lambda} \circ F\gamma^c \boldsymbol{v} \cdot \nabla^S \gamma^s \mathbf{n} \cdot \nabla\gamma\big) dv \\ &- \int_{V_\infty} \nabla \cdot \boldsymbol{\lambda\lambda} \circ F\gamma^c \boldsymbol{v} \cdot \nabla^S \gamma^s \mathbf{n} \cdot \nabla\gamma dv \\ &- \int_{V_\infty} \boldsymbol{\lambda} \cdot \nabla\boldsymbol{\lambda} \circ F\gamma^c \boldsymbol{v} \cdot \nabla^S \gamma^s \mathbf{n} \cdot \nabla\gamma dv \\ &- \int_{V_\infty} \boldsymbol{\lambda} \circ F\boldsymbol{\lambda} \cdot \nabla\gamma^c \boldsymbol{v} \cdot \nabla^S \gamma^s \mathbf{n} \cdot \nabla\gamma dv \\ &- \int_{V_\infty} \boldsymbol{\lambda} \circ F\gamma^c \boldsymbol{\lambda} \cdot \nabla(\boldsymbol{v} \cdot \nabla^S \gamma^s) \mathbf{n} \cdot \nabla\gamma dv \\ &- \int_{V_\infty} \boldsymbol{\lambda} \circ F\gamma^c \boldsymbol{v} \cdot \nabla^S \gamma^s \boldsymbol{\lambda} \cdot \nabla(\mathbf{n} \cdot \nabla\gamma) dv\end{aligned}$$

By Equation (41), we have

$$\begin{aligned}\int_C \nabla^c \circ F dc = &- \int_{V_\infty} (\boldsymbol{\lambda\lambda} \cdot \nabla + \boldsymbol{vv} \cdot \nabla + \mathbf{nn} \cdot \nabla) \cdot \boldsymbol{\lambda\lambda} \circ F\gamma^c \boldsymbol{v} \cdot \nabla^S \gamma^s \mathbf{n} \cdot \nabla\gamma dv \\ &- \int_{V_\infty} \boldsymbol{\lambda} \cdot \nabla^c \boldsymbol{\lambda} \circ F\gamma^c \boldsymbol{v} \cdot \nabla^S \gamma^s \mathbf{n} \cdot \nabla\gamma dv \\ &- \int_{\infty} \boldsymbol{\lambda} \circ F\boldsymbol{\lambda} \cdot \nabla\gamma^c \boldsymbol{v} \cdot \nabla^S \gamma^s \mathbf{n} \cdot \nabla\gamma dv \\ &- \int_{V_\infty} \boldsymbol{\lambda} \circ F\gamma^c \boldsymbol{\lambda} \cdot \big(\nabla\boldsymbol{v} \cdot \nabla^S \gamma^s + \nabla\nabla^S \gamma^s \cdot \boldsymbol{v}\big) \mathbf{n} \cdot \nabla\gamma dv \\ &- \int_{V_\infty} \boldsymbol{\lambda} \circ F\gamma^c \boldsymbol{v} \cdot \nabla^S \gamma^s \boldsymbol{\lambda} \cdot (\nabla\mathbf{n} \cdot \nabla\gamma + \nabla\nabla\gamma \cdot \mathbf{n}) dv\end{aligned}$$

By Equations (195), (200) and the definition of ∇^S, we have

$$\begin{aligned}\int_C \nabla^c \circ F dc = &- \int_{V_\infty} \boldsymbol{v} \cdot \nabla\boldsymbol{\lambda} \cdot \boldsymbol{v\lambda} \circ F\gamma^c \boldsymbol{v} \cdot \nabla^S \gamma^s \mathbf{n} \cdot \nabla\gamma dv \\ &- \int_{V_\infty} \mathbf{n} \cdot \nabla\boldsymbol{\lambda} \cdot \mathbf{n}\boldsymbol{\lambda} \circ F\gamma^c \boldsymbol{v} \cdot \nabla^S \gamma^s \mathbf{n} \cdot \nabla\gamma dv \\ &- \int_{V_\infty} \boldsymbol{\lambda} \cdot \nabla^c \boldsymbol{\lambda} \circ F\gamma^c \boldsymbol{v} \cdot \nabla^S \gamma^s \mathbf{n} \cdot \nabla\gamma dv \\ &- \int_{V_\infty} \boldsymbol{\lambda} \circ F\boldsymbol{\lambda} \cdot \nabla\gamma^c \boldsymbol{v} \cdot \nabla^S \gamma^s \mathbf{n} \cdot \nabla\gamma dv \\ &- \int_{V_\infty} \boldsymbol{\lambda} \circ F\gamma^c \boldsymbol{\lambda} \cdot \nabla\nabla^S \gamma^s \cdot \boldsymbol{v}\mathbf{n} \cdot \nabla\gamma dv \\ &- \int_{V_\infty} \boldsymbol{\lambda} \circ F\gamma^c \boldsymbol{v} \cdot \nabla^S \gamma^s \boldsymbol{\lambda} \cdot \nabla\nabla\gamma \cdot \mathbf{n} dv\end{aligned} \tag{463}$$

The last two terms of the right side of Equation (463) can be rearranged into

$$\int_{V_\infty} \boldsymbol{\lambda} \circ F\gamma^{c}\boldsymbol{\lambda} \cdot \nabla\nabla^{S}\gamma^{s} \cdot \boldsymbol{\nu}\mathbf{n} \cdot \nabla\gamma dv = \int_{V_\infty} F \circ \nabla^{c}\nabla^{S}\gamma^{s} \cdot \boldsymbol{\nu}\mathbf{n} \cdot \nabla\gamma\gamma^{c} dv \tag{464}$$

$$\int_{V_\infty} \boldsymbol{\lambda} \circ F\gamma^{c}\boldsymbol{\lambda} \cdot \nabla\nabla\gamma \cdot \boldsymbol{n}\boldsymbol{\nu} \cdot \nabla^{S}\gamma^{s} dv = \int_{V_\infty} F \circ \nabla^{c}\nabla\gamma \cdot \mathbf{n}\gamma^{c}\boldsymbol{\nu} \cdot \nabla^{S}\gamma^{s} dv \tag{465}$$

Equations (216), (258) and (261) allow us to write Equations (464) and (465) as

$$\int_{V_\infty} \boldsymbol{\lambda} \circ F\gamma^{c}\boldsymbol{\lambda} \cdot \nabla\nabla^{S}\gamma^{s} \cdot \boldsymbol{\nu}\mathbf{n} \cdot \nabla\gamma dv = -\int_{V_\infty} F \circ \boldsymbol{\lambda}\boldsymbol{\nu} \cdot \nabla\boldsymbol{\lambda} \cdot \nabla\gamma^{s}\mathbf{n} \cdot \nabla\gamma\gamma^{c} dv \tag{466}$$

$$\int_{V_\infty} \boldsymbol{\lambda} \circ F\gamma^{c}\boldsymbol{\nu} \cdot \nabla^{S}\gamma^{s}\boldsymbol{\lambda} \cdot \nabla\nabla\gamma \cdot \mathbf{n} dv = -\int_{V_\infty} F \circ \boldsymbol{\lambda}\mathbf{n} \cdot \nabla\boldsymbol{\lambda} \cdot \nabla\gamma\gamma^{c}\boldsymbol{\nu} \cdot \nabla^{S}\gamma^{s} dv \tag{467}$$

Substituting Equations (466) and (467) into Equation (463) yields

$$\begin{aligned}
\int_C \nabla^{c} \circ F dc = & -\int_{V_\infty} \boldsymbol{\nu} \cdot \nabla\boldsymbol{\lambda} \cdot \boldsymbol{\nu}\boldsymbol{\lambda} \circ F\gamma^{c}\boldsymbol{\nu} \cdot \nabla^{S}\gamma^{s}\mathbf{n} \cdot \nabla\gamma dv \\
& -\int_{V_\infty} \mathbf{n} \cdot \nabla\boldsymbol{\lambda} \cdot \mathbf{n}\boldsymbol{\lambda} \circ F\gamma^{c}\boldsymbol{\nu} \cdot \nabla^{S}\gamma^{s}\mathbf{n} \cdot \nabla\gamma dv \\
& -\int_{V_\infty} \boldsymbol{\lambda} \cdot \nabla^{c}\boldsymbol{\lambda} \circ F\gamma^{c}\boldsymbol{\nu} \cdot \nabla^{S}\gamma^{s}\mathbf{n} \cdot \nabla\gamma dv \\
& -\int_{V_\infty} \boldsymbol{\lambda} \circ F\boldsymbol{\lambda} \cdot \nabla\gamma^{c}\boldsymbol{\nu} \cdot \nabla^{S}\gamma^{s}\mathbf{n} \cdot \nabla\gamma dv \\
& +\int_{V_\infty} \boldsymbol{\lambda} \circ F\boldsymbol{\nu} \cdot \nabla\boldsymbol{\lambda} \cdot \nabla\gamma^{s}\mathbf{n} \cdot \nabla\gamma\gamma^{c} dv \\
& +\int_{V_\infty} \boldsymbol{\lambda} \circ F\mathbf{n} \cdot \nabla\boldsymbol{\lambda} \cdot \nabla\gamma\gamma^{c}\boldsymbol{\nu} \cdot \nabla^{S}\gamma^{s} dv \\
= & -\int_C (\boldsymbol{\lambda} \cdot \nabla^{c}\boldsymbol{\lambda}) \circ F dc + (\mathbf{e} \circ F)|_{\text{ends}}
\end{aligned} \tag{468}$$

in which we have used Equations (165)–(167) and (275). By substituting Equation (468) into Equation (469), we finally obtain

$$\int_C \nabla \circ F dc = \int_C \nabla^{S} \circ F dc - \int_C (\boldsymbol{\lambda} \cdot \nabla^{c}\boldsymbol{\lambda}) \circ F dc + (\mathbf{e} \circ F)|_{\text{ends}}$$

Remark. Equation (459) contains the following three formulas:

Gradient

$$\int_C \nabla f dc = \int_C \nabla^{S} f dc - \int_C (\boldsymbol{\lambda} \cdot \nabla^{c}\boldsymbol{\lambda}) f dc + \mathbf{e} f|_{\text{ends}} \tag{469}$$

Divergence

$$\int_C \nabla \cdot \mathbf{f}dc = \int_C \nabla^S \cdot \mathbf{f}dc + \int_C (\nabla^S \cdot \boldsymbol{\lambda})\boldsymbol{\lambda} \cdot \mathbf{f}dc - \int_C (\boldsymbol{\lambda} \cdot \nabla^c \boldsymbol{\lambda}) \cdot \mathbf{f}dc + (\mathbf{e} \cdot \mathbf{f})|_{\text{ends}} \quad (470)$$

Curl

$$\int_C \nabla \times \mathbf{f}dc = \int_C \nabla^S \times \mathbf{f}dc - \int_C (\boldsymbol{\lambda} \cdot \nabla^c \boldsymbol{\lambda}) \times \mathbf{f}dc + (\mathbf{e} \times \mathbf{f})|_{\text{ends}} \quad (471)$$

4.3.2 Integration of spatial operators over a straight line fixed in space

This group of integration theorems involves integrations of $(\partial f/\partial t)|_{\mathbf{x}}$, ∇f, $\nabla \cdot \mathbf{f}$ and $\nabla \times \mathbf{f}$ over a straight line L fixed in space (Figure 22). Here $f(\mathbf{x}, t)$ and $\mathbf{f}(\mathbf{x}, t)$ are microscale scalar and vector functions of position vector $\mathbf{x}$ and time t.

In this group of integration theorems, L is the straight line segment of integration. Λ is the unit vector tangent to segment L and $\mathbf{e}$ the particular unit vector which is tangent to the end points of segment L with positive outward direction. $\mathbf{n}^*$ is the unit vector at the ends of the L which is normal to the surface of a volume that is pierced by the straight line. ∇^S is the surficial operator such that $\nabla^S = \nabla - \Lambda\Lambda \cdot \nabla$, $\mathbf{w}$ the velocity of the endpoints of the line L such that $\mathbf{w} = \Lambda\Lambda \cdot \mathbf{w}$, $\mathbf{f}^c$ a vector tangent to L such that $\mathbf{f}^c = \Lambda\Lambda \cdot \mathbf{f}$ and $\mathbf{f}^s$ a vector normal to L such that $\mathbf{f}^s = \mathbf{f} - \mathbf{f}^c$.

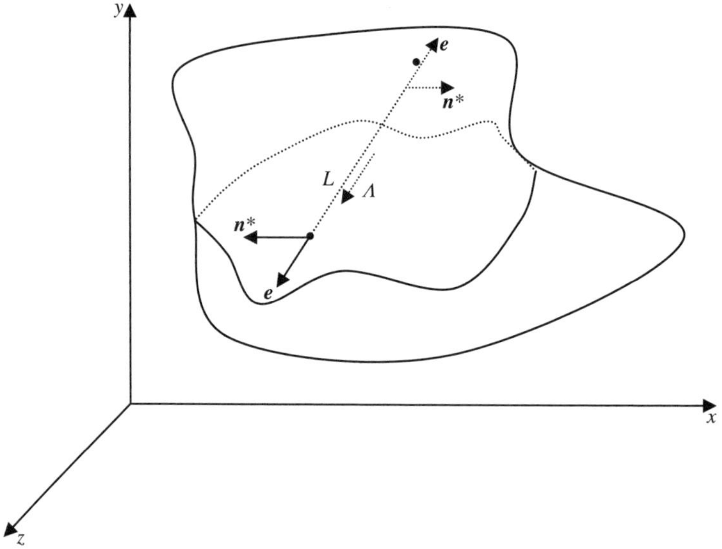

Figure 22 A straight line segment L whose endpoints are the points of intersection of the line with the surface of a volume ($\mathbf{n}^*$ *is* the outward unit normal of the surface; Λ unit tangent of L; and $\mathbf{e}$ unit vector tangent to L and pointing outward from L at the endpoints of L)

Theorem 23.

$$\int_L \left.\frac{\partial f}{\partial t}\right|_{\mathbf{x}} \mathrm{d}L = \frac{\partial}{\partial t}\int_L f\mathrm{d}L - \left.\left(\frac{\mathbf{n}^* \cdot \mathbf{w}}{\mathbf{e} \cdot \mathbf{n}^*} f\right)\right|_{\text{ends}} \tag{472}$$

Proof. By defining an indicator function $\gamma(\mathbf{x})$ such that $\gamma = 1$ within V and $\gamma = 0$ outside of V, we obtain

$$\int_L \left.\frac{\partial f}{\partial t}\right|_{\mathbf{x}} \mathrm{d}L = \int_{L_\infty} \left.\frac{\partial f}{\partial t}\right|_{\mathbf{x}} \gamma \mathrm{d}L \tag{473}$$

An application of the chain rule to the right side of Equation (473) yields

$$\begin{aligned}\int_L \left.\frac{\partial f}{\partial t}\right|_{\mathbf{x}} \mathrm{d}L &= \int_{L_\infty} \left.\frac{\partial (f\gamma)}{\partial t}\right|_{\mathbf{x}} \mathrm{d}L - \int_{L_\infty} f\left.\frac{\partial \gamma}{\partial t}\right|_{\mathbf{x}} \mathrm{d}L \\ &= \frac{\partial}{\partial t}\int_L f\mathrm{d}L + \int_{L_\infty} f\mathbf{w}\cdot\nabla\gamma\mathrm{d}L\end{aligned} \tag{474}$$

in which we have used Equation (180). By Equation (304), we have

$$\int_L \left.\frac{\partial f}{\partial t}\right|_{\mathbf{x}} \mathrm{d}L = \frac{\partial}{\partial t}\int_C f\mathrm{d}L - \left.\left(\frac{\mathbf{n}^* \cdot \mathbf{w}f}{\mathbf{e} \cdot \mathbf{n}^*}\right)\right|_{\text{ends}}$$

Theorem 24.

$$\int_L \nabla f \mathrm{d}L = \nabla^{\mathrm{S}} \int_L f\,\mathrm{d}L + \left.\left(\frac{\mathbf{n}^* f}{\mathbf{e} \cdot \mathbf{n}^*}\right)\right|_{\text{ends}} \tag{475}$$

Proof. Define an indicator function $\gamma(\mathbf{x})$ such that $\gamma = 1$ within V and $\gamma = 0$ outside of V. Therefore

$$\int_L \nabla f \mathrm{d}L = \int_{L_\infty} \nabla f \gamma \mathrm{d}L \tag{476}$$

An application of the chain rule to the right side of Equation (476) yields

$$\begin{aligned}\int_L \nabla f \mathrm{d}L &= \int_{L_\infty} \nabla(f\gamma)\mathrm{d}L - \int_{L_\infty} f\nabla\gamma\mathrm{d}L \\ &= \int_{L_\infty} \nabla^{\mathrm{S}}(f\gamma)\mathrm{d}L + \int_{L_\infty} \nabla^{\mathrm{c}}(f\gamma)\mathrm{d}L - \int_{L_\infty} f\nabla\gamma\mathrm{d}L \\ &= \nabla^{\mathrm{S}}\int_L f\mathrm{d}L - \int_{L_\infty} f\nabla\gamma\mathrm{d}L\end{aligned} \tag{477}$$

in which we have used Equation (246) and $\nabla^{\mathrm{S}} + \nabla^{\mathrm{c}} = \nabla$. By Equation (308), we obtain

$$\int_L \nabla f \mathrm{d}L = \nabla^{\mathrm{S}} \int_L f\mathrm{d}L + \left.\left(\frac{\mathbf{n}^* f}{\mathbf{e} \cdot \mathbf{n}^*}\right)\right|_{\text{ends}}$$

Theorem 25.

$$\int_L \nabla\cdot\mathbf{f}dL = \nabla^S\cdot\int_L f^s dL + \int_L \nabla^S\cdot\Lambda\Lambda\cdot\mathbf{f}^c dL + \left(\frac{\mathbf{n}^*\cdot\mathbf{f}}{\mathbf{e}\cdot\mathbf{n}^*}\right)\Bigg|_{\text{ends}} \tag{478}$$

Proof. By Equation (72), we have

$$\nabla\cdot\mathbf{f} = \nabla^S\cdot\mathbf{f}^s + \nabla^c\cdot\mathbf{f}^c + \nabla\cdot\Lambda\mathbf{f}\cdot\Lambda - \Lambda\cdot\nabla\Lambda\cdot\mathbf{f} \tag{479}$$

Thus

$$\int_L \nabla\cdot\mathbf{f}dL = \int_L \nabla^S\cdot\mathbf{f}^s dL + \int_L \nabla^c\cdot\mathbf{f}^c dL + \int_L \nabla\cdot\Lambda\mathbf{f}\cdot\Lambda dL - \int_L \Lambda\cdot\nabla\Lambda\cdot\mathbf{f}dL \tag{480}$$

Note that

$$\nabla^c\cdot\Lambda = \Lambda(\Lambda\cdot\nabla)\cdot\Lambda = \Lambda\cdot\nabla\Lambda\cdot\Lambda = 0$$

Therefore

$$\nabla\cdot\Lambda = \nabla^S\cdot\Lambda \tag{481}$$

Also,

$$\Lambda\cdot\nabla\Lambda = \Lambda\cdot\nabla^c\Lambda \tag{482}$$

Along the segment L, Λ is a constant vector so that

$$\Lambda\cdot\nabla^c\Lambda = 0 \tag{483}$$

Equations (481)–(483) allow us to rewrite Equation (480) as

$$\int_L \nabla\cdot\mathbf{f}dL = \int_L \nabla^S\cdot\mathbf{f}^s dL + \int_L \nabla^c\cdot\mathbf{f}^c dL + \int_L \nabla^S\cdot\Lambda\mathbf{f}^c\cdot\Lambda dL \tag{484}$$

Define an indicator function $\gamma(\mathbf{x})$ such that $\gamma = 1$ within V and $\gamma = 0$ outside of V. Hence

$$\int_L \nabla\cdot\mathbf{f}dL = \int_{L_\infty} \nabla^S\cdot\mathbf{f}^s\gamma dL + \int_{L_\infty} \nabla^c\cdot\mathbf{f}^c\gamma dL + \int_L \nabla^S\cdot\Lambda\mathbf{f}^c\cdot\Lambda dL \tag{485}$$

An application of the chain rule to the first two terms on the right side of Equation (485) yields

$$\begin{aligned}\int_L \nabla\cdot\mathbf{f}dL &= \int_{L_\infty} \nabla^S\cdot(\mathbf{f}^s\gamma)dL - \int_{L_\infty} \mathbf{f}^s\cdot\nabla^S\gamma dL + \int_{L_\infty} \nabla^c\cdot(\mathbf{f}^c\gamma)dL \\ &\quad - \int_{L_\infty} \mathbf{f}^c\cdot\nabla^c\gamma dL + \int_L \nabla^S\cdot\Lambda\mathbf{f}^c\cdot\Lambda dL \\ &= \nabla^S\cdot\int_{L_\infty} \mathbf{f}^s\gamma dL - \int_{L_\infty} \mathbf{f}^s\cdot\nabla\gamma dL - \int_{L_\infty} \mathbf{f}^c\cdot\nabla^c\gamma dL \\ &\quad + \int_L \nabla^S\cdot\Lambda\mathbf{f}^c\cdot\Lambda dL\end{aligned} \tag{486}$$

In deriving Equation (486), Equation (247) and

$$\mathbf{f}^s\cdot\nabla = \mathbf{f}^s\cdot\nabla^c + \mathbf{f}^s\cdot\nabla^S = \mathbf{f}^s\cdot\nabla^S$$

$$\mathbf{f}^c\cdot\nabla = \mathbf{f}^c\cdot\nabla^S + \mathbf{f}\cdot\nabla^c = \mathbf{f}^c\cdot\nabla^c$$

have been used. We have also used the fact that γ has the same role as γ^c. Rearranging Equation (486) yields

$$\int_L \nabla \cdot \mathbf{f} dL = \nabla^S \cdot \int_L \mathbf{f}^s dL - \int_{L_\infty} \mathbf{f} \cdot \nabla \gamma dL + \int_L \nabla^c \cdot \boldsymbol{\Lambda} \mathbf{f}^c \cdot \boldsymbol{\Lambda} dL \tag{487}$$

By using Equation (304) and noting that $\nabla^S \cdot \boldsymbol{\Lambda} = \nabla^c \cdot \boldsymbol{\Lambda}$, we obtain

$$\int_L \nabla \cdot \mathbf{f} dL = \nabla^S \cdot \int_L \mathbf{f}^s dL + \int_L \nabla^S \cdot \boldsymbol{\Lambda}\boldsymbol{\Lambda} \cdot \mathbf{f}^c dL + \left.\left(\frac{\mathbf{f} \cdot \mathbf{n}^*}{\mathbf{e} \cdot \mathbf{n}^*}\right)\right|_{\text{ends}}$$

Theorem 26.

$$\int_L \nabla \times \mathbf{f} dL = \nabla^S \times \int_L f dL + \left.\left(\frac{\mathbf{n}^* \times \mathbf{f}}{\mathbf{e} \cdot \mathbf{n}^*}\right)\right|_{\text{ends}} \tag{488}$$

Proof. Defining an indicator function $\gamma(\mathbf{x})$ such that $\gamma = 1$ within V and $\gamma = 0$ outside of V, we obtain

$$\int_L \nabla \times \mathbf{f} dL = \int_{L_\infty} \nabla \times \mathbf{f}\gamma dL \tag{489}$$

An application of the chain rule to the right side of Equation (489) yields

$$\begin{aligned} \int_L \nabla \times \mathbf{f} dL &= \int_{L_\infty} \nabla \times (\mathbf{f}\gamma) dL - \int_{L_\infty} \nabla\gamma \times \mathbf{f} dL \\ &= \int_{L_\infty} \nabla^S \times (\mathbf{f}\gamma) dL + \int_{L_\infty} \nabla^c \times (\mathbf{f}\gamma) dL - \int_{L_\infty} \nabla\gamma \times \mathbf{f} dL \\ &= \nabla^S \times \int_{L_\infty} \mathbf{f}\gamma dL - \int_{L_\infty} \nabla\gamma \times \mathbf{f} dL \end{aligned} \tag{490}$$

in which we have used Equation (248) and $\nabla = \nabla^S + \nabla^c$; the role of γ is also the same as that of γ^c. By Equation (310), we conclude

$$\int_L \nabla \times \mathbf{f} dL = \nabla^S \times \int_L \mathbf{f} dL + \left.\left(\frac{\mathbf{n}^* \times \mathbf{f}}{\mathbf{e} \cdot \mathbf{n}^*}\right)\right|_{\text{ends}}$$

4.3.3 Curvilineal integration of surficial operators

This group of integration theorems involves integrations of $\int_V (\partial f/\partial t)|_{\mathbf{u}}$, $\nabla^S f$, $\nabla^S \cdot \mathbf{f}$, and $\nabla^S \times \mathbf{f}$ over a curve C (Figure 23). Here $f(\mathbf{x}, t)$ and $\mathbf{f}(\mathbf{x}, t)$ are microscale scalar and vector functions of position vector $\mathbf{x}$ and time t. Generally, the curve C can deform and translate with time.

In these theorems C is the curve of integration. $\boldsymbol{\lambda}$ is the unit vector tangent to C and $\mathbf{e}$ the particular unit vector that is tangent to C at its endpoints with positive outward direction. $\mathbf{n}$ is the unit vector normal to the surface containing curve C and $\boldsymbol{\nu}$ the unit vector normal to C and tangent to the surface containing C. ∇^S is the surficial operator such that $\nabla^S = \nabla - \mathbf{nn} \cdot \nabla$ and ∇^ν is the curvilineal operator along the curve normal to C and in the surface containing C such that $\nabla^\nu = \nabla^S - \boldsymbol{\lambda\lambda} \cdot \nabla^S$. $\mathbf{w}$ is the velocity of C, $\mathbf{w}^s$ the velocity of C in the $\boldsymbol{\nu}$ direction

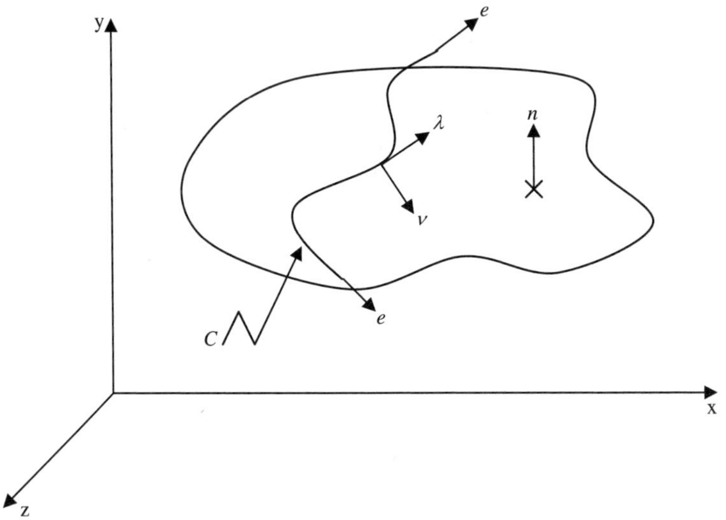

Figure 23 Simple integration curve of C liying in a surface (λ is unit tangent of C; **ν** unit normal of C and tangent to the surface; **n** unit normal of the surface; and **e** unit vector tangent to C and pointing outward from C at the endpoints of C).

such that $\mathbf{w}^s = \boldsymbol{\nu}\boldsymbol{\nu}\cdot\mathbf{w}$ and $\mathbf{w}^\lambda$ is the velocity of C in the $\boldsymbol{\lambda}$ direction such that $\mathbf{w}^\lambda = \boldsymbol{\lambda}\boldsymbol{\lambda}\cdot\mathbf{w}$. $\mathbf{f}^n$ is a vector normal to C and to the surface containing C such that $\mathbf{f}^n = \mathbf{nn}\cdot\mathbf{f}$, $\mathbf{f}^s$ is a surficial vector such that $\mathbf{f}^s = \mathbf{f} - \mathbf{f}^n$, $\mathbf{f}^\lambda$ is a vector tangent to C such that $\mathbf{f}^\lambda = \boldsymbol{\lambda}\boldsymbol{\lambda}\cdot\mathbf{f}$ and $\mathbf{f}^\nu$ is a vector normal to C such that $\mathbf{f}^\nu = \boldsymbol{\nu}\boldsymbol{\nu}\cdot\mathbf{f}$.

Theorem 27.

$$\int_C \nabla^S f dc = \int_C \nabla^\nu f dc - \int_C \boldsymbol{\lambda}\cdot\nabla\boldsymbol{\lambda} f dc + (\mathbf{e}f)|_{\text{ends}} \tag{491}$$

Proof. Extend the surface S to form a closed surface S_∞. Denote the volume enclosed by S_∞ as V and the whole three-dimensional space as V_∞.

Define three indicator functions $\gamma^c(\mathbf{l})$, $\gamma^s(\mathbf{u})$ and $\gamma(\mathbf{x})$ such that $\gamma^c = 1$ on the curve C and $\gamma^c = 0$ elsewhere, $\gamma^s = 1$ on the surface S and $\gamma^s = 0$ elsewhere and $\gamma = 1$ within V but $\gamma = 0$ outside of V. Therefore

$$\begin{aligned}\int_C \nabla^S f dc &= \int_C \nabla^\nu f dc + \int_C \nabla^\lambda f dc \\ &= \int_C \nabla^\nu f dc + \int_{V_\infty} \nabla^\lambda f \gamma^c \boldsymbol{\nu}\cdot\nabla^S\gamma^s\mathbf{n}\cdot\nabla\gamma dv\end{aligned} \tag{492}$$

in which we used $\nabla^S = \nabla^\nu + \nabla^\lambda$. An application of the chain rule in the second term yields

$$\begin{aligned}
&\int_{V_\infty} \boldsymbol{\lambda\lambda}\cdot\nabla f\gamma^{\mathrm{c}}\boldsymbol{v}\cdot\nabla^{\mathrm{S}}\gamma^{\mathrm{s}}\mathbf{n}\cdot\nabla\gamma dv\\
&\quad=\int_{V_\infty}\nabla\cdot(\boldsymbol{\lambda\lambda}f\gamma^{\mathrm{c}}\boldsymbol{v}\cdot\nabla^{\mathrm{S}}\gamma^{\mathrm{s}}\mathbf{n}\cdot\nabla\gamma)dv-\int_{V_\infty}\nabla\cdot\boldsymbol{\lambda\lambda}f\gamma^{\mathrm{c}}\boldsymbol{v}\cdot\nabla^{\mathrm{S}}\gamma^{\mathrm{s}}\mathbf{n}\cdot\nabla\gamma dv\\
&\qquad-\int_{V_\infty}\boldsymbol{\lambda}\cdot\nabla\boldsymbol{\lambda}f\gamma^{\mathrm{c}}\boldsymbol{v}\cdot\nabla^{\mathrm{S}}\gamma^{\mathrm{s}}\mathbf{n}\cdot\nabla\gamma dv-\int_{V_\infty}\boldsymbol{\lambda}f\boldsymbol{\lambda}\cdot\nabla\gamma^{\mathrm{c}}\boldsymbol{v}\cdot\nabla^{\mathrm{S}}\gamma^{\mathrm{s}}\mathbf{n}\cdot\nabla\gamma dv\\
&\qquad-\int_{V_\infty}\boldsymbol{\lambda}f\gamma^{\mathrm{c}}\boldsymbol{\lambda}\cdot\nabla(\boldsymbol{v}\cdot\nabla^{\mathrm{S}}\gamma^{\mathrm{s}})\mathbf{n}\cdot\nabla\gamma dv-\int_{V_\infty}\boldsymbol{\lambda}f\gamma^{\mathrm{c}}\boldsymbol{v}\cdot\nabla^{\mathrm{S}}\gamma^{\mathrm{s}}\boldsymbol{\lambda}\cdot\nabla(\mathbf{n}\cdot\nabla\gamma)dv
\end{aligned}$$

By Equations (41) and (231), we have

$$\begin{aligned}
&\int_{V_\infty} \boldsymbol{\lambda\lambda}\cdot\nabla f\gamma^{\mathrm{c}}\boldsymbol{v}\cdot\nabla^{\mathrm{S}}\gamma^{\mathrm{s}}\mathbf{n}\cdot\nabla\gamma dv\\
&\quad=-\int_{V_\infty}\boldsymbol{v}\cdot\nabla\boldsymbol{\lambda}\cdot\boldsymbol{v\lambda}f\gamma^{\mathrm{c}}\boldsymbol{v}\cdot\nabla^{\mathrm{S}}\gamma^{\mathrm{s}}\mathbf{n}\cdot\nabla\gamma dv-\int_{V_\infty}\boldsymbol{n}\cdot\nabla\boldsymbol{\lambda}\cdot\boldsymbol{n\lambda}f\gamma^{\mathrm{c}}\boldsymbol{v}\cdot\nabla^{\mathrm{S}}\gamma^{\mathrm{s}}\mathbf{n}\cdot\nabla\gamma dv\\
&\qquad-\int_{V_\infty}\boldsymbol{\lambda}\cdot\nabla\boldsymbol{\lambda}f\gamma^{\mathrm{c}}\boldsymbol{v}\cdot\nabla^{\mathrm{S}}\gamma^{\mathrm{s}}\mathbf{n}\cdot\nabla\gamma dv-\int_{V_\infty}\boldsymbol{\lambda}f\boldsymbol{\lambda}\cdot\nabla\gamma^{\mathrm{c}}\boldsymbol{v}\cdot\nabla^{\mathrm{S}}\gamma^{\mathrm{s}}\mathbf{n}\cdot\nabla\gamma dv\\
&\qquad-\int_{V_\infty}\boldsymbol{\lambda}f\gamma^{\mathrm{c}}\boldsymbol{\lambda}\cdot(\nabla\boldsymbol{v}\cdot\nabla^{\mathrm{S}}\gamma^{\mathrm{s}}+\nabla\nabla^{\mathrm{S}}\gamma^{\mathrm{s}}\cdot\boldsymbol{v})\mathbf{n}\cdot\nabla\gamma dv\\
&\qquad-\int_{V_\infty}\boldsymbol{\lambda}f\gamma^{\mathrm{c}}\boldsymbol{v}\cdot\nabla^{\mathrm{S}}\gamma^{\mathrm{s}}\boldsymbol{\lambda}\cdot(\nabla\mathbf{n}\cdot\nabla\gamma+\nabla\nabla\gamma\cdot\mathbf{n})dv
\end{aligned}$$

By Equations (195) and (200),

$$\begin{aligned}
&\int_{V_\infty} \boldsymbol{\lambda\lambda}\cdot\nabla f\gamma^{\mathrm{c}}\boldsymbol{v}\cdot\nabla^{\mathrm{S}}\gamma^{\mathrm{s}}\mathbf{n}\cdot\nabla\gamma dv\\
&\quad=-\int_{V_\infty}\boldsymbol{v}\cdot\nabla\boldsymbol{\lambda}\cdot\boldsymbol{v\lambda}f\gamma^{\mathrm{c}}\boldsymbol{v}\cdot\nabla^{\mathrm{S}}\gamma^{\mathrm{s}}\mathbf{n}\cdot\nabla\gamma dv-\int_{V_\infty}\boldsymbol{n}\cdot\nabla\boldsymbol{\lambda}\cdot\boldsymbol{n\lambda}f\gamma^{\mathrm{c}}\boldsymbol{v}\cdot\nabla^{\mathrm{S}}\gamma^{\mathrm{s}}\mathbf{n}\cdot\nabla\gamma dv\\
&\qquad-\int_{V_\infty}\boldsymbol{\lambda}\cdot\nabla\boldsymbol{\lambda}f\gamma^{\mathrm{c}}\boldsymbol{v}\cdot\nabla^{\mathrm{S}}\gamma^{\mathrm{s}}\mathbf{n}\cdot\nabla\gamma dv-\int_{V_\infty}\boldsymbol{\lambda}f\boldsymbol{\lambda}\cdot\nabla\gamma^{\mathrm{c}}\boldsymbol{v}\cdot\nabla^{\mathrm{S}}\gamma^{\mathrm{s}}\mathbf{n}\cdot\nabla\gamma dv\\
&\qquad-\int_{V_\infty}\boldsymbol{\lambda}f\gamma^{\mathrm{c}}\boldsymbol{\lambda}\cdot\nabla\nabla^{\mathrm{S}}\gamma^{\mathrm{s}}\cdot\boldsymbol{v}\mathbf{n}\cdot\nabla\gamma dv-\int_{V_\infty}\boldsymbol{\lambda}f\gamma^{\mathrm{c}}\boldsymbol{v}\cdot\nabla^{\mathrm{S}}\gamma^{\mathrm{s}}\boldsymbol{\lambda}\cdot\nabla\nabla\gamma\cdot\mathbf{n}dv
\end{aligned}\tag{493}$$

Equations (216), (309) and (261) allow us to rewrite Equation (493) as

$$\begin{aligned}
&\int_{V_\infty} \boldsymbol{\lambda\lambda}\cdot\nabla f\gamma^{\mathrm{c}}\boldsymbol{v}\cdot\nabla^{\mathrm{S}}\gamma^{\mathrm{s}}\mathbf{n}\cdot\nabla\gamma dv\\
&\quad=-\int_{V_\infty}\boldsymbol{v}\cdot\nabla\boldsymbol{\lambda}\cdot\boldsymbol{v\lambda}f\gamma^{\mathrm{c}}\boldsymbol{v}\cdot\nabla^{\mathrm{S}}\gamma^{\mathrm{s}}\mathbf{n}\cdot\nabla\gamma dv-\int_{V_\infty}\boldsymbol{n}\cdot\nabla\boldsymbol{\lambda}\cdot\boldsymbol{n\lambda}f\gamma^{\mathrm{c}}\boldsymbol{v}\cdot\nabla^{\mathrm{S}}\gamma^{\mathrm{s}}\mathbf{n}\cdot\nabla\gamma dv\\
&\qquad-\int_{V_\infty}\boldsymbol{\lambda}\cdot\nabla\boldsymbol{\lambda}f\gamma^{\mathrm{c}}\boldsymbol{v}\cdot\nabla^{\mathrm{S}}\gamma^{\mathrm{s}}\mathbf{n}\cdot\nabla\gamma dv-\int_{V_\infty}\boldsymbol{\lambda}f\boldsymbol{\lambda}\cdot\nabla\gamma^{\mathrm{c}}\boldsymbol{v}\cdot\nabla^{\mathrm{S}}\gamma^{\mathrm{s}}\mathbf{n}\cdot\nabla\gamma dv\\
&\qquad+\int_{V_\infty}f\gamma^{\mathrm{c}}\boldsymbol{\lambda v}\cdot\nabla\boldsymbol{\lambda}\cdot\nabla\gamma^{\mathrm{s}}\mathbf{n}\cdot\nabla\gamma dv+\int_{V_\infty}\boldsymbol{\lambda}f\gamma^{\mathrm{c}}\mathbf{n}\cdot\nabla\boldsymbol{\lambda}\cdot\nabla\gamma\boldsymbol{v}\cdot\nabla^{\mathrm{S}}\gamma^{\mathrm{s}}dv\\
&\quad=-\int_{V_\infty}\boldsymbol{\lambda}\cdot\nabla\boldsymbol{\lambda}f\gamma^{\mathrm{c}}\boldsymbol{v}\cdot\nabla^{\mathrm{S}}\gamma^{\mathrm{s}}\mathbf{n}\cdot\nabla\gamma dv-\int_{V_\infty}\boldsymbol{\lambda}f\boldsymbol{\lambda}\cdot\nabla\gamma^{\mathrm{c}}\boldsymbol{v}\cdot\nabla^{\mathrm{S}}\gamma^{\mathrm{s}}\mathbf{n}\cdot\nabla\gamma dv
\end{aligned}\tag{494}$$

By Equations (165) and (275),

$$\int_{V_\infty} \boldsymbol{\lambda\lambda} \cdot \nabla f \gamma^c \boldsymbol{v} \cdot \nabla^S \gamma^s \mathbf{n} \cdot \nabla \gamma dv = -\int_C \boldsymbol{\lambda} \cdot \nabla \boldsymbol{\lambda} f dc + (\mathbf{e}f)|_{\text{ends}} \tag{495}$$

Substituting Equation (494) into Equation (492) finally leads to

$$\int_C \nabla^S f dc = \int_C \nabla^v f dc - \int_C \boldsymbol{\lambda} \cdot \nabla \boldsymbol{\lambda} f dc + (\mathbf{e}f)|_{\text{ends}} \tag{496}$$

Theorem 28.

$$\int_C \nabla^S \cdot \mathbf{f} dc = \int_C \nabla^v \cdot \mathbf{f}^v dc - \int_C (\boldsymbol{\lambda} \cdot \nabla \boldsymbol{\lambda}) \cdot \mathbf{f} dc - \int_C (\boldsymbol{v} \cdot \nabla \boldsymbol{v}) \cdot \mathbf{f} dc + (\mathbf{e} \cdot \mathbf{f}^\lambda)|_{\text{ends}} \tag{497}$$

Proof. By $\nabla^S = \nabla^v + \nabla^\lambda$, $\mathbf{f} = \mathbf{f}^\lambda + \mathbf{f}^v + \mathbf{f}^n$, we obtain

$$\begin{aligned}
\int_C \nabla^S \cdot \mathbf{f} dc &= \int_C \nabla^v \cdot \mathbf{f} dc + \int_C \nabla^\lambda \cdot \mathbf{f} dc \\
&= \int_C \nabla^v \cdot \mathbf{f}^v dc + \int_C \nabla^v \cdot \mathbf{f}^\lambda dc + \int_C \nabla^v \cdot \mathbf{f}^n dc + \int_C \nabla^\lambda \cdot \mathbf{f} dc \\
&= \int_C \nabla^v \cdot \mathbf{f}^v dc + \int_C (\boldsymbol{vv} \cdot \nabla) \cdot \mathbf{f}^\lambda dc \\
&\quad + \int_C (\boldsymbol{vv} \cdot \nabla) \cdot \mathbf{f}^n dc + \int_C \nabla^\lambda \cdot \mathbf{f} dc \\
&= \int_C \nabla^v \cdot \mathbf{f}^v dc + \int_C (\boldsymbol{vv} \cdot \nabla) \cdot (\boldsymbol{\lambda} \mathbf{f} \cdot \boldsymbol{\lambda}) dc \\
&\quad + \int_C (\boldsymbol{vv} \cdot \nabla) \cdot (\mathbf{nf} \cdot \mathbf{n}) dc + \int_C \nabla^\lambda \cdot \mathbf{f} dc \\
&= \int_C \nabla^v \cdot \mathbf{f}^v dc + \int_C \boldsymbol{v} \cdot \nabla \boldsymbol{\lambda} \cdot \boldsymbol{v} \mathbf{f} \cdot \boldsymbol{\lambda} + \int_C \boldsymbol{v} \cdot \boldsymbol{\lambda v} \cdot \nabla (\mathbf{f} \cdot \boldsymbol{\lambda}) dc \\
&\quad + \int_C \boldsymbol{v} \cdot \nabla \mathbf{n} \cdot \boldsymbol{v} \mathbf{f} \cdot \mathbf{n} dc + \int_C \boldsymbol{v} \cdot \mathbf{n} \boldsymbol{v} \cdot \nabla (\mathbf{f} \cdot \mathbf{n}) dc + \int_C \nabla^\lambda \cdot \mathbf{f} dc \\
&= \int_C \nabla^v \cdot \mathbf{f}^v dc - \int_C \boldsymbol{v} \cdot \nabla \boldsymbol{v} \cdot \boldsymbol{\lambda} \mathbf{f} \cdot \boldsymbol{\lambda} dc \\
&\quad - \int_C \boldsymbol{v} \cdot \nabla \boldsymbol{v} \cdot \mathbf{nf} \cdot \mathbf{n} dc + \int_C \nabla^\lambda \cdot \mathbf{f} dc \\
&= \int_C \nabla^v \cdot \mathbf{f}^v - \int_C \boldsymbol{v} \cdot \nabla \boldsymbol{v} \cdot (\boldsymbol{\lambda} \mathbf{f} \cdot \boldsymbol{\lambda} + \mathbf{nf} \cdot \mathbf{n}) dc + \int_C \nabla^\lambda \cdot \mathbf{f} dc \\
&= \int_C \nabla^v \cdot \mathbf{f}^v dc - \int_C \boldsymbol{v} \cdot \nabla \boldsymbol{v} \cdot \mathbf{f} dc + \int_C \nabla^\lambda \cdot \mathbf{f} dc
\end{aligned} \tag{498}$$

in which we have used Equations (16), (31), (23) and (24).

Define three indicator functions $\gamma^s(\mathbf{u})$, $\gamma^c(\mathbf{l})$ and $\gamma(\mathbf{x})$ such that $\gamma^c = 1$ on the curve C but $\gamma^c = 0$ elsewhere, $\gamma^s = 1$ on the surface S but $\gamma^s = 0$ elsewhere and $\gamma = 1$ within V but $\gamma = 0$ outside of V. The third term on the right side of

Equation (498) is thus

$$\int_C \nabla^\lambda \cdot \mathbf{f} dc = \int_{V_\infty} \nabla^\lambda \cdot \mathbf{f}\gamma^c \boldsymbol{v} \cdot \nabla^S \gamma^s \mathbf{n} \cdot \nabla\gamma dv$$
$$= \int_{V_\infty} (\boldsymbol{\lambda\lambda} \cdot \nabla) \cdot \mathbf{f}\gamma^c \boldsymbol{v} \cdot \nabla^S \gamma^s \mathbf{n} \cdot \nabla\gamma dv$$
$$= \int_{V_\infty} \boldsymbol{\lambda} \cdot \nabla \mathbf{f} \cdot \boldsymbol{\lambda}\gamma^c \boldsymbol{v} \cdot \nabla^S \gamma^s \mathbf{n} \cdot \nabla\gamma dv \tag{499}$$

An application of the chain rule to the right side of Equation (499) yields

$$\int_C \nabla^\lambda \cdot \mathbf{f} dc$$
$$= \int_{V_\infty} \nabla \cdot (\boldsymbol{\lambda\lambda} \cdot \mathbf{f}\gamma^c \boldsymbol{v} \cdot \nabla^S \gamma^s \mathbf{n} \cdot \nabla\gamma) dv - \int_{V_\infty} \nabla \cdot \boldsymbol{\lambda\lambda} \cdot \mathbf{f}\gamma^c \boldsymbol{v} \cdot \nabla^S \gamma^s \mathbf{n} \cdot \nabla\gamma dv$$
$$- \int_{V_\infty} \boldsymbol{\lambda} \cdot \nabla\boldsymbol{\lambda} \cdot \mathbf{f}\gamma^c \boldsymbol{v} \cdot \nabla^S \gamma^s \mathbf{n} \cdot \nabla\gamma dv - \int_{V_\infty} \boldsymbol{\lambda} \cdot \mathbf{f}\boldsymbol{\lambda} \cdot \nabla\gamma^c \boldsymbol{v} \cdot \nabla^S \gamma^s \mathbf{n} \cdot \nabla\gamma dv$$
$$- \int_{V_\infty} \boldsymbol{\lambda} \cdot \mathbf{f}\gamma^c \boldsymbol{\lambda} \cdot \nabla(\boldsymbol{v} \cdot \nabla^S \gamma^s) \mathbf{n} \cdot \nabla\gamma dv - \int_{V_\infty} \boldsymbol{\lambda} \cdot \mathbf{f}\gamma^c \boldsymbol{v} \cdot \nabla^S \gamma^s \boldsymbol{\lambda} \cdot \nabla(\mathbf{n} \cdot \nabla\gamma) dv$$

By Equations (41) and (231), we therefore obtain

$$\int_C \nabla^\lambda \cdot \mathbf{f} dc$$
$$= - \int_{V_\infty} (\boldsymbol{vv} \cdot \nabla) \cdot \boldsymbol{\lambda\lambda} \cdot \mathbf{f}\gamma^c \boldsymbol{v} \cdot \nabla^S \gamma^s \mathbf{n} \cdot \nabla\gamma dv - \int_{V_\infty} (\boldsymbol{\lambda\lambda} \cdot \nabla) \cdot \boldsymbol{\lambda\lambda} \cdot \mathbf{f}\gamma^c \boldsymbol{v} \cdot \nabla^S \gamma^s \mathbf{n} \cdot \nabla\gamma dv$$
$$- \int_{V_\infty} (\boldsymbol{nn} \cdot \nabla) \cdot \boldsymbol{\lambda\lambda} \cdot \mathbf{f}\gamma^c \boldsymbol{v} \cdot \nabla^S \gamma^s \mathbf{n} \cdot \nabla\gamma dv - \int_{V_\infty} \boldsymbol{\lambda} \cdot \nabla\boldsymbol{\lambda} \cdot \mathbf{f}\gamma^c \boldsymbol{v} \cdot \nabla^S \gamma^s \mathbf{n} \cdot \nabla\gamma dv$$
$$- \int_{V_\infty} \boldsymbol{\lambda} \cdot \mathbf{f}\boldsymbol{\lambda} \cdot \nabla\gamma^c \boldsymbol{v} \cdot \nabla^S \gamma^s \mathbf{n} \cdot \nabla\gamma dv - \int_{V_\infty} \boldsymbol{\lambda} \cdot \mathbf{f}\gamma^c \boldsymbol{\lambda} \cdot (\nabla\boldsymbol{v} \cdot \nabla^S \gamma^s + \nabla\nabla^S \gamma^s \cdot \boldsymbol{v}) \mathbf{n} \cdot \nabla\gamma dv$$
$$- \int_{V_\infty} \boldsymbol{\lambda} \cdot \mathbf{f}\gamma^c \boldsymbol{v} \cdot \nabla^S \gamma^s \boldsymbol{\lambda} \cdot (\nabla\mathbf{n} \cdot \nabla\gamma + \nabla\nabla\gamma \cdot \mathbf{n}) dv$$

By Equations (21), (195), (200), and (231),

$$\int_C \nabla^\lambda \cdot \mathbf{f} dc$$
$$= - \int_{V_\infty} \boldsymbol{v} \cdot \nabla\boldsymbol{\lambda} \cdot \boldsymbol{v\lambda} \cdot \mathbf{f}\gamma^c \boldsymbol{v} \cdot \nabla^S \gamma^s \mathbf{n} \cdot \nabla\gamma dv - \int_{V_\infty} \boldsymbol{n} \cdot \nabla\boldsymbol{\lambda} \cdot \boldsymbol{n\lambda} \cdot \mathbf{f}\gamma^c \boldsymbol{v} \cdot \nabla^S \gamma^s \mathbf{n} \cdot \nabla\gamma dv$$
$$- \int_{V_\infty} \boldsymbol{\lambda} \cdot \nabla\boldsymbol{\lambda} \cdot \mathbf{f}\gamma^c \boldsymbol{v} \cdot \nabla^S \gamma^s \mathbf{n} \cdot \nabla\gamma dv - \int_{V_\infty} \boldsymbol{\lambda} \cdot \mathbf{f}\boldsymbol{\lambda} \cdot \nabla\gamma^c \boldsymbol{v} \cdot \nabla^S \gamma^s \mathbf{n} \cdot \nabla\gamma dv$$
$$- \int_{V_\infty} \gamma^c \mathbf{f} \cdot \nabla^c \nabla\gamma^s \cdot \boldsymbol{v}\mathbf{n} \cdot \nabla\gamma dv - \int_{V_\infty} \gamma^c \boldsymbol{v} \cdot \nabla^S \gamma^s \mathbf{f} \cdot \nabla^c \nabla\gamma \cdot \mathbf{n} dv \tag{500}$$

in which we have used Equation (216). Also, by Equations (166) and (275),

$$\int_C \nabla^\lambda \cdot \mathbf{f}dc = -\int_C \boldsymbol{v} \cdot \nabla\boldsymbol{\lambda} \cdot \boldsymbol{v}\boldsymbol{\lambda} \cdot \mathbf{f}dc - \int_C \mathbf{n} \cdot \nabla\boldsymbol{\lambda} \cdot \mathbf{n}\boldsymbol{\lambda} \cdot \mathbf{f}dc - \int_C \boldsymbol{\lambda} \cdot \nabla\boldsymbol{\lambda} \cdot \mathbf{f}dc$$
$$+ (\mathbf{e} \cdot \mathbf{f})|_{\text{ends}} + \int_C \mathbf{f} \cdot \boldsymbol{\lambda}\boldsymbol{v} \cdot \nabla\boldsymbol{\lambda} \cdot \boldsymbol{v}dc + \int_C \mathbf{f} \cdot \boldsymbol{\lambda}\mathbf{n} \cdot \nabla\boldsymbol{\lambda} \cdot \mathbf{n}dc$$
$$= -\int_C \boldsymbol{\lambda} \cdot \nabla\boldsymbol{\lambda} \cdot \mathbf{f}dc + (\mathbf{e} \cdot \mathbf{f}^\lambda)|_{\text{ends}} \quad (501)$$

Substituting Equation (501) into Equation (498) finally leads to:

$$\int_C \nabla^S \cdot \mathbf{f}dc = \int_C \nabla^v \cdot \mathbf{f}^v dc - \int_C (\boldsymbol{\lambda} \cdot \nabla\boldsymbol{\lambda}) \cdot \mathbf{f}dc - \int_C (\boldsymbol{v} \cdot \nabla\boldsymbol{v}) \cdot \mathbf{f}dc + (\mathbf{e} \cdot \mathbf{f}^\lambda)|_{\text{ends}}$$

Theorem 29.

$$\int_C \nabla^S \times \mathbf{f}dc = \int_C \nabla^v \times (\mathbf{f}^\lambda + \mathbf{f}^n)dc - \int_C (\boldsymbol{v} \cdot \nabla\boldsymbol{v}) \times \mathbf{f}^v dc$$
$$- \int_C (\boldsymbol{\lambda} \cdot \nabla\boldsymbol{\lambda}) \times \mathbf{f}dc + [\mathbf{e} \times (\mathbf{f}^n + \mathbf{f}^v)]|_{\text{ends}} \quad (502)$$

Proof. Note that

$$\nabla^S = \nabla^v + \nabla^\lambda$$

Also,

$$\mathbf{f} = \mathbf{f}^\lambda + \mathbf{f}^n + \mathbf{f}^v$$

Thus

$$\int_C \nabla^S \times \mathbf{f}dc = \int_C \nabla^v \times \mathbf{f}dc + \int_C \nabla^\lambda \times \mathbf{f}dc$$
$$= \int_C \nabla^v \times (\mathbf{f}^v + \mathbf{f}^\lambda + \mathbf{f}^n)dc + \int_C \nabla^\lambda \times \mathbf{f}dc$$
$$= \int_C \nabla^v \times (\mathbf{f}^\lambda + \mathbf{f}^n)dc + \int_C \nabla^v \times \mathbf{f}^v dc + \int_C \nabla^\lambda \times \mathbf{f}dc \quad (503)$$

The second term on the right side of Equation (503) is

$$\int_C \nabla^v \times \mathbf{f}^v dc = \int_C (\boldsymbol{v}\boldsymbol{v} \cdot \nabla) \times (\mathbf{f} \cdot \boldsymbol{v}\boldsymbol{v})dc$$
$$= \int_C \boldsymbol{v} \times (\boldsymbol{v} \cdot \nabla\boldsymbol{v})\mathbf{f} \cdot \boldsymbol{v}dc + \int_C \boldsymbol{v} \times \boldsymbol{v}\boldsymbol{v} \cdot \nabla(\mathbf{f} \cdot \boldsymbol{v})dc$$
$$= -\int_C \boldsymbol{v} \cdot \nabla\boldsymbol{v} \times \boldsymbol{v}\mathbf{f} \cdot \boldsymbol{v}dc$$
$$= -\int_C \boldsymbol{v} \cdot \nabla\boldsymbol{v} \times \mathbf{f}^v dc \quad (504)$$

Define three indicator function $\gamma^c(\mathbf{l})$, γ^s $(\mathbf{u})$ and $\gamma(\mathbf{x})$ such that $\gamma^c = 1$ on the curve C and $\gamma^c = 0$ elsewhere, $\gamma^s = 1$ on the surface S and $\gamma^s = 0$ elsewhere and

$\gamma = 1$ within V but $\gamma = 0$ outside of V. The third term on the right side of Equation (503) is thus

$$\begin{aligned}\int_C \nabla^\lambda \times \mathbf{f}dc &= \int_{V_\infty} \nabla^\lambda \times \mathbf{f}\gamma^c \boldsymbol{v} \cdot \nabla^S \gamma^s \mathbf{n} \cdot \nabla\gamma dv \\ &= \int_{V_\infty} (\boldsymbol{\lambda\lambda} \cdot \nabla) \times \mathbf{f}\gamma^c \boldsymbol{v} \cdot \nabla^S \gamma^s \mathbf{n} \cdot \nabla\gamma dv \\ &= \int_{V_\infty} \boldsymbol{\lambda} \times (\boldsymbol{\lambda} \cdot \nabla\mathbf{f})\gamma^c \boldsymbol{v} \cdot \nabla^S \gamma^s \mathbf{n} \cdot \nabla\gamma dv \end{aligned} \quad (505)$$

Applying Equation (41) and the chain rule to the right side of Equation (505) yields

$$\begin{aligned}&\int_C \nabla^\lambda \times \mathbf{f}dc \\ &= \int_{V_\infty} \nabla \cdot (\boldsymbol{\lambda\lambda} \times \mathbf{f}\gamma^c \boldsymbol{v} \cdot \nabla^S \gamma^s \mathbf{n} \cdot \nabla\gamma)dv - \int_{V_\infty} \nabla \cdot \boldsymbol{\lambda\lambda} \times \mathbf{f}\gamma^c \boldsymbol{v} \cdot \nabla^S \gamma^s \mathbf{n} \cdot \nabla\gamma dv \\ &\quad - \int_{V_\infty} \boldsymbol{\lambda} \cdot \nabla\boldsymbol{\lambda} \times \mathbf{f}\gamma^c \boldsymbol{v} \cdot \nabla^S \gamma^s \mathbf{n} \cdot \nabla\gamma dv - \int_{V_\infty} \boldsymbol{\lambda} \times \mathbf{f}\boldsymbol{\lambda} \cdot \nabla\gamma^c \boldsymbol{v} \cdot \nabla^S \gamma^s \mathbf{n} \cdot \nabla\gamma dv \\ &\quad - \int_{V_\infty} \boldsymbol{\lambda} \times \mathbf{f}\gamma^c \boldsymbol{\lambda} \cdot \nabla(\boldsymbol{v} \cdot \nabla^S \gamma^s)\mathbf{n} \cdot \nabla\gamma dv - \int_{V_\infty} \boldsymbol{\lambda} \times \mathbf{f}\gamma^c \boldsymbol{v} \cdot \nabla^S \gamma^s \boldsymbol{\lambda} \cdot \nabla(\mathbf{n} \cdot \nabla\gamma)dv \end{aligned} \quad (506)$$

By Equations (41) and (231), we have

$$\begin{aligned}&\int_C \nabla^\lambda \times \mathbf{f}dc \\ &= -\int_{V_\infty} (\boldsymbol{vv} \cdot \nabla) \cdot \boldsymbol{\lambda\lambda} \times \mathbf{f}\gamma^c \boldsymbol{v} \cdot \nabla^S \gamma^s \mathbf{n} \cdot \nabla\gamma dv - \int_{V_\infty} (\boldsymbol{\lambda\lambda} \cdot \nabla) \cdot \boldsymbol{\lambda\lambda} \times \mathbf{f}\gamma^c \boldsymbol{v} \cdot \nabla^S \gamma^s \mathbf{n} \cdot \nabla\gamma dv \\ &\quad - \int_{V_\infty} (\boldsymbol{nn} \cdot \nabla) \cdot \boldsymbol{\lambda\lambda} \times \mathbf{f}\gamma^c \boldsymbol{v} \cdot \nabla^S \gamma^s \mathbf{n} \cdot \nabla\gamma dv - \int_{V_\infty} \boldsymbol{\lambda} \cdot \nabla\boldsymbol{\lambda} \times \mathbf{f}\gamma^c \boldsymbol{v} \cdot \nabla^S \gamma^s \mathbf{n} \cdot \nabla\gamma dv \\ &\quad - \int_{V_\infty} \boldsymbol{\lambda} \times \mathbf{f}\boldsymbol{\lambda} \cdot \nabla\gamma^c \boldsymbol{v} \cdot \nabla^S \gamma^s \mathbf{n} \cdot \nabla\gamma dv + \int_{V_\infty} \gamma^c \mathbf{f} \times \boldsymbol{\lambda\lambda} \cdot (\nabla\boldsymbol{v} \cdot \nabla^S \gamma^s + \nabla\nabla^S \gamma^s \cdot \boldsymbol{v})\mathbf{n} \cdot \nabla\gamma dv \\ &\quad + \int_{V_\infty} \gamma^c \mathbf{f} \times \boldsymbol{\lambda\lambda} \cdot (\nabla\mathbf{n} \cdot \nabla\gamma + \nabla\nabla\gamma \cdot \mathbf{n})\boldsymbol{v} \cdot \nabla^S \gamma^s dv \end{aligned} \quad (507)$$

By Equations (41), (195) and (200), we also have

$$\begin{aligned}&\int_C \nabla^\lambda \times \mathbf{f}dc \\ &= -\int_{V_\infty} \boldsymbol{v} \cdot \nabla\boldsymbol{\lambda} \cdot \boldsymbol{v\lambda} \times \mathbf{f}\gamma^c \boldsymbol{v} \cdot \nabla^S \gamma^s \mathbf{n} \cdot \nabla\gamma dv - \int_{V_\infty} \boldsymbol{n} \cdot \nabla\boldsymbol{\lambda} \cdot \mathbf{n}\boldsymbol{\lambda} \times \mathbf{f}\gamma^c \boldsymbol{v} \cdot \nabla^S \gamma^s \mathbf{n} \cdot \nabla\gamma dv \\ &\quad - \int_{V_\infty} \boldsymbol{\lambda} \cdot \nabla\boldsymbol{\lambda} \times \mathbf{f}\gamma^c \boldsymbol{v} \cdot \nabla^S \gamma^s \mathbf{n} \cdot \nabla\gamma dv - \int_{V_\infty} \boldsymbol{\lambda} \times \mathbf{f}\boldsymbol{\lambda} \cdot \nabla\gamma^c \boldsymbol{v} \cdot \nabla^S \gamma^s \mathbf{n} \cdot \nabla\gamma dv \\ &\quad + \int_{V_\infty} \gamma^c \mathbf{f} \times \nabla^c \nabla^S \gamma^s \cdot \boldsymbol{v}\mathbf{n} \cdot \nabla\gamma dv + \int_{V_\infty} \gamma^c \mathbf{f} \times \nabla^c \nabla\gamma \cdot \mathbf{n}\boldsymbol{v} \cdot \nabla^S \gamma^s dv \end{aligned} \quad (508)$$

By Equations (216), (258) and (261),

$$\begin{aligned}&\int_C \nabla^\lambda \times \mathbf{f}\mathrm{d}c\\ &= -\int_{V_\infty} \boldsymbol{v}\cdot\nabla\boldsymbol{\lambda}\cdot\boldsymbol{v}\boldsymbol{\lambda}\times\mathbf{f}\gamma^c\boldsymbol{v}\cdot\nabla^S\gamma^s\mathbf{n}\cdot\nabla\gamma\mathrm{d}v - \int_{V_\infty}\mathbf{n}\cdot\nabla\boldsymbol{\lambda}\cdot\mathbf{n}\boldsymbol{\lambda}\times\mathbf{f}\gamma^c\boldsymbol{v}\cdot\nabla^S\gamma^s\mathbf{n}\cdot\nabla\gamma\mathrm{d}v\\ &\quad -\int_{V_\infty}\boldsymbol{\lambda}\cdot\nabla\boldsymbol{\lambda}\times\mathbf{f}\gamma^c\boldsymbol{v}\cdot\nabla^S\gamma^s\mathbf{n}\cdot\nabla\gamma\mathrm{d}v - \int_{V_\infty}\boldsymbol{\lambda}\times\mathbf{f}\boldsymbol{\lambda}\cdot\nabla\gamma^c\boldsymbol{v}\cdot\nabla^S\gamma^s\mathbf{n}\cdot\nabla\gamma\mathrm{d}v\\ &\quad -\int_{V_\infty}\gamma^c\mathbf{f}\times\boldsymbol{\lambda}\boldsymbol{v}\cdot\nabla\boldsymbol{\lambda}\cdot\nabla^S\gamma^s\mathbf{n}\cdot\nabla\gamma\mathrm{d}v - \int_{V_\infty}\gamma^c\mathbf{f}\times\boldsymbol{\lambda}\mathbf{n}\cdot\nabla\boldsymbol{\lambda}\cdot\nabla\gamma\boldsymbol{v}\cdot\nabla^S\gamma^s\mathrm{d}v\end{aligned} \tag{509}$$

Applying Equations (165) and (275) leads to

$$\begin{aligned}\int_C \nabla^\lambda\times\mathbf{f}\mathrm{d}c &= -\int_C \boldsymbol{v}\cdot\nabla\boldsymbol{\lambda}\cdot\boldsymbol{v}\boldsymbol{\lambda}\times\mathbf{f}\mathrm{d}c - \int_C\mathbf{n}\cdot\nabla\boldsymbol{\lambda}\cdot\mathbf{n}\boldsymbol{\lambda}\times\mathbf{f}\mathrm{d}c\\ &\quad -\int_C\boldsymbol{\lambda}\cdot\nabla\boldsymbol{\lambda}\times\mathbf{f}\mathrm{d}c + (\mathbf{e}\times\mathbf{f})|_{\text{ends}} - \int_C\mathbf{f}\times\boldsymbol{\lambda}\boldsymbol{v}\cdot\nabla\boldsymbol{\lambda}\cdot\boldsymbol{v}\mathrm{d}c\\ &\quad -\int_C\mathbf{f}\times\boldsymbol{\lambda}\mathbf{n}\cdot\nabla\boldsymbol{\lambda}\cdot\mathbf{n}\mathrm{d}c\\ &= -\int_C\boldsymbol{\lambda}\cdot\nabla\boldsymbol{\lambda}\times\mathbf{f}\mathrm{d}c + (\mathbf{e}\times\mathbf{f})|_{\text{ends}}\end{aligned} \tag{510}$$

Substituting Equations (510) and (504) into (503) finally yields

$$\begin{aligned}\int_C\nabla^S\times\mathbf{f}\mathrm{d}c &= \int_C\nabla^\nu\times(\mathbf{f}^\lambda+\mathbf{f}^n)\mathrm{d}c - \int_C(\boldsymbol{v}\cdot\nabla\boldsymbol{v})\times\mathbf{f}^\nu\mathrm{d}c\\ &\quad -\int_C(\boldsymbol{\lambda}\cdot\nabla\boldsymbol{\lambda})\times\mathbf{f}\mathrm{d}c + [\mathbf{e}\times(\mathbf{f}^n+\mathbf{f}^\nu)]|_{\text{ends}}\end{aligned} \tag{511}$$

in which we have used $\mathbf{e}\times\mathbf{f}=0$.

Theorem 30.

$$\int_C\left.\frac{\partial f}{\partial t}\right|_{\mathbf{u}}\mathrm{d}c = \frac{\partial}{\partial t}\int_C f\mathrm{d}c - \int_C\mathbf{w}^\nu\cdot\nabla^\nu f\mathrm{d}c + \int_C(\boldsymbol{\lambda}\cdot\nabla\boldsymbol{\lambda})\cdot\mathbf{w}f\mathrm{d}c - (\mathbf{e}\cdot\mathbf{w}^\lambda f)|_{\text{ends}} \tag{512}$$

Proof. By Equation (130), we have

$$\left.\frac{\partial f}{\partial t}\right|_{\mathbf{u}} = \left.\frac{\partial f}{\partial t}\right|_{\mathbf{x}} + \mathbf{w}\cdot\mathbf{n}\mathbf{n}\cdot\nabla f \tag{513}$$

Thus

$$\int_C\left.\frac{\partial f}{\partial t}\right|_{\mathbf{u}}\mathrm{d}c = \int_C\left.\frac{\partial f}{\partial t}\right|_{\mathbf{x}}\mathrm{d}c + \int_C\mathbf{w}\cdot\mathbf{n}\mathbf{n}\cdot\nabla f\mathrm{d}c \tag{514}$$

Applying Equation (275) in the first integral on the right side of Equation (514) yields

$$\int_C\left.\frac{\partial f}{\partial t}\right|_{\mathbf{u}}\mathrm{d}c = \int_{V_\infty}\left.\frac{\partial f}{\partial t}\right|_{\mathbf{x}}\gamma^c(\boldsymbol{v}\cdot\nabla^S\gamma^s)(\mathbf{n}\cdot\nabla\gamma)\mathrm{d}v + \int_C\mathbf{w}\cdot\mathbf{n}\mathbf{n}\cdot\nabla f\mathrm{d}c \tag{515}$$

where the notation used for the indicator functions and for the unit normal vectors is the same as that in Theorem 29.

The chain rule is now applied to obtain

$$\begin{aligned}
&\int_{V_\infty} \left.\frac{\partial f}{\partial t}\right|_{\mathbf{x}} \gamma^{\mathrm{c}}(\mathbf{v}\cdot\nabla^{\mathrm{S}}\gamma^{\mathrm{s}})(\mathbf{n}\cdot\nabla\gamma)\mathrm{d}v \\
&= \int_{V_\infty} \frac{\partial[f\gamma^{\mathrm{c}}(\mathbf{v}\cdot\nabla\gamma^{\mathrm{s}})(\mathbf{n}\cdot\nabla\gamma)]}{\partial t}\mathrm{d}v - \int_{V_\infty} f\frac{\partial\gamma^{\mathrm{c}}}{\partial t}(\mathbf{v}\cdot\nabla\gamma^{\mathrm{s}})(\mathbf{n}\cdot\nabla\gamma)\mathrm{d}v \\
&- \int_{V_\infty} f\gamma^{\mathrm{c}}\left(\frac{\partial\mathbf{v}}{\partial t}\cdot\nabla\gamma^{\mathrm{s}}\right)(\mathbf{n}\cdot\nabla\gamma)\mathrm{d}v - \int_{V_\infty} f\gamma^{\mathrm{c}}\left(\mathbf{v}\cdot\nabla\frac{\partial\gamma^{\mathrm{s}}}{\partial t}\right)(\mathbf{n}\cdot\nabla\gamma)\mathrm{d}v \\
&- \int_{V_\infty} f\gamma^{\mathrm{c}}\mathbf{v}\cdot\nabla\gamma^{\mathrm{s}}\left(\frac{\partial\mathbf{n}}{\partial t}\cdot\nabla\gamma\right)\mathrm{d}v - \int_{V_\infty} f\gamma^{\mathrm{c}}(\mathbf{v}\cdot\nabla\gamma^{\mathrm{s}})\left(\mathbf{n}\cdot\nabla\frac{\partial\gamma}{\partial t}\right)\mathrm{d}v \qquad (516)
\end{aligned}$$

The order of integration and differentiation may be interchanged in the first integral because the limits of integration are independent of time. Equation (275) may be applied to convert the integral over V_∞ to an integral over C. In doing this, the partial derivative must be converted to a total derivative so that the curve is followed as it moves in space. The first term on the right side of Equation (516) can be rearranged into

$$\int_{V_\infty} \frac{\partial[f\gamma^{\mathrm{c}}(\mathbf{v}\cdot\nabla\gamma^{\mathrm{s}})(\mathbf{n}\cdot\nabla\gamma)]}{\partial t}\mathrm{d}v = \frac{\partial}{\partial t}\int_{V_\infty} f\gamma^{\mathrm{c}}(\mathbf{v}\cdot\nabla\gamma^{\mathrm{s}})(\mathbf{n}\cdot\nabla\gamma) = \frac{\mathrm{d}}{\mathrm{d}t}\int_C f\mathrm{d}c \qquad (517)$$

The time derivatives of the curvilineal indicator function γ^{c} in the second term on the right side of Equation (516) can be written in terms of its gradient by Equation (229). Also by Equation (275), we obtain

$$\begin{aligned}
-\int_{V_\infty} \frac{\partial\gamma^{\mathrm{c}}}{\partial t}(\mathbf{v}\cdot\nabla\gamma^{\mathrm{s}})(\mathbf{n}\cdot\nabla\gamma)\mathrm{d}v &= \int_{V_\infty} f\mathbf{w}\cdot\nabla\gamma^{\mathrm{c}}(\mathbf{v}\cdot\nabla\gamma^{\mathrm{s}})(\mathbf{n}\cdot\nabla\gamma)\mathrm{d}v \\
&= \int_{C_\infty} f\mathbf{w}\cdot\nabla\gamma^{\mathrm{c}}\mathrm{d}c = -f\mathbf{w}\cdot\mathbf{e}|_{\mathrm{ends}}
\end{aligned}$$

where $\mathbf{e}$ is shown in Figure 23.

By Equations (194) and (199), both the third and the fifth integrals on the right side of Equation (516) are zero. Equations (224) and (180) may be used, respectively, to replace the time derivative of γ^{s} in the fourth term and the time derivative of γ in the sixth term by their gradients. Equation (516) becomes:

$$\begin{aligned}
\int_{V_\infty} \frac{\partial f}{\partial t}\gamma^{\mathrm{c}}(\mathbf{v}\cdot\nabla\gamma^{\mathrm{s}})(\mathbf{n}\cdot\nabla\gamma)\mathrm{d}v &= \frac{\mathrm{d}}{\mathrm{d}t}\int_C f\mathrm{d}c - f\mathbf{w}\cdot\mathbf{e}|_{\mathrm{ends}} \\
&+ \int_{V_\infty} f\gamma^{\mathrm{c}}[\mathbf{v}\cdot\nabla(\mathbf{w}\cdot\nabla\gamma^{\mathrm{s}})](\mathbf{n}\cdot\nabla\mathbf{v})\mathrm{d}v \qquad (518) \\
&+ \int_{V_\infty} f\gamma^{\mathrm{c}}(\mathbf{v}\cdot\nabla\gamma^{\mathrm{s}})[\mathbf{n}\cdot\nabla(\mathbf{w}\cdot\nabla\gamma)]\mathrm{d}v
\end{aligned}$$

Further manipulation is required for the third and fourth terms on the right side of this equation to convert them back to integrals over C. Here we discuss the manipulation of the third integral in detail. For the fourth integral the procedure is very similar.

Applying the chain rule to the third integral to obtain:

$$\begin{aligned}\int_{V_\infty} f\gamma^{\mathrm{c}}[\boldsymbol{v}\cdot\nabla(\mathbf{w}\cdot\nabla\gamma^{\mathrm{s}})][\mathbf{n}\cdot\nabla\gamma]\mathrm{d}v &= \int_{V_\infty}\nabla\cdot[\boldsymbol{v}f\gamma^{\mathrm{c}}(\mathbf{w}\cdot\nabla\gamma^{\mathrm{s}})(\mathbf{n}\cdot\nabla\gamma)]\mathrm{d}v\\ &\quad-\int_{V_\infty}\nabla\cdot[\boldsymbol{v}f\gamma^{\mathrm{c}}(\mathbf{n}\cdot\nabla\gamma)](\mathbf{w}\cdot\nabla\gamma^{\mathrm{s}})\mathrm{d}v\end{aligned}\tag{519}$$

By Equation (231), the first term on the right side of this equation is zero. Applying the chain rule to the second term leads Equation (519) to

$$\begin{aligned}&\int_{V_\infty} f\gamma^{\mathrm{c}}[\boldsymbol{v}\cdot\nabla(\mathbf{w}\cdot\nabla\gamma^{\mathrm{s}})][\mathbf{n}\cdot\nabla\gamma]\mathrm{d}v\\ &\quad=-\int_{V_\infty}\nabla\cdot\boldsymbol{v}f\gamma^{\mathrm{c}}(\mathbf{n}\cdot\nabla\gamma)(\mathbf{w}\cdot\nabla\gamma^{\mathrm{s}})\mathrm{d}v-\int_{V_\infty}\boldsymbol{v}\cdot\nabla f\gamma^{\mathrm{c}}\mathbf{n}\cdot\nabla\gamma\mathbf{w}\cdot\nabla\gamma^{\mathrm{s}}\mathrm{d}v\\ &\quad-\int_{V_\infty} f\boldsymbol{v}\cdot\nabla\gamma^{\mathrm{c}}\mathbf{n}\cdot\nabla\gamma^{\mathrm{c}}\mathbf{w}\cdot\nabla\gamma^{\mathrm{s}}\mathrm{d}v-\int_{V_\infty} f\gamma^{\mathrm{c}}(\boldsymbol{v}\cdot\nabla\mathbf{n})\cdot\nabla\gamma\mathbf{w}\cdot\nabla\gamma^{\mathrm{s}}\mathrm{d}v\\ &\quad-\int_{V_\infty} f\gamma^{\mathrm{c}}\mathbf{n}\cdot[\nabla\nabla\gamma\cdot\boldsymbol{v}](\mathbf{w}\cdot\nabla\gamma^{\mathrm{s}})\mathrm{d}v\end{aligned}\tag{520}$$

The first two integrals on the right side may be readily transformed to integrals over a curve using Equation (275). The third integral is zero because $\boldsymbol{v}\cdot\nabla\gamma^{\mathrm{c}}$ is zero by Equation (215). The fourth integral is also zero because by Equation (195) $\nabla\mathbf{n}\cdot\nabla\gamma=0$. The fifth integral may be simplified by applying Equation (249) and $\nabla\nabla\gamma\cdot\boldsymbol{v}=-\nabla\,\boldsymbol{v}\cdot\nabla\gamma$ first and then transforming the resulting expression to an integral over a curve using Equation (275). After all these manipulations, Equation (520) becomes

$$\begin{aligned}\int_{V_\infty} f\gamma^{\mathrm{c}}[\boldsymbol{v}\cdot\nabla(\mathbf{w}\cdot\nabla\gamma^{\mathrm{s}})][\mathbf{n}\cdot\nabla\gamma]\mathrm{d}v &= -\int_C\nabla\cdot\boldsymbol{v}f\mathbf{w}\cdot\boldsymbol{v}\mathrm{d}c-\int_C\mathbf{w}\cdot\boldsymbol{v}\boldsymbol{v}\cdot\nabla f\mathrm{d}c\\ &\quad+\int_C f(\mathbf{n}\cdot\nabla\boldsymbol{v}\cdot\mathbf{n})\mathbf{w}\cdot\boldsymbol{v}\mathrm{d}c\end{aligned}\tag{521}$$

Similarly, the last term in Equation (518) is

$$\begin{aligned}\int_{V_\infty} f\gamma^{\mathrm{c}}[\boldsymbol{v}\cdot\nabla\gamma^{\mathrm{s}}][\mathbf{n}\cdot\nabla(\mathbf{w}\cdot\nabla\gamma)]\mathrm{d}v &= -\int_C(\nabla\cdot\mathbf{n})f\mathbf{w}\cdot\mathbf{n}\mathrm{d}c-\int_C\mathbf{w}\cdot\mathbf{n}\mathbf{n}\cdot\nabla f\mathrm{d}c\\ &\quad+\int_C f(\boldsymbol{v}\cdot\nabla\mathbf{n}\cdot\boldsymbol{v})\mathbf{w}\cdot\mathbf{n}\mathrm{d}c\end{aligned}\tag{522}$$

Substituting these last two equations into Equation (518) and regrouping terms yields

$$\begin{aligned}&\int_{V_\infty}\left.\frac{\partial f}{\partial t}\right|_{\mathbf{x}}\gamma^{\mathrm{c}}[\boldsymbol{v}\cdot\nabla\gamma^{\mathrm{s}}][\mathbf{n}\cdot\nabla\gamma]\mathrm{d}v\\ &\quad=\frac{\mathrm{d}}{\mathrm{d}t}\int_C f\mathrm{d}c-f\mathbf{w}\cdot e|_{\mathrm{ends}}-\int_C(\nabla\cdot\mathbf{n}-\boldsymbol{v}\cdot\nabla\mathbf{n}\cdot\boldsymbol{v})f\mathbf{w}\cdot\mathbf{n}\mathrm{d}c\\ &\quad-\int_C(\nabla\cdot\boldsymbol{v}-\mathbf{n}\cdot\nabla\boldsymbol{v}\cdot\mathbf{n})f\mathbf{w}\cdot\boldsymbol{v}\mathrm{d}c-\int_C(\mathbf{w}\cdot\boldsymbol{v}\boldsymbol{v}+\mathbf{w}\cdot\mathbf{n}\mathbf{n})\cdot\nabla f\mathrm{d}c\end{aligned}\tag{523}$$

By Equation (53),

$$\nabla \cdot \mathbf{n} = \boldsymbol{\lambda} \cdot \nabla \mathbf{n} \cdot \boldsymbol{\lambda} + \mathbf{n} \cdot \nabla \mathbf{n} \cdot \mathbf{n} + \boldsymbol{\nu} \cdot \nabla \mathbf{n} \cdot \boldsymbol{\nu} \tag{524}$$

$$\nabla \cdot \boldsymbol{\nu} = \boldsymbol{\lambda} \cdot \nabla \boldsymbol{\nu} \cdot \boldsymbol{\lambda} + \mathbf{n} \cdot \nabla \boldsymbol{\nu} \cdot \mathbf{n} + \boldsymbol{\nu} \cdot \nabla \boldsymbol{\nu} \cdot \boldsymbol{\nu} \tag{525}$$

By Equation (21),

$$\nabla \mathbf{n} \cdot \mathbf{n} = \nabla \boldsymbol{\nu} \cdot \boldsymbol{\nu} = 0$$

Furthermore, the velocity of the curve is

$$\mathbf{w} = \mathbf{w} \cdot \boldsymbol{\lambda\lambda} + \mathbf{w} \cdot \boldsymbol{\nu\nu} + \mathbf{w} \cdot \mathbf{nn} \tag{526}$$

Substituting these expressions into Equation (523) yields

$$\begin{aligned} &\int_{V_\infty} \left.\frac{\partial f}{\partial t}\right|_{\mathbf{x}} \gamma^{c}(\boldsymbol{\nu} \cdot \nabla^{S}\gamma^{s})(\mathbf{n} \cdot \nabla \gamma) \mathrm{d}v \\ &= \frac{\mathrm{d}}{\mathrm{d}t}\int_C f \mathrm{d}c - (f\mathbf{w} \cdot \mathbf{e})|_{\text{ends}} - \int_C \boldsymbol{\lambda} \cdot \nabla \mathbf{n} \cdot \boldsymbol{\lambda} f \mathbf{w} \cdot \mathbf{n} \mathrm{d}c \\ &- \int_C \boldsymbol{\lambda} \cdot \nabla \boldsymbol{\nu} \cdot \boldsymbol{\lambda} f \mathbf{w} \cdot \boldsymbol{\nu} \mathrm{d}c - \int_C \mathbf{w} \cdot \nabla f \mathrm{d}c + \int_C \mathbf{w} \cdot \boldsymbol{\lambda\lambda} \cdot \nabla f \mathrm{d}c \end{aligned} \tag{527}$$

By Equations (25) and (24), we have

$$\nabla \mathbf{n} \cdot \boldsymbol{\lambda} = -\nabla \boldsymbol{\lambda} \cdot \mathbf{n}$$

and

$$\nabla \boldsymbol{\nu} \cdot \boldsymbol{\lambda} = -\nabla \boldsymbol{\lambda} \cdot \boldsymbol{\nu}$$

Equation (527) may thus be rearranged into

$$\begin{aligned} \int_{V_\infty} \left.\frac{\partial f}{\partial t}\right|_{\mathbf{x}} \gamma^{c}[\boldsymbol{\nu} \cdot \nabla \gamma^{s}][\mathbf{n} \cdot \nabla \gamma] \mathrm{d}v &= \frac{\mathrm{d}}{\mathrm{d}t}\int_C f \mathrm{d}c - f\mathbf{w} \cdot \mathbf{e}|_{\text{ends}} + \int_C (\boldsymbol{\lambda} \cdot \nabla \boldsymbol{\lambda}) \cdot \mathbf{w} f \mathrm{d}c \\ &- \int_C \mathbf{w} \cdot \nabla f \mathrm{d}c + \int_C \mathbf{w} \cdot \boldsymbol{\lambda\lambda} \cdot \nabla f \mathrm{d}c \end{aligned} \tag{528}$$

where we have used

$$\boldsymbol{\lambda} \cdot \nabla \boldsymbol{\lambda} \cdot (\mathbf{nn} \cdot \mathbf{w} + \boldsymbol{\nu\nu} \cdot \mathbf{w}) = \boldsymbol{\lambda} \cdot \nabla \boldsymbol{\lambda} \cdot \mathbf{w}$$

Finally, substituting Equation (528) into Equation (515) and rearranging terms yield

$$\int_C \left.\frac{\partial f}{\partial t}\right|_{\mathbf{u}} \mathrm{d}c = \frac{\mathrm{d}}{\mathrm{d}t}\int_C f \mathrm{d}c - \int_C \mathbf{w}^{\nu} \cdot \nabla^{\nu} f \, \mathrm{d}c + \int_C \boldsymbol{\lambda} \cdot \nabla \boldsymbol{\lambda} \cdot \mathbf{w} f \mathrm{d}c - (\mathbf{e} \cdot \mathbf{w} f)|_{\text{ends}} \tag{529}$$

Let

$$\mathbf{w} \cdot \boldsymbol{\nu\nu} = \mathbf{w}^{\nu}$$

$$\boldsymbol{\nu\nu} \cdot \nabla = \nabla^{\nu}$$

Then

$$\mathbf{w} \cdot \boldsymbol{\nu\nu} \cdot \nabla = \mathbf{w}^{\nu} \cdot \nabla^{\nu}$$

Thus Equation (529) can be written as

$$\int_C \left.\frac{\partial f}{\partial t}\right|_{\mathbf{u}} \mathrm{d}c = \frac{\partial}{\partial t}\int_C f \mathrm{d}c - \int_C \mathbf{w}^{\nu} \cdot \nabla^{\nu} \mathbf{f} \mathbf{d} \mathbf{c} + \int_C (\boldsymbol{\lambda} \cdot \nabla \boldsymbol{\lambda}) \cdot \mathbf{w} f \mathrm{d}c - (\mathbf{e} \cdot \mathbf{w}^{\lambda} f)|_{\text{ends}}$$

4.3.4 Integration of surficial operators over a straight line fixed on a planar surface

This group of integration theorems involves integrations of $(\partial f/\partial t)|_{\mathbf{u}}$, $\nabla^{S} f$, $\nabla^{S} \cdot \mathbf{f}$ and $\nabla^{S} \times \mathbf{f}$ over a straight line L fixed on a planar surface (Figure 24). Here $f(\mathbf{u}, t)$ and $\mathbf{f}(\mathbf{u}, t)$ are microscale scalar and vector functions of surficial position vector $\mathbf{x}$ and time t.

In these theorems L is the straight-line segment of integration, $\boldsymbol{\Lambda}$ the unit vector tangent to segment L, and $\mathbf{e}$ the unit vector tangent to segment L at the endpoints with positive outward direction. $\mathbf{N}$ is the unit vector normal to the plane containing segment L, $\mathbf{n}^*$ the unit vector at the intersection of the ends of the segment L with the curve bounding the plane containing L that is normal to the curve and tangent to the plane. ∇^{S} is the surficial operator such that $\nabla^{S} = \nabla - \mathbf{NN} \cdot \nabla$, ∇^{ν} the curvilineal operator along the curve normal to L and in the surface containing L such that $\nabla^{\nu} = \nabla^{S} - \boldsymbol{\Lambda\Lambda} \cdot \nabla^{S}$. $\mathbf{w}$ is the velocity of the endpoints of L such that $\mathbf{w} = \boldsymbol{\Lambda\Lambda} \cdot \mathbf{w}$. $\mathbf{f}^{\lambda}$ is the vector tangent to L such that $\mathbf{f}^{\lambda} = \boldsymbol{\Lambda\Lambda} \cdot \mathbf{f}$, $\mathbf{f}^{\mathrm{n}}$ is the vector normal to L and normal to the plane containing L such that $\mathbf{f}^{\mathrm{n}} = \mathbf{NN} \cdot \mathbf{f}$ and $\mathbf{f}^{\nu}$ is the vector normal to L and tangent to the plane containing L such that $\mathbf{f}^{\nu} = \boldsymbol{\nu\nu} \cdot \mathbf{f}$.

Theorem 31.

$$\int_L \left.\frac{\partial f}{\partial t}\right|_{\mathbf{u}} \mathrm{d}L = \frac{\partial}{\partial t}\int_L f \mathrm{d}L - \left.\left(\frac{\mathbf{n}^* \cdot \mathbf{w} f}{\mathbf{e} \cdot \mathbf{n}^*}\right)\right|_{\text{ends}} \tag{530}$$

Proof. By Equation (137), we have

$$\int_L \left.\frac{\partial f}{\partial t}\right|_{\mathbf{u}} \mathrm{d}L = \int_L \left.\frac{\partial f}{\partial t}\right|_{1} \mathrm{d}L - \int_L (\mathbf{w} - \boldsymbol{\Lambda} \cdot \mathbf{w}\boldsymbol{\Lambda}) \cdot \nabla^{S} f \mathrm{d}L \tag{531}$$

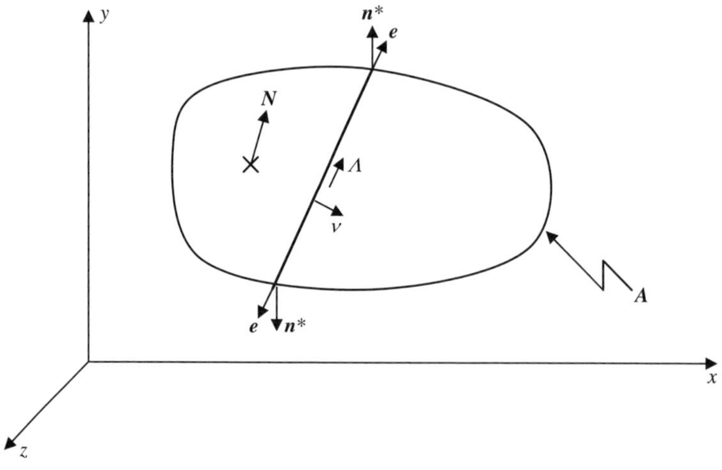

Figure 24 Straight line segment L liying in plane A ($\boldsymbol{\Lambda}$ is unit tangent of L; $\boldsymbol{\nu}$ unit normal of L and tangent to A; $\mathbf{N}$ unit normal of A; $\mathbf{n}^*$ unit tangent of A and normal to the boundary curve of A; and $\mathbf{e}$ unit tangent of L and pointing outward from L at the endpoints of L).

Note here that

$$\mathbf{w} = \Lambda \cdot \mathbf{w}\Lambda$$

Thus Equation (531) becomes

$$\int_L \left.\frac{\partial f}{\partial t}\right|_{\mathbf{u}} \mathrm{d}L = \int_L \left.\frac{\partial f}{\partial t}\right|_{\mathbf{l}} \mathrm{d}L \tag{532}$$

Assume that the straight line segment L is on the plane A (Figure 24). By defining an indicator function $\gamma^{\mathrm{s}}(\mathbf{u})$ such that $\gamma^{\mathrm{s}} = 1$ on the plane A and $\gamma^{\mathrm{s}} = 0$ elsewhere, we obtain

$$\int_L \left.\frac{\partial f}{\partial t}\right|_{\mathbf{u}} \mathrm{d}L = \int_L \left.\frac{\partial f}{\partial t}\right|_{\mathbf{l}} \gamma^{\mathrm{s}} \mathrm{d}L \tag{533}$$

An application of the chain rule to the right side of Equation (533) yields

$$\int_L \left.\frac{\partial f}{\partial t}\right|_{\mathbf{u}} \mathrm{d}L = \int_{L_\infty} \left.\frac{\partial (f\gamma^{\mathrm{s}})}{\partial t}\right|_{\mathbf{l}} \mathrm{d}L - \int_{L_\infty} f\left.\frac{\partial \gamma^{\mathrm{s}}}{\partial t}\right|_{\mathbf{l}} \mathrm{d}L \tag{534}$$

As the limits of the integral are independent of t, the first term on the right side of Equation (534) becomes

$$\int_{L_\infty} \left.\frac{\partial (f\gamma^{\mathrm{s}})}{\partial t}\right|_{\mathbf{l}} \mathrm{d}L = \frac{\partial}{\partial t}\int_{L_\infty} f\gamma^{\mathrm{s}} \mathrm{d}L = \frac{\partial}{\partial t}\int_L f \mathrm{d}L \tag{535}$$

By Equation (224), the second term of the right side of Equation (534) is

$$\int_{L_\infty} f\left.\frac{\partial \gamma^{\mathrm{s}}}{\partial t}\right|_{\mathbf{l}} \mathrm{d}L = -\int_{L_\infty} f\mathbf{w} \cdot \nabla \gamma^{\mathrm{s}} \mathrm{d}L \tag{536}$$

by Equation (299),

$$\int_{L_\infty} f\left.\frac{\partial \gamma^{\mathrm{s}}}{\partial t}\right|_{\mathbf{l}} \mathrm{d}L = \left.\frac{\mathbf{n}^* \cdot \mathbf{w}f}{\mathbf{e} \cdot \mathbf{n}^*}\right|_{\mathrm{ends}} \tag{537}$$

Equations (535) and (536) finally allow us to rewrite Equation (534) as

$$\int_L \left.\frac{\partial f}{\partial t}\right|_{\mathbf{u}} \mathrm{d}L = \frac{\partial}{\partial t}\int_L f\mathrm{d}L - \left.\left(\frac{\mathbf{n}^* \cdot \mathbf{w}f}{\mathbf{e} \cdot \mathbf{n}^*}\right)\right|_{\mathrm{ends}}$$

Theorem 32.

$$\int_L \nabla^{\mathrm{S}} f \mathrm{d}L = \nabla^{v} \int_L f\mathrm{d}L + \left.\left(\frac{\mathbf{n}^* f}{\mathbf{e} \cdot \mathbf{n}^*}\right)\right|_{\mathrm{ends}} \tag{538}$$

Proof. Assume that the straight-line segment L is on the plane A (Figure 24). Define an indicator function $\gamma^{\mathrm{s}}(\mathbf{u})$ such that $\gamma^{\mathrm{s}} = 1$ on the plane A and $\gamma^{\mathrm{s}} = 0$ elsewhere. We have

$$\int_L \nabla^{\mathrm{S}} f \mathrm{d}L = \int_{L_\infty} \nabla^{\mathrm{S}} f \gamma^{\mathrm{s}} \mathrm{d}L \tag{539}$$

An application of the chain rule to the right side of Equation (439) yields

$$\int_L \nabla^S f \mathrm{d}L = \int_{L_\infty} \nabla^S (f\gamma^s)\mathrm{d}L - \int_{L_\infty} f\nabla^S \gamma^s \mathrm{d}L \tag{540}$$

By Equation (298), the second term of the right side of Equation (540) may be written as

$$\int_{L_\infty} f\nabla^S \gamma^s \mathrm{d}L = -\frac{f\mathbf{n}^*}{\mathbf{e}\cdot\mathbf{n}^*}\bigg|_{\text{ends}} \tag{541}$$

By the definition of the operator ∇^S, the first term of the right-hand side of Equation (540) is

$$\int_{L_\infty} \nabla^S (f\gamma^s)\mathrm{d}L = \int_{L_\infty} \nabla^v (f\gamma^s)\mathrm{d}L + \int_{L_\infty} \nabla^\lambda (f\gamma^s)\mathrm{d}L \tag{542}$$

Since the operator ∇^v is independent of the integration variable, we may take it out of the integration sign so that

$$\int_{L_\infty} \nabla^v (f\gamma^s)\mathrm{d}L = \nabla^v \int_{L_\infty} f\gamma^s \mathrm{d}L \tag{543}$$

The definition of the indicator function γ^s allows us to rewrite Equation (543)

$$\int_{L_\infty} \nabla^v (f\gamma^s)\mathrm{d}L = \nabla^v \int_L f\mathrm{d}L \tag{544}$$

The second term of the right side of Equation (542) may be written as

$$\int_{L_\infty} \nabla^\lambda (f\gamma^s)\mathrm{d}L = \int_{L_\infty} \nabla^c (f\gamma^s)\mathrm{d}L \tag{545}$$

where $\nabla^c = \nabla^\lambda = \mathit{\Lambda}\mathit{\Lambda}\cdot\nabla$. By using Equation (246) and the fact that the role of γ^s here is the same as that of γ^c, we obtain

$$\int_{L_\infty} \nabla^\lambda (f\gamma^s)\mathrm{d}L = 0 \tag{546}$$

Substituting Equations (544) and (546) into Equation (542) yields

$$\int_{L_\infty} \nabla^S (f\gamma^s)\mathrm{d}L = \nabla^v \int_L f\mathrm{d}L \tag{547}$$

Substituting Equations (541) and (547) into Equation (540) finally leads to

$$\int_L \nabla^S f\mathrm{d}L = \nabla^v \int_L f\mathrm{d}c + \left(\frac{\mathbf{n}^* f}{\mathbf{e}\cdot\mathbf{n}^*}\right)\bigg|_{\text{ends}}$$

Theorem 33.

$$\int_L \nabla^S \cdot \mathbf{f}\mathrm{d}L = \nabla^v \cdot \int_L \mathbf{f}^v \mathrm{d}L + \int_L (\nabla^v \cdot \mathit{\Lambda})\mathit{\Lambda}\cdot \mathbf{f}^\lambda \mathrm{d}L + \left(\frac{\mathbf{n}^*\cdot\mathbf{f}}{\mathbf{e}\cdot\mathbf{n}^*}\right)\bigg|_{\text{ends}} \tag{548}$$

Proof. Assume that the straight-line segment L is on a plane A (Figure 2.7). Define an indicator function $\gamma^s(\mathbf{u})$ such that $\gamma^s = 1$ on the plane A and $\gamma^s = 0$

elsewhere. We obtain

$$\int_L \nabla^S \cdot \mathbf{f} dL = \int_{L_\infty} \nabla^S \cdot \mathbf{f}\gamma^s dL \tag{549}$$

An application of the chain rule to the right side of Equation (549) yields

$$\int_L \nabla^S \cdot \mathbf{f} dL = -\int_{L_\infty} \nabla^S \cdot (\mathbf{f}\gamma^s) dL - \int_{L_\infty} \mathbf{f} \cdot \nabla^S \gamma^s dL \tag{550}$$

By Equation (299), the second term of the right side of Equation (550) becomes

$$\int_{L_\infty} \mathbf{f} \cdot \nabla^S \gamma^s dL = -\left(\frac{\mathbf{n}^* \cdot \mathbf{f}}{\mathbf{e} \cdot \mathbf{n}^*}\right)\bigg|_{\text{ends}} \tag{551}$$

By the definition of ∇^S, the first term of the right side of Equation (550) is

$$\int_{L_\infty} \nabla^S \cdot (\mathbf{f}\gamma^s) dL = \int_{L_\infty} \nabla^\lambda \cdot (\mathbf{f}\gamma^s) dL + \int_{L_\infty} \nabla^v \cdot (\mathbf{f}\gamma^s) dL \tag{552}$$

By making use of Equation (247) to the first term of the right side of Equation (552) and noting that the role of γ^s here is the same as that of γ^c, we obtain

$$\int_{L_\infty} \nabla^\lambda \cdot (\mathbf{f}\gamma^s) dL = 0 \tag{553}$$

The second term of the right side of Equation (552) is

$$\begin{aligned}\int_{L_\infty} \nabla^v \cdot (\mathbf{f}\gamma^s) dL &= \int_{L_\infty} \nabla^v \cdot (\mathbf{f}^v\gamma^s + \mathbf{f}^\lambda\gamma^s) dL \\ &= \int_{L_\infty} \nabla^v \cdot (\mathbf{f}^v\gamma^s) dL + \int_{L_\infty} \nabla^v \cdot (\mathbf{f}^\lambda\gamma^s) dL \end{aligned} \tag{554}$$

Because ∇^v is independent of the integration variables, we have

$$\int_{L_\infty} \nabla^v \cdot (\mathbf{f}^v\gamma^s) dL = \nabla^v \cdot \int_{L_\infty} \mathbf{f}^v\gamma^s dL = \nabla^v \cdot \int_L \mathbf{f}^v dL \tag{555}$$

By the definition of the function $\mathbf{f}^\lambda$, we obtain

$$\begin{aligned}\int_{L_\infty} \nabla^v \cdot (\mathbf{f}^\lambda\gamma^s) dL &= \int_{L_\infty} \nabla^v \cdot (\boldsymbol{\Lambda}\boldsymbol{\Lambda} \cdot \mathbf{f}\gamma^s) dL \\ &= \int_{L_\infty} \nabla^v \cdot \boldsymbol{\Lambda}\boldsymbol{\Lambda} \cdot \mathbf{f}\gamma^s dL + \int_{L_\infty} \boldsymbol{\Lambda} \cdot \nabla^v (\boldsymbol{\Lambda} \cdot \mathbf{f}\gamma^s) dL \end{aligned} \tag{556}$$

Since $\boldsymbol{\Lambda} \cdot \nabla^v = 0$,

$$\int_{L_\infty} \nabla^v \cdot (\mathbf{f}^\lambda\gamma^s) dL = \int_{L_\infty} \nabla^v \cdot \boldsymbol{\Lambda}\boldsymbol{\Lambda} \cdot \mathbf{f}\gamma^s dL \tag{557}$$

by the definition of the indicator function γ^s,

$$\int_{L_\infty} \nabla^v \cdot (\mathbf{f}^\lambda\gamma^s) dL = \int_L \nabla^v \cdot \boldsymbol{\Lambda}\boldsymbol{\Lambda} \cdot \mathbf{f} dL \tag{558}$$

Substituting Equations (555) and (558) into Equation (554) yields

$$\int_{L_\infty} \nabla^v \cdot (\mathbf{f}\gamma^s) dL = \nabla^v \cdot \int_L \mathbf{f}^v dL + \int_L \nabla^v \cdot \boldsymbol{\Lambda}\boldsymbol{\Lambda} \cdot \mathbf{f}^\lambda dL \tag{559}$$

Combining Equations (553) and (559) allows us to rewrite Equation (552) as

$$\int_{L_\infty} \nabla^S \cdot (\mathbf{f}\gamma^s)dL = \nabla^v \cdot \int_L \mathbf{f}^v dL + \int_L \nabla^v \cdot \boldsymbol{\Lambda}\boldsymbol{\Lambda} \cdot \mathbf{f}^\lambda dL \tag{560}$$

Finally, substituting Equations (551) and (560) into Equation (550) leads to

$$\int_L \nabla^S \cdot \mathbf{f}dL = \nabla^v \cdot \int_L \mathbf{f}^\lambda dL + \int_L (\nabla^v \cdot \boldsymbol{\Lambda})\boldsymbol{\Lambda} \cdot \mathbf{f}^\lambda dL + \left(\frac{\mathbf{n}^* \cdot \mathbf{f}}{\mathbf{e} \cdot \mathbf{n}^*}\right)\Bigg|_{\text{ends}}$$

Theorem 34.

$$\int_L \nabla^S \times \mathbf{f}dL = \nabla^v \times \int_L (\mathbf{f}^\lambda + \mathbf{f}^n)dc + \int_L (\nabla^v \cdot \boldsymbol{\Lambda})\boldsymbol{\Lambda} \times \mathbf{f}^v dL + \left(\frac{\mathbf{n}^* \times \mathbf{f}}{\mathbf{e} \cdot \mathbf{n}^*}\right)\Bigg|_{\text{ends}} \tag{561}$$

Proof. Assume that the straight line segment L is on the plane A (Figure 2.7). Define an indicator function $\gamma^s(\mathbf{u})$ such that $\gamma^s = 1$ on the plane A and $\gamma^s = 0$ elsewhere. We obtain

$$\int_L \nabla^S \times \mathbf{f}dL = \int_{L_\infty} \nabla^S \times \mathbf{f}\gamma^s dL \tag{562}$$

An application of the chain rule to the right side of Equation (562) yields

$$\int_L \nabla^S \times \mathbf{f}dL = \int_{L_\infty} \nabla^S \times (\mathbf{f}\gamma^s)dL + \int_{L_\infty} \mathbf{f} \times \nabla\gamma^s dL \tag{563}$$

By a similar derivation as that for Equation (300), the second term of the right side of Equation (563) becomes

$$\int_{L_\infty} \mathbf{f} \times \nabla\gamma^s dL = \left(\frac{\mathbf{n}^* \times \mathbf{f}}{\mathbf{e} \cdot \mathbf{n}^*}\right)\Bigg|_{\text{ends}} \tag{564}$$

Note that

$$\nabla^v = \nabla^S - \boldsymbol{\Lambda}\,\boldsymbol{\Lambda} \cdot \nabla^S$$

The first term of the right side of Equation (563) is

$$\begin{aligned}\int_{L_\infty} \nabla^S \times (\mathbf{f}\gamma^s)dL &= \int_{L_\infty} (\nabla^v + \boldsymbol{\Lambda}\,\boldsymbol{\Lambda} \cdot \nabla^S) \times (\mathbf{f}\gamma^s)dL \\ &= \int_{L_\infty} \nabla^v \times (\mathbf{f}\gamma^s)dL + \int_{L_\infty} (\boldsymbol{\Lambda}\,\boldsymbol{\Lambda} \cdot \nabla^S) \times (\mathbf{f}\gamma^s)dL \end{aligned} \tag{565}$$

By

$$(\boldsymbol{\Lambda}\,\boldsymbol{\Lambda} \cdot \nabla^S) \times \mathbf{f}^\lambda = 0$$

$$(\boldsymbol{\Lambda}\boldsymbol{\Lambda} \cdot \nabla^S) \times \mathbf{f}^n = 0$$

we obtain

$$\begin{aligned}\int_{L_\infty} (\boldsymbol{\Lambda}\boldsymbol{\Lambda} \cdot \nabla^S) \times (\mathbf{f}\gamma^s)dL &= \int_{L_\infty} (\boldsymbol{\Lambda}\boldsymbol{\Lambda} \cdot \nabla^S) \times (\mathbf{f}^v\gamma^s)dL = \int_{L_\infty} \nabla^S \cdot (\boldsymbol{\Lambda}\boldsymbol{\Lambda} \times \mathbf{f}^v\gamma^s)dL \\ &= \int_{L_\infty} \nabla^S \cdot \boldsymbol{\Lambda}\boldsymbol{\Lambda} \times \mathbf{f}^v\gamma^s dL + \int_{L_\infty} \boldsymbol{\Lambda} \cdot \nabla^S(\boldsymbol{\Lambda} \times \mathbf{f}^v\gamma^s)dL \\ &= \int_{L_\infty} \nabla^S \cdot \boldsymbol{\Lambda}\boldsymbol{\Lambda} \times \mathbf{f}^v\gamma^s dL + \int_{L_\infty} \nabla^\lambda(\boldsymbol{\Lambda} \times \mathbf{f}^v\gamma^s)dL\end{aligned}$$

By using Equation (246) and noting that the role of γ^s here is the same as that of γ^c,

$$\int_{L_\infty} (\boldsymbol{\Lambda\Lambda} \cdot \nabla^S) \times (\mathbf{f}\gamma^s)\mathrm{d}L = \int_{L_\infty} \nabla^S \cdot \boldsymbol{\Lambda\Lambda} \times \mathbf{f}^\nu \gamma^s \mathrm{d}L$$
$$= \int_L \nabla^\nu \cdot \boldsymbol{\Lambda\Lambda} \times \mathbf{f}^\nu \mathrm{d}L \quad (566)$$

By Equation (248) and

$$\int_{L_\infty} \nabla^\nu \times (\mathbf{f}\gamma^s)\mathrm{d}L = \int_{L_\infty} \nabla^\nu \times [(\mathbf{f}^\nu + \mathbf{f}^\mathrm{n} + \mathbf{f}^\lambda)\gamma^s]\mathrm{d}L$$

we obtain

$$\int_{L_\infty} \nabla^\nu \times (\mathbf{f}\gamma^s)\mathrm{d}L = \int_{L_\infty} \nabla^\nu \times [(\mathbf{f}^\mathrm{n} + \mathbf{f}^\lambda)\gamma^s]\mathrm{d}L = \nabla^\nu \times \int_{L_\infty} (\mathbf{f}^\mathrm{n} + \mathbf{f}^\lambda)\gamma^s \mathrm{d}L$$
$$= \nabla^\nu \times \int_L (\mathbf{f}^\mathrm{n} + \mathbf{f}^\lambda)\mathrm{d}L \quad (567)$$

Substituting Equations (566) and (567) into Equation (565) yields

$$\int_{L_\infty} \nabla^S \times (\mathbf{f}\gamma^s)\mathrm{d}L = \int_L \nabla^\nu \cdot \boldsymbol{\Lambda\Lambda} \times \mathbf{f}^\nu \mathrm{d}L + \nabla^\nu \times \int_L (\mathbf{f}^\mathrm{n} + \mathbf{f}^\lambda)\mathrm{d}L \quad (568)$$

Finally, substituting Equations (564) and (568) into Equation (563) concludes

$$\int_L \nabla^S \times \mathbf{f}\mathrm{d}L = \nabla^\nu \times \int_L (\mathbf{f}^\mathrm{n} + \mathbf{f}^\lambda)\mathrm{d}L + \int_L \nabla^\nu \cdot \boldsymbol{\Lambda\Lambda} \times \mathbf{f}^\nu \mathrm{d}L + \left(\frac{\mathbf{n}^* \times \mathbf{f}}{\mathbf{e} \cdot \mathbf{n}^*}\right)\bigg|_{\mathrm{ends}}$$

4.3.5 Curvilineal integration of curvilineal operators

This group of integration theorems involves integrations of $(\partial f/\partial t)|_\mathbf{l}$, $\nabla^c f$, $\nabla^c \cdot \mathbf{f}$ and $\nabla^c \times \mathbf{f}$ over a curve C (Figure 21). Here $f(\mathbf{l}, t)$ and $\mathbf{f}(\mathbf{l}, t)$ are microscale scalar and vector functions of curvilineal position vector $\mathbf{l}$ and time t. Generally, the curve C can deform and translate with time.

Theorem 35.

$$\int_C \frac{\partial f}{\partial t}\bigg|_\mathbf{l} \mathrm{d}c = \frac{\mathrm{d}}{\mathrm{d}t}\int_C f\mathrm{d}c + \int_C (\boldsymbol{\lambda} \cdot \nabla^c \boldsymbol{\lambda}) \cdot \mathbf{w}^s f\mathrm{d}c - (\mathbf{e} \cdot \mathbf{w}^c f)|_{\mathrm{ends}} \quad (569)$$

where C is the integration curve, $\boldsymbol{\lambda}$ the unit vector tangent to curve C, and $\mathbf{e}$ the particular unit vector which is tangent to C at its endpoints with positive outward direction from the curve. ∇^c is the curvilineal operator such that $\nabla^c = \boldsymbol{\lambda\lambda} \cdot \nabla$, $\mathbf{w}$ the velocity of C, and $\mathbf{w}^c$ the velocity of C tangent to C such that $\mathbf{w}^c = \boldsymbol{\lambda\lambda} \cdot \mathbf{w}$. $\mathbf{w}^s$ is the velocity of C normal to C such that $\mathbf{w}^s = \mathbf{w} - \mathbf{w}^c$.

Proof. Define three indicator functions $\gamma(\mathbf{x})$, $\gamma^c(\mathbf{l})$ and $\gamma^s(\mathbf{u})$ such that

$$\int_C \frac{\partial f}{\partial t}\bigg|_\mathbf{l} \mathrm{d}c = \int_{V_\infty} \frac{\partial f}{\partial t}\bigg|_\mathbf{l} \gamma^c \boldsymbol{v} \cdot \nabla^S \gamma^s \mathbf{n} \cdot \nabla\gamma \mathrm{d}v \quad (570)$$

By Equation (134),

$$\int_C \frac{\partial f}{\partial t}\bigg|_\mathbf{l} \mathrm{d}c = \int_{V_\infty} \frac{\partial f}{\partial t}\bigg|_\mathbf{x} \gamma^c \boldsymbol{v} \cdot \nabla^S \gamma^s \mathbf{n} \cdot \nabla\gamma \mathrm{d}v + \int_{V_\infty} \mathbf{w}^s \cdot \nabla f \gamma^c \boldsymbol{v} \cdot \nabla^S \gamma^s \mathbf{n} \cdot \nabla\gamma \mathrm{d}v \quad (571)$$

where $\mathbf{w}^{\mathrm{s}} = \mathbf{w} - \mathbf{w} \cdot \boldsymbol{\lambda}\boldsymbol{\lambda}$. An application of the chain rule to the two terms of the right-hand side of Equation (571) yields

$$\begin{aligned}
\int_C \left.\frac{\partial f}{\partial t}\right|_1 \mathrm{d}c = & \int_{V_\infty} \left.\frac{\partial(f\gamma^{\mathrm{c}}\boldsymbol{v} \cdot \nabla^{\mathrm{S}}\gamma^{\mathrm{c}}\mathbf{n} \cdot \nabla\gamma)}{\partial t}\right|_{\mathbf{x}} \mathrm{d}v - \int_{V_\infty} f\left.\frac{\partial \gamma^{\mathrm{c}}}{\partial t}\right|_{\mathbf{x}} \boldsymbol{v} \cdot \nabla^{\mathrm{S}}\gamma^{\mathrm{s}}\mathbf{n} \cdot \nabla\gamma \mathrm{d}v \\
& - \int_{V_\infty} f\gamma^{\mathrm{c}}\left(\left.\frac{\partial \boldsymbol{v}}{\partial t}\right|_{\mathbf{x}} \cdot \nabla^{\mathrm{S}}\gamma^{\mathrm{s}} + \boldsymbol{v} \cdot \nabla^{\mathrm{S}}\left.\frac{\partial \gamma^{\mathrm{s}}}{\partial t}\right|_{\mathbf{x}}\right)\mathbf{n} \cdot \nabla\gamma \mathrm{d}v \\
& - \int_{V_\infty} f\gamma^{\mathrm{c}}\boldsymbol{v} \cdot \nabla^{\mathrm{S}}\gamma^{\mathrm{s}}\left(\left.\frac{\partial \mathbf{n}}{\partial t}\right|_{\mathbf{x}} \cdot \nabla\gamma + \mathbf{n} \cdot \nabla\left.\frac{\partial \gamma}{\partial t}\right|_{\mathbf{x}}\right)\mathrm{d}v \\
& + \int_{V_\infty} \nabla \cdot (\mathbf{w}^{\mathrm{s}} f\gamma^{\mathrm{c}}\boldsymbol{v} \cdot \nabla^{\mathrm{S}}\gamma^{\mathrm{s}}\mathbf{n} \cdot \nabla\gamma)\mathrm{d}v \\
& - \int_{V_\infty} \nabla \cdot \mathbf{w}^{\mathrm{s}} f\gamma^{\mathrm{c}}\boldsymbol{v} \cdot \nabla^{\mathrm{S}}\gamma^{\mathrm{s}}\mathbf{n} \cdot \nabla\gamma \mathrm{d}v \\
& - \int_{V_\infty} f\mathbf{w}^{\mathrm{s}} \cdot \nabla\gamma^{\mathrm{c}}\boldsymbol{v} \cdot \nabla^{\mathrm{S}}\gamma^{\mathrm{s}}\mathbf{n} \cdot \nabla\gamma \mathrm{d}v \\
& - \int_{V_\infty} f\gamma^{\mathrm{c}}\mathbf{w}^{\mathrm{s}} \cdot (\nabla\boldsymbol{v} \cdot \nabla^{\mathrm{S}}\gamma^{\mathrm{s}} + \nabla\nabla^{\mathrm{S}}\gamma^{\mathrm{s}} \cdot \boldsymbol{v})\mathbf{n} \cdot \nabla\gamma \mathrm{d}v \\
& - \int_{V_\infty} f\gamma^{\mathrm{c}}\boldsymbol{v} \cdot \nabla^{\mathrm{S}}\gamma^{\mathrm{s}}\mathbf{w}^{\mathrm{s}} \cdot (\nabla\mathbf{n} \cdot \nabla\gamma + \nabla\nabla\gamma \cdot \mathbf{n})\mathrm{d}v
\end{aligned} \tag{572}$$

As the integration limits are independent of t, the differentiation with respect to time t can be exchanged with the integration over the volume V_∞ in the first term on the right-hand side of Equation (578).

By Equation (229),

$$\left.\frac{\partial \gamma^{\mathrm{c}}}{\partial t}\right|_{\mathbf{x}} = \left.\frac{\partial \gamma^{\mathrm{c}}}{\partial t}\right|_1 = -\mathbf{w} \cdot \nabla\gamma^{\mathrm{c}} \tag{573}$$

By Equations (194) and (199),

$$\left.\frac{\partial \boldsymbol{v}}{\partial t}\right|_{\mathbf{x}} \cdot \nabla^{\mathrm{S}}\gamma^{\mathrm{s}} = 0 \tag{574}$$

$$\left.\frac{\partial \mathbf{n}}{\partial t}\right|_{\mathbf{x}} \cdot \nabla\gamma = 0 \tag{575}$$

By Equations (180) and (224),

$$\left.\frac{\partial \gamma^{\mathrm{s}}}{\partial t}\right|_{\mathbf{x}} = -\mathbf{w} \cdot \nabla\gamma^{\mathrm{s}} \tag{576}$$

$$\left.\frac{\partial \gamma}{\partial t}\right|_{\mathbf{x}} = -\mathbf{w} \cdot \nabla\gamma \tag{577}$$

By Equation (213), the fifth term on the right side of Equation (572) is zero. Therefore, Equation (572) becomes

$$\begin{aligned}\int_C \frac{\partial f}{\partial t}\bigg|_1 \mathrm{d}c = {} & \frac{\partial}{\partial t}\int_{V_\infty} f\gamma^c \boldsymbol{v}\cdot\nabla^S\gamma^s\mathbf{n}\cdot\nabla\gamma \mathrm{d}v + \int_{V_\infty} f\mathbf{w}\cdot\nabla^c\gamma^c\boldsymbol{v}\cdot\nabla^S\gamma^s\mathbf{n}\cdot\nabla\gamma\mathrm{d}v \\ & + \int_{V_\infty} f\gamma^c\boldsymbol{v}\cdot\nabla^S(\mathbf{w}\cdot\nabla\gamma^s)\mathbf{n}\cdot\nabla\gamma\mathrm{d}v + \int_{V_\infty} f\gamma^c\boldsymbol{v}\cdot\nabla^S\gamma^s\mathbf{n}\cdot\nabla(\mathbf{w}\cdot\nabla\gamma)\mathrm{d}v \\ & - \int_{V_\infty}\nabla\cdot\mathbf{w}^s f\gamma^c\boldsymbol{v}\cdot\nabla^S\gamma^s\mathbf{n}\cdot\nabla\gamma\mathrm{d}v - \int_{V_\infty} f\mathbf{w}^s\cdot\nabla^c\gamma^c\boldsymbol{v}\cdot\nabla^S\gamma^s\mathbf{n}\cdot\nabla\gamma\mathrm{d}v \\ & - \int_{V_\infty} f\gamma^c\mathbf{w}^s\cdot(\nabla\boldsymbol{v}\cdot\nabla^S\gamma^s + \nabla\nabla^S\gamma^s\cdot\boldsymbol{v})\mathbf{n}\cdot\nabla\gamma\mathrm{d}v \\ & - \int_{V_\infty} f\gamma^c\boldsymbol{v}\cdot\nabla^S\gamma^s\mathbf{w}^s\cdot(\nabla\mathbf{n}\cdot\nabla\gamma + \nabla\nabla\gamma\cdot\mathbf{n})\mathrm{d}v \end{aligned} \tag{578}$$

By Equations (195), (200) and $\boldsymbol{v}\cdot\nabla^S = \boldsymbol{v}\cdot\nabla$, Equation (578) becomes

$$\begin{aligned}\int_C \frac{\partial f}{\partial t}\bigg|_1 \mathrm{d}c = {} & \frac{\partial}{\partial t}\int_{V_\infty} f\gamma^c \boldsymbol{v}\cdot\nabla^S\gamma^s\mathbf{n}\cdot\nabla\gamma \mathrm{d}v + \int_{V_\infty} f\mathbf{w}\cdot\nabla^c\gamma^c\boldsymbol{v}\cdot\nabla^S\gamma^s\mathbf{n}\cdot\nabla\gamma\mathrm{d}v \\ & - \int_{V_\infty} f\mathbf{w}^s\cdot\nabla^c\gamma^c\boldsymbol{v}\cdot\nabla^S\gamma^s\mathbf{n}\cdot\nabla\gamma\mathrm{d}v + \int_{V_\infty} f\gamma^c\boldsymbol{v}\cdot\nabla^S\mathbf{w}\cdot\nabla\gamma^s\mathbf{n}\cdot\nabla\gamma\mathrm{d}v \\ & + \int_{V_\infty} f\gamma^c\boldsymbol{v}\cdot\nabla^S\nabla\gamma^s\cdot\mathbf{w}\mathbf{n}\cdot\nabla\gamma\mathrm{d}v + \int_{V_\infty} f\gamma^c\boldsymbol{v}\cdot\nabla^S\gamma^s\mathbf{n}\cdot\nabla\mathbf{w}\cdot\nabla\gamma\mathrm{d}v \\ & + \int_{V_\infty} f\gamma^c\boldsymbol{v}\cdot\nabla^S\gamma^s\mathbf{n}\cdot\nabla\nabla\gamma\cdot\mathbf{w}\mathrm{d}v - \int_{V_\infty}\nabla\cdot\mathbf{w}^s f\gamma^c\boldsymbol{v}\cdot\nabla^S\gamma^s\mathbf{n}\cdot\nabla\gamma\mathrm{d}v \\ & - \int_{V_\infty} f\gamma^c\mathbf{w}^s\cdot\nabla\nabla^S\gamma^s\cdot\boldsymbol{v}\mathbf{n}\cdot\nabla\gamma\mathrm{d}v - \int_{V_\infty} f\gamma^c\boldsymbol{v}\cdot\nabla^S\gamma^s\mathbf{w}^s\cdot\nabla\nabla\gamma\cdot\mathbf{n}\mathrm{d}v \end{aligned} \tag{579}$$

By $\mathbf{w}-\mathbf{w}^s = \mathbf{w}^c$, $\boldsymbol{v}\cdot\nabla^S\mathbf{w}\cdot\nabla\gamma^s = \boldsymbol{v}\cdot\nabla\mathbf{w}\cdot\boldsymbol{v}\boldsymbol{v}\cdot\nabla\gamma^s$, $\mathbf{n}\cdot\nabla\mathbf{w}\cdot\nabla\gamma = \mathbf{n}\cdot\nabla\mathbf{w}\cdot\mathbf{n}\mathbf{n}\cdot\nabla\gamma$, $\nabla^S\gamma^s = \nabla\gamma^s$, we have

$$\begin{aligned}\int_C \frac{\partial f}{\partial t}\bigg|_1 \mathrm{d}c = {} & \frac{\partial}{\partial t}\int_{V_\infty} f\gamma^c \boldsymbol{v}\cdot\nabla^S\gamma^s\mathbf{n}\cdot\nabla\gamma \mathrm{d}v + \int_{V_\infty} f\mathbf{w}\cdot\nabla^c\gamma^c\boldsymbol{v}\cdot\nabla^S\gamma^s\mathbf{n}\cdot\nabla\gamma\mathrm{d}v \\ & + \int_{V_\infty} f\gamma^c(\boldsymbol{v}\cdot\nabla\mathbf{w}\cdot\boldsymbol{v} + \mathbf{n}\cdot\nabla\mathbf{w}\cdot\mathbf{n})\boldsymbol{v}\cdot\nabla\gamma^s\mathbf{n}\cdot\nabla\gamma\mathrm{d}v \\ & + \int_{V_\infty} f\gamma^c\boldsymbol{v}\cdot\nabla\nabla\gamma^s\cdot\mathbf{w}\mathbf{n}\cdot\nabla\gamma\mathrm{d}v \\ & + \int_{V_\infty} f\gamma^c\mathbf{n}\cdot\nabla\nabla\gamma\cdot\mathbf{w}\boldsymbol{v}\cdot\nabla^S\gamma^s\mathrm{d}v \\ & - \int_{V_\infty}\nabla\cdot\mathbf{w}^s f\gamma^c\boldsymbol{v}\cdot\nabla^S\gamma^s\mathbf{n}\cdot\nabla\gamma\mathrm{d}v - \int_{V_\infty} f\gamma^c\mathbf{w}^s\cdot\nabla\nabla^S\gamma^s\cdot\boldsymbol{v}\mathbf{n}\cdot\nabla\gamma\mathrm{d}v \\ & - \int_{V_\infty} f\gamma^c\boldsymbol{v}\cdot\nabla^S\gamma^s\mathbf{w}^s\cdot\nabla\nabla\gamma\cdot\mathbf{n}\mathrm{d}v \end{aligned} \tag{580}$$

By Equation (53),

$$\nabla \cdot \mathbf{w} = \boldsymbol{\lambda} \cdot \nabla \mathbf{w} \cdot \boldsymbol{\lambda} + \boldsymbol{\nu} \cdot \nabla \mathbf{w} \cdot \boldsymbol{\nu} + \mathbf{n} \cdot \nabla \mathbf{w} \cdot \mathbf{n} \tag{581}$$

Thus Equation (580) can be rearranged into

$$\begin{aligned}\int_C \frac{\partial f}{\partial t}\bigg|_1 \mathrm{d}c &= \frac{\partial}{\partial t}\int_{V_\infty} f\gamma^{\mathrm{c}}\boldsymbol{\nu} \cdot \nabla^{\mathrm{S}}\gamma^{\mathrm{s}}\mathbf{n} \cdot \nabla\gamma \mathrm{d}v + \int_{V_\infty} f\mathbf{w}^{\mathrm{c}} \cdot \nabla^{\mathrm{c}}\gamma^{\mathrm{c}}\boldsymbol{\nu} \cdot \nabla^{\mathrm{S}}\gamma^{\mathrm{s}}\mathbf{n} \cdot \nabla\gamma \mathrm{d}v \\ &+ \int_{V_\infty} f\gamma^{\mathrm{c}}\nabla \cdot \mathbf{w}^{\mathrm{c}}\boldsymbol{\nu} \cdot \nabla^{\mathrm{S}}\gamma^{\mathrm{s}}\mathbf{n} \cdot \nabla\gamma \mathrm{d}v + \int_{V_\infty} f\gamma^{\mathrm{c}}\mathbf{w}^{\mathrm{c}} \cdot \nabla\nabla\gamma^{\mathrm{s}} \cdot \boldsymbol{\nu}\mathbf{n} \cdot \nabla\gamma \mathrm{d}v \\ &+ \int_{V_\infty} f\gamma^{\mathrm{c}}\mathbf{w}^{\mathrm{c}} \cdot \nabla\nabla\gamma \cdot \mathbf{n}\boldsymbol{\nu} \cdot \nabla^{\mathrm{S}}\gamma^{\mathrm{s}} \mathrm{d}v - \int_{V_\infty} f\gamma^{\mathrm{c}}\boldsymbol{\lambda} \cdot \nabla\mathbf{w} \cdot \boldsymbol{\lambda}\boldsymbol{\nu} \cdot \nabla^{\mathrm{S}}\gamma^{\mathrm{s}} \mathrm{d}v\end{aligned} \tag{582}$$

By Equations (258), (261) and

$$\begin{aligned}\nabla \cdot \mathbf{w}^{\mathrm{c}} &= \nabla \cdot (\boldsymbol{\lambda}\boldsymbol{\lambda} \cdot \mathbf{w}) = \nabla \cdot \boldsymbol{\lambda}\boldsymbol{\lambda} \cdot \mathbf{w} + \boldsymbol{\lambda} \cdot \nabla(\boldsymbol{\lambda} \cdot \mathbf{w}) \\ &= \nabla \cdot \boldsymbol{\lambda}\boldsymbol{\lambda} \cdot \mathbf{w} + \boldsymbol{\lambda} \cdot \nabla\boldsymbol{\lambda} \cdot \mathbf{w} + \boldsymbol{\lambda} \cdot \nabla\mathbf{w} \cdot \boldsymbol{\lambda}\end{aligned}$$

$$\begin{aligned}\mathbf{w}^{\mathrm{c}} \cdot \nabla\nabla\gamma^{\mathrm{s}} \cdot \boldsymbol{\nu} &= \mathbf{w} \cdot \nabla^{\mathrm{c}}\nabla\gamma^{\mathrm{s}} \cdot \boldsymbol{\nu} \\ \mathbf{w}^{\mathrm{c}} \cdot \nabla\nabla\gamma \cdot \mathbf{n} &= \mathbf{w} \cdot \nabla^{\mathrm{c}}\nabla\gamma \cdot \mathbf{n}\end{aligned}$$

we obtain

$$\begin{aligned}\int_C \frac{\partial f}{\partial t}\bigg|_1 \mathrm{d}c &= \frac{\partial}{\partial t}\int_{V_\infty} f\gamma^{\mathrm{c}}\boldsymbol{\nu} \cdot \nabla^{\mathrm{S}}\gamma^{\mathrm{s}}\mathbf{n} \cdot \nabla\gamma \mathrm{d}v + \int_{V_\infty} f\mathbf{w}^{\mathrm{c}} \cdot \nabla^{\mathrm{c}}\gamma^{\mathrm{c}}\boldsymbol{\nu} \cdot \nabla^{\mathrm{S}}\gamma^{\mathrm{s}}\mathbf{n} \cdot \nabla\gamma \mathrm{d}v \\ &+ \int_{V_\infty} f\gamma^{\mathrm{c}}\nabla \cdot \boldsymbol{\lambda}\boldsymbol{\lambda} \cdot \mathbf{w}\boldsymbol{\nu} \cdot \nabla^{\mathrm{S}}\gamma^{\mathrm{s}}\mathbf{n} \cdot \nabla\gamma \mathrm{d}v \\ &+ \int_{V_\infty} f\gamma^{\mathrm{c}}\boldsymbol{\lambda} \cdot \nabla^{\mathrm{c}}\boldsymbol{\lambda} \cdot \mathbf{w}^{\mathrm{s}}\boldsymbol{\nu} \cdot \nabla^{\mathrm{S}}\gamma^{\mathrm{s}}\mathbf{n} \cdot \nabla\gamma \mathrm{d}v \\ &- \int_{V_\infty} f\gamma^{\mathrm{c}}\mathbf{w} \cdot \boldsymbol{\lambda}\boldsymbol{\nu} \cdot \nabla\boldsymbol{\lambda} \cdot \nabla\gamma^{\mathrm{s}}\mathbf{n} \cdot \nabla\gamma \mathrm{d}v - \int_{V_\infty} f\gamma^{\mathrm{c}}\mathbf{w} \cdot \boldsymbol{\lambda}\mathbf{n} \cdot \nabla\boldsymbol{\lambda} \cdot \nabla\gamma\boldsymbol{\nu} \cdot \nabla^{\mathrm{S}}\gamma^{\mathrm{s}} \mathrm{d}v \\ &- \int_{V_\infty} f\gamma^{\mathrm{c}}\mathbf{w} \cdot \boldsymbol{\lambda}\mathbf{n} \cdot \nabla\boldsymbol{\lambda} \cdot \nabla\gamma\boldsymbol{\nu} \cdot \nabla^{\mathrm{S}}\gamma^{\mathrm{s}} \mathrm{d}v\end{aligned} \tag{583}$$

By Equations (275) and (165), we have

$$\begin{aligned}\int_C \frac{\partial f}{\partial t}\bigg|_1 \mathrm{d}c &= \frac{\partial}{\partial t}\int_C f \mathrm{d}c - (\mathbf{e} \cdot \mathbf{w}^{\mathrm{s}} f)|_{\mathrm{ends}} + \int_C \nabla \cdot \boldsymbol{\lambda} f\boldsymbol{\lambda} \cdot \mathbf{w} \mathrm{d}c \\ &+ \int_C f\boldsymbol{\lambda} \cdot \nabla^{\mathrm{c}}\boldsymbol{\lambda} \cdot \mathbf{w}^{\mathrm{s}} \mathrm{d}c - \int_C f\mathbf{w} \cdot \boldsymbol{\lambda}\boldsymbol{\nu} \cdot \nabla\boldsymbol{\lambda} \cdot \boldsymbol{\nu} - \int_C f\mathbf{w} \cdot \boldsymbol{\lambda}\mathbf{n} \cdot \nabla\boldsymbol{\lambda} \cdot \mathbf{n} \mathrm{d}c\end{aligned} \tag{584}$$

By Equation (53),

$$\begin{aligned}\int_C f\boldsymbol{\lambda} \cdot \mathbf{w}\nabla \cdot \boldsymbol{\lambda} \mathrm{d}c &= \int_C f\boldsymbol{\lambda} \cdot \mathbf{w}(\boldsymbol{\lambda} \cdot \nabla\boldsymbol{\lambda} \cdot \boldsymbol{\lambda} + \boldsymbol{\nu} \cdot \nabla\boldsymbol{\lambda} \cdot \boldsymbol{\nu} + \mathbf{n} \cdot \nabla\boldsymbol{\lambda} \cdot \mathbf{n}) \mathrm{d}c \\ &= \int_C f\boldsymbol{\lambda} \cdot \mathbf{w}\boldsymbol{\nu} \cdot \nabla\boldsymbol{\lambda} \cdot \boldsymbol{\nu} \mathrm{d}c + \int_C f\boldsymbol{\lambda} \cdot \mathbf{w}\mathbf{n} \cdot \nabla\boldsymbol{\lambda} \cdot \mathbf{n} \mathrm{d}c\end{aligned} \tag{585}$$

Substituting this equation into Equation (584) finally leads to:

$$\int_C \frac{\partial f}{\partial t}\bigg|_1 \mathrm{d}c = \frac{\partial}{\partial t}\int_C f\mathrm{d}c + \int_C (\boldsymbol{\lambda}\cdot\nabla^c\boldsymbol{\lambda})\cdot\mathbf{w}^s f\mathrm{d}c - (\mathbf{e}\cdot\mathbf{w}^c f)|_{\text{ends}}$$

Theorem 36.

$$\int_C \nabla^c \circ F\mathrm{d}c = -\int_C (\boldsymbol{\lambda}\cdot\nabla^c\boldsymbol{\lambda})\circ F\mathrm{d}c + (\mathbf{e}\circ F)|_{\text{ends}} \tag{586}$$

where $\nabla^c \circ F$ can be $\nabla^c f$, $\nabla^c\cdot\mathbf{f}$ or $\nabla^c\times\mathbf{f}$

Proof. By the three indicator functions γ^c (**l**), γ^s(**u**) and γ(**x**) defined previously, we obtain

$$\int_C \nabla^c \circ F\mathrm{d}c = \int_{V_\infty} \nabla^c \circ F\gamma^c \boldsymbol{v}\cdot\nabla^S\gamma^s\mathbf{n}\cdot\nabla\gamma\mathrm{d}v \tag{587}$$

An application of the chain rule to the right-hand side of Equation (587) yields

$$\begin{aligned}\int_C \nabla^c \circ F\mathrm{d}c = &\int_{V_\infty} \nabla\cdot(\boldsymbol{\lambda}\boldsymbol{\lambda}\circ F\gamma^c\boldsymbol{v}\cdot\nabla^S\gamma^s\mathbf{n}\cdot\nabla\gamma)\mathrm{d}v - \int_{V_\infty}\nabla\cdot\boldsymbol{\lambda}\boldsymbol{\lambda}\circ F\gamma^c\boldsymbol{v}\cdot\nabla^S\gamma^s\mathbf{n}\cdot\nabla\gamma\mathrm{d}v\\ &-\int_{V_\infty}\boldsymbol{\lambda}\cdot\nabla\boldsymbol{\lambda}\circ F\gamma^c\boldsymbol{v}\cdot\nabla^S\gamma^s\mathbf{n}\cdot\nabla\gamma\mathrm{d}v - \int_{V_\infty}\boldsymbol{\lambda}\circ F\boldsymbol{\lambda}\cdot\nabla\gamma^c\boldsymbol{v}\cdot\nabla^S\gamma^s\mathbf{n}\cdot\nabla\gamma\mathrm{d}v\\ &-\int_{V_\infty}\boldsymbol{\lambda}\circ F\gamma^c\boldsymbol{\lambda}\cdot\nabla(\boldsymbol{v}\cdot\nabla^S\gamma^s)\mathbf{n}\cdot\nabla\gamma\mathrm{d}v\\ &-\int_{V_\infty}\boldsymbol{\lambda}\circ F\gamma^c\boldsymbol{v}\cdot\nabla^S\gamma^s\boldsymbol{\lambda}\cdot\nabla(\mathbf{n}\cdot\nabla\gamma)\mathrm{d}v\end{aligned} \tag{588}$$

By Equations (195), (200) and

$$\nabla(\boldsymbol{v}\cdot\nabla^S\gamma^s) = \nabla\boldsymbol{v}\cdot\nabla^S\gamma^s + \nabla\nabla^S\gamma^s\cdot\boldsymbol{v} \tag{589}$$

$$\nabla(\mathbf{n}\cdot\nabla\gamma) = \nabla\mathbf{n}\cdot\nabla\gamma + \nabla\nabla\gamma\cdot\mathbf{n} \tag{590}$$

the last two terms of Equation (588) are equivalent to

$$\int_{V_\infty}\boldsymbol{\lambda}\circ F\gamma^c\boldsymbol{\lambda}\cdot\nabla(\nabla^S\gamma^s\cdot\boldsymbol{v})\mathbf{n}\cdot\nabla\gamma\mathrm{d}v = \int_{V_\infty}\boldsymbol{\lambda}\circ F\gamma^c\boldsymbol{\lambda}\cdot\nabla\nabla^S\gamma^s\cdot\boldsymbol{v}\mathbf{n}\cdot\nabla\gamma\mathrm{d}v \tag{591}$$

$$\int_{V_\infty}\boldsymbol{\lambda}\circ F\gamma^c\nabla^S\gamma^s\cdot\boldsymbol{v}\boldsymbol{\lambda}\cdot\nabla(\mathbf{n}\cdot\nabla\gamma)\mathrm{d}v = \int_{V_\infty}\boldsymbol{\lambda}\circ F\gamma^c\boldsymbol{\lambda}\cdot\nabla\nabla\gamma\cdot\mathbf{n}\boldsymbol{v}\cdot\nabla^S\gamma^s\mathrm{d}v \tag{592}$$

By Equation (216) and $\boldsymbol{\lambda}\boldsymbol{\lambda}\cdot\nabla = \nabla^c$, Equations (591) and (592) become

$$\int_{V_\infty}\boldsymbol{\lambda}\circ F\gamma^c\boldsymbol{\lambda}\cdot\nabla(\nabla^S\gamma^s\cdot\boldsymbol{v})\mathbf{n}\cdot\nabla\gamma\mathrm{d}v = -\int_{V_\infty}\gamma^c F\circ\nabla^c\nabla\gamma^s\cdot\boldsymbol{v}\mathbf{n}\cdot\gamma\mathrm{d}v \tag{593}$$

$$\int_{V_\infty}\boldsymbol{\lambda}\circ F\gamma^c\nabla^S\gamma^s\cdot\boldsymbol{v}\boldsymbol{\lambda}\cdot\nabla(\mathbf{n}\cdot\nabla\gamma)\mathrm{d}v = -\int_{V_\infty}\gamma^c F\circ\nabla^c\nabla\gamma\cdot\mathbf{n}\boldsymbol{v}\cdot\nabla^S\gamma^s\mathrm{d}v \tag{594}$$

Equations (258) and (261) enable us to write Equations (593) and (594) as

$$\int_{V_\infty} \boldsymbol{\lambda} \circ F\gamma^{c}\boldsymbol{\lambda} \cdot \nabla(\nabla^{S}\gamma^{s} \cdot \boldsymbol{v})\mathbf{n} \cdot \nabla\gamma dv = \int_{V_\infty} \gamma^{c}F \circ \boldsymbol{\lambda}\boldsymbol{v} \cdot \nabla\boldsymbol{\lambda} \cdot \nabla\gamma^{s}\mathbf{n} \cdot \nabla\gamma dv \tag{595}$$

$$\int_{V_\infty} \boldsymbol{\lambda} \circ F\gamma^{c}\nabla^{S}\gamma^{s} \cdot \boldsymbol{v}\boldsymbol{\lambda} \cdot \nabla(\mathbf{n} \cdot \nabla\gamma)dv = \int_{V_\infty} \gamma^{c}F \circ \boldsymbol{\lambda}\mathbf{n} \cdot \nabla\boldsymbol{\lambda} \cdot \nabla\gamma\boldsymbol{v} \cdot \nabla^{S}\gamma^{s}dv \tag{596}$$

Making use of Equation (231) and substituting Equations (595) and (596) into Equation (588) yields

$$\begin{aligned}\int_C \nabla^{c} \circ F dc = &-\int_{V_\infty} \nabla \cdot \boldsymbol{\lambda}\boldsymbol{\lambda} \circ F\gamma^{c}\boldsymbol{v} \cdot \nabla^{S}\gamma^{s}\mathbf{n} \cdot \nabla\gamma dv - \int_{V_\infty} \boldsymbol{\lambda} \cdot \nabla\boldsymbol{\lambda} \circ F\gamma^{c}\boldsymbol{v} \cdot \nabla^{S}\gamma^{s}\mathbf{n} \cdot \nabla\gamma dv \\ &- \int_{V_\infty} \boldsymbol{\lambda} \circ F\boldsymbol{\lambda} \cdot \nabla\gamma^{c}\boldsymbol{v} \cdot \nabla^{S}\gamma^{s}\mathbf{n} \cdot \nabla\gamma dv - \int_{V_\infty} \gamma^{c}F \circ \boldsymbol{\lambda}\boldsymbol{v} \cdot \nabla\boldsymbol{\lambda} \cdot \nabla\gamma^{s}\mathbf{n} \cdot \nabla\gamma dv \\ &- \int_{V_\infty} \gamma^{c}F \circ \boldsymbol{\lambda}\mathbf{n} \cdot \nabla\boldsymbol{\lambda} \cdot \nabla\gamma\boldsymbol{v} \cdot \nabla^{S}\gamma^{s}dv\end{aligned} \tag{597}$$

By Equations (275) and (165), Equations (597) leads to

$$\begin{aligned}\int_C \nabla^{c} \circ F dc = &-\int_C \nabla \cdot \boldsymbol{\lambda}\boldsymbol{\lambda} \circ F dc - \int_C \boldsymbol{\lambda} \cdot \nabla^{c}\boldsymbol{\lambda} \circ F dc + (\mathbf{e} \circ F)|_{\text{ends}} \\ &- \int_C F \circ \boldsymbol{\lambda}\boldsymbol{v} \cdot \nabla\boldsymbol{\lambda} \cdot \boldsymbol{v} dc - \int_C F \circ \boldsymbol{\lambda}\mathbf{n} \cdot \nabla\boldsymbol{\lambda} \cdot \mathbf{n} dc\end{aligned} \tag{598}$$

By Equation (53), we have

$$\begin{aligned}\nabla \cdot \boldsymbol{\lambda} &= \boldsymbol{\lambda} \cdot \nabla\boldsymbol{\lambda} \cdot \boldsymbol{\lambda} + \mathbf{n} \cdot \nabla\boldsymbol{\lambda} \cdot \mathbf{n} + \boldsymbol{v} \cdot \nabla\boldsymbol{\lambda} \cdot \boldsymbol{v} \\ &= \mathbf{n} \cdot \nabla\boldsymbol{\lambda} \cdot \mathbf{n} + \boldsymbol{v} \cdot \nabla\boldsymbol{\lambda} \cdot \boldsymbol{v}\end{aligned}$$

Therefore, Equation (598) becomes

$$\int_C \nabla^{c} \circ F dc = -\int_C (\boldsymbol{\lambda} \cdot \nabla^{c}\boldsymbol{\lambda}) \circ F dc + (\mathbf{e} \circ F)|_{\text{ends}}$$

Remark. Equation (586) contains the following three formulas:

Gradient

$$\int_C \nabla^{c} f dc = -\int_C (\boldsymbol{\lambda} \cdot \nabla^{c}\boldsymbol{\lambda}) f dc + (\mathbf{e}f)|_{\text{ends}} \tag{599}$$

Divergence

$$\int_C \nabla^{c} \cdot \mathbf{f} dc = -\int_C (\boldsymbol{\lambda} \cdot \nabla^{c}\boldsymbol{\lambda}) \cdot \mathbf{f} dc + (\mathbf{e} \cdot \mathbf{f})|_{\text{ends}} \tag{600}$$

Curl

$$\int_C \nabla^{c} \times \mathbf{f} dc = -\int_C (\boldsymbol{\lambda} \cdot \nabla^{c}\boldsymbol{\lambda}) \times \mathbf{f} dc + (\mathbf{e} \times \mathbf{f})|_{\text{ends}} \tag{601}$$

4.3.6 Integration of curvilineal operators over a straight line fixed in space

This group of integration theorems involves integrations of $(\partial f/\partial t)|_1$, $\nabla^c f$, $\nabla^c \cdot \mathbf{f}$ and $\nabla^c \times \mathbf{f}$ over a straight line L fixed in space (Figures 22 and 24). Here $f(\mathbf{x}, t)$ and $\mathbf{f}(\mathbf{x}, t)$ are microscale scalar and vector functions of position vector $\mathbf{l}$ and time t.

Theorem 37.

$$\int_L \left.\frac{\partial f}{\partial t}\right|_1 \mathrm{d}c = \frac{\partial}{\partial t}\int_L f\mathrm{d}c - (\mathbf{e}\cdot\mathbf{w})f|_{\text{ends}} \tag{602}$$

where L is the straight line segment of integration. $\boldsymbol{\Lambda}$ is the unit vector tangent to L, $\mathbf{e}$ the particular unit vector which is tangent to L at its endpoints with positive outward direction from the line segment, and $\mathbf{w}$ the velocity of the endpoints of L such that $\mathbf{w} = \boldsymbol{\Lambda}\boldsymbol{\Lambda}\cdot\mathbf{w}$.

Proof. By defining an indicator function $\gamma^c(\mathbf{l})$ such that $\gamma^c = 1$ on the straight-line segment L and $\gamma^c = 0$ elsewhere,

$$\int_L \left.\frac{\partial f}{\partial t}\right|_1 \mathrm{d}L = \int_{L_\infty} \left.\frac{\partial f}{\partial t}\right|_1 \gamma^c \mathrm{d}L \tag{603}$$

Applying the chain rule to the right-hand side of Equation (603) yields

$$\int_L \left.\frac{\partial f}{\partial t}\right|_1 \mathrm{d}c = \int_{L_\infty} \left.\frac{\partial (f\gamma^c)}{\partial t}\right|_1 \mathrm{d}L - \int_{L_\infty} f\left.\frac{\partial \gamma^c}{\partial t}\right|_1 \mathrm{d}L \tag{604}$$

The integration operator can exchange with the differentiation operator with respect to the time t in the first term on the right side of Equation (604), due to the independency of the integration domain. By using Equation (229) for the second term, Equation (604) becomes

$$\int_L \left.\frac{\partial f}{\partial t}\right|_1 \mathrm{d}L = \frac{\partial}{\partial t}\int_L f\mathrm{d}L + \int_{L_\infty} f\mathbf{w}\cdot\nabla^c\gamma^c \mathrm{d}L \tag{605}$$

By Equation (166), we finally obtain

$$\int_L \left.\frac{\partial f}{\partial t}\right|_1 \mathrm{d}L = \frac{\partial}{\partial t}\int_L f\mathrm{d}c - (\mathbf{e}\cdot\mathbf{w}f)|_{\text{ends}} \tag{606}$$

Theorem 38.

$$\int_L \nabla^c \circ F\mathrm{d}L = (\mathbf{e}\circ F)|_{\text{ends}} \tag{607}$$

where $\nabla^c \circ F$ can be $\nabla^c f$, $\nabla^c \cdot \mathbf{f}$ or $\nabla^c \times \mathbf{f}$. ∇^c is the curvilineal operator such that $\nabla^c = \boldsymbol{\Lambda}\boldsymbol{\Lambda}\cdot\nabla$.

Proof. By defining an indicator function $\gamma^c(\mathbf{l})$ such that $\gamma^c = 1$ on the straight line segment L and $\gamma^c = 0$ elsewhere,

$$\int_L \nabla^c \circ F\mathrm{d}L = \int_{L_\infty} \nabla^c \circ F\gamma^c \mathrm{d}L \tag{608}$$

An application of the chain rule to the right side of Equation (608) yields

$$\int_L \nabla^c \circ F\mathrm{d}L = \int_{L_\infty} \nabla^c \circ (F\gamma)\mathrm{d}L - \int_{L_\infty} \nabla^c \gamma^c \circ F\mathrm{d}c \tag{609}$$

By Equations (165)–(167) and Equations (246)–(248), Equation (609) leads to

$$\int_L \nabla^c \circ F\mathrm{d}L = (\mathbf{e} \circ F)|_{\text{ends}} \tag{610}$$

Remark. Equation (607) contains the following three formulars:

Gradient

$$\int_L \nabla^c f\mathrm{d}c = (\mathbf{e}f)|_{\text{ends}} \tag{611}$$

Divergence

$$\int_L \nabla^c \cdot \mathbf{f}\mathrm{d}L = (\mathbf{e} \cdot \mathbf{f})|_{\text{ends}} \tag{612}$$

Curl

$$\int_L \nabla^c \times \mathbf{f}\mathrm{d}L = (\mathbf{e} \times \mathbf{f})|_{\text{ends}} \tag{613}$$

5. AVERAGING THEOREMS

We develop 33 averaging theorems that relate the integral of a derivative of a microscale function to the derivative of that integral. These theorems are powerful tools for the scaling among microscales, macroscales, and megascales.

In developing averaging theorems, the integration is made over a region within an REV. These REVs are constructed to reflect the different scales of integration for each coordinate direction. Although the averaging region is contained within a volume, the actual integration may occur over curves, surfaces, or subvolumes within that REV.

In the derivation of averaging theorems, we often use the following four indicator functions:

(1) $\gamma(\mathbf{x}, \xi)$ ($\gamma = 1$ within the REV and $\gamma = 0$ elsewhere);
(2) $\gamma^\alpha(\mathbf{x}+\xi, t)$ ($\gamma^\alpha = 1$ in the portion V_α occupied by the α-phase within the REV and $\gamma^\alpha = 0$ elsewhere);
(3) $\gamma^s(\mathbf{u}(\mathbf{x}+\xi), t)$ ($\gamma^s = 1$ on the surface $S_{\alpha\beta}$ within the REV between the α-phase and all others and $\gamma^s = 0$ elsewhere); and
(4) $\gamma^c(\mathbf{l}(\mathbf{x}+\xi), t)$ ($\gamma^c = 1$ along the boundary curve $C_{\alpha\beta\varepsilon}$ of $S_{\alpha\beta}$ and $\gamma^c = 0$ elsewhere).

5.1 Three macroscopic dimensions

In this section we develop 11 averaging theorems regarding spatial integration of spatial operators, surficial integration of surficial operators, and curvilineal integration of curvilineal operators within a REV of 3 macroscopic dimensions. The appropriate REV is a spherical volume with a radius independent of position and time (Figures 25–27).

5.1.1 Spatial operator theorems

The REV used for this group of averaging theorems is shown in Figure 25.

Theorem 39.

$$\int_{V_\alpha} \left.\frac{\partial f}{\partial t}\right|_{\mathbf{x}} dv = \frac{\partial}{\partial t}\int_{V_\alpha} f dv - \int_{S_{\alpha\beta}} \mathbf{n}\cdot\mathbf{w} f ds \tag{614}$$

where V_α is the portion of the REV occupied by the α-phase. $S_{\alpha\beta}$ is the surface within the REV between the α-phase and all the other phases. $\mathbf{n}$ is the unit vector normal to $S_{\alpha\beta}$ pointing outward from the α-phase, $\mathbf{w}$ the velocity of $S_{\alpha\beta}$ and f a spatial function $f(\mathbf{x}, t)$ defined in the α-phase.

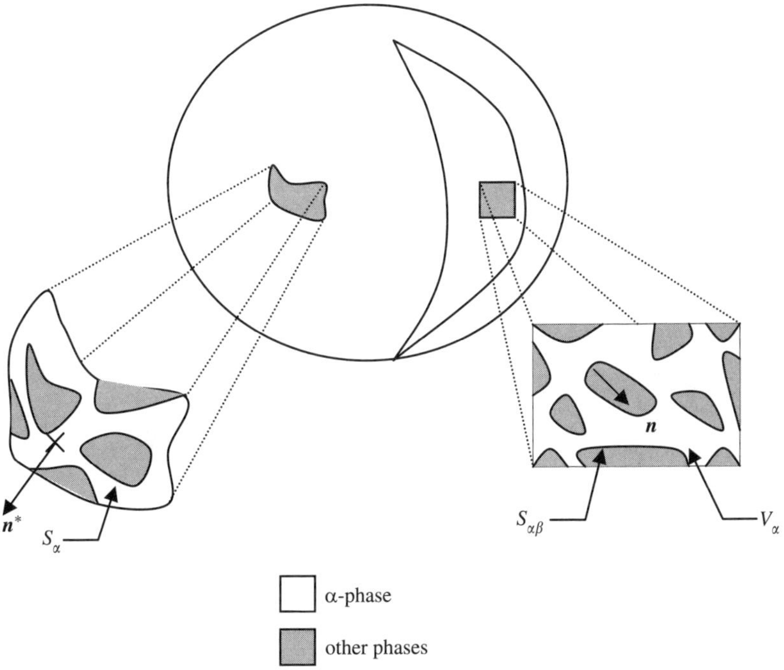

Figure 25 Spherical REV (independent of space and time; wedge removed for illustrative purposes).

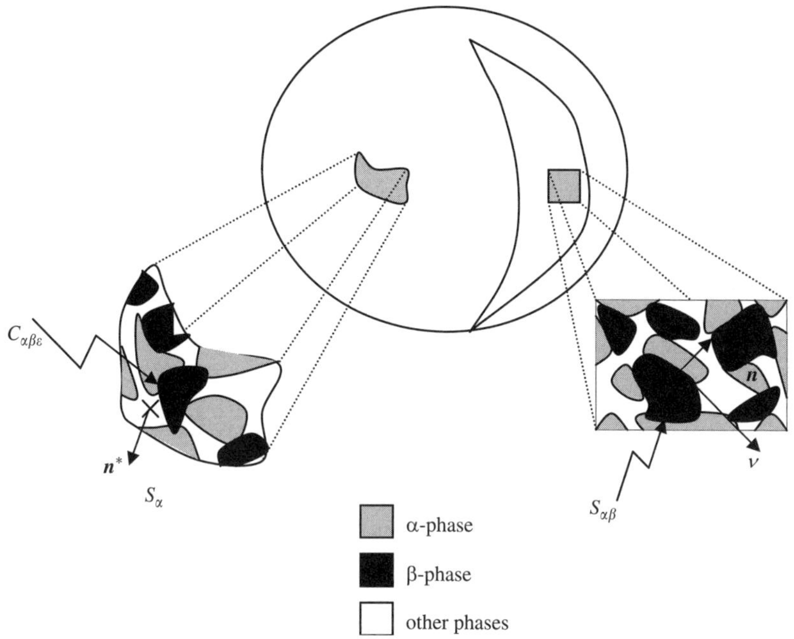

Figure 26 Spherical REV (independent of space and time; wedge removed for illustrative purposes).

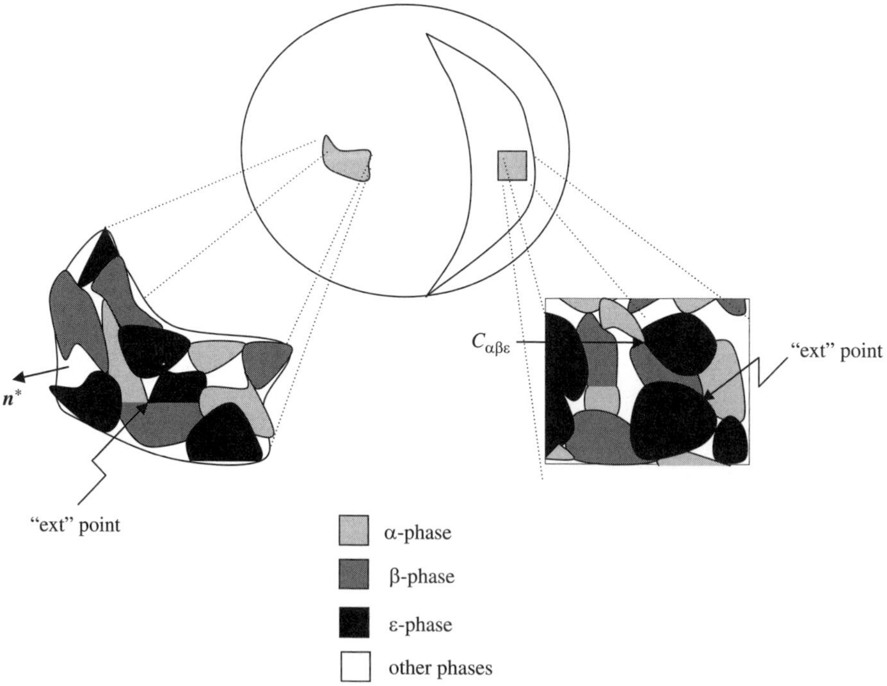

Figure 27 Spherical REV (independent of space and time; wedge removed for illustrative purposes).

Proof. By the definitions of the indicator functions $\gamma^\alpha(\mathbf{x}+\xi, t)$ and $\gamma(\mathbf{x}, \xi)$, we obtain

$$\int_{V_\alpha} \left.\frac{\partial f}{\partial t}\right|_{\mathbf{x}} dv = \int_{V_\infty} \left.\frac{\partial f}{\partial t}\right|_{\mathbf{x}} \gamma\gamma^\alpha dv \tag{615}$$

An application of the chain rule to the right-hand side of Equation (615) yields

$$\int_{V_\alpha} \left.\frac{\partial f}{\partial t}\right|_{\mathbf{x}} dv = \int_{V_\infty} \left.\frac{\partial (f\gamma\gamma^\alpha)}{\partial t}\right|_{\mathbf{x}} dv - \int_{V_\infty} f \left.\frac{\partial (\gamma\gamma^\alpha)}{\partial t}\right|_{\mathbf{x}} dv \tag{616}$$

Note that the integration domain in the first integral on the right side of Equation (616) is independent of the time t. Therefore,

$$\int_{V_\infty} \frac{\partial (f\gamma\gamma^\alpha)}{\partial t} dv = \frac{\partial}{\partial t}\int_{V_\infty} f\gamma\gamma^\alpha dv \tag{617}$$

Again using the definitions of the indicator functions γ and γ^α,

$$\int_{V_\infty} \frac{\partial (f\gamma\gamma^\alpha)}{\partial t} dv = \frac{\partial}{\partial t}\int_{V_\alpha} f dv \tag{618}$$

By the definitions of γ and γ^α and Equation (180) with γ replaced by $\gamma\gamma^\alpha$, the second integral on the right-hand side of Equation (616) becomes

$$\int_{V_\infty} f \left.\frac{\partial (\gamma\gamma^\alpha)}{\partial t}\right|_{\mathbf{x}} dv = -\int_{V_\infty} f\mathbf{w}\cdot\nabla(\gamma\gamma^\alpha) dv \tag{619}$$

Using the definitions of γ and γ^α again and Equation (186), we have

$$\int_{V_\infty} f \left.\frac{\partial (\gamma\gamma^\alpha)}{\partial t}\right|_{\mathbf{x}} dv = \int_{S_{\alpha\beta}} f\mathbf{w}\cdot\mathbf{n} ds \tag{620}$$

Substituting Equations (618) and (620) into Equation (616) finally yields

$$\int_{V_\alpha} \left.\frac{\partial f}{\partial t}\right|_{\mathbf{x}} dv = \frac{\partial}{\partial t}\int_{V_\alpha} f dv - \int_{S_{\alpha\beta}} \mathbf{n}\cdot\mathbf{w} f ds$$

Theorem 40.

$$\int_{V_\alpha} \nabla \circ F dv = \nabla \circ \int_{V_\alpha} F dv + \int_{S_{\alpha\beta}} \mathbf{n}\circ F ds \tag{621}$$

where $\nabla \circ F$ can be any of the three operators: ∇f, $\nabla\cdot\mathbf{f}$ and $\nabla\times\mathbf{f}$. $f(\mathbf{x}, t)$, and $\mathbf{f}(\mathbf{x}, t)$ are spatial scalar and vector functions of position vector $\mathbf{x}$ and time t defined in α-phase, respectively. ∇ in the integrand is the microscopic spatial operator, $\nabla = \nabla_\xi$. ∇ operating on the integral is the macroscopic spatial operator, $\nabla = \nabla\mathbf{x}$.

Proof. By the definitions of the indicator functions $\gamma^\alpha(\mathbf{x}+\xi, t)$ and $\gamma(\mathbf{x}, \xi)$, we have

$$\int_{V_\alpha} \nabla \circ F dv = \int_{V_\infty} \nabla \circ F\gamma\gamma^\alpha dv \tag{622}$$

An application of the chain rule to the right side of Equation (622) yields

$$\int_{V_\alpha} \nabla \circ F dv = \int_{V_\infty} \nabla \circ (F\gamma\gamma^\alpha) dv - \int_{V_\infty} \nabla(\gamma\gamma^\alpha) \circ F dv \tag{623}$$

Because the integration domain is independent of the integration variables, the first integral on the right side of Equation (623) may be written as

$$\int_{V_\infty} \nabla \circ (F\gamma\gamma^\alpha) dv = \nabla \circ \int_{V_\infty} F\gamma\gamma^\alpha dv$$

which is, by the definitions of the indicator functions γ^α and γ,

$$\int_{V_\infty} \nabla \circ (F\gamma\gamma^\alpha) dv = \nabla \circ \int_{V_\alpha} F dv \tag{624}$$

By Equations (185)–(187), the second integral on the right side of Equation (623) becomes

$$\int_{V_\infty} \nabla(\gamma\gamma^\alpha) \circ F dv = - \int_{S_{\alpha\beta}} \mathbf{n} \circ F ds \tag{625}$$

Substituting Equations (624) and (625) into Equation (623) finally yields

$$\int_{V_\alpha} \nabla \circ F dv = \nabla \circ \int_{V_\alpha} F dv + \int_{S_{\alpha\beta}} \mathbf{n} \circ F ds$$

Remark. Equation (621) contains the following three formulas:

$$\int_{V_\alpha} \nabla f dv = \nabla \int_{V_\alpha} f dv + \int_{S_{\alpha\beta}} \mathbf{n} f ds \tag{626}$$

$$\int_{V_\alpha} \nabla \cdot \mathbf{f} dv = \nabla \cdot \int_{V_\alpha} \mathbf{f} dv + \int_{S_{\alpha\beta}} \mathbf{n} \cdot \mathbf{f} ds \tag{627}$$

$$\int_{V_\alpha} \nabla \times \mathbf{f} dv = \nabla \times \int_{V_\alpha} \mathbf{f} dv + \int_{S_{\alpha\beta}} \mathbf{n} \times \mathbf{f} ds \tag{628}$$

Theorem 40 has the following alternative form:

Theorem 41.

$$\int_{V_\alpha} \nabla \circ F dv = \int_{S_\alpha} \mathbf{n}^* \circ F ds + \int_{S_{\alpha\beta}} \mathbf{n} \circ F ds \tag{629}$$

where S_α is the portion of the external boundary of the REV that intersects the α-phase. $\mathbf{n}^*$ is the unit vector normal to the external boundary of the REV pointing outward from the REV.

Proof. By the definitions of the indicator functions of $\gamma^\alpha(\mathbf{x} + \xi, t)$ and $\gamma(\mathbf{x}, \xi)$, we obtain

$$\int_{V_\alpha} \nabla \circ F dv = \int_{V_\infty} \nabla \circ F\gamma\gamma^\alpha dv \tag{630}$$

An application of the chain rule to the right-hand side of Equation (630) yields

$$\int_{V_\alpha} \nabla \circ F dv = \int_{V_\infty} \nabla \circ (F\gamma\gamma^\alpha) dv - \int_{V_\infty} \nabla\gamma \circ F\gamma^\alpha dv - \int_{V_\infty} \nabla\gamma^\alpha \circ F\gamma dv \tag{631}$$

By Equations (230)–(232), it becomes

$$\int_{V_\alpha} \nabla \circ F dv = -\int_{V_\infty} \nabla\gamma \circ F\gamma^\alpha dv - \int_{V_\infty} \nabla\gamma^\alpha \circ F\gamma dv \tag{632}$$

By Equations (185)–(187) and the definitions of the indicator functions γ and γ^α, Equation (632) may be written as

$$\int_{V_\alpha} \nabla \circ F dv = \int_{S_\alpha} \mathbf{n}^* \circ F ds + \int_{S_{\alpha\beta}} \mathbf{n} \circ F ds$$

Remark 1. Equation (629) contains the following three formulas:

Gradient

$$\int_{V_\alpha} \nabla f dv = \int_{S_\alpha} \mathbf{n}^* f ds + \int_{S_{\alpha\beta}} \mathbf{n} f ds \tag{633}$$

Divergence

$$\int_{V_\alpha} \nabla \cdot \mathbf{f} dv = \int_{S_\alpha} \mathbf{n}^* \cdot \mathbf{f} ds + \int_{S_{\alpha\beta}} \mathbf{n} \cdot \mathbf{f} ds \tag{634}$$

Curl

$$\int_{V_\alpha} \nabla \times \mathbf{f} dv = \int_{S_\alpha} \mathbf{n}^* \times \mathbf{f} ds + \int_{S_{\alpha\beta}} \mathbf{n} \times \mathbf{f} ds \tag{635}$$

Remark 2. By comparing Theorems 40 and 41, we readily obtain

$$\int_{S_\alpha} \mathbf{n}^* \circ F ds = \nabla \circ \int_{V_\alpha} F dv$$

5.1.2 Surficial operator theorems

The REV used for this group of averaging theorems is shown in Figure 26.

Theorem 42.

$$\begin{aligned} \int_{S_{\alpha\beta}} \left.\frac{\partial f}{\partial t}\right|_{\mathbf{u}} ds = {} & \frac{\partial}{\partial t} \int_{S_{\alpha\beta}} f ds + \nabla \cdot \int_{S_{\alpha\beta}} \mathbf{w}^c f ds \\ & - \int_{S_{\alpha\beta}} (\nabla^S \cdot \mathbf{n}) \mathbf{n} \cdot \mathbf{w}^c f ds - \int_{C_{\alpha\beta\varepsilon}} \boldsymbol{\nu} \cdot \mathbf{w}^s f dc \end{aligned} \tag{636}$$

where $S_{\alpha\beta}$ is the surface within the REV between the α- and β-phases. $C_{\alpha\beta\varepsilon}$ is the boundary curve of $S_{\alpha\beta}$ within the REV that is also the location where the α- and β-phases meet other phases. $\mathbf{n}$ is the unit vector normal to $S_{\alpha\beta}$ and $\boldsymbol{\nu}$ the unit vector normal on the $C_{\alpha\beta\varepsilon}$ curve pointing positive outward from $S_{\alpha\beta}$ such that

$\mathbf{n}\cdot\boldsymbol{\nu}=0$. ∇^S is the microscopic surficial operator such that $\nabla^S=\nabla\xi-\mathbf{nn}\cdot\nabla\xi$ and ∇ operating on the integral is the macroscopic spatial operator, $\nabla=\nabla_\mathbf{x}$. $\mathbf{w}$ is the velocity of $S_{\alpha\beta}$, $\mathbf{w}^c$ is the velocity of $S_{\alpha\beta}$ normal to $S_{\alpha\beta}$ such that $\mathbf{w}^c=\mathbf{nn}\cdot\mathbf{w}$, $\mathbf{w}^s$ is the velocity of $S_{\alpha\beta}$ tangent to $S_{\alpha\beta}$ such that $\mathbf{w}^s=\mathbf{w}-\mathbf{w}^c$ and f is a surficial function $f(\mathbf{u},t)$ defined on $S_{\alpha\beta}$.

Proof. By the definitions of the three indicator functions $\gamma(x,\xi)$, $\gamma^s(\mathbf{u}(\mathbf{x}+\xi),t)$, and $\gamma^\alpha(\mathbf{x}+\xi,t)$, we have

$$\int_{S_{\alpha\beta}}\left.\frac{\partial f}{\partial t}\right|_\mathbf{u}\mathrm{d}s=-\int_{V_\infty}\left.\frac{\partial f}{\partial t}\right|_\mathbf{u}\gamma\gamma^s\mathbf{n}\cdot\nabla\gamma^\alpha\mathrm{d}v \tag{637}$$

Applying Equation (130) yields

$$\int_{S_{\alpha\beta}}\left.\frac{\partial f}{\partial t}\right|_\mathbf{u}\mathrm{d}s=-\int_{V_\infty}\left.\frac{\partial f}{\partial t}\right|_\mathbf{x}\gamma\gamma^s\mathbf{n}\cdot\nabla\gamma^\alpha\mathrm{d}v-\int_{V_\infty}\mathbf{w}\cdot\mathbf{nn}\cdot\nabla f\gamma\gamma^s\mathbf{n}\cdot\nabla\gamma^\alpha\mathrm{d}v \tag{638}$$

An application of the chain rule to the two integrals on the right-hand side of Equation (638) leads to

$$\begin{aligned}\int_{S_{\alpha\beta}}\left.\frac{\partial f}{\partial t}\right|_\mathbf{u}\mathrm{d}s=&-\int_{V_\infty}\frac{\partial}{\partial t}(f\gamma\gamma^s\mathbf{n}\cdot\nabla\gamma^\alpha)\mathrm{d}v+\int_{V_\infty}f\left.\frac{\partial\gamma}{\partial t}\right|_\mathbf{x}\gamma^s\mathbf{n}\cdot\nabla\gamma^\alpha\mathrm{d}v\\&+\int_{V_\infty}f\gamma\left.\frac{\partial\gamma^s}{\partial t}\right|_\mathbf{x}\mathbf{n}\cdot\nabla\gamma^\alpha\mathrm{d}v+\int_{V_\infty}f\gamma\gamma^s\left(\frac{\partial\mathbf{n}}{\partial t}\cdot\nabla\gamma^\alpha+\mathbf{n}\cdot\nabla\frac{\partial\gamma^\alpha}{\partial t}\right)\mathrm{d}v\\&-\int_{V_\infty}\nabla\cdot(\mathbf{nw}\cdot\mathbf{n}f\gamma\gamma^s\mathbf{n}\cdot\nabla\gamma^\alpha)\mathrm{d}v+\int_{V_\infty}\nabla\cdot\mathbf{nw}\cdot\mathbf{n}f\gamma\gamma^s\mathbf{n}\cdot\nabla\gamma^\alpha\mathrm{d}v\\&+\int_{V_\infty}\mathbf{n}\cdot\nabla\mathbf{w}\cdot\mathbf{n}f\gamma\gamma^s\mathbf{n}\cdot\nabla\gamma^\alpha\mathrm{d}v+\int_{V_\infty}\mathbf{n}\cdot\nabla\mathbf{n}\cdot\mathbf{w}f\gamma\gamma^s\mathbf{n}\cdot\nabla\gamma^\alpha\mathrm{d}v\\&+\int_{V_\infty}\mathbf{w}\cdot\mathbf{n}f\mathbf{n}\cdot\nabla\gamma\gamma^s\mathbf{n}\cdot\nabla\gamma^\alpha\mathrm{d}v+\int_{V_\infty}\mathbf{w}\cdot\mathbf{nn}\cdot\nabla\gamma^s f\gamma\mathbf{n}\cdot\nabla\gamma^\alpha\mathrm{d}v\\&+\int_{V_\infty}\mathbf{w}\cdot\mathbf{n}f\gamma\gamma^s\mathbf{n}\cdot(\nabla\mathbf{n}\cdot\nabla\gamma^\alpha+\nabla\nabla\gamma^\alpha\cdot\mathbf{n})\mathrm{d}v\end{aligned} \tag{639}$$

Since the integration regions are constants for the first and fifth terms on the right side of Equation (639), the operators $\partial/\partial t$ and ∇ can be interchanged with the integration sign. Also applying Equations (180), (194), (195), (200), and (224), we obtain

$$\begin{aligned}\int_{S_{\alpha\beta}}\left.\frac{\partial f}{\partial t}\right|_\mathbf{u}\mathrm{d}s=&-\frac{\partial}{\partial t}\int_{V_\infty}f\gamma\gamma^s\mathbf{n}\cdot\nabla\gamma^\alpha\mathrm{d}v-\int_{V_\infty}f\mathbf{w}\cdot\nabla\gamma\gamma^s\mathbf{n}\cdot\nabla\gamma^\alpha\mathrm{d}v\\&-\int_{V_\infty}f\gamma\mathbf{w}\cdot\nabla\gamma^s\mathbf{n}\cdot\nabla\gamma^\alpha\mathrm{d}v-\int_{V_\infty}f\gamma\gamma^s\mathbf{n}\cdot\nabla(\mathbf{w}\cdot\nabla\gamma^\alpha)\mathrm{d}v\\&-\nabla\cdot\int_{V_\infty}\mathbf{nw}\cdot\mathbf{n}f\gamma\gamma^s\mathbf{n}\cdot\nabla\gamma^\alpha\mathrm{d}v+\int_{V_\infty}\nabla\cdot\mathbf{nw}\cdot\mathbf{n}f\gamma\gamma^s\mathbf{n}\cdot\nabla\gamma^\alpha\mathrm{d}v\\&+\int_{V_\infty}\mathbf{n}\cdot\nabla\mathbf{w}\cdot\mathbf{n}f\gamma\gamma^s\mathbf{n}\cdot\nabla\gamma^\alpha\mathrm{d}v+\int_{V_\infty}\mathbf{n}\cdot\nabla\mathbf{n}\cdot\mathbf{w}f\gamma\gamma^s\mathbf{n}\cdot\nabla\gamma^\alpha\mathrm{d}v\\&+\int_{V_\infty}\mathbf{w}\cdot\mathbf{n}f\mathbf{n}\cdot\nabla\gamma\gamma^s\mathbf{n}\cdot\nabla\gamma^\alpha\mathrm{d}v+\int_{V_\infty}\mathbf{w}\cdot\mathbf{nn}\cdot\nabla\gamma^s f\gamma\mathbf{n}\cdot\nabla\gamma^\alpha\mathrm{d}v\\&+\int_{V_\infty}\mathbf{w}\cdot\mathbf{n}f\gamma\gamma^s\mathbf{n}\cdot\nabla\nabla\gamma^\alpha\cdot\mathbf{n}\mathrm{d}v\end{aligned} \tag{640}$$

By making use of the relation,

$$\nabla(\mathbf{w}\cdot\nabla\gamma^{\alpha}) = \nabla\mathbf{w}\cdot\nabla\gamma^{\alpha} + \nabla\nabla\gamma^{\alpha}\cdot\mathbf{w}$$

and combining the second and tenth terms on the right side of Equation (640), we have

$$\begin{aligned}\int_{S_{\alpha\beta}} \left.\frac{\partial f}{\partial t}\right|_{\mathbf{u}} ds = &-\frac{\partial}{\partial t}\int_{V_\infty} f\gamma\gamma^{s}\mathbf{n}\cdot\nabla\gamma^{\alpha}dv - \int_{V_\infty} f\gamma\mathbf{w}\cdot\nabla^{S}\gamma^{s}\mathbf{n}\cdot\nabla\gamma^{\alpha}dv\\ &-\int_{V_\infty} f\gamma\gamma^{s}\mathbf{n}\cdot\nabla\nabla\gamma^{\alpha}\cdot\mathbf{w}dv - \nabla\cdot\int_{V_\infty}\mathbf{nw}\cdot\mathbf{n}f\gamma\gamma^{s}\mathbf{n}\cdot\nabla\gamma^{\alpha}dv\\ &+\int_{V_\infty}\nabla\cdot\mathbf{nw}\cdot\mathbf{n}f\gamma\gamma^{s}\mathbf{n}\cdot\nabla\gamma^{\alpha}dv + \int_{V_\infty}\mathbf{n}\cdot\nabla\mathbf{n}\cdot\mathbf{w}f\gamma\gamma^{s}\mathbf{n}\cdot\nabla\gamma^{\alpha}dv\\ &+\int_{V_\infty}\mathbf{w}\cdot\mathbf{n}f\gamma\gamma^{s}\mathbf{n}\cdot\nabla\nabla\gamma^{\alpha}\cdot\mathbf{n}dv \end{aligned}\tag{641}$$

By combining the third and the last terms on the right side of Equation (641) and applying Equation (256),

$$\begin{aligned}\int_{S_{\alpha\beta}} \left.\frac{\partial f}{\partial t}\right|_{\mathbf{u}} ds = &-\frac{\partial}{\partial t}\int_{V_\infty} f\gamma\gamma^{s}\mathbf{n}\cdot\nabla\gamma^{\alpha}dv - \int_{V_\infty} f\gamma\mathbf{w}^{s}\cdot\nabla^{S}\gamma^{s}\mathbf{n}\cdot\nabla\gamma^{\alpha}dv\\ &+\int_{V_\infty}\mathbf{n}\cdot\nabla\mathbf{n}\cdot\mathbf{w}f\gamma\gamma^{s}\mathbf{n}\cdot\nabla\gamma^{\alpha}dv - \nabla\cdot\int_{V_\infty}\mathbf{nw}\cdot\mathbf{n}f\gamma\gamma^{s}\mathbf{n}\cdot\nabla\gamma^{\alpha}dv\\ &+\int_{V_\infty}\nabla\cdot\mathbf{nw}\cdot\mathbf{n}f\gamma\gamma^{s}\mathbf{n}\cdot\nabla\gamma^{\alpha}dv - \int_{V_\infty}\mathbf{n}\cdot\nabla\mathbf{n}\cdot\mathbf{w}f\gamma\gamma^{s}\mathbf{n}\cdot\nabla\gamma^{\alpha}dv \end{aligned}\tag{642}$$

By Equations (318), (184), and (275), Equation (642) becomes

$$\int_{S_{\alpha\beta}} \left.\frac{\partial f}{\partial t}\right|_{\mathbf{u}} = \frac{\partial}{\partial t}\int_{S_{\alpha\beta}} f ds + \nabla\cdot\int_{S_{\alpha\beta}}\mathbf{w}^{c}f ds - \int_{S_{\alpha\beta}}(\nabla^{S}\cdot\mathbf{n})\mathbf{n}\cdot\mathbf{w}^{c}f ds - \int_{C_{\alpha\beta\varepsilon}} f\boldsymbol{v}\cdot\mathbf{w}^{s}dc$$

Theorem 43.

$$\int_{S_{\alpha\beta}}\nabla^{S}\circ F ds = \nabla\circ\int_{S_{\alpha\beta}} F ds - \nabla\cdot\int_{S_{\alpha\beta}}\mathbf{nn}\circ F ds + \int_{S_{\alpha\beta}}\nabla^{S}\cdot\mathbf{nn}\circ F ds + \int_{C_{\alpha\beta\varepsilon}}\boldsymbol{v}\circ F ds \tag{643}$$

where $\nabla\circ F$ can be any of the three operators: $\nabla^{S}f$, $\nabla^{S}\cdot\mathbf{f}$, and $\nabla^{S}\times\mathbf{f}$. $f(\mathbf{u}, t)$ and $\mathbf{f}(\mathbf{u}, t)$ are scalar and vector functions of surficial position vector $\mathbf{u}$ and time t defined on $S_{\alpha\beta}$. $\mathbf{n}^{*}$ is the unit vector normal to the external boundary of the REV pointing outward.

Proof. By the definitions of the indicator functions $\gamma(\mathbf{x}, \xi)$, $\gamma^{s}(\mathbf{u}(\mathbf{x}+\xi), t)$, and $\gamma^{\alpha}(\mathbf{x}+\xi)$, we have

$$\int_{S_{\alpha\beta}}\nabla^{S}\circ F ds = -\int_{V_\infty}\nabla^{S}\circ F\gamma\gamma^{s}\mathbf{n}\cdot\nabla\gamma^{\alpha}dv \tag{644}$$

By $\nabla^{S} = \nabla - \mathbf{nn}\cdot\nabla$, Equation (644) may be written as

$$\begin{aligned}\int_{S_{\alpha\beta}}\nabla^{S}\circ F ds &= -\int_{V_\infty}\nabla\circ F\gamma\gamma^{s}\mathbf{n}\cdot\nabla\gamma^{\alpha}dv + \int_{V_\infty}\mathbf{nn}\cdot\nabla\circ F\gamma\gamma^{s}\mathbf{n}\cdot\nabla\gamma^{\alpha}dv\\ &= -\int_{V_\infty}\nabla\circ F\gamma\gamma^{s}\mathbf{n}\cdot\nabla\gamma^{\alpha}dv + \int_{V_\infty}\mathbf{n}\circ(\mathbf{n}\cdot\nabla F)\gamma\gamma^{s}\mathbf{n}\cdot\nabla\gamma^{\alpha}dv \end{aligned}\tag{645}$$

By applying the chain rule to the two terms on the right-hand side of Equation (645), we obtain

$$\begin{aligned}\int_{S_{\alpha\beta}} \nabla^{S} \circ F \mathrm{d}s = & -\int_{V_\infty} \nabla \circ (F\gamma\gamma^{s}\mathbf{n}\cdot\nabla\gamma^{\alpha})\mathrm{d}v + \int_{V_\infty} \nabla\gamma \circ F\gamma^{s}\mathbf{n}\cdot\nabla\gamma^{\alpha}\mathrm{d}v \\ & + \int_{V_\infty} \nabla\gamma^{s} \circ F\gamma\mathbf{n}\cdot\nabla\gamma^{\alpha}\mathrm{d}v + \int_{V_\infty} \nabla(\mathbf{n}\cdot\nabla\gamma^{\alpha}) \circ F\gamma\gamma^{s}\mathrm{d}v \\ & + \int_{V_\infty} \nabla\cdot(\mathbf{nn}\circ F\gamma\gamma^{s}\mathbf{n}\cdot\nabla\gamma^{\alpha})\mathrm{d}v - \int_{V_\infty} \nabla\cdot\mathbf{nn}\circ F\gamma\gamma^{s}\mathbf{n}\cdot\nabla\gamma^{\alpha}\mathrm{d}v \\ & - \int_{V_\infty} (\mathbf{n}\cdot\nabla\mathbf{n})\circ F\gamma\gamma^{s}\mathbf{n}\cdot\nabla\gamma^{\alpha}\mathrm{d}v - \int_{V_\infty} \mathbf{n}\circ F\mathbf{n}\cdot\nabla\gamma\gamma^{s}\mathbf{n}\cdot\nabla\gamma^{\alpha}\mathrm{d}v \\ & - \int_{V_\infty} \mathbf{n}\circ F\gamma\mathbf{n}\cdot\nabla\gamma^{s}\mathbf{n}\cdot\nabla\gamma^{\alpha}\mathrm{d}v - \int_{V_\infty} \mathbf{n}\circ F\gamma\gamma^{s}\mathbf{n}\cdot\nabla(\mathbf{n}\cdot\nabla\gamma^{\alpha})\mathrm{d}v \end{aligned} \tag{646}$$

Note that the integration domains are independent of the integration variables in the first and fifth terms on the right side of Equation (646). The operator ∇ and the integration sign can be interchanged. Note also that $S_{\alpha\beta}$ is within the REV. By Equation (181), the second term cancels the eighth term. Thus Equation (646) becomes

$$\begin{aligned}\int_{S_{\alpha\beta}} \nabla^{S} \circ F \mathrm{d}s = & -\nabla\circ\int_{V_\infty} F\gamma\gamma^{s}\mathbf{n}\cdot\nabla\gamma^{\alpha}\mathrm{d}v + \int_{V_\infty} \nabla\gamma^{s}\circ F\gamma\mathbf{n}\cdot\nabla\gamma^{\alpha}\mathrm{d}v \\ & + \int_{V_\infty} (\nabla\mathbf{n}\cdot\nabla\gamma^{\alpha} + \nabla\nabla\gamma^{\alpha}\cdot\mathbf{n})\circ F\gamma\gamma^{s}\mathrm{d}v \\ & + \nabla\cdot\int_{V_\infty} \mathbf{nn}\circ F\gamma\gamma^{s}\mathbf{n}\cdot\nabla\gamma^{\alpha}\mathrm{d}v - \int_{V_\infty} \nabla\cdot\mathbf{nn}\circ F\gamma\gamma^{s}\mathbf{n}\cdot\nabla\gamma^{\alpha}\mathrm{d}v \\ & - \int_{V_\infty} (\mathbf{n}\cdot\nabla\mathbf{n})\circ F\gamma\gamma^{s}\mathbf{n}\cdot\nabla\gamma^{\alpha}\mathrm{d}v - \int_{V_\infty} \mathbf{n}\circ F\gamma\mathbf{n}\cdot\nabla\gamma^{s}\mathbf{n}\cdot\nabla\gamma^{\alpha}\mathrm{d}v \\ & - \int_{V_\infty} \mathbf{n}\circ F\gamma\gamma^{s}\mathbf{n}\cdot(\nabla\mathbf{n}\cdot\nabla\gamma^{\alpha} + \nabla\nabla\gamma^{\alpha}\cdot\mathbf{n})\mathrm{d}v \end{aligned} \tag{647}$$

By combining the second and seventh terms on the right side of Equation (647) and applying Equation (195), we have

$$\begin{aligned}\int_{S_{\alpha\beta}} \nabla^{S} \circ F \mathrm{d}s = & -\nabla\circ\int_{V_\infty} F\gamma\gamma^{s}\mathbf{n}\cdot\nabla\gamma^{\alpha}\mathrm{d}v + \int_{V_\infty} \nabla^{S}\gamma^{s}\circ F\gamma\mathbf{n}\cdot\nabla\gamma^{\alpha}\mathrm{d}v \\ & + \nabla\cdot\int_{V_\infty} \mathbf{nn}\circ F\gamma\gamma^{s}\mathbf{n}\cdot\nabla\gamma^{\alpha}\mathrm{d}v - \int_{V_\infty} \nabla\cdot\mathbf{nn}\circ F\gamma\gamma^{s}\mathbf{n}\cdot\nabla\gamma^{\alpha}\mathrm{d}v \\ & + \int_{V_\infty} (\nabla\nabla\gamma^{\alpha}\cdot\mathbf{n})\circ F\gamma\gamma^{s}\mathrm{d}v \\ & - \int_{V_\infty} \mathbf{n}\circ F\gamma\gamma^{s}\mathbf{n}\cdot(\nabla\nabla\gamma^{\alpha}\cdot\mathbf{n})\mathrm{d}v \\ & - \int_{V_\infty} \mathbf{n}\cdot\nabla\mathbf{n}\circ F\gamma\gamma^{s}\mathbf{n}\cdot\nabla\gamma^{\alpha}\mathrm{d}v \end{aligned} \tag{648}$$

By combining the fifth and sixth terms on the right side of Equation (648) and applying Equation (256) and $\nabla - \mathbf{nn} \cdot \nabla = \nabla^S$, we have

$$\begin{aligned}\int_{S_{\alpha\beta}} \nabla^S \circ F dv = & -\nabla \circ \int_{V_\infty} F\gamma\gamma^s \mathbf{n} \cdot \nabla\gamma^\alpha dv + \int_{V_\infty} \nabla^S \gamma^s \circ F\gamma\mathbf{n} \cdot \nabla\gamma^\alpha dv \\ & + \nabla \cdot \int_{V_\infty} \mathbf{nn} \circ F\gamma\gamma^s \mathbf{n} \cdot \nabla\gamma^\alpha dv - \int_{V_\infty} \nabla \cdot \mathbf{nn} \circ F\gamma\gamma^s \mathbf{n} \cdot \nabla\gamma^\alpha dv \\ & + \int_{V_\infty} (\mathbf{n} \cdot \nabla\mathbf{n}) \circ F\gamma\gamma^s \mathbf{n} \cdot \nabla\gamma^\alpha dv - \int_{V_\infty} \mathbf{n} \cdot \nabla\mathbf{n} \circ F\gamma\gamma^s \mathbf{n} \cdot \nabla\gamma^\alpha dv \\ = & -\nabla \circ \int_{V_\infty} f\gamma\gamma^s \mathbf{n} \cdot \nabla\gamma^\alpha dv + \int_{V_\infty} \nabla^S \gamma^s \circ F\gamma\mathbf{n} \cdot \nabla\gamma^\alpha dv \\ & + \nabla \cdot \int_{V_\infty} \mathbf{nn} \circ F\gamma\gamma^s \mathbf{n} \cdot \nabla\gamma^\alpha dv - \int_{V_\infty} \nabla \cdot \mathbf{nn} \circ F\gamma\gamma^s \mathbf{n} \cdot \nabla\gamma^\alpha dv \qquad (649)\end{aligned}$$

Equations (138), (183), and (275) allow us to rewrite Equation (649) as

$$\int_{S_{\alpha\beta}} \nabla^S \circ F dv = \nabla \circ \int_{S_{\alpha\beta}} F ds - \nabla \cdot \int_{S_{\alpha\beta}} \mathbf{nn} \circ F ds + \int_{S_{\alpha\beta}} \nabla^S \cdot \mathbf{nn} \circ F ds + \int_{C_{\alpha\beta\varepsilon}} \boldsymbol{\nu} \circ F dc$$

Remark. Equation (643) contains the following three formulas:

Gradient

$$\int_{S_{\alpha\beta}} \nabla^S f dv = \nabla \int_{S_{\alpha\beta}} f ds - \nabla \cdot \int_{S_{\alpha\beta}} \mathbf{nn} f ds + \int_{S_{\alpha\beta}} (\nabla^S \cdot \mathbf{n})\mathbf{n} f ds + \int_{C_{\alpha\beta\varepsilon}} \boldsymbol{\nu} f dc \qquad (650)$$

Divergence

$$\begin{aligned}\int_{S_{\alpha\beta}} \nabla^S \cdot \mathbf{f} ds = & \nabla \cdot \int_{S_{\alpha\beta}} \mathbf{f} ds - \nabla \cdot \int_{S_{\alpha\beta}} \mathbf{nn} \cdot \mathbf{f} ds \\ & + \int_{S_{\alpha\beta}} (\nabla^S \cdot \mathbf{n})\mathbf{n} \cdot \mathbf{f} ds + \int_{C_{\alpha\beta\varepsilon}} \boldsymbol{\nu} \cdot \mathbf{f} ds \qquad (651)\end{aligned}$$

Curl

$$\begin{aligned}\int_{S_{\alpha\beta}} \nabla^S \times \mathbf{f} ds = & \nabla \times \int_{S_{\alpha\beta}} \mathbf{f} ds - \nabla \cdot \int_{S_{\alpha\beta}} \mathbf{nn} \times \mathbf{f} ds \\ & + \int_{S_{\alpha\beta}} (\nabla^S \cdot \mathbf{n})\mathbf{n} \times \mathbf{f} ds + \int_{C_{\alpha\beta\varepsilon}} \boldsymbol{\nu} \times \mathbf{f} ds \qquad (652)\end{aligned}$$

Theorem 42 has the following alternative form:

Theorem 44.

$$\begin{aligned}\int_{S_{\alpha\beta}} \left.\frac{\partial f}{\partial t}\right|_{\mathbf{u}} ds = & \frac{\partial}{\partial t} \int_{S_{\alpha\beta}} f ds - \int_{S_{\alpha\beta}} (\nabla^S \cdot \mathbf{n})\mathbf{n} \cdot \mathbf{w}^c f ds + \int_{C_{\alpha\beta}} \frac{\mathbf{n}^* \cdot \mathbf{w} f}{\boldsymbol{\nu} \cdot \mathbf{n}^*} dc \\ & - \int_{C_{\alpha\beta}} \boldsymbol{\nu} \cdot \mathbf{w}^s f dc - \int_{C_{\alpha\beta\varepsilon}} \boldsymbol{\nu} \cdot \mathbf{w}^s f dc \qquad (653)\end{aligned}$$

where $\mathbf{n}^*$ is the unit vector normal to the external boundary of the REV that is pointing outward from the REV and $C_{\alpha\beta}$ the curve of intersection of the $S_{\alpha\beta}$ surface with the external surface of the REV.

Proof. By the definitions of the indicator functions $\gamma(\mathbf{x}, \xi)$, $\gamma^s(\mathbf{u}(\mathbf{x}+\xi), t)$, and $\gamma^\alpha(\mathbf{x}+\xi, t)$, we have

$$\int_{S_{\alpha\beta}} \left.\frac{\partial f}{\partial t}\right|_{\mathbf{u}} \mathrm{d}s = -\int_{V_\infty} \left.\frac{\partial f}{\partial t}\right|_{\mathbf{u}} \gamma\gamma^s \mathbf{n}\cdot\nabla\gamma^\alpha \mathrm{d}v \tag{654}$$

By Equation (130), Equation (654) becomes

$$\int_{S_{\alpha\beta}} \left.\frac{\partial f}{\partial t}\right|_{\mathbf{u}} \mathrm{d}s = -\int_{V_\infty} \left.\frac{\partial f}{\partial t}\right|_{\mathbf{x}} \gamma\gamma^s \mathbf{n}\cdot\nabla\gamma^\alpha \mathrm{d}v - \int_{V_\infty} \mathbf{w}\cdot\mathbf{n}\mathbf{n}\cdot\nabla f\gamma\gamma^s \mathbf{n}\cdot\nabla\gamma^\alpha \mathrm{d}v \tag{655}$$

An application of the chain rule to both the terms on the right-hand side of Equation (655) yields

$$\begin{aligned}\int_{S_{\alpha\beta}} \left.\frac{\partial f}{\partial t}\right|_{\mathbf{u}} \mathrm{d}s = & -\int_{V_\infty} \frac{\partial(f\gamma\gamma^s \mathbf{n}\cdot\nabla\gamma^\alpha)}{\partial t}\mathrm{d}v + \int_{V_\infty} f\left.\frac{\partial(\gamma\gamma^s)}{\partial t}\right|_{\mathbf{x}} \mathbf{n}\cdot\nabla\gamma^\alpha \mathrm{d}v \\ & + \int_{V_\infty} f\gamma\gamma^s \frac{\partial \mathbf{n}}{\partial t}\cdot\nabla\gamma^\alpha \mathrm{d}v + \int_{V_\infty} f\gamma\gamma^s \mathbf{n}\cdot\nabla\frac{\partial\gamma^\alpha}{\partial t}\mathrm{d}v \\ & - \int_{V_\infty} \nabla\cdot(\mathbf{n}\mathbf{w}\cdot\mathbf{n}f\gamma\gamma^s \mathbf{n}\cdot\nabla\gamma^\alpha)\mathrm{d}v \\ & + \int_{V_\infty} \nabla\cdot\mathbf{n}\mathbf{w}\cdot\mathbf{n}f\gamma\gamma^s \mathbf{n}\cdot\nabla\gamma^\alpha \mathrm{d}v \\ & + \int_{V_\infty} \mathbf{n}\cdot\nabla\mathbf{w}\cdot\mathbf{n}f\gamma\gamma^s \mathbf{n}\cdot\nabla\gamma^\alpha \mathrm{d}v \\ & + \int_{V_\infty} \mathbf{n}\cdot\nabla\mathbf{n}\cdot\mathbf{w}f\gamma\gamma^s \mathbf{n}\cdot\nabla\gamma^\alpha \mathrm{d}v \\ & + \int_{V_\infty} \mathbf{w}\cdot\mathbf{n}f\mathbf{n}\cdot\nabla(\gamma\gamma^s)\mathbf{n}\cdot\nabla\gamma^\alpha \mathrm{d}v \\ & + \int_{V_\infty} \mathbf{w}\cdot\mathbf{n}f\gamma\gamma^s \mathbf{n}\cdot(\nabla\mathbf{n}\cdot\nabla\gamma^\alpha + \nabla\nabla\gamma^\alpha\cdot\mathbf{n})\mathrm{d}v\end{aligned} \tag{656}$$

As the integration regions are independent of the integration variables for the first term on the right side of Equation (656), the differentiation with respect to the time t can be interchanged with the integration operator. This allows us to rewrite it as

$$\int_{V_\infty} \frac{\partial(f\gamma\gamma^s \mathbf{n}\cdot\nabla\gamma^\alpha)}{\partial t}\mathrm{d}v = \frac{\partial}{\partial t}\int_{V_\infty} f\gamma\gamma^s \mathbf{n}\cdot\nabla\gamma^\alpha \mathrm{d}v \tag{657}$$

Because γ is independent of time, we can write the second term on the right side of Equation (656) as

$$\int_{V_\infty} f\left.\frac{\partial(\gamma\gamma^s)}{\partial t}\right|_{\mathbf{x}} \mathbf{n}\cdot\nabla\gamma^\alpha \mathrm{d}v = \int_{V_\infty} f\gamma\left.\frac{\partial\gamma^s}{\partial t}\right|_{\mathbf{x}} \mathbf{n}\cdot\nabla\gamma^\alpha \mathrm{d}v \tag{658}$$

Applying Equation (224) yields

$$\int_{V_\infty} f\frac{\partial(\gamma\gamma^{\mathrm{s}})}{\partial t}\bigg|_{\mathbf{x}}\mathbf{n}\cdot\nabla\gamma^\alpha \mathrm{d}v = -\int_{V_\infty} f\gamma\mathbf{w}\cdot\nabla\gamma^{\mathrm{s}}\mathbf{n}\cdot\nabla\gamma^\alpha\mathrm{d}v$$

$$= -\int_{V_\infty} f\mathbf{w}\cdot\nabla(\gamma\gamma^{\mathrm{s}})\mathbf{n}\cdot\nabla\gamma^\alpha\mathrm{d}v$$

$$+\int_{V_\infty} f\mathbf{w}\cdot\nabla\gamma\gamma^{\mathrm{s}}\mathbf{n}\cdot\nabla\gamma^\alpha\mathrm{d}v \quad (659)$$

By Equation (194), the third integral is zero on the right side of Equation (656). By Equation (180), the fourth term on the right side of Equation (656) is

$$\int_{V_\infty} f\gamma\gamma^{\mathrm{s}}\mathbf{n}\cdot\nabla\frac{\partial\gamma^\alpha}{\partial t}\mathrm{d}v = -\int_{V_\infty} f\gamma\gamma^{\mathrm{s}}\mathbf{n}\cdot\nabla\mathbf{w}\cdot\nabla\gamma^\alpha\mathrm{d}v - \int_{V_\infty} f\gamma\gamma^{\mathrm{s}}\mathbf{n}\cdot\nabla\nabla\gamma^\alpha\cdot\mathbf{w}\mathrm{d}v \quad (660)$$

By Equation (231), the fifth term is equal to zero on the right side of Equation (656). By Equation (195), we may rewrite the last term on the right side of Equation (656) as

$$\int_{V_\infty}\mathbf{w}\cdot\mathbf{n}f\gamma\gamma^{\mathrm{s}}\mathbf{n}\cdot(\nabla\mathbf{n}\cdot\nabla\gamma^\alpha + \nabla\nabla\gamma^\alpha\cdot\mathbf{n})\mathrm{d}v = \int_{V_\infty}\mathbf{w}\cdot\mathbf{n}f\gamma\gamma^{\mathrm{s}}\mathbf{n}\cdot\nabla\nabla\gamma^\alpha\cdot\mathbf{n}\mathrm{d}v \quad (661)$$

Equations (657)–(660) enable us to conclude that

$$\int_{S_{\alpha\beta}}\frac{\partial f}{\partial t}\bigg|_{\mathbf{u}}\mathrm{d}s = -\frac{\partial}{\partial t}\int_{V_\infty} f\gamma\gamma^{\mathrm{s}}\mathbf{n}\cdot\nabla\gamma^\alpha\mathrm{d}v - \int_{V_\infty} f\mathbf{w}\cdot\nabla(\gamma\gamma^{\mathrm{s}})\mathbf{n}\cdot\nabla\gamma^\alpha\mathrm{d}v$$

$$+\int_{V_\infty} f\mathbf{w}\cdot\nabla\gamma\gamma^{\mathrm{s}}\mathbf{n}\cdot\nabla\gamma^\alpha\mathrm{d}v - \int_{V_\infty} f\gamma\gamma^{\mathrm{s}}\mathbf{n}\cdot\nabla\mathbf{w}\cdot\nabla\gamma^\alpha\mathrm{d}v$$

$$-\int_{V_\infty} f\gamma\gamma^{\mathrm{s}}\mathbf{n}\cdot\nabla\nabla\gamma^\alpha\cdot\mathbf{w}\mathrm{d}v + \int_{V_\infty}\nabla\cdot\mathbf{n}\mathbf{w}\cdot\mathbf{n}f\gamma\gamma^{\mathrm{s}}\mathbf{n}\cdot\nabla\gamma^\alpha\mathrm{d}v$$

$$+\int_{V_\infty}\mathbf{n}\cdot\nabla\mathbf{w}\cdot\mathbf{n}f\gamma\gamma^{\mathrm{s}}\mathbf{n}\cdot\nabla\gamma^\alpha\mathrm{d}v + \int_{V_\infty}\mathbf{n}\cdot\nabla\mathbf{n}\cdot\mathbf{w}f\gamma\gamma^{\mathrm{s}}\mathbf{n}\cdot\nabla\gamma^\alpha\mathrm{d}v$$

$$+\int_{V_\infty}\mathbf{w}\cdot\mathbf{n}f\mathbf{n}\cdot\nabla(\gamma\gamma^{\mathrm{s}})\mathbf{n}\cdot\nabla\gamma^\alpha\mathrm{d}v + \int_{V_\infty}\mathbf{w}\cdot\mathbf{n}f\gamma\gamma^{\mathrm{s}}\mathbf{n}\cdot\nabla\nabla\gamma^\alpha\cdot\mathbf{n}\mathrm{d}v \quad (662)$$

Regrouping the terms on the right side of Equation (662) yields

$$\int_{S_{\alpha\beta}}\frac{\partial f}{\partial t}\bigg|_{\mathbf{u}}\mathrm{d}s = -\frac{\partial}{\partial t}\int_{V_\infty} f\gamma\gamma^{\mathrm{s}}\mathbf{n}\cdot\nabla\gamma^\alpha\mathrm{d}v - \int_{V_\infty} f\mathbf{w}\cdot\nabla^{\mathrm{S}}(\gamma\gamma^{\mathrm{s}})\mathbf{n}\cdot\nabla\gamma^\alpha\mathrm{d}v$$

$$+\int_{V_\infty} f\mathbf{w}\cdot\nabla\gamma\gamma^{\mathrm{s}}\mathbf{n}\cdot\nabla\gamma^\alpha\mathrm{d}v - \int_{V_\infty} f\gamma\gamma^{\mathrm{s}}\mathbf{n}\cdot\nabla\mathbf{w}\cdot\nabla\gamma^\alpha\mathrm{d}v$$

$$-\int_{V_\infty} f\gamma\gamma^{\mathrm{s}}\mathbf{w}\cdot\nabla^{\mathrm{S}}\nabla\gamma^\alpha\cdot\mathbf{n}\mathrm{d}v + \int_{V_\infty}\nabla\cdot\mathbf{n}\mathbf{w}\cdot\mathbf{n}f\gamma\gamma^{\mathrm{s}}\mathbf{n}\cdot\nabla\gamma^\alpha\mathrm{d}v$$

$$+\int_{V_\infty}\mathbf{n}\cdot\nabla\mathbf{w}\cdot\mathbf{n}f\gamma\gamma^{\mathrm{s}}\mathbf{n}\cdot\nabla\gamma^\alpha\mathrm{d}v$$

$$+\int_{V_\infty}\mathbf{n}\cdot\nabla\mathbf{n}\cdot\mathbf{w}f\gamma\gamma^{\mathrm{s}}\mathbf{n}\cdot\nabla\gamma^\alpha\mathrm{d}v \quad (663)$$

Applying Equation (256) to the fifth term on the right-hand side of Equation (663) leads to

$$\int_{S_{\alpha\beta}} \frac{\partial f}{\partial t}\Big|_{\mathbf{u}} ds = -\frac{\partial}{\partial t}\int_{V_\infty} f\gamma\gamma^s \mathbf{n}\cdot\nabla\gamma^\alpha dv - \int_{V_\infty} f\mathbf{w}\cdot\nabla^S(\gamma\gamma^s)\mathbf{n}\cdot\nabla\gamma^\alpha dv$$
$$+ \int_{V_\infty} f\mathbf{w}\cdot\nabla\gamma\gamma^s\mathbf{n}\cdot\nabla\gamma^\alpha dv - \int_{V_\infty} f\gamma\gamma^s\mathbf{n}\cdot\nabla\mathbf{w}\cdot\nabla\gamma^\alpha dv$$
$$- \int_{V_\infty} f\gamma\gamma^s\mathbf{n}\cdot\nabla\mathbf{n}\cdot\mathbf{w}\mathbf{n}\cdot\nabla\gamma^\alpha dv + \int_{V_\infty} \nabla\cdot\mathbf{n}\mathbf{w}\cdot\mathbf{n}f\gamma\gamma^s\mathbf{n}\cdot\nabla\gamma^\alpha dv$$
$$+ \int_{V_\infty} \mathbf{n}\cdot\nabla\mathbf{w}\cdot\mathbf{n}f\gamma\gamma^s\mathbf{n}\cdot\nabla\gamma^\alpha dv + \int_{V_\infty} \mathbf{n}\cdot\nabla\mathbf{n}\cdot\mathbf{w}f\gamma\gamma^s\mathbf{n}\cdot\nabla\gamma^\alpha dv \quad (664)$$

By the definitions of the indicator functions γ^s and γ^α and Equation (272), the third term on the right side of Equation (664) is

$$\int_{V_\infty} f\mathbf{w}\cdot\nabla\gamma\gamma^s\mathbf{n}\cdot\nabla\gamma^\alpha dv = -\int_{S_{\alpha\beta}} f\mathbf{w}\cdot\nabla\gamma ds \quad (665)$$

which becomes, by Equation (284)

$$\int_{V_\infty} f\mathbf{w}\cdot\nabla\gamma\gamma^s\mathbf{n}\cdot\nabla\gamma^\alpha dv = \int_{C_{\alpha\beta}} f\frac{\mathbf{w}\cdot\mathbf{n}^*}{\mathbf{v}\cdot\mathbf{n}^*}dc \quad (666)$$

Substituting Equation (666) into Equation (664) and applying Equations (272) and (275) yields

$$\int_{S_{\alpha\beta}} \frac{\partial f}{\partial t}\Big|_{\mathbf{u}} ds = \frac{\partial}{\partial t}\int_{S_{\alpha\beta}} f ds - \int_{C_{\alpha\beta}} \mathbf{v}\cdot\mathbf{w}f dc - \int_{C_{\alpha\beta\varepsilon}} \mathbf{v}\cdot\mathbf{w}f dc$$
$$+ \int_{C_{\alpha\beta}} \frac{\mathbf{n}^*\cdot\mathbf{w}}{\mathbf{v}\cdot\mathbf{n}^*} dc - \int_{S_{\alpha\beta}} \nabla\cdot\mathbf{n}\mathbf{n}\cdot\mathbf{w}f dc \quad (667)$$

Since $\mathbf{v}\cdot\mathbf{w} = \mathbf{v}\cdot\mathbf{w}^s$, $\nabla\cdot\mathbf{n} = \nabla^S\cdot\mathbf{n}$, and $\mathbf{n}\cdot\mathbf{w} = \mathbf{n}\cdot\mathbf{w}^c$, we finally have

$$\int_{S_{\alpha\beta}} \frac{\partial f}{\partial t}\Big|_{\mathbf{u}} ds = \frac{\partial}{\partial t}\int_{S_{\alpha\beta}} f ds - \int_{S_{\alpha\beta}} (\nabla^S\cdot\mathbf{n})\mathbf{n}\cdot\mathbf{w}^c f dc + \int_{C_{\alpha\beta}} \frac{\mathbf{n}^*\cdot\mathbf{w}}{\mathbf{v}\cdot\mathbf{n}^*} dc$$
$$- \int_{C_{\alpha\beta}} \mathbf{v}\cdot\mathbf{w}^s f dc - \int_{C_{\alpha\beta\varepsilon}} \mathbf{v}\cdot\mathbf{w}^s f dc$$

Theorem 43 has the following alternative form:

Theorem 45.

$$\int_{S_{\alpha\beta}} \nabla^S \circ F ds = \int_{S_{\alpha\beta}} (\nabla^S\cdot\mathbf{n})\mathbf{n}\circ F ds + \int_{C_{\alpha\beta}} \mathbf{v}\circ F dc + \int_{C_{\alpha\beta\varepsilon}} \mathbf{v}\circ F dc \quad (668)$$

where $\nabla^S \circ F$ can be any of the three operators: $\nabla^S f$, $\nabla^S\cdot\mathbf{f}$, and $\nabla^S\times\mathbf{f}$. $f(\mathbf{u}, t)$ and $\mathbf{f}(\mathbf{u}, t)$ are scalar and vector functions of surficial position vector $\mathbf{u}$ and time t defined on $S_{\alpha\beta}$. $\mathbf{n}^*$ is the unit vector normal to the external boundary of the REV that is pointing outward from the REV.

Proof. By the definitions of the indicator functions $\gamma\ (\mathbf{x}, \xi)$, $\gamma^{\alpha}(\mathbf{x}+\xi, t)$, and $\gamma^{s}(\mathbf{u}(\mathbf{x}+\xi), t)$, we have

$$\int_{S_{\alpha\beta}} \nabla^{S} \circ F ds = -\int_{V_{\infty}} \nabla^{S} \circ F\gamma\gamma^{s}\mathbf{n}\cdot\nabla\gamma^{\alpha} dv \tag{669}$$

Since $\nabla^{S} = \nabla - \mathbf{nn}\cdot\nabla$, Equation (669) may be written as

$$\int_{S_{\alpha\beta}} \nabla^{S} \circ F ds = -\int_{V_{\infty}} \nabla \circ F\gamma\gamma^{s}\mathbf{n}\cdot\nabla\gamma^{\alpha} dv + \int_{V_{\infty}} \mathbf{n}\circ(\mathbf{n}\cdot\nabla F)\gamma\gamma^{s}\mathbf{n}\cdot\nabla\gamma^{\alpha} dv \tag{670}$$

An application of the chain rule to the two integrals on the right side of Equation (670) yields

$$\begin{aligned}\int_{S_{\alpha\beta}} \nabla^{S} \circ F ds = & -\int_{V_{\infty}} \nabla\circ(F\gamma\gamma^{s}\mathbf{n}\cdot\nabla\gamma^{\alpha})dv + \int_{V_{\infty}} \nabla(\gamma\gamma^{s})\circ F\mathbf{n}\cdot\nabla\gamma^{\alpha} dv \\ & + \int_{V_{\infty}} \gamma\gamma^{s}(\nabla\mathbf{n}\cdot\nabla\gamma^{\alpha} + \nabla\nabla\gamma^{\alpha}\cdot\mathbf{n})\circ F dv \\ & + \int_{V_{\infty}} \nabla\cdot(\mathbf{nn}\circ F\gamma\gamma^{s}\mathbf{n}\cdot\nabla\gamma^{\alpha})dv - \int_{V_{\infty}} \nabla\cdot\mathbf{nn}\circ F\gamma\gamma^{s}\mathbf{n}\cdot\nabla\gamma^{\alpha} dv \\ & - \int_{V_{\infty}} \mathbf{n}\cdot\nabla\mathbf{n}\circ F\gamma\gamma^{s}\mathbf{n}\cdot\nabla\gamma^{\alpha} dv - \int_{V_{\infty}} \mathbf{n}\circ F\mathbf{n}\cdot\nabla(\gamma\gamma^{s})\mathbf{n}\cdot\nabla\gamma^{\alpha} dv \\ & - \int_{V_{\infty}} \mathbf{n}\circ F\gamma\gamma^{s}\mathbf{n}\cdot(\nabla\mathbf{n}\cdot\nabla\gamma^{\alpha} + \nabla\nabla\gamma^{\alpha}\cdot\mathbf{n})dv\end{aligned} \tag{671}$$

By Equations (230)–(232), both the first and the fourth terms on the right side of Equation (671) are zero. Furthermore, by combining the second and seventh terms using $\nabla - \mathbf{nn}\cdot\nabla = \nabla^{S}$ and applying Equation (195), we obtain

$$\begin{aligned}\int_{S_{\alpha\beta}} \nabla^{S} \circ F ds = & \int_{V_{\infty}} \nabla^{S}(\gamma\gamma^{s})\circ F\mathbf{n}\cdot\nabla\gamma^{\alpha} dv + \int_{V_{\infty}} \gamma\gamma^{s}(\nabla\nabla\gamma^{\alpha}\cdot\mathbf{n})\circ F dv \\ & - \int_{V_{\infty}} \nabla\cdot\mathbf{nn}\circ F\gamma\gamma^{s}\mathbf{n}\cdot\nabla\gamma^{\alpha} dv - \int_{V_{\infty}} \mathbf{n}\cdot\nabla\mathbf{n}\circ F\gamma\gamma^{s}\mathbf{n}\cdot\nabla\gamma^{\alpha} dv \\ & - \int_{V_{\infty}} \mathbf{n}\circ F\gamma\gamma^{s}\mathbf{n}\cdot\nabla\nabla\gamma^{\alpha}\cdot\mathbf{n} dv\end{aligned} \tag{672}$$

By combining the second and fifth terms, we have

$$\begin{aligned}\int_{S_{\alpha\beta}} \nabla^{S} \circ F ds = & \int_{V_{\infty}} \nabla^{S}(\gamma\gamma^{s})\circ F\mathbf{n}\cdot\nabla\gamma^{\alpha} dv + \int_{V_{\infty}} \gamma\gamma^{s}(\nabla^{S}\nabla\gamma^{\alpha}\cdot\mathbf{n})\circ F dv \\ & - \int_{V_{\infty}} \nabla\cdot\mathbf{nn}\circ F\gamma\gamma^{s}\mathbf{n}\cdot\nabla\gamma^{\alpha} dv \\ & - \int_{V_{\infty}} \mathbf{n}\cdot\nabla\mathbf{n}\circ F\gamma\gamma^{s}\mathbf{n}\cdot\nabla\gamma^{\alpha} dv\end{aligned} \tag{673}$$

Applying Equation (256) yields

$$\begin{aligned}\int_{S_{\alpha\beta}} \nabla^S \circ F ds &= \int_{V_\infty} \nabla^S(\gamma\gamma^s) \circ F\mathbf{n} \cdot \nabla\gamma^\alpha dv + \int_{V_\infty} \gamma\gamma^s(\mathbf{n} \cdot \nabla\mathbf{n} \circ F)\mathbf{n} \cdot \nabla\gamma^\alpha dv \\ &\quad - \int_{V_\infty} \nabla \cdot \mathbf{nn} \circ F\gamma\gamma^s\mathbf{n} \cdot \nabla\gamma^\alpha dv - \int_{V_\infty} \mathbf{n} \cdot \nabla\mathbf{n} \circ F\gamma\gamma^s\mathbf{n} \cdot \nabla\gamma^\alpha dv \\ &= \int_{V_\infty} \nabla^S(\gamma\gamma^s) \circ F\mathbf{n} \cdot \nabla\gamma^\alpha dv - \int_{V_\infty} \nabla \cdot \mathbf{nn} \circ F\gamma\gamma^s\mathbf{n} \cdot \nabla\gamma^\alpha dv \end{aligned} \tag{674}$$

By Equations (272) and (275), Equation (674) finally becomes

$$\int_{S_{\alpha\beta}} \nabla^S \circ F ds = \int_{S_{\alpha\beta}} (\nabla^S \cdot \mathbf{n})\mathbf{n} \circ F ds + \int_{C_{\alpha\beta}} \boldsymbol{v} \circ F dc + \int_{C_{\alpha\beta\varepsilon}} \boldsymbol{v} \circ F dc$$

Remark 1. Equation (668) contains the following three formulas:

Gradient

$$\int_{S_{\alpha\beta}} \nabla^S f ds = \int_{S_{\alpha\beta}} (\nabla^S \cdot \mathbf{n})\mathbf{n} f ds + \int_{C_{\alpha\beta}} \boldsymbol{v} f dc + \int_{C_{\alpha\beta\varepsilon}} \boldsymbol{v} f dc \tag{675}$$

Divergence

$$\int_{S_{\alpha\beta}} \nabla^S \cdot \mathbf{f} ds = \int_{S_{\alpha\beta}} (\nabla^S \cdot \mathbf{n})\mathbf{n} \cdot \mathbf{f}^c ds + \int_{C_{\alpha\beta}} \boldsymbol{v} \cdot \mathbf{f}^s dc + \int_{C_{\alpha\beta\varepsilon}} \boldsymbol{v} \cdot \mathbf{f}^s dc \tag{676}$$

Curl

$$\int_{S_{\alpha\beta}} \nabla^S \times \mathbf{f} ds = \int_{S_{\alpha\beta}} (\nabla^S \cdot \mathbf{n})\mathbf{n} \times \mathbf{f} ds + \int_{C_{\alpha\beta}} \boldsymbol{v} \times \mathbf{f} dc + \int_{C_{\alpha\beta\varepsilon}} \boldsymbol{v} \times \mathbf{f} dc \tag{677}$$

Remark 2. By comparing with Theorem 43, we have

$$\nabla \int_{S_{\alpha\beta}} f ds - \nabla \cdot \int_{S_{\alpha\beta}} \mathbf{nn} f ds = \int_{C_{\alpha\beta}} \boldsymbol{v} f dc$$

$$\nabla \cdot \int_{S_{\alpha\beta}} \mathbf{f}^s ds = \int_{C_{\alpha\beta}} \boldsymbol{v} \cdot \mathbf{f}^s dc$$

$$\nabla \times \int_{S_{\alpha\beta}} \mathbf{f} ds - \nabla \cdot \int_{S_{\alpha\beta}} \mathbf{nn} \times \mathbf{f}^s ds = \int_{C_{\alpha\beta}} v \times \mathbf{f} dc$$

5.1.3 Curvilineal operator theorems

The REV used for this group of averaging theorems is shown in Figure 27.

Theorem 46.

$$\begin{aligned}\int_{C_{\alpha\beta\varepsilon}} \left.\frac{\partial f}{\partial t}\right|_1 dc &= \frac{\partial}{\partial t} \int_{C_{\alpha\beta\varepsilon}} f dc + \nabla \cdot \int_{C_{\alpha\beta\varepsilon}} \mathbf{w}^s f dc \\ &\quad + \int_{C_{\alpha\beta\varepsilon}} \boldsymbol{\lambda} \cdot \nabla^c \boldsymbol{\lambda} \cdot \mathbf{w}^c f dc - \sum_{\text{pts}} \mathbf{e} \cdot \mathbf{w}^c f \end{aligned} \tag{678}$$

where $C_{\alpha\beta\varepsilon}$ is the contact curve within the REV that is the location where the α-, β-, and ε-phases meet. pts refers to the end points of the $C_{\alpha\beta\varepsilon}$ curve where the α-, β-, and ε-phases meet a fourth phase. $\boldsymbol{\lambda}$ is the unit vector tangent to curve $C_{\alpha\beta\varepsilon}$ and $\mathbf{e}$ the unit vector tangent to $C_{\alpha\beta\varepsilon}$ at the endpoints that has positive outward direction from the curve such that $\boldsymbol{\lambda}\cdot\mathbf{e} = \pm 1$. ∇^{c} is the microscopic curvilineal operator such that $\nabla^{c} = \boldsymbol{\lambda}\boldsymbol{\lambda}\cdot\nabla\xi$ and ∇ outside the integral is the macroscopic spatial operator $\nabla = \nabla_{\mathbf{x}}$. $\mathbf{w}$ is the velocity of $C_{\alpha\beta\varepsilon}$, $\mathbf{w}^{c}$ the velocity of $C_{\alpha\beta\varepsilon}$ tangent to $C_{\alpha\beta\varepsilon}$ such that $\mathbf{w}^{c} = \boldsymbol{\lambda}\boldsymbol{\lambda}\cdot\mathbf{w}$, $\mathbf{w}^{s}$ the velocity of $C_{\alpha\beta\varepsilon}$ normal to $C_{\alpha\beta\varepsilon}$ such that $\mathbf{w}^{s} = \mathbf{w} - \mathbf{w}^{c}$, and f a curvilineal function $f(\mathbf{l}, t)$ defined on $C_{\alpha\beta\varepsilon}$.

Proof. By Equation (275), we obtain

$$\int_{C_{\alpha\beta\varepsilon}} \left.\frac{\partial f}{\partial t}\right|_{\mathbf{l}} \mathrm{d}c = \int_{V_\infty} \left.\frac{\partial f}{\partial t}\right|_{\mathbf{l}} \gamma\gamma^{c}\boldsymbol{v}\cdot\nabla^{S}\gamma^{s}\mathbf{n}\cdot\nabla\gamma^{\alpha}\mathrm{d}v \tag{679}$$

By Equation (269),

$$\begin{aligned}\int_{C_{\alpha\beta\varepsilon}} \left.\frac{\partial f}{\partial t}\right|_{\mathbf{l}} \mathrm{d}c = &\int_{V_\infty} \left.\frac{\partial f}{\partial t}\right|_{\mathbf{x}} \gamma\gamma^{c}\boldsymbol{v}\cdot\nabla^{S}\gamma^{s}\mathbf{n}\cdot\nabla\gamma^{\alpha}\mathrm{d}v \\ &+ \int_{V_\infty} \mathbf{w}^{s}\cdot\nabla f\gamma\gamma^{c}\boldsymbol{v}\cdot\nabla^{S}\gamma^{s}\mathbf{n}\cdot\nabla\gamma^{\alpha}\mathrm{d}v\end{aligned} \tag{680}$$

An application of the chain rule to the terms on the right side of Equation (680) yields

$$\begin{aligned}\int_{C_{\alpha\beta\varepsilon}} \left.\frac{\partial f}{\partial t}\right|_{\mathbf{l}} \mathrm{d}c = &\int_{V_\infty} \frac{\partial \gamma}{\partial t}(f\gamma\gamma^{c}\boldsymbol{v}\cdot\nabla^{S}\gamma^{s}\mathbf{n}\cdot\nabla\gamma^{\alpha})\mathrm{d}v - \int_{V_\infty} f\left.\frac{\partial \gamma}{\partial t}\right|_{\mathbf{x}} \gamma^{c}\boldsymbol{v}\cdot\nabla^{S}\gamma^{s}\mathbf{n}\cdot\nabla\gamma^{\alpha}\mathrm{d}v \\ &- \int_{V_\infty} f\left.\frac{\partial \gamma^{c}}{\partial t}\right|_{\mathbf{x}} \gamma\boldsymbol{v}\cdot\nabla^{S}\gamma^{s}\mathbf{n}\cdot\nabla\gamma^{\alpha}\mathrm{d}v - \int_{V_\infty} f\gamma\gamma^{c}\frac{\partial \boldsymbol{v}}{\partial t}\cdot\nabla^{S}\gamma^{s}\mathbf{n}\cdot\nabla\gamma^{\alpha}\mathrm{d}v \\ &- \int_{V_\infty} f\gamma\gamma^{c}\boldsymbol{v}\cdot\nabla^{S}\left.\frac{\partial \gamma^{s}}{\partial t}\right|_{\mathbf{x}} \mathbf{n}\cdot\nabla\gamma^{\alpha}\mathrm{d}v - \int_{V_\infty} f\gamma\gamma^{c}\boldsymbol{v}\cdot\nabla^{S}\gamma^{s}\left.\frac{\partial \mathbf{n}}{\partial t}\right|_{\mathbf{x}}\cdot\nabla\gamma^{\alpha}\mathrm{d}v \\ &- \int_{V_\infty} f\gamma\gamma^{c}\boldsymbol{v}\cdot\nabla^{S}\gamma^{s}\mathbf{n}\cdot\nabla\left.\frac{\partial \gamma^{\alpha}}{\partial t}\right|_{\mathbf{x}}\mathrm{d}v \\ &+ \int_{V_\infty} \nabla\cdot(\mathbf{w}^{s}f\gamma\gamma^{c}\boldsymbol{v}\cdot\nabla^{S}\gamma^{s}\mathbf{n}\cdot\nabla\gamma^{\alpha})\mathrm{d}v \\ &- \int_{V_\infty} \nabla\cdot\mathbf{w}^{s}f\gamma\gamma^{c}\boldsymbol{v}\cdot\nabla^{S}\gamma^{s}\mathbf{n}\cdot\nabla\gamma^{\alpha}\mathrm{d}v \\ &- \int_{V_\infty} \mathbf{w}^{s}\cdot\nabla\gamma f\gamma^{c}\boldsymbol{v}\cdot\nabla^{S}\gamma^{s}\mathbf{n}\cdot\nabla\gamma^{\alpha}\mathrm{d}v \\ &- \int_{V_\infty} f\gamma\mathbf{w}^{s}\cdot\nabla\gamma^{c}\boldsymbol{v}\cdot\nabla^{S}\gamma^{s}\mathbf{n}\cdot\nabla\gamma^{\alpha}\mathrm{d}v \\ &- \int_{V_\infty} f\gamma\gamma^{c}\mathbf{w}^{s}\cdot(\nabla\boldsymbol{v}\cdot\nabla^{S}\gamma^{s} + \nabla\nabla^{S}\gamma^{s}\cdot\boldsymbol{v})\mathbf{n}\cdot\nabla\gamma^{\alpha}\mathrm{d}v \\ &- \int_{V_\infty} f\gamma\gamma^{c}\boldsymbol{v}\cdot\nabla^{S}\gamma^{s}\mathbf{w}^{s}\cdot(\nabla\mathbf{n}\cdot\nabla\gamma^{\alpha} + \nabla\nabla\gamma^{\alpha}\cdot\mathbf{n})\mathrm{d}v\end{aligned} \tag{681}$$

The gradient operator in the first and eighth terms on the right-hand side of Equation (681) can be interchanged with the integration operator because the integration domains are independent of the integration variables. Note that the curve $C_{\alpha\beta\varepsilon}$ is within the REV. By Equations (180), (199), (200), (224), and (229), we thus have

$$
\begin{aligned}
&\int_{C_{\alpha\beta\varepsilon}} z\frac{\partial f}{\partial t}\Big|_1 \mathrm{d}c \\
&= \frac{\partial}{\partial t}\int_{V_\infty} f\gamma\gamma^{\mathrm{c}}\boldsymbol{v}\cdot\nabla^{\mathrm{S}}\gamma^{\mathrm{s}}\mathbf{n}\cdot\nabla\gamma^{\alpha}\mathrm{d}v \\
&\quad+\int_{V_\infty} f\gamma\mathbf{w}\cdot\nabla\gamma^{\mathrm{c}}\boldsymbol{v}\cdot\nabla^{\mathrm{S}}\gamma^{\mathrm{s}}\mathbf{n}\cdot\nabla\gamma^{\alpha}\mathrm{d}v+\int_{V_\infty} f\gamma\gamma^{\mathrm{c}}\boldsymbol{v}\cdot\nabla(\mathbf{w}\cdot\nabla\gamma^{\mathrm{s}})\mathbf{n}\cdot\nabla\gamma^{\alpha}\mathrm{d}v \\
&\quad+\int_{V_\infty} f\gamma\gamma^{\mathrm{c}}\boldsymbol{v}\cdot\nabla^{\mathrm{S}}\gamma^{\mathrm{s}}\mathbf{n}\cdot\nabla(\mathbf{w}\cdot\nabla\gamma^{\alpha})\mathrm{d}v+\nabla\cdot\int_{V_\infty}\mathbf{w}^{\mathrm{s}}f\gamma\gamma^{\mathrm{c}}\boldsymbol{v}\cdot\nabla^{\mathrm{S}}\gamma^{\mathrm{s}}\mathbf{n}\cdot\nabla\gamma^{\alpha}\mathrm{d}v \\
&\quad-\int_{V_\infty}\nabla\cdot(\mathbf{w}-\boldsymbol{\lambda}\boldsymbol{\lambda}\cdot\mathbf{w})f\gamma\gamma^{\mathrm{c}}\boldsymbol{v}\cdot\nabla^{\mathrm{S}}\gamma^{\mathrm{s}}\mathbf{n}\cdot\nabla\gamma^{\alpha}\mathrm{d}v \\
&\quad-\int_{V_\infty} f\gamma\mathbf{w}^{\mathrm{s}}\cdot\nabla\gamma^{\mathrm{c}}\boldsymbol{v}\cdot\nabla^{\mathrm{S}}\gamma^{\mathrm{s}}\mathbf{n}\cdot\nabla\gamma^{\alpha}\mathrm{d}v-\int_{V_\infty}\gamma\gamma^{\mathrm{c}}\mathbf{w}^{\mathrm{c}}\cdot\nabla\nabla\gamma^{\mathrm{s}}\cdot\boldsymbol{v}\mathbf{n}\cdot\nabla\gamma^{\alpha}\mathrm{d}v \\
&\quad-\int_{V_\infty} f\gamma\gamma^{\mathrm{c}}\boldsymbol{v}\cdot\nabla^{\mathrm{S}}\gamma^{\mathrm{s}}\mathbf{w}^{\mathrm{s}}\cdot\nabla\nabla\gamma^{\alpha}\cdot\mathbf{n}\mathrm{d}v=\frac{\partial}{\partial t}\int_{V_\infty} f\gamma\gamma^{\mathrm{c}}\boldsymbol{v}\cdot\nabla^{\mathrm{S}}\gamma^{\mathrm{s}}\mathbf{n}\cdot\nabla\gamma^{\alpha}\mathrm{d}v \\
&\quad+\int_{V_\infty} f\gamma\mathbf{w}^{\mathrm{c}}\cdot\nabla\gamma^{\mathrm{c}}\boldsymbol{v}\cdot\nabla^{\mathrm{S}}\gamma^{\mathrm{s}}\mathbf{n}\cdot\nabla\gamma^{\alpha}\mathrm{d}v+\int_{V_\infty} f\gamma\gamma^{\mathrm{c}}\boldsymbol{v}\cdot\nabla\mathbf{w}\cdot\nabla\gamma^{\mathrm{s}}\mathbf{n}\cdot\nabla\gamma^{\alpha}\mathrm{d}v \\
&\quad+\int_{V_\infty} f\gamma\gamma^{\mathrm{c}}\boldsymbol{v}\cdot\nabla\nabla\gamma^{\mathrm{s}}\cdot\mathbf{w}\mathbf{n}\cdot\nabla\gamma^{\alpha}\mathrm{d}v+\int_{V_\infty} f\gamma\gamma^{\mathrm{c}}\boldsymbol{v}\cdot\nabla^{\mathrm{S}}\gamma^{\mathrm{s}}\mathbf{n}\cdot\nabla\mathbf{w}\cdot\nabla\gamma^{\alpha}\mathrm{d}v \\
&\quad+\int_{V_\infty} f\gamma\gamma^{\mathrm{c}}\boldsymbol{v}\cdot\nabla^{\mathrm{S}}\gamma^{\mathrm{s}}\mathbf{n}\cdot\nabla\nabla\gamma^{\alpha}\cdot\mathbf{w}\mathrm{d}v+\nabla\cdot\int_{V_\infty}\mathbf{w}^{\mathrm{s}}f\gamma\gamma^{\mathrm{c}}\boldsymbol{v}\cdot\nabla^{\mathrm{S}}\gamma^{\mathrm{s}}\mathbf{n}\cdot\nabla\gamma^{\alpha}\mathrm{d}v \\
&\quad-\int_{V_\infty}\nabla\cdot\mathbf{w}f\gamma\gamma^{\mathrm{c}}\boldsymbol{v}\cdot\nabla^{\mathrm{S}}\gamma^{\mathrm{s}}\mathbf{n}\cdot\nabla\gamma^{\alpha}\mathrm{d}v+\int_{V_\infty}\nabla\cdot\boldsymbol{\lambda}\boldsymbol{\lambda}\cdot\mathbf{w}f\gamma\gamma^{\mathrm{c}}\boldsymbol{v}\cdot\nabla^{\mathrm{S}}\gamma^{\mathrm{s}}\mathbf{n}\cdot\nabla\gamma^{\alpha}\mathrm{d}v \\
&\quad+\int_{V_\infty}\boldsymbol{\lambda}\cdot\nabla\boldsymbol{\lambda}\cdot\mathbf{w}f\gamma\gamma^{\mathrm{c}}\boldsymbol{v}\cdot\nabla^{\mathrm{S}}\gamma^{\mathrm{s}}\mathbf{n}\cdot\nabla\gamma^{\alpha}\mathrm{d}v+\int_{V_\infty}\boldsymbol{\lambda}\cdot\nabla\mathbf{w}\cdot\boldsymbol{\lambda}f\gamma\gamma^{\mathrm{c}}\boldsymbol{v}\cdot\nabla^{\mathrm{S}}\gamma^{\mathrm{s}}\mathbf{n}\cdot\nabla\gamma^{\alpha}\mathrm{d}v \\
&\quad-\int_{V_\infty} f\gamma\gamma^{\mathrm{c}}\mathbf{w}^{\mathrm{s}}\cdot\nabla\nabla\gamma^{\mathrm{s}}\cdot\boldsymbol{v}\mathbf{n}\cdot\nabla\gamma^{\alpha}\mathrm{d}v-\int_{V_\infty} f\gamma\gamma^{\mathrm{c}}\boldsymbol{v}\cdot\nabla^{\mathrm{S}}\gamma^{\mathrm{s}}\mathbf{w}^{\mathrm{s}}\cdot\nabla\nabla\gamma^{\alpha}\cdot\mathbf{n}\mathrm{d}v
\end{aligned}
\tag{682}
$$

A rearrangement of terms yields

$$
\begin{aligned}
\int_{C_{\alpha\beta\varepsilon}}\frac{\partial f}{\partial t}\Big|_1 \mathrm{d}c =&\frac{\partial}{\partial t}\int_{V_\infty} f\gamma\gamma^{\mathrm{c}}\boldsymbol{v}\cdot\nabla^{\mathrm{S}}\gamma^{\mathrm{s}}\mathbf{n}\cdot\nabla\gamma^{\alpha}\mathrm{d}v \\
&+\int_{V_\infty} f\gamma\mathbf{w}^{\mathrm{c}}\cdot\nabla\gamma^{\mathrm{c}}\boldsymbol{v}\cdot\nabla^{\mathrm{S}}\gamma^{\mathrm{s}}\mathbf{n}\cdot\nabla\gamma^{\alpha}\mathrm{d}v \\
&+\int_{V_\infty} f\gamma\gamma^{\mathrm{c}}\boldsymbol{v}\cdot\nabla\mathbf{w}\cdot\nabla\gamma^{\mathrm{s}}\mathbf{n}\cdot\nabla\gamma^{\alpha}\mathrm{d}v
\end{aligned}
$$

$$+\int_{V_\infty} f\gamma\gamma^c \mathbf{w}^c\cdot\nabla\nabla\gamma^s\cdot\boldsymbol{v}\mathbf{n}\cdot\nabla\gamma^\alpha dv$$
$$+\int_{V_\infty} f\gamma\gamma^c \boldsymbol{v}\cdot\nabla^S\gamma^s\mathbf{n}\cdot\nabla\mathbf{w}\cdot\nabla\gamma^\alpha dv$$
$$+\int_{V_\infty} f\gamma\gamma^c \mathbf{w}^c\cdot\nabla\nabla\gamma^\alpha\cdot\mathbf{n}\boldsymbol{v}\cdot\nabla^S\gamma^s dv$$
$$+\nabla\cdot\int_{V_\infty}\mathbf{w}^s f\gamma\gamma^c\boldsymbol{v}\cdot\nabla^S\gamma^s\mathbf{n}\cdot\nabla\gamma^\alpha dv$$
$$-\int_{V_\infty}\nabla\cdot\mathbf{w}f\gamma\gamma^c\boldsymbol{v}\cdot\nabla^S\gamma^s\mathbf{n}\cdot\nabla\gamma^\alpha dv$$
$$+\int_{V_\infty}\nabla\cdot\boldsymbol{\lambda}\boldsymbol{\lambda}\cdot\mathbf{w}f\gamma\gamma^c\boldsymbol{v}\cdot\nabla^S\gamma^s\mathbf{n}\cdot\nabla\gamma^\alpha dv$$
$$+\int_{V_\infty}\boldsymbol{\lambda}\cdot\nabla\boldsymbol{\lambda}\cdot\mathbf{w}f\gamma\gamma^c\boldsymbol{v}\cdot\nabla^S\gamma^s\mathbf{n}\cdot\nabla\gamma^\alpha dv$$
$$+\int_{V_\infty}\boldsymbol{\lambda}\cdot\nabla\mathbf{w}\cdot\boldsymbol{\lambda}f\gamma\gamma^c\boldsymbol{v}\cdot\nabla^S\gamma^s\mathbf{n}\cdot\nabla\gamma^\alpha dv \quad (683)$$

in which we used $\mathbf{w}–\mathbf{w}^s = \mathbf{w}^c$. By Equations (53), (258), and (261),

$$\int_{C_{\alpha\beta\varepsilon}}\left.\frac{\partial f}{\partial t}\right|_1 dc$$
$$=\frac{\partial}{\partial t}\int_{V_\infty} f\gamma\gamma^c\boldsymbol{v}\cdot\nabla^S\gamma^s\mathbf{n}\cdot\nabla\gamma^\alpha dv+\int_{V_\infty} f\gamma\mathbf{w}^c\cdot\nabla\gamma^c\boldsymbol{v}\cdot\nabla^S\gamma^s\mathbf{n}\cdot\nabla\gamma^\alpha dv$$
$$+\int_{V_\infty} f\gamma\gamma^c\boldsymbol{v}\cdot\nabla\mathbf{w}\cdot\nabla\gamma^s\mathbf{n}\cdot\nabla\gamma^\alpha dv-\int_{V_\infty} f\gamma\gamma^c\mathbf{w}\cdot\boldsymbol{\lambda}\boldsymbol{v}\cdot\nabla\boldsymbol{\lambda}\cdot\nabla\gamma^s\mathbf{n}\cdot\nabla\gamma^\alpha dv$$
$$+\int_{V_\infty} f\gamma\gamma^c\boldsymbol{v}\cdot\nabla^S\gamma^s\mathbf{n}\cdot\nabla\mathbf{w}\cdot\nabla\gamma^\alpha dv-\int_{V_\infty} f\gamma\gamma^c\mathbf{w}\cdot\boldsymbol{\lambda}\mathbf{n}\cdot\nabla\boldsymbol{\lambda}\cdot\nabla\gamma^\alpha\boldsymbol{v}\cdot\nabla^S\gamma^s dv$$
$$+\nabla\cdot\int_{V_\infty}\mathbf{w}^s f\gamma\gamma^c\boldsymbol{v}\cdot\nabla^S\gamma^s\mathbf{n}\cdot\nabla\gamma^\alpha dv-\int_{V_\infty}\boldsymbol{\lambda}\cdot\nabla\mathbf{w}\cdot\boldsymbol{\lambda}f\gamma\gamma^c\boldsymbol{v}\cdot\nabla^S\gamma^s\mathbf{n}\cdot\nabla\gamma^\alpha dv$$
$$-\int_{V_\infty}\mathbf{n}\cdot\nabla\mathbf{w}\cdot\mathbf{n}f\gamma\gamma^c\boldsymbol{v}\cdot\nabla^S\gamma^s\mathbf{n}\cdot\nabla\gamma^\alpha dv-\int_{V_\infty}\boldsymbol{v}\cdot\nabla\mathbf{w}\cdot f\gamma\gamma^c\boldsymbol{v}\cdot\nabla^S\gamma^s\mathbf{n}\cdot\nabla\gamma^\alpha dv$$
$$+\int_{V_\infty}\boldsymbol{v}\cdot\nabla\boldsymbol{\lambda}\cdot\boldsymbol{v}\boldsymbol{\lambda}\cdot\mathbf{w}f\gamma\gamma^c\boldsymbol{v}\cdot\nabla^S\gamma^s\mathbf{n}\cdot\nabla\gamma^\alpha dv$$
$$+\int_{V_\infty}\boldsymbol{n}\cdot\nabla\boldsymbol{\lambda}\cdot\boldsymbol{n}\boldsymbol{\lambda}\cdot\mathbf{w}f\gamma\gamma^c\boldsymbol{v}\cdot\nabla^S\gamma^s\mathbf{n}\cdot\nabla\gamma^\alpha dv$$
$$+\int_{V_\infty}\boldsymbol{\lambda}\cdot\nabla\boldsymbol{\lambda}\cdot w f\gamma\gamma^c\boldsymbol{v}\cdot\nabla^S\gamma^s\mathbf{n}\cdot\nabla\gamma^\alpha dv$$
$$+\int_{V_\infty}\boldsymbol{\lambda}\cdot\nabla w\cdot\boldsymbol{\lambda}f\gamma\gamma^c\boldsymbol{v}\cdot\nabla^S\gamma^s\mathbf{n}\cdot\nabla\gamma^\alpha dv \quad (684)$$

By Equations (275) and (166), we finally obtain

$$\int_{C_{\alpha\beta\varepsilon}} \frac{\partial f}{\partial t}\bigg|_{1} dc = \frac{\partial}{\partial t}\int_{C_{\alpha\beta\varepsilon}} f dc + \nabla\cdot\int_{C_{\alpha\beta\varepsilon}} \mathbf{w}^{s} f dc + \int_{C_{\alpha\beta\varepsilon}} \boldsymbol{\lambda}\cdot\nabla^{c}\boldsymbol{\lambda}\cdot\mathbf{w}^{s} f dc - \sum_{\text{pts}} \mathbf{e}\cdot\mathbf{w}^{c} f$$

Theorem 47.

$$\int_{C_{\alpha\beta\varepsilon}} \nabla^{c}\circ F dc = \nabla\cdot\int_{C_{\alpha\beta\varepsilon}} \boldsymbol{\lambda}\boldsymbol{\lambda}\circ F dc - \int_{C_{\alpha\beta\varepsilon}} (\boldsymbol{\lambda}\cdot\nabla^{c}\boldsymbol{\lambda})\circ F dc + \sum_{\text{pts}} \mathbf{e}\circ F \tag{685}$$

where $\nabla^{c}\circ F$ can be any of the three operators: $\nabla^{c} f$, $\nabla^{c}\cdot\mathbf{f}$, and $\nabla^{c}\times\mathbf{f}$. $f(\mathbf{l}, t)$ and $\mathbf{f}(\mathbf{u}, t)$ are scalar and vector functions of curvilineal position vector $\mathbf{u}$ and time t defined on $C_{\alpha\beta\varepsilon}$, respectively.

Proof. Define the indicator function $\lambda^{c}(\mathbf{l}(\mathbf{x}+\boldsymbol{\xi}), t)$ such that $\gamma^{c} = 1$ on $C_{\alpha\beta\varepsilon}$ and $\gamma^{c} = 0$ elsewhere. By Equation (275), we obtain

$$\int_{C_{\alpha\beta\varepsilon}} \nabla^{c}\circ F dc = \int_{V_{\infty}} \boldsymbol{\lambda}\circ(\boldsymbol{\lambda}\cdot\nabla F)\gamma\gamma^{c}\boldsymbol{v}\cdot\nabla^{S}\gamma^{s}\mathbf{n}\cdot\nabla\gamma^{\alpha} dv \tag{686}$$

An application of the chain rule to the right-hand side of Equation (686) yields

$$\begin{aligned}
\int_{C_{\alpha\beta\varepsilon}} \nabla^{c}\circ F dc = & \int_{V_{\infty}} \nabla\cdot(\boldsymbol{\lambda}\boldsymbol{\lambda}\circ F\gamma\gamma^{c}\boldsymbol{v}\cdot\nabla^{S}\gamma^{s}\mathbf{n}\cdot\nabla\gamma^{\alpha}) dv \\
& - \int_{V_{\infty}} \nabla\cdot\boldsymbol{\lambda}\boldsymbol{\lambda}\circ F\gamma\gamma^{c}\boldsymbol{v}\cdot\nabla^{S}\gamma^{s}\mathbf{n}\cdot\nabla\gamma^{\alpha} dv \\
& - \int_{V_{\infty}} \boldsymbol{\lambda}\cdot\nabla\boldsymbol{\lambda}\circ F\gamma\gamma^{c}\boldsymbol{v}\cdot\nabla^{S}\gamma^{s}\mathbf{n}\cdot\nabla\gamma^{\alpha} dv \\
& - \int_{V_{\infty}} \boldsymbol{\lambda}\circ F\nabla\gamma\cdot\boldsymbol{\lambda}\gamma^{c}\boldsymbol{v}\cdot\nabla^{S}\gamma^{s}\mathbf{n}\cdot\nabla\gamma^{\alpha} dv \\
& - \int_{V_{\infty}} \boldsymbol{\lambda}\circ F\gamma\boldsymbol{\lambda}\cdot\nabla\gamma^{c}\boldsymbol{v}\cdot\nabla^{S}\gamma^{s}\mathbf{n}\cdot\nabla\gamma^{\alpha} dv \\
& - \int_{V_{\infty}} \boldsymbol{\lambda}\circ F\gamma\gamma^{c}\boldsymbol{\lambda}\cdot(\nabla\boldsymbol{v}\cdot\nabla^{S}\gamma^{s} \\
& + \nabla\nabla^{S}\gamma^{s}\cdot\boldsymbol{v})\mathbf{n}\cdot\nabla\gamma^{\alpha} dv \\
& - \int_{V_{\infty}} \boldsymbol{\lambda}\circ F\gamma\gamma^{c}\boldsymbol{\lambda}\cdot(\nabla\mathbf{n}\cdot\nabla\gamma^{\alpha} \\
& + \nabla\nabla\gamma^{\alpha}\cdot\mathbf{n})\boldsymbol{v}\cdot\nabla^{S}\gamma^{s} dv
\end{aligned} \tag{687}$$

The operator ∇ can be interchanged with the integration operator for the first integral on the right side of Equation (687) because its integration domain is independent of the integration variables. Since the curve $C_{\alpha\beta\varepsilon}$ is within the REV, the fourth term is zero by the definition of the indicator function γ.

By Equations (195), (200), and (216), we have

$$\begin{aligned}
&\int_{C_{\alpha\beta\varepsilon}} \nabla^{c} \circ F dc \\
&= \nabla \cdot \int_{V_\infty} \boldsymbol{\lambda}\boldsymbol{\lambda} \circ F\gamma\gamma^{c}\boldsymbol{v} \cdot \nabla^{S}\gamma^{s}\mathbf{n} \cdot \nabla\gamma^{\alpha} dv - \int_{V_\infty} \nabla \cdot \boldsymbol{\lambda}\boldsymbol{\lambda} \circ F\gamma\gamma^{c}\boldsymbol{v} \cdot \nabla^{S}\gamma^{s}\mathbf{n} \cdot \nabla\gamma^{\alpha} dv \\
&\quad - \int_{V_\infty} \boldsymbol{\lambda} \cdot \nabla\boldsymbol{\lambda} \circ F\gamma\gamma^{c}\boldsymbol{v} \cdot \nabla^{S}\gamma^{s}\mathbf{n} \cdot \nabla\gamma^{\alpha} dv - \int_{V_\infty} \boldsymbol{\lambda} \circ F\gamma\boldsymbol{\lambda} \cdot \nabla\gamma^{c}\boldsymbol{v} \cdot \nabla^{S}\gamma^{s}\mathbf{n} \cdot \nabla\gamma^{\alpha} dv \\
&\quad - \int_{V_\infty} \boldsymbol{\lambda} \circ F\gamma\gamma^{c}\boldsymbol{\lambda} \cdot \nabla\nabla^{S}\gamma^{s} \cdot \boldsymbol{v}\mathbf{n} \cdot \nabla\gamma^{\alpha} dv - \int_{V_\infty} \boldsymbol{\lambda} \circ F\gamma\gamma^{c}\boldsymbol{v} \cdot \nabla^{S}\gamma^{s}\mathbf{n} \cdot \nabla\nabla\gamma^{\alpha} \cdot \boldsymbol{\lambda} dv
\end{aligned} \tag{688}$$

By Equation (53), we have

$$\begin{aligned}
\nabla \cdot \boldsymbol{\lambda} &= \boldsymbol{\lambda} \cdot \nabla\boldsymbol{\lambda} \cdot \boldsymbol{\lambda} + \mathbf{n} \cdot \nabla\boldsymbol{\lambda} \cdot \mathbf{n} + \boldsymbol{v} \cdot \nabla\boldsymbol{\lambda} \cdot \boldsymbol{v} \\
&= \mathbf{n} \cdot \nabla\boldsymbol{\lambda} \cdot \mathbf{n} + \boldsymbol{v} \cdot \nabla\boldsymbol{\lambda} \cdot \boldsymbol{v}
\end{aligned} \tag{689}$$

By Equations (258), (261), and (689),

$$\begin{aligned}
\int_{C_{\alpha\beta\varepsilon}} \nabla \circ F dc = \nabla \cdot &\int_{V_\infty} \boldsymbol{\lambda}\boldsymbol{\lambda} \circ F\gamma\gamma^{c}\boldsymbol{v} \cdot \nabla^{S}\gamma^{s}\mathbf{n} \cdot \nabla\gamma^{\alpha} dv \\
&- \int_{V_\infty} \mathbf{n} \cdot \nabla\boldsymbol{\lambda} \cdot \mathbf{n}\boldsymbol{\lambda} \circ F\gamma\gamma^{c}\boldsymbol{v} \cdot \nabla^{S}\gamma^{s}\mathbf{n} \cdot \nabla\gamma^{\alpha} dv \\
&- \int_{V_\infty} \boldsymbol{v} \cdot \nabla\boldsymbol{\lambda} \cdot \boldsymbol{v}\boldsymbol{\lambda} \circ F\gamma\gamma^{c}\boldsymbol{v} \cdot \nabla^{S}\gamma^{s}\mathbf{n} \cdot \nabla\gamma^{\alpha} dv \\
&- \int_{V_\infty} \boldsymbol{\lambda} \cdot \nabla\boldsymbol{\lambda} \circ F\gamma\gamma^{c}\boldsymbol{v} \cdot \nabla^{S}\gamma^{s}\mathbf{n} \cdot \nabla\gamma^{\alpha} dv \\
&- \int_{V_\infty} \boldsymbol{\lambda} \circ F\gamma\boldsymbol{\lambda} \cdot \nabla\gamma^{c}\boldsymbol{v} \cdot \nabla^{S}\gamma^{s}\mathbf{n} \cdot \nabla\gamma^{\alpha} dv \\
&- \int_{V_\infty} F \circ \boldsymbol{\lambda}\gamma\gamma^{c}\boldsymbol{v} \cdot \nabla\boldsymbol{\lambda} \cdot \nabla\gamma^{s}\mathbf{n} \cdot \nabla\gamma^{\alpha} dv \\
&- \int_{V_\infty} F \circ \boldsymbol{\lambda}\gamma\gamma^{c}\mathbf{n} \cdot \nabla\boldsymbol{\lambda} \cdot \nabla\gamma^{\alpha}\boldsymbol{v} \cdot \nabla^{S}\gamma^{s} dv
\end{aligned} \tag{690}$$

By Equations (165)–(167) and (275), Equation (690) finally leads to

$$\int_{C_{\alpha\beta\varepsilon}} \nabla^{c} \circ F dc = \nabla \cdot \int_{C_{\alpha\beta\varepsilon}} \boldsymbol{\lambda}\boldsymbol{\lambda} \circ F dc - \int_{C_{\alpha\beta\varepsilon}} (\boldsymbol{\lambda} \cdot \nabla^{c}\boldsymbol{\lambda}) \circ F dc + \sum_{\text{pts}} \mathbf{e} \circ F$$

Remark. Equation (685) contains the following three formulas:

Gradient

$$\int_{C_{\alpha\beta\varepsilon}} \nabla^{c} f dc = \nabla \cdot \int_{C_{\alpha\beta\varepsilon}} \boldsymbol{\lambda}\boldsymbol{\lambda} f dc - \int_{C_{\alpha\beta\varepsilon}} \boldsymbol{\lambda} \cdot \nabla^{c}\boldsymbol{\lambda} f dc + \sum_{\text{pts}} \mathbf{e} f \tag{691}$$

Divergence

$$\int_{C_{\alpha\beta\varepsilon}} \nabla^{c}\cdot \mathbf{f}dc = \nabla\cdot \int_{C_{\alpha\beta\varepsilon}} \boldsymbol{\lambda\lambda}\cdot \mathbf{f}dc - \int_{C_{\alpha\beta\varepsilon}} \boldsymbol{\lambda}\cdot\nabla^{c}\boldsymbol{\lambda}\cdot \mathbf{f}dc + \sum_{\text{pts}} \mathbf{e}\cdot\mathbf{f} \tag{692}$$

Curl

$$\int_{C_{\alpha\beta\varepsilon}} \nabla^{c}\times \mathbf{f}dc = \nabla\cdot \int_{C_{\alpha\beta\varepsilon}} \boldsymbol{\lambda\lambda}\times \mathbf{f}dc - \int_{C_{\alpha\beta\varepsilon}} \boldsymbol{\lambda}\cdot\nabla^{c}\boldsymbol{\lambda}\times \mathbf{f}dc + \sum_{\text{pts}} \mathbf{e}\times\mathbf{f} \tag{693}$$

Theorem 46 has the following alternative form:

Theorem 48.

$$\begin{aligned}\int_{C_{\alpha\beta\varepsilon}} \left.\frac{\partial f}{\partial t}\right|_{1} dc = \frac{\partial}{\partial t}\int_{C_{\alpha\beta\varepsilon}} f dc + \int_{C_{\alpha\beta\varepsilon}} \boldsymbol{\lambda}\cdot\nabla^{c}\boldsymbol{\lambda}\cdot\mathbf{w}^{s} f dc + \sum_{\text{ext}} \frac{\mathbf{n}^{*}\cdot\mathbf{w}f}{\mathbf{e}\cdot\mathbf{n}^{*}} \\ - \sum_{\text{ext}} \mathbf{e}\cdot\mathbf{w}^{c} f - \sum_{\text{pts}} \mathbf{e}\cdot\mathbf{w}^{c} f\end{aligned} \tag{694}$$

where ext refers to the points of intersection of the $C_{\alpha\beta\varepsilon}$ curve with the external surface of the REV. $\mathbf{n}^{*}$ is the unit vector normal to the external boundary of the REV that is pointing outward from the REV.

Proof. By Equation (275), we obtain

$$\int_{C_{\alpha\beta\varepsilon}} \left.\frac{\partial f}{\partial t}\right|_{1} dc = \int_{V_\infty} \left.\frac{\partial f}{\partial t}\right|_{1} \gamma\gamma^{c}\boldsymbol{v}\cdot\nabla^{S}\gamma^{s}\mathbf{n}\cdot\nabla\gamma^{\alpha} dv \tag{695}$$

which is, Equation (134),

$$\int_{C_{\alpha\beta\varepsilon}} \left.\frac{\partial f}{\partial t}\right|_{1} dc = \int_{V_\infty} \left.\frac{\partial f}{\partial t}\right|_{\mathbf{x}} \gamma\gamma^{c}\boldsymbol{v}\cdot\nabla^{S}\gamma^{s}\mathbf{n}\cdot\nabla\gamma^{\alpha} dv + \int_{V_\infty} \mathbf{w}^{s}\cdot\nabla f\gamma\gamma^{c}\boldsymbol{v}\cdot\nabla^{S}\gamma^{s}\mathbf{n}\cdot\nabla\gamma^{\alpha} dv \tag{696}$$

An application of the chain rule to the right side of Equation (696) yields

$$\begin{aligned}&\int_{C_{\alpha\beta\varepsilon}} \left.\frac{\partial f}{\partial t}\right|_{1} dc \\ &= \int_{V_\infty} \frac{\partial(f\gamma\gamma^{c}\boldsymbol{v}\cdot\nabla^{S}\gamma^{s}\mathbf{n}\cdot\nabla\gamma^{\alpha})}{\partial t} dv - \int_{V_\infty} f\frac{\partial(\gamma\gamma^{c})}{\partial t}\boldsymbol{v}\cdot\nabla^{S}\gamma^{s}\mathbf{n}\cdot\nabla\gamma^{\alpha} dv \\ &\quad - \int_{V_\infty} f\gamma\gamma^{c}\frac{\partial\boldsymbol{v}}{\partial t}\cdot\nabla^{S}\gamma^{s}\mathbf{n}\cdot\nabla\gamma^{\alpha} dv - \int_{V_\infty} f\gamma\gamma^{c}\boldsymbol{v}\cdot\nabla\frac{\partial\gamma^{s}}{\partial t}\mathbf{n}\cdot\nabla\gamma^{\alpha} dv \\ &\quad - \int_{V_\infty} f\gamma\gamma^{c}\boldsymbol{v}\cdot\nabla^{S}\gamma^{s}\frac{\partial\mathbf{n}}{\partial t}\cdot\nabla\gamma^{\alpha} dv - \int_{V_\infty} f\gamma\gamma^{c}\boldsymbol{v}\cdot\nabla^{S}\gamma^{s}\mathbf{n}\cdot\nabla\frac{\partial\gamma^{\alpha}}{\partial t} dv \\ &\quad + \int_{V_\infty} \nabla\cdot(\mathbf{w}^{s} f\gamma\gamma^{c}\boldsymbol{v}\cdot\nabla^{S}\gamma^{s}\mathbf{n}\cdot\nabla\gamma^{\alpha}) dv - \int_{V_\infty} \nabla\cdot\mathbf{w}^{s} f\gamma\gamma^{c}\boldsymbol{v}\cdot\nabla^{S}\gamma^{s}\mathbf{n}\cdot\nabla\gamma^{\alpha} dv \\ &\quad - \int_{V_\infty} f\mathbf{w}^{s}\cdot\nabla(\gamma\gamma^{c})\boldsymbol{v}\cdot\nabla^{S}\gamma^{s}\mathbf{n}\cdot\nabla\gamma^{\alpha} dv - \int_{V_\infty} f\gamma\gamma^{c}\mathbf{w}^{s}\cdot\nabla\boldsymbol{v}\cdot\nabla^{S}\gamma^{s}\mathbf{n}\cdot\nabla\gamma^{\alpha} dv \\ &\quad - \int_{V_\infty} f\gamma\gamma^{c}\mathbf{w}^{s}\cdot\nabla\nabla\gamma^{s}\cdot\boldsymbol{v}\mathbf{n}\cdot\nabla\gamma^{\alpha} dv - \int_{V_\infty} f\gamma\gamma^{c}\boldsymbol{v}\cdot\nabla^{S}\gamma^{s}\mathbf{w}^{s}\cdot\nabla\mathbf{n}\cdot\nabla\gamma^{\alpha} dv \\ &\quad - \int_{V_\infty} f\gamma\gamma^{c}\boldsymbol{v}\cdot\nabla^{S}\gamma^{s}\mathbf{w}^{s}\cdot\nabla\nabla\gamma^{\alpha}\cdot\mathbf{n} dv\end{aligned} \tag{697}$$

Differentiation with respect to the time t can be interchanged with the integration operator in the first integral on the right-hand side of Equation (697) because its integration domain is independent of t. Because γ is independent of t, for the second term we have

$$\frac{\partial(\gamma\gamma^{c})}{\partial t} = \gamma\frac{\partial\gamma^{c}}{\partial t}$$

By Equation (231), the seventh term is zero. By Equations (180), (194), (195), (199), (200), and (224), we obtain

$$\begin{aligned}
\int_{C_{\alpha\beta\varepsilon}} \frac{\partial f}{\partial t}\bigg|_{1} \mathrm{d}c
&= \frac{\partial}{\partial t}\int_{V_\infty} f\gamma\gamma^{c}\mathbf{v}\cdot\nabla^{S}\gamma^{s}\mathbf{n}\cdot\nabla\gamma^{\alpha}\mathrm{d}v + \int_{V_\infty} f\gamma\mathbf{w}\cdot\nabla\gamma^{c}\mathbf{v}\cdot\nabla^{S}\gamma^{s}\mathbf{n}\cdot\nabla\gamma^{\alpha}\mathrm{d}v \\
&+ \int_{V_\infty} f\gamma\gamma^{c}\mathbf{v}\cdot\nabla\mathbf{w}\cdot\nabla\gamma^{s}\mathbf{n}\cdot\nabla\gamma^{\alpha}\mathrm{d}v + \int_{V_\infty} f\gamma\gamma^{c}\mathbf{v}\cdot\nabla\nabla\gamma^{s}\cdot\mathbf{w}\mathbf{n}\cdot\nabla\gamma^{\alpha}\mathrm{d}v \\
&+ \int_{V_\infty} f\gamma\gamma^{c}\mathbf{v}\cdot\nabla^{S}\gamma^{s}\mathbf{n}\cdot\nabla\mathbf{w}\cdot\nabla\gamma^{\alpha}\mathrm{d}v + \int_{V_\infty} f\gamma\gamma^{c}\mathbf{v}\cdot\nabla^{S}\gamma^{s}\mathbf{n}\cdot\nabla\nabla\gamma^{\alpha}\cdot\mathbf{w}\mathrm{d}v \\
&- \int_{V_\infty} \nabla\cdot\mathbf{w}^{s} f\gamma\gamma^{c}\mathbf{v}\cdot\nabla^{S}\gamma^{s}\mathbf{n}\cdot\nabla\gamma^{\alpha}\mathrm{d}v - \int_{V_\infty} f\mathbf{w}^{s}\cdot\nabla(\gamma\gamma^{c})\mathbf{v}\cdot\nabla^{S}\gamma^{s}\mathbf{n}\cdot\nabla^{\alpha}\mathrm{d}v \\
&- \int_{V_\infty} f\gamma\gamma^{c}\mathbf{w}^{s}\cdot\nabla\mathbf{v}\cdot\nabla^{S}\gamma^{s}\mathbf{n}\cdot\nabla\gamma^{\alpha}\mathrm{d}v - \int_{V_\infty} f\gamma\gamma^{c}\mathbf{w}^{s}\cdot\nabla\nabla\gamma^{s}\cdot\mathbf{v}\mathbf{n}\cdot\nabla\gamma^{\alpha}\mathrm{d}v \\
&- \int_{V_\infty} f\gamma\gamma^{c}\mathbf{v}\cdot\nabla^{S}\gamma^{s}\mathbf{w}^{s}\cdot\nabla\mathbf{n}\cdot\nabla\gamma^{\alpha}\mathrm{d}v - \int_{V_\infty} f\gamma\gamma^{c}\mathbf{v}\cdot\nabla^{S}\gamma^{s}\mathbf{w}^{s}\cdot\nabla\nabla\gamma^{\alpha}\cdot\mathbf{n}\mathrm{d}v
\end{aligned} \tag{698}$$

Regrouping the terms on the right side of Equation (698) yields

$$\begin{aligned}
\int_{C_{\alpha\beta\varepsilon}} \frac{\partial f}{\partial t}\bigg|_{1} \mathrm{d}v
&= \frac{\partial}{\partial t}\int_{V_\infty} f\gamma\gamma^{c}\mathbf{v}\cdot\nabla^{S}\gamma^{s}\mathbf{n}\cdot\nabla\gamma^{\alpha}\mathrm{d}v + \int_{V_\infty} f\mathbf{w}^{c}\cdot\nabla(\gamma\gamma^{c})\mathbf{v}\cdot\nabla^{S}\gamma^{s}\mathbf{n}\cdot\nabla\gamma^{\alpha}\mathrm{d}v \\
&- \int_{V_\infty} f\mathbf{w}\cdot\nabla\gamma\gamma^{c}\mathbf{v}\cdot\nabla^{S}\gamma^{s}\mathbf{n}\cdot\nabla\gamma^{\alpha}\mathrm{d}v + \int_{V_\infty} f\gamma\gamma^{c}\mathbf{v}\cdot\nabla\nabla\gamma^{s}\cdot\mathbf{w}^{c}\mathbf{n}\cdot\nabla\gamma^{\alpha}\mathrm{d}v \\
&+ \int_{V_\infty} f\gamma\gamma^{c}\mathbf{v}\cdot\nabla\mathbf{w}\cdot\nabla\gamma^{s}\mathbf{n}\cdot\nabla\gamma^{\alpha}\mathrm{d}v + \int_{V_\infty} f\gamma\gamma^{c}\mathbf{v}\cdot\nabla^{S}\gamma^{s}\mathbf{n}\cdot\nabla\mathbf{w}\cdot\nabla\gamma^{\alpha}\mathrm{d}v \\
&+ \int_{V_\infty} f\gamma\gamma^{c}\mathbf{v}\cdot\nabla^{S}\gamma^{s}\mathbf{n}\cdot\nabla\nabla\gamma^{\alpha}\cdot\mathbf{w}^{c}\mathrm{d}v - \int_{V_\infty} \nabla\cdot\mathbf{w} f\gamma\gamma^{c}\mathbf{v}\cdot\nabla^{S}\gamma^{s}\mathbf{n}\cdot\nabla\gamma^{\alpha}\mathrm{d}v \\
&+ \int_{V_\infty} \nabla\cdot(\boldsymbol{\lambda}\boldsymbol{\lambda}\cdot\mathbf{w}) f\gamma\gamma^{c}\mathbf{v}\cdot\nabla^{S}\gamma^{s}\mathbf{n}\cdot\nabla\gamma^{\alpha}\mathrm{d}v \\
&- \int_{V_\infty} f\gamma\gamma^{c}\mathbf{w}^{s}\cdot\nabla\mathbf{v}\cdot\nabla^{S}\gamma^{s}\mathbf{n}\cdot\nabla\gamma^{\alpha}\mathrm{d}v - \int_{V_\infty} f\gamma\gamma^{c}\mathbf{w}^{s}\cdot\nabla\mathbf{n}\cdot\nabla\gamma^{\alpha}\mathbf{v}\cdot\nabla^{S}\gamma^{s}\mathrm{d}v
\end{aligned} \tag{699}$$

in which we used $\mathbf{w}^{s} = \mathbf{w} - \boldsymbol{\lambda}\boldsymbol{\lambda}\cdot\mathbf{w}$. By Equation (53),

$$\nabla\cdot\mathbf{w} = \boldsymbol{\lambda}\cdot\nabla\mathbf{w}\cdot\boldsymbol{\lambda} + \mathbf{n}\cdot\nabla\mathbf{w}\cdot\mathbf{n} + \mathbf{v}\cdot\nabla\mathbf{w}\cdot\mathbf{v} \tag{700}$$

By Equations (195) and (200), we have

$$\begin{aligned}
&\int_{C_{\alpha\beta\varepsilon}} \frac{\partial f}{\partial t}\Big|_1 \mathrm{d}v \\
&= \frac{\partial}{\partial t}\int_{V_\infty} f\gamma\gamma^{\mathrm{c}}\boldsymbol{v}\cdot\nabla^{\mathrm{S}}\gamma^{\mathrm{s}}\mathbf{n}\cdot\nabla\gamma^{\alpha}\mathrm{d}v + \int_{V_\infty} f\mathbf{w}^{\mathrm{c}}\cdot\nabla(\gamma\gamma^{\mathrm{c}})\boldsymbol{v}\cdot\nabla^{\mathrm{S}}\gamma^{\mathrm{s}}\mathbf{n}\cdot\nabla\gamma^{\alpha}\mathrm{d}v \\
&\quad - \int_{V_\infty} f\mathbf{w}\cdot\nabla\gamma\gamma^{\mathrm{c}}\boldsymbol{v}\cdot\nabla^{\mathrm{S}}\gamma^{\mathrm{s}}\mathbf{n}\cdot\nabla^{\alpha}\mathrm{d}v + \int_{V_\infty} f\gamma\gamma^{\mathrm{c}}\boldsymbol{v}\cdot\nabla\nabla\gamma^{\mathrm{s}}\cdot\mathbf{w}^{\mathrm{c}}\mathbf{n}\cdot\nabla\gamma^{\alpha}\mathrm{d}v \\
&\quad + \int_{V_\infty} f\gamma\gamma^{\mathrm{c}}\boldsymbol{v}\cdot\nabla\mathbf{w}\cdot\nabla\gamma^{\mathrm{s}}\mathbf{n}\cdot\nabla\gamma^{\alpha}\mathrm{d}v + \int_{V_\infty} f\gamma\gamma^{\mathrm{c}}\boldsymbol{v}\cdot\nabla^{\mathrm{S}}\gamma^{\mathrm{s}}\mathbf{n}\cdot\nabla\mathbf{w}\cdot\nabla\gamma^{\alpha}\mathrm{d}v \\
&\quad + \int_{V_\infty} f\gamma\gamma^{\mathrm{c}}\boldsymbol{v}\cdot\nabla^{\mathrm{S}}\gamma^{\mathrm{s}}\mathbf{n}\cdot\nabla\nabla\gamma^{\alpha}\cdot\mathbf{w}^{\mathrm{c}}\mathrm{d}v \\
&\quad - \int_{V_\infty} \boldsymbol{\lambda}\cdot\nabla\mathbf{w}\cdot\boldsymbol{\lambda} f\gamma\gamma^{\mathrm{c}}\boldsymbol{v}\cdot\nabla^{\mathrm{S}}\gamma^{\mathrm{s}}\mathbf{n}\cdot\nabla\gamma^{\alpha}\mathrm{d}v \\
&\quad - \int_{V_\infty} \mathbf{n}\cdot\nabla\mathbf{w}\cdot\mathbf{n} f\gamma\gamma^{\mathrm{c}}\boldsymbol{v}\cdot\nabla^{\mathrm{S}}\gamma^{\mathrm{s}}\mathbf{n}\cdot\nabla\gamma^{\alpha}\mathrm{d}v \\
&\quad - \int_{V_\infty} \boldsymbol{v}\cdot\nabla\mathbf{w}\cdot\boldsymbol{v} f\gamma\gamma^{\mathrm{c}}\boldsymbol{v}\cdot\nabla^{\mathrm{S}}\gamma^{\mathrm{s}}\mathbf{n}\cdot\nabla\gamma^{\alpha}\mathrm{d}v \\
&\quad + \int_{V_\infty} \nabla\cdot\boldsymbol{\lambda}\boldsymbol{\lambda}\cdot\mathbf{w} f\gamma\gamma^{\mathrm{c}}\boldsymbol{v}\cdot\nabla^{\mathrm{S}}\gamma^{\mathrm{s}}\mathbf{n}\cdot\nabla\gamma^{\alpha}\mathrm{d}v \\
&\quad + \int_{V_\infty} \boldsymbol{\lambda}\cdot\nabla\boldsymbol{\lambda}\cdot\mathbf{w} f\gamma\gamma^{\mathrm{c}}\boldsymbol{v}\cdot\nabla^{\mathrm{S}}\gamma^{\mathrm{s}}\mathbf{n}\cdot\nabla\gamma^{\alpha}\mathrm{d}v \\
&\quad + \int_{V_\infty} \boldsymbol{\lambda}\cdot\nabla\mathbf{w}\cdot\boldsymbol{\lambda} f\gamma\gamma^{\mathrm{c}}\boldsymbol{v}\cdot\nabla^{\mathrm{S}}\gamma^{\mathrm{s}}\mathbf{n}\cdot\nabla\gamma^{\alpha}\mathrm{d}v
\end{aligned} \tag{701}$$

Applying Equations (53), (258), and (261), yields

$$\begin{aligned}
\int_{C_{\alpha\beta\varepsilon}} \frac{\partial f}{\partial t}\Big|_1 \mathrm{d}v &= \frac{\partial}{\partial t}\int_{V_\infty} f\gamma\gamma^{\mathrm{c}}\boldsymbol{v}\cdot\nabla^{\mathrm{S}}\gamma^{\mathrm{s}}\mathbf{n}\cdot\nabla\gamma^{\alpha}\mathrm{d}v \\
&\quad + \int_{V_\infty} f\mathbf{w}^{\mathrm{c}}\cdot\nabla(\gamma\gamma^{\mathrm{c}})\boldsymbol{v}\cdot\nabla^{\mathrm{S}}\gamma^{\mathrm{s}}\mathbf{n}\cdot\nabla\gamma^{\alpha}\mathrm{d}v \\
&\quad - \int_{V_\infty} f\mathbf{w}\cdot\nabla\gamma\gamma^{\mathrm{c}}\boldsymbol{v}\cdot\nabla^{\mathrm{S}}\gamma^{\mathrm{s}}\mathbf{n}\cdot\nabla\gamma^{\alpha}\mathrm{d}v \\
&\quad + \int_{V_\infty} \boldsymbol{\lambda}\cdot\nabla\boldsymbol{\lambda}\cdot\mathbf{w} f\gamma\gamma^{\mathrm{c}}\boldsymbol{v}\cdot\nabla^{\mathrm{S}}\gamma^{\mathrm{s}}\mathbf{n}\cdot\nabla\gamma^{\alpha}\mathrm{d}v
\end{aligned} \tag{702}$$

By Equations (166), (275), and (308), Equation (702) becomes

$$\begin{aligned}
\int_{C_{\alpha\beta\varepsilon}} \frac{\partial f}{\partial t}\Big|_1 \mathrm{d}v &= \frac{\partial}{\partial t}\int_{C_{\alpha\beta\varepsilon}} f\mathrm{d}c + \int_{C_{\alpha\beta\varepsilon}} \boldsymbol{\lambda}\cdot\nabla^{\mathrm{c}}\boldsymbol{\lambda}\cdot\mathbf{w}^{\mathrm{s}} f\mathrm{d}c \\
&\quad + \sum_{\mathrm{ext}} \frac{\mathbf{n}^*\cdot\mathbf{w}}{\mathbf{e}\cdot\mathbf{n}^*} f - \sum_{\mathrm{ext}} \mathbf{e}\cdot\mathbf{w}^{\mathrm{c}} f - \sum_{\mathrm{pts}} \mathbf{e}\cdot\mathbf{w}^{\mathrm{c}} f
\end{aligned}$$

in which we have used $\boldsymbol{\lambda}\cdot\nabla\boldsymbol{\lambda}\cdot\mathbf{w} = \boldsymbol{\lambda}\cdot\nabla^{\mathrm{c}}\boldsymbol{\lambda}\cdot\mathbf{w} = \boldsymbol{\lambda}\cdot\nabla^{\mathrm{c}}\boldsymbol{\lambda}\cdot\mathbf{w}^{\mathrm{s}}$.

Theorem 47 has the following alternative form:

Theorem 49.

$$\int_{C_{\alpha\beta\varepsilon}} \nabla^{c} \circ F dc = -\int_{C_{\alpha\beta\varepsilon}} (\boldsymbol{\lambda} \cdot \nabla^{c}\boldsymbol{\lambda}) \circ F dc + \sum_{\text{ext}} \mathbf{e} \circ F + \sum_{\text{pts}} \mathbf{e} \circ F \tag{703}$$

where $\nabla^{c} \circ F$ can be any of the three operators: $\nabla^{c} f$, $\nabla^{c} \cdot \mathbf{f}$, and $\nabla^{c} \times \mathbf{f}$. $f(\mathbf{l}, t)$ and $\mathbf{f}(\mathbf{l}, t)$ are scalar and vector functions of curvilineal position vector $\mathbf{l}$ and time t defined on $C_{\alpha\beta\varepsilon}$.

Proof. By Equation (275), we obtain

$$\int_{C_{\alpha\beta\varepsilon}} \nabla^{c} \circ F dc = \int_{V_{\infty}} \boldsymbol{\lambda} \circ (\boldsymbol{\lambda} \cdot \nabla F)\gamma\gamma^{c}\boldsymbol{v} \cdot \nabla^{S}\gamma^{s}\mathbf{n} \cdot \nabla\gamma^{\alpha} dv \tag{704}$$

An application of the chain rule to the right-hand side of Equation (704) yields

$$\begin{aligned}\int_{C_{\alpha\beta\varepsilon}} \nabla^{c} \circ F dc = &\int_{V_{\infty}} \nabla \cdot (\boldsymbol{\lambda}\boldsymbol{\lambda} \circ F\gamma\gamma^{c}\boldsymbol{v} \cdot \nabla^{S}\gamma^{s}\mathbf{n} \cdot \nabla\gamma^{\alpha}) dv \\ &- \int_{V_{\infty}} \nabla \cdot \boldsymbol{\lambda}\boldsymbol{\lambda} \circ F\gamma\gamma^{c}\boldsymbol{v} \cdot \nabla^{S}\gamma^{s}\mathbf{n} \cdot \nabla\gamma^{\alpha} dv \\ &- \int_{V_{\infty}} \boldsymbol{\lambda} \cdot \nabla\boldsymbol{\lambda} \circ F\gamma\gamma^{c}\boldsymbol{v} \cdot \nabla^{S}\gamma^{s}\mathbf{n} \cdot \nabla\gamma^{\alpha} dv \\ &- \int_{V_{\infty}} \boldsymbol{\lambda} \circ F\boldsymbol{\lambda} \cdot \nabla(\gamma\gamma^{c})\boldsymbol{v} \cdot \nabla^{S}\gamma^{s}\mathbf{n} \cdot \nabla\gamma^{\alpha} dv \\ &- \int_{V_{\infty}} \boldsymbol{\lambda} \circ F\gamma\gamma^{c}\boldsymbol{\lambda} \cdot (\nabla\boldsymbol{v} \cdot \nabla^{S}\gamma^{s} + \nabla\nabla^{S}\gamma^{s} \cdot \boldsymbol{v})\mathbf{n} \cdot \nabla\gamma^{\alpha} dv \\ &- \int_{V_{\infty}} \boldsymbol{\lambda} \circ F\gamma\gamma^{c}\boldsymbol{\lambda} \cdot (\nabla\mathbf{n} \cdot \nabla\gamma^{\alpha} + \nabla\nabla\gamma^{\alpha} \cdot \mathbf{n})\boldsymbol{v} \cdot \nabla^{S}\gamma^{s} dv\end{aligned} \tag{705}$$

By Equation (231), the first integral on the right-hand side of Equation (705) is zero. By Equations (195), (200), and (216),

$$\begin{aligned}\int_{C_{\alpha\beta\varepsilon}} \nabla^{c} \circ F = &-\int_{V_{\infty}} \nabla \cdot \boldsymbol{\lambda}\boldsymbol{\lambda} \circ F\gamma\gamma^{c}\boldsymbol{v} \cdot \nabla^{S}\gamma^{s}\mathbf{n} \cdot \nabla\gamma^{\alpha} dv \\ &- \int_{V_{\infty}} \boldsymbol{\lambda} \cdot \nabla\boldsymbol{\lambda} \circ F\gamma\gamma^{c}\boldsymbol{v} \cdot \nabla^{S}\gamma^{s}\mathbf{n} \cdot \nabla\gamma^{\alpha} dv \\ &- \int_{V_{\infty}} \boldsymbol{\lambda} \circ F\nabla(\gamma\gamma^{c}) \cdot \boldsymbol{\lambda}\boldsymbol{v} \cdot \nabla^{S}\gamma^{s}\mathbf{n} \cdot \nabla\gamma^{\alpha} dv \\ &+ \int_{V_{\infty}} \gamma\gamma^{c} F \circ \nabla^{c}\nabla\gamma^{s} \cdot \boldsymbol{v}\mathbf{n} \cdot \nabla\gamma^{\alpha} dv \\ &+ \int_{V_{\infty}} \gamma\gamma^{c} F \circ \nabla^{c}\nabla\gamma^{\alpha} \cdot \mathbf{n}\boldsymbol{v} \cdot \nabla^{S}\gamma^{s} dv\end{aligned} \tag{706}$$

By Equation (53),

$$\begin{aligned}\nabla \cdot \boldsymbol{\lambda} &= \boldsymbol{\lambda} \cdot \nabla\boldsymbol{\lambda} \cdot \boldsymbol{\lambda} + \mathbf{n} \cdot \nabla\boldsymbol{\lambda} \cdot \mathbf{n} + \boldsymbol{v} \cdot \nabla\boldsymbol{\lambda} \cdot \boldsymbol{v} \\ &= \mathbf{n} \cdot \nabla\boldsymbol{\lambda} \cdot \mathbf{n} + \boldsymbol{v} \cdot \nabla \cdot \boldsymbol{v}.\end{aligned} \tag{707}$$

By Equations (707), (258), and (261), we have

$$\begin{aligned}\int_{C_{\alpha\beta\varepsilon}} \nabla^{c} \circ F \mathrm{d}c = & -\int_{V_\infty} \mathbf{n} \cdot \nabla\boldsymbol{\lambda} \cdot \mathbf{n}\boldsymbol{\lambda} \circ F\gamma\gamma^{c}\boldsymbol{v} \cdot \nabla^{S}\gamma^{s}\mathbf{n} \cdot \nabla\gamma^{\alpha}\mathrm{d}v \\ & -\int_{V_\infty} \boldsymbol{v} \cdot \nabla\boldsymbol{\lambda} \cdot \boldsymbol{v}\boldsymbol{\lambda} \circ F\gamma\gamma^{c}\boldsymbol{v} \cdot \nabla^{S}\gamma^{s}\mathbf{n} \cdot \nabla\gamma^{\alpha}\mathrm{d}v \\ & -\int_{V_\infty} \boldsymbol{\lambda} \cdot \nabla\boldsymbol{\lambda} \circ F\gamma\gamma^{c}\boldsymbol{v} \cdot \nabla^{S}\gamma^{s}\mathbf{n} \cdot \nabla\gamma^{\alpha}\mathrm{d}v \\ & -\int_{V_\infty} \boldsymbol{\lambda} \circ F\boldsymbol{\lambda} \cdot \nabla(\gamma^{c}\gamma)\boldsymbol{v} \cdot \nabla^{S}\gamma^{s}\mathbf{n} \cdot \nabla\gamma^{\alpha}\mathrm{d}v \\ & -\int_{V_\infty} F \circ \boldsymbol{\lambda}\gamma\gamma^{c}\boldsymbol{v} \cdot \nabla\boldsymbol{\lambda} \cdot \nabla\gamma^{s}\mathbf{n} \cdot \nabla\gamma^{\alpha}\mathrm{d}v \\ & -\int_{V_\infty} F \circ \boldsymbol{\lambda}\gamma\gamma^{c}\mathbf{n} \cdot \nabla\boldsymbol{\lambda} \cdot \nabla\gamma^{\alpha}\boldsymbol{v} \cdot \nabla^{S}\gamma^{s}\mathrm{d}v \end{aligned} \tag{708}$$

By Equations (165)–(167) and (275), Equation (708) becomes

$$\int_{C_{\alpha\beta\varepsilon}} \nabla^{c} \circ F \mathrm{d}c = \int_{C_{\alpha\beta\varepsilon}} (\boldsymbol{\lambda} \cdot \nabla^{c}\boldsymbol{\lambda}) \circ F \mathrm{d}c + \sum_{\text{ext}} \mathbf{e} \circ F + \sum_{\text{pts}} \mathbf{e} \circ F$$

Remark 1. Equation (703) contains the following three formulas:

Gradient

$$\int_{C_{\alpha\beta\varepsilon}} \nabla^{c}\mathbf{f}\mathrm{d}c = -\int_{C_{\alpha\beta\varepsilon}} (\boldsymbol{\lambda} \cdot \nabla^{c}\boldsymbol{\lambda})f\mathrm{d}c + \sum_{\text{ext}} \mathbf{e}f + \sum_{\text{pts}} \mathbf{e}f \tag{709}$$

Divergence

$$\int_{C_{\alpha\beta\varepsilon}} \nabla^{c} \cdot \mathbf{f}\mathrm{d}c = -\int_{C_{\alpha\beta\varepsilon}} (\boldsymbol{\lambda} \cdot \nabla^{c}\boldsymbol{\lambda}) \cdot \mathbf{f}\mathrm{d}c + \sum_{\text{ext}} \mathbf{e} \cdot \mathbf{f} + \sum_{\text{pts}} \mathbf{e} \cdot \mathbf{f} \tag{710}$$

Curl

$$\int_{C_{\alpha\beta\varepsilon}} \nabla^{c} \times \mathbf{f}\mathrm{d}c = -\int_{C_{\alpha\beta\varepsilon}} (\boldsymbol{\lambda} \cdot \nabla^{c}\boldsymbol{\lambda}) \times \mathbf{f}\mathrm{d}c + \sum_{\text{ext}} \mathbf{e} \times \mathbf{f} + \sum_{\text{pts}} \mathbf{e} \times \mathbf{f} \tag{711}$$

Remark 2. By comparing Theorems 47 and 49, we obtain

$$\nabla \cdot \int_{C_{\alpha\beta\varepsilon}} \boldsymbol{\lambda}\boldsymbol{\lambda}f\mathrm{d}c = \sum_{\text{ext}} \mathbf{e}f$$

$$\nabla \cdot \int_{C_{\alpha\beta\varepsilon}} \mathbf{f}^{c}\mathrm{d}c = \sum_{\text{ext}} \mathbf{e} \cdot \mathbf{f}^{c}$$

$$\nabla \cdot \int_{C_{\alpha\beta\varepsilon}} \boldsymbol{\lambda}\boldsymbol{\lambda} \times \mathbf{f}^{s}\mathrm{d}c = \sum_{\text{ext}} \mathbf{e} \times \mathbf{f}^{s}$$

5.2 Two macroscopic dimensions

In this section we develop 11 averaging theorems regarding spatial integration of spatial operators, surficial integration of surficial operators and curvilineal integration of curvilineal operators within a REV of two macroscopic dimensions and one megascopic dimension. The appropriate REV is a circular cylinder of diameter D, which is independent of position and time (Figure 28–30). The axis orientation of the cylinder is also independent of time but may vary with position. However, this variation with position is such that the curvature radius of the surface orthogonal to the axis of the cylinder as a function of position must be much smaller than the diameter of the cylinder such that $|\nabla^S \cdot \boldsymbol{\Lambda}|D \ll 1$. The volume of the REV may change with time and position due to changes in its length as determined by the intersection of the cylinder with the megascopic boundaries of the system. The theorems presented in this section will contain macroscopic derivatives in the two directions perpendicular to the axis of the

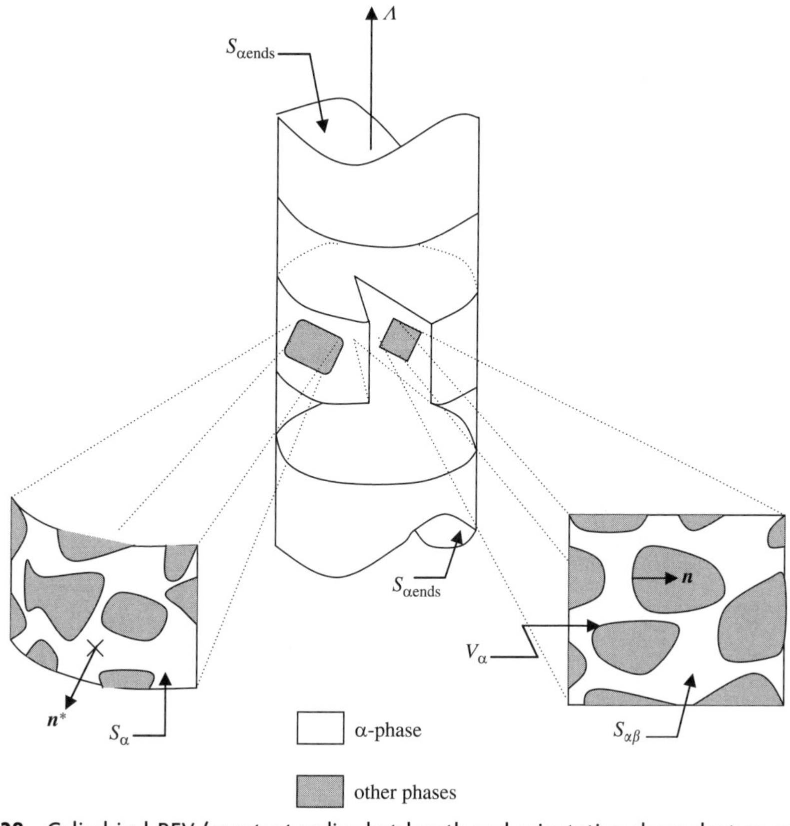

Figure 28 Cylindrical REV (constant radius but length and orientation dependent on space and time; wedge removed for illustrative purposes).

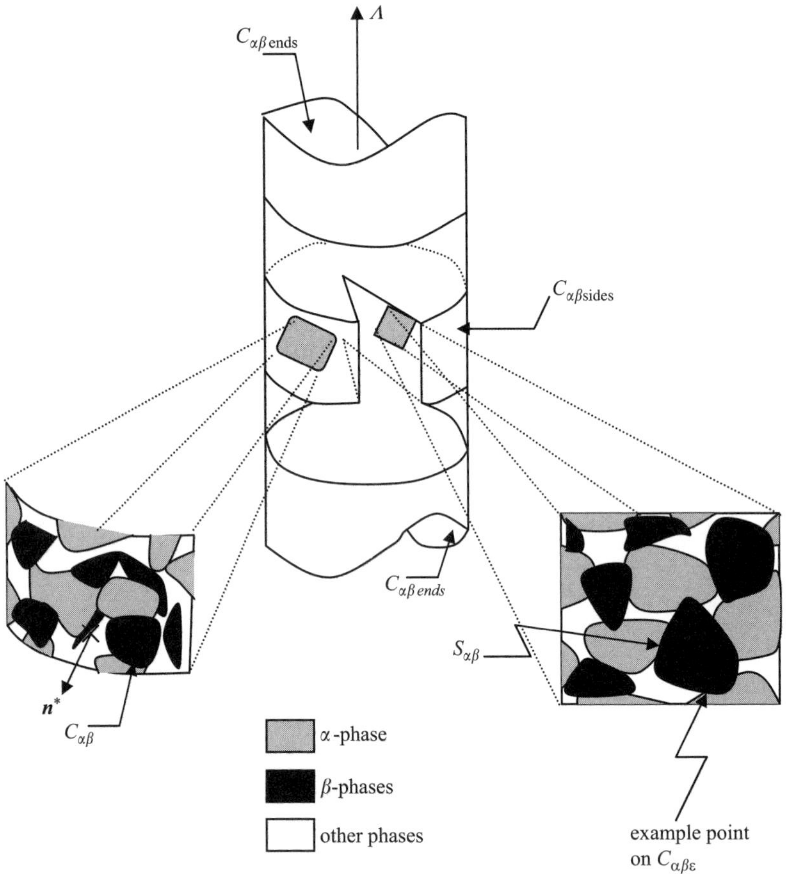

Figure 29 Cylindrical REV (constant radius but length and orientation dependent on space and time; wedge removed for illustrative purposes).

cylindrical REV and flux terms over the cylinder ends where the REV intersects the megascopic system boundaries.

5.2.1 Spatial operator theorems

The REV used for this group of averaging theorems is shown in Figure 28.

Theorem 50.

$$\int_{V_\alpha} \frac{\partial f}{\partial t}\bigg|_{\mathbf{x}} \mathrm{d}v = \frac{\partial}{\partial t}\int_{V_\alpha} f \mathrm{d}v - \int_{S_{\alpha\beta}} \mathbf{n}\cdot\mathbf{w} f \mathrm{d}s - \int_{S_{\alpha\text{ends}}} \mathbf{n}^*\cdot\mathbf{w} f \mathrm{d}s \tag{712}$$

where V_α is the portion of the REV occupied by the α-phase, $S_{\alpha\beta}$ the surface within the REV between the α-phase and all other phases, $S_{\alpha\text{ends}}$ the portion of the end boundaries of the REV that is occupied by the α-phases, $\mathbf{n}^*$ the unit vector

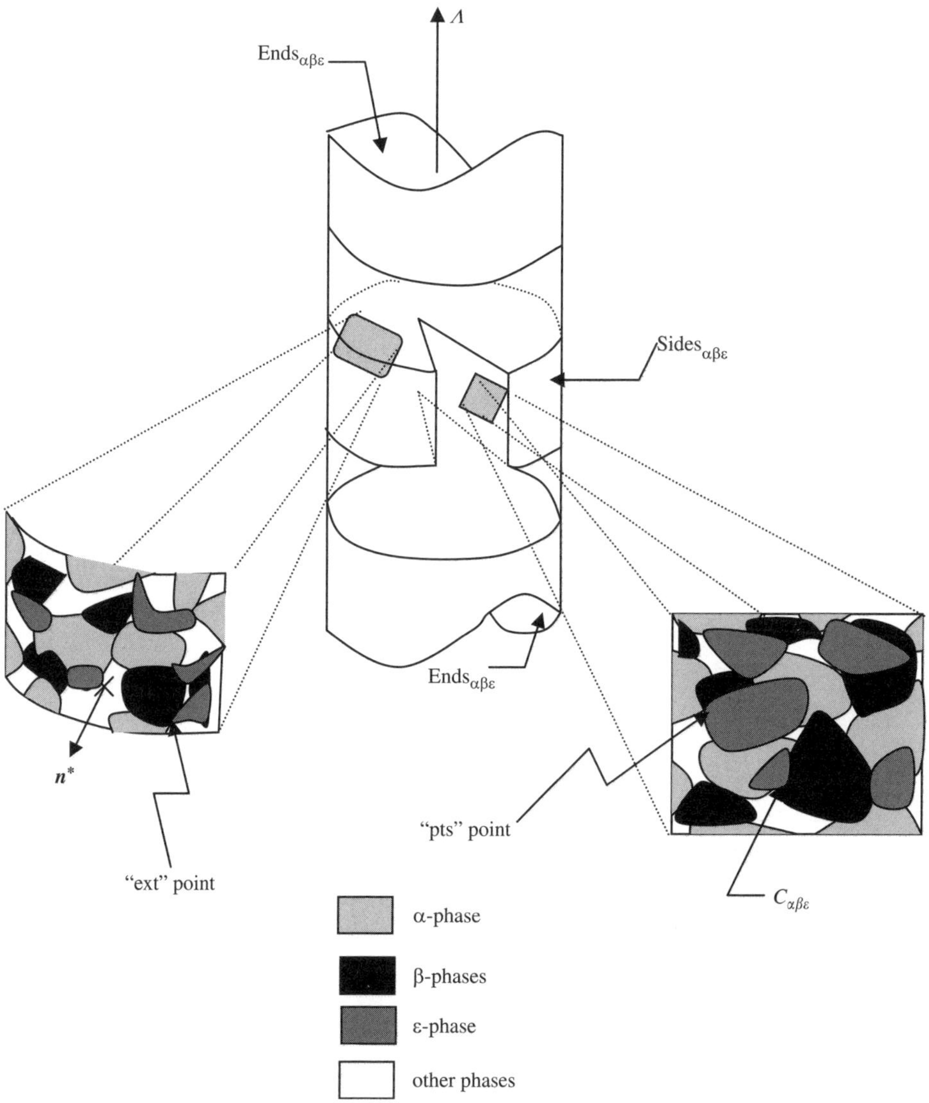

Figure 30 Cylindrical REV (constant radius but length and orientation dependent on space and time; wedge removed for illustrative purposes).

normal to the external boundary of the REV that is pointing outward from the REV, and $\mathbf{n}$ the unit vector normal to the $S_{\alpha\beta}$ surface that is pointing outward from the α-phase. $\boldsymbol{\Lambda}$ is the unit vector tangent to the axis of the REV, $\mathbf{w}$ the velocity of either $S_{\alpha\beta}$ or $S_{\alpha\text{ends}}$, and f a spatial function $f(\mathbf{x}, t)$ defined in the α-phase.

Proof. By the indicator function $\gamma^{\alpha}(\mathbf{x}+\xi, t)$, we have

$$\int_{V_\alpha} \frac{\partial f}{\partial t}\bigg|_{\mathbf{x}} \mathrm{d}v = \int_{V} \frac{\partial f}{\partial t}\bigg|_{\mathbf{x}} \gamma^{\alpha} \mathrm{d}v \tag{713}$$

The integral on the right side of this equation can be changed to an integral over all space by using the indicator function $\gamma(\mathbf{x}+\xi, t)$ to obtain

$$\int_{V_\alpha} \frac{\partial f}{\partial t}\bigg|_{\mathbf{x}} \mathrm{d}v = \int_{V_\infty} \frac{\partial f}{\partial t}\bigg|_{\mathbf{x}} \gamma^{\alpha}\gamma \mathrm{d}v. \tag{714}$$

An application of the chain rule to the right side of Equation (714) yields

$$\int_{V_\alpha} \frac{\partial f}{\partial t}\bigg|_{\mathbf{x}} \mathrm{d}v = \int_{V_\infty} \frac{\partial (f\gamma\gamma^{\alpha})}{\partial t}\bigg|_{\mathbf{x}} \mathrm{d}v - \int_{V_\infty} f\frac{\partial \gamma^{\alpha}}{\partial t}\bigg|_{\mathbf{x}} \gamma \mathrm{d}v - \int_{V_\infty} f\gamma^{\alpha}\frac{\partial \gamma}{\partial t}\bigg|_{\mathbf{x}} \mathrm{d}v \tag{715}$$

Because the boundaries of V_∞ are independent of time, differentiation and integration can be interchanged in the first term on the right side of Equation (715). Then the definitions of the indicator functions may be used to convert the derivative of an integral over V_∞ to the derivative of an integral over V_α.

$$\int_{V_\infty} \frac{\partial (f\gamma^{\alpha}\gamma)}{\partial t}\bigg|_{\mathbf{x}} \mathrm{d}v = \frac{\partial}{\partial t}\int_{V_\infty} f\gamma^{\alpha}\gamma \mathrm{d}v = \frac{\partial}{\partial t}\int_{V_\alpha} f \mathrm{d}v \tag{716}$$

Substitution of Equation (716) into Equation (715) leads to

$$\int_{V_\alpha} \frac{\partial f}{\partial t}\bigg|_{\mathbf{x}} \mathrm{d}v = \frac{\partial}{\partial t}\int_{V_\alpha} f \mathrm{d}v - \int_{V_\infty} f\frac{\partial \gamma^{\alpha}}{\partial t}\bigg|_{\mathbf{x}} \gamma \mathrm{d}v - \int_{V_\infty} f\gamma^{\alpha}\frac{\partial \gamma}{\partial t}\bigg|_{\mathbf{x}} \mathrm{d}v \tag{717}$$

By Equation (180), Equation (717) becomes

$$\int_{V_\alpha} \frac{\partial f}{\partial t}\bigg|_{\mathbf{x}} \mathrm{d}v = \frac{\partial}{\partial t}\int_{V_\alpha} f \mathrm{d}v + \int_{V_\infty} f\mathbf{w}\cdot\nabla\gamma^{\alpha}\gamma \mathrm{d}v + \int_{V_\infty} f\gamma^{\alpha}\mathbf{w}\cdot\nabla\gamma \mathrm{d}v \tag{718}$$

By the definitions of γ and γ^{α} together with Equation (186), we have

$$\int_{V_\alpha} \frac{\partial f}{\partial t}\bigg|_{\mathbf{x}} \mathrm{d}v = \frac{\partial}{\partial t}\int_{V_\alpha} f \mathrm{d}v - \int_{S_{\alpha\beta}} f\mathbf{n}\cdot\mathbf{w}\mathrm{d}s - \int_{S_{\alpha\mathrm{ends}}} \mathbf{n}^{*}\cdot\mathbf{w}f\mathrm{d}s$$

Theorem 51.

$$\int_{V_\alpha} \nabla\circ F\mathrm{d}v = \nabla^{S}\circ\int_{V_\alpha} F\mathrm{d}v + \int_{S_{\alpha\beta}} \mathbf{n}\circ F\mathrm{d}s + \int_{S_{\alpha\mathrm{ends}}} \mathbf{n}^{*}\circ F\mathrm{d}s \tag{719}$$

where $\nabla\circ F$ can be any of the three operators: ∇f, $\nabla\cdot\mathbf{f}$, and $\nabla\times\mathbf{f}$. $f(\mathbf{x}, t)$ and $\mathbf{f}(\mathbf{x}, t)$ are spatial scalar and vector functions of position vector $\mathbf{x}$ and time t defined in the α-phase. ∇ is the microscopic spatial operator, $\nabla = \nabla_\xi$ and ∇^{S} the two-dimensional macroscopic operator in the direction orthogonal to the axis of the REV such that $\nabla^{S} = \nabla_{\mathbf{x}} - \boldsymbol{\Lambda\Lambda}\cdot\nabla_{\mathbf{x}}$.

Proof. The indicator functions $\gamma(\mathbf{x}, \boldsymbol{\xi})$ and $\gamma^\alpha(\mathbf{x}+\boldsymbol{\xi}, t)$ can be used to convert the integral over V_α to an integral over V_∞.

$$\int_{V_\alpha} \nabla \circ F dv = \int_{V_\infty} (\nabla \circ F)\gamma\gamma^\alpha dv \tag{720}$$

An application of the chain rule to the right side of Equation (720) yields

$$\int_{V_\infty} \nabla \circ F dv = \int_{V_\infty} \nabla \circ (F\gamma\gamma^\alpha) dv - \int_{V_\infty} \nabla \gamma^\alpha \circ F\gamma dv - \int_{V_\infty} \nabla \gamma \circ F\gamma^\alpha dv \tag{721}$$

By the definitions of γ and γ^α and Equations (185)–(187), the last two terms in this expression are converted to surface integrals over the α–β interface and the edge of the region of interest. Therefore, Equation (721) becomes

$$\int_{V_\alpha} \nabla \circ F dv = \int_{V_\infty} \nabla \circ (F\gamma\gamma^\alpha) dv + \int_{S_{\alpha\beta}} \mathbf{n} \circ F dv + \int_{S_{\alpha ends}} \mathbf{n}^* \circ F ds \tag{722}$$

Further manipulations will be only for the first term on the right side of Equation (722).

The gradient operator may be decomposed into a macroscopic part and a microscopic part such that

$$\nabla = \nabla^S + \nabla^c \tag{723}$$

By substituting Equation (723) into Equation (722), we obtain

$$\int_{V_\infty} \nabla \circ (F\gamma\gamma^\alpha) dv = \int_{V_\infty} \nabla^S \circ (F\gamma\gamma^\alpha) dv + \int_{V_\infty} \nabla^c \circ (F\gamma\gamma^\alpha) dv$$
$$+ \int_{S_{\alpha\beta}} \mathbf{n} \circ F dv + \int_{S_{\alpha ends}} \mathbf{n}^* \circ F ds \tag{724}$$

In the first term on the right, the operator ∇^S may be moved outside of the integral and the indicator functions may be eliminated to return the domain of integration to V_α. The second integral in Equation (724) is zero because the integration results in evaluation of the integrand at the boundary of V_∞ where γ is zero. Thus Equation (724) becomes

$$\int_{V_\infty} \nabla \circ (F\gamma\gamma^\alpha) dv = \nabla^S \circ \int_{V_\alpha} F dv + \int_{S_{\alpha\beta}} \mathbf{n} \circ F dv + \int_{S_{\alpha ends}} \mathbf{n}^* \circ F ds$$

Remark. Equation (719) contains the following three formulas:

Gradient

$$\int_{V_\alpha} \nabla f dv = \nabla^S \int_{V_\alpha} f dv + \int_{S_{\alpha\beta}} \mathbf{n} f ds + \int_{\alpha ends} \mathbf{n}^* f ds \tag{725}$$

Divergence

$$\int_{V_\alpha} \nabla \cdot \mathbf{f} dv = \nabla^S \cdot \int_{V_\alpha} \mathbf{f} dv + \int_{S_{\alpha\beta}} \mathbf{n} \cdot \mathbf{f} ds + \int_{\alpha ends} \mathbf{n}^* \cdot \mathbf{f} ds \tag{726}$$

Curl

$$\int_{V_\alpha} \nabla \times \mathbf{f} dv = \nabla^S \times \int_{V_\alpha} \mathbf{f} dv + \int_{S_{\alpha\beta}} \mathbf{n} \times \mathbf{f} ds + \int_{\alpha \mathrm{ends}} \mathbf{n}^* \times \mathbf{f} ds \tag{727}$$

Theorem 51 has the following alternative forms:

Theorem 52.

$$\int_{V_\alpha} \nabla \circ F dv = \int_{S_\alpha} \mathbf{n}^* \circ F ds + \int_{S_{\alpha\beta}} \mathbf{n} \circ F ds \tag{728}$$

where S_α is the portion of the external boundary of the REV that intersects the α-phase, including the end boundaries $S_{\alpha \mathrm{ends}}$.

Proof. The indicator functions $\gamma(\mathbf{x}, \xi)$ and $\gamma^\alpha(\mathbf{x} + \xi, t)$ may be employed to rewrite the integral under study as follows:

$$\int_{V_\alpha} \nabla \circ F dv = \int_{V_\infty} \nabla \circ F \gamma \gamma^\alpha dv \tag{729}$$

An application of the chain rule to the right side of Equation (729) yields

$$\int_{V_\infty} \nabla \circ F dv = \int_{V_\infty} \nabla \circ (F \gamma \gamma^\alpha) dv - \int_{V_\infty} \nabla \gamma \circ F \gamma^\alpha dv - \int_{V_\infty} \nabla \gamma^\alpha \circ F \gamma dv \tag{730}$$

By Equations (230)–(232), the first integral on the right side of Equation (730) is zero. By an application of Equations (185), (186), and (187) to the last two integrals on the right side of Equation (730), we obtain

$$\int_{V_\alpha} \nabla \circ F dv = \int_{S_\alpha} \mathbf{n}^* \circ F ds + \int_{S_{\alpha\beta}} \mathbf{n} \circ F ds$$

Remark. Equation (728) contains the following three formulas:

Gradient

$$\int_{V_\alpha} \nabla f dv = \int_{S_\alpha} \mathbf{n}^* f ds + \int_{S_{\alpha\beta}} \mathbf{n} f ds \tag{731}$$

Divergence

$$\int_{V_\alpha} \nabla \cdot \mathbf{f} dv = \int_{S_\alpha} \mathbf{n}^* \cdot \mathbf{f} ds + \int_{S_{\alpha\beta}} \mathbf{n} \cdot \mathbf{f} ds \tag{732}$$

Curl

$$\int_{V_\alpha} \nabla \times \mathbf{f} dv = \int_{S_\alpha} \mathbf{n}^* \times \mathbf{f} ds + \int_{S_{\alpha\beta}} \mathbf{n} \times \mathbf{f} ds \tag{733}$$

5.2.2 Surficial operator theorems

The REV used for this group of averaging theorems is shown in Figure 29.

Theorem 53.

$$\int_{S_{\alpha\beta}} \frac{\partial f}{\partial t}\bigg|_{\mathbf{u}} ds = \frac{\partial}{\partial t}\int_{S_{\alpha\beta}} f ds + \nabla^S \cdot \int_{S_{\alpha\beta}} \mathbf{w}^c f ds - \int_{S_{\alpha\beta}} (\nabla^S \cdot \mathbf{n})\mathbf{n}\cdot \mathbf{w}^c f ds$$
$$- \int_{C_{\alpha\beta\varepsilon}} \boldsymbol{\nu}\cdot \mathbf{w}^s f dc - \int_{C_{\alpha\beta \mathrm{ends}}} \boldsymbol{\nu}\cdot \mathbf{w}^s f dc \tag{734}$$

where $S_{\alpha\beta}$ is the surface between the α- and β-phases, $C_{\alpha\beta\varepsilon}$ the boundary curve of $S_{\alpha\beta}$ within the REV that also is the location where the α- and β-phases meet a third phase, $C_{\alpha\beta\mathrm{ends}}$ the curve at the ends of the REV formed by the intersection of $S_{\alpha\beta}$ with the ends of the cylinder, $\mathbf{n}$ the unit vector normal to the $S_{\alpha\beta}$ surface, and $\boldsymbol{\nu}$ the unit vector tangent to the $S_{\alpha\beta}$ surface and normal to $C_{\alpha\beta\varepsilon}$ and $C_{\alpha\beta\mathrm{ends}}$ such that $\mathbf{n}\cdot\boldsymbol{\nu} = 0$. $\boldsymbol{\Lambda}$ is the unit vector tangent to the axis of the REV. ∇ is the microscopic spatial operator such that $\nabla = \nabla_\xi$ and ∇^S the microscopic surficial operator, $\nabla^S = \nabla_\xi - \mathbf{nn}\cdot\nabla_\xi$. ∇^S is the two-dimensional macroscopic spatial operator in the directions orthogonal to the axis of the REV, $\nabla^S = \nabla_\mathbf{x} - \boldsymbol{\Lambda}\boldsymbol{\Lambda}\cdot\nabla_\mathbf{x}$. $\mathbf{w}$ is the velocity of $S_{\alpha\beta}$, $\mathbf{w}^c$ the velocity of $S_{\alpha\beta}$ normal to $S_{\alpha\beta}$ such that $\mathbf{w}^c = \mathbf{nn}\cdot\mathbf{w}$, $\mathbf{w}^s$ the velocity of $S_{\alpha\beta}$ tangent to $S_{\alpha\beta}$ such that $\mathbf{w}^s = \mathbf{w} - \mathbf{w}^c$, and f a surficial function $f(\mathbf{u}, t)$ defined on $S_{\alpha\beta}$.

Proof. By the definitions of the indicator functions $\gamma(\mathbf{x}, \boldsymbol{\xi})$, $\gamma^s(\mathbf{u}(\mathbf{x}+\boldsymbol{\xi}), t)$, and $\gamma^\alpha(\mathbf{x}+\boldsymbol{\xi}, t)$, we obtain

$$\int_{S_{\alpha\beta}} \frac{\partial f}{\partial t}\bigg|_{\mathbf{u}} ds = -\int_{V_\infty} \frac{\partial f}{\partial t}\bigg|_{\mathbf{u}} \gamma\gamma^s \mathbf{n}\cdot\nabla\gamma^\alpha dv \tag{735}$$

which is, by Equation (130),

$$\int_{S_{\alpha\beta}} \frac{\partial f}{\partial t}\bigg|_{\mathbf{u}} ds = -\int_{V_\infty} \frac{\partial f}{\partial t}\bigg|_{\mathbf{x}} \gamma\gamma^s \mathbf{n}\cdot\nabla\gamma^\alpha dv - \int_{V_\infty} \mathbf{w}\cdot\mathbf{nn}\cdot\nabla f\gamma\gamma^s\mathbf{n}\cdot\nabla\gamma^\alpha dv \tag{736}$$

An application of the chain rule to the right side of Equation (736) yields

$$\int_{S_{\alpha\beta}} \frac{\partial f}{\partial t}\bigg|_{\mathbf{u}} ds = -\int_{V_\infty} \frac{\partial (f\gamma\gamma^s\mathbf{n}\cdot\nabla\gamma^\alpha)}{\partial t} dv + \int_{V_\infty} f\frac{\partial(\gamma\gamma^s)}{\partial t}\mathbf{n}\cdot\nabla\gamma^\alpha dv$$
$$+ \int_{V_\infty} f\gamma\gamma^s \frac{\partial \mathbf{n}}{\partial t}\cdot\nabla\gamma^\alpha dv + \int_{V_\infty} f\gamma\gamma^s\mathbf{n}\cdot\nabla\frac{\partial\gamma^\alpha}{\partial t} dv$$
$$- \int_{V_\infty} \nabla\cdot(\mathbf{nn}\cdot\mathbf{w}f\gamma\gamma^s\mathbf{n}\cdot\nabla\gamma^\alpha) dv$$
$$+ \int_{V_\infty} \nabla\cdot\mathbf{nn}\cdot\mathbf{w}f\gamma\gamma^s\mathbf{n}\cdot\nabla\gamma^\alpha dv$$
$$+ \int_{V_\infty} \mathbf{n}\cdot\nabla\mathbf{n}\cdot\mathbf{n}f\gamma\gamma^s\mathbf{n}\cdot\nabla\gamma^\alpha dv$$
$$+ \int_{V_\infty} \mathbf{n}\cdot\nabla\mathbf{w}\cdot\mathbf{n}f\gamma\gamma^s\mathbf{n}\cdot\nabla\gamma^\alpha dv$$
$$+ \int_{V_\infty} \mathbf{n}\cdot\mathbf{w}f\mathbf{n}\cdot\nabla(\gamma\gamma^s)\mathbf{n}\cdot\nabla\gamma^\alpha dv$$
$$+ \int_{V_\infty} \mathbf{n}\cdot\mathbf{w}f\gamma\gamma^s\mathbf{n}\cdot(\nabla\mathbf{n}\cdot\nabla\gamma^\alpha + \nabla\nabla\gamma^\alpha\cdot\mathbf{n}) dv \tag{737}$$

Differentiation with respect to time t may be interchanged with the integration in the first term on the right side of Equation (737) because its integration domain is independent of time t. By Equation (224), the second term becomes

$$\int_{V_\infty} f\frac{\partial(\gamma\gamma^s)}{\partial t}\mathbf{n}\cdot\nabla\gamma^\alpha dv = -\int_{V_\infty} f\mathbf{w}\cdot\nabla(\gamma\gamma^s)\mathbf{n}\cdot\nabla\gamma^\alpha dv \tag{738}$$

By Equation (194), the third term on the right side of Equation (737) is zero. By Equation (180), the fourth term on the right side of Equation (737) becomes

$$\int_{V_\infty} f\gamma\gamma^s\mathbf{n}\cdot\nabla\frac{\partial\gamma^\alpha}{\partial t}dv = -\int_{V_\infty} f\gamma\gamma^s\mathbf{n}\cdot\nabla\mathbf{w}\cdot\nabla\gamma^\alpha dv - \int_{V_\infty} f\gamma\gamma^s\mathbf{n}\cdot\nabla\nabla\gamma^\alpha\cdot\mathbf{w}dv \tag{739}$$

Similar to the proof of Theorem 51, the fifth integral on the right side of Equation (737) may be rewritten as

$$\int_{V_\infty} \nabla\cdot(\mathbf{nn}\cdot\mathbf{w}f\gamma\gamma^s\mathbf{n}\cdot\nabla\gamma^\alpha) = \nabla^S\cdot\int_{V_\infty}\mathbf{w}^c f\gamma\gamma^s\mathbf{n}\cdot\nabla\gamma^\alpha dv \tag{740}$$

By Equation (138) and $\mathbf{n}\cdot\mathbf{w} = \mathbf{n}\cdot\mathbf{w}^c$, the sixth term on the right side of Equation (737) is

$$\int_{V_\infty} \nabla\cdot\mathbf{nn}\cdot\mathbf{w}f\gamma\gamma^s\mathbf{n}\cdot\nabla\gamma^\alpha dv = \int_{V_\infty}\nabla^S\cdot\mathbf{nn}\cdot\mathbf{w}^c f\gamma\gamma^s\mathbf{n}\cdot\nabla\gamma^\alpha dv \tag{741}$$

By Equation (195), the last integral on the right side of Equation (737) may be written as

$$\int_{V_\infty}\mathbf{n}\cdot\mathbf{w}f\gamma\gamma^s\mathbf{n}\cdot(\nabla\nabla\gamma^\alpha\cdot\mathbf{n}+\nabla\mathbf{n}\cdot\nabla\gamma^\alpha)dv = \int_{V_\infty}\gamma\gamma^s f\mathbf{w}\cdot\mathbf{nn}\cdot\nabla\nabla\gamma^\alpha\cdot\mathbf{n}dv \tag{742}$$

By applying these to the right side terms of Equation (737) and regrouping,

$$\begin{aligned}\int_{S_{\alpha\beta}}\frac{\partial f}{\partial t}\bigg|_{\mathbf{u}}ds = &-\frac{\partial}{\partial t}\int_{V_\infty} f\gamma\gamma^s\mathbf{n}\cdot\nabla\gamma^\alpha dv - \int_{V_\infty} f\mathbf{w}\cdot\nabla^S(\gamma\gamma^s)\mathbf{n}\cdot\nabla\gamma^\alpha dv\\ &-\int_{V_\infty} f\gamma\gamma^s\mathbf{n}\cdot\nabla\mathbf{w}\cdot\nabla\gamma^\alpha dv + \int_{V_\infty}\mathbf{n}\cdot\nabla\mathbf{w}\cdot\mathbf{n}f\gamma\gamma^s\mathbf{n}\cdot\nabla\gamma^\alpha dv\\ &-\int_{V_\infty} f\gamma\gamma^s\mathbf{w}\cdot\nabla^S\nabla\gamma^\alpha\cdot\mathbf{n}dv - \nabla^S\cdot\int_{V_\infty}\mathbf{nn}\cdot\mathbf{w}f\gamma\gamma^s\mathbf{n}\cdot\nabla\gamma^\alpha dv\\ &+\int_{V_\infty}\nabla^S\cdot\mathbf{nn}\cdot\mathbf{w}^c f\gamma\gamma^s\mathbf{n}\cdot\nabla\gamma^\alpha dv + \int_{V_\infty}\mathbf{n}\cdot\nabla\mathbf{n}\cdot\mathbf{w}f\gamma\gamma^s\mathbf{n}\cdot\nabla\gamma^\alpha dv\end{aligned} \tag{743}$$

Applying Equation (256) to the fifth integral on the right side of Equation (743) yields

$$\int_{V_\infty} f\gamma\gamma^s\mathbf{w}\cdot\nabla^S\nabla\gamma^\alpha\cdot\mathbf{n}dv = \int_{V_\infty}\mathbf{n}\cdot\nabla\mathbf{n}\cdot\mathbf{w}f\gamma\gamma^s\mathbf{n}\cdot\nabla\gamma^\alpha dv \tag{744}$$

By the definition of the indicator function γ^α, the third term cancels the fourth term. With these manipulations together with application of Equations (275) and (272) to the terms on the right side of Equation (743), we finally obtain

$$\int_{S_{\alpha\beta}} \frac{\partial f}{\partial t}\bigg|_{\mathbf{u}} ds = \frac{\partial}{\partial t}\int_{S_{\alpha\beta}} f ds + \nabla^S \cdot \int_{S_{\alpha\beta}} \mathbf{w}^c f ds - \int_{S_{\alpha\beta}} (\nabla^S \cdot \mathbf{n})\mathbf{n} \cdot \mathbf{w}^c f ds$$
$$- \int_{C_{\alpha\beta\varepsilon}} \boldsymbol{\nu} \cdot \mathbf{w}^s f ds - \int_{C_{\alpha\beta\text{ends}}} \boldsymbol{\nu} \cdot \mathbf{w}^s f dc$$

Theorem 54.

$$\int_{S_{\alpha\beta}} \nabla^S \circ F ds = \nabla^S \circ \int_{S_{\alpha\beta}} F ds - \nabla^S \cdot \int_{S_{\alpha\beta}} \mathbf{nn} \circ F ds + \int_{S_{\alpha\beta}} \nabla^S \cdot \mathbf{nn} \circ F ds$$
$$+ \int_{C_{\alpha\beta\varepsilon}} \boldsymbol{\nu} \circ F dc + \int_{C_{\alpha\beta\text{ends}}} \boldsymbol{\nu} \circ F dc \tag{745}$$

where $\nabla^S \circ F$ can be any of the three operators: $\nabla^S f$, $\nabla^S \cdot \mathbf{f}$, and $\nabla^S \times \mathbf{f}$. $f(\mathbf{u}, t)$ and $\mathbf{f}(\mathbf{u}, t)$ are scalar and vector functions of surficial position vector $\mathbf{u}$ and time t defined on $S_{\alpha\beta}$, respectively. $\mathbf{n}^*$ is the unit vector normal to the external boundary of the REV that is pointing outward from the REV.

Proof. By the definitions of the indicator functions $\gamma(\mathbf{x}, \xi)$, $\gamma^s(\mathbf{u}(\mathbf{x} + \xi), t)$ and $\gamma^\alpha(\mathbf{x} + \xi, t)$, we obtain

$$\int_{S_{\alpha\beta}} \nabla^S \circ F ds = - \int_{V_\infty} \nabla^S \circ F \gamma\gamma^s \mathbf{n} \cdot \nabla\gamma^\alpha dv \tag{746}$$

Note that $\nabla^S = \nabla - \mathbf{nn} \cdot \nabla$. Thus,

$$\int_{S_{\alpha\beta}} \nabla^S \circ F ds = - \int_{V_\infty} \nabla \circ F\gamma\gamma^s \mathbf{n} \cdot \nabla\gamma^\alpha dv + \int_{V_\infty} (\mathbf{nn} \cdot \nabla) \circ F\gamma\gamma^s \mathbf{n} \cdot \nabla\gamma^\alpha dv \tag{747}$$

An application of the chain rule to the right side of Equation (747) yields

$$\int_{S_{\alpha\beta}} \nabla^S \circ F ds$$
$$= - \int_{V_\infty} \nabla \circ (F\gamma\gamma^s \mathbf{n} \cdot \nabla\gamma^\alpha) dv + \int_{V_\infty} \nabla(\gamma\gamma^s) \circ F\mathbf{n} \cdot \nabla\gamma^\alpha dv$$
$$+ \int_{V_\infty} \gamma\gamma^s (\nabla\mathbf{n} \cdot \nabla\gamma^\alpha + \nabla\nabla\gamma^\alpha \cdot \mathbf{n}) dv + \int_{V_\infty} \nabla \cdot (\mathbf{nn} \circ F\gamma\gamma^s \mathbf{n} \cdot \nabla\gamma^\alpha) dv$$
$$- \int_{V_\infty} \nabla \cdot \mathbf{nn} \circ F\gamma\gamma^s \mathbf{n} \cdot \nabla\gamma^\alpha dv - \int_{V_\infty} \mathbf{n} \cdot \nabla\mathbf{n} \circ F\gamma\gamma^s \mathbf{n} \cdot \nabla\gamma^\alpha dv$$
$$- \int_{V_\infty} \mathbf{n} \circ F\mathbf{n} \cdot \nabla(\gamma\gamma^s)\mathbf{n} \cdot \nabla\gamma^\alpha dv - \int_{V_\infty} \mathbf{n} \circ F\mathbf{n} \cdot (\nabla\mathbf{n} \cdot \nabla\gamma^\alpha + \nabla\nabla\gamma^\alpha \cdot \mathbf{n}) dv \tag{748}$$

Applying Equation (195) to the third and last integrals on the right side of Equation (748) and regrouping the other terms, Equation (748) becomes

$$\int_{S_{\alpha\beta}} \nabla^S \circ F ds = -\int_{V_\infty} \nabla \circ (F\gamma\gamma^s \mathbf{n} \cdot \nabla\gamma^\alpha) dv + \int_{V_\infty} \nabla^S(\gamma\gamma^s) \circ F\mathbf{n} \cdot \nabla\gamma^\alpha dv$$
$$+ \int_{V_\infty} \gamma\gamma^s(\nabla^S\nabla\gamma^\alpha \cdot \mathbf{n}) \circ F dv + \int_{V_\infty} \nabla \cdot (\mathbf{nn} \circ F\gamma\gamma^s \mathbf{n} \cdot \nabla\gamma^\alpha) dv$$
$$- \int_{V_\infty} \nabla \cdot \mathbf{nn} \circ F\gamma\gamma^s \mathbf{n} \cdot \nabla\gamma^\alpha dv - \int_{V_\infty} \mathbf{n} \cdot \nabla\mathbf{n} \circ F\gamma\gamma^s \mathbf{n} \cdot \nabla\gamma^\alpha dv \quad (749)$$

By $\nabla = \nabla^S + \nabla^c$, we can rewrite the first and fourth terms on the right side of Equation (749) as

$$\int_{V_\infty} \nabla \circ (F\gamma\gamma^s \mathbf{n} \cdot \nabla\gamma^\alpha) dv = \int_{V_\infty} \nabla^S \circ (F\gamma\gamma^s \mathbf{n} \cdot \nabla\gamma^\alpha) dv$$
$$+ \int_{V_\infty} \nabla^c \circ (F\gamma\gamma^s \mathbf{n} \cdot \nabla\gamma^\alpha) dv \quad (750)$$

$$\int_{V_\infty} \nabla \cdot (\mathbf{nn} \circ F\gamma\gamma^s \mathbf{n} \cdot \nabla\gamma^\alpha) dv = \int_{V_\infty} \nabla^S \cdot (\mathbf{nn} \circ F\gamma\gamma^s \mathbf{n} \cdot \nabla\gamma^\alpha) dv$$
$$+ \int_{V_\infty} \nabla^c \cdot (\mathbf{nn} \circ F\gamma\gamma^s \mathbf{n} \cdot \nabla\gamma^\alpha) dv \quad (751)$$

The order of the gradient operators ∇^S and the integration may be interchanged because the operator ∇^S is with respect to the macroscopic variable while the integration is with respect to the microscopic variable for the first term on the right side of Equations (750) and (751). By the definitions of the indicator functions γ and γ^s, the second term on the right side of Equations (750) and (751) is zero. Therefore, Equations (750) and (751) become

$$\int_{V_\infty} \nabla \circ (F\gamma\gamma^s \mathbf{n} \cdot \nabla\gamma^\alpha) dv = \nabla^S \circ \int_{V_\infty} F\gamma\gamma^s \mathbf{n} \cdot \nabla\gamma^\alpha dv \quad (752)$$

$$\int_{V_\infty} \nabla \cdot (\mathbf{nn} \circ F\gamma\gamma^s \mathbf{n} \cdot \nabla\gamma^\alpha) dv = \nabla^S \cdot \int_{V_\infty} \mathbf{nn} \circ F\gamma\gamma^s \mathbf{n} \cdot \nabla\gamma^\alpha dv \quad (753)$$

By the definitions of the indicator functions on the right side of Equations (752) and (753), we have

$$\int_{V_\infty} \nabla \circ (F\gamma\gamma^s \mathbf{n} \cdot \nabla\gamma^\alpha) dv = -\nabla^S \circ \int_{S_{\alpha\beta}} F ds \quad (754)$$

$$\int_{V_\infty} \nabla \cdot (\mathbf{nn} \circ F\gamma\gamma^s \mathbf{n} \cdot \nabla\gamma^\alpha) dv = -\nabla^S \cdot \int_{S_{\alpha\beta}} \mathbf{nn} \circ F ds \quad (755)$$

By substituting Equations (754) and (755) into Equation (749), we obtain

$$\int_{S_{\alpha\beta}} \nabla^S \circ F ds = \nabla^S \circ \int_{S_{\alpha\beta}} F ds - \nabla^S \cdot \int_{S_{\alpha\beta}} \mathbf{nn} \circ F ds$$
$$+ \int_{V_\infty} \gamma\gamma^s(\nabla^S\nabla\gamma^\alpha \cdot \mathbf{n}) \circ F dv + \int_{V_\infty} \nabla^S(\gamma\gamma^s) \circ F\mathbf{n} \cdot \nabla\gamma^\alpha dv$$
$$- \int_{V_\infty} \nabla \cdot \mathbf{nn} \circ F\gamma\gamma^s \mathbf{n} \cdot \nabla\gamma^\alpha dv - \int_{V_\infty} \mathbf{n} \cdot \nabla\mathbf{n} \circ F\gamma\gamma^s \mathbf{n} \cdot \nabla\gamma^\alpha dv \quad (756)$$

By Equation (256), the third term on the right side of Equation (756) cancels the last term. Applying Equation (275) to the fourth term on the right side of Equation (756) and Equations (138) and (272) to the fifth term, Equation (756) becomes

$$\int_{S_{\alpha\beta}} \nabla^S \circ F ds = \nabla^S \circ \int_{S_{\alpha\beta}} F ds - \nabla^S \cdot \int_{S_{\alpha\beta}} \mathbf{nn} \circ F ds$$
$$+ \int_{S_{\alpha\beta}} \nabla^S \cdot \mathbf{nn} \circ F ds + \int_{C_{\alpha\beta\varepsilon}} \boldsymbol{\nu} \circ F dc + \int_{C_{\alpha\beta \text{ends}}} \boldsymbol{\nu} \circ F dc$$

Remark. Equation (745) contains the following three formulas:

Gradient

$$\int_{S_{\alpha\beta}} \nabla^S f ds = \nabla^S \int_{S_{\alpha\beta}} f ds - \nabla^S \cdot \int_{S_{\alpha\beta}} \mathbf{nn} f ds + \int_{S_{\alpha\beta}} \nabla^S \cdot \mathbf{nn} f ds$$
$$+ \int_{C_{\alpha\beta\varepsilon}} \boldsymbol{\nu} f dc + \int_{C_{\alpha\beta \text{ends}}} \boldsymbol{\nu} f dc \tag{757}$$

Divergence

$$\int_{S_{\alpha\beta}} \nabla^S \cdot \mathbf{f} ds = \nabla^S \cdot \int_{S_{\alpha\beta}} \mathbf{f} ds - \nabla^S \cdot \int_{S_{\alpha\beta}} \mathbf{nnf} ds + \int_{S_{\alpha\beta}} \nabla^S \cdot \mathbf{nn} \cdot \mathbf{f} ds$$
$$+ \int_{C_{\alpha\beta\varepsilon}} \boldsymbol{\nu} \cdot \mathbf{f} dc + \int_{C_{\alpha\beta \text{ends}}} \boldsymbol{\nu} \cdot \mathbf{f} dc \tag{758}$$

Curl

$$\int_{S_{\alpha\beta}} \nabla^S \times \mathbf{f} ds = \nabla^S \times \int_{S_{\alpha\beta}} \mathbf{f} ds - \nabla^S \cdot \int_{S_{\alpha\beta}} \mathbf{nn} \times \mathbf{f} ds + \int_{S_{\alpha\beta}} (\nabla^S \cdot \mathbf{n})\mathbf{n} \times \mathbf{f} ds$$
$$+ \int_{C_{\alpha\beta\varepsilon}} \boldsymbol{\nu} \times \mathbf{f} dc + \int_{C_{\alpha\beta \text{ends}}} \boldsymbol{\nu} \times \mathbf{f} dc \tag{759}$$

Theorem 53 has the following alternative form:

Theorem 55.

$$\int_{S_{\alpha\beta}} \left.\frac{\partial f}{\partial t}\right|_{\mathbf{u}} ds = \frac{\partial}{\partial t} \int_{S_{\alpha\beta}} f ds - \int_{S_{\alpha\beta}} \nabla^S \cdot \mathbf{nn} \cdot \mathbf{w}^c f ds$$
$$- \int_{C_{\alpha\beta}} \mathbf{w}^s \cdot \boldsymbol{\nu} f dc - \int_{C_{\alpha\beta\varepsilon}} f \mathbf{w}^s \cdot \boldsymbol{\nu} ds \tag{760}$$

where $C_{\alpha\beta}$ is the curve of intersection of $S_{\alpha\beta}$ with the external surface of the REV.

Proof. By the definition of the indicator functions $\gamma\ (\mathbf{x}, \xi)$, $\gamma^s(\mathbf{u}(\mathbf{x}+\xi), t)$, and $\gamma^\alpha(\mathbf{x}+\xi, t)$, we have

$$\int_{S_{\alpha\beta}} \left.\frac{\partial f}{\partial t}\right|_{\mathbf{u}} ds = -\int_{V_\infty} \left.\frac{\partial f}{\partial t}\right|_{\mathbf{u}} \gamma\gamma^s \mathbf{n} \cdot \nabla\gamma^\alpha dv \tag{761}$$

which is, by Equation (130),

$$\int_{S_{\alpha\beta}} \left.\frac{\partial f}{\partial t}\right|_{\mathbf{u}} ds = -\int_{V_\infty} \left(\left.\frac{\partial f}{\partial t}\right|_{\mathbf{x}} + \mathbf{w} \cdot \mathbf{nn} \cdot \nabla f \right) \gamma\gamma^s \mathbf{n} \cdot \nabla\gamma^\alpha dv \tag{762}$$

An application of the chain rule to the right side of Equation (762) yields

$$\begin{aligned}
&\int_{S_{\alpha\beta}} \frac{\partial f}{\partial t}\bigg|_{\mathbf{u}} ds \\
&= -\int_{V_\infty} \frac{\partial}{\partial t}(f\gamma\gamma^s \mathbf{n}\cdot\nabla\gamma^\alpha)dv + \int_{V_\infty} f\frac{\partial(\gamma\gamma^s)}{\partial t}\mathbf{n}\cdot\nabla\gamma^\alpha dv \\
&+ \int_{V_\infty} f\gamma\gamma^s\left(\frac{\partial \mathbf{n}}{\partial t}\cdot\nabla\gamma^\alpha + \mathbf{n}\cdot\nabla\frac{\partial\gamma^\alpha}{\partial t}\right)dv - \int_{V_\infty} \nabla\cdot(\mathbf{nn}\cdot\mathbf{w}f\gamma\gamma^s\mathbf{n}\cdot\nabla\gamma^\alpha)dv \\
&+ \int_{V_\infty} \nabla^S\cdot\mathbf{nn}\cdot\mathbf{w}^c f\gamma\gamma^s\mathbf{n}\cdot\nabla\gamma^\alpha dv + \int_{V_\infty} \mathbf{n}\cdot\nabla\mathbf{n}\cdot\mathbf{w}f\gamma\gamma^s\mathbf{n}\cdot\nabla\gamma^\alpha dv \\
&+ \int_{V_\infty} \mathbf{n}\cdot\nabla\mathbf{w}\cdot\mathbf{n}f\gamma\gamma^s\mathbf{n}\cdot\nabla\gamma^\alpha dv + \int_{V_\infty} \mathbf{n}\cdot\mathbf{w}f\mathbf{n}\cdot\nabla(\gamma\gamma^s)\mathbf{n}\cdot\nabla\gamma^\alpha dv \\
&+ \int_{V_\infty} \mathbf{n}\cdot\mathbf{w}f\gamma\gamma^s\mathbf{n}\cdot(\nabla\mathbf{n}\cdot\nabla\gamma^\alpha + \nabla\nabla\gamma^\alpha\cdot\mathbf{n})dv
\end{aligned} \tag{763}$$

Because the integration variables do not include the time t, the differentiation sign $\partial/\partial t$ may be factored out from the integration sign in the first integral on the right side of Equation (763). By Equation (231), the fourth integral is zero on the right side of Equation (763). Note that the velocity at REV boundary surface is zero. With Equation (224) applied to the second integral on the right side of Equation (763), we obtain

$$\begin{aligned}
&\int_{V_\infty} f\frac{\partial(\gamma\gamma^s)}{\partial t}\mathbf{n}\cdot\nabla\gamma^\alpha dv + \int_{V_\infty} \mathbf{n}\cdot\mathbf{w}f\mathbf{n}\cdot\nabla(\gamma\gamma^s)\mathbf{n}\cdot\nabla\gamma^\alpha dv \\
&= -\int_{V_\infty} f\mathbf{w}\cdot\nabla(\gamma\gamma^s)\mathbf{n}\cdot\nabla\gamma^\alpha dv + \int_{V_\infty} \mathbf{n}\cdot\mathbf{w}f\mathbf{n}\cdot\nabla(\gamma\gamma^s)\mathbf{n}\cdot\nabla\gamma^\alpha dv \\
&= -\int_{V_\infty} f\mathbf{w}\cdot\nabla^S(\gamma\gamma^s)\mathbf{n}\cdot\nabla\gamma^\alpha dv
\end{aligned} \tag{764}$$

By Equations (180) and (194),

$$\begin{aligned}
&\int_{V_\infty} f\gamma\gamma^s\left(\frac{\partial \mathbf{n}}{\partial t}\cdot\nabla\gamma^\alpha + \mathbf{n}\cdot\nabla\frac{\partial\gamma^\alpha}{\partial t}\right)dv \\
&= -\int_{V_\infty} f\gamma\gamma^s\mathbf{n}\cdot\nabla\mathbf{w}\cdot\nabla\gamma^\alpha dv - \int_{V_\infty} f\gamma\gamma^s\mathbf{n}\cdot\nabla\nabla\gamma^\alpha\cdot\mathbf{w}dv
\end{aligned} \tag{765}$$

By Equation (195), we have

$$\int_{V_\infty} \mathbf{n}\cdot\mathbf{w}f\gamma\gamma^s\mathbf{n}\cdot(\nabla\mathbf{n}\cdot\nabla\gamma^\alpha + \nabla\nabla\gamma^\alpha\cdot\mathbf{n})dv = \int_{V_\infty} f\gamma\gamma^s\mathbf{w}\cdot\mathbf{nn}\cdot\nabla\nabla\gamma^\alpha\cdot\mathbf{n}dv \tag{766}$$

Thus Equation (763) becomes

$$\begin{aligned}
\int_{S_{\alpha\beta}} \frac{\partial f}{\partial t}\bigg|_{\mathbf{u}} ds &= -\frac{\partial}{\partial t}\int_{V_\infty} f\gamma\gamma^s\mathbf{n}\cdot\nabla\gamma^\alpha dv - \int_{V_\infty} f\gamma\mathbf{w}\cdot\nabla^S\gamma^s\mathbf{n}\cdot\nabla\gamma^\alpha dv \\
&- \int_{V_\infty} \gamma^s f\mathbf{w}\cdot\nabla^S\gamma\mathbf{n}\cdot\nabla\gamma^\alpha dv - \int_{V_\infty} f\gamma\gamma^s\mathbf{w}\cdot\nabla^S\nabla\gamma^\alpha\cdot\mathbf{n}dv \\
&+ \int_{V_\infty} \nabla^S\cdot\mathbf{nn}\cdot\mathbf{w}^c f\gamma\gamma^s\mathbf{n}\cdot\nabla\gamma^\alpha dv + \int_{V_\infty} \mathbf{n}\cdot\nabla\mathbf{n}\cdot\mathbf{w}f\gamma\gamma^s\mathbf{n}\cdot\nabla\gamma^\alpha dv
\end{aligned} \tag{767}$$

By Equation (256)

$$\int_{V_\infty} f\gamma\gamma^{s}\mathbf{w}\cdot\nabla^{S}\nabla\gamma^{\alpha}\cdot\mathbf{n}dv = \int_{V_\infty} f\gamma\gamma^{s}\mathbf{n}\cdot\nabla\mathbf{n}\cdot\mathbf{wn}\cdot\nabla\gamma^{\alpha}dv \tag{768}$$

Substituting Equation (768) into Equation (767) yields

$$\int_{S_{\alpha\beta}} \frac{\partial f}{\partial t}\bigg|_{\mathbf{u}} ds = -\frac{\partial}{\partial t}\int_{V_\infty} f\gamma\gamma^{s}\mathbf{n}\cdot\nabla\gamma^{\alpha}dv - \int_{V_\infty} f\gamma\mathbf{w}\cdot\nabla^{S}\gamma^{s}\mathbf{n}\cdot\nabla\gamma^{\alpha}dv$$
$$- \int_{V_\infty} f\gamma^{s}\mathbf{w}\cdot\nabla^{S}\gamma\mathbf{n}\cdot\nabla\gamma^{\alpha}dv + \int_{V_\infty} f\gamma\gamma^{s}\nabla^{S}\cdot\mathbf{nn}\cdot\mathbf{w}^{c}\mathbf{n}\cdot\nabla\gamma^{\alpha}dv \tag{769}$$

By Equations (186), (272), and (275), we obtain

$$\int_{S_{\alpha\beta}} \frac{\partial f}{\partial t}\bigg|_{\mathbf{u}} ds = \frac{\partial}{\partial t}\int_{S_{\alpha\beta}} f ds - \int_{S_{\alpha\beta}} \nabla^{S}\cdot\mathbf{nn}\cdot\mathbf{w}^{c} f ds - \int_{C_{\alpha\beta}} \mathbf{w}^{s}\cdot\boldsymbol{\nu} f dc - \int_{C_{\alpha\beta\varepsilon}} f\mathbf{w}^{s}\cdot\boldsymbol{\nu} ds$$

Theorem 54 has the following alternative form:

Theorem 56.

$$\int_{S_{\alpha\beta}} \nabla^{S}\circ F ds = \int_{S_{\alpha\beta}} \nabla^{S}\cdot\mathbf{nn}\circ F ds + \int_{C_{\alpha\beta}} \boldsymbol{\nu}\circ F ds + \int_{C_{\alpha\beta\varepsilon}} \boldsymbol{\nu}\circ F dc \tag{770}$$

Proof. By the indicator functions γ $(\mathbf{x}, \xi)$, $\gamma^{s}(\mathbf{u}(\mathbf{x}, \xi), t)$, and $\gamma^{\alpha}(\mathbf{x}+\xi, t)$, the integral under study can be expressed as

$$\int_{S_{\alpha\beta}} \nabla^{S}\circ F ds = -\int_{V_\infty} (\nabla - \mathbf{nn}\cdot\nabla)\circ F\gamma\gamma^{s}\mathbf{n}\cdot\nabla\gamma^{\alpha}dv \tag{771}$$

An application of the chain rule to the right-hand side of Equation (771) yields

$$\int_{S_{\alpha\beta}} \nabla^{S}\circ F ds$$
$$= -\int_{V_\infty} \nabla\circ(F\gamma\gamma^{s}\mathbf{n}\cdot\nabla\gamma^{\alpha})dv + \int_{V_\infty} \nabla(\gamma\gamma^{s})\circ F\mathbf{n}\cdot\nabla\gamma^{\alpha}dv$$
$$+ \int_{V_\infty} \gamma\gamma^{s}(\nabla\mathbf{n}\cdot\nabla\gamma^{\alpha} + \nabla\nabla\gamma^{\alpha}\cdot\mathbf{n})\circ F dv + \int_{V_\infty} \nabla\cdot(\mathbf{nn}\circ F\gamma\gamma^{s}\mathbf{n}\cdot\nabla\gamma^{\alpha})dv$$
$$- \int_{V_\infty} \nabla\cdot\mathbf{nn}\circ F\gamma\gamma^{s}\mathbf{n}\cdot\nabla\gamma^{\alpha}dv - \int_{V_\infty} \mathbf{n}\cdot\nabla\mathbf{n}\circ F\gamma\gamma^{s}\mathbf{n}\cdot\nabla\gamma^{\alpha}dv$$
$$- \int_{V_\infty} \mathbf{n}\circ F\mathbf{n}\cdot\nabla(\gamma\gamma^{s})\mathbf{n}\cdot\nabla\gamma^{\alpha}dv$$
$$- \int_{V_\infty} \mathbf{n}\circ F\gamma\gamma^{s}\mathbf{n}\cdot(\nabla\mathbf{n}\cdot\nabla\gamma^{\alpha} + \nabla\nabla\gamma^{\alpha}\cdot\mathbf{n})dv \tag{772}$$

With Equations (230)–(232) applied to the first and fourth integrals, respectively, we conclude that the two integrals are zero. By Equation (195), we have

$$\int_{V_\infty} \gamma\gamma^{s}(\nabla\mathbf{n}\cdot\nabla\gamma^{\alpha} + \nabla\nabla\gamma^{\alpha}\cdot\mathbf{n})\circ F dv = \int_{V_\infty} \gamma\gamma^{s}(\nabla\nabla\gamma^{\alpha}\cdot\mathbf{n})\circ F dv \tag{773}$$

$$\int_{V_\infty} \mathbf{n}\circ F\gamma\gamma^{s}\mathbf{n}\cdot(\nabla\mathbf{n}\cdot\nabla\gamma^{\alpha} + \nabla\nabla\gamma^{\alpha}\cdot\mathbf{n})dv = \int_{V_\infty} \gamma\gamma^{s}(\nabla\nabla\gamma^{\alpha}\cdot\mathbf{n})\cdot\mathbf{nn}\circ F dv \tag{774}$$

Finally, regrouping the terms on the right side of Equation (772) yields

$$\int_{S_{\alpha\beta}} \nabla^{S} \circ F ds = \int_{V_\infty} \nabla^{S}(\gamma\gamma^{s}) \circ F\mathbf{n} \cdot \nabla\gamma^{\alpha} dv + \int_{V_\infty} \gamma\gamma^{s}(\nabla^{S}\nabla\gamma^{\alpha} \cdot \mathbf{n}) \circ F dv - \int_{V_\infty} \nabla^{S} \cdot \mathbf{nn} \circ F\gamma\gamma^{s}\mathbf{n} \cdot \nabla\gamma^{\alpha} dv - \int_{V_\infty} \mathbf{n} \cdot \nabla\mathbf{n} \circ F\gamma\gamma^{s}\mathbf{n} \cdot \nabla\gamma^{\alpha} dv \tag{775}$$

By Equation (256), the second integral on the right side of Equation (775) becomes

$$\int_{V_\infty} \gamma\gamma^{s}(\nabla^{S}\nabla\gamma^{\alpha} \cdot \mathbf{n}) \circ F dv = \int_{V_\infty} \gamma\gamma^{s}\mathbf{n} \cdot \nabla\mathbf{n} \circ F\mathbf{n} \cdot \nabla\gamma^{\alpha} dv \tag{776}$$

Substituting Equation (776) into Equation (775) yields

$$\int_{S_{\alpha\beta}} \nabla^{S} \circ F ds = \int_{V_\infty} (\nabla^{S}\gamma\gamma^{s} + \gamma\nabla^{S}\gamma^{s}) \circ F\mathbf{n} \cdot \nabla\gamma^{\alpha} dv - \int_{V_\infty} \nabla^{S} \cdot \mathbf{nn} \circ F\gamma\gamma^{s}\mathbf{n} \cdot \nabla\gamma^{\alpha} dv \tag{777}$$

With Equations (275) and (272) applied to the first and second integrals, respectively, we conclude that

$$\int_{S_{\alpha\beta}} \nabla^{S} \circ F ds = \int_{S_{\alpha\beta}} \nabla^{S} \cdot \mathbf{nn} \circ F ds + \int_{C_{\alpha\beta}} \mathbf{v} \circ F dc + \int_{C_{\alpha\beta\varepsilon}} \mathbf{v} \circ F dc$$

Remark. Equation (770) contains the following three formulas:

Gradient

$$\int_{S_{\alpha\beta}} \nabla^{S} f ds = \int_{S_{\alpha\beta}} (\nabla^{S} \cdot \mathbf{n})\mathbf{n} f ds + \int_{C_{\alpha\beta}} \mathbf{v} f ds + \int_{C_{\alpha\beta\varepsilon}} \mathbf{v} f dc \tag{778}$$

Divergence

$$\int_{S_{\alpha\beta}} \nabla^{S} \cdot \mathbf{f} ds = \int_{S_{\alpha\beta}} \nabla^{S} \cdot \mathbf{nn} \cdot \mathbf{f} ds + \int_{C_{\alpha\beta}} \mathbf{v} \cdot \mathbf{f} dc + \int_{C_{\alpha\beta\varepsilon}} \mathbf{v} \cdot \mathbf{f} dc \tag{779}$$

Curl

$$\int_{S_{\alpha\beta}} \nabla^{S} \times \mathbf{f} ds = \int_{S_{\alpha\beta}} \nabla^{S} \cdot \mathbf{nn} \times \mathbf{f} ds + \int_{C_{\alpha\beta}} \mathbf{v} \times \mathbf{f} dc + \int_{C_{\alpha\beta\varepsilon}} \mathbf{v} \times \mathbf{f} dc \tag{780}$$

5.2.3 Curvilineal operator theorems

The REV used for this group of averaging theorems is shown in Figure 30.

Theorem 57.

$$\int_{C_{\alpha\beta\varepsilon}} \left.\frac{\partial f}{\partial t}\right|_{1} dc = \frac{\partial}{\partial t}\int_{C_{\alpha\beta\varepsilon}} f dc + \nabla^{S} \cdot \int_{C_{\alpha\beta\varepsilon}} \mathbf{w}^{s} f dc + \int_{C_{\alpha\beta\varepsilon}} (\boldsymbol{\lambda} \cdot \nabla^{c}\boldsymbol{\lambda}) \cdot \mathbf{w}^{s} f dc - \sum_{\text{pts}} \mathbf{e} \cdot \mathbf{w}^{c} f - \sum_{\text{ends}_{\alpha\beta\varepsilon}} \mathbf{e} \cdot \mathbf{w}^{c} f \tag{781}$$

where $C_{\alpha\beta\varepsilon}$ is the contact curve within the REV that exists as the location where the α-, β-, and ε-phases meet. pts refers to the end points of the $C_{\alpha\beta\varepsilon}$ curve within the REV where the α-, β-, and ε-phases meet a fourth phase. ends$_{\alpha\beta\varepsilon}$ is the points at the ends of the REV formed by the intersection of the bounding surface of the cylinder with $C_{\alpha\beta\varepsilon}$. $\mathbf{n}^*$ is the unit vector normal to the external boundary of the REV that is pointing outward from the REV, $\boldsymbol{\lambda}$ the unit vector tangent to $C_{\alpha\beta\varepsilon}$, and $\mathbf{e}$ the unit vector tangent to $C_{\alpha\beta\varepsilon}$ at pts and ext pointing outward from the curve such that $\mathbf{e}\cdot\boldsymbol{\lambda} = \pm 1$. ∇^{c} is the microscopic curvilineal operator, $\nabla^{c} = \boldsymbol{\lambda\lambda}\cdot\nabla_{\xi}$, $\mathbf{w}$ the velocity of $C_{\alpha\beta\varepsilon}$, and $\mathbf{w}^{c}$ the velocity of $C_{\alpha\beta\varepsilon}$ tangent to $C_{\alpha\beta\varepsilon}$ such that $\mathbf{w}^{c} = \boldsymbol{\lambda\lambda}\cdot\mathbf{w}$. $\mathbf{w}^{s}$ is the velocity of $C_{\alpha\beta\varepsilon}$ normal to $C_{\alpha\beta\varepsilon}$ such that $\mathbf{w}^{S} = \mathbf{w}-\mathbf{w}^{c}$ and f a curvilineal function $f(\mathbf{l}, t)$ defined on $C_{\alpha\beta\varepsilon}$.

Proof. By Equation (134), the integral under study may be expressed as

$$\int_{C_{\alpha\beta\varepsilon}} \left.\frac{\partial f}{\partial t}\right|_{\mathbf{l}} \mathrm{d}c = \int_{C_{\alpha\beta\varepsilon}} \left.\frac{\partial f}{\partial t}\right|_{\mathbf{x}} + \mathbf{w}^{s}\cdot\nabla f \mathrm{d}c \tag{782}$$

By the indicator functions $\gamma(\mathbf{x}, \xi)$, $\gamma^{s}(\mathbf{u}(\mathbf{x}+\xi), t)$, $\gamma^{\alpha}(\mathbf{x}+\xi, t)$, and $\gamma^{c}(\mathbf{l}(\mathbf{x}+\xi), t)$, the integral on the right-hand side of Equation (782) is equivalent to

$$\begin{aligned}\int_{C_{\alpha\beta\varepsilon}} \left.\frac{\partial f}{\partial t}\right|_{\mathbf{l}} \mathrm{d}c = &\int_{V_\infty} \left.\frac{\partial f}{\partial t}\right|_{\mathbf{x}} \gamma\gamma^{c}\mathbf{v}\cdot\nabla\gamma^{s}\mathbf{n}\cdot\nabla\gamma^{\alpha}\mathrm{d}v \\ &+ \int_{V_\infty} \mathbf{w}^{s}\cdot\nabla f\gamma\gamma^{c}\mathbf{v}\cdot\nabla\gamma^{s}\mathbf{n}\cdot\nabla\gamma^{\alpha}\mathrm{d}v\end{aligned} \tag{783}$$

An application of the chain rule to the integrals on the right side of Equation (783) yields

$$\begin{aligned}\int_{C_{\alpha\beta\varepsilon}} \left.\frac{\partial f}{\partial t}\right|_{\mathbf{l}} \mathrm{d}c = &\int_{V_\infty} \frac{\partial}{\partial t}(f\gamma\gamma^{c}\mathbf{v}\cdot\nabla\gamma^{s}\mathbf{n}\cdot\nabla\gamma^{\alpha})\mathrm{d}v - \int_{V_\infty} f\frac{\partial(\gamma\gamma^{c})}{\partial t}\mathbf{v}\cdot\nabla\gamma^{s}\mathbf{n}\cdot\nabla\gamma^{\alpha}\mathrm{d}v \\ &- \int_{V_\infty} f\gamma\gamma^{c}\left(\frac{\partial\mathbf{v}}{\partial t}\cdot\nabla\gamma^{s} + \mathbf{v}\cdot\frac{\partial\gamma^{s}}{\partial t}\right)\mathbf{n}\cdot\nabla\gamma^{\alpha}\mathrm{d}v \\ &- \int_{V_\infty} f\gamma\gamma^{c}\mathbf{v}\cdot\nabla\gamma^{s}\left(\frac{\partial\mathbf{n}}{\partial t}\cdot\nabla\gamma^{\alpha} + \mathbf{n}\cdot\nabla\frac{\partial\gamma^{\alpha}}{\partial t}\right)\mathrm{d}v \\ &+ \int_{V_\infty} \nabla\cdot(\mathbf{w}^{s}f\gamma\gamma^{c}\mathbf{v}\cdot\nabla\gamma^{s}\mathbf{n}\cdot\nabla\gamma^{\alpha})\mathrm{d}v \\ &- \int_{V_\infty} \nabla\cdot\mathbf{w}^{s}f\gamma\gamma^{c}\mathbf{v}\cdot\nabla\gamma^{s}\mathbf{n}\cdot\nabla\gamma^{\alpha}\mathrm{d}v \\ &- \int_{V_\infty} f\mathbf{w}^{s}\nabla(\gamma\gamma^{c})\mathbf{v}\cdot\nabla\gamma^{s}\mathbf{n}\cdot\nabla\gamma^{\alpha}\mathrm{d}v \\ &- \int_{V_\infty} f\gamma\gamma^{c}\mathbf{w}^{s}\cdot(\nabla\mathbf{v}\cdot\nabla\gamma^{s} + \nabla\nabla\gamma^{s}\cdot\mathbf{v})\mathbf{n}\cdot\nabla\gamma^{\alpha}\mathrm{d}v \\ &- \int_{V_\infty} f\gamma\gamma^{c}\mathbf{v}\cdot\nabla\gamma^{s}\mathbf{w}^{s}\cdot(\nabla\mathbf{n}\cdot\nabla\gamma^{\alpha} + \nabla\nabla\gamma^{\alpha}\cdot\mathbf{n})\mathrm{d}v\end{aligned} \tag{784}$$

Because the integration domain is independent of time t, the differentiation $\partial/\partial t$ can be moved out of the integration sign for the first integral on the right side of Equation (784). By $\nabla = \nabla^S + \nabla^c$, the fifth integral on the right side of Equation (784) becomes

$$\int_{V_\infty} \nabla \cdot (\mathbf{w}^s f\gamma\gamma^c \boldsymbol{v} \cdot \nabla\gamma^s \mathbf{n} \cdot \nabla\gamma^\alpha) dv = \nabla^S \cdot \int_{V_\infty} \mathbf{w}^s f\gamma\gamma^c \boldsymbol{v} \cdot \nabla\gamma^s \mathbf{n} \cdot \nabla\gamma^\alpha dv \tag{785}$$

By Equations (227), (224), and (229),

$$\frac{\partial(\gamma\gamma^c)}{\partial t} = -\mathbf{w} \cdot \nabla(\gamma\gamma^c), \quad \frac{\partial\gamma^\alpha}{\partial t} = -\mathbf{w} \cdot \nabla\gamma^\alpha, \quad \frac{\partial\gamma^s}{\partial t} = -\mathbf{w} \cdot \nabla\gamma^s$$

By applying these formulas and Equations (195) and (199) to the integrals on the right side of Equation (784), we obtain

$$\begin{aligned}
\int_{C_{\alpha\beta\varepsilon}} \frac{\partial f}{\partial t}\Big|_1 dc
&= \frac{\partial}{\partial t}\int_{V_\infty} f\gamma\gamma^c \boldsymbol{v} \cdot \nabla\gamma^s \mathbf{n} \cdot \nabla\gamma^\alpha dv + \int_{V_\infty} \mathbf{w} \cdot \nabla(\gamma\gamma^c) f \boldsymbol{v} \cdot \nabla\gamma^s \mathbf{n} \cdot \nabla\gamma^\alpha dv \\
&+ \int_{V_\infty} f\gamma\gamma^c \boldsymbol{v} \cdot \nabla\mathbf{w} \cdot \nabla\gamma^s \mathbf{n} \cdot \nabla\gamma^\alpha dv + \int_{V_\infty} f\gamma\gamma^c \boldsymbol{v} \cdot \nabla\nabla\gamma^s \cdot \mathbf{w}\mathbf{n} \cdot \nabla\gamma^\alpha dv \\
&+ \int_{V_\infty} f\gamma\gamma^c \boldsymbol{v} \cdot \nabla\gamma^s \mathbf{n} \cdot \nabla\mathbf{w} \cdot \nabla\gamma^\alpha dv + \int_{V_\infty} f\gamma\gamma^c \boldsymbol{v} \cdot \nabla\gamma^s \mathbf{n} \cdot \nabla\nabla\gamma^\alpha \cdot \mathbf{w} dv \\
&+ \nabla^S \cdot \int_{V_\infty} \mathbf{w}^s f\gamma\gamma^c \boldsymbol{v} \cdot \nabla\gamma^s \mathbf{n} \cdot \nabla\gamma^\alpha dv - \int_{V_\infty} \nabla \cdot \mathbf{w}^s f\gamma\gamma^c \boldsymbol{v} \cdot \nabla\gamma^s \mathbf{n} \cdot \nabla\gamma^\alpha dv \\
&- \int_{V_\infty} f\mathbf{w}^s \cdot \nabla(\gamma\gamma^c) \boldsymbol{v} \cdot \nabla\gamma^s \mathbf{n} \cdot \nabla\gamma^\alpha dv - \int_{V_\infty} f\gamma\gamma^c \mathbf{w}^s \cdot \nabla\nabla\gamma^s \cdot \boldsymbol{v}\mathbf{n} \cdot \nabla\gamma^\alpha dv \\
&- \int_{V_\infty} f\gamma\gamma^c \boldsymbol{v} \cdot \nabla\gamma^s \mathbf{w}^s \cdot \nabla\nabla\gamma^\alpha \cdot \mathbf{n} dv
\end{aligned} \tag{786}$$

Regrouping the terms on the right side of Equation (786) yields

$$\begin{aligned}
\int_{C_{\alpha\beta\varepsilon}} \frac{\partial f}{\partial t}\Big|_1 dc
&= \frac{\partial}{\partial t}\int_{V_\infty} f\gamma\gamma^c \boldsymbol{v} \cdot \nabla\gamma^s \mathbf{n} \cdot \nabla\gamma^\alpha dv + \nabla^S \cdot \int_{V_\infty} \mathbf{w}^s f\gamma\gamma^c \boldsymbol{v} \cdot \nabla\gamma^s \mathbf{n} \cdot \nabla\gamma^\alpha dv \\
&+ \int_{V_\infty} f\gamma\gamma^c \boldsymbol{v} \cdot \nabla\mathbf{w} \cdot \nabla\gamma^s \mathbf{n} \cdot \nabla\gamma^\alpha dv + \int_{V_\infty} f\gamma\gamma^c \boldsymbol{v} \cdot \nabla\gamma^s \mathbf{n} \cdot \nabla\mathbf{w} \cdot \nabla\gamma^\alpha dv \\
&+ \int_{V_\infty} f\gamma\gamma^c \mathbf{w} \cdot \nabla^c\nabla\gamma^s \cdot \boldsymbol{v}\mathbf{n} \cdot \nabla\gamma^\alpha dv + \int_{V_\infty} f\gamma\gamma^c \boldsymbol{v} \cdot \nabla\gamma^s \mathbf{w} \cdot \nabla^c\nabla\gamma^\alpha \cdot \mathbf{n} dv \\
&- \int_{V_\infty} \nabla \cdot (\mathbf{w} - \boldsymbol{\lambda}\boldsymbol{\lambda} \cdot \mathbf{w}) f\gamma\gamma^c \boldsymbol{v} \cdot \nabla\gamma^s \mathbf{n} \cdot \nabla\gamma^\alpha dv \\
&+ \int_{V_\infty} f\mathbf{w} \cdot \nabla^c(\gamma\gamma^c) \boldsymbol{v} \cdot \nabla\gamma^s \mathbf{n} \cdot \nabla\gamma^\alpha dv
\end{aligned} \tag{787}$$

By Equation (53), we have

$$\nabla \cdot \mathbf{w} = \boldsymbol{\lambda} \cdot \nabla\mathbf{w} \cdot \boldsymbol{\lambda} + \boldsymbol{v} \cdot \nabla\mathbf{w} \cdot \boldsymbol{v} + \mathbf{n} \cdot \nabla\mathbf{w} \cdot \mathbf{n}$$

$$\nabla \cdot (\boldsymbol{\lambda\lambda} \cdot \mathbf{w}) = \boldsymbol{v} \cdot \nabla\boldsymbol{\lambda} \cdot \boldsymbol{v\lambda} \cdot \mathbf{w} + \mathbf{n} \cdot \nabla\boldsymbol{\lambda} \cdot \mathbf{n}\boldsymbol{\lambda} \cdot \mathbf{w} \\ + \boldsymbol{\lambda} \cdot \nabla^{c}\boldsymbol{\lambda} \cdot \mathbf{w}^{c} + \boldsymbol{\lambda} \cdot \nabla\mathbf{w} \cdot \boldsymbol{\lambda}$$

Applying these together with Equations (261) and (258) to the fifth and sixth terms leads to

$$\begin{aligned}\int_{C_{\alpha\beta\varepsilon}} \left.\frac{\partial f}{\partial t}\right|_{1} \mathrm{d}c = {} & \frac{\partial}{\partial t}\int_{V_\infty} f\gamma\gamma^{c}\boldsymbol{v} \cdot \nabla\gamma^{s}\mathbf{n} \cdot \nabla\gamma^{\alpha}\mathrm{d}v + \nabla^{S} \cdot \int_{V_\infty} \mathbf{w}^{s} f\gamma\gamma^{c}\boldsymbol{v} \cdot \nabla\gamma^{s}\mathbf{n} \cdot \gamma^{\alpha}\mathrm{d}v \\ & + \int_{V_\infty} f\gamma\gamma^{c}\boldsymbol{v} \cdot \nabla\mathbf{w} \cdot \nabla\gamma^{s}\mathbf{n} \cdot \nabla\gamma^{\alpha}\mathrm{d}v \\ & + \int_{V_\infty} f\gamma\gamma^{c}\boldsymbol{v} \cdot \nabla\gamma^{s}\mathbf{n} \cdot \nabla\mathbf{w} \cdot \nabla\gamma^{\alpha}\mathrm{d}v \\ & - \int_{V_\infty} f\gamma\gamma^{c}\mathbf{w} \cdot \boldsymbol{\lambda v} \cdot \nabla\boldsymbol{\lambda} \cdot \nabla\gamma^{s}\mathbf{n} \cdot \nabla\gamma^{\alpha}\mathrm{d}v \\ & - \int_{V_\infty} f\gamma\gamma^{c}\mathbf{w} \cdot \boldsymbol{\lambda}\mathbf{n} \cdot \nabla\boldsymbol{\lambda} \cdot \nabla\gamma^{\alpha}\boldsymbol{v} \cdot \nabla\gamma^{s}\mathrm{d}v \\ & - \int_{V_\infty} \boldsymbol{\lambda} \cdot \nabla\mathbf{w} \cdot \boldsymbol{\lambda} f\gamma\gamma^{c}\boldsymbol{v} \cdot \nabla\gamma^{s}\mathbf{n} \cdot \nabla\gamma^{\alpha}\mathrm{d}v \\ & - \int_{V_\infty} \boldsymbol{n} \cdot \nabla\mathbf{w} \cdot \boldsymbol{n} f\gamma\gamma^{c}\boldsymbol{v} \cdot \nabla\gamma^{s}\mathbf{n} \cdot \nabla\gamma^{\alpha}\mathrm{d}v \\ & - \int_{V_\infty} \boldsymbol{v} \cdot \nabla\mathbf{w} \cdot \boldsymbol{v} f\gamma\gamma^{c}\boldsymbol{v} \cdot \nabla\gamma^{s}\mathbf{n} \cdot \nabla\gamma^{\alpha}\mathrm{d}v \\ & + \int_{V_\infty} \boldsymbol{v} \cdot \nabla\boldsymbol{\lambda} \cdot \boldsymbol{v} f\gamma\gamma^{c}\boldsymbol{v} \cdot \nabla\gamma^{s}\mathbf{n} \cdot \nabla\gamma^{\alpha}\boldsymbol{\lambda} \cdot \mathbf{w}\mathrm{d}v \\ & + \int_{V_\infty} \boldsymbol{n} \cdot \nabla\mathbf{w} \cdot \boldsymbol{n} f\gamma\gamma^{c}\boldsymbol{v} \cdot \nabla\gamma^{s}\mathbf{n} \cdot \nabla\gamma^{\alpha}\boldsymbol{\lambda} \cdot \mathbf{w}\mathrm{d}v \\ & + \int_{V_\infty} \boldsymbol{\lambda} \cdot \nabla^{c}\boldsymbol{\lambda} \cdot \mathbf{w}^{s} f\gamma\gamma^{c}\boldsymbol{v} \cdot \nabla\gamma^{s}\mathbf{n} \cdot \nabla\gamma^{\alpha}\mathrm{d}v \\ & + \int_{V_\infty} \boldsymbol{\lambda} \cdot \nabla\mathbf{w} \cdot \boldsymbol{\lambda} f\gamma\gamma^{c}\boldsymbol{v} \cdot \nabla\gamma^{s}\mathbf{n} \cdot \nabla\gamma^{\alpha}\mathrm{d}v \\ & + \int_{V_\infty} f\mathbf{w} \cdot \nabla^{c}(\gamma\gamma^{c})\boldsymbol{v} \cdot \nabla\gamma^{s}\mathbf{n} \cdot \nabla\gamma^{\alpha}\mathrm{d}v \end{aligned} \tag{788}$$

Equation (788) can be further simplified to

$$\begin{aligned}\int_{C_{\alpha\beta\varepsilon}} \left.\frac{\partial f}{\partial t}\right|_{1} \mathrm{d}c = {} & \frac{\partial}{\partial t}\int_{V_\infty} f\gamma\gamma^{c}\boldsymbol{v} \cdot \nabla\gamma^{s}\mathbf{n} \cdot \nabla\gamma^{\alpha}\mathrm{d}v \\ & + \nabla^{S} \cdot \int_{V_\infty} \mathbf{w}^{s} f\gamma\gamma^{c}\boldsymbol{v} \cdot \nabla\gamma^{s}\mathbf{n} \cdot \nabla\gamma^{\alpha}\mathrm{d}v \\ & + \int_{V_\infty} \boldsymbol{\lambda} \cdot \nabla^{c}\boldsymbol{\lambda} \cdot \mathbf{w}^{s} f\gamma\gamma^{c}\boldsymbol{v} \cdot \nabla\gamma^{s}\mathbf{n} \cdot \nabla\gamma^{\alpha}\mathrm{d}v \\ & + \int_{V_\infty} f\mathbf{w} \cdot \nabla^{c}(\gamma\gamma^{c})\boldsymbol{v} \cdot \nabla\gamma^{s}\mathbf{n} \cdot \nabla\gamma^{\alpha}\mathrm{d}v \end{aligned} \tag{789}$$

By Equations (275) and (166), we obtain

$$\int_{C_{\alpha\beta\varepsilon}} \frac{\partial f}{\partial t}\bigg|_{\mathbf{l}} \mathrm{d}c = \frac{\partial}{\partial t}\int_{C_{\alpha\beta\varepsilon}} f\mathrm{d}c + \nabla^{S}\cdot\int_{C_{\alpha\beta\varepsilon}} \mathbf{w}^{s} f\mathrm{d}c + \int_{C_{\alpha\beta\varepsilon}} (\boldsymbol{\lambda}\cdot\nabla^{c}\boldsymbol{\lambda})\cdot\mathbf{w}^{s} f\mathrm{d}c$$
$$-\sum_{\text{pts}} \mathbf{e}\cdot\mathbf{w}^{c} f - \sum_{\text{ends}_{\alpha\beta\varepsilon}} \mathbf{e}\cdot\mathbf{w}^{c} f$$

Theorem 58.

$$\int_{C_{\alpha\beta\varepsilon}} \nabla^{c}\circ F\mathrm{d}c = \nabla^{S}\cdot\int_{C_{\alpha\beta\varepsilon}} \boldsymbol{\lambda}\boldsymbol{\lambda}\circ F\mathrm{d}c - \int_{C_{\alpha\beta\varepsilon}} (\boldsymbol{\lambda}\cdot\nabla^{c}\boldsymbol{\lambda})\circ F\mathrm{d}c$$
$$+\sum_{\text{pts}} \mathbf{e}\circ F + \sum_{\text{ends}_{\alpha\beta\varepsilon}} \mathbf{e}\circ F \tag{790}$$

where $\nabla^{c}\circ F$ can be any of the three operators: $\nabla^{c} f$, $\nabla^{c}\cdot\mathbf{f}$, and $\nabla^{c}\times\mathbf{f}$. $f(\mathbf{l}, t)$ and $\mathbf{f}(\mathbf{l}, t)$ are scalar and vector functions of curvilineal position vector $\mathbf{l}$ and time t defined on $C_{\alpha\beta\varepsilon}$.

Proof. The indicator functions $\gamma(\mathbf{x}, \xi)$, $\gamma^{\alpha}(\mathbf{u}(\mathbf{x}+\xi), t)$, $\gamma^{s}(\mathbf{u}(\mathbf{x}+\xi), t)$, and $\gamma^{c}(\mathbf{l}(\mathbf{x}+\xi), t)$ allow us to rewrite the integral under study as

$$\int_{C_{\alpha\beta\varepsilon}} \nabla^{c}\circ F\mathrm{d}c = \int_{V_\infty} (\boldsymbol{\lambda}\boldsymbol{\lambda}\cdot\nabla)\circ F\gamma\gamma^{c}\mathbf{n}\cdot\nabla\gamma^{\alpha}\boldsymbol{v}\cdot\nabla\gamma^{s}\mathrm{d}v \tag{791}$$

An application of the chain rule to the right side of Equation (791) yields

$$\int_{C_{\alpha\beta\varepsilon}} \nabla^{c}\circ F\mathrm{d}c = \int_{V_\infty} \nabla\cdot(\boldsymbol{\lambda}\boldsymbol{\lambda}\circ F\gamma\gamma^{c}\mathbf{n}\cdot\nabla\gamma^{\alpha}\boldsymbol{v}\cdot\nabla\gamma^{s})\mathrm{d}v$$
$$-\int_{V_\infty} \nabla\cdot\boldsymbol{\lambda}\boldsymbol{\lambda}\circ F\gamma\gamma^{c}\mathbf{n}\cdot\nabla\gamma^{\alpha}\boldsymbol{v}\cdot\nabla\gamma^{s}\mathrm{d}v$$
$$-\int_{V_\infty} \boldsymbol{\lambda}\cdot\nabla\boldsymbol{\lambda}\circ F\gamma\gamma^{c}\mathbf{n}\cdot\nabla\gamma^{\alpha}\boldsymbol{v}\cdot\nabla\gamma^{s}\mathrm{d}v$$
$$-\int_{V_\infty} \boldsymbol{\lambda}\circ F\boldsymbol{\lambda}\cdot\nabla(\gamma\gamma^{c})\mathbf{n}\cdot\nabla\gamma^{\alpha}\boldsymbol{v}\cdot\nabla\gamma^{s}\mathrm{d}v$$
$$-\int_{V_\infty} \boldsymbol{\lambda}\circ F\gamma\gamma^{c}\boldsymbol{\lambda}\cdot(\nabla\boldsymbol{v}\cdot\nabla\gamma^{s} + \nabla\nabla\gamma^{s}\cdot\boldsymbol{v})\mathbf{n}\cdot\nabla\gamma^{\alpha}\mathrm{d}v$$
$$-\int_{V_\infty} \boldsymbol{\lambda}\circ F\gamma\gamma^{c}\boldsymbol{v}\cdot\nabla\gamma^{s}\boldsymbol{\lambda}\cdot(\nabla\mathbf{n}\cdot\nabla\gamma^{\alpha} + \nabla\nabla\gamma^{\alpha}\cdot\mathbf{n})\mathrm{d}v \tag{792}$$

By a similar approach as in the derivation of Equation (785), we can obtain

$$\int_{V_\infty} \nabla\cdot(\boldsymbol{\lambda}\boldsymbol{\lambda}\circ F\gamma\gamma^{c}\mathbf{n}\cdot\nabla\gamma^{\alpha}\boldsymbol{v}\cdot\nabla\gamma^{s})\mathrm{d}v = \nabla^{S}\cdot\int_{V_\infty} \boldsymbol{\lambda}\boldsymbol{\lambda}\circ F\gamma\gamma^{c}\mathbf{n}\cdot\nabla\gamma^{\alpha}\boldsymbol{v}\cdot\nabla\gamma^{s}\mathrm{d}v \tag{793}$$

By Equations (21) and (53),

$$\nabla \cdot \boldsymbol{\lambda} = \mathbf{n} \cdot \nabla\boldsymbol{\lambda} \cdot \mathbf{n} + \boldsymbol{v} \cdot \nabla\boldsymbol{\lambda} \cdot \boldsymbol{v} \tag{794}$$

Substituting Equation (794) into the second integral on the right side of Equation (792) yields

$$\begin{aligned}\int_{V_\infty} \nabla \cdot \boldsymbol{\lambda}\boldsymbol{\lambda} \circ F\gamma\gamma^{\mathrm{c}}\boldsymbol{v} \cdot \nabla\gamma^{\mathrm{s}}\mathbf{n} \cdot \nabla\gamma^{\alpha}\mathrm{d}v &= \int_{V_\infty} \boldsymbol{\lambda} \circ F\mathbf{n} \cdot \nabla\boldsymbol{\lambda} \cdot \mathbf{n}\gamma\gamma^{\mathrm{c}}\boldsymbol{v} \cdot \nabla\gamma^{\mathrm{s}}\mathbf{n} \cdot \nabla\gamma^{\alpha}\mathrm{d}v \\ &+ \int_{V_\infty} \boldsymbol{\lambda} \circ F\boldsymbol{v} \cdot \nabla\boldsymbol{\lambda} \cdot \boldsymbol{v}\gamma\gamma^{\mathrm{c}}\boldsymbol{v} \cdot \nabla\gamma^{\mathrm{s}}\mathbf{n} \cdot \nabla\gamma^{\alpha}\mathrm{d}v\end{aligned} \tag{795}$$

Equation (200) enables us to obtain

$$\begin{aligned}\int_{V_\infty} \boldsymbol{\lambda} \circ F\gamma\gamma^{\mathrm{c}}\boldsymbol{\lambda} \cdot (\nabla\boldsymbol{v} \cdot \nabla\gamma^{\mathrm{s}} + \nabla\nabla\gamma^{\mathrm{s}} \cdot \boldsymbol{v})\mathbf{n} \cdot \nabla\gamma^{\alpha}\mathrm{d}v &= \int_{V_\infty} \boldsymbol{\lambda} \circ F\gamma\gamma^{\mathrm{c}}\boldsymbol{\lambda} \cdot \nabla\nabla\gamma^{\mathrm{s}} \cdot \boldsymbol{v}\mathbf{n} \cdot \nabla\gamma^{\alpha}\mathrm{d}v\end{aligned} \tag{796}$$

By Equation (195), the last term on the right side of Equation (792) becomes

$$\begin{aligned}\int_{V_\infty} \boldsymbol{\lambda} \circ F\gamma\gamma^{\mathrm{c}}\boldsymbol{v} \cdot \nabla\gamma^{\mathrm{s}}\boldsymbol{\lambda} \cdot (\nabla\mathbf{n} \cdot \nabla\gamma^{\alpha} + \nabla\nabla\gamma^{\alpha} \cdot \mathbf{n})\mathrm{d}v &= \int_{V_\infty} \boldsymbol{\lambda} \circ F\gamma\gamma^{\mathrm{c}}\boldsymbol{v} \cdot \nabla\gamma^{\mathrm{s}}\boldsymbol{\lambda} \cdot \nabla\nabla\gamma^{\alpha} \cdot \mathbf{n}\mathrm{d}v\end{aligned} \tag{797}$$

Substituting Equations (793), (795)–(797) into the right side of Equation (792) and regrouping the resulting terms yields

$$\begin{aligned}\int_{C_{\alpha\beta\varepsilon}} \nabla^{\mathrm{c}} \circ F\mathrm{d}c = \nabla^{\mathrm{S}} \cdot \int_{V_\infty} \boldsymbol{\lambda}\boldsymbol{\lambda} \circ F\gamma\gamma^{\mathrm{c}}\mathbf{n} \cdot \nabla\gamma^{\alpha}\boldsymbol{v} \cdot \nabla\gamma^{\mathrm{s}}\mathrm{d}s \\ - \int_{V_\infty} \boldsymbol{\lambda} \circ F\mathbf{n} \cdot \nabla\boldsymbol{\lambda} \cdot \mathbf{n}\gamma\gamma^{\mathrm{c}}\boldsymbol{v} \cdot \nabla\gamma^{\mathrm{s}}\mathbf{n} \cdot \nabla\gamma^{\alpha}\mathrm{d}v \\ - \int_{V_\infty} \boldsymbol{\lambda} \circ F\boldsymbol{v} \cdot \nabla\boldsymbol{\lambda} \cdot \boldsymbol{v}\gamma\gamma^{\mathrm{c}}\boldsymbol{v} \cdot \nabla\gamma^{\mathrm{s}}\mathbf{n} \cdot \nabla\gamma^{\alpha}\mathrm{d}v \\ - \int_{V_\infty} (\boldsymbol{\lambda} \cdot \nabla^{\mathrm{c}}\boldsymbol{\lambda}) \circ F\gamma\gamma^{\mathrm{c}}\mathbf{n} \cdot \nabla\gamma^{\alpha}\boldsymbol{v} \cdot \nabla\gamma^{\mathrm{s}}\mathrm{d}v \\ - \int_{V_\infty} \nabla^{\mathrm{c}}(\gamma\gamma^{\mathrm{c}}) \circ F\mathbf{n} \cdot \nabla\gamma^{\alpha}\boldsymbol{v} \cdot \nabla\gamma^{\mathrm{s}}\mathrm{d}v \\ - \int_{V_\infty} \gamma\gamma^{\mathrm{c}}\nabla^{\mathrm{c}}\nabla\gamma^{\mathrm{s}} \cdot \boldsymbol{v} \circ F\mathbf{n} \cdot \nabla\gamma^{\alpha}\mathrm{d}v \\ - \int_{V_\infty} \gamma\gamma^{\mathrm{c}}\nabla^{\mathrm{c}}\nabla\gamma^{\alpha} \cdot \mathbf{n} \circ F\boldsymbol{v} \cdot \nabla\gamma^{\mathrm{s}}\mathrm{d}v\end{aligned} \tag{798}$$

By Equations (258) and (261), this equation becomes

$$\begin{aligned}
\int_{C_{\alpha\beta\varepsilon}} \nabla^{c} \circ F dc = \nabla^{S} \cdot \int_{V_\infty} & \boldsymbol{\lambda\lambda} \circ F\gamma\gamma^{c}\mathbf{n} \cdot \nabla\gamma^{\alpha}\boldsymbol{v} \cdot \nabla\gamma^{s} ds \\
& - \int_{V_\infty} \boldsymbol{\lambda} \circ F\mathbf{n} \cdot \nabla\boldsymbol{\lambda} \cdot \mathbf{n}\gamma\gamma^{c}\boldsymbol{v} \cdot \nabla\gamma^{s}\mathbf{n} \cdot \nabla\gamma^{\alpha} dv \\
& - \int_{V_\infty} \boldsymbol{\lambda} \circ F\boldsymbol{v} \cdot \nabla\boldsymbol{\lambda} \cdot \boldsymbol{v}\gamma\gamma^{c}\boldsymbol{v} \cdot \nabla\gamma^{s}\mathbf{n} \cdot \nabla\gamma^{\alpha} dv \\
& - \int_{V_\infty} (\boldsymbol{\lambda} \cdot \nabla^{c}\boldsymbol{\lambda}) \circ F\gamma\gamma^{c}\mathbf{n} \cdot \nabla\gamma^{\alpha}\boldsymbol{v} \cdot \nabla\gamma^{s} dv \\
& - \int_{V_\infty} \nabla^{c}(\gamma\gamma^{c}) \circ F\mathbf{n} \cdot \nabla\gamma^{\alpha}\boldsymbol{v} \cdot \nabla\gamma^{s} dv \\
& + \int_{V_\infty} \gamma\gamma^{c}\boldsymbol{\lambda} \circ F\boldsymbol{v} \cdot \nabla\boldsymbol{\lambda} \cdot \nabla\gamma^{s}\mathbf{n} \cdot \nabla\gamma^{\alpha} dv \\
& + \int_{V_\infty} \gamma\gamma^{c}\boldsymbol{\lambda} \circ F\boldsymbol{v} \cdot \nabla\gamma^{s}\mathbf{n} \cdot \nabla\boldsymbol{\lambda} \cdot \nabla\gamma^{\alpha} dv
\end{aligned}$$

By Equations (165)–(167) and (275), we finally obtain

$$\begin{aligned}
\int_{C_{\alpha\beta\varepsilon}} \nabla^{c} \circ F dc = \nabla^{S} \cdot \int_{C_{\alpha\beta\varepsilon}} & \boldsymbol{\lambda\lambda} \circ F dc - \int_{C_{\alpha\beta\varepsilon}} (\boldsymbol{\lambda} \cdot \nabla^{c}\boldsymbol{\lambda}) \circ F dc \\
& + \sum_{\text{pts}} \mathbf{e} \circ F + \sum_{\text{ends}_{\alpha\beta\varepsilon}} \mathbf{e} \circ F
\end{aligned}$$

Remark. Equation (790) contains the following three formulas:

Gradient

$$\int_{C_{\alpha\beta\varepsilon}} \nabla^{c} f dc = -\int_{C_{\alpha\beta\varepsilon}} (\boldsymbol{\lambda} \cdot \nabla^{c}\boldsymbol{\lambda}) f dc + \sum_{\text{pts}} \mathbf{e}f + \sum_{\text{ends}_{\alpha\beta\varepsilon}} \mathbf{e}f + \nabla^{S} \cdot \int_{C_{\alpha\beta\varepsilon}} \boldsymbol{\lambda\lambda} f dc \tag{799}$$

Divergence

$$\begin{aligned}
\int_{C_{\alpha\beta\varepsilon}} \nabla^{c} \cdot \mathbf{f} dc = & -\int_{C_{\alpha\beta\varepsilon}} (\boldsymbol{\lambda} \cdot \nabla^{c}\boldsymbol{\lambda}) \cdot \mathbf{f} dc + \sum_{\text{pts}} \mathbf{e} \cdot \mathbf{f} \\
& + \sum_{\text{end}_{\alpha\beta\varepsilon}} \mathbf{e} \cdot \mathbf{f} + \nabla^{S} \cdot \int_{C_{\alpha\beta\varepsilon}} \boldsymbol{\lambda\lambda} \cdot \mathbf{f} dc
\end{aligned} \tag{800}$$

Curl

$$\begin{aligned}
\int_{C_{\alpha\beta\varepsilon}} \nabla^{c} \times \mathbf{f} dc = & -\int_{C_{\alpha\beta\varepsilon}} (\boldsymbol{\lambda} \cdot \nabla^{c}\boldsymbol{\lambda}) \times \mathbf{f} dc + \sum_{\text{pts}} \mathbf{e} \times \mathbf{f} + \sum_{\text{end}_{\alpha\beta\varepsilon}} \mathbf{e} \times \mathbf{f} \\
& + \nabla^{S} \cdot \int_{C_{\alpha\beta\varepsilon}} \boldsymbol{\lambda\lambda} \times \mathbf{f} dc
\end{aligned} \tag{801}$$

Theorem 57 has the following alternative form:

Theorem 59.

$$\int_{C_{\alpha\beta\varepsilon}} \left.\frac{\partial f}{\partial t}\right|_{1} \mathrm{d}c = \frac{\partial}{\partial t}\int_{C_{\alpha\beta\varepsilon}} f\mathrm{d}c + \int_{C_{\alpha\beta\varepsilon}} \boldsymbol{\lambda}\cdot\nabla^{\mathrm{c}}\boldsymbol{\lambda}\cdot\mathbf{w}^{\mathrm{c}} f\mathrm{d}c - \sum_{\mathrm{pts}} \mathbf{e}\cdot\mathbf{w}^{\mathrm{s}} f - \sum_{\mathrm{ext}} \mathbf{e}\cdot\mathbf{w}^{\mathrm{c}} f \quad (802)$$

where ext refers to the points of intersection of the $C_{\alpha\beta\varepsilon}$ curve with the external boundary of the REV. pts refers to the end points of the $C_{\alpha\beta\varepsilon}$ curve within the REV where the α-, β-, and ε-phases meet a fourth phase.

Proof. By Equation (134), the integral under study may be expressed as

$$\int_{C_{\alpha\beta\varepsilon}} \left.\frac{\partial f}{\partial t}\right|_{1} \mathrm{d}c = \int_{C_{\alpha\beta\varepsilon}} \left.\frac{\partial f}{\partial t}\right|_{\mathbf{x}} + \mathbf{w}^{\mathrm{s}}\cdot\nabla f\mathrm{d}c \quad (803)$$

By the definitions of the indicator functions $\gamma\ (\mathbf{x},\boldsymbol{\xi})$, $\gamma^{\mathrm{s}}(\mathbf{u}(\mathbf{x}+\boldsymbol{\xi}),t)$, $\gamma^{\alpha}(\mathbf{x}+\boldsymbol{\xi},t)$, and $\gamma^{\mathrm{c}}(\mathbf{l}(\mathbf{x}+\boldsymbol{\xi}),t)$, the integral on the right-hand side of Equation (803) may be written as

$$\int_{C_{\alpha\beta\varepsilon}} \left.\frac{\partial f}{\partial t}\right|_{1} \mathrm{d}c = \int_{V_\infty} \left.\frac{\partial f}{\partial t}\right|_{\mathbf{x}} \gamma\gamma^{\mathrm{c}}\mathbf{v}\cdot\nabla\gamma^{\mathrm{s}}\mathbf{n}\cdot\nabla\gamma^{\alpha}\mathrm{d}v + \int_{V_\infty} \mathbf{w}^{\mathrm{s}}\cdot\nabla f\gamma\gamma^{\mathrm{c}}\mathbf{v}\cdot\nabla\gamma^{\mathrm{s}}\mathbf{n}\cdot\nabla\gamma^{\alpha}\mathrm{d}v \quad (804)$$

An application of the chain rule to the integrals on the right side of Equation (804) yields

$$\begin{aligned}
\int_{C_{\alpha\beta\varepsilon}} \left.\frac{\partial f}{\partial t}\right|_{1} \mathrm{d}c = & \int_{V_\infty} \frac{\partial}{\partial t}(f\gamma\gamma^{\mathrm{c}}\mathbf{v}\cdot\nabla\gamma^{\mathrm{s}}\mathbf{n}\cdot\nabla\gamma^{\alpha})\mathrm{d}v \\
& - \int_{V_\infty} f\left(\frac{\partial\gamma}{\partial t}\gamma^{\mathrm{c}} + \gamma\frac{\partial\gamma^{\mathrm{c}}}{\partial t}\right)\mathbf{v}\cdot\nabla\gamma^{\mathrm{s}}\mathbf{n}\cdot\nabla\gamma^{\alpha}\mathrm{d}v \\
& - \int_{V_\infty} f\gamma\gamma^{\mathrm{c}}\left(\frac{\partial\mathbf{v}}{\partial t}\cdot\nabla\gamma^{\mathrm{s}} + \mathbf{v}\cdot\nabla\frac{\partial\gamma^{\mathrm{s}}}{\partial t}\right)\mathbf{n}\cdot\nabla\gamma^{\alpha}\mathrm{d}v \\
& - \int_{V_\infty} f\gamma\gamma^{\mathrm{c}}\mathbf{v}\cdot\nabla\gamma^{\mathrm{s}}\left(\frac{\partial\mathbf{n}}{\partial t}\cdot\nabla\gamma^{\alpha} + \mathbf{n}\cdot\nabla\frac{\partial\gamma^{\alpha}}{\partial t}\right)\mathrm{d}v \\
& + \int_{V_\infty} \nabla\cdot(\mathbf{w}^{\mathrm{s}} f\gamma\gamma^{\mathrm{c}}\mathbf{v}\cdot\nabla\gamma^{\mathrm{s}}\mathbf{n}\cdot\nabla\gamma^{\alpha})\mathrm{d}v \\
& - \int_{V_\infty} \nabla\cdot\mathbf{w}^{\mathrm{s}} f\gamma\gamma^{\mathrm{c}}\mathbf{v}\cdot\nabla\gamma^{\mathrm{s}}\mathbf{n}\cdot\nabla\gamma^{\alpha}\mathrm{d}v \\
& - \int_{V_\infty} f\mathbf{w}^{\mathrm{s}}\cdot\nabla(\gamma\gamma^{\mathrm{c}})\mathbf{v}\cdot\nabla\gamma^{\mathrm{s}}\mathbf{n}\cdot\nabla\gamma^{\alpha}\mathrm{d}v \\
& - \int_{V_\infty} f\gamma\gamma^{\mathrm{c}}\mathbf{w}^{\mathrm{s}}\cdot(\nabla\mathbf{v}\cdot\nabla\gamma^{\mathrm{s}} + \nabla\nabla\gamma^{\mathrm{s}}\cdot\mathbf{v})\mathbf{n}\cdot\nabla\gamma^{\alpha}\mathrm{d}v \\
& - \int_{V_\infty} f\gamma\gamma^{\mathrm{c}}\mathbf{v}\cdot\nabla\gamma^{\mathrm{s}}\mathbf{w}^{\mathrm{s}}\cdot(\nabla\mathbf{n}\cdot\nabla\gamma^{\alpha} + \nabla\nabla\gamma^{\alpha}\cdot\mathbf{n})\mathrm{d}v \quad (805)
\end{aligned}$$

The differentiation $\partial/\partial t$ may be moved out of the integration sign in the first integral on the right side of Equation (805). By Equation (231), the fifth term on the right side of Equation (805) is zero. By Equations (180), (224), and (229),

$$\frac{\partial \gamma}{\partial t} = -\mathbf{w} \cdot \nabla \gamma, \quad \frac{\partial \gamma^{\mathrm{c}}}{\partial t} = -\mathbf{w} \cdot \nabla \gamma^{\mathrm{c}}$$
$$\frac{\partial \gamma^{\mathrm{s}}}{\partial t} = -\mathbf{w} \cdot \nabla \gamma^{\mathrm{s}}, \quad \frac{\partial \gamma^{\alpha}}{\partial t} = -\mathbf{w} \cdot \nabla \gamma^{\alpha}$$

With these manipulations and Equations (195) and (200) applied to the integrals on the right side of Equation (805), we obtain

$$\begin{aligned}
&\int_{C_{\alpha\beta\varepsilon}} \left.\frac{\partial f}{\partial t}\right|_1 \mathrm{d}c \\
&= \frac{\partial}{\partial t}\int_{V_\infty} f\gamma\gamma^{\mathrm{c}}\boldsymbol{\nu} \cdot \nabla\gamma^{\mathrm{s}}\mathbf{n} \cdot \nabla\gamma^{\alpha}\mathrm{d}v + \int_{V_\infty} f\mathbf{w} \cdot \nabla(\gamma\gamma^{\mathrm{c}})\boldsymbol{\nu} \cdot \nabla\gamma^{\mathrm{s}}\mathbf{n} \cdot \nabla\gamma^{\alpha}\mathrm{d}v \\
&+ \int_{V_\infty} f\gamma\gamma^{\mathrm{c}}\boldsymbol{\nu} \cdot \nabla\mathbf{w} \cdot \nabla\gamma^{\mathrm{s}}\mathbf{n} \cdot \nabla\gamma^{\alpha}\mathrm{d}v + \int_{V_\infty} f\gamma\gamma^{\mathrm{c}}\boldsymbol{\nu} \cdot \nabla\nabla\gamma^{\mathrm{s}} \cdot \mathbf{w}\mathbf{n} \cdot \nabla\gamma^{\alpha}\mathrm{d}v \\
&+ \int_{V_\infty} f\gamma\gamma^{\mathrm{c}}\boldsymbol{\nu} \cdot \nabla\gamma^{\mathrm{s}}\mathbf{n} \cdot \nabla\mathbf{w} \cdot \nabla\gamma^{\alpha}\mathrm{d}v + \int_{V_\infty} f\gamma\gamma^{\mathrm{c}}\boldsymbol{\nu} \cdot \nabla\gamma^{\mathrm{s}}\mathbf{n} \cdot \nabla\nabla\gamma^{\alpha} \cdot \mathbf{w}\mathrm{d}v \\
&- \int_{V_\infty} \nabla \cdot \mathbf{w}^{\mathrm{s}} f\gamma\gamma^{\mathrm{c}}\boldsymbol{\nu} \cdot \nabla\gamma^{\mathrm{s}}\mathbf{n} \cdot \nabla\gamma^{\alpha}\mathrm{d}v - \int_{V_\infty} f\mathbf{w}^{\mathrm{s}} \cdot \nabla(\gamma\gamma^{\mathrm{c}})\boldsymbol{\nu} \cdot \nabla\gamma^{\mathrm{s}}\mathbf{n} \cdot \nabla\gamma^{\alpha}\mathrm{d}v \\
&- \int_{V_\infty} f\gamma\gamma^{\mathrm{c}}\boldsymbol{\nu} \cdot \nabla\gamma^{\mathrm{s}}\mathbf{w}^{\mathrm{s}} \cdot \nabla\nabla\gamma^{\alpha} \cdot \mathbf{n}\mathrm{d}v - \int_{V_\infty} f\gamma\gamma^{\mathrm{c}}\mathbf{w}^{\mathrm{s}} \cdot \nabla\nabla\gamma^{\mathrm{s}} \cdot \boldsymbol{\nu}\mathbf{n} \cdot \nabla\gamma^{\alpha}\mathrm{d}v
\end{aligned} \tag{806}$$

Regrouping the right side terms of Equation (806) yields

$$\begin{aligned}
&\int_{C_{\alpha\beta\varepsilon}} \left.\frac{\partial f}{\partial t}\right|_1 \mathrm{d}c \\
&= \frac{\partial}{\partial t}\int_{V_\infty} f\gamma\gamma^{\mathrm{c}}\boldsymbol{\nu} \cdot \nabla\gamma^{\mathrm{s}}\mathbf{n} \cdot \nabla\gamma^{\alpha}\mathrm{d}v + \int_{V_\infty} f\gamma\gamma^{\mathrm{c}}\boldsymbol{\nu} \cdot \nabla\mathbf{w} \cdot \nabla\gamma^{\mathrm{s}}\mathbf{n} \cdot \nabla\gamma^{\alpha}\mathrm{d}v \\
&+ \int_{V_\infty} f\gamma\gamma^{\mathrm{c}}\boldsymbol{\nu} \cdot \nabla\gamma^{\mathrm{s}}\mathbf{n} \cdot \nabla\mathbf{w} \cdot \nabla\gamma^{\alpha}\mathrm{d}v + \int_{V_\infty} f\gamma\gamma^{\mathrm{c}}\mathbf{w} \cdot \nabla^{\mathrm{c}}\nabla\gamma^{\mathrm{s}} \cdot \boldsymbol{\nu}\mathbf{n} \cdot \nabla\gamma^{\alpha}\mathrm{d}v \\
&+ \int_{V_\infty} f\gamma\gamma^{\mathrm{c}}\boldsymbol{\nu} \cdot \nabla\gamma^{\mathrm{s}}\mathbf{w} \cdot \nabla^{\mathrm{c}}\nabla\gamma^{\alpha} \cdot \mathbf{n}\mathrm{d}v \\
&- \int_{V_\infty} \nabla \cdot (\mathbf{w} - \boldsymbol{\lambda}\boldsymbol{\lambda} \cdot \mathbf{w}) f\gamma\gamma^{\mathrm{c}}\boldsymbol{\nu} \cdot \nabla\gamma^{\mathrm{s}}\mathbf{n} \cdot \nabla\gamma^{\alpha}\mathrm{d}v \\
&+ \int_{V_\infty} f\mathbf{w} \cdot \nabla^{\mathrm{c}}(\gamma\gamma^{\mathrm{c}})\boldsymbol{\nu} \cdot \nabla\gamma^{\mathrm{s}}\mathbf{n} \cdot \nabla\gamma^{\alpha}\mathrm{d}v
\end{aligned} \tag{807}$$

By Equation (53),

$$\nabla \cdot \mathbf{w} = \boldsymbol{\lambda} \cdot \nabla\mathbf{w} \cdot \boldsymbol{\lambda} + \boldsymbol{\nu} \cdot \nabla\mathbf{w} \cdot \boldsymbol{\nu} + \mathbf{n} \cdot \nabla\mathbf{w} \cdot \mathbf{n}$$
$$\nabla \cdot (\boldsymbol{\lambda}\boldsymbol{\lambda} \cdot \mathbf{w}) = \boldsymbol{\nu} \cdot \nabla\boldsymbol{\lambda} \cdot \boldsymbol{\nu}\boldsymbol{\lambda} \cdot \mathbf{w} + \mathbf{n} \cdot \nabla\boldsymbol{\lambda} \cdot \mathbf{n}\boldsymbol{\lambda} \cdot \mathbf{w} + \boldsymbol{\lambda} \cdot \nabla^{\mathrm{c}}\boldsymbol{\lambda} \cdot \mathbf{w}^{\mathrm{s}} + \boldsymbol{\lambda} \cdot \nabla\mathbf{w} \cdot \boldsymbol{\lambda}$$

With these equations and Equations (258) and (261), we obtain

$$
\begin{aligned}
&\int_{C_{\alpha\beta\varepsilon}} \frac{\partial f}{\partial t}\bigg|_1 \mathrm{d}c \\
&= \frac{\partial}{\partial t}\int_{V_\infty} f\gamma\gamma^{\mathrm{c}}\boldsymbol{\nu}\cdot\nabla\gamma^{\mathrm{s}}\mathbf{n}\cdot\nabla\gamma^{\alpha}\mathrm{d}v + \int_{V_\infty} f\gamma\gamma^{\mathrm{c}}\boldsymbol{\nu}\cdot\nabla\mathbf{w}\cdot\nabla\gamma^{\mathrm{s}}\mathbf{n}\cdot\nabla\gamma^{\alpha}\mathrm{d}v \\
&\quad + \int_{V_\infty} f\gamma\gamma^{\mathrm{c}}\boldsymbol{\nu}\cdot\nabla\gamma^{\mathrm{s}}\mathbf{n}\cdot\nabla\mathbf{w}\cdot\nabla\gamma^{\alpha}\mathrm{d}v - \int_{V_\infty} f\gamma\gamma^{\mathrm{c}}\mathbf{w}\cdot\boldsymbol{\lambda}\boldsymbol{\nu}\cdot\nabla\boldsymbol{\lambda}\cdot\nabla\gamma^{\mathrm{s}}\mathbf{n}\cdot\nabla\gamma^{\alpha}\mathrm{d}v \\
&\quad - \int_{V_\infty} f\gamma\gamma^{\mathrm{c}}\mathbf{w}\cdot\boldsymbol{\lambda}\mathbf{n}\cdot\nabla\boldsymbol{\lambda}\cdot\nabla\gamma^{\alpha}\boldsymbol{\nu}\cdot\nabla\gamma^{\mathrm{s}}\mathrm{d}v \\
&\quad - \int_{V_\infty} \boldsymbol{\lambda}\cdot\nabla\mathbf{w}\cdot\boldsymbol{\lambda}f\gamma\gamma^{\mathrm{c}}\boldsymbol{\nu}\cdot\nabla\gamma^{\mathrm{c}}\mathbf{n}\cdot\nabla\gamma^{\alpha}\mathrm{d}v \\
&\quad - \int_{V_\infty} \mathbf{n}\cdot\nabla\mathbf{w}\cdot\mathbf{n}f\gamma\gamma^{\mathrm{c}}\boldsymbol{\nu}\cdot\nabla\gamma^{\mathrm{s}}\mathbf{n}\cdot\nabla\gamma^{\alpha}\mathrm{d}v \\
&\quad - \int_{V_\infty} \boldsymbol{\nu}\cdot\nabla\mathbf{w}\cdot\boldsymbol{\nu}f\gamma\gamma^{\mathrm{c}}\boldsymbol{\nu}\cdot\nabla\gamma^{\mathrm{s}}\mathbf{n}\cdot\gamma^{\alpha}\mathrm{d}v \\
&\quad + \int_{V_\infty} \boldsymbol{\nu}\cdot\nabla\boldsymbol{\lambda}\cdot\boldsymbol{\nu}\boldsymbol{\lambda}\cdot\nabla\gamma^{\mathrm{s}}\mathbf{n}\cdot\nabla\gamma^{\alpha}f\gamma\gamma^{\mathrm{c}}\boldsymbol{\lambda}\cdot\mathbf{w}\mathrm{d}v \\
&\quad + \int_{V_\infty} \mathbf{n}\cdot\nabla\boldsymbol{\lambda}\cdot\mathbf{n}\boldsymbol{\lambda}\cdot\mathbf{w}f\gamma\gamma^{\mathrm{c}}\boldsymbol{\nu}\cdot\nabla\gamma^{\mathrm{s}}\mathbf{n}\cdot\nabla\gamma^{\alpha}\mathrm{d}v \\
&\quad + \int_{V_\infty} \boldsymbol{\lambda}\cdot\nabla^{\mathrm{c}}\boldsymbol{\lambda}\cdot\mathbf{w}^{\mathrm{s}}f\gamma\gamma^{\mathrm{c}}\boldsymbol{\nu}\cdot\nabla\gamma^{\mathrm{s}}\mathbf{n}\cdot\nabla\gamma^{\alpha}\mathrm{d}v \\
&\quad + \int_{V_\infty} \boldsymbol{\lambda}\cdot\nabla\mathbf{w}\cdot\boldsymbol{\lambda}f\gamma\gamma^{\mathrm{c}}\boldsymbol{\nu}\cdot\nabla\gamma^{\mathrm{s}}\mathbf{n}\cdot\nabla\gamma^{\alpha}\mathrm{d}v \\
&\quad + \int_{V_\infty} f\mathbf{w}\cdot\nabla^{\mathrm{c}}(\gamma\gamma^{\mathrm{c}})\boldsymbol{\nu}\cdot\nabla\gamma^{\mathrm{s}}\mathbf{n}\cdot\nabla\gamma^{\alpha}\mathrm{d}v
\end{aligned} \tag{808}
$$

Note also that the second, third, fourth, and fifth terms cancel the eighth, seventh, ninth, and tenth terms, respectively. Therefore, Equation (808) becomes

$$
\begin{aligned}
\int_{C_{\alpha\beta\varepsilon}} \frac{\partial f}{\partial t}\bigg|_1 \mathrm{d}c &= \frac{\partial}{\partial t}\int_{V_\infty} f\gamma\gamma^{\mathrm{c}}\boldsymbol{\nu}\cdot\nabla\gamma^{\mathrm{s}}\mathbf{n}\cdot\nabla\gamma^{\alpha}\mathrm{d}v \\
&\quad + \int_{V_\infty} \boldsymbol{\lambda}\cdot\nabla^{\mathrm{c}}\boldsymbol{\lambda}\cdot\mathbf{w}^{\mathrm{s}}f\gamma\gamma^{\mathrm{c}}\boldsymbol{\nu}\cdot\nabla\gamma^{\mathrm{s}}\mathbf{n}\cdot\nabla\gamma^{\alpha}\mathrm{d}v \\
&\quad + \int_{V_\infty} f\mathbf{w}\cdot\nabla^{\mathrm{c}}(\gamma\gamma^{\mathrm{c}})\boldsymbol{\nu}\cdot\nabla\gamma^{\mathrm{s}}\mathbf{n}\cdot\nabla\gamma^{\alpha}\mathrm{d}v
\end{aligned} \tag{809}
$$

By Equations (275) and (166), we finally obtain

$$
\begin{aligned}
\int_{C_{\alpha\beta\varepsilon}} \frac{\partial f}{\partial t}\bigg|_1 \mathrm{d}c &= \frac{\partial}{\partial t}\int_{C_{\alpha\beta\varepsilon}} f\mathrm{d}c + \int_{C_{\alpha\beta\varepsilon}} \boldsymbol{\lambda}\cdot\nabla^{\mathrm{c}}\boldsymbol{\lambda}\cdot\mathbf{w}^{\mathrm{s}}f\mathrm{d}c \\
&\quad - \sum_{\mathrm{pts}} \mathbf{e}\cdot\mathbf{w}^{\mathrm{c}}f - \sum_{\mathrm{ext}} \mathbf{e}\cdot\mathbf{w}^{\mathrm{c}}f
\end{aligned}
$$

Theorem 58 has another alternative form:

Theorem 60.

$$\int_{C_{\alpha\beta\varepsilon}} \nabla^{c} \circ F dc = -\int_{C_{\alpha\beta\varepsilon}} (\boldsymbol{\lambda} \cdot \nabla^{c}\boldsymbol{\lambda}) \circ F dc + \sum_{\text{pts}} \mathbf{e} \circ F + \sum_{\text{ext}} \mathbf{e} \circ F \tag{810}$$

Proof. By the indicator functions $\gamma(\mathbf{x}, \xi)$, $\gamma^{s}(\mathbf{u}(\mathbf{x}+\xi), t)$, $\gamma^{\alpha}(\mathbf{x}+\xi, t)$, and $\gamma^{c}(\mathbf{l}(\mathbf{x}+\xi), t)$, the integral under study can be written as

$$\int_{C_{\alpha\beta\varepsilon}} \nabla^{c} \circ F dc = \int_{V_{\infty}} \nabla^{c} \circ F \boldsymbol{v} \cdot \nabla\gamma^{s}\mathbf{n} \cdot \nabla\gamma^{\alpha}\gamma\gamma^{c} dv \tag{811}$$

An application of the chain rule to the right side of Equation (811) yields

$$\begin{aligned}\int_{C_{\alpha\beta\varepsilon}} \nabla^{c} \circ F dc = & \int_{V_{\infty}} \nabla \cdot (\boldsymbol{\lambda\lambda} \circ F \boldsymbol{v} \cdot \nabla\gamma^{s}\mathbf{n} \cdot \nabla\gamma^{\alpha}\gamma\gamma^{c}) dv \\ & - \int_{V_{\infty}} \nabla \cdot \boldsymbol{\lambda\lambda} \circ F \boldsymbol{v} \cdot \nabla\gamma^{s}\mathbf{n} \cdot \nabla\gamma^{\alpha}\gamma\gamma^{c} dv \\ & - \int_{V_{\infty}} \boldsymbol{\lambda} \cdot \nabla\boldsymbol{\lambda} \circ F \boldsymbol{v} \cdot \nabla\gamma^{s}\mathbf{n} \cdot \nabla\gamma^{\alpha}\gamma\gamma^{c} dv \\ & - \int_{V_{\infty}} \boldsymbol{\lambda} \circ F \boldsymbol{\lambda} \cdot (\nabla\boldsymbol{v} \cdot \nabla\gamma^{s} + \nabla\nabla\gamma^{s} \cdot \boldsymbol{v})\mathbf{n} \cdot \nabla\gamma^{\alpha}\gamma\gamma^{c} dv \\ & - \int_{V_{\infty}} \boldsymbol{\lambda} \circ F \boldsymbol{v} \cdot \nabla\gamma^{s}\boldsymbol{\lambda} \cdot (\nabla\mathbf{n} \cdot \nabla\gamma^{\alpha} + \nabla\nabla\gamma^{\alpha} \cdot \mathbf{n})\gamma\gamma^{c} dv \\ & - \int_{V_{\infty}} \boldsymbol{\lambda} \circ F \boldsymbol{v} \cdot \nabla\gamma^{s}\mathbf{n} \cdot \nabla\gamma^{\alpha}\boldsymbol{\lambda} \cdot \nabla(\gamma\gamma^{c}) dv\end{aligned} \tag{812}$$

By Equation (231), the first integral on the right side of Equation (812) is zero. By Equations (21) and (53),

$$\begin{aligned}\nabla \cdot \boldsymbol{\lambda} &= \boldsymbol{\lambda} \cdot \nabla\boldsymbol{\lambda} \cdot \boldsymbol{\lambda} + \mathbf{n} \cdot \nabla\boldsymbol{\lambda} \cdot \mathbf{n} + \boldsymbol{v} \cdot \nabla\boldsymbol{\lambda} \cdot \boldsymbol{v} \\ &= \mathbf{n} \cdot \nabla\boldsymbol{\lambda} \cdot \mathbf{n} + \boldsymbol{v} \cdot \nabla\boldsymbol{\lambda} \cdot \boldsymbol{v}\end{aligned} \tag{813}$$

Equation (813) allows us to rewrite the second integral on the right side of Equation (812) as

$$\begin{aligned}& \int_{V_{\infty}} \nabla \cdot \boldsymbol{\lambda\lambda} \circ F \boldsymbol{v} \cdot \nabla\gamma^{s}\mathbf{n} \cdot \nabla\gamma^{\alpha}\gamma\gamma^{c} dc \\ & = \int_{V_{\infty}} \mathbf{n} \cdot \nabla\boldsymbol{\lambda} \cdot \mathbf{n}\boldsymbol{\lambda} \circ F \boldsymbol{v} \cdot \nabla\gamma^{s}\mathbf{n} \cdot \nabla\gamma^{\alpha}\gamma\gamma^{c} dv \\ & \quad + \int_{V_{\infty}} \boldsymbol{v} \cdot \nabla\boldsymbol{\lambda} \cdot \boldsymbol{v}\boldsymbol{\lambda} \circ F \boldsymbol{v} \cdot \nabla\gamma^{s}\mathbf{n} \cdot \nabla\gamma^{\alpha}\gamma\gamma^{c} dv\end{aligned} \tag{814}$$

Since $\boldsymbol{\lambda} \cdot \nabla = \boldsymbol{\lambda} \cdot \nabla^{c}$, the third integral on the right side of Equation (812) becomes

$$\int_{V_{\infty}} \boldsymbol{\lambda} \cdot \nabla\boldsymbol{\lambda} \circ F \boldsymbol{v} \cdot \nabla\gamma^{s}\mathbf{n} \cdot \nabla\gamma^{\alpha}\gamma\gamma^{c} dv = \int_{V_{\infty}} \boldsymbol{\lambda} \cdot \nabla^{c}\boldsymbol{\lambda} \circ F \boldsymbol{v} \cdot \nabla\gamma^{s}\mathbf{n} \cdot \nabla\gamma^{\alpha}\gamma\gamma^{c} dv \tag{815}$$

With Equations (200) and (261) applied to the fourth integral on the right side of Equation (812), we obtain

$$\int_{V_\infty} \boldsymbol{\lambda} \circ F\boldsymbol{\lambda} \cdot (\nabla \boldsymbol{v} \cdot \nabla \gamma^{\mathrm{s}} + \nabla\nabla\gamma^{\mathrm{s}} \cdot \boldsymbol{v})\mathbf{n} \cdot \nabla\gamma^{\alpha}\gamma\gamma^{\mathrm{c}} \mathrm{d}v$$
$$= -\int_{V_\infty} \boldsymbol{\lambda} \circ F\boldsymbol{v} \cdot \nabla\boldsymbol{\lambda} \cdot \nabla\gamma^{\mathrm{s}}\mathbf{n} \cdot \nabla\gamma^{\alpha}\gamma\gamma^{\mathrm{c}} \mathrm{d}v \tag{816}$$

Similarly, we have the fifth term on the right side of Equation (812),

$$\int_{V_\infty} \boldsymbol{\lambda} \circ F\boldsymbol{\lambda} \cdot (\nabla \mathbf{n} \cdot \nabla \gamma^{\alpha} + \nabla\nabla\gamma^{\alpha} \cdot \mathbf{n})\boldsymbol{v} \cdot \nabla\gamma^{\mathrm{s}}\gamma\gamma^{\mathrm{c}} \mathrm{d}c$$
$$= -\int_{V_\infty} \boldsymbol{\lambda} \circ F\mathbf{n} \cdot \nabla\boldsymbol{\lambda} \cdot \nabla\gamma^{\alpha}\boldsymbol{v} \cdot \nabla\gamma^{\mathrm{s}}\gamma\gamma^{\mathrm{s}} \mathrm{d}v \tag{817}$$

The last term on the right-hand side of Equation (812) may be rewritten as

$$\int_{V_\infty} \boldsymbol{\lambda} \circ F\boldsymbol{v} \cdot \nabla\gamma^{\mathrm{s}}\mathbf{n} \cdot \nabla\gamma^{\alpha}\boldsymbol{\lambda} \cdot \nabla(\gamma\gamma^{\mathrm{c}})\mathrm{d}v = \int_{V_\infty} \nabla^{\mathrm{c}}(\gamma\gamma^{\mathrm{c}}) \circ F\boldsymbol{v} \cdot \nabla\gamma^{\mathrm{s}}\mathbf{n} \cdot \nabla\gamma^{\alpha}\mathrm{d}v \tag{818}$$

By Equations (165)–(167) and (275),

$$\int_{V_\infty} \boldsymbol{\lambda} \circ F\boldsymbol{v} \cdot \nabla\gamma^{\mathrm{s}}\mathbf{n} \cdot \nabla\gamma^{\alpha}\boldsymbol{\lambda} \cdot \nabla(\gamma\gamma^{\mathrm{c}})\mathrm{d}v = -\sum_{\mathrm{pts}} \mathbf{e} \circ F - \sum_{\mathrm{ext}} \mathbf{e} \circ F \tag{819}$$

Substituting Equations (814)–(819) into Equation (812) yields

$$\begin{aligned}\int_{C_{\alpha\beta\varepsilon}} \nabla^{\mathrm{c}} \circ F\mathrm{d}c = &-\int_{V_\infty} \mathbf{n} \cdot \nabla\boldsymbol{\lambda} \cdot \mathbf{n}\boldsymbol{\lambda} \circ F\boldsymbol{v} \cdot \nabla\gamma^{\mathrm{s}}\mathbf{n} \cdot \nabla\gamma^{\alpha}\gamma\gamma^{\mathrm{c}}\mathrm{d}v \\ &-\int_{V_\infty} \boldsymbol{v} \cdot \nabla\boldsymbol{\lambda} \cdot \boldsymbol{v}\boldsymbol{\lambda} \circ F\boldsymbol{v} \cdot \nabla\gamma^{\mathrm{s}}\mathbf{n} \cdot \nabla\gamma^{\alpha}\gamma\gamma^{\mathrm{c}}\mathrm{d}v \\ &-\int_{V_\infty} \boldsymbol{\lambda} \cdot \nabla^{\mathrm{c}}\boldsymbol{\lambda} \circ F\boldsymbol{v} \cdot \nabla\gamma^{\mathrm{s}}\mathbf{n} \cdot \nabla\gamma^{\alpha}\gamma\gamma^{\mathrm{c}}\mathrm{d}v \\ &+\int_{V_\infty} \boldsymbol{\lambda} \circ F\boldsymbol{v} \cdot \nabla\boldsymbol{\lambda} \cdot \nabla\gamma^{\mathrm{s}}\mathbf{n} \cdot \nabla\gamma^{\alpha}\gamma\gamma^{\mathrm{c}}\mathrm{d}v \\ &+\int_{V_\infty} \boldsymbol{\lambda} \circ F\mathbf{n} \cdot \nabla\boldsymbol{\lambda} \cdot \nabla\gamma^{\alpha}\boldsymbol{v} \cdot \nabla\gamma^{\mathrm{s}}\gamma\gamma^{\mathrm{c}}\mathrm{d}v \\ &+\sum_{\mathrm{pts}} \mathbf{e} \circ F - \sum_{\mathrm{ext}} \mathbf{e} \circ F\end{aligned} \tag{820}$$

By Equation (275),

$$\begin{aligned}\int_{C_{\alpha\beta\varepsilon}} \nabla^{\mathrm{c}} \circ F = &-\int_{C_{\alpha\beta\varepsilon}} \mathbf{n} \cdot \nabla\boldsymbol{\lambda} \cdot \mathbf{n}\boldsymbol{\lambda} \circ F\mathrm{d}c - \int_{C_{\alpha\beta\varepsilon}} \boldsymbol{v} \cdot \nabla\boldsymbol{\lambda} \cdot \boldsymbol{v}\boldsymbol{\lambda} \circ F\mathrm{d}c \\ &-\int_{C_{\alpha\beta\varepsilon}} \boldsymbol{\lambda} \cdot \nabla^{\mathrm{c}}\boldsymbol{\lambda} \circ F\mathrm{d}c + \int_{C_{\alpha\beta\varepsilon}} \boldsymbol{v} \cdot \nabla\boldsymbol{\lambda} \cdot \boldsymbol{v}\boldsymbol{\lambda} \circ F\mathrm{d}c \\ &+\int_{C_{\alpha\beta\varepsilon}} \boldsymbol{\lambda} \circ F\mathbf{n} \cdot \nabla\boldsymbol{\lambda} \cdot \mathbf{n}\mathrm{d}c + \sum_{\mathrm{pts}} \mathbf{e} \circ F + \sum_{\mathrm{ext}} \mathbf{e} \circ F\end{aligned} \tag{821}$$

Finally, we obtain

$$\int_{C_{\alpha\beta\varepsilon}} \nabla^{c} \circ F = -\int_{C_{\alpha\beta\varepsilon}} (\boldsymbol{\lambda} \cdot \nabla^{c}\boldsymbol{\lambda}) \circ F dc + \sum_{\text{pts}} \mathbf{e} \circ F + \sum_{\text{ext}} \mathbf{e} \circ F$$

Remark. Equation (810) contains the following three formulas:

Gradient

$$\int_{C_{\alpha\beta\varepsilon}} \nabla^{c} f dc = -\int_{C_{\alpha\beta\varepsilon}} \boldsymbol{\lambda} \cdot \nabla^{c}\boldsymbol{\lambda} f dc + \sum_{\text{pts}} \mathbf{e}f + \sum_{\text{ext}} \mathbf{e}f \tag{822}$$

Divergence

$$\int_{C_{\alpha\beta\varepsilon}} \nabla^{c} \cdot \mathbf{f} dc = -\int_{C_{\alpha\beta\varepsilon}} \boldsymbol{\lambda} \cdot \nabla^{c}\boldsymbol{\lambda} \cdot \mathbf{f} dc + \sum_{\text{pts}} \mathbf{e} \cdot \mathbf{f} + \sum_{\text{ext}} \mathbf{e} \cdot \mathbf{f} \tag{823}$$

Curl

$$\int_{C_{\alpha\beta\varepsilon}} \nabla^{c} \times \mathbf{f} dc = -\int_{C_{\alpha\beta\varepsilon}} (\boldsymbol{\lambda} \cdot \nabla^{c}\boldsymbol{\lambda}) \times \mathbf{f} dc + \sum_{\text{pts}} \mathbf{e} \times \mathbf{f} + \sum_{\text{ext}} \mathbf{e} \times \mathbf{f} \tag{824}$$

5.3 One macroscopic dimension

In this section we develop 11 averaging theorems regarding spatial integration of spatial operators, surficial integration of surficial operators, and curvilineal integration of curvilineal operators within an REV of one macroscopic dimension and two megascopic dimensions. The appropriate REV is a slab of finite thickness, D, with parallel faces (Figures 31–33). The faces of the REV intersect the megascopic boundary of the system of interest along their edges. The macroscopic direction is perpendicular to the faces of the slab. The two megascopic directions are tangent to the slab faces. The thickness of the slab is constant in time and space but the orientation of this REV may vary with position as long as $|\mathbf{N} \cdot \nabla_{\mathbf{x}}\mathbf{N}|D \leq 1$. The volume of the effective REV may not be constant due to changes in the intersection of the slab with the physical megascopic boundaries of a system that does not have a constant cross-section or whose cross-section changes with time. The theorems presented in this section will contain macroscopic spatial derivatives in the direction perpendicular to the slab face and flux terms along the edges of the slab.

5.3.1 Spatial operator theorems

The REV used for this group of averaging theorems is shown in Figure 31.

Theorem 61.

$$\int_{V_\alpha} \left.\frac{\partial f}{\partial t}\right|_{\mathbf{x}} dv = \frac{\partial}{\partial t}\int_{V_\alpha} f dv - \int_{S_{\alpha\beta}} \mathbf{n} \cdot \mathbf{w}^{c} f ds - \int_{S_{\alpha\text{edges}}} \mathbf{n}^{*} \cdot \mathbf{w} f ds \tag{825}$$

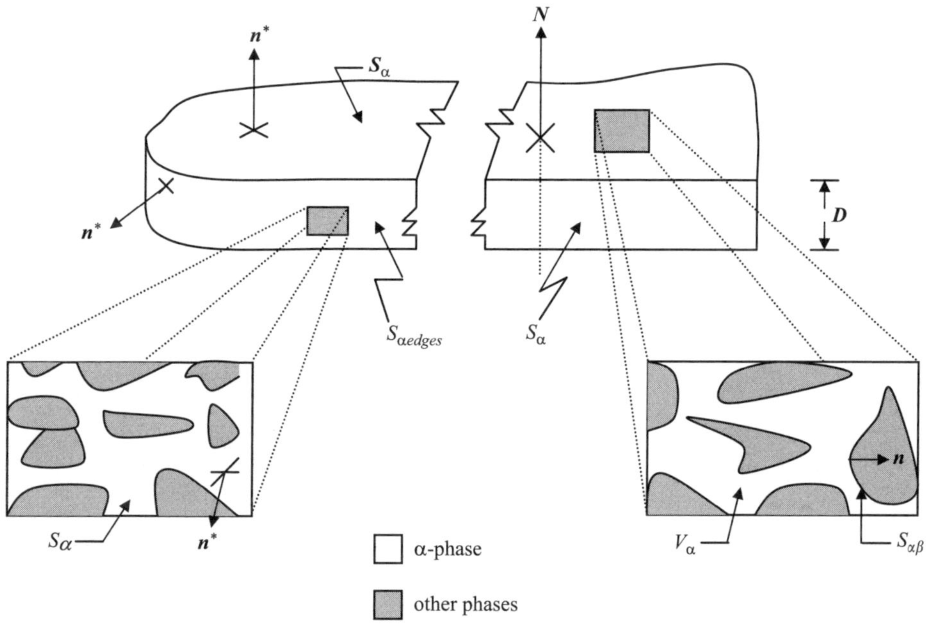

Figure 31 Slab REV (constant thickness but extent and orientation dependent on space and time; wedge removed for illustrative purposes).

where V_α is the portion of the REV occupied by the α-phase. $S_{\alpha\beta}$ is the surface within the REV between the α-phase and all other phases. $S_{\alpha\text{edges}}$ is the portion of the edge boundaries of the REV that is occupied by the α-phase. $\mathbf{n}^*$ is the unit vector normal to the external boundary of the REV that is pointing outward from the REV. $\mathbf{n}$ is the unit vector normal to the $S_{\alpha\beta}$ surface that is pointing outward from the α-phase. $\mathbf{N}$ is the unit vector normal to the face of the slab REV. ∇^C is the one-dimensional macroscopic spatial operator in the direction normal to the face of the slab REV such that $\nabla^C = \mathbf{NN} \cdot \nabla_\mathbf{x}$. $\mathbf{w}$ is the velocity of either $S_{\alpha\beta}$ or $S_{\alpha\text{edges}}$, $\mathbf{w}^c$ the velocity of $S_{\alpha\beta}$ normal to $S_{\alpha\beta}$ such that $\mathbf{w}^c = \mathbf{nn} \cdot \mathbf{w}$. f is a spatial function $f(\mathbf{x}, t)$ defined in the α-phase and ∇ the microscopic spatial operator $\nabla = \nabla_\xi$.

Proof. The integral over the α-phase in the REV of the time derivative of a spatial function f can be written in terms of the integral over the REV by making use of the indicator function $\gamma^\alpha(\mathbf{x} + \xi, t)$ as

$$\int_{V_\alpha} \frac{\partial f}{\partial t}\bigg|_\mathbf{x} \mathrm{d}v = \int_V \frac{\partial f}{\partial t} \gamma^\alpha \mathrm{d}v \tag{826}$$

The integral on the right side of this equation can be further changed to an integral over all space by using the indicator function $\gamma(\mathbf{x}+\xi)$ to obtain

$$\int_{V_\alpha} \frac{\partial f}{\partial t}\bigg|_\mathbf{x} \mathrm{d}v = \int_{V_\infty} \frac{\partial f}{\partial t}\bigg|_\mathbf{x} \gamma^\alpha \gamma \mathrm{d}v \tag{827}$$

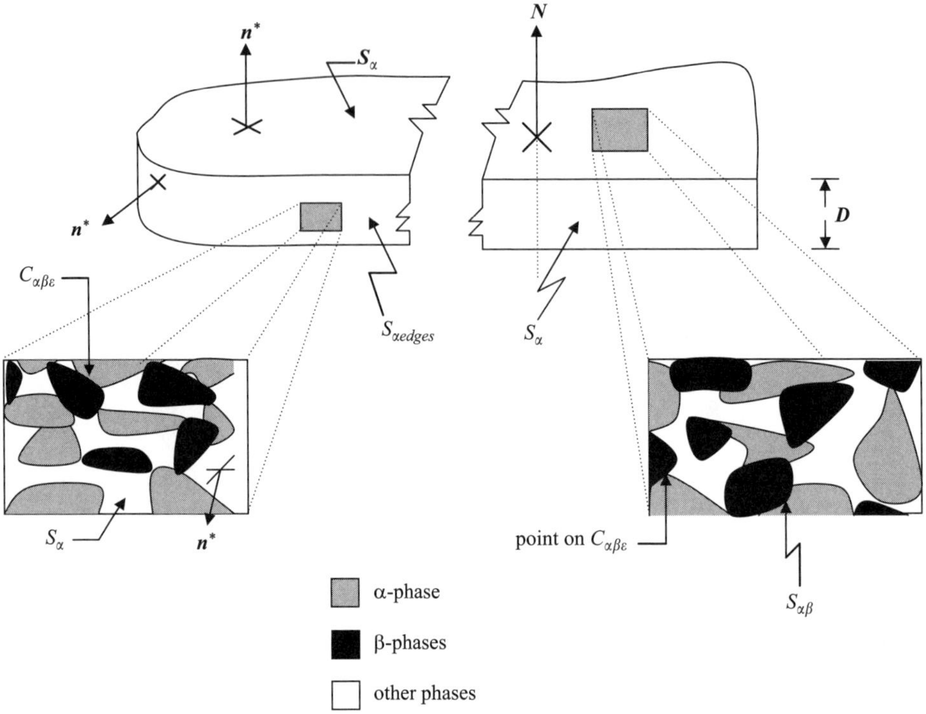

Figure 32 Slab REV (constant thickness but extent and orientation dependent on space and time; wedge removed for illustrative purposes).

The time derivative is the partial derivative of f evaluated at a fixed position in space. An application of the chain rule to the right-hand side of Equation (827) gives:

$$\int_{V_\alpha} \frac{\partial f}{\partial t}\bigg|_{\mathbf{x}} \mathrm{d}v = \int_{V_\infty} \frac{\partial (f\gamma^\alpha \gamma)}{\partial t}\bigg|_{\mathbf{x}} \mathrm{d}v - \int_{V_\infty} f \frac{\partial \gamma^\alpha}{\partial t}\bigg|_{\mathbf{x}} \gamma \mathrm{d}v - \int_{V_\infty} f\gamma^\alpha \frac{\partial \gamma}{\partial t}\bigg|_{\mathbf{x}} \mathrm{d}v \tag{828}$$

Because the boundaries of V_∞ are independent of time, differentiation and integration can be interchanged in the first term on the right side of Equation (828). Then the definitions of the indicator functions may be used to convert the derivative of an integral over V_∞ to the derivative of an integral over V_α:

$$\int_{V_\infty} \frac{\partial (f\gamma\gamma^\alpha)}{\partial t}\bigg|_{\mathbf{x}} \mathrm{d}v = \frac{\partial}{\partial t}\int_{V_\infty} f\gamma^\alpha \gamma \mathrm{d}v = \frac{\partial}{\partial t}\int_{V_\infty} f\gamma^\alpha \mathrm{d}v = \frac{\partial}{\partial t}\int_{V_\alpha} f \mathrm{d}v \tag{829}$$

Note that the time derivative on the right side of this equation is evaluated by holding the position of the REV fixed. Substitution of Equation (829) into Equation (826) yields:

$$\int_{V_\alpha} \frac{\partial f}{\partial t}\bigg|_{\mathbf{x}} \mathrm{d}v = \frac{\partial}{\partial t}\int_{V_\alpha} f \mathrm{d}v - \int_{V_\infty} f \frac{\partial \gamma^\alpha}{\partial t}\bigg|_{\mathbf{x}} \gamma \mathrm{d}v - \int_{V_\infty} f\gamma^\alpha \frac{\partial \gamma}{\partial t}\bigg|_{\mathbf{x}} \mathrm{d}v \tag{830}$$

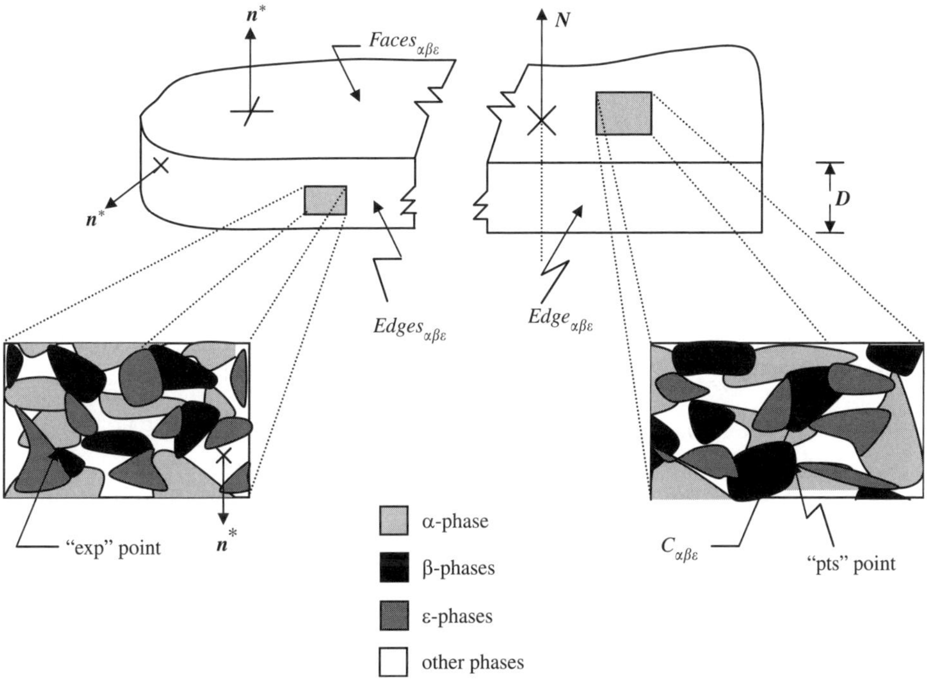

Figure 33 Slab REV (constant thickness but extent and orientation dependent on space and time; wedge removed for illustrative purposes).

Equation (180) indicates that $(\partial\gamma/\partial t)|_{\mathbf{x}} = -\mathbf{w}\cdot\nabla\gamma$, where $\mathbf{w}$ is the velocity of the boundary of the REV. Similarly, the indicator function identifying the α-phase satisfies the identity $(\partial\gamma^{\alpha}/\partial t)|_{\mathbf{x}} = -\mathbf{w}\cdot\nabla\gamma^{\alpha}$, where $\mathbf{w}$ is the velocity of the interface between the α-phase and all other phases. With these applied, Equation (830) becomes

$$\int_{V_\alpha}\frac{\partial f}{\partial t}\bigg|_{\mathbf{x}}dv = \frac{\partial}{\partial t}\int_{V_\alpha} f dv + \int_{V_\infty} f(\mathbf{w}\cdot\nabla\gamma^{\alpha})\gamma dv + \int_{V_\infty} f\gamma^{\alpha}\mathbf{w}\cdot\nabla\gamma dv \tag{831}$$

The indicator functions may be eliminated by restating the integrals over the surface $S_{\alpha\beta}$ and $S_{\alpha\text{edges}}$. Thus we obtain

$$\int_{V_\alpha}\frac{\partial f}{\partial t}\bigg|_{\mathbf{x}}dv = \frac{\partial}{\partial t}\int_{V_\alpha} f dv - \int_{S_{\alpha\beta}}\mathbf{n}\cdot\mathbf{w}f ds - \int_{S_{\alpha\text{edges}}}\mathbf{n}^*\cdot\mathbf{w}f ds$$

Theorem 62.

$$\int_{V_\alpha}\nabla\circ F dv = \nabla^{C}\circ\int_{V_\alpha} F dv + \int_{S_{\alpha\beta}}\mathbf{n}\circ F ds + \int_{S_{\alpha\text{edges}}}\mathbf{n}^*\circ F ds \tag{832}$$

where $\nabla\circ F$ can be any of the three operators: ∇f, $\nabla\mathbf{f}$ and $\nabla\times\mathbf{f}$. $f(\mathbf{x}, t)$, and $\mathbf{f}(\mathbf{x}, t)$ are spatial scalar and vector functions of position vector $\mathbf{x}$ and time t defined in the α-phase.

Proof. The indicator functions $\gamma(\mathbf{x}, \xi)$ and $\gamma^{\alpha}(\mathbf{x}+\xi, t)$ can be used to convert the integral over V_α to an integral over V_∞

$$\int_{V_\infty} \nabla \circ F dv = \int_{V_\infty} (\nabla \circ F)\gamma\gamma^{\alpha} dv \tag{833}$$

An application of the chain rule to the right side of Equation (833) yields

$$\int_{V_\alpha} \nabla \circ F dv = \int_{V_\infty} \nabla \circ (F\gamma\gamma^{\alpha}) dv - \int_{V_\infty} \nabla\gamma^{\alpha} \circ F\gamma dv - \int_{V_\infty} \nabla\gamma \circ F\gamma^{\alpha} dv \tag{834}$$

The last two terms in this expression are converted to surface integrals over the α–β interface and the edge of the region of interest, respectively, by Equations (185)–(187). This results in

$$\int_{V_\alpha} \nabla \circ F dv = \int_{V_\infty} \nabla \circ (F\gamma\gamma^{\alpha}) dv + \int_{S_{\alpha\beta}} \mathbf{n} \circ F ds + \int_{S_{\alpha \text{edges}}} \mathbf{n}^* \circ F ds \tag{835}$$

Since $\nabla = \nabla^C + \nabla^S$, the first term on the right side of Equation (835) becomes

$$\int_{V_\infty} \nabla \circ (F\gamma\gamma^{\alpha}) dv = \int_{V_\infty} \nabla^C \circ (F\gamma\gamma^{\alpha}) dv + \int_{V_\infty} \nabla^S \circ (F\gamma\gamma^{\alpha}) dv \tag{836}$$

In the first term on the right, the operator ∇^c may be moved outside of the integral and the indicator functions eliminated to return the domain of integration to V_∞. That is

$$\int_{V_\infty} \nabla \circ (F\gamma\gamma^{\alpha}) dv = \nabla^c \circ \int_{V_\alpha} F dv \tag{837}$$

In deriving this equation, the definitions of γ and γ^{α} together with Equations (233), (240), and (245) have been used. Finally, substitution of this expression into Equation (835) leads to

$$\int_{V_\alpha} \nabla \circ F dv = \nabla^C \circ \int_{V_\alpha} F dv + \int_{S_{\alpha\beta}} \mathbf{n} \circ F ds + \int_{S_{\alpha \text{edges}}} \mathbf{n}^* \circ F ds$$

Remark. Equation (832) contains the following three formulas:

Gradient

$$\int_{V_\alpha} \nabla f dv = \nabla^C \int_{V_\alpha} f dv + \int_{S_{\alpha\beta}} \mathbf{n} f ds + \int_{S_{\alpha \text{edges}}} \mathbf{n}^* f ds \tag{838}$$

Divergence

$$\int_{V_\alpha} \nabla \cdot \mathbf{f} dv = \nabla^C \cdot \int_{V_\alpha} \mathbf{f} dv + \int_{S_{\alpha\beta}} \mathbf{n} \cdot \mathbf{f} ds + \int_{S_{\alpha \text{edges}}} \mathbf{n}^* \cdot \mathbf{f} ds \tag{839}$$

Curl

$$\int_{V_\alpha} \nabla \times \mathbf{f} dv = \nabla^C \times \int_{V_\alpha} \mathbf{f} dv + \int_{S_{\alpha\beta}} \mathbf{n} \times \mathbf{f} ds + \int_{S_{\alpha \text{edges}}} \mathbf{n}^* \times \mathbf{f} ds \tag{840}$$

Theorem 62 has an alternative form given by:

Theorem 63.

$$\int_{V_\alpha} \nabla \circ F dv = \int_{S_\alpha} \mathbf{n}^* \circ F ds + \int_{S_{\alpha\beta}} \mathbf{n} \circ F ds \tag{841}$$

where S_α is the portion of the external boundary of the REV that intersects the α-phase, including the portion of the edge surface $S_{\alpha\text{edges}}$.

Proof. The indicator functions $\gamma(\mathbf{x}, \xi)$ and $\gamma^\alpha(\mathbf{x} + \xi, t)$ can be used to rewrite the integral under study as follows

$$\int_{V_\alpha} \nabla \circ F dv = \int_{V_\infty} \nabla \circ F\gamma\gamma^\alpha dv \tag{842}$$

An application of the chain rule to the right side of Equation (842) yields

$$\int_{V_\alpha} \nabla \circ F dv = \int_{V_\infty} \nabla \circ (F\gamma\gamma^\alpha) dv - \int_{V_\infty} \nabla\gamma \circ F\gamma^\alpha dv - \int_{V_\infty} \nabla\gamma^\alpha \circ F\gamma dv \tag{843}$$

The definition of the indicator function $\gamma(\mathbf{x}, \xi)$ enables us to conclude that the first integral on the right side of Equation (843) is vanished. With the application of Equations (185)–(187) to the last two integrals on the right side of Equation (843), we obtain

$$\int_{V_\alpha} \nabla \circ F dv = \int_{S_\alpha} \mathbf{n}^* \circ F ds + \int_{S_{\alpha\beta}} \mathbf{n} \circ F ds$$

Remark. Equation (841) contains the following three formulas:

Gradient

$$\int_{V_\alpha} \nabla f dv = \int_{S_\alpha} \mathbf{n}^* f ds + \int_{S_{\alpha\beta}} \mathbf{n} f ds \tag{844}$$

Divergence

$$\int_{V_\alpha} \nabla \cdot \mathbf{f} dv = \int_{S_\alpha} \mathbf{n}^* \cdot \mathbf{f} ds + \int_{S_{\alpha\beta}} \mathbf{n} \cdot \mathbf{f} ds \tag{845}$$

Curl

$$\int_{V_\alpha} \nabla \times \mathbf{f} dv = \int_{S_\alpha} \mathbf{n}^* \times \mathbf{f} ds + \int_{S_{\alpha\beta}} \mathbf{n} \times \mathbf{f} ds \tag{846}$$

5.3.2 Surficial operator theorems

The REV used for this group of averaging theorems is shown in Figure 32.

Theorem 64.

$$\int_{S_{\alpha\beta}} \left.\frac{\partial f}{\partial t}\right|_{\mathbf{u}} ds = \frac{\partial}{\partial t} \int_{S_{\alpha\beta}} f ds + \nabla^C \cdot \int_{S_{\alpha\beta}} \mathbf{w}^c f ds - \int_{S_{\alpha\beta}} \nabla^S \cdot \mathbf{nn} \cdot \mathbf{w}^c f ds$$
$$- \int_{C_{\alpha\beta\varepsilon}} \boldsymbol{\nu} \cdot \mathbf{w}^s f dc - \int_{C_{\alpha\beta\text{edge}}} \boldsymbol{\nu} \cdot \mathbf{w}^s f dc \tag{847}$$

where $S_{\alpha\beta}$ is the surface within the slab REV between the α- and the β-phases. $C_{\alpha\beta\varepsilon}$ is the boundary curve of $S_{\alpha\beta}$ within the REV that also is the location where the α- and β-phases meet a third phase. $C_{\alpha\beta\text{edge}}$ is the curve at the edge of the REV formed by the intersection of $S_{\alpha\beta}$ with the edge of the slab. $\mathbf{n}^*$ is the unit vector normal to the external boundary of the REV that is pointing outward from the REV. $\mathbf{n}$ is the unit vector normal to the $S_{\alpha\beta}$ surface. $\boldsymbol{\nu}$ is the unit vector tangent to the $S_{\alpha\beta}$ surface and normal to $C_{\alpha\beta\varepsilon}$ and $C_{\alpha\beta\text{edge}}$ such that $\mathbf{n}\cdot\boldsymbol{\nu}=0$. $\mathbf{N}$ is the unit vector normal to the face of the slab REV. ∇^{S} is the microscopic surficial operator $\nabla^{S}=\nabla_{\xi}-\mathbf{nn}\cdot\nabla_{\xi}$. ∇^{C} is the one-dimensional macroscopic spatial operator in the direction normal to the face of the slab REV such that $\nabla^{C}=\mathbf{NN}\cdot\nabla_{\mathbf{x}}$. $\mathbf{w}$ is the velocity of either $S_{\alpha\beta}$ or $S_{\alpha\beta\text{edge}}$, $\mathbf{w}^{c}$ the velocity of $S_{\alpha\beta}$ normal to $S_{\alpha\beta}$ such that $\mathbf{w}^{c}=\mathbf{nn}\cdot\mathbf{w}$, $\mathbf{w}^{s}$ is the velocity of $S_{\alpha\beta}$ tangent to $S_{\alpha\beta}$ such that $\mathbf{w}^{s}=\mathbf{w}-\mathbf{w}^{c}$ and f is a surficial function $f(\mathbf{u},t)$ defined on $S_{\alpha\beta}$.

Proof. By Equation (130), the integral under study is rewritten as

$$\int_{S_{\alpha\beta}}\left.\frac{\partial f}{\partial t}\right|_{\mathbf{u}}\mathrm{d}s=\int_{S_{\alpha\beta}}\left.\frac{\partial f}{\partial t}\right|_{\mathbf{x}}+\mathbf{w}\cdot\mathbf{nn}\cdot\nabla f\mathrm{d}s \tag{848}$$

By the definitions of the indicator functions $\gamma(\mathbf{x},\xi)$, $\gamma^{s}(\mathbf{u}(\mathbf{x}+\xi),t)$ and $\gamma^{\alpha}(\mathbf{x}+\xi,t)$, Equation (848) may be expressed as

$$\int_{S_{\alpha\beta}}\left.\frac{\partial f}{\partial t}\right|_{\mathbf{u}}\mathrm{d}s=-\int_{V_{\infty}}\frac{\partial f}{\partial t}\gamma\gamma^{s}\mathbf{n}\cdot\nabla\gamma^{\alpha}\mathrm{d}v-\int_{V_{\infty}}\mathbf{w}\cdot\mathbf{nn}\cdot\nabla f\gamma\gamma^{s}\mathbf{n}\cdot\nabla\gamma^{\alpha}\mathrm{d}v \tag{849}$$

An application of the chain rule to the two integrals on the right side of Equation (849), yields

$$\begin{aligned}\int_{S_{\alpha\beta}}\left.\frac{\partial f}{\partial t}\right|_{\mathbf{u}}\mathrm{d}s=&-\int_{V_{\infty}}\frac{\partial}{\partial t}(f\gamma\gamma^{s}\mathbf{n}\cdot\nabla\gamma^{\alpha})\mathrm{d}v-\int_{V_{\infty}}\nabla\cdot(\mathbf{nn}\cdot\mathbf{w}f\gamma\gamma^{s}\mathbf{n}\cdot\nabla\gamma^{\alpha})\mathrm{d}v\\&+\int_{V_{\infty}}f\frac{\partial(\gamma\gamma^{s})}{\partial t}\mathbf{n}\cdot\nabla\gamma^{\alpha}\mathrm{d}v+\int_{V_{\infty}}f\gamma\gamma^{s}\frac{\partial\mathbf{n}}{\partial t}\cdot\nabla\gamma^{\alpha}\mathrm{d}v\\&+\int_{V_{\infty}}f\gamma\gamma^{s}\mathbf{n}\cdot\nabla\frac{\partial\gamma^{\alpha}}{\partial t}\mathrm{d}v+\int_{V_{\infty}}\nabla\cdot\mathbf{nn}\cdot\mathbf{w}f\gamma\gamma^{s}\mathbf{n}\cdot\nabla\gamma^{\alpha}\mathrm{d}v\\&+\int_{V_{\infty}}\mathbf{n}\cdot\nabla\mathbf{n}\cdot\mathbf{w}f\gamma\gamma^{s}\mathbf{n}\cdot\nabla\gamma^{\alpha}\mathrm{d}v+\int_{V_{\infty}}\mathbf{n}\cdot\nabla\mathbf{w}\cdot\mathbf{n}f\gamma\gamma^{s}\mathbf{n}\cdot\nabla\gamma^{\alpha}\mathrm{d}v\\&+\int_{V_{\infty}}\mathbf{n}\cdot\mathbf{w}f\mathbf{n}\cdot\nabla(\gamma\gamma^{s})\mathbf{n}\cdot\nabla\gamma^{\alpha}\mathrm{d}v\\&+\int_{V_{\infty}}\mathbf{n}\cdot\mathbf{w}f\gamma\gamma^{s}\mathbf{n}\cdot(\nabla\mathbf{n}\cdot\nabla\gamma^{\alpha}+\nabla\nabla\gamma^{\alpha}\cdot\mathbf{n})\mathrm{d}v\end{aligned} \tag{850}$$

In the first term on the right side of Equation (850), the order of integration and differentiation with respect to time t can be interchanged because the integration domain is independent of time t. Since $\nabla=\nabla^{C}+\nabla^{S}$, the second term on the right side of Equation (850) becomes

$$\int_{V_\infty} \nabla \cdot (\mathbf{nn} \cdot \mathbf{w} f \gamma \gamma^{\mathrm{s}} \mathbf{n} \cdot \nabla \gamma^{\alpha}) \mathrm{d}v = \int_{V_\infty} \nabla^{\mathrm{C}} \cdot (\mathbf{nn} \cdot \mathbf{w} f \gamma \gamma^{\mathrm{s}} \mathbf{n} \cdot \nabla \gamma^{\alpha}) \mathrm{d}v + \int_{V_\infty} \nabla^{\mathrm{S}} \cdot (\mathbf{nn} \cdot \mathbf{w} f \gamma \gamma^{\mathrm{s}} \mathbf{n} \cdot \nabla \gamma^{\alpha}) \mathrm{d}v \tag{851}$$

Because the operator ∇^{c} is with respect to the macroscopic variables, it may be moved out from the integration. By the definition of the indicator function γ and Equation (240), the second term on the right side of Equation (851) vanishes. Therefore, Equation (851) becomes

$$\int_{V_\infty} \nabla \cdot (\mathbf{nn} \cdot \mathbf{w} f \gamma \gamma^{\mathrm{s}} \mathbf{n} \cdot \nabla \gamma^{\alpha}) \mathrm{d}v = \nabla^{\mathrm{C}} \cdot \int_{V_\infty} \mathbf{nn} \cdot \mathbf{w} f \gamma \gamma^{\mathrm{s}} \mathbf{n} \cdot \nabla \gamma^{\alpha} \mathrm{d}v \tag{852}$$

By Equation (223), the third term on the right side of Equation (850) may be written as

$$\int_{V_\infty} f \frac{\partial(\gamma \gamma^{\mathrm{s}})}{\partial t} \mathbf{n} \cdot \nabla \gamma^{\alpha} \mathrm{d}v = - \int_{V_\infty} f \mathbf{w} \cdot \nabla(\gamma \gamma^{\mathrm{s}}) \mathbf{n} \cdot \nabla \gamma^{\alpha} \mathrm{d}v \tag{853}$$

With Equations (194) and (195) applied to the fourth and the last terms on the right side of Equation (850), we obtain

$$\int_{V_\infty} f \gamma \gamma^{\mathrm{s}} \frac{\partial \mathbf{n}}{\partial t} \cdot \nabla \gamma^{\alpha} \mathrm{d}v = 0 \tag{854}$$

$$\int_{V_\infty} \mathbf{n} \cdot \mathbf{w} f \gamma \gamma^{\mathrm{s}} \mathbf{n} \cdot (\nabla \mathbf{n} \cdot \nabla \gamma^{\alpha} + \nabla \nabla \gamma^{\alpha} \cdot \mathbf{n}) \mathrm{d}v = \int_{V_\infty} \mathbf{n} \cdot \mathbf{w} f \gamma \gamma^{\mathrm{s}} \mathbf{n} \cdot \nabla \nabla \gamma^{\alpha} \cdot \mathbf{n} \mathrm{d}v \tag{855}$$

By Equation (180), the fifth integral on the right side of Equation (850) becomes

$$\int_{V_\infty} f \gamma \gamma^{\mathrm{s}} \mathbf{n} \cdot \nabla \frac{\partial \gamma^{\alpha}}{\partial t} \mathrm{d}v = - \int_{V_\infty} f \gamma \gamma^{\mathrm{s}} \mathbf{n} \cdot \nabla(\mathbf{w} \cdot \nabla \gamma^{\alpha}) \mathrm{d}v = - \int_{V_\infty} f \gamma \gamma^{\mathrm{s}} \mathbf{n} \cdot \nabla \mathbf{w} \cdot \nabla \gamma^{\alpha} \mathrm{d}v - \int_{V_\infty} f \gamma \gamma^{\mathrm{s}} \mathbf{n} \cdot \nabla \nabla \gamma^{\alpha} \cdot \mathbf{w} \mathrm{d}v \tag{856}$$

By substituting Equations (851)–(855) into the right side of Equation (850) and with some regrouping, we obtain

$$\begin{aligned} \int_{S_{\alpha\beta}} \left. \frac{\partial f}{\partial t} \right|_{\mathbf{u}} \mathrm{d}s = & - \frac{\partial}{\partial t} \int_{V_\infty} f \gamma \gamma^{\mathrm{s}} \mathbf{n} \cdot \nabla \gamma^{\alpha} \mathrm{d}v - \nabla^{\mathrm{c}} \cdot \int_{V_\infty} \mathbf{nn} \cdot \mathbf{w} f \gamma \gamma^{\mathrm{s}} \mathbf{n} \cdot \nabla \gamma^{\alpha} \mathrm{d}v \\ & - \int_{V_\infty} f \mathbf{w} \cdot \nabla^{\mathrm{S}}(\gamma \gamma^{\mathrm{s}}) \mathbf{n} \cdot \nabla \gamma^{\alpha} \mathrm{d}v - \int_{V_\infty} f \gamma \gamma^{\mathrm{s}} \mathbf{n} \cdot \nabla \mathbf{w} \cdot \nabla \gamma^{\alpha} \mathrm{d}v \\ & - \int_{V_\infty} f \gamma \gamma^{\mathrm{s}} \mathbf{w} \cdot \nabla^{\mathrm{S}} \nabla \gamma^{\alpha} \cdot \mathbf{n} \mathrm{d}v + \int_{V_\infty} \nabla \cdot \mathbf{nn} \cdot \mathbf{w} f \gamma \gamma^{\mathrm{s}} \mathbf{n} \cdot \nabla \gamma^{\alpha} \mathrm{d}v \\ & + \int_{V_\infty} \mathbf{n} \cdot \nabla \mathbf{n} \cdot \mathbf{w} f \gamma \gamma^{\mathrm{s}} \mathbf{n} \cdot \nabla \gamma^{\alpha} \mathrm{d}v + \int_{V_\infty} \mathbf{n} \cdot \nabla \mathbf{w} \cdot \mathbf{n} f \gamma \gamma^{\mathrm{s}} \mathbf{n} \cdot \nabla \gamma^{\alpha} \mathrm{d}v \end{aligned} \tag{857}$$

By Equation (255),

$$\int_{V_\infty} f\gamma\gamma^s \mathbf{w}\cdot\nabla^S\nabla\gamma^\alpha\cdot\mathbf{n}dv = \int_{V_\infty} f\gamma\gamma^s\mathbf{n}\cdot\nabla\mathbf{n}\cdot\mathbf{w}\mathbf{n}\cdot\nabla\gamma^\alpha dv \tag{858}$$

By the definitions of $C_{\alpha\beta\varepsilon}$, $C_{\alpha\beta\text{edge}}$, γ, γ^s, and Equation (275), we have

$$\begin{aligned}\int_{S_{\alpha\beta}} \left.\frac{\partial f}{\partial t}\right|_{\mathbf{u}} ds = &-\frac{\partial}{\partial t}\int_{V_\infty} f\gamma\gamma^s\mathbf{n}\cdot\nabla\gamma^\alpha dv - \nabla^C\cdot\int_{V_\infty}\mathbf{nn}\cdot\mathbf{w}f\gamma\gamma^c\mathbf{n}\cdot\nabla\gamma^\alpha dv\\ &-\int_{C_{\alpha\beta\text{edge}}}\mathbf{w}\cdot\nu dc - \int_{C_{\alpha\beta\varepsilon}}\mathbf{w}\cdot\nu dc - \int_{V_\infty} f\gamma\gamma^s\mathbf{n}\cdot\nabla\mathbf{w}\cdot\nabla\gamma^\alpha dv\\ &-\int_{V_\infty} f\gamma\gamma^s\mathbf{n}\cdot\nabla\mathbf{n}\cdot\mathbf{w}dv + \int_{V_\infty}\nabla\cdot\mathbf{nn}\cdot\mathbf{w}f\gamma\gamma^s\mathbf{n}\cdot\nabla\gamma^\alpha dv\\ &+\int_{V_\infty}\mathbf{n}\cdot\nabla\mathbf{n}\cdot\mathbf{w}f\gamma\gamma^s\mathbf{n}\cdot\nabla\gamma^\alpha dv + \int_{V_\infty}\mathbf{n}\cdot\nabla\mathbf{w}\cdot\mathbf{n}f\gamma\gamma^s\mathbf{n}\cdot\nabla\gamma^\alpha dv\end{aligned} \tag{859}$$

Note also that

$$\nu\cdot\mathbf{w} = \nu\cdot\mathbf{w}^s$$
$$\nabla\cdot\mathbf{n} = \nabla^S\cdot\mathbf{n}$$

By Equations (271) and (136), Equation (859) finally becomes

$$\begin{aligned}\int_{S_{\alpha\beta}} \left.\frac{\partial f}{\partial t}\right|_{\mathbf{u}} ds = &\frac{\partial}{\partial t}\int_{S_{\alpha\beta}} f ds + \nabla^C\cdot\int_{S_{\alpha\beta}}\mathbf{w}^c f ds - \int_{S_{\alpha\beta}}\nabla^S\cdot\mathbf{nn}\cdot\mathbf{w}^c f ds\\ &-\int_{C_{\alpha\beta\varepsilon}}\nu\cdot\mathbf{w}^s f dc - \int_{C_{\alpha\beta\text{edge}}}\nu\cdot\mathbf{w}^s f dc\end{aligned}$$

Theorem 65.

$$\begin{aligned}\int_{S_{\alpha\beta}}\nabla^S\circ F ds = &\nabla^C\circ\int_{S_{\alpha\beta}} F ds - \nabla^C\cdot\int_{S_{\alpha\beta}}\mathbf{nn}\circ F ds\\ &+\int_{S_{\alpha\beta}}\nabla^S\cdot\mathbf{nn}\circ F ds + \int_{C_{\alpha\beta\varepsilon}}\nu\circ F dc + \int_{C_{\alpha\beta\text{edge}}}\nu\circ F dc\end{aligned} \tag{860}$$

where $\nabla^S\circ F$ can be any of the three operators: $\nabla^S f$, $\nabla^S\cdot\mathbf{f}$, and $\nabla^S\times\mathbf{f}$. $f(\mathbf{u},t)$ and $\mathbf{f}(\mathbf{u},t)$ are scalar and vector functions of surficial position vector $\mathbf{u}$ and time t defined on $S_{\alpha\beta}$.

Proof. The indicator functions $\gamma(\mathbf{x},\ \xi)$, $\gamma^s(\mathbf{u}(\mathbf{x}+\xi),t)$ and $\gamma^\alpha(\mathbf{x}+\xi,t)$ may be employed to convert the integration under study to the integration over the whole space

$$\int_{S_{\alpha\beta}}\nabla^S\circ F ds = -\int_{V_\infty}\nabla^S\circ F\gamma\gamma^s\mathbf{n}\cdot\nabla\gamma^\alpha dv \tag{861}$$

By $\nabla^S = \nabla - \mathbf{nn}\cdot\nabla$, we have

$$\int_{S_{\alpha\beta}}\nabla^S\circ F ds = -\int_{V_\infty}\nabla\circ F\gamma\gamma^s\mathbf{n}\cdot\nabla\gamma^\alpha dv + \int_{V_\infty}(\mathbf{nn}\cdot\nabla)\circ F\gamma\gamma^s\mathbf{n}\cdot\nabla\gamma^\alpha dv \tag{862}$$

An application of the chain rule to the terms on the right side of Equation (862) yields

$$\begin{aligned}\int_{S_{\alpha\beta}} \nabla^{S} \circ F ds \\ = &- \int_{V_\infty} \nabla \circ (F\gamma\gamma^{s}\mathbf{n} \cdot \nabla\gamma^{\alpha}) dv + \int_{V_\infty} \nabla \cdot (\mathbf{nn} \circ F\gamma\gamma^{s}\mathbf{n} \cdot \nabla\gamma^{\alpha}) dv \\ &+ \int_{V_\infty} \nabla(\gamma\gamma^{s}) \circ F\mathbf{n} \cdot \nabla\gamma^{\alpha} dv + \int_{V_\infty} \gamma\gamma^{s}(\nabla\mathbf{n} \cdot \nabla\gamma^{\alpha} + \nabla\nabla\gamma^{\alpha} \cdot \mathbf{n}) \circ F dv \\ &- \int_{V_\infty} \nabla \cdot \mathbf{nn} \circ F\gamma\gamma^{s}\mathbf{n} \cdot \nabla\gamma^{\alpha} dv - \int_{V_\infty} \mathbf{n} \cdot \nabla\mathbf{n} \circ F\gamma\gamma^{s}\mathbf{n} \cdot \nabla\gamma^{\alpha} dv \\ &- \int_{V_\infty} \mathbf{n} \circ F\mathbf{n} \cdot \nabla(\gamma\gamma^{s})\mathbf{n} \cdot \nabla\gamma^{\alpha} dv \\ &- \int_{V_\infty} \mathbf{n} \circ F\gamma\gamma^{s}\mathbf{n} \cdot (\nabla\mathbf{n} \cdot \nabla\gamma^{\alpha} + \nabla\nabla\gamma^{\alpha} \cdot \mathbf{n}) dv \qquad (863)\end{aligned}$$

With Equation (195) applied to the fourth and eighth terms on the right side of Equation (863), Equation (138) to the fifth term and regrouping the remaining terms, Equation (863) becomes

$$\begin{aligned}\int_{S_{\alpha\beta}} \nabla^{S} \circ F ds = &- \int_{V_\infty} \nabla \circ (F\gamma\gamma^{s}\mathbf{n} \cdot \nabla\gamma^{\alpha}) dv + \int_{V_\infty} \nabla \cdot (\mathbf{nn} \circ F\gamma\gamma^{s}\mathbf{n} \cdot \nabla\gamma^{\alpha}) dv \\ &+ \int_{V_\infty} \nabla^{S} \cdot (\gamma\gamma^{s}) \circ F\mathbf{n} \cdot \nabla\gamma^{\alpha} dv + \int_{V_\infty} \gamma\gamma^{s}\nabla^{S}\nabla\gamma^{\alpha} \cdot \mathbf{n} \circ F dv \\ &- \int_{V_\infty} \nabla^{S} \cdot \mathbf{nn} \circ F\gamma\gamma^{s}\mathbf{n} \cdot \nabla\gamma^{\alpha} dv \\ &- \int_{V_\infty} \mathbf{n} \cdot \nabla\mathbf{n} \circ F\gamma\gamma^{s}\mathbf{n} \cdot \nabla\gamma^{\alpha} dv \qquad (864)\end{aligned}$$

By a similar derivation as that for Theorem 64 (Equation (847)), we have

$$\int_{V_\infty} \nabla \circ (F\gamma\gamma^{s}\mathbf{n} \cdot \nabla\gamma^{\alpha}) dv = \nabla^{C} \circ \int_{V_\infty} F\gamma\gamma^{s}\mathbf{n} \cdot \nabla\gamma^{\alpha} dv \qquad (865)$$

$$\int_{V_\infty} \nabla \cdot (\mathbf{nn} \circ F\mathbf{n} \cdot \nabla\gamma^{\alpha}\gamma\gamma^{s}) dv = \nabla^{C} \cdot \int_{V_\infty} \mathbf{nn} \circ F\mathbf{n} \cdot \nabla\gamma^{\alpha}\gamma\gamma^{s} dv \qquad (866)$$

By Equation (275), the third term on the right side of Equation (864) becomes

$$\int_{V_\infty} \nabla^{S}(\gamma\gamma^{s}) \circ F\mathbf{n} \cdot \nabla\gamma^{\alpha} dv = \int_{C_{\alpha\beta\mathrm{edge}}} \boldsymbol{\nu} \circ F dc + \int_{C_{\alpha\beta\varepsilon}} \boldsymbol{\nu} \circ F dc \qquad (867)$$

Applying Equation (255) to the fourth term on the right side of Equation (864) leads to

$$\int_{V_\infty} \gamma\gamma^{s}(\nabla^{S}\nabla\gamma^{\alpha} \cdot \mathbf{n}) \circ F dv = \int_{V_\infty} \gamma\gamma^{s}\mathbf{n} \cdot \nabla\mathbf{n} \circ F\mathbf{n} \cdot \nabla\gamma^{\alpha} dv \qquad (868)$$

Thus this term cancels the last term on the right side of Equation (864).

With these manipulations applied to the terms on the right side of Equation (864), we obtain

$$\int_{S_{\alpha\beta}} \nabla^S \circ F ds = -\nabla^C \circ \int_{V_\infty} F\gamma\gamma^s \mathbf{n} \cdot \nabla\gamma^\alpha dv + \nabla^C \cdot \int_{V_\infty} \mathbf{nn} \circ F\mathbf{n} \cdot \nabla\gamma^\alpha\gamma\gamma^s dv$$
$$+ \int_{C_{\alpha\beta\text{edge}}} \boldsymbol{v} \circ F dc + \int_{C_{\alpha\beta\varepsilon}} \boldsymbol{v} \circ F dc - \int_{V_\infty} \nabla^S \cdot \mathbf{nn} \circ F\gamma\gamma^s \mathbf{n} \cdot \nabla\gamma^\alpha dv \quad (869)$$

By Equation (186), Equation (869) finally leads to

$$\int_{S_{\alpha\beta}} \nabla^S \circ F ds = \nabla^C \circ \int_{S_{\alpha\beta}} F ds - \nabla^C \cdot \int_{S_{\alpha\beta}} \mathbf{nn} \circ F ds$$
$$+ \int_{S_{\alpha\beta}} \nabla^S \cdot \mathbf{nn} \circ F ds + \int_{C_{\alpha\beta\varepsilon}} \boldsymbol{v} \circ F dc + \int_{C_{\alpha\beta\text{edge}}} \boldsymbol{v} \circ F dc$$

Remark. Equation (860) contains the following three formulas:

Gradient

$$\int_{S_{\alpha\beta}} \nabla^S f ds = \nabla^C \int_{S_{\alpha\beta}} f ds - \nabla^C \cdot \int_{S_{\alpha\beta}} \mathbf{nn} f ds + \int_{S_{\alpha\beta}} (\nabla^S \cdot \mathbf{n})\mathbf{n} f ds$$
$$+ \int_{C_{\alpha\beta\varepsilon}} \boldsymbol{v} f dc + \int_{C_{\alpha\beta\text{edge}}} \boldsymbol{v} f dc \quad (870)$$

Divergence

$$\int_{S_{\alpha\beta}} \nabla^S \cdot \mathbf{f} ds = \nabla^C \cdot \int_{S_{\alpha\beta}} \mathbf{f} ds + \int_{S_{\alpha\beta}} \nabla^S \cdot \mathbf{nn} \cdot \mathbf{f} ds + \int_{C_{\alpha\beta\varepsilon}} v \cdot \mathbf{f} dc + \int_{C_{\alpha\beta\text{edge}}} v \cdot \mathbf{f} dc \quad (871)$$

Curl

$$\int_{S_{\alpha\beta}} \nabla^S \times \mathbf{f} ds = \nabla^C \times \int_{S_{\alpha\beta}} \mathbf{f} ds - \nabla^C \cdot \int_{S_{\alpha\beta}} \mathbf{nn} \times \mathbf{f} ds + \int_{S_{\alpha\beta}} \nabla^S \cdot \mathbf{nn} \times \mathbf{f} ds$$
$$+ \int_{C_{\alpha\beta\varepsilon}} \boldsymbol{v} \times \mathbf{f} dc + \int_{C_{\alpha\beta\text{edge}}} \boldsymbol{v} \times \mathbf{f} dc \quad (872)$$

Theorem 64 has the following alternative form:

Theorem 66.

$$\int_{S_{\alpha\beta}} \left.\frac{\partial f}{\partial t}\right|_{\mathbf{u}} ds = \frac{\partial}{\partial t} \int_{S_{\alpha\beta}} f ds - \int_{S_{\alpha\beta}} (\nabla^S \cdot \mathbf{n})\mathbf{n} \cdot \mathbf{w}^c f ds - \int_{C_{\alpha\beta\varepsilon}} \boldsymbol{v} \cdot \mathbf{w}^s f dc$$
$$- \int_{C_{\alpha\beta}} \boldsymbol{v} \cdot \mathbf{w}^s f dc \quad (873)$$

where $C_{\alpha\beta\varepsilon}$ is the boundary curve of $S_{\alpha\beta}$ within the REV that also is the location where the α- and β-phases meet at third phase.

Proof. By the definition of the indicator functions $\gamma(\mathbf{x}, \xi)$, $\gamma^s(\mathbf{u}(\mathbf{x}+\xi),t)$ and $\gamma^\alpha(\mathbf{x}+\xi,t)$ we have

$$\int_{S_{\alpha\beta}} \left.\frac{\partial f}{\partial t}\right|_{\mathbf{u}} \mathrm{d}s = -\int_{V_\infty} \left.\frac{\partial f}{\partial t}\right|_{\mathbf{u}} \gamma\gamma^s \mathbf{n}\cdot\nabla\gamma^\alpha \mathrm{d}v \tag{874}$$

which is, by Equation (130),

$$\int_{S_{\alpha\beta}} \left.\frac{\partial f}{\partial t}\right|_{\mathbf{u}} \mathrm{d}s = -\int_{V_\infty} \left(\left.\frac{\partial f}{\partial t}\right|_{\mathrm{x}} + \mathbf{w}\cdot\mathbf{nn}\cdot\nabla f\right)\gamma\gamma^s \mathbf{n}\cdot\nabla\gamma^\alpha \mathrm{d}v \tag{875}$$

An application of the chain rule to the right side of Equation (875) yields

$$\begin{aligned}\int_{S_{\alpha\beta}} \left.\frac{\partial f}{\partial t}\right|_{\mathbf{u}} \mathrm{d}s = &-\int_{V_\infty} \frac{\partial}{\partial t}(f\gamma\gamma^s\mathbf{n}\cdot\nabla\gamma^\alpha)\mathrm{d}v + \int_{V_\infty} f\frac{\partial(\gamma\gamma^s)}{\partial t}\mathbf{n}\cdot\nabla\gamma^\alpha\mathrm{d}v \\ &+\int_{V_\infty} f\gamma\gamma^s\left(\frac{\partial\mathbf{n}}{\partial t}\cdot\nabla\gamma^\alpha + \mathbf{n}\cdot\nabla\frac{\partial\gamma^\alpha}{\partial t}\right)\mathrm{d}v \\ &-\int_{V_\infty} \nabla\cdot(\mathbf{nn}\cdot\mathbf{w}f\gamma\gamma^s\mathbf{n}\cdot\nabla\gamma^\alpha)\mathrm{d}v \\ &+\int_{V_\infty} \nabla^S\cdot\mathbf{nn}\cdot\mathbf{w}^c f\gamma\gamma^s\mathbf{n}\cdot\nabla\gamma^\alpha\mathrm{d}v \\ &+\int_{V_\infty} \mathbf{n}\cdot\nabla\mathbf{n}\cdot\mathbf{w}f\gamma\gamma^s\mathbf{n}\cdot\nabla\gamma^\alpha\mathrm{d}v \\ &+\int_{V_\infty} \mathbf{n}\cdot\nabla\mathbf{w}\cdot\mathbf{n}f\gamma\gamma^s\mathbf{n}\cdot\nabla\gamma^\alpha\mathrm{d}v \\ &+\int_{V_\infty} \mathbf{n}\cdot\mathbf{w}f\mathbf{n}\cdot\nabla(\gamma\gamma^s)\mathbf{n}\cdot\nabla\gamma^\alpha\mathrm{d}v \\ &+\int_{V_\infty} \mathbf{n}\cdot\mathbf{w}f\gamma\gamma^s\mathbf{n}\cdot(\nabla\mathbf{n}\cdot\nabla\gamma^\alpha + \nabla\nabla\gamma^\alpha\cdot\mathbf{n})\mathrm{d}v\end{aligned} \tag{876}$$

Because the integration variables do not involve the time t, the differentiation $\partial/\partial t$ may be moved out from the integration in the first integral on the right side of Equation (876). By Equation (230), the fourth integral is zero on the right side of Equation (876). After applying Equation (223) to the second term on the right side of Equation (876), it can be combined with the eight term. Note also that the velocity of the REV faces is zero; thus we obtain

$$\begin{aligned}&\int_{V_\infty} f\frac{\partial(\gamma\gamma^s)}{\partial t}\mathbf{n}\cdot\nabla\gamma^\alpha\mathrm{d}v + \int_{V_\infty} \mathbf{n}\cdot\mathbf{w}f\mathbf{n}\cdot\nabla(\gamma\gamma^s)\mathbf{n}\cdot\nabla\gamma^\alpha\mathrm{d}v \\ &= -\int_{V_\infty} f\mathbf{w}\cdot\nabla(\gamma\gamma^s)\mathbf{n}\cdot\nabla\gamma^\alpha\mathrm{d}v + \int_{V_\infty} \mathbf{n}\cdot\mathbf{w}f\mathbf{n}\cdot\nabla(\gamma\gamma^s)\mathbf{n}\cdot\nabla\gamma^\alpha\mathrm{d}v \\ &= -\int_{V_\infty} f\mathbf{w}\cdot\nabla^S(\gamma\gamma^s)\mathbf{n}\cdot\nabla\gamma^\alpha\mathrm{d}v\end{aligned} \tag{877}$$

By Equations (180) and (194),

$$\int_{V_\infty} f\gamma\gamma^s\left(\frac{\partial \mathbf{n}}{\partial t}\cdot\nabla\gamma^\alpha+\mathbf{n}\cdot\nabla\frac{\partial\gamma^\alpha}{\partial t}\right)dv = -\int_{V_\infty} f\gamma\gamma^s\mathbf{n}\cdot\nabla\mathbf{w}\cdot\nabla\gamma^\alpha dv - \int_{V_\infty} f\gamma\gamma^s\mathbf{n}\cdot\nabla\nabla\gamma^\alpha\cdot\mathbf{w}dv \quad (878)$$

By Equation (195), we have

$$\int_{V_\infty} \mathbf{n}\cdot\mathbf{w}f\gamma\gamma^s\mathbf{n}\cdot(\nabla\mathbf{n}\cdot\nabla\gamma^\alpha\cdot\nabla\nabla\gamma^\alpha\cdot\mathbf{n})dv = \int_{V_\infty} f\gamma\gamma^s\mathbf{w}\cdot\mathbf{nn}\cdot\nabla\nabla\gamma^\alpha\cdot\mathbf{n}dv \quad (879)$$

Thus Equation (876) becomes

$$\int_{S_{\alpha\beta}}\left.\frac{\partial f}{\partial t}\right|_{\mathbf{u}}ds = -\frac{\partial}{\partial t}\int_{V_\infty} f\gamma\gamma^s\mathbf{n}\cdot\nabla\gamma^\alpha dv - \int_{V_\infty} f\gamma\mathbf{w}\cdot\nabla^S\gamma^s\mathbf{n}\cdot\nabla\gamma^\alpha dv - \int_{V_\infty}\gamma^s f\mathbf{w}\cdot\nabla^S\gamma\mathbf{n}\cdot\nabla\gamma^\alpha dv - \int_{V_\infty} f\gamma\gamma^s\mathbf{w}\cdot\nabla^S\nabla\gamma^\alpha\cdot\mathbf{n}dv + \int_{V_\infty}\nabla^S\cdot\mathbf{nn}\cdot\mathbf{w}^c f\gamma\gamma^s\mathbf{n}\cdot\nabla\gamma^\alpha dv + \int_{V_\infty}\mathbf{n}\cdot\nabla\mathbf{n}\cdot\mathbf{w}f\gamma\gamma^s\mathbf{n}\cdot\nabla\gamma^\alpha dv \quad (880)$$

By Equation (720),

$$\int_{V_\infty} f\gamma\gamma^s\mathbf{w}\cdot\nabla^S\nabla\gamma^\alpha\cdot\mathbf{n}dv = \int_{V_\infty} f\gamma\gamma^s\mathbf{n}\cdot\nabla\mathbf{n}\cdot\mathbf{wn}\cdot\nabla\gamma^\alpha dv \quad (881)$$

Substituting Equation (881) into Equation (880) yields

$$\int_{S_{\alpha\beta}}\left.\frac{\partial f}{\partial t}\right|_{\mathbf{u}}ds = -\frac{\partial}{\partial t}\int_{V_\infty} f\gamma\gamma^s\mathbf{n}\cdot\nabla\gamma^\alpha dv - \int_{V_\infty} f\gamma\mathbf{w}\cdot\nabla^S\gamma^s\mathbf{n}\cdot\nabla\gamma^\alpha dv - \int_{V_\infty} f\gamma^s\mathbf{w}\cdot\nabla^S\gamma\mathbf{n}\cdot\nabla\gamma^\alpha dv + \int_{V_\infty} f\gamma\gamma^s\nabla^S\cdot\mathbf{nn}\cdot\mathbf{w}^c\mathbf{n}\cdot\nabla\gamma^\alpha dv \quad (882)$$

By Equations (275) and (271), we obtain

$$\int_{S_{\alpha\beta}}\left.\frac{\partial f}{\partial t}\right|_{\mathbf{u}}ds = \frac{\partial}{\partial t}\int_{S_{\alpha\beta}} fds - \int_{S_{\alpha\beta}}(\nabla^S\cdot\mathbf{n})\mathbf{n}\cdot\mathbf{w}^c fds - \int_{C_{\alpha\beta}}\boldsymbol{\nu}\cdot\mathbf{w}^s fdc - \int_{C_{\alpha\beta\varepsilon}} f\boldsymbol{\nu}\cdot\mathbf{w}^s ds$$

Theorem 65 has the following alternative form:

Theorem 67.

$$\int_{S_{\alpha\beta}}\nabla^S\circ Fds = \int_{S_{\alpha\beta}}(\nabla^S\cdot\mathbf{n})\mathbf{n}\circ Fds + \int_{C_{\alpha\beta}}\boldsymbol{\nu}\circ Fds + \int_{C_{\alpha\beta\varepsilon}}\boldsymbol{\nu}\circ Fdc \quad (883)$$

Proof. By the definition of the indicator functions $\gamma(\mathbf{x},\,\xi)$, $\gamma^s(\mathbf{u}(\mathbf{x}+\xi),\,t)$ and $\gamma^\alpha(\mathbf{x}+\xi,\,t)$ the integral under study can be expressed as

$$\int_{S_{\alpha\beta}}\nabla^S\circ Fds = -\int_{V_\infty}(\nabla-\mathbf{nn}\cdot\nabla)\circ F\gamma\gamma^s\mathbf{n}\cdot\nabla\gamma^\alpha dv \quad (884)$$

An application of the chain rule to the right side of Equation (884) yields

$$\begin{aligned}\int_{S_{\alpha\beta}} \nabla^S \circ F ds = &-\int_{V_\infty} \nabla \circ (F\gamma\gamma^s \mathbf{n} \cdot \nabla\gamma^\alpha) dv + \int_{V_\infty} \nabla(\gamma\gamma^s) \circ F\mathbf{n} \cdot \nabla\gamma^\alpha dv \\ &+ \int_{V_\infty} \gamma\gamma^s (\nabla\mathbf{n} \cdot \nabla\gamma^\alpha + \nabla\nabla\gamma^\alpha \cdot \mathbf{n}) \circ F dv \\ &+ \int_{V_\infty} \nabla \cdot (\mathbf{nn} \circ F\gamma\gamma^s \mathbf{n} \cdot \nabla\gamma^\alpha) dv - \int_{V_\infty} \nabla \cdot \mathbf{nn} \circ F\gamma\gamma^s \mathbf{n} \cdot \nabla\gamma^\alpha dv \\ &- \int_{V_\infty} \mathbf{n} \cdot \nabla\mathbf{n} \circ F\gamma\gamma^s \mathbf{n} \cdot \nabla\gamma^\alpha dv - \int_{V_\infty} \mathbf{n} \circ F\mathbf{n} \cdot \nabla(\gamma\gamma^s) \mathbf{n} \cdot \nabla\gamma^\alpha dv \\ &- \int_{V_\infty} \mathbf{n} \circ F\gamma\gamma^s \mathbf{n} \cdot (\nabla\mathbf{n} \cdot \nabla\gamma^\alpha \cdot \nabla\nabla\gamma^\alpha \cdot \mathbf{n}) dv \end{aligned} \tag{885}$$

With the application of Equations (229)–(231) respectively, we conclude that the first and fourth integral are zero. By Equation (195),

$$\int_{V_\infty} \gamma\gamma^s (\nabla\mathbf{n} \cdot \nabla\gamma^\alpha + \nabla\nabla\gamma^\alpha \cdot \mathbf{n}) \circ F dv = \int_{V_\infty} \gamma\gamma^s (\nabla\nabla\gamma^\alpha \cdot \mathbf{n}) \circ F dv \tag{886}$$

$$\int_{V_\infty} \mathbf{n} \circ F\gamma\gamma^s \mathbf{n} \cdot (\nabla\mathbf{n} \cdot \nabla\gamma^\alpha + \nabla\nabla\gamma^\alpha \cdot \mathbf{n}) dv = \int_{V_\infty} \gamma\gamma^s (\nabla\nabla\gamma^\alpha \cdot \mathbf{n}) \cdot \mathbf{nn} \circ F dv \tag{887}$$

Finally, regrouping the terms on the right side of Equation (885) yields

$$\begin{aligned}\int_{S_{\alpha\beta}} \nabla^S \circ F ds = &\int_{V_\infty} \nabla^S(\gamma\gamma^s) \circ F\mathbf{n} \cdot \nabla\gamma^\alpha dv + \int_{V_\infty} \gamma\gamma^s (\nabla^S\nabla\gamma^\alpha \cdot \mathbf{n}) \circ F dv \\ &- \int_{V_\infty} \nabla^S \cdot \mathbf{nn} \circ F\gamma\gamma^s \mathbf{n} \cdot \nabla\gamma^\alpha dv \\ &- \int_{V_\infty} \mathbf{n} \cdot \nabla\mathbf{n} \circ F\gamma\gamma^s \mathbf{n} \cdot \nabla\gamma^\alpha dv \end{aligned} \tag{888}$$

By Equation (255), the second integral on the right side of Equation (888) becomes

$$\int_{V_\infty} \gamma\gamma^s (\nabla^S\nabla\gamma^\alpha \cdot \mathbf{n}) \circ F dv = \int_{V_\infty} \gamma\gamma^s \mathbf{n} \cdot \nabla\mathbf{n} \circ F\mathbf{n} \cdot \nabla\gamma^\alpha dv \tag{889}$$

Substituting Equation (889) into Equation (888) yields

$$\begin{aligned}\int_{S_{\alpha\beta}} \nabla^S \circ F ds = &\int_{V_\infty} (\nabla^S\gamma\gamma^s + \gamma\nabla^S\gamma^s) \circ F\mathbf{n} \cdot \nabla\gamma^\alpha dv \\ &- \int_{V_\infty} \nabla^S \cdot \mathbf{nn} \circ F\gamma\gamma^s \mathbf{n} \cdot \nabla\gamma^\alpha dv \end{aligned} \tag{890}$$

With Equations (275) and (271) applied to the first and second integrals, respectively, we finally have

$$\int_{S_{\alpha\beta}} \nabla^S \circ F ds = \int_{S_{\alpha\beta}} \nabla^S \cdot \mathbf{nn} \circ F + \int_{C_{\alpha\beta\varepsilon}} \mathbf{v} \circ F dc + \int_{C_{\alpha\beta\varepsilon}} \mathbf{v} \circ F dc$$

Remark. Equation (883) contains the following three formulas:

Gradient

$$\int_{S_{\alpha\beta}} \nabla^{S} f ds = \int_{S_{\alpha\beta}} (\nabla^{S} \cdot \mathbf{n})\mathbf{n} f ds + \int_{C_{\alpha\beta}} \nu f ds + \int_{C_{\alpha\beta\varepsilon}} \nu f dc \tag{891}$$

Divergence

$$\int_{S_{\alpha\beta}} \nabla^{S} \cdot \mathbf{f} ds = \int_{S_{\alpha\beta}} (\nabla^{S} \cdot \mathbf{n})\mathbf{n} \cdot \mathbf{f} ds + \int_{C_{\alpha\beta}} \mathbf{v} \cdot \mathbf{f} ds + \int_{C_{\alpha\beta\varepsilon}} \mathbf{v} \cdot \mathbf{f} dc \tag{892}$$

Curl

$$\int_{S_{\alpha\beta}} \nabla^{S} \times \mathbf{f} ds = \int_{S_{\alpha\beta}} (\nabla^{S} \cdot \mathbf{n})\mathbf{n} \times \mathbf{f} ds + \int_{C_{\alpha\beta}} \mathbf{v} \times \mathbf{f} ds + \int_{C_{\alpha\beta\varepsilon}} \mathbf{v} \times \mathbf{f} dc \tag{893}$$

5.3.3 Curvilineal operator theorems

The REV used for this group of averaging theorems is shown in Figure 33.

Theorem 68.

$$\int_{C_{\alpha\beta\varepsilon}} \left.\frac{\partial f}{\partial t}\right|_{\mathbf{l}} dc = \frac{\partial}{\partial t} \int_{C_{\alpha\beta\varepsilon}} f dc + \nabla^{C} \cdot \int_{C_{\alpha\beta\varepsilon}} \mathbf{w}^{s} f dc + \int_{C_{\alpha\beta\varepsilon}} \boldsymbol{\lambda} \cdot \nabla^{c} \boldsymbol{\lambda} \cdot \mathbf{w}^{s} f dc$$
$$- \sum_{\text{pts}} \mathbf{e} \cdot \mathbf{w}^{c} f - \sum_{\text{edge}_{\alpha\beta\varepsilon}} \mathbf{e} \cdot \mathbf{w}^{c} f \tag{894}$$

where $C_{\alpha\beta\varepsilon}$ is the contact curve within the slab REV that is the location where the α-, β- and ε-phases meet. pts refers to the end points of the $C_{\alpha\beta\varepsilon}$ curve within the REV where the α-, β- and ε-phases meet a fourth phase. $\text{edge}_{\alpha\beta\varepsilon}$ is the points at the edge of the REV formed by the intersection of bounding surface of the slab with $C_{\alpha\beta\varepsilon}$. $\mathbf{n}^*$ is the unit vector normal to the external boundary of the REV that is pointing outward from the REV. $\boldsymbol{\lambda}$ is the unit vector tangent to $C_{\alpha\beta\varepsilon}$, $\mathbf{e}$ the unit vector tangent to $C_{\alpha\beta\varepsilon}$ at pts and $\text{edge}_{\alpha\beta\varepsilon}$ that is pointing outward from the curve such that $\mathbf{e} \cdot \boldsymbol{\lambda} = \pm 1$ and $\mathbf{N}$ the unit vector normal to the face of the slab REV. ∇^{c} is the microscopic curvilineal operator such that $\nabla^{c} = \boldsymbol{\lambda}\boldsymbol{\lambda} \cdot \nabla_{\xi}$ and ∇^{C} the one-dimensional macroscopic spatial operator in the direction normal to the face of the slab REV such that $\nabla^{C} = \mathbf{NN} \cdot \nabla_{\mathbf{x}}$. $\mathbf{w}$ is the velocity of $C_{\alpha\beta\varepsilon}$ and $\mathbf{w}^{c}$ the velocity of $C_{\alpha\beta\varepsilon}$ such that $\mathbf{w}^{s} = \mathbf{w} - \mathbf{w}^{c}$ and f is a curvilineal function $f(\mathbf{l}, t)$ defined on $C_{\alpha\beta\varepsilon}$.

Proof. By Equation (133), the integral under study may be expressed as

$$\int_{C_{\alpha\beta\varepsilon}} \left.\frac{\partial f}{\partial t}\right|_{\mathbf{l}} dc = \int_{C_{\alpha\beta\varepsilon}} \left.\frac{\partial f}{\partial t}\right|_{\mathbf{x}} + \mathbf{w}^{s} \cdot \nabla f dc \tag{895}$$

By the definitions of the indicator functions $\gamma(\mathbf{x}, \xi)$, $\gamma^{s}(\mathbf{u}(\mathbf{x}+\xi), t)$, $\gamma^{\alpha}(\mathbf{x}+\xi, t)$, and $\gamma^{c}(\mathbf{l}(\mathbf{x}+\xi), t)$, the integral on the right-hand side of Equation (895) may be expressed as

$$\int_{C_{\alpha\beta\varepsilon}} \left.\frac{\partial f}{\partial t}\right|_{\mathbf{l}} \mathrm{d}c = \int_{V_\infty} \left.\frac{\partial f}{\partial t}\right|_{\mathbf{x}} \gamma\gamma^{c}\boldsymbol{v}\cdot\nabla\gamma^{s}\mathbf{n}\cdot\nabla\gamma^{\alpha}\mathrm{d}v + \int_{V_\infty} \mathbf{w}^{s}\cdot\nabla f\gamma\gamma^{c}\boldsymbol{v}\cdot\nabla\gamma^{s}\mathbf{n}\cdot\nabla\gamma^{\alpha}\mathrm{d}v \tag{896}$$

An application of the chain rule to the integrals on the right side of Equation (896) yields

$$\begin{aligned}
\int_{C_{\alpha\beta\varepsilon}} \left.\frac{\partial f}{\partial t}\right|_{\mathbf{l}} \mathrm{d}c = & \int_{V_\infty} \frac{\partial}{\partial t}(f\gamma\gamma^{c}\boldsymbol{v}\cdot\nabla\gamma^{s}\mathbf{n}\cdot\nabla\gamma^{\alpha}\mathrm{d}v) \\
& - \int_{V_\infty} f\left(\frac{\partial\gamma}{\partial t}\gamma^{c} + \gamma\frac{\partial\gamma^{c}}{\partial t}\right)\boldsymbol{v}\cdot\nabla\gamma^{s}\mathbf{n}\cdot\nabla\gamma^{\alpha}\mathrm{d}v \\
& - \int_{V_\infty} f\gamma\gamma^{c}\left(\frac{\partial\boldsymbol{v}}{\partial t}\cdot\nabla\gamma^{s} + \boldsymbol{v}\cdot\nabla\frac{\partial\gamma^{s}}{\partial t}\right)\mathbf{n}\cdot\nabla\gamma^{\alpha}\mathrm{d}v \\
& - \int_{V_\infty} f\gamma\gamma^{c}\boldsymbol{v}\cdot\nabla\gamma^{s}\left(\frac{\partial\mathbf{n}}{\partial t}\cdot\nabla\gamma^{\alpha} + \mathbf{n}\cdot\nabla\frac{\partial\gamma^{\alpha}}{\partial t}\right)\mathrm{d}v \\
& + \int_{V\infty} \nabla\cdot(\mathbf{w}^{s}f\gamma\gamma^{c}\boldsymbol{v}\cdot\nabla\gamma^{s}\mathbf{n}\cdot\nabla\gamma^{\alpha})\mathrm{d}v \\
& - \int_{V_\infty} \nabla\cdot\mathbf{w}^{s}f\gamma\gamma^{c}\boldsymbol{v}\cdot\nabla\gamma^{s}\mathbf{n}\cdot\nabla\gamma^{\alpha}\mathrm{d}v \\
& - \int_{V_\infty} f\mathbf{w}^{s}\cdot\nabla(\gamma\gamma^{c})\boldsymbol{v}\cdot\nabla\gamma^{s}\mathbf{n}\cdot\nabla\gamma^{\alpha}\mathrm{d}v \\
& - \int_{V_\infty} f\gamma\gamma^{c}\mathbf{w}^{s}\cdot(\nabla\boldsymbol{v}\cdot\nabla\gamma^{s} + \nabla\nabla\gamma^{s}\cdot\boldsymbol{v})\mathbf{n}\cdot\nabla\gamma^{\alpha}\mathrm{d}v \\
& - \int_{V_\infty} f\gamma\gamma^{c}\boldsymbol{v}\cdot\nabla\gamma^{s}\mathbf{w}^{s}\cdot(\nabla\mathbf{n}\cdot\nabla\gamma^{\alpha} + \nabla\nabla\gamma^{\alpha}\cdot\mathbf{n})\mathrm{d}v
\end{aligned} \tag{897}$$

Because the integration domain is independent of the time t, the differentiation $\partial/\partial t$ can be moved out from the integration in the first integral on the right side of Equation (897). Similarly, the order of the integration and the operation ∇ may be interchanged in the firth integral on the right side of Equation (897). By Equations (180), (223), and (228),

$$\frac{\partial\gamma}{\partial t} = -\mathbf{w}\cdot\nabla\gamma, \quad \frac{\partial\gamma^{c}}{\partial t} = -\mathbf{w}\cdot\nabla\gamma^{c}$$
$$\frac{\partial\gamma^{s}}{\partial t} = -\mathbf{w}\cdot\nabla\gamma^{s}, \quad \frac{\partial\gamma^{\alpha}}{\partial t} = -\mathbf{w}\cdot\nabla\gamma^{\alpha}$$

With these relations and Equations (195) and (200) applied to the integrals on the right side of Equation (897), we obtain

$$
\begin{aligned}
&\int_{C_{\alpha\beta\varepsilon}} \frac{\partial f}{\partial t}\Big|_1 \mathrm{d}c \\
&= \frac{\partial}{\partial t}\int_{V_\infty} f\gamma\gamma^{\mathrm{c}}\boldsymbol{v}\cdot\nabla\gamma^{\mathrm{s}}\mathbf{n}\cdot\nabla\gamma^{\alpha}\mathrm{d}v \\
&+ \int_{V_\infty} f(\mathbf{w}\cdot\nabla\gamma\gamma^{\mathrm{c}} + \gamma\mathbf{w}\cdot\nabla\gamma^{\mathrm{c}})\boldsymbol{v}\cdot\nabla\gamma^{\mathrm{s}}\mathbf{n}\cdot\nabla\gamma^{\alpha}\mathrm{d}v \\
&+ \int_{V_\infty} f\gamma\gamma^{\mathrm{c}}\boldsymbol{v}\cdot\nabla\mathbf{w}\cdot\nabla\gamma^{\mathrm{s}}\mathbf{n}\cdot\nabla\gamma^{\alpha}\mathrm{d}v + \int_{V_\infty} f\gamma\gamma^{\mathrm{c}}\boldsymbol{v}\cdot\nabla\nabla\gamma^{\mathrm{s}}\cdot\mathbf{w}\mathbf{n}\cdot\nabla\gamma^{\alpha}\mathrm{d}v \\
&+ \int_{V_\infty} f\gamma\gamma^{\mathrm{c}}\boldsymbol{v}\cdot\nabla\gamma^{\mathrm{s}}\mathbf{n}\cdot\nabla\mathbf{w}\cdot\nabla\gamma^{\alpha}\mathrm{d}v + \int_{V_\infty} f\gamma\gamma^{\mathrm{c}}\boldsymbol{v}\cdot\nabla\gamma^{\mathrm{s}}\mathbf{n}\cdot\nabla\nabla\gamma^{\alpha}\cdot\mathbf{w}\mathrm{d}v \\
&+ \nabla^{\mathrm{C}}\cdot\int_{V_\infty} \mathbf{w}^{\mathrm{s}} f\gamma\gamma^{\mathrm{c}}\boldsymbol{v}\cdot\nabla\gamma^{\mathrm{s}}\mathbf{n}\cdot\nabla\gamma^{\alpha}\mathrm{d}v - \int_{V_\infty} \nabla\cdot\mathbf{w}^{\mathrm{s}} f\gamma\gamma^{\mathrm{c}}\boldsymbol{v}\cdot\nabla\gamma^{\mathrm{s}}\mathbf{n}\cdot\nabla\gamma^{\alpha}\mathrm{d}v \\
&- \int_{v_\infty} f\mathbf{w}^{\mathrm{s}}\cdot(\nabla\gamma\gamma^{\mathrm{c}} + \nabla\gamma^{\mathrm{c}}\gamma)\boldsymbol{v}\cdot\nabla\gamma^{\mathrm{s}}\mathbf{n}\cdot\nabla\gamma^{\alpha}\mathrm{d}v \\
&- \int_{V_\infty} f\gamma\gamma^{\mathrm{c}}\mathbf{w}^{\mathrm{s}}\cdot\nabla\nabla\gamma^{\mathrm{s}}\cdot\boldsymbol{v}\mathbf{n}\cdot\nabla\gamma^{\alpha}\mathrm{d}v - \int_{V_\infty} f\gamma\gamma^{\mathrm{c}}\boldsymbol{v}\cdot\nabla\gamma^{\mathrm{s}}\mathbf{w}^{\mathrm{s}}\cdot\nabla\nabla\gamma^{\alpha}\cdot\mathbf{n}\mathrm{d}v
\end{aligned}
\tag{898}
$$

Regrouping the terms on the right side of Equation (898) yields

$$
\begin{aligned}
&\int_{C_{\alpha\beta\varepsilon}} \frac{\partial f}{\partial t}\Big|_1 \mathrm{d}c \\
&= \frac{\partial}{\partial t}\int_{V_\infty} f\gamma\gamma^{\mathrm{c}}\boldsymbol{v}\cdot\nabla\gamma^{\mathrm{s}}\mathbf{n}\cdot\nabla\gamma^{\alpha}\mathrm{d}v + \nabla^{\mathrm{C}}\cdot\int_{V_\infty} \mathbf{w}^{\mathrm{s}} f\gamma\gamma^{\mathrm{c}}\boldsymbol{v}\cdot\nabla\gamma^{\mathrm{s}}\mathbf{n}\cdot\nabla\gamma^{\alpha}\mathrm{d}v \\
&+ \int_{V_\infty} f\gamma\gamma^{\mathrm{c}}\boldsymbol{v}\cdot\nabla\mathbf{w}\cdot\nabla\gamma^{\mathrm{s}}\mathbf{n}\cdot\nabla\gamma^{\alpha}\mathrm{d}v \\
&+ \int_{V_\infty} f\gamma\gamma^{\mathrm{c}}\boldsymbol{v}\cdot\nabla\gamma^{\mathrm{s}}\mathbf{n}\cdot\nabla\mathbf{w}\cdot\nabla\gamma^{\alpha}\mathrm{d}v \\
&+ \int_{V_\infty} f\gamma\gamma^{\mathrm{c}}\mathbf{w}\cdot\nabla^{\mathrm{c}}\nabla\gamma^{\mathrm{s}}\cdot\boldsymbol{v}\mathbf{n}\cdot\nabla\gamma^{\alpha}\mathrm{d}v \\
&+ \int_{V_\infty} f\gamma\gamma^{\mathrm{c}}\boldsymbol{v}\cdot\nabla\gamma^{\mathrm{s}}\mathbf{w}\cdot\nabla^{\mathrm{c}}\nabla\gamma^{\alpha}\cdot\mathbf{n}\mathrm{d}v \\
&- \int_{V_\infty} \nabla\cdot(\mathbf{w}-\boldsymbol{\lambda}\boldsymbol{\lambda}\cdot\mathbf{w}) f\gamma\gamma^{\mathrm{c}}\boldsymbol{v}\cdot\nabla\gamma^{\mathrm{s}}\mathbf{n}\cdot\nabla\gamma^{\alpha}\mathrm{d}v \\
&+ \int_{V_\infty} f\mathbf{w}\cdot\nabla^{\mathrm{c}}(\gamma\gamma^{\mathrm{c}})\boldsymbol{v}\cdot\nabla\gamma^{\mathrm{s}}\mathbf{n}\cdot\nabla\gamma^{\alpha}\mathrm{d}v
\end{aligned}
\tag{899}
$$

By Equation (53),

$$
\nabla\cdot\mathbf{w} = \boldsymbol{\lambda}\cdot\nabla\mathbf{w}\cdot\boldsymbol{\lambda} + \boldsymbol{v}\cdot\nabla\mathbf{w}\cdot\boldsymbol{v} + \mathbf{n}\cdot\nabla\mathbf{w}\cdot\mathbf{n}
$$

$$
\nabla\cdot(\boldsymbol{\lambda}\boldsymbol{\lambda}\cdot\mathbf{w}) = \boldsymbol{v}\cdot\nabla\boldsymbol{\lambda}\cdot\boldsymbol{v}\boldsymbol{\lambda}\cdot\mathbf{w} + \mathbf{n}\cdot\nabla\boldsymbol{\lambda}\cdot\mathbf{n}\boldsymbol{\lambda}\cdot\mathbf{w} + \boldsymbol{\lambda}\cdot\nabla^{\mathrm{c}}\boldsymbol{\lambda}\cdot\mathbf{w}^{\mathrm{s}} + \boldsymbol{\lambda}\cdot\nabla\mathbf{w}\cdot\boldsymbol{\lambda}
$$

With these equations and Equations (255) and (257), we obtain

$$\begin{aligned}
\int_{C_{\alpha\beta\varepsilon}} \left.\frac{\partial f}{\partial t}\right|_1 \mathrm{d}c = {} & \frac{\partial}{\partial t}\int_{V_\infty} f\gamma\gamma^{\mathrm{c}}\boldsymbol{v}\cdot\nabla\gamma^{\mathrm{s}}\mathbf{n}\cdot\nabla\gamma^{\alpha}\mathrm{d}v \\
& + \nabla^{\mathrm{C}}\cdot\int_{V_\infty} \mathbf{w}^{\mathrm{s}} f\gamma\gamma^{\mathrm{c}}\boldsymbol{v}\cdot\nabla\gamma^{\mathrm{s}}\mathbf{n}\cdot\nabla\gamma^{\alpha}\mathrm{d}v \\
& + \int_{V_\infty} f\gamma\gamma^{\mathrm{c}}\boldsymbol{v}\cdot\nabla\mathbf{w}\cdot\nabla\gamma^{\mathrm{s}}\mathbf{n}\cdot\nabla\gamma^{\alpha}\mathrm{d}v \\
& + \int_{V_\infty} f\gamma\gamma^{\mathrm{c}}\boldsymbol{v}\cdot\nabla\gamma^{\mathrm{s}}\mathbf{n}\cdot\nabla\mathbf{w}\cdot\nabla\gamma^{\alpha}\mathrm{d}v \\
& - \int_{V_\infty} f\gamma\gamma^{\mathrm{c}}\mathbf{w}\cdot\boldsymbol{\lambda}\boldsymbol{v}\cdot\nabla\boldsymbol{\lambda}\cdot\nabla\gamma^{\mathrm{s}}\mathbf{n}\cdot\nabla\gamma^{\alpha}\mathrm{d}v \\
& - \int_{V_\infty} f\gamma\gamma^{\mathrm{c}}\mathbf{w}\cdot\boldsymbol{\lambda}\mathbf{n}\cdot\nabla\boldsymbol{\lambda}\cdot\nabla\gamma^{\alpha}\boldsymbol{v}\cdot\nabla\gamma^{\mathrm{s}}\mathrm{d}v \\
& - \int_{V_\infty} \mathbf{n}\cdot\nabla\mathbf{w}\cdot\mathbf{n} f\gamma\gamma^{\mathrm{c}}\boldsymbol{v}\cdot\nabla\gamma^{\mathrm{s}}\mathbf{n}\cdot\nabla\gamma^{\alpha}\mathrm{d}v \\
& - \int_{V_\infty} \boldsymbol{\lambda}\cdot\nabla\mathbf{w}\cdot\boldsymbol{\lambda} f\gamma\gamma^{\mathrm{c}}\boldsymbol{v}\cdot\nabla\gamma^{\mathrm{s}}\mathbf{n}\cdot\nabla\gamma^{\alpha}\mathrm{d}v \\
& - \int_{V_\infty} \boldsymbol{v}\cdot\nabla\mathbf{w}\cdot\boldsymbol{v} f\gamma\gamma^{\mathrm{c}}\boldsymbol{v}\cdot\nabla\gamma^{\mathrm{s}}\mathbf{n}\cdot\nabla\gamma^{\alpha}\mathrm{d}v \\
& + \int_{V_\infty} \gamma\gamma^{\mathrm{c}}\boldsymbol{v}\cdot\nabla\boldsymbol{\lambda}\cdot\boldsymbol{v}\boldsymbol{\lambda}\cdot\mathbf{w}\boldsymbol{v}\cdot\nabla\gamma^{\mathrm{s}}\mathbf{n}\cdot\nabla\gamma^{\alpha}\mathrm{d}v \\
& + \int_{V_\infty} \mathbf{n}\cdot\nabla\boldsymbol{\lambda}\cdot\mathbf{n}\boldsymbol{\lambda}\cdot\mathbf{w} f\gamma\gamma^{\mathrm{c}}\boldsymbol{v}\cdot\nabla\gamma^{\mathrm{s}}\mathbf{n}\cdot\nabla\gamma^{\alpha}\mathrm{d}v \\
& + \int_{V_\infty} \boldsymbol{\lambda}\cdot\nabla^{\mathrm{c}}\boldsymbol{\lambda}\cdot\mathbf{w}^{\mathrm{s}} f\gamma\gamma^{\mathrm{c}}\boldsymbol{v}\cdot\nabla\gamma^{\mathrm{s}}\mathbf{n}\cdot\nabla\gamma^{\alpha}\mathrm{d}v \\
& + \int_{V_\infty} \boldsymbol{\lambda}\cdot\nabla\mathbf{w}\cdot\boldsymbol{\lambda} f\gamma\gamma^{\mathrm{c}}\boldsymbol{v}\cdot\nabla\gamma^{\mathrm{s}}\mathbf{n}\cdot\nabla\gamma^{\alpha}\mathrm{d}v \\
& + \int_{V_\infty} f\mathbf{w}\cdot\nabla^{\mathrm{c}}(\gamma\gamma^{\mathrm{c}})\boldsymbol{v}\cdot\nabla\gamma^{\mathrm{s}}\mathbf{n}\cdot\nabla\gamma^{\alpha}\mathrm{d}v
\end{aligned} \tag{900}$$

Thus Equation (900) may be simplified to

$$\begin{aligned}
\int_{C_{\alpha\beta\varepsilon}} \left.\frac{\partial f}{\partial t}\right|_1 \mathrm{d}c = {} & \frac{\partial}{\partial t} f\gamma\gamma^{\mathrm{c}}\boldsymbol{v}\cdot\nabla\gamma^{\mathrm{s}}\mathbf{n}\cdot\nabla\gamma^{\alpha}\mathrm{d}v \\
& + \nabla^{\mathrm{C}}\cdot\int_{V_\infty} \mathbf{w}^{\mathrm{s}} f\gamma\gamma^{\mathrm{c}}\boldsymbol{v}\cdot\nabla\gamma^{\mathrm{s}}\mathbf{n}\cdot\nabla\gamma^{\alpha}\mathrm{d}v \\
& + \int_{V_\infty} \boldsymbol{\lambda}\cdot\nabla^{\mathrm{c}}\boldsymbol{\lambda}\cdot\mathbf{w}^{\mathrm{s}} f\gamma\gamma^{\mathrm{c}}\boldsymbol{v}\cdot\nabla\gamma^{\mathrm{s}}\mathbf{n}\cdot\nabla\gamma^{\alpha}\mathrm{d}v \\
& + \int_{V_\infty} f\mathbf{w}\cdot\nabla^{\mathrm{c}}(\gamma\gamma^{\mathrm{c}})\boldsymbol{v}\cdot\nabla\gamma^{\mathrm{s}}\mathbf{n}\cdot\nabla\gamma^{\alpha}\mathrm{d}v
\end{aligned} \tag{901}$$

By Equations (275) and (166), we finally obtain

$$\int_{C_{\alpha\beta\varepsilon}} \frac{\partial f}{\partial t}\bigg|_1 dc = \frac{\partial}{\partial t}\int_{C_{\alpha\beta\varepsilon}} f dc + \nabla^C \cdot \int_{C_{\alpha\beta\varepsilon}} \mathbf{w}^s f dc + \int_{C_{\alpha\beta\varepsilon}} \boldsymbol{\lambda}\cdot\nabla^c\boldsymbol{\lambda}\cdot\mathbf{w}^s f dc - \sum_{\text{pts}} \mathbf{e}\cdot\mathbf{w}^c f - \sum_{\text{edge}\alpha\beta\varepsilon} \mathbf{e}\cdot\mathbf{w}^c f$$

Theorem 69.

$$\int_{C_{\alpha\beta\varepsilon}} \nabla^C \circ F dc = \nabla^C \cdot \int_{C_{\alpha\beta\varepsilon}} \boldsymbol{\lambda}\boldsymbol{\lambda} \circ F dc - \int_{C_{\alpha\beta\varepsilon}} (\boldsymbol{\lambda}\cdot\nabla^C\boldsymbol{\lambda}) \circ F dc + \sum_{\text{pts}} \mathbf{e}\circ F + \sum_{\text{edge}\alpha\beta\varepsilon} \mathbf{e}\circ F \quad (902)$$

Proof. By the definitions of the indicator functions $\gamma(\mathbf{x},\xi)$, $\gamma^s(\mathbf{u}(\mathbf{x}+\xi),t)$, $\gamma^\alpha(\mathbf{x}+\xi,t)$ and $\gamma^c(1(\mathbf{x}+\xi),t)$, the integral under study can be written as

$$\int_{C_{\alpha\beta\varepsilon}} \nabla^c \circ F dc = \int_{V_\infty} \nabla^c \circ F \boldsymbol{v}\cdot\nabla\gamma^s \mathbf{n}\cdot\nabla\gamma^\alpha\gamma\gamma^c dv \quad (903)$$

An application of the chain rule to the right side of Equation (903) yields

$$\begin{aligned}\int_{C_{\alpha\beta\varepsilon}} \nabla^c \circ F dc = &\int_{V_\infty} \nabla\cdot(\boldsymbol{\lambda}\boldsymbol{\lambda}\circ F\boldsymbol{v}\cdot\nabla\gamma^s\mathbf{n}\cdot\nabla\gamma^\alpha\gamma\gamma^c)dv \\ &- \int_{V_\infty} \nabla\cdot\boldsymbol{\lambda}\boldsymbol{\lambda}\circ F\boldsymbol{v}\cdot\nabla\gamma^s\mathbf{n}\cdot\nabla\gamma^\alpha\gamma\gamma^c dv \\ &- \int_{V_\infty} \boldsymbol{\lambda}\cdot\nabla\boldsymbol{\lambda}\circ F\boldsymbol{v}\cdot\nabla\gamma^s\mathbf{n}\cdot\nabla\gamma^\alpha\gamma\gamma^c dv \\ &- \int_{V_\infty} \boldsymbol{\lambda}\circ F\boldsymbol{\lambda}\cdot(\nabla\boldsymbol{v}\cdot\nabla\gamma^s + \nabla\nabla\gamma^s\cdot\boldsymbol{v})\mathbf{n}\cdot\nabla\gamma^\alpha\gamma\gamma^c dv \\ &- \int_{V_\infty} \boldsymbol{\lambda}\circ F\boldsymbol{v}\cdot\nabla\gamma^s\boldsymbol{\lambda}\cdot(\nabla\mathbf{n}\cdot\nabla\gamma^\alpha + \nabla\nabla\gamma^\alpha\cdot\mathbf{n})\gamma\gamma^c dv \\ &- \int_{V_\infty} \boldsymbol{\lambda}\circ F\boldsymbol{v}\cdot\nabla\gamma^s\mathbf{n}\cdot\nabla\gamma^\alpha\boldsymbol{\lambda}\cdot\nabla(\gamma\gamma^c)dv \end{aligned} \quad (904)$$

Since $\nabla = \nabla^C + \nabla^S$, the first integral on the right side of Equation (904) may be written as

$$\int_{V_\infty} \nabla\cdot(\boldsymbol{\lambda}\boldsymbol{\lambda}\circ F\boldsymbol{v}\cdot\nabla\gamma^s\mathbf{n}\cdot\nabla\gamma^\alpha\gamma\gamma^c)dv = \int_{V_\infty} \nabla^C\cdot(\boldsymbol{\lambda}\boldsymbol{\lambda}\circ F\boldsymbol{v}\cdot\nabla\gamma^s\mathbf{n}\cdot\nabla\gamma^\alpha\gamma\gamma^c)dv + \int_{V_\infty} \nabla^S\cdot(\boldsymbol{\lambda}\boldsymbol{\lambda}\circ F\boldsymbol{v}\cdot\nabla\gamma^s\mathbf{n}\cdot\nabla\gamma^\alpha\gamma\gamma^c)dv \quad (905)$$

In the first integral on the right side of Equation (905), the operator ∇^c can be moved out from the integration because the operator is with respect to the macroscopic variables. By Equation (239), the second integral on the right side of Equation (904) is zero. Thus Equation (905) becomes

$$\int_{V_\infty} \nabla\cdot(\boldsymbol{\lambda}\boldsymbol{\lambda}\circ F\boldsymbol{v}\cdot\nabla\gamma^s\mathbf{n}\cdot\nabla\gamma^\alpha\gamma\gamma^c)dv = \nabla^C\cdot\int_{V_\infty} \boldsymbol{\lambda}\boldsymbol{\lambda}\circ F\boldsymbol{v}\cdot\nabla\gamma^s\mathbf{n}\cdot\nabla\gamma^\alpha\gamma\gamma^c dv \quad (906)$$

By Equations (21) and (53),

$$\nabla \cdot \boldsymbol{\lambda} = \boldsymbol{\lambda} \cdot \nabla\boldsymbol{\lambda} \cdot \boldsymbol{\lambda} + \mathbf{n} \cdot \nabla\boldsymbol{\lambda} \cdot \mathbf{n} + \boldsymbol{v} \cdot \nabla\boldsymbol{\lambda} \cdot \boldsymbol{v} = \mathbf{n} \cdot \nabla\boldsymbol{\lambda} \cdot \mathbf{n} + \boldsymbol{v} \cdot \nabla\boldsymbol{\lambda} \cdot \boldsymbol{v} \tag{907}$$

Equation (907) enables us to rewrite the second integral on the right side of Equation (904) as

$$\int_{V_\infty} \nabla \cdot \boldsymbol{\lambda}\boldsymbol{\lambda} \circ F\boldsymbol{v} \cdot \nabla\gamma^{\mathrm{s}}\mathbf{n} \cdot \nabla\gamma^{\alpha}\gamma\gamma^{\mathrm{c}}dv = \int_{V_\infty} \mathbf{n} \cdot \nabla\boldsymbol{\lambda} \cdot \mathbf{n}\boldsymbol{\lambda} \circ F\boldsymbol{v} \cdot \nabla\gamma^{\mathrm{s}}\mathbf{n} \cdot \nabla\gamma^{\alpha}\gamma\gamma^{\mathrm{c}}dv + \int_{V_\infty} \boldsymbol{v} \cdot \nabla\boldsymbol{\lambda} \cdot \boldsymbol{v}\boldsymbol{\lambda} \circ F\boldsymbol{v} \cdot \nabla\gamma^{\mathrm{s}}\mathbf{n} \cdot \nabla\gamma^{\alpha}\gamma\gamma^{\mathrm{c}}dv \tag{908}$$

Since $\boldsymbol{\lambda} \cdot \nabla = \boldsymbol{\lambda} \cdot \nabla^{\mathrm{c}}$, the third integral on the right side of Equation (904) becomes

$$\int_{V_\infty} \boldsymbol{\lambda} \cdot \nabla\boldsymbol{\lambda} \circ F\boldsymbol{v} \cdot \nabla\gamma^{\mathrm{s}}\mathbf{n} \cdot \nabla\gamma^{\alpha}\gamma\gamma^{\mathrm{c}}dv = \int_{V_\infty} \boldsymbol{\lambda} \cdot \nabla^{\mathrm{c}}\boldsymbol{\lambda} \circ F\boldsymbol{v} \cdot \nabla\gamma^{\mathrm{s}}\mathbf{n} \cdot \nabla\gamma^{\alpha}\gamma\gamma^{\mathrm{c}}dv \tag{909}$$

With Equations (200) and (257) applied to the fourth integral on the right side of Equation (904), we obtain

$$\int_{V_\infty} \boldsymbol{\lambda} \circ F\,\boldsymbol{\lambda} \cdot (\nabla\boldsymbol{v} \cdot \nabla\gamma^{\mathrm{s}} + \nabla\nabla\gamma^{\mathrm{s}} \cdot \boldsymbol{v})\mathbf{n} \cdot \nabla\gamma^{\alpha}\gamma\gamma^{\mathrm{c}}dv = -\int_{V_\infty} \boldsymbol{\lambda} \circ F\boldsymbol{v} \cdot \nabla\boldsymbol{\lambda} \cdot \nabla\gamma^{\mathrm{s}}\mathbf{n} \cdot \nabla\gamma^{\alpha}\gamma\gamma^{\mathrm{c}}dv \tag{910}$$

Similarly, we have the fifth term on the right side of Equation (904),

$$\int_{V_\infty} \boldsymbol{\lambda} \circ F\,\boldsymbol{\lambda} \cdot (\nabla\mathbf{n} \cdot \nabla\gamma^{\alpha} + \nabla\nabla\gamma^{\alpha} \cdot \mathbf{n})\boldsymbol{v} \cdot \nabla\gamma^{\mathrm{s}}\gamma\gamma^{\mathrm{c}}dv = -\int_{V_\infty} \boldsymbol{\lambda} \circ F\mathbf{n} \cdot \nabla\boldsymbol{\lambda} \cdot \nabla\gamma^{\alpha}\boldsymbol{v} \cdot \nabla\gamma^{\mathrm{s}}\gamma\gamma^{\mathrm{c}}dv \tag{911}$$

The last term on the right-hand side of Equation (904) may be rewritten as

$$\int_{V_\infty} \boldsymbol{\lambda} \circ F\boldsymbol{v} \cdot \nabla\gamma^{\mathrm{s}}\mathbf{n} \cdot \nabla\gamma^{\alpha}\boldsymbol{\lambda} \cdot \nabla(\gamma\gamma^{\mathrm{c}})dv = \int_{V_\infty} \nabla^{\mathrm{c}}(\gamma\gamma^{\mathrm{c}}) \circ F\boldsymbol{v} \cdot \nabla\gamma^{\mathrm{s}}\mathbf{n} \cdot \nabla\gamma^{\alpha}dv \tag{912}$$

By Equations (165)–(167) and (275),

$$\int_{V_\infty} \boldsymbol{\lambda} \circ F\boldsymbol{v} \cdot \nabla\gamma^{\mathrm{s}}\mathbf{n} \cdot \nabla\gamma^{\alpha}\boldsymbol{\lambda} \cdot \nabla(\gamma\gamma^{\mathrm{c}})dv = -\sum_{\mathrm{pts}} \mathbf{e} \circ F - \sum_{\mathrm{edge}\alpha\beta\varepsilon} \mathbf{e} \circ F \tag{913}$$

Substituting Equations (904), (908)–(911), and (913) into Equation (904) yields

$$\begin{aligned}
\int_{C_{\alpha\beta\varepsilon}} \nabla^{\mathrm{c}} \circ F dc
&= \nabla^{\mathrm{C}} \cdot \int_{V_\infty} \boldsymbol{\lambda}\boldsymbol{\lambda} \circ F\boldsymbol{v} \cdot \nabla\gamma^{\mathrm{s}}\mathbf{n} \cdot \nabla\gamma^{\alpha}\gamma\gamma^{\mathrm{c}}dv - \int_{V_\infty} \mathbf{n} \cdot \nabla\boldsymbol{\lambda} \cdot \mathbf{n}\boldsymbol{\lambda} \circ F\boldsymbol{v} \cdot \nabla\gamma^{\mathrm{s}}\mathbf{n} \cdot \nabla\gamma^{\alpha}\gamma\gamma^{\mathrm{c}}dv \\
&\quad - \int_{V_\infty} \boldsymbol{v} \cdot \nabla\boldsymbol{\lambda} \cdot \boldsymbol{v}\boldsymbol{\lambda} \circ F\boldsymbol{v} \cdot \nabla\gamma^{\mathrm{s}}\mathbf{n} \cdot \nabla\gamma^{\alpha}\gamma\gamma^{\mathrm{c}}dv - \int_{V_\infty} \boldsymbol{\lambda} \cdot \nabla^{\mathrm{c}}\boldsymbol{\lambda} \circ F\boldsymbol{v} \cdot \nabla\gamma^{\mathrm{s}}\mathbf{n} \cdot \nabla\gamma^{\alpha}\gamma\gamma^{\mathrm{c}}dv \\
&\quad + \int_{V_\infty} \boldsymbol{\lambda} \circ F\boldsymbol{v} \cdot \nabla\boldsymbol{\lambda} \cdot \nabla\gamma^{\mathrm{s}}\mathbf{n} \cdot \nabla\gamma^{\alpha}\gamma\gamma^{\mathrm{c}}dv + \int_{V_\infty} \boldsymbol{\lambda} \circ F\mathbf{n} \cdot \nabla\boldsymbol{\lambda} \cdot \nabla\gamma^{\alpha}\boldsymbol{v} \cdot \nabla\gamma^{\mathrm{s}}\gamma\gamma^{\mathrm{c}}dv \\
&\quad + \sum_{\mathrm{pts}} \mathbf{e} \circ F + \sum_{\mathrm{edge}\alpha\beta\varepsilon} \mathbf{e} \circ F
\end{aligned} \tag{914}$$

By Equation (275),

$$\begin{aligned}\int_{C_{\alpha\beta\varepsilon}} \nabla^{C} \circ F dc &= \nabla^{C}. \int_{C_{\alpha\beta\varepsilon}} \boldsymbol{\lambda\lambda} \circ F dc - \int_{C_{\alpha\beta\varepsilon}} \mathbf{n} \cdot \nabla\boldsymbol{\lambda} \cdot \mathbf{n}\boldsymbol{\lambda} \circ F dc \\ &- \int_{C_{\alpha\beta\varepsilon}} \boldsymbol{v} \cdot \nabla\boldsymbol{\lambda} \cdot \boldsymbol{v}\boldsymbol{\lambda} \circ F dc - \int_{C_{\alpha\beta\varepsilon}} (\boldsymbol{\lambda} \cdot \nabla^{c}\boldsymbol{\lambda}) \circ F dc \\ &+ \int_{C_{\alpha\beta\varepsilon}} \boldsymbol{\lambda} \circ F\boldsymbol{v} \cdot \nabla\boldsymbol{\lambda} \cdot \boldsymbol{v} dc + \int_{C_{\alpha\beta\varepsilon}} \boldsymbol{\lambda} \circ F\mathbf{n} \cdot \nabla\boldsymbol{\lambda} \cdot \mathbf{n} dc \\ &+ \sum_{\text{pts}} \mathbf{e} \circ F + \sum_{\text{edge}\alpha\beta\varepsilon} \mathbf{e} \circ F \end{aligned} \tag{915}$$

Rearranging Equation (915) finally yields

$$\int_{C_{\alpha\beta\varepsilon}} \nabla^{C} \circ F dc = \nabla^{C} \cdot \int_{C_{\alpha\beta\varepsilon}} \boldsymbol{\lambda\lambda} \circ F dc - \int_{C_{\alpha\beta\varepsilon}} (\boldsymbol{\lambda} \cdot \nabla^{C}\boldsymbol{\lambda}) \circ F dc + \sum_{\text{pts}} \mathbf{e} \circ F + \sum_{\text{edge}\alpha\beta\varepsilon} \mathbf{e} \circ F$$

Remark. Equation (902) contains the following three formulas:

Gradient

$$\int_{C_{\alpha\beta\varepsilon}} \nabla^{C} f dc = \nabla^{C} \cdot \int_{C_{\alpha\beta\varepsilon}} \boldsymbol{\lambda\lambda} f dc - \int_{C_{\alpha\beta\varepsilon}} (\boldsymbol{\lambda} \cdot \nabla^{C}\boldsymbol{\lambda}) f dc + \sum_{\text{pts}} \mathbf{e} f + \sum_{\text{edge}\alpha\beta\varepsilon} \mathbf{e} f \tag{916}$$

Divergence

$$\int_{C_{\alpha\beta\varepsilon}} \nabla^{c} \cdot \mathbf{f} dc = \nabla^{C} \cdot \int_{C_{\alpha\beta\varepsilon}} \mathbf{f} dc - \int_{C_{\alpha\beta\varepsilon}} (\boldsymbol{\lambda} \cdot \nabla^{c}\boldsymbol{\lambda}) \cdot \mathbf{f} dc + \sum_{\text{pts}} \mathbf{e} \cdot \mathbf{f} + \sum_{\text{edge}\alpha\beta\varepsilon} \mathbf{e} \cdot \mathbf{f} \tag{917}$$

Curl

$$\begin{aligned}\int_{C_{\alpha\beta\varepsilon}} \nabla^{C} \times \mathbf{f} dc &= \nabla^{C} \cdot \int_{C_{\alpha\beta\varepsilon}} \boldsymbol{\lambda\lambda} \times \mathbf{f} dc - \int_{C_{\alpha\beta\varepsilon}} (\boldsymbol{\lambda} \cdot \nabla^{C}\boldsymbol{\lambda}) \times \mathbf{f} dc \\ &+ \sum_{\text{pts}} \mathbf{e} \times \mathbf{f} + \sum_{\text{edge}\alpha\beta\varepsilon} \mathbf{e} \times \mathbf{f} \end{aligned} \tag{918}$$

Theorem 68 also has the following alternative form:

Theorem 70.

$$\int_{C_{\alpha\beta\varepsilon}} \left.\frac{\partial f}{\partial t}\right|_{\mathbf{l}} dc = \frac{\partial}{\partial t} \int_{C_{\alpha\beta\varepsilon}} f dc + \int_{C_{\alpha\beta\varepsilon}} \boldsymbol{\lambda} \cdot \nabla^{c}\boldsymbol{\lambda} \cdot \mathbf{w}^{s} f dc - \sum_{\text{pts}} \mathbf{e} \cdot \mathbf{w}^{c} f - \sum_{\text{ext}} \mathbf{e} \cdot \mathbf{w}^{c} f \tag{919}$$

where ext refers to the points of intersection of the curve $C_{\alpha\beta\varepsilon}$ with the external boundary of the REV.

Proof. By Equation (133), the integral under study may be expressed as

$$\int_{C_{\alpha\beta\varepsilon}} \left.\frac{\partial f}{\partial t}\right|_{\mathbf{l}} dc = \int_{C_{\alpha\beta\varepsilon}} \left.\frac{\partial f}{\partial t}\right|_{\mathbf{x}} + \mathbf{w}^{s} \cdot \nabla f dc \tag{920}$$

By the definitions of the indicator functions $\gamma(\mathbf{x}, \xi)$, $\gamma^{s}(\mathbf{u}(\mathbf{x} + \xi), t)$, $\gamma^{\alpha}(\mathbf{x} + \xi, t)$ and $\gamma^{c}(\mathbf{1}(\mathbf{x} + \xi), t)$, the integral of the right-hand side of Equation (920) may be expressed as

$$\int_{C_{\alpha\beta\varepsilon}} \left.\frac{\partial f}{\partial t}\right|_{\mathbf{l}} \mathrm{d}c = \int_{V_\infty} \left.\frac{\partial f}{\partial t}\right|_{\mathbf{x}} \gamma\gamma^{\mathrm{c}}\mathbf{v}\cdot\nabla\gamma^{\mathrm{s}}\mathbf{n}\cdot\nabla\gamma^{\alpha}\mathrm{d}v + \int_{V_\infty} \mathbf{w}^{\mathrm{s}}\cdot\nabla f\gamma\gamma^{\mathrm{c}}\mathbf{v}\cdot\nabla\gamma^{\mathrm{s}}\mathbf{n}\cdot\nabla\gamma^{\alpha}\mathrm{d}v \quad (921)$$

An application of the chain rule to the integrals on the right side of Equation (921) yields

$$\begin{aligned}
&\int_{C_{\alpha\beta\varepsilon}} \left.\frac{\partial f}{\partial t}\right|_{\mathbf{l}} \mathrm{d}c \\
&= \int_{V_\infty} \frac{\partial}{\partial t}(f\gamma\gamma^{\mathrm{c}}\mathbf{v}\cdot\nabla\gamma^{\mathrm{s}}\mathbf{n}\cdot\nabla\gamma^{\alpha})\mathrm{d}v - \int_{V_\infty} f\left(\frac{\partial\gamma}{\partial t}\gamma^{\mathrm{c}} + \gamma\frac{\partial\gamma^{\mathrm{c}}}{\partial t}\right)\mathbf{v}\cdot\nabla\gamma^{\mathrm{s}}\mathbf{n}\cdot\nabla\gamma^{\alpha}\mathrm{d}v \\
&\quad - \int_{V_\infty} f\gamma\gamma^{\mathrm{c}}\left(\frac{\partial\mathbf{v}}{\partial t}\cdot\nabla\gamma^{\mathrm{s}} + \mathbf{v}\cdot\nabla\frac{\partial\gamma^{\mathrm{s}}}{\partial t}\right)\mathbf{n}\cdot\nabla\gamma^{\alpha}\mathrm{d}v \\
&\quad - \int_{V_\infty} f\gamma\gamma^{\mathrm{c}}\mathbf{v}\cdot\nabla\gamma^{\mathrm{s}}\left(\frac{\partial\mathbf{n}}{\partial t}\cdot\nabla\gamma^{\alpha} + \mathbf{n}\cdot\nabla\frac{\partial\gamma^{\alpha}}{\partial t}\right)\mathrm{d}v \\
&\quad + \int_{V_\infty} \nabla\cdot(\mathbf{w}^{\mathrm{s}}f\gamma\gamma^{\mathrm{c}}\mathbf{v}\cdot\nabla\gamma^{\mathrm{s}}\mathbf{n}\cdot\nabla\gamma^{\alpha})\mathrm{d}v - \int_{V_\infty} \nabla\cdot\mathbf{w}^{\mathrm{s}}f\gamma\gamma^{\mathrm{c}}\mathbf{v}\cdot\nabla\gamma^{\mathrm{s}}\mathbf{n}\cdot\nabla\gamma^{\alpha}\mathrm{d}v \\
&\quad - \int_{V_\infty} f\mathbf{w}^{\mathrm{s}}\cdot\nabla(\gamma\gamma^{\mathrm{c}})\mathbf{v}\cdot\nabla\gamma^{\mathrm{s}}\mathbf{n}\cdot\nabla\gamma^{\alpha}\mathrm{d}v \\
&\quad - \int_{V_\infty} f\gamma\gamma^{\mathrm{c}}\mathbf{w}^{\mathrm{s}}\cdot(\nabla\mathbf{v}\cdot\nabla\gamma^{\mathrm{s}} + \nabla\nabla\gamma^{\mathrm{s}}\cdot\mathbf{v})\mathbf{n}\cdot\nabla\gamma^{\alpha}\mathrm{d}v \\
&\quad - \int_{V_\infty} f\gamma\gamma^{\mathrm{c}}\mathbf{v}\cdot\nabla\gamma^{\mathrm{s}}\mathbf{w}^{\mathrm{s}}\cdot(\nabla\mathbf{n}\cdot\nabla\gamma^{\alpha} + \nabla\nabla\gamma^{\alpha}\cdot\mathbf{n})\mathrm{d}v \qquad (922)
\end{aligned}$$

In the first integral on the right side of Equation (922), the differentiation $\partial/\partial t$ may be moved out from the integration. By Equation (231), the fifth integral is zero. By Equations (180), (223), and (228),

$$\frac{\partial\gamma}{\partial t} = -\mathbf{w}\cdot\nabla\gamma, \quad \frac{\partial\gamma^{\mathrm{c}}}{\partial t} = -\mathbf{w}\cdot\nabla\gamma^{\mathrm{c}}$$
$$\frac{\partial\gamma^{\mathrm{s}}}{\partial t} = -\mathbf{w}\cdot\nabla\gamma^{\mathrm{s}}, \quad \frac{\partial\gamma^{\alpha}}{\partial t} = -\mathbf{w}\cdot\nabla\gamma^{\alpha}$$

With these and Equations (194), (195), (199), and (200) applied to the integrals on the right side of Equation (922), we obtain

$$\begin{aligned}
&\int_{C_{\alpha\beta\varepsilon}} \left.\frac{\partial f}{\partial t}\right|_{\mathbf{l}} \mathrm{d}c \\
&= \frac{\partial}{\partial t}\int_{V_\infty} f\gamma\gamma^{\mathrm{c}}\mathbf{v}\cdot\nabla\gamma^{\mathrm{s}}\mathbf{n}\cdot\nabla\gamma^{\alpha}\mathrm{d}v + \int_{V_\infty} f\mathbf{w}\cdot\nabla(\gamma\gamma^{\mathrm{c}})\mathbf{v}\cdot\nabla\gamma^{\mathrm{s}}\mathbf{n}\cdot\nabla\gamma^{\alpha}\mathrm{d}v \\
&\quad + \int_{V_\infty} f\gamma\gamma^{\mathrm{c}}\mathbf{v}\cdot\nabla\mathbf{w}\cdot\nabla\gamma^{\mathrm{s}}\mathbf{n}\cdot\nabla\gamma^{\alpha}\mathrm{d}v + \int_{V_\infty} f\gamma\gamma^{\mathrm{c}}\mathbf{v}\cdot\nabla\nabla\gamma^{\mathrm{s}}\cdot\mathbf{w}\mathbf{n}\cdot\nabla\gamma^{\alpha}\mathrm{d}v \\
&\quad + \int_{V_\infty} f\gamma\gamma^{\mathrm{c}}\mathbf{v}\cdot\nabla\gamma^{\mathrm{s}}\mathbf{n}\cdot\nabla\mathbf{w}\cdot\nabla\gamma^{\alpha}\mathrm{d}v + \int_{V_\infty} f\gamma\gamma^{\mathrm{c}}\mathbf{v}\cdot\nabla\gamma^{\mathrm{s}}\mathbf{n}\cdot\nabla\nabla\gamma^{\alpha}\cdot\mathbf{w}\mathrm{d}v
\end{aligned}$$

$$
\begin{aligned}
&- \int_{V_\infty} \nabla \cdot \mathbf{w}^{s} f\gamma\gamma^{c}\boldsymbol{v} \cdot \nabla\gamma^{s}\mathbf{n} \cdot \nabla\gamma^{\alpha} dv - \int_{V_\infty} f\mathbf{w}^{s} \cdot \nabla(\gamma\gamma^{c})\boldsymbol{v} \cdot \nabla\gamma^{s}\mathbf{n} \cdot \nabla\gamma^{\alpha} dv \\
&- \int_{V_\infty} f\gamma\gamma^{c}\boldsymbol{v} \cdot \nabla\gamma^{s}\mathbf{w}^{s} \cdot \nabla\nabla\gamma^{\alpha} \cdot \mathbf{n} dv - \int_{V_\infty} f\gamma\gamma^{c}\mathbf{w}^{s} \cdot \nabla\nabla\gamma^{s} \cdot \boldsymbol{v}\mathbf{n} \cdot \nabla\gamma^{\alpha} dv
\end{aligned}
\tag{923}
$$

Regrouping the terms on the right side of Equation (923) yields

$$
\begin{aligned}
\int_{C_{\alpha\beta\varepsilon}} \frac{\partial f}{\partial t}\bigg|_1 dc = {} & \frac{\partial}{\partial t}\int_{V_\infty} f\gamma\gamma^{c}\boldsymbol{v} \cdot \nabla\gamma^{s}\mathbf{n} \cdot \nabla\gamma^{\alpha} dv + \int_{V_\infty} f\gamma\gamma^{c}\boldsymbol{v} \cdot \nabla\mathbf{w} \cdot \nabla\gamma^{s}\mathbf{n} \cdot \nabla\gamma^{\alpha} dv \\
& + \int_{V_\infty} f\gamma\gamma^{c}\boldsymbol{v} \cdot \nabla\gamma^{s}\mathbf{n} \cdot \nabla\mathbf{w} \cdot \nabla\gamma^{\alpha} dv + \int_{V_\infty} f\gamma\gamma^{c}\mathbf{w} \cdot \nabla^{c}\nabla\gamma^{s} \cdot \boldsymbol{v}\mathbf{n} \cdot \nabla\gamma^{\alpha} dv \\
& + \int_{V} f\gamma\gamma^{c}\boldsymbol{v} \cdot \nabla\gamma^{s}\mathbf{w} \cdot \nabla^{c}\nabla\gamma^{\alpha} \cdot \mathbf{n} dv \\
& - \int_{V_\infty} \nabla \cdot (\mathbf{w} - \boldsymbol{\lambda}\boldsymbol{\lambda} \cdot \mathbf{w}) f\gamma\gamma^{c}\boldsymbol{v} \cdot \nabla\gamma^{s}\mathbf{n} \cdot \nabla\gamma^{\alpha} dv \\
& + \int_{V_\infty} f\mathbf{w} \cdot \nabla^{c}(\gamma\gamma^{c})\boldsymbol{v} \cdot \nabla\gamma^{s}\mathbf{n} \cdot \nabla\gamma^{\alpha} dv
\end{aligned}
\tag{924}
$$

By Equation (53), we have

$$
\begin{aligned}
\nabla \cdot \mathbf{w} &= \boldsymbol{\lambda} \cdot \nabla\mathbf{w} \cdot \boldsymbol{\lambda} + \boldsymbol{v} \cdot \nabla\mathbf{w} \cdot \boldsymbol{v} + \mathbf{n} \cdot \nabla\mathbf{w} \cdot \mathbf{n} \\
\nabla \cdot (\boldsymbol{\lambda}\boldsymbol{\lambda} \cdot \mathbf{w}) &= \boldsymbol{v} \cdot \nabla\boldsymbol{\lambda} \cdot \boldsymbol{v}\boldsymbol{\lambda} \cdot \mathbf{w} + \mathbf{n} \cdot \nabla\boldsymbol{\lambda} \cdot \mathbf{n}\boldsymbol{\lambda} \cdot \mathbf{w} + \boldsymbol{\lambda} \cdot \nabla^{c}\boldsymbol{\lambda} \cdot \mathbf{w}^{s} + \boldsymbol{\lambda} \cdot \nabla\mathbf{w} \cdot \boldsymbol{\lambda}
\end{aligned}
$$

By Equations (257) and (260), we thus obtain

$$
\begin{aligned}
\int_{C_{\alpha\beta\varepsilon}} \frac{\partial f}{\partial t}\bigg|_1 dc = {} & \frac{\partial}{\partial t}\int_{V_\infty} f\gamma\gamma^{c}\boldsymbol{v} \cdot \nabla\gamma^{s}\mathbf{n} \cdot \nabla\gamma^{\alpha} dv \\
& + \int_{V_\infty} f\gamma\gamma^{c}\boldsymbol{v} \cdot \nabla\mathbf{w} \cdot \nabla\gamma^{s}\mathbf{n} \cdot \nabla\gamma^{\alpha} dv \\
& + \int_{V_\infty} f\gamma\gamma^{c}\boldsymbol{v} \cdot \nabla\gamma^{s}\mathbf{n} \cdot \nabla\mathbf{w}.\nabla\gamma^{\alpha} dv \\
& - \int_{V_\infty} f\gamma\gamma^{c}\mathbf{w} \cdot \boldsymbol{\lambda}\boldsymbol{v} \cdot \nabla\boldsymbol{\lambda} \cdot \nabla\gamma^{s}\mathbf{n} \cdot \nabla\gamma^{\alpha} dv \\
& - \int_{V_\infty} f\gamma\gamma^{c}\mathbf{w} \cdot \boldsymbol{\lambda}\mathbf{n} \cdot \nabla\boldsymbol{\lambda} \cdot \nabla\gamma^{\alpha}\boldsymbol{v} \cdot \nabla\gamma^{s} dv \\
& - \int_{V_\infty} \boldsymbol{\lambda} \cdot \nabla\mathbf{w} \cdot \boldsymbol{\lambda} f\gamma\gamma^{c}\boldsymbol{v} \cdot \nabla\gamma^{s}\mathbf{n} \cdot \nabla\gamma^{\alpha} dv \\
& - \int_{V_\infty} \mathbf{n} \cdot \nabla\mathbf{w} \cdot \mathbf{n} f\gamma\gamma^{c}\boldsymbol{v} \cdot \nabla\gamma^{s}\mathbf{n} \cdot \nabla\gamma^{\alpha} dv \\
& - \int_{V_\infty} \boldsymbol{v} \cdot \nabla\mathbf{w} \cdot \boldsymbol{v} f\gamma\gamma^{c}\boldsymbol{v} \cdot \nabla\gamma^{s}\mathbf{n} \cdot \nabla\gamma^{\alpha} dv \\
& + \int_{V_\infty} \boldsymbol{v} \cdot \nabla\boldsymbol{\lambda} \cdot \boldsymbol{v}\boldsymbol{\lambda} \cdot \nabla\gamma^{s}\mathbf{n} \cdot \nabla\gamma^{\alpha} f\gamma\gamma^{c}\boldsymbol{\lambda} \cdot \mathbf{w} dv
\end{aligned}
$$

$$+\int_{V_\infty} \mathbf{n}\cdot\nabla\boldsymbol{\lambda}\cdot\mathbf{n}\boldsymbol{\lambda}\cdot\mathbf{w}f\gamma\gamma^c\boldsymbol{v}\cdot\nabla\gamma^s\mathbf{n}\cdot\nabla\gamma^\alpha dv$$
$$+\int_{V_\infty} \boldsymbol{\lambda}\cdot\nabla^c\boldsymbol{\lambda}\cdot\mathbf{w}^s f\gamma\gamma^c\boldsymbol{v}\cdot\nabla\gamma^s\mathbf{n}\cdot\nabla\gamma^\alpha dv$$
$$+\int_{V_\infty} \boldsymbol{\lambda}\cdot\nabla\mathbf{w}\cdot\boldsymbol{\lambda}f\gamma\gamma^c\boldsymbol{v}\cdot\nabla\gamma^s\mathbf{n}\cdot\nabla\gamma^\alpha dv$$
$$+\int_{V_\infty} f\mathbf{w}\cdot\nabla^c(\gamma\gamma^c)\boldsymbol{v}\cdot\nabla\gamma^s\mathbf{n}\cdot\nabla\gamma^\alpha dv \tag{925}$$

which may be implied to

$$\int_{C_{\alpha\beta\varepsilon}} \left.\frac{\partial f}{\partial t}\right|_1 dc = \frac{\partial}{\partial t}\int_{V_\infty} f\gamma\gamma^c\boldsymbol{v}\cdot\nabla\gamma^s\mathbf{n}\cdot\nabla\gamma^\alpha dv + \int_{V_\infty} \boldsymbol{\lambda}\cdot\nabla^c\boldsymbol{\lambda}\cdot\mathbf{w}^s f\gamma\gamma^c\boldsymbol{v}\cdot\nabla\gamma^s\mathbf{n}\cdot\nabla\gamma^\alpha dv$$
$$+\int_{V_\infty} f\mathbf{w}\cdot\nabla^c(\gamma\gamma^c)\boldsymbol{v}\cdot\nabla\gamma^s\mathbf{n}\cdot\nabla\gamma^\alpha dv \tag{926}$$

By Equations (275), (165), (166), and (167), we finally have

$$\int_{C_{\alpha\beta\varepsilon}} \left.\frac{\partial f}{\partial t}\right|_1 dc = \frac{\partial}{\partial t}\int_{C_{\alpha\beta\varepsilon}} f dc + \int_{C_{\alpha\beta\varepsilon}} \boldsymbol{\lambda}\cdot\nabla^c\boldsymbol{\lambda}\cdot\mathbf{w}^s f dc - \sum_{\text{pts}} \mathbf{e}\cdot\mathbf{w}^c f - \sum_{\text{ext}} \mathbf{e}\cdot\mathbf{w}^c f$$

Theorem 69 also has the following alternative form

Theorem 71.

$$\int_{C_{\alpha\beta\varepsilon}} \nabla^c\circ F dc = -\int_{C_{\alpha\beta\varepsilon}} (\boldsymbol{\lambda}\cdot\nabla^c\boldsymbol{\lambda})\circ F dc + \sum_{\text{pts}} \mathbf{e}\circ F + \sum_{\text{ext}} \mathbf{e}\circ F \tag{927}$$

Proof. By the indicator functions $\gamma(\mathbf{x},\boldsymbol{\xi})$, $\gamma^s(\mathbf{u}(\mathbf{x}+\boldsymbol{\xi}),t)$, $\gamma^\alpha(\mathbf{x}+\boldsymbol{\xi},t)$, and $\gamma^c(1(\mathbf{x}+\boldsymbol{\xi}),t)$, the integral under study can be written as

$$\int_{C_{\alpha\beta\varepsilon}} \nabla^c\circ F dc = \int_{V_\infty} \nabla^c\circ F\boldsymbol{v}\cdot\nabla\gamma^s\mathbf{n}\cdot\nabla\gamma^\alpha\gamma\gamma^c dv \tag{928}$$

An application of the chain rule to the right side of Equation (928) yields

$$\int_{C_{\alpha\beta\varepsilon}} \nabla^c\circ F dc = \int_{V_\infty} \nabla\cdot(\boldsymbol{\lambda}\boldsymbol{\lambda}\circ F\boldsymbol{v}\cdot\nabla\gamma^s\mathbf{n}\cdot\nabla\gamma^\alpha\gamma\gamma^c)dv$$
$$-\int_{V_\infty} \nabla\cdot\boldsymbol{\lambda}\boldsymbol{\lambda}\circ F\boldsymbol{v}\cdot\nabla\gamma^s\mathbf{n}\cdot\nabla\gamma^\alpha\gamma\gamma^c dv$$
$$-\int_{V_\infty} \boldsymbol{\lambda}\cdot\nabla\boldsymbol{\lambda}\circ F\boldsymbol{v}\cdot\nabla\gamma^s\mathbf{n}\cdot\nabla\gamma^\alpha\gamma\gamma^c dv$$
$$-\int_{V_\infty} \boldsymbol{\lambda}\circ F\,\boldsymbol{\lambda}\cdot(\nabla\boldsymbol{v}\cdot\nabla\gamma^s + \nabla\nabla\gamma^s\cdot\boldsymbol{v})\mathbf{n}\cdot\nabla\gamma^\alpha\gamma\gamma^c dv$$
$$-\int_{V_\infty} \boldsymbol{\lambda}\circ F\boldsymbol{v}\cdot\nabla\gamma^s\boldsymbol{\lambda}\cdot(\nabla\mathbf{n}\cdot\nabla\gamma^\alpha + \nabla\nabla\gamma^\alpha\cdot\mathbf{n})\gamma\gamma^c dv$$
$$-\int_{V_\infty} \boldsymbol{\lambda}\circ F\boldsymbol{v}\cdot\nabla\gamma^s\mathbf{n}\cdot\nabla\gamma^\alpha\boldsymbol{\lambda}\cdot\nabla(\gamma\gamma^c)dv \tag{929}$$

By Equation (231), the first integral on the right side of Equation (929) is zero. By Equations (21) and (53),

$$\nabla \cdot \boldsymbol{\lambda} = \boldsymbol{\lambda} \cdot \nabla \boldsymbol{\lambda} \cdot \boldsymbol{\lambda} + \mathbf{n} \cdot \nabla \boldsymbol{\lambda} \cdot \mathbf{n} + \boldsymbol{v} \cdot \nabla \boldsymbol{\lambda} \cdot \boldsymbol{v} = \mathbf{n} \cdot \nabla \boldsymbol{\lambda} \cdot \mathbf{n} + \boldsymbol{v} \cdot \nabla \boldsymbol{\lambda} \cdot \boldsymbol{v} \tag{930}$$

Equation (930) allows us to rewrite the second integral on the right side of Equation (929) as

$$\int_{V_\infty} \nabla \cdot \boldsymbol{\lambda}\boldsymbol{\lambda} \circ F\boldsymbol{v} \cdot \nabla\gamma^{\mathrm{s}}\mathbf{n} \cdot \nabla\gamma^{\alpha}\gamma\gamma^{\mathrm{c}}\mathrm{d}v = \int_{V_\infty} \mathbf{n} \cdot \nabla \boldsymbol{\lambda} \cdot \mathbf{n}\boldsymbol{\lambda} \circ F\boldsymbol{v} \cdot \nabla\gamma^{\mathrm{s}}\mathbf{n} \cdot \nabla\gamma^{\alpha}\gamma\gamma^{\mathrm{c}}\mathrm{d}v + \int_{V_\infty} \boldsymbol{v} \cdot \nabla \boldsymbol{\lambda} \cdot \boldsymbol{v}\boldsymbol{\lambda} \circ F\boldsymbol{v} \cdot \nabla\gamma^{\mathrm{s}}\mathbf{n} \cdot \nabla\gamma^{\alpha}\gamma\gamma^{\mathrm{c}}\mathrm{d}v \tag{931}$$

Since $\boldsymbol{\lambda} \cdot \nabla = \boldsymbol{\lambda} \cdot \nabla^{\mathrm{c}}$, the third integral on the right side of Equation (929) becomes

$$\int_{V_\infty} \boldsymbol{\lambda} \cdot \nabla \boldsymbol{\lambda} \circ F\boldsymbol{v} \cdot \nabla\gamma^{\mathrm{s}}\mathbf{n} \cdot \nabla\gamma^{\alpha}\gamma\gamma^{\mathrm{c}}\mathrm{d}v = \int_{V_\infty} \boldsymbol{\lambda} \cdot \nabla^{\mathrm{c}} \boldsymbol{\lambda} \circ F\boldsymbol{v} \cdot \nabla\gamma^{\mathrm{s}}\mathbf{n} \cdot \nabla\gamma^{\alpha}\gamma\gamma^{\mathrm{c}}\mathrm{d}v \tag{932}$$

Applying Equations (200) and (261) to the fourth integral on the right side of Equation (929), we obtain

$$\int_{V_\infty} \boldsymbol{\lambda} \circ F\,\boldsymbol{\lambda} \cdot (\nabla \boldsymbol{v} \cdot \nabla\gamma^{\mathrm{s}} + \nabla\nabla\gamma^{\mathrm{s}} \cdot \boldsymbol{v})\mathbf{n} \cdot \nabla\gamma^{\alpha}\gamma\gamma^{\mathrm{c}}\mathrm{d}v = -\int_{V_\infty} \boldsymbol{\lambda} \circ F\boldsymbol{v} \cdot \nabla \boldsymbol{\lambda} \cdot \nabla\gamma^{\mathrm{s}}\mathbf{n} \cdot \nabla\gamma^{\alpha}\gamma\gamma^{\mathrm{c}}\mathrm{d}v \tag{933}$$

Similarly, we have the fifth term on the right side of Equation (929),

$$\int_{V_\infty} \boldsymbol{\lambda} \circ F\,\boldsymbol{\lambda} \cdot (\nabla \mathbf{n} \cdot \nabla\gamma^{\alpha} + \nabla\nabla\gamma^{\alpha} \cdot \mathbf{n})\boldsymbol{v} \cdot \nabla\gamma^{\mathrm{s}}\gamma\gamma^{\mathrm{c}}\mathrm{d}v = -\int_{V_\infty} \boldsymbol{\lambda} \circ F\mathbf{n} \cdot \nabla \boldsymbol{\lambda} \cdot \nabla\gamma^{\alpha}\boldsymbol{v} \cdot \nabla\gamma^{\mathrm{s}}\gamma\gamma^{\mathrm{c}}\mathrm{d}v \tag{934}$$

The last term on the right-hand side of Equation (930) may be rewritten as

$$\int_{V_\infty} \boldsymbol{\lambda} \circ F\boldsymbol{v} \cdot \nabla\gamma^{\mathrm{s}}\mathbf{n} \cdot \nabla\gamma^{\alpha}\boldsymbol{\lambda} \cdot \nabla(\gamma\gamma^{\mathrm{c}})\mathrm{d}v = \int_{V_\infty} \nabla^{\mathrm{c}}(\gamma\gamma^{\mathrm{c}}) \circ F\boldsymbol{v} \cdot \nabla\gamma^{\mathrm{s}}\mathbf{n} \cdot \nabla\gamma^{\alpha}\mathrm{d}v \tag{935}$$

By Equations (165)–(167) and (275),

$$\int_{V_\infty} \boldsymbol{\lambda} \circ F\boldsymbol{v} \cdot \nabla\gamma^{\mathrm{s}}\mathbf{n} \cdot \nabla\gamma^{\alpha}\boldsymbol{\lambda} \cdot \nabla(\gamma\gamma^{\mathrm{c}})\mathrm{d}v = -\sum_{\mathrm{pts}} \mathbf{e} \circ F - \sum_{\mathrm{ext}} \mathbf{e} \circ F \tag{936}$$

Substituting Equations (931)–(934) and (936) into Equation (929) yields

$$\begin{aligned}
\int_{C_{\alpha\beta\varepsilon}} \nabla^{\mathrm{c}} \circ F\mathrm{d}c = &-\int_{V_\infty} \mathbf{n} \cdot \nabla \boldsymbol{\lambda} \cdot \mathbf{n}\boldsymbol{\lambda} \circ F\boldsymbol{v} \cdot \nabla\gamma^{\mathrm{s}}\mathbf{n} \cdot \nabla\gamma^{\alpha}\gamma\gamma^{\mathrm{c}}\mathrm{d}v \\
&-\int_{V_\infty} \boldsymbol{v} \cdot \nabla \boldsymbol{\lambda} \cdot \boldsymbol{v}\boldsymbol{\lambda} \circ F\boldsymbol{v} \cdot \nabla\gamma^{\mathrm{s}}\mathbf{n} \cdot \nabla\gamma^{\alpha}\gamma\gamma^{\mathrm{c}}\mathrm{d}v \\
&-\int_{V_\infty} \boldsymbol{\lambda} \cdot \nabla^{\mathrm{c}} \boldsymbol{\lambda} \circ F\boldsymbol{v} \cdot \nabla\gamma^{\mathrm{s}}\mathbf{n} \cdot \nabla\gamma^{\alpha}\gamma\gamma^{\mathrm{c}}\mathrm{d}v \\
&+\int_{V_\infty} \boldsymbol{\lambda} \circ F\boldsymbol{v} \cdot \nabla \boldsymbol{\lambda} \cdot \nabla\gamma^{\mathrm{s}}\mathbf{n} \cdot \nabla\gamma^{\alpha}\gamma\gamma^{\mathrm{c}}\mathrm{d}v \\
&+\int_{V_\infty} \boldsymbol{\lambda} \circ F\mathbf{n} \cdot \nabla \boldsymbol{\lambda} \cdot \nabla\gamma^{\alpha}\boldsymbol{v} \cdot \nabla\gamma^{\mathrm{s}}\gamma\gamma^{\mathrm{c}}\mathrm{d}v + \sum_{\mathrm{pts}} \mathbf{e} \circ F + \sum_{\mathrm{ext}} \mathbf{e} \circ F
\end{aligned} \tag{937}$$

By Equation (275), we finally obtain

$$\int_{C_{\alpha\beta\varepsilon}} \nabla^c \circ F dc = -\int_{C_{\alpha\beta\varepsilon}} (\boldsymbol{\lambda} \cdot \nabla^c \boldsymbol{\lambda}) \circ F dc + \sum_{\text{pts}} \mathbf{e} \circ F + \sum_{\text{ext}} \mathbf{e} \circ F$$

Remark. Equation (927) contains the following three formulas:

Gradient

$$\int_{C_{\alpha\beta\varepsilon}} \nabla^c f dc = -\int_{C_{\alpha\beta\varepsilon}} (\boldsymbol{\lambda} \cdot \nabla^c \boldsymbol{\lambda}) f dc + \sum_{\text{pts}} \mathbf{e} f + \sum_{\text{ext}} \mathbf{e} f \tag{938}$$

Divergence

$$\int_{C_{\alpha\beta\varepsilon}} \nabla^c \cdot \mathbf{f} dc = -\int_{C_{\alpha\beta\varepsilon}} (\boldsymbol{\lambda} \cdot \nabla^c \boldsymbol{\lambda}) \cdot \mathbf{f} dc + \sum_{\text{pts}} \mathbf{e} \cdot \mathbf{f} + \sum_{\text{ext}} \mathbf{e} \cdot \mathbf{f} \tag{939}$$

Curl

$$\int_{C_{\alpha\beta\varepsilon}} \nabla^c \times \mathbf{f} dc = -\int_{C_{\alpha\beta\varepsilon}} (\boldsymbol{\lambda} \cdot \nabla^c \boldsymbol{\lambda}) \times \mathbf{f} dc + \sum_{\text{pts}} \mathbf{e} \times \mathbf{f} + \sum_{\text{ext}} \mathbf{e} \times \mathbf{f} \tag{940}$$

6. APPLICATIONS IN TRANSPORT-PHENOMENA MODELING AND SCALING

Multiscale theorems provide simple and powerful tools for modeling multiscale phenomena at various scales and for interchanging among those scales. We show this by some sample applications of the theorems in modeling and scaling transport phenomena.

6.1 Introduction

The term *transport phenomena* is used to describe processes in which mass, momentum, energy, and entropy move about in matter. The subject may be treated from a molecular point view (molecular scale l_{mo}; kinetic theory), from a microscopic point of view (microscale l_{mi}; continuum mechanics), from a macroscopic point of view (macroscale l_{ma}; continuum mechanics), or from a megascopic point of view (megascale l_{me}; equipment description). Transport phenomena arise in many disciplines, such as chemical engineering, civil engineering, electrical engineering, mechanical engineering, metallurgical engineering, biological engineering, biomedical engineering, nuclear engineering, space engineering, biotechnology, microelectronics, nanotechnology, polymer science, mathematics, physics, and chemistry. Evolution of transport phenomena has been thus very rapid and extensive. Fully half of the publications in chemical engineering journals have, for example, something to do with transport phenomena (Bird, 2004).

Transport phenomena are governed and modeled by conservation or balance laws of mass, momentum, energy, and entropy. For some systems it may be desirable to perform this balancing (modeling) at a microscopic point in space; for others, it may be desirable to consider a point on a surface; or sometimes a large volume may be the appropriate scale at which to apply the balance equation. Therefore we are often required to model transport processes at various scales and to interchange among those scales (model-upscaling and model-downscaling). The multiscale theorems in Sections 4 and 5 provide a convenient set of rules that facilitate such modeling and model-scaling, as demonstrated in Section 1.3. We further demonstrate this applicability for modeling single-phase turbulent flow (Sections 6.2 and 6.3), heat conduction in two-phase systems (Section 6.4), transport in porous and multiphase systems (Sections 6.5–6.7) and for developing the thermodynamically constrained averaging theory (TCAT) approach for modeling flow and transport phenomena in multiscale porous-medium systems (Section 6.8). The readers are referred to Drew and Passman (1999), Whitaker (1999), Rajagopal and Tao (1995), Slattery (1972, 1999), Slattery et al. (2007), Joseph (1990), Joseph and Renardy (1993), and Truesdell and Rajagopal (2000) for some basic important applications of multiscale theorems in transport phenomena.

Whereas mass, momentum, energy, and entropy transfer developed independently as branches of classical physics long ago (mass transfer, fluid mechanics, heat transfer, and thermodynamics), their unified study has recently found its place as one of the fundamental engineering sciences (transport phenomena). Although less than half a century old, this approach continues to grow and find applications in new fields such as biotechnology, microelectronics, nanotechnology, and polymer science. The primary reason for the unified study of mass, momentum, energy, and entropy transfer is the fact that, in many industrial processes and in nature, the transport of all four entities occurs simultaneously. Also, the four areas share the same kinds of equations and approaches for attacking problems. In this section, we will mainly use this unified approach to study transport phenomena.

6.2 Energy budget equations in turbulent flows

The conventional method of studying turbulence decomposes turbulent transport problems into a large scale for simulations and a small scale for modeling (Davidson, 2004; Lesieur et al., 2005; Piquet, 1999; Pope, 2000; Sagaut, 2006a; Wang, 1997a, 2002). The large-scale equations are obtained from the microscale conservation equations using the ensemble averaging in classical Reynolds average simulations (RAS) and using filtering in large eddy simulations (LES) (Davidson, 2004; Gatski et al., 1996; Lesieur et al., 2005; Piquet, 1999; Pope, 2000; Sagaut, 2006a; Van den Akker, 2006; Wang, 2007). During averaging or filtering, attention is devoted exclusively to obtaining large-scale equations without taking account of geometrical structures and interactions between the large- and small-scale eddies within the averaging domain.

Information regarding such geometrical structures and interactions is actually very important for understanding the interaction between transports at two different scales and in particular for modeling the effect of small-scale transport on large-scale transport. An omission of such information in averaging or filtering has resulted in the very difficult task of modeling Reynolds stress, turbulent energy flux and turbulent mass flux in the RAS (Churchill, 2001; Davidson, 2004; Gatski et al., 1996; Piquet, 1999; Pope, 2000; Wang, 1997, 2002), and of modeling subgrid scale (SGS) stress, SGS energy flux and SGS mass flux in LES (Frisch, 1995; Gatski et al., 1996; Lesieur et al., 2005; Sagaut, 2006a; Wang, 2007).

The multiscale theorems which are presented in the present work are obtained by averaging over an integration domain or over volumes, surfaces, or curves contained within an averaging volume. During averaging, geometrical structures of and interactions among various components (phases, interfaces, common curves, common points, etc.) within the integration or averaging domain have been considered. The application of these theorems to transport phenomena involving turbulence enables us to develop a new strategy of modeling the relationship between transports at different scales in turbulent flows. We demonstrate this here by establishing scale-by-scale energy budget equations in *real space*. We will develop macroscale balance equations of turbulent eddies for various geometrical and topological structures in the next section. These results provide a new tool for examining turbulence both theoretically and computationally.

Turbulent energy cascade is normally studied in wave number space (Frisch, 1995). Here we apply some averaging theorems to develop the scale-by-scale energy budget equations in real space.

It is generally accepted that turbulent flows satisfy conservation of mass and momentum, where for incompressible fluids the former reduces to the divergence free condition of the velocity field and the latter reads

$$\rho\frac{\partial U_j}{\partial t}+\rho U_i\frac{\partial U_j}{\partial x_i}=\frac{\partial \tau_{ij}}{\partial x_i} \tag{941}$$

where $i, j = 1, 2, 3$, the conventional Einstein's summation has been used, ρ is the density, $\mathbf{U} = [U_1, U_2, U_3]^T$ the velocity vector, $\mathbf{x} = [x_1, x_2, x_3]^T$ the position vector and τ_{ij} is the stress tensor. Multiplying both sides of Equation (941) by U_j yields

$$\rho U_j\frac{\partial U_j}{\partial t}+\rho U_iU_j\frac{\partial U_j}{\partial x_i}=U_j\frac{\partial \tau_{ij}}{\partial x_i} \tag{942}$$

This equation can be reformulated as

$$\rho\frac{DE}{Dt}-\frac{\partial(U_j\tau_{ij})}{\partial x_i}=-S_{ij}\tau_{ij} \tag{943}$$

where $E=\frac{1}{2}U_jU_j$, $DE/Dt=(\partial E/\partial t)+U_i(\partial E/\partial x_i)$, and $S_{ij}=\frac{1}{2}((\partial U_i/\partial x_j)+(\partial U_j/\partial x_i))$ is the rate-of-strain tensor. For Newtonian fluids,

$$\tau_{ij}=-p\delta_{ij}+2\rho\nu S_{ij}$$

where $\delta_{ij} = \{^{1,\ i=j}_{0,\ i\neq j}$, p is the pressure and v the viscosity. Therefore we have (Pope, 2000)

$$\frac{DE}{Dt} + \nabla \cdot \mathbf{T}r = -2vS_{ij}S_{ij} \tag{944}$$

in which the divergence free condition of the velocity field has been used, and

$$Tr_i = \frac{U_i p}{\rho} - 2vS_{ij}U_j$$

Turbulent flows consist of eddies at various scales. The Richardson energy cascade shows that the largest eddies attain energy from the mean flow, break down, and then transfer the energy to smaller eddies (Frisch, 1995). The smaller eddies break down further and transfer the energy to an even smaller one. This process will continue until the length of the small eddies reaches the dissipation length η at which these smallest eddies will convert the energy to heat.

Consider one group of such eddies with length scale l and velocity scale U_l, named l-group for convenience. The volume occupied by the l-group eddies is denoted by V_l. This group of eddies draws energy from its neighboring group of eddies with a relatively larger length scale L and velocity scale U_L, named L-group, through the interface S_{Ll} between the two groups. The L-group eddies obtain energy from a neighboring group of eddies with a relatively larger scale through their interface S_L. At the same time, the l-group eddies pass energy to a neighboring group of eddies with a relatively smaller scale through their interface S_l.

Integrating Equation (942) over the volume V_L yields

$$\int_{V_L} \frac{\partial E}{\partial t} dv + \int_{V_L} \mathbf{U} \cdot \nabla E dv + \int_{V_L} \nabla \cdot \mathbf{T}r dv = -2 \int_{V_L} vS_{ij}S_{ij} dv \tag{945}$$

Theorems 39 and 41 (Equations (614) and (634)) enable us to rewrite Equation (945) as

$$\frac{\partial}{\partial t} \int_{V_L} E dv + \int_{S_{lL}} \mathbf{n} \cdot ((\mathbf{U} - \mathbf{w})E + \mathbf{T}r) ds - \int_{S_L} \mathbf{n}^* \cdot ((\mathbf{U} - \mathbf{w})E + \mathbf{T}r) ds$$
$$= -2 \int_{V_L} vS_{ij}S_{ij} dv \tag{946}$$

where $\mathbf{w}$ is the velocity of S_{Ll} and $\mathbf{n}$ and $\mathbf{n}^*$ the unit normal vectors pointing outward from S_{Ll} and S_L, respectively. Mass conservation requires that $\mathbf{U} = \mathbf{w}$ on the surfaces S_{Ll} and S_L. Therefore we have

$$\frac{\partial}{\partial t} \int_{V_L} E dv + \int_{S_{lL}} \mathbf{n} \cdot \mathbf{T}r ds - \int_{S_L} \mathbf{n}^* \cdot \mathbf{T}r ds = -2v \int_{V_L} S_{ij}S_{ij} dv \tag{947}$$

By introducing the notation,

$$\bar{E} = \frac{1}{V_L} \int_{V_L} E dv \tag{948}$$

we obtain

$$\frac{\partial \bar{E}}{\partial t} + Tr_{Ll} - Tr_L = -2\nu \frac{1}{V_L} \int_{V_L} S_{ij} S_{ij} \mathrm{d}v \tag{949}$$

where

$$Tr_{Ll} = \frac{1}{V_L} \int_{S_{lL}} \mathbf{n} \cdot \mathbf{T} r \mathrm{d}s \tag{950}$$

and

$$Tr_L = \frac{1}{V_L} \int_{S_L} \mathbf{n}^* \cdot \mathbf{T} r \mathrm{d}s \tag{951}$$

Tr_{Ll} represents the energy flux rate from the L-group eddies to the l-group eddies, whereas Tr_L stands for the energy flux rate from the neighboring group of eddies with a relatively larger scale to the L-group eddies.

Similarly, integrating Equation (944) over V_l and applying Theorems 39 and 41 yields

$$\frac{\partial}{\partial t} \int_{V_l} E \mathrm{d}v - \int_{S_{Ll}} \mathbf{n} \cdot \mathbf{T} r \mathrm{d}s + \int_{S_l} \mathbf{n}^{**} \cdot \mathbf{T} r \mathrm{d}s = -2\nu \int_{V_l} S_{ij} S_{ij} \mathrm{d}v \tag{952}$$

where $\mathbf{n}^{**}$ is the unit normal vector of S_l. By Equation (948), we have

$$\frac{\partial}{\partial t} \bar{E} - Tr_{lL} + Tr_l = -2\nu \frac{1}{V_l} \int_{V_l} S_{ij} S_{ij} \mathrm{d}v \tag{953}$$

where

$$Tr_{lL} = \frac{1}{V_l} \int_{S_{Ll}} \mathbf{n} \cdot \mathbf{T} r \mathrm{d}s \tag{954}$$

and

$$Tr_l = \frac{1}{V_l} \int_{S_l} \mathbf{n}^{**} \cdot \mathbf{T} r \mathrm{d}s \tag{955}$$

Here Tr_{lL} is the energy flux rate received by the l-group eddies from the L-group eddies, and Tr_l the energy flux rate from the l-group eddies to its neighboring group of eddies with a relatively smaller scale.

Now consider a group of eddies of length scale η at which the dissipation process becomes dominant. This group of eddies draws energy from its neighboring l-group eddies with the scale l located in the inertial range. The interface between these two groups of eddies is denoted by $S_{l\eta}$.V_η is used to represent the volume occupied by the η-group of eddies. By integrating Equation (944) over V_η and applying Theorems 39 and 41, we have

$$\frac{\partial}{\partial t} \int_{V_\eta} E \mathrm{d}v - \int_{S_{l\eta}} \mathbf{n}^{l\eta} \cdot \mathbf{T} r \mathrm{d}s = -2\nu \int_{V_\eta} S_{ij} S_{ij} \mathrm{d}V \tag{956}$$

where $\mathbf{n}^{l\eta}$ is the unit normal vector of $S_{l\eta}$. This equation can be rewritten as

$$\frac{\partial}{\partial t} \bar{E} - Tr_{l\eta} = -2\nu \frac{1}{V_\eta} \int_{V_\eta} S_{ij} S_{ij} \mathrm{d}V \tag{957}$$

By the definition of the dissipation rate ε_t (Frisch, 1995), Equation (957) becomes

$$\frac{\partial}{\partial t}\bar{E} - Tr_{l\eta} = -\varepsilon_t \tag{958}$$

Therefore, the averaging theorems enable us to establish the scale-by-scale energy budget equations in real space (Equations (949), (953), and (958)). Equations (950), (951), (954), and (955) also provide us with the energy flux rates among various groups of eddies.

6.3 Macroscale equations of turbulence eddies

Turbulence flows consist of eddies at various scales and with various geometrical and topological structures (Davidson, 2004; Piquet, 1999; Pope, 2000). Some eddies possess a three-dimensional spatial structure; others demonstrate a two-dimensional thin layer structure; and still others behave like a one-dimensional filament. The classical RAS and LES ignore these structural differences and the interactions among them in developing large-scale equations, resulting in the very difficult task of Reynolds modeling or SGS modeling (Bonn et al., 1993; Brachet, 1990; Brachet, 1991; Douady and Couder, 1993; Douady et al., 1991; Hosokawa and Yamamoto, 1990; Kerr, 1985; She et al., 1991; Vincent and Meneguzzi, 1991). To consider the coupling of fine and coarse scales, multiscale methods have recently been developed in computational science to simulate turbulent flows (Hughes, 1995; Hughes et al., 1998; Hughes et al., 2000; Hughes et al., 2001; Sagaut et al., 2006). Nonlinear dynamical theory has also been applied to study turbulence with special attention to the geometrical and topological structures of turbulent flows. By applying fractal geometry theory, for example, the Kolmogorov turbulence theory has been extended to include the phenomenon of intermittency (Frisch, 1995). Bifurcation and chaos theory has also been employed to study the topological structures of turbulent flows (Moffatt, 2001a, b; Moffatt, 2002).

In this section, we apply the multiscale theorems to develop macroscale equations of turbulence eddies of different structures. Since the multiscale theorems contain information regarding eddy structures of geometrical and topological types and the interactions among eddies of various structures, the macroscale equations developed are endowed with this important information for understanding and modeling turbulence. In light of the multistage theorems and to take account of the geometrical features of eddies and interaction among eddies, we view turbulence as a system that consists of different 'phases'; every phase has its own characteristic length scale and velocity scale; the interaction among different 'phases' takes place on the interfaces among them to reflect transports of mass, momentum, energy, and entropy among the eddies of different scales.

6.3.1 Three-dimensional spatial eddies

In this subsection, we apply the averaging method and the multistage theorems to develop macroscale equations for three-dimensional spatial eddies.

The starting point for averaging is the general microscale conservation equation for the property ψ (Equation (1)). The volume of the *REV* is denoted by δV. In the *REV*, the volume of one 'phase' of a characteristic length l and a velocity scale u_l is denoted by δV^l. S^{lk} stands for the interface between this 'phase' and all other phases. Integrating Equation (1) over δV^l yields

$$\int_{\delta V^l}\left[\frac{\partial(\rho\psi)}{\partial t}+\nabla\cdot(\rho\mathbf{v}\psi)-\nabla\cdot\mathbf{i}-\rho f\right]\mathrm{d}V=\int_{\delta V^l}\rho G\mathrm{d}V \tag{959}$$

where the ∇ in the integrand is the microscopic spatial operator. By applying Theorems 39 and 40 (Equations (614) and (627)), this equation becomes

$$\begin{aligned}&\frac{\partial}{\partial t}\int_{\delta V^l}\rho\psi\mathrm{d}V-\int_{S^{lk}}\mathbf{n}\cdot\mathbf{w}\rho\psi\mathrm{d}S+\nabla\cdot\int_{\delta V^l}\rho\mathbf{v}\psi\mathrm{d}V+\int_{S^{lk}}\mathbf{n}\cdot(\rho\mathbf{v}\psi)\mathrm{d}S\\&-\nabla\cdot\int_{\delta V^l}\mathbf{i}\mathrm{d}V-\int_{S^{lk}}\mathbf{n}\cdot\mathbf{i}\mathrm{d}S-\int_{\delta V^l}\rho\mathbf{f}\mathrm{d}V-\int_{\delta V^l}\rho G\mathrm{d}V=0\end{aligned} \tag{960}$$

where $\mathbf{n}$ is the unit vector normal to S^{lk}, ∇ the macroscopic spatial operator, and $\mathbf{w}$ the velocity of S^{lk}. After introducing the notation,

$$\langle F\rangle=\frac{1}{\delta V^l}\int_{\delta V^l}F\mathrm{d}V,\ \bar{F}=\frac{1}{\langle\rho\rangle\delta V^l}\int_{\delta V^l}\rho F\mathrm{d}V$$

Equation (960) becomes

$$\begin{aligned}&\frac{\partial}{\partial t}(\langle\rho\rangle\delta V^l\bar{\psi})+\nabla\cdot(\langle\rho\rangle\delta V^l\overline{\mathbf{v}\psi}-\delta V^l\langle\mathbf{i}\rangle)-\langle\rho\rangle\delta V^l(\bar{f}+\bar{G})\\&+\int_{S^{lk}}\mathbf{n}\cdot(\rho\mathbf{v}\psi)-\mathbf{n}\cdot\mathbf{w}\rho\psi-\mathbf{n}\cdot\mathbf{i}\mathrm{d}S=0\end{aligned} \tag{961}$$

By introducing

$$\hat{\psi}=\psi-\bar{\psi},\hat{\mathbf{v}}=\mathbf{v}-\bar{\mathbf{v}}$$

and applying the identities

$$\overline{\bar{\mathbf{v}}\hat{\psi}}=0,\ \ \overline{\hat{\mathbf{v}}\bar{\psi}}=0$$

we obtain

$$\begin{aligned}&\frac{\partial}{\partial t}(\langle\rho\rangle\delta V^l\bar{\psi})+\nabla\cdot(\langle\rho\rangle\delta V^l\bar{\mathbf{v}}\bar{\psi})+\nabla\cdot[\langle\rho\rangle\delta V^l\overline{\hat{\mathbf{v}}\hat{\psi}}-\delta V^l\langle\mathbf{i}\rangle]\\&-\langle\rho\rangle\delta V^l(\bar{f}+\bar{G})+\int_{S^{lk}}\mathbf{n}\cdot(\rho\mathbf{v}\psi)-\mathbf{n}\cdot\mathbf{w}\rho\psi-\mathbf{n}\cdot\mathbf{i}\mathrm{d}S=0\end{aligned}$$

By introducing the notation

$$\varepsilon^l=\frac{\delta V^l}{\delta V}$$

we finally obtain the macroscale balance equation of eddies with length scale l and velocity scale u_l

$$\frac{\partial}{\partial t}(\langle\rho\rangle\varepsilon^l\bar{\psi})+\nabla\cdot(\langle\rho\rangle\varepsilon^l\bar{\mathbf{v}}\bar{\psi})+\nabla\cdot[\langle\rho\rangle\varepsilon^l\overline{\hat{\mathbf{v}}\hat{\psi}}-\varepsilon^l\langle\mathbf{i}\rangle]-\langle\rho\rangle\varepsilon^l\bar{f}=\langle\rho\rangle\varepsilon^l\bar{G}+\hat{e}^l\bar{\psi}+\hat{I}^l \tag{962}$$

where

$$\hat{e}^l = \frac{1}{\delta V}\int_{S^{lk}} \rho \mathbf{n}\cdot(\mathbf{w}-\mathbf{v})\mathrm{d}S$$

$$\hat{I}^l = \frac{1}{\delta V}\int_{S^{lk}} \mathbf{n}\cdot[\mathbf{i}+\rho(\mathbf{w}-\mathbf{v})\hat{\psi}]\mathrm{d}\mathbf{S}$$

By making an appropriate choice of ψ, $\mathbf{i}$, f, and G in Table 2, we can obtain the specific macroscale equations that are listed in Table 6.

6.3.2 Two-dimensional surficial eddies

This subsection is on the application of the multiscale theorems in deriving macroscale equations for eddies with a thin layer structure that can be approximated by a surface. Let l and u_l stand for the characteristic length and velocity scales of such eddies. S^{lk} is the surface region (also area) occupied by these eddies in a REV of volume δV. C^{lkj} represents the contact line between these eddies and all others. Integrating Equation (1) over S^{lk} yields

$$\int_{S^{lk}} \left[\frac{\partial(\rho\psi)}{\partial t} + \nabla\cdot(\rho\mathbf{v}\psi) - \nabla\cdot\mathbf{i} - \rho f - \rho G\right]\mathrm{d}S = 0 \tag{963}$$

By applying Theorems 42 and 43 (Equations (636) and (642)) and $\nabla = \nabla^{\mathrm{S}} + \nabla^{\mathrm{c}}$, this equation becomes

$$\begin{aligned}
&\frac{\partial}{\partial t}\int_{S^{lk}} \rho\psi \mathrm{d}S + \nabla\cdot\int_{S^{lk}} \mathbf{w}^{\mathrm{c}}\rho\psi\mathrm{d}S - \int_{S^{lk}} \nabla^{\mathrm{S}}\cdot\mathbf{nn}\cdot\mathbf{w}^{\mathrm{c}}\rho\psi\mathrm{d}S \\
&- \int_{C^{lkj}} \boldsymbol{\nu}\cdot\mathbf{w}^{\mathrm{s}}\rho\psi\mathrm{d}C + \nabla\cdot\int_{S^{lk}} \rho\mathbf{v}\psi\mathrm{d}S + \int_{S^{lk}} \nabla^{\mathrm{c}}\cdot(\rho\mathbf{v}\psi)\mathrm{d}S \\
&- \nabla\cdot\int_{S^{lk}} \mathbf{nn}\cdot(\rho\mathbf{v}\psi)\mathrm{d}S + \int_{S^{lk}} \nabla^{\mathrm{S}}\cdot\mathbf{nn}\cdot(\rho\mathbf{v}\psi)\mathrm{d}S + \int_{C^{lkg}} \boldsymbol{\nu}\cdot(\rho\mathbf{v}\psi)\mathrm{d}C \\
&- \nabla\cdot\int_{S^{lk}} \mathbf{i}\mathrm{d}S - \int_{S^{lk}} \nabla^{\mathrm{c}}\cdot\mathbf{i}\mathrm{d}S + \nabla\cdot\int_{S^{lk}} \mathbf{nn}\cdot\mathbf{i}\mathrm{d}S - \int_{S^{lk}} \nabla^{\mathrm{S}}\cdot\mathbf{nn}\cdot\mathbf{i}\mathrm{d}S \\
&- \int_{C^{lkj}} \boldsymbol{\nu}\cdot\mathbf{i}\mathrm{d}C - \int_{S^{lk}} \rho(f+G)\mathrm{d}S = 0
\end{aligned} \tag{964}$$

where $\mathbf{n}$ is the unit normal vector of $S^{lk}, \boldsymbol{\nu}$ the unit normal vector of C^{lkj} positive outward from S^{lk} such that $\mathbf{n}\cdot\boldsymbol{\nu} = 0$, $\nabla^{\mathrm{c}} = \mathbf{nn}\cdot\nabla$, $\nabla^{\mathrm{S}} = \nabla - \nabla^{\mathrm{c}}$, $\mathbf{w}$ is the velocity of S^{lk}, $\mathbf{w}^{\mathrm{c}} = \mathbf{nn}\cdot\mathbf{w}$, and $\mathbf{w}^{\mathrm{s}} = \mathbf{w} - \mathbf{w}^{\mathrm{c}}$. After introducing the notation,

$$\langle F\rangle = \frac{1}{S^{lk}}\int_{S^{lk}} F\mathrm{d}S$$

$$\bar{F} = \frac{1}{\langle\rho\rangle S^{lk}}\int_{S^{lk}} \rho F\mathrm{d}S$$

Table 6 Macroscale balance equations for 3D spatial eddies

Conservation law	Equation
Mass	$\frac{\partial}{\partial t}(\langle\rho\rangle\varepsilon^l) + \nabla\cdot(\langle\rho\rangle\varepsilon^l\bar{\mathbf{v}}) = \hat{e}^l$ $\hat{e}^l = \frac{1}{\delta V}\int_{S^{lk}} \rho\mathbf{n}\cdot(\mathbf{w}-\mathbf{v})dS$
Momentum	$\frac{\partial}{\partial t}(\langle\rho\rangle\varepsilon^l\bar{\mathbf{v}}) + \nabla\cdot(\langle\rho\rangle\varepsilon^l\bar{\mathbf{v}}\bar{\mathbf{v}}) + \nabla\cdot[\langle\rho\rangle\varepsilon^l\overline{\hat{\mathbf{v}}\hat{\mathbf{v}}} - \varepsilon^l\langle\mathbf{t}\rangle]$ $- \langle\rho\rangle\varepsilon^l\bar{\mathbf{g}} = \hat{e}^l\bar{\mathbf{v}} + \hat{I}^l$ $\hat{e}^l = \frac{1}{\delta V}\int_{S^{lk}} \rho\mathbf{n}\cdot(\mathbf{w}-\mathbf{v})dS$ $\hat{I}^l = \frac{1}{\delta V}\int_{S^{lk}} \mathbf{n}\cdot[\mathbf{i}+\rho(\mathbf{w}-\mathbf{v})\hat{\mathbf{v}}]dS$
Angular momentum	$\frac{\partial}{\partial t}(\langle\rho\rangle\varepsilon^l\overline{\mathbf{r}\times\mathbf{v}}) + \nabla\cdot(\langle\rho\rangle\varepsilon^l\bar{\mathbf{v}}\overline{\mathbf{r}\times\mathbf{v}}) + \nabla\cdot[\langle\rho\rangle\varepsilon^l\overline{\hat{\mathbf{v}}\widehat{\mathbf{r}\times\mathbf{v}}}$ $- \varepsilon^l\langle\mathbf{r}\times\mathbf{t}\rangle] - \langle\rho\rangle\varepsilon^l\overline{\mathbf{r}\times\mathbf{g}} = \hat{e}^l\overline{\mathbf{r}\times\mathbf{v}} + \hat{I}^l$ $\hat{e}^l = \frac{1}{\delta V}\int_{S^{lk}} \rho\mathbf{n}\cdot(\mathbf{w}-\mathbf{v})dS,$ $\hat{I}^l = \frac{1}{\delta V}\int_{S^{lk}} \mathbf{n}\cdot[\mathbf{r}\times\mathbf{t}+\rho(\mathbf{w}-\mathbf{v})\widehat{\mathrm{r}\times\mathrm{v}}]dS$
Energy	$\frac{\partial}{\partial t}(\langle\rho\rangle\varepsilon^l\overline{E}) + \nabla\cdot(\langle\rho\rangle\varepsilon^l\bar{\mathbf{v}}\overline{E}) + \nabla\cdot[\langle\rho\rangle\varepsilon^l\overline{\hat{\mathbf{v}}\hat{E}}$ $- \varepsilon^l\langle\mathbf{t}\cdot\mathbf{v}+\mathbf{q}\rangle] - \langle\rho\rangle\varepsilon^l\overline{(\mathbf{g}\cdot\mathbf{v}+h)} = \hat{e}^l\overline{E} + \hat{I}^l$ $\hat{e}^l = \frac{1}{\delta V}\int_{S^{lk}} \rho\mathbf{n}\cdot(\mathbf{w}-\mathbf{v})dS$ $\hat{I}^l = \frac{1}{\delta V}\int_{S^{lk}} \mathbf{n}\cdot[\mathbf{i}+\rho(\mathbf{w}-\mathbf{v})\hat{E}]dS$ $E = e + \frac{1}{2}v^2$
Entropy	$\frac{\partial}{\partial t}(\langle\rho\rangle\varepsilon^l\bar{\eta}) + \nabla\cdot(\langle\rho\rangle\varepsilon^l\overline{\mathbf{v}\eta}) + \nabla\cdot[\langle\rho\rangle\varepsilon^l\overline{\hat{\mathbf{v}}\hat{\eta}}$ $- \varepsilon^l\langle\phi\rangle] - \langle\rho\rangle\varepsilon^l\overline{\mathbf{b}} = \langle\rho\rangle\varepsilon^l\bar{\Lambda} + \hat{e}^l\bar{\eta} + \hat{I}^l$ $\hat{e}^l = \frac{1}{\delta V}\int_{S^{lk}} \rho\mathbf{n}\cdot(\mathbf{w}-\mathbf{v})dS$ $\hat{I}^l = \frac{1}{\delta V}\int_{S^{lk}} \mathbf{n}\cdot[\phi+\rho(\mathbf{w}-\mathbf{v})\hat{\eta}]dS$

Equation (964) becomes

$$\begin{aligned}
&\frac{\partial}{\partial t}(\langle\rho\rangle S^{lk}\bar{\psi}) + \nabla\cdot(\langle\rho\rangle S^{lk}\overline{\mathbf{v}\psi}) + \nabla\cdot(\langle\rho\rangle S^{lk}\overline{\mathbf{w}^{c}\psi}) \\
&- \overline{\nabla^{S}\cdot\mathbf{nn}\cdot\mathbf{w}^{c}\psi}\langle\rho\rangle S^{lk} + \langle\nabla^{c}\cdot(\rho\mathbf{v}\psi)\rangle S^{lk} \\
&- \nabla\cdot\overline{(\mathbf{nn}\cdot(\mathbf{v}\psi)}\langle\rho\rangle S^{lk}) + \overline{\nabla^{S}\cdot\mathbf{nn}\cdot(\mathbf{v}\psi)}\langle\rho\rangle S^{lk} - \nabla\cdot(\langle\mathbf{i}\rangle S^{lk}) \\
&- \langle\nabla^{c}\cdot\mathbf{i}\rangle S^{lk} + \nabla\cdot(\langle\mathbf{nn}\cdot\mathbf{i}\rangle S^{lk}) - \langle\nabla^{S}\cdot\mathbf{nn}\cdot\mathbf{i}\rangle S^{lk} - \langle\rho\rangle S^{lk}\bar{f} \\
&= \langle\rho\rangle S^{lkj}\bar{G} + \int_{C^{lkj}}\mathbf{v}\cdot(\mathbf{w}^{s}-\mathbf{v})\rho\psi dC + \int_{C^{lkj}}\mathbf{v}\cdot\mathbf{i}dC
\end{aligned} \tag{965}$$

By introducing the notations

$$\hat{\psi} = \psi - \bar{\psi}, \quad \hat{\mathbf{v}} = \mathbf{v} - \bar{\mathbf{v}}, \quad \varepsilon^{lk} = \frac{S^{lk}}{\delta V}$$

and applying the identities

$$\overline{\bar{\mathbf{v}}\hat{\psi}} = \overline{\hat{\mathbf{v}}\bar{\psi}} = 0$$

we obtain the macroscale balance equation for two-dimensional surficial eddies in turbulent flows

$$\begin{aligned}
&\frac{\partial}{\partial t}(\langle\rho\rangle\varepsilon^{lk}\bar{\psi}) + \nabla\cdot(\langle\rho\rangle\varepsilon^{lk}\bar{\mathbf{v}}\bar{\psi}) + \nabla\cdot[\langle\rho\rangle\varepsilon^{lk} \\
&\overline{\hat{\mathbf{v}}\hat{\psi}} + \langle\rho\rangle\varepsilon^{lk}\overline{\mathbf{w}^{c}\psi} - \langle\rho\rangle\varepsilon^{lk}\overline{\mathbf{nn}\cdot(\mathbf{v}\psi)} - \langle\mathbf{i}\rangle\varepsilon^{lk} \\
&+ \langle\mathbf{nn}\cdot\mathbf{i}\rangle\varepsilon^{lk}] - \overline{\nabla^{S}\cdot\mathbf{nn}\cdot\mathbf{w}^{c}\psi}\langle\rho\rangle\varepsilon^{lk} + \langle\nabla^{c}\cdot(\rho\mathbf{v}\psi)\rangle\varepsilon^{lk} \\
&+ \overline{\nabla^{S}\cdot\mathbf{nn}\cdot(\mathbf{v}\psi)}\langle\rho\rangle\varepsilon^{lk} - \langle\nabla^{c}\cdot\mathbf{i}\rangle\varepsilon^{lk} - \langle\nabla^{S}\cdot\mathbf{nn}\cdot\mathbf{i}\rangle\varepsilon^{lk} \\
&- \langle\rho\rangle\varepsilon^{lk}\bar{f} = \langle\rho\rangle\varepsilon^{lk}\bar{G} + \hat{e}^{l}\bar{\psi} + \hat{I}^{l}
\end{aligned} \tag{966}$$

where

$$\hat{e}^{l} = \frac{1}{\delta V}\int_{C^{lkj}}\mathbf{v}\cdot(\mathbf{w}^{s}-\mathbf{v})\rho dC$$

$$\hat{I}^{l} = \frac{1}{\delta V}\int_{C^{lkj}}\mathbf{v}\cdot[\mathbf{i}+(\mathbf{w}^{s}-\mathbf{v})\rho\hat{\psi}]dC$$

By making an appropriate choice of ψ, $\mathbf{i}$, f, and G in Table 2, we can obtain the specific macroscale equations that are listed in Table 7.

6.3.3 One-dimensional vortex filaments

Much effort has been made on the study of vortex filaments in turbulent flows: their existence, birth, death, proliferation, structure, interaction with other eddies, and their role in organizing large-scale structures (Bonn et al., 1993; Brachet, 1990; Brachet, 1991; Douady and Couder, 1993; Douady et al., 1991; Hosokawa and Yamamoto, 1990; Kerr, 1985; She et al., 1991; Vincent and Meneguzzi, 1991). Both

Table 7 Macroscale balance equations for 2D surficial eddies

Conservation law	Equation
Mass	$\frac{\partial}{\partial t}(\langle\rho\rangle\varepsilon^{lk}) + \nabla\cdot(\langle\rho\rangle\varepsilon^{lk}\bar{\mathbf{v}}) + \nabla\cdot[\langle\rho\rangle\varepsilon^{lk}\overline{\mathbf{w}^c}) - \langle\rho\rangle\varepsilon^{lk}\overline{\mathbf{nn}\cdot\mathbf{v}}] - \overline{\nabla^S\cdot\mathbf{nn}\cdot\mathbf{w}^c}\langle\rho\rangle\varepsilon^{lk} + \langle\nabla^c\cdot(\rho\mathbf{v})\rangle\varepsilon^{lk} + \overline{\nabla^S\cdot\mathbf{nn}\cdot\mathbf{v}}\langle\rho\rangle\varepsilon^{lk} = \hat{e}^l$ $\hat{e}^l = \frac{1}{\delta V}\int_{C^{lkj}} \mathbf{\nu}\cdot(\mathbf{w}^s - \mathbf{v})\rho dc$
Momentum	$\frac{\partial}{\partial t}(\langle\rho\rangle\varepsilon^{lk}\bar{\mathbf{v}}) + \nabla\cdot(\langle\rho\rangle\varepsilon^{lk}\bar{\mathbf{v}}\bar{\mathbf{v}}) + \nabla\cdot[\langle\rho\rangle\varepsilon^{lk}\overline{\hat{\mathbf{v}}\hat{\mathbf{v}}} + \langle\rho\rangle\varepsilon^{lk}\overline{\mathbf{w}^c\mathbf{v}} - \langle\rho\rangle\varepsilon^{lk}\overline{\mathbf{nn}\cdot(\mathbf{vv})} - \langle\mathbf{t}\rangle\varepsilon^{lk} + \langle\mathbf{nn}\cdot\mathbf{t}\rangle\varepsilon^{lk}] - \overline{\nabla^S\cdot\mathbf{nn}\times\mathbf{w}^c\mathbf{v}}\langle\rho\rangle\varepsilon^{lk} + \langle\nabla^c\cdot(\rho\mathbf{vv})\rangle\varepsilon^{lk} + \overline{\nabla^S\cdot\mathbf{nn}\cdot(\mathbf{vv})}\langle\rho\rangle\varepsilon^{lk} - \langle\nabla^c\cdot\mathbf{t}\rangle\varepsilon^{lk} - \langle\nabla^S\cdot\mathbf{nn}\cdot\mathbf{t}\rangle\varepsilon^{lk} - \langle\rho\rangle\varepsilon^{lk}\bar{\mathbf{g}} = \hat{e}^l\bar{\mathbf{v}} + \hat{I}^l$ $\hat{e}^l = \frac{\mathbf{1}}{\delta V}\int_{C^{lkj}} \mathbf{\nu}\cdot(\mathbf{w}^s - \mathbf{v})\rho dC$ $\hat{I}^l = \frac{1}{\delta V}\int_{C^{lkj}} \mathbf{\nu}\cdot[\mathbf{t} + (\mathbf{w}^s - \mathbf{v})\rho\hat{\mathbf{v}}]dC$
Angular momentum	$\frac{\partial}{\partial t}(\langle\rho\rangle\varepsilon^{lk}\overline{\mathbf{r}\times\mathbf{v}}) + \nabla\cdot(\langle\rho\rangle\varepsilon^{lk}\overline{\bar{\mathbf{v}}\mathbf{r}\times\mathbf{v}}) + \nabla\cdot[\langle\rho\rangle\varepsilon^{lk}\overline{\hat{\mathbf{v}}\widehat{\mathbf{r}\times\mathbf{v}}} + \langle\rho\rangle\varepsilon^{lk}\overline{\mathbf{w}^c\mathbf{r}\times\mathbf{v}} - \langle\rho\rangle\varepsilon^{lk}\overline{\mathbf{nn}\cdot(\mathbf{vr}\times\mathbf{v})} - \langle\mathbf{r}\times\mathbf{t}\rangle\varepsilon^{lk} + \langle\mathbf{nn}\cdot(\mathbf{r}\times\mathbf{t})\rangle\varepsilon^{lk}] - \overline{\nabla^S\cdot\mathbf{nn}\cdot\mathbf{w}^c\mathbf{r}\times\mathbf{v}}\langle\rho\rangle\varepsilon^{lk} + \langle\nabla^c\cdot(\rho\mathbf{vr}\times\mathbf{v})\rangle\varepsilon^{lk} + \overline{\nabla^S\cdot\mathbf{nn}\cdot(\mathbf{vr}\times\mathbf{v})}\langle\rho\rangle\varepsilon^{lk} - \langle\nabla^c\cdot(\mathbf{r}\times\mathbf{t})\rangle\varepsilon^{lk} - \langle\nabla^S\cdot\mathbf{nn}\cdot(\mathbf{r}\times\mathbf{t})\rangle\varepsilon^{lk} - \langle\rho\rangle\varepsilon^{lk}\overline{\mathbf{r}\times\mathbf{g}} = \hat{e}^l\overline{\mathbf{r}\times\mathbf{v}} + \hat{I}^l$ $\hat{e}^l = \frac{1}{\delta V}\int_{C^{lkj}} \mathbf{\nu}\cdot(\mathbf{w}^s - \mathbf{v})\rho dC,$ $\hat{I}^l = \frac{1}{\delta V}\int_{C^{lkj}} \mathbf{\nu}\cdot[\mathbf{i} + (\mathbf{w} - \mathbf{v})\rho\widehat{\mathbf{r}\times\mathbf{v}}]dC$

Table 7 *(Continued)*

Conservation law	Equation
Energy	$\frac{\partial}{\partial t}(\langle\rho\rangle\varepsilon^{lk}\bar{E}) + \nabla\cdot(\langle\rho\rangle\varepsilon^{lk}\bar{\mathbf{v}}\bar{E} + \nabla\cdot[\langle\rho\rangle\varepsilon^{lk}\overline{\hat{\mathbf{v}}\hat{E}} + \langle\rho\rangle\varepsilon^{lk}\overline{\mathbf{w}^c E} - \langle\rho\rangle\varepsilon^{lk}\overline{\mathbf{nn}\cdot(\mathbf{v}E)} - \langle\mathbf{t}\cdot\mathbf{v}+\mathbf{q}\rangle\varepsilon^{lk} + \langle\mathbf{nn}\cdot(\mathbf{t}\cdot\mathbf{v}+\mathbf{q})\rangle\varepsilon^{lk}] - \overline{\nabla^S\cdot\mathbf{nn}\cdot\mathbf{w}^c E}\langle\rho\rangle\varepsilon^{lk} + \langle\nabla^c\cdot(\rho\mathbf{v}E)\rangle\varepsilon^{lk} + \overline{\nabla^S\cdot\mathbf{nn}\cdot(\mathbf{v}E)}\langle\rho\rangle\varepsilon^{lk} - \langle\nabla^c\cdot(\mathbf{t}\cdot\mathbf{v} + \mathbf{q})\rangle\varepsilon^{lk} - \langle\nabla^S\cdot\mathbf{nn}\cdot(\mathbf{t}\cdot\mathbf{v}+\mathbf{q})\rangle\varepsilon^{lk} - \langle\rho\rangle\varepsilon^{lk}\overline{(\mathbf{g}\cdot\mathbf{v}+h)} = \hat{e}^l\bar{E} + \hat{I}^l$ $\hat{e}^l = \frac{1}{\delta V}\int_{C^{lkj}} \mathbf{v}\cdot(\mathbf{w}^s - \mathbf{v})\rho dC,$ $\hat{I}^l = \frac{1}{\delta V}\int_{C^{lkj}} \mathbf{v}\cdot[\mathbf{i} + (\mathbf{w}^s - \mathbf{v})\rho\hat{E}]dC$ $E = e + \frac{1}{2}v^2$
Entropy	$\frac{\partial}{\partial t}(\langle\rho\rangle\varepsilon^{lk}\bar{\eta}) + \nabla\cdot(\langle\rho\rangle\varepsilon^{lk}\bar{\mathbf{v}}\bar{\eta}) + \nabla\cdot[\langle\rho\rangle\varepsilon^{lk}\overline{\hat{\mathbf{v}}\hat{\eta}} + \langle\rho\rangle\varepsilon^{lk}\overline{\mathbf{w}^c\eta} - \langle\rho\rangle\varepsilon^{lk}\overline{\mathbf{nn}\cdot(\mathbf{v}\eta)} - \langle\phi\rangle\varepsilon^{lk} + \langle\mathbf{nn}\cdot\phi\rangle\varepsilon^{lk}] - \overline{\nabla^S\cdot\mathbf{nn}\cdot\mathbf{w}^c\eta}\langle\rho\rangle\varepsilon^{lk} + \langle\nabla^c\cdot(\rho\mathbf{v}\eta)\rangle\varepsilon^{lk} - \overline{\nabla^S\cdot\mathbf{nn}\cdot v\eta}\langle\rho\rangle\varepsilon^{lk} + \langle\nabla^c\cdot\phi\rangle\varepsilon^{lk} - \langle\nabla^c\cdot\mathbf{nn}\cdot\phi\rangle\varepsilon^{lk} - \langle\rho\rangle\varepsilon^{lk}\bar{b} = \langle\rho\rangle\varepsilon^{lk}\bar{\Lambda} + \hat{e}\bar{\eta} + \hat{I}$ $\hat{e} = \frac{1}{\delta V}\int_{C^{lkj}} \mathbf{v}\cdot(\mathbf{w}^s - \mathbf{v})\rho dC,$ $\hat{I} = \frac{1}{\delta V}\int_{C^{lkj}} \mathbf{v}\cdot[\phi + (\mathbf{w}^s - \mathbf{v})\rho\hat{\eta}]dC$

numerical and experimental studies show that the vortex filaments are actually tubes with an approximately circular cross-section. Their diameter is of the order of Kolmogorov dissipation scale η, and their length is some where between the Taylor scale and the integral scale l_0[701,719]. As $\eta \ll l_0$, it is not unreasonable to approximate the filaments using mathematical curves. In this subsection, we apply the averaging method and the multiscale theorems to develop macroscale equations for these one-dimensional vortex filaments.

Integrating Equation (1) over a filament C^l yields

$$\int_{C^l}\left[\frac{\partial(\rho\psi)}{\partial t} + \nabla\cdot(\rho\mathbf{v}\psi) - \nabla\cdot\mathbf{i} - \rho f - \rho G\right]\mathrm{d}C = 0 \tag{967}$$

Applying Theorems 46 and 47 (Equations (678) and (685)) and $\nabla = \nabla^S + \nabla^c$, this equation becomes

$$\begin{aligned}
&\frac{\partial}{\partial t}\int_{C^l} \rho\psi dC + \nabla\cdot\int_{C^l} \mathbf{w}^s\rho\psi dC \\
&+ \int_{C^l} \boldsymbol{\lambda}\cdot\nabla^c\boldsymbol{\lambda}\cdot\mathbf{w}^s\rho\psi dC - \sum_{\text{pts}} \mathbf{e}\cdot\mathbf{w}^c\rho\psi \\
&+ \int_{C^l} \nabla^S\cdot(\rho\mathbf{v}\psi)dC + \nabla\cdot\int_{C^l}(\rho\mathbf{v}\psi)^c dC \\
&- \int_{C^l} \boldsymbol{\lambda}\cdot\nabla^c\boldsymbol{\lambda}\cdot(\rho\mathbf{v}\psi)^s dC + \sum_{\text{pts}} \mathbf{e}\cdot(\rho\mathbf{v}\psi)^c \\
&- \int_{C^l} \nabla^S\cdot\mathbf{i}dC - \nabla\cdot\int_{C^l}\mathbf{i}^c + \int_{C^l}\boldsymbol{\lambda}\cdot\nabla^c\boldsymbol{\lambda}\cdot\mathbf{i}^s dC \\
&- \sum_{\text{pts}}\mathbf{e}\cdot\mathbf{i}^c - \int_{C^l}\rho f dC - \int_{C^l}\rho G dC = 0
\end{aligned} \tag{968}$$

where C^l is the filament and also stands for its length, pts refers to the endpoints of C^l, $\boldsymbol{\lambda}$ the unit vector tangent to C^l, $\mathbf{e}$ the unit vector tangent to C^l at its endpoints that is positive in the outward direction from C^l such that $\boldsymbol{\lambda}\cdot\mathbf{e} = \pm 1$, ∇^c the microscopic curvilinear operator such that $\nabla^c = \boldsymbol{\lambda}\boldsymbol{\lambda}\cdot\nabla$, $\mathbf{w}$ the velocity of C^l, $\mathbf{w}^c$ the tangent velocity component of C^l such that $\mathbf{w}^c = \boldsymbol{\lambda}\boldsymbol{\lambda}\cdot\mathbf{w}$, and $\mathbf{w}^s = \mathbf{w} - \mathbf{w}^c$. By introducing the notation

$$\langle F\rangle = \frac{1}{C^l}\int_{C^l} F dC$$

$$\bar{F} = \frac{1}{\langle\rho\rangle C^l}\int_{C^l}\rho F dC$$

we have

$$\begin{aligned}
&\frac{\partial}{\partial t}(\langle\rho\rangle C^l\bar{\psi}) + \nabla\cdot(\langle\rho\rangle C^l\overline{\mathbf{w}^s\psi}) + \langle\rho\rangle C^l\overline{\boldsymbol{\lambda}\cdot\nabla^c\boldsymbol{\lambda}\cdot\mathbf{w}^s\psi} \\
&- \sum_{\text{pts}}\mathbf{e}\cdot\mathbf{w}^c\rho\psi + \langle\nabla\cdot(\rho\mathbf{v}\psi)\rangle C^l + \nabla\cdot(\langle\rho\rangle C^l\overline{\mathbf{v}\psi}) \\
&- \nabla\cdot(\langle\rho\rangle C^l\overline{(\mathbf{v}\psi)^s}) - \langle\rho\rangle C^l\overline{\boldsymbol{\lambda}\cdot\nabla^c\boldsymbol{\lambda}\cdot(\mathbf{v}\psi)^s} \\
&+ \sum_{\text{pts}}\mathbf{e}\cdot(\rho\mathbf{v}\psi)^c - \langle\nabla^S\cdot\mathbf{i}\rangle C^l - \nabla\cdot(\langle\mathbf{i}^c\rangle C^l) \\
&+ \langle\boldsymbol{\lambda}\cdot\nabla^c\boldsymbol{\lambda}\cdot\mathbf{i}^s\rangle C^l - \sum_{\text{pts}}\mathbf{e}\cdot\mathbf{i}^c \\
&- f\langle\rho\rangle C^l - G\langle\rho\rangle C^l = 0
\end{aligned} \tag{969}$$

By introducing

$$\hat{\mathbf{v}} = \mathbf{v} - \bar{\mathbf{v}}, \hat{\psi} = \psi - \overline{\psi}, \varepsilon^l = \frac{C^l}{\delta V}$$

and applying the identities

$$\overline{\bar{\mathbf{v}}\hat{\psi}} = \overline{\hat{\mathbf{v}}\bar{\psi}} = 0$$

we obtain the macroscale balance equation for vortex filaments.

$$\begin{aligned}
&\frac{\partial}{\partial t}(\langle\rho\rangle\varepsilon^l\bar{\psi}) + \nabla\cdot(\langle\rho\rangle\varepsilon^l\bar{\mathbf{v}}\bar{\psi}) + \nabla\cdot[\langle\rho\rangle\varepsilon^l\overline{\hat{\mathbf{v}}\hat{\psi}} \\
&+ \langle\rho\rangle\varepsilon^l\overline{\mathbf{w}^{\mathrm{s}}\psi} - \langle\rho\rangle\varepsilon^l\overline{(\mathbf{v}\psi)^{\mathrm{s}}} - \langle\mathbf{i}^{\mathrm{c}}\rangle\varepsilon^l] + \langle\rho\rangle\varepsilon^l\overline{\boldsymbol{\lambda}\cdot\nabla^{\mathrm{c}}\boldsymbol{\lambda}\cdot\mathbf{w}^{\mathrm{s}}\psi} \\
&+ \langle\nabla^{\mathrm{S}}\cdot(\rho\mathbf{v}\psi)\rangle\varepsilon^l - \langle\rho\rangle\varepsilon^l\overline{\boldsymbol{\lambda}\cdot\nabla^{\mathrm{c}}\boldsymbol{\lambda}\cdot(\mathbf{v}\psi)^{\mathrm{s}}} - \langle\nabla^{\mathrm{S}}\cdot\mathbf{i}\rangle\varepsilon^l \\
&+ \langle\boldsymbol{\lambda}\cdot\nabla^{\mathrm{c}}\boldsymbol{\lambda}\cdot\mathbf{i}^{\mathrm{s}}\rangle\varepsilon^l - \bar{f}\langle\rho\rangle\varepsilon^l = \bar{G}\langle\rho\rangle\varepsilon^l + \hat{e}\bar{\psi} + \hat{I}
\end{aligned} \tag{970}$$

where

$$\hat{e} = \frac{1}{\delta V}\sum_{\text{pts}}\mathbf{e}\cdot(\mathbf{w}^{\mathrm{c}} - \mathbf{v}^{\mathrm{c}})\rho$$

$$\hat{I} = \frac{1}{\delta V}\sum_{\text{pts}}\mathbf{e}\cdot[\mathbf{i}^{\mathrm{c}} + (\mathbf{w}^{\mathrm{c}} - \mathbf{v}^{\mathrm{c}})\rho\hat{\psi}]$$

By making an appropriate choice of ψ, $\mathbf{i}$, f, and G in Table 2, we can obtain the specific macroscale equations that are listed in Table 8.

6.4 Heat conduction in two-phase systems

In this section, we show the application of the multiscale theorems in developing one- and two-equation macroscale models for heat conduction in two-phase systems. These macroscale models form the foundation of examining heat conduction in two-phase systems and are tools for seeking physical insights into the process. We demonstrate this by developing, based on the two-equation macroscale model, the equivalence of dual-phase-lagging heat conduction and Fourier heat conduction in two-phase systems subject to a lack of local thermal equilibrium.

6.4.1 One- and two-equation models

The microscale model for heat conduction in two-phase systems is well known. It consists of the field equation and the constitutive equation. The field equation comes from the conservation of energy (the first law of thermodynamics). The commonly used constitutive equation is the Fourier law of heat conduction for the relation between the temperature gradient ∇T and the heat flux density vector $\mathbf{q}$ (Wang, 1994).

For transport in two-phase systems, the macroscale is a phenomenological scale that is much larger than the microscale of pores and grains and much smaller than the system length scale. Interest in the macroscale rather than the microscale comes from the fact that a prediction at the microscale is complicated due to the complex microscale geometry of two-phase systems such as porous media, and also because we are usually more interested in large scales of transport for practical applications. Existence of such a macroscale description

Table 8 Macroscale balance equations for vortex filaments

Conservation law	Equation
Mass	$\frac{\partial}{\partial t}(\langle\rho\rangle\varepsilon^l) + \nabla\cdot(\langle\rho\rangle\varepsilon^l\bar{\mathbf{v}}) + \nabla\cdot[\langle\rho\rangle\varepsilon^l\overline{\mathbf{w}^s} - \langle\rho\rangle\varepsilon^l\overline{\mathbf{v}^s}] + \langle\rho\rangle\varepsilon^l\overline{\boldsymbol{\lambda}\cdot\nabla^c\boldsymbol{\lambda}\cdot\mathbf{w}^s} + \langle\nabla^S\cdot(\rho\mathbf{v})\rangle\varepsilon^l - \langle\rho\rangle\varepsilon^l\overline{\boldsymbol{\lambda}\cdot\nabla^c\boldsymbol{\lambda}\cdot\mathbf{v}^s} = \hat{e}$ $\hat{e} = \frac{1}{\delta V}\sum_{\text{pts}}\mathbf{e}\cdot(\mathbf{w}^c - \mathbf{v}^c)\rho$
Momentum	$\frac{\partial}{\partial t}(\langle\rho\rangle\varepsilon^l\bar{\mathbf{v}}) + \nabla\cdot(\langle\rho\rangle\varepsilon^l\bar{\mathbf{v}}\bar{\mathbf{v}}) + \nabla\cdot[\langle\rho\rangle\varepsilon^l\overline{\hat{\mathbf{v}}\hat{\mathbf{v}}} + \langle\rho\rangle\varepsilon^l\overline{\mathbf{w}^s\mathbf{v}} - \langle\rho\rangle\varepsilon^l\overline{(\mathbf{v}\mathbf{v})^s} - \langle\mathbf{t}^c\rangle\varepsilon^l] + \langle\rho\rangle\varepsilon^l\overline{\boldsymbol{\lambda}\cdot\nabla^c\boldsymbol{\lambda}\cdot\mathbf{w}^s\mathbf{v}} + \langle\nabla^S\cdot(\rho\mathbf{v}\mathbf{v})\rangle\varepsilon^l - \langle\rho\rangle\varepsilon^l\overline{\boldsymbol{\lambda}\cdot\nabla^c\boldsymbol{\lambda}\cdot(\mathbf{v}\mathbf{v})^s} - \langle\nabla^S\cdot\mathbf{t}\rangle\varepsilon^l + \langle\boldsymbol{\lambda}\cdot\nabla^c\boldsymbol{\lambda}\cdot\mathbf{t}^s\rangle\varepsilon^l - \bar{f}\langle\rho\rangle\varepsilon^l = G\langle\rho\rangle\varepsilon^l + \hat{e}\bar{\mathbf{v}} + \hat{I}$ $\hat{e} = \frac{1}{\delta V}\sum_{\text{pts}}\mathbf{e}\cdot(\mathbf{w}^c - \mathbf{v}^c)\rho,$ $\hat{I} = \frac{1}{\delta V}\sum_{\text{pts}}\mathbf{e}\cdot[\mathbf{t}^c + (\mathbf{w}^c - \mathbf{v}^c)\rho\hat{\mathbf{v}}]$
Angular momentum	$\frac{\partial}{\partial t}(\langle\rho\rangle\varepsilon^l\overline{\mathbf{r}\times\mathbf{v}}) + \nabla\cdot(\langle\rho\rangle\varepsilon^l\bar{\mathbf{v}}\overline{\mathbf{r}\times\mathbf{v}}) + \nabla\cdot[\langle\rho\rangle\varepsilon^l\overline{\hat{\mathbf{v}}\widehat{\mathbf{r}\times\mathbf{v}}} + \langle\rho\rangle\varepsilon^l\overline{\mathbf{w}^s\mathbf{r}\times\mathbf{v}} - \langle\rho\rangle\varepsilon^l\overline{(\mathbf{v}\mathbf{r}\times\mathbf{v})^s} - \langle(\mathbf{r}\times\mathbf{t})^c\rangle\varepsilon^l] + \langle\rho\rangle\varepsilon^l\overline{\boldsymbol{\lambda}\cdot\nabla^c\boldsymbol{\lambda}\cdot\mathbf{w}^s\mathbf{r}\times\mathbf{v}} + \langle\nabla^S\cdot(\rho\mathbf{v}\mathbf{r}\times\mathbf{v})\rangle\varepsilon^l - \langle\rho\rangle\varepsilon^l\overline{\boldsymbol{\lambda}\cdot\nabla^c\boldsymbol{\lambda}\cdot(\mathbf{v}\mathbf{r}\times\mathbf{v})^s} - \langle\nabla^S\cdot(\mathbf{r}\times\mathbf{t})\rangle\varepsilon^l + \langle\boldsymbol{\lambda}\cdot\nabla^c\boldsymbol{\lambda}\cdot(\mathbf{r}\times\mathbf{t})^s\rangle\varepsilon^l - \overline{\mathbf{r}\times\mathbf{g}}\langle\rho\rangle\varepsilon^l = \hat{e}\overline{\mathbf{r}\times\mathbf{v}} + \hat{I}$ $\hat{e} = \frac{1}{\delta V}\sum_{\text{pts}}\mathbf{e}\cdot(\mathbf{w}^c - \mathbf{v}^c)\rho,$ $\hat{I} = \frac{1}{\delta V}\sum_{\text{pts}}\mathbf{e}\cdot[(\mathbf{r}\times\mathbf{t})^c + (\mathbf{w}^c - \mathbf{v}^c)\rho\widehat{\mathbf{r}\times\mathbf{v}}]$
Energy	$\frac{\partial}{\partial t}(\langle\rho\rangle\varepsilon^l\bar{E}) + \nabla\cdot(\langle\rho\rangle\varepsilon^l\bar{\mathbf{v}}\bar{E}) + \nabla\cdot[\langle\rho\rangle\varepsilon^l\overline{\hat{\mathbf{v}}\hat{E}} + \langle\rho\rangle\varepsilon^l\overline{\mathbf{w}^s E} - \langle\rho\rangle\varepsilon^l\overline{(\mathbf{v}E)^s} - \langle(\mathbf{t}\cdot\mathbf{v}+\mathbf{q})^c\rangle\varepsilon^l] + \langle\rho\rangle\varepsilon^l\overline{\boldsymbol{\lambda}\cdot\nabla^c\boldsymbol{\lambda}\cdot\mathbf{w}^s E} + \langle\nabla^S\cdot(\rho\mathbf{v}E)\rangle\varepsilon^l - \langle\rho\rangle\varepsilon^l\overline{\boldsymbol{\lambda}\cdot\nabla^c\boldsymbol{\lambda}\cdot(\mathbf{v}E)^s} - \langle\nabla^S\cdot(\mathbf{t}\cdot\mathbf{v}+\mathbf{q})\rangle\varepsilon^l + \langle\boldsymbol{\lambda}\cdot\nabla^c\boldsymbol{\lambda}\cdot(\mathbf{t}\cdot\mathbf{v}+\mathbf{q})^s\rangle\varepsilon^l - \overline{(\mathbf{g}\cdot\mathbf{v}+h)}\langle\rho\rangle\varepsilon^l = \hat{e}\bar{E} + \hat{I}$

Table 8 (*Continued*)

Conservation law	Equation
	$\hat{e} = \frac{1}{\delta V}\sum_{\text{pts}} \mathbf{e}\cdot(\mathbf{w}^c - \mathbf{v}^c)\rho,$ $\hat{I} = \frac{1}{\delta V}\sum_{\text{pts}} \mathbf{e}\cdot[(\mathbf{t}\cdot\mathbf{v}+\mathbf{q})^c + (\mathbf{w}^c - \mathbf{v}^c)\rho\hat{E}]$ $E = e + \frac{1}{2}v^2$
Entropy	$\frac{\partial}{\partial t}(\langle\rho\rangle\varepsilon^l\bar{\eta}) + \nabla\cdot(\langle\rho\rangle\varepsilon^l\overline{\mathbf{v}\eta}) + \nabla\cdot[\langle\rho\rangle\varepsilon^l\overline{\hat{\mathbf{v}}\hat{\eta}} + \langle\rho\rangle\varepsilon^l\overline{\mathbf{w}^s\eta} - \langle\rho\rangle\varepsilon^l\overline{(\mathbf{v}\eta)^s} - \langle\phi^c\rangle\varepsilon^l] + \langle\rho\rangle\varepsilon^l\overline{\boldsymbol{\lambda}\cdot\nabla^c\boldsymbol{\lambda}\cdot\mathbf{w}^s\eta} + \langle\nabla^S\cdot(\rho\mathbf{v}\eta)\rangle\varepsilon^l - \langle\rho\rangle\varepsilon^l\overline{\boldsymbol{\lambda}\cdot\nabla^c\boldsymbol{\lambda}\cdot(\mathbf{v}\eta)^s} - \langle\nabla^S\cdot\phi\rangle\varepsilon^l + \langle\boldsymbol{\lambda}\cdot\nabla^c\boldsymbol{\lambda}\cdot\phi^s\rangle\varepsilon^l - \bar{b}\langle\rho\rangle\varepsilon^l = \bar{\Lambda}\langle\rho\rangle\varepsilon^l + \hat{e}\bar{\eta} + \hat{I}$ $\hat{e} = \frac{1}{\delta V}\sum_{\text{pts}} \mathbf{e}\cdot(\mathbf{w}^c - \mathbf{v}^c)\rho,$ $\hat{I} = \frac{1}{\delta V}\sum_{\text{pts}} \mathbf{e}\cdot[\phi^c + (\mathbf{w}^c - \mathbf{v}^c)\rho\hat{\eta}]$

equivalent to the microscale behavior requires a good separation of length scales and has been well discussed in Auriault (1991).

To develop a macroscale model of transport in two-phase systems, the method of volume averaging starts with a microscale description (Wang, 2000; Whitaker, 1999). Both conservation and constitutive equations are introduced at the microscale. The resulting microscale field equations are then averaged over an REV, the smallest differential volume resulting in statistically meaningful local averaging properties, to obtain the macroscale field equations. In the process of averaging, the *averaging theorems* are used to convert integrals of gradient, divergence, curl, and partial time derivatives of a function into some combination of gradient, divergence, curl, and partial time derivatives of integrals of the function and integrals over the boundary of the REV (Wang, 2000; Whitaker, 1999). The readers are referred to Wang (2000) and Whitaker (1999) for the details of the method of volume averaging and to Wang (2000) for a summary of the other methods of obtaining macroscale models.

Quintard and Whitaker (1993) use the method of volume averaging to develop one- and two-equation macroscale models for heat conduction in two-phase systems. First, they define the microscale problem by the first law of thermodynamics and the Fourier law of heat conduction (Figure 34)

$$(\rho c)_\beta \frac{\partial T_\beta}{\partial t} = \nabla \cdot (k_\beta \nabla T_\beta), \text{ in the } \beta\text{-phase} \tag{971}$$

$$(\rho c)_\sigma \frac{\partial T_\sigma}{\partial t} = \nabla \cdot (k_\sigma \nabla T_\sigma), \text{ in the } \sigma\text{-phase} \tag{972}$$

$$T_\beta = T_\sigma, \text{ at the } \beta - \sigma \text{ interface } A_{\beta\sigma} \tag{973}$$

$$\mathbf{n}_{\beta\sigma} \cdot k_\beta \nabla T_\beta = \mathbf{n}_{\beta\sigma} \cdot k_\sigma \nabla T_\sigma, \text{ at the } \beta - \sigma \text{ interface } A_{\beta\sigma} \tag{974}$$

Here ρ, c, and k are the density, specific heat and thermal conductivity, respectively. Subscripts β and σ refer to the β- and σ-phases, respectively. $A_{\beta\sigma}$ represents the area of the β–σ interface contained in the REV; $\mathbf{n}_{\beta\sigma}$ is the outward-directed surface normal from the β-phase toward the σ-phase and $\mathbf{n}_{\beta\sigma} = -\mathbf{n}_{\sigma\beta}$ (Figure 34). To be thorough, Quintard and Whitaker (1993) have also specified the initial conditions and the boundary conditions at the entrances and exits of the REV; however, we need not do so for our discussion.

Next Quintard and Whitaker (1993) apply the superficial averaging process to Equations (971) and (972) to obtain,

$$\frac{1}{V_{\text{REV}}} \int_{V_\beta} (\rho c)_\beta \frac{\partial T_\beta}{\partial t} \mathrm{d}V = \frac{1}{V_{\text{REV}}} \int_{V_\beta} \nabla \cdot (k_\beta \nabla T_\beta) \mathrm{d}V \tag{975}$$

and

$$\frac{1}{V_{\text{REV}}} \int_{V_\sigma} (\rho c)_\sigma \frac{\partial T_\sigma}{\partial t} \mathrm{d}V = \frac{1}{V_{\text{REV}}} \int_{V_\sigma} \nabla \cdot (k_\sigma \nabla T_\sigma) \mathrm{d}V \tag{976}$$

where V_{REV}, V_β, and V_σ are the volumes of the REV, β-phase in REV and σ-phase in REV, respectively. We should note that the superficial temperature is evaluated at the centroid of the REV, whereas the phase temperature is evaluated throughout the REV. Neglecting variations of (ρc) within the REV and

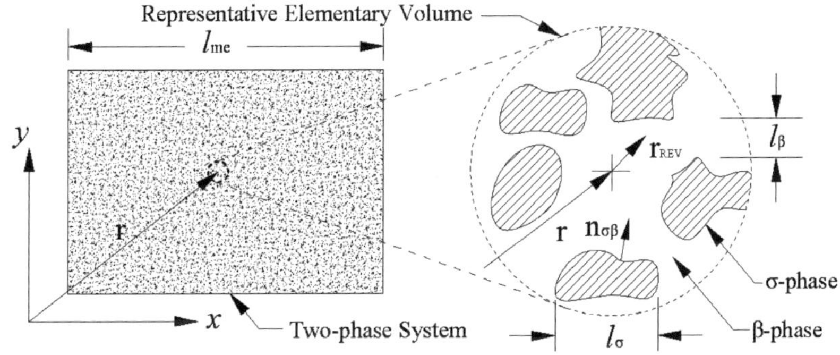

Figure 34 Rigid two-phase system.

considering the system to be rigid so that V_β and V_σ are not functions of time, the volume-averaged form of Equations (971) and (972) are

$$(\rho c)_\beta \frac{\partial \langle T_\beta \rangle}{\partial t} = \langle \nabla \cdot (k_\beta \nabla T_\beta) \rangle \tag{977}$$

and

$$(\rho c)_\sigma \frac{\partial \langle T_\sigma \rangle}{\partial t} = \langle \nabla \cdot (k_\sigma \nabla T_\sigma) \rangle \tag{978}$$

where angle brackets indicate superficial quantities such as

$$\langle T_\beta \rangle = \frac{1}{V_{\mathrm{REV}}} \int_{V_\beta} T_\beta \mathrm{d}V$$

and

$$\langle T_\sigma \rangle = \frac{1}{V_{\mathrm{REV}}} \int_{V_\sigma} T_\sigma \mathrm{d}V$$

The superficial average, however, is an unsuitable variable because it can yield erroneous results. For example, if the temperature of the β-phase were constant, the superficial average would differ from it (Quintard and Whitaker, 1993). However, intrinsic phase averages do not have this shortcoming. These averages are defined by

$$\langle T_\beta \rangle^\beta = \frac{1}{V_\beta} \int_{V_\beta} T_\beta \mathrm{d}V \tag{979}$$

and

$$\langle T_\sigma \rangle^\sigma = \frac{1}{V_\sigma} \int_{V_\sigma} T_\sigma \mathrm{d}V \tag{980}$$

Also, intrinsic averages are related to superficial averages by

$$\langle T_\beta \rangle = \varepsilon_\beta \langle T_\beta \rangle^\beta \tag{981}$$

and

$$\langle T_\sigma \rangle = \varepsilon_\sigma \langle T_\sigma \rangle^\sigma \tag{982}$$

where ε_β and ε_σ are the volume fractions of the β- and σ-phases with $\varepsilon_\beta = \varphi, \varepsilon_\sigma = 1 - \varphi$ with a constant porosity φ for a rigid two-phase system.

Quintard and Whitaker (1993) substitute Equations (981) and (982) into Equations (977) and (978) to obtain

$$\varepsilon_\beta (\rho c)_\beta \frac{\partial \langle T_\beta \rangle^\beta}{\partial t} = \langle \nabla \cdot (k_\beta \nabla T_\beta) \rangle \tag{983}$$

and

$$\varepsilon_\sigma (\rho c)_\sigma \frac{\partial \langle T_\sigma \rangle^\sigma}{\partial t} = \langle \nabla \cdot (k_\sigma \nabla T_\sigma) \rangle \tag{984}$$

Next Quintard and Whitaker (1993) apply the spatial averaging theorem (Theorem 40, Equation (626)) to Equations (983) and (984) and neglect variations of physical properties within the REV. The result is

$$\underbrace{\varepsilon_\beta(\rho c)_\beta \frac{\partial \langle T_\beta \rangle^\beta}{\partial t}}_{\text{accumulation}} = \underbrace{\nabla \cdot \left\{ k_\beta \left[\varepsilon_\beta \nabla \langle T_\beta \rangle^\beta + \langle T_\beta \rangle^\beta \nabla \varepsilon_\beta + \frac{1}{V_{\text{REV}}} \int_{A_{\beta\sigma}} \mathbf{n}_{\beta\sigma} T_\beta \mathrm{d}A \right] \right\}}_{\text{conduction}}$$
$$+ \underbrace{\frac{1}{V_{\text{REV}}} \int_{A_{\beta\sigma}} \mathbf{n}_{\beta\sigma} \cdot k_\beta \nabla T_\beta \mathrm{d}A}_{\text{interfacial flux}} \tag{985}$$

and

$$\underbrace{\varepsilon_\sigma(\rho c)_\sigma \frac{\partial \langle T_\sigma \rangle^\sigma}{\partial t}}_{\text{accumulation}} = \underbrace{\nabla \cdot \left\{ k_\sigma \left[\varepsilon_\sigma \nabla \langle T_\sigma \rangle^\sigma + \langle T_\sigma \rangle^\sigma \nabla \varepsilon_\sigma + \frac{1}{V_{\text{REV}}} \int_{A_{\beta\sigma}} \mathbf{n}_{\beta\sigma} T_\sigma \mathrm{d}A \right] \right\}}_{\text{conduction}}$$
$$+ \underbrace{\frac{1}{V_{\text{REV}}} \int_{A_{\beta\sigma}} \mathbf{n}_{\beta\sigma} \cdot k_\sigma \nabla T_\sigma \mathrm{d}A}_{\text{interfacial flux}} \tag{986}$$

By introducing the spatial decompositions $T_\beta = \langle T_\beta \rangle^\beta + \tilde{T}_\beta$ and $T_\sigma = \langle T_\sigma \rangle^\sigma + \tilde{T}_\sigma$ and by applying scaling arguments and Theorem 40 (Equation (626)), Equations (985) and (986) are simplified into (Quintard and Whitaker, 1993)

$$\varepsilon_\beta(\rho c)_\beta \frac{\partial \langle T_\beta \rangle^\beta}{\partial t} = \nabla \cdot \left\{ k_\beta \left[\varepsilon_\beta \nabla \langle T_\beta \rangle^\beta + \frac{1}{V_{\text{REV}}} \int_{A_{\beta\sigma}} \mathbf{n}_{\beta\sigma} \tilde{T}_\beta \mathrm{d}A \right] \right\}$$
$$+ \frac{1}{V_{\text{REV}}} \int_{A_{\beta\sigma}} \mathbf{n}_{\beta\sigma} \cdot k_\beta \nabla \langle T_\beta \rangle^\beta \mathrm{d}A + \frac{1}{V_{\text{REV}}} \int_{A_{\beta\sigma}} \mathbf{n}_{\beta\sigma} \cdot k_\beta \nabla \tilde{T}_\beta \mathrm{d}A \tag{987}$$

and

$$\varepsilon_\sigma(\rho c)_\sigma \frac{\partial \langle T_\sigma \rangle^\sigma}{\partial t} = \nabla \cdot \left\{ k_\sigma \left[\varepsilon_\sigma \nabla \langle T_\sigma \rangle^\sigma + \frac{1}{V_{\text{REV}}} \int_{A_{\beta\sigma}} \mathbf{n}_{\sigma\beta} \tilde{T}_\sigma \mathrm{d}A \right] \right\}$$
$$+ \frac{1}{V_{\text{REV}}} \int_{A_{\beta\sigma}} \mathbf{n}_{\sigma\beta} \cdot k_\sigma \nabla \langle T_\sigma \rangle^\sigma \mathrm{d}A + \frac{1}{V_{\text{REV}}} \int_{A_{\beta\sigma}} \mathbf{n}_{\sigma\beta} \cdot k_\sigma \nabla \tilde{T}_\sigma \mathrm{d}A \tag{988}$$

After developing the closure for $\tilde{T}_\beta$ and $\tilde{T}_\sigma$ (see Quintard and Whitaker, 1993; Whitaker, 1999; for details), Quintard and Whitaker (1993) obtain a two-equation model

$$\varepsilon_\beta(\rho c)_\beta \frac{\partial \langle T_\beta \rangle^\beta}{\partial t} = \nabla \cdot \{ \mathbf{K}_{\beta\beta} \cdot \nabla \langle T_\beta \rangle^\beta + \mathbf{K}_{\beta\sigma} \cdot \nabla \langle T_\sigma \rangle^\sigma \} + h a_v (\langle T_\sigma \rangle^\sigma - \langle T_\beta \rangle^\beta) \tag{989}$$

and

$$\varepsilon_\sigma(\rho c)_\sigma \frac{\partial \langle T_\sigma \rangle^\sigma}{\partial t} = \nabla \cdot \{ \mathbf{K}_{\sigma\sigma} \cdot \nabla \langle T_\sigma \rangle^\sigma + \mathbf{K}_{\sigma\beta} \cdot \nabla \langle T_\beta \rangle^\beta \} - h a_v (\langle T_\sigma \rangle^\sigma - \langle T_\beta \rangle^\beta) \tag{990}$$

where h and a_v come from modeling of the interfacial flux and are the film heat transfer coefficient and the interfacial area per unit volume, respectively.

$\mathbf{K}_{\beta\beta}$, $\mathbf{K}_{\sigma\sigma}$, $\mathbf{K}_{\beta\sigma}$ and $\mathbf{K}_{\sigma\beta}$ are the effective thermal conductivity tensors, and the coupled thermal conductivity tensors are equal

$$\mathbf{K}_{\beta\sigma} = \mathbf{K}_{\sigma\beta}$$

When the system is isotropic and the physical properties of the two phases are constant, Equations (989) and (990) reduce to

$$\gamma_\beta \frac{\partial \langle T_\beta \rangle^\beta}{\partial t} = k_\beta \Delta \langle T_\beta \rangle^\beta + k_{\beta\sigma} \Delta \langle T_\sigma \rangle^\sigma + ha_v(\langle T_\sigma \rangle^\sigma - \langle T_\beta \rangle^\beta) \tag{991}$$

and

$$\gamma_\sigma \frac{\partial \langle T_\sigma \rangle^\sigma}{\partial t} = k_\sigma \Delta \langle T_\sigma \rangle^\sigma + k_{\sigma\beta} \Delta \langle T_\beta \rangle^\beta + ha_v(\langle T_\sigma \rangle^\sigma - \langle T_\beta \rangle^\beta) \tag{992}$$

where $\gamma_\beta = \varphi(\rho c)_\beta$ and $\gamma_\sigma = (1-\varphi)(\rho c)_\sigma$ are the β- and σ-phases effective thermal capacities, respectively, φ the porosity, k_β and k_σ the effective thermal conductivities of the β- and σ-phases, respectively, and $k_{\beta\sigma} = k_{\sigma\beta}$ is the cross effective thermal conductivity of the two phases.

The one-equation model is valid whenever the two temperatures $\langle T_\beta \rangle^\beta$ and $\langle T_\sigma \rangle^\sigma$ are sufficiently close to each other so that

$$\langle T_\beta \rangle^\beta = \langle T_\sigma \rangle^\sigma = \langle T \rangle \tag{993}$$

This local thermal equilibrium is valid when any one of the following three conditions occurs (Quintard and Whitaker, 1993; Whitaker, 1999): (1) either ε_β or ε_σ tends to zero, (2) the difference in the β- and σ-phases physical properties tends to zero, and (3) the square of the ratio of length scales $(l_{\beta\sigma}/L)^2$ tends to zero (e.g. steady, one-dimensional heat conduction). Here $l_{\beta\sigma}^2 = [\varepsilon_\beta \varepsilon_\sigma(\varepsilon_\beta k_\sigma + \varepsilon_\sigma k_\beta)]/(ha_v)$ and $L = L_T L_{T1}$ with L_T and L_{T1} as the characteristic lengths of $\nabla\langle T \rangle$ and $\nabla\nabla\langle T \rangle$, respectively, such that $\nabla\langle T \rangle = O(\Delta\langle T \rangle / L_T)$ and $\nabla\nabla\langle T \rangle = O(\Delta\langle T \rangle / L_{T1} L_T)$.

When the local thermal equilibrium is valid, (Quintard and Whitaker, 1993) add Equations (989) and (990) to obtain a one-equation model

$$\langle \rho \rangle C \frac{\partial \langle T \rangle}{\partial t} = \nabla \cdot [\mathbf{K}_{\text{eff}} \cdot \nabla \langle T \rangle] \tag{994}$$

Here $\langle \rho \rangle$ is the spatial average density defined by

$$\langle \rho \rangle = \varepsilon_\beta \rho_\beta + \varepsilon_\sigma \rho_\sigma \tag{995}$$

and C is the mass-fraction-weighted thermal capacity given by

$$C = \frac{\varepsilon_\beta (\rho c)_\beta + \varepsilon_\sigma (\rho c)_\sigma}{\varepsilon_\beta \rho_\beta + \varepsilon_\sigma \rho_\sigma} \tag{996}$$

The effective thermal conductivity tensor is

$$\mathbf{K}_{\text{eff}} = \mathbf{K}_{\beta\beta} + 2\mathbf{K}_{\beta\sigma} + \mathbf{K}_{\sigma\sigma} \tag{997}$$

The choice between the one-equation model and the two-equation model has been well discussed in Whitaker (1999) and Quintard and Whitaker (1993). They have also developed methods of determining the effective thermal conductivity tensor $\mathbf{K}_{\text{eff}}$ in the one-equation model and the four coefficients

$\mathbf{K}_{\beta\beta}$, $\mathbf{K}_{\beta\sigma} = \mathbf{K}_{\sigma\beta}$, $\mathbf{K}_{\sigma\sigma}$, and ha_v in the two-equation model. Their studies suggest that the coupling coefficients are on the order of the smaller of $\mathbf{K}_{\beta\beta}$ and $\mathbf{K}_{\sigma\sigma}$. Therefore, the coupled conductive terms should not be omitted in any detailed two-equation model of heat conduction processes. When the principle of local thermal equilibrium is not valid, the commonly used two-equation model in the literature is the one without the coupled conductive terms (Glatzmaier and Ramirez, 1988)

$$\varepsilon_\beta(\rho c)_\beta \frac{\partial \langle T_\beta \rangle^\beta}{\partial t} = \nabla \cdot (\mathbf{K}_{\beta\beta} \nabla \cdot \langle T_\beta \rangle^\beta) + ha_v(\langle T_\sigma \rangle^\sigma - \langle T_\beta \rangle^\beta) \tag{998}$$

and

$$\varepsilon_\sigma(\rho c)_\sigma \frac{\partial \langle T_\sigma \rangle^\sigma}{\partial t} = \nabla \cdot (\mathbf{K}_{\sigma\sigma} \cdot \nabla \langle T_\sigma \rangle^\sigma) - ha_v(\langle T_\sigma \rangle^\sigma - \langle T_\beta \rangle^\beta) \tag{999}$$

On the basis of the above analysis, we now know that the coupled conductive terms $\mathbf{K}_{\beta\sigma} \cdot \nabla\langle T_\sigma \rangle^\sigma$ and $\mathbf{K}_{\sigma\beta} \cdot \nabla\langle T_\beta \rangle^\beta$ cannot be discarded in the exact representation of the two-equation model. However, we could argue that Equations (998) and (999) represent a reasonable approximation of Equations (989) and (990) for a heat conduction process in which $\nabla\langle T_\beta \rangle^\beta$ and $\nabla\langle T_\sigma \rangle^\sigma$ are *sufficiently close* to each other. Under these circumstances $\mathbf{K}_{\beta\beta}$ in Equation (998) would be given by $\mathbf{K}_{\beta\beta} + \mathbf{K}_{\beta\sigma}$ while $\mathbf{K}_{\sigma\sigma}$ in Equation (999) should be interpreted as $\mathbf{K}_{\sigma\beta} + \mathbf{K}_{\sigma\sigma}$. This limitation of Equations (998) and (999) is believed to be the reason behind the paradox of heat conduction in porous media subject to lack of local thermal equilibrium analyzed in Vadasz (2005b). For an isotropic system with constant physical properties of the two phases, Equations (998) and (999) reduce to the traditional formulation of heat conduction in two-phase systems (Bejan, 2004; Bejan et al., 2004; Nield and Bejan, 2006; Vadasz, 2005b)

$$\gamma_\beta \frac{\partial \langle T_\beta \rangle^\beta}{\partial t} = k_{e\beta} \Delta \langle T_\beta \rangle^\beta + ha_v(\langle T_\sigma \rangle^\sigma - \langle T_\beta \rangle^\beta) \tag{1000}$$

and

$$\gamma_\sigma \frac{\partial \langle T_\sigma \rangle^\sigma}{\partial t} = k_{e\sigma} \Delta \langle T_\sigma \rangle^\beta - ha_v(\langle T_\sigma \rangle^\sigma - \langle T_\beta \rangle^\beta) \tag{1001}$$

where we introduce the *equivalent* effective thermal conductivities $k_{e\beta} = k_\beta + k_{\beta\sigma}$ and $k_{e\sigma} = k_\sigma + k_{\sigma\beta}$ for the β- and σ-phases, respectively, to take the above note into account. To describe the thermal energy exchange between solid and gas phases in casting sand, Tzou (1997) has also directly postulated Equations (1000) and (1001) (using k_β and k_σ rather than $k_{e\beta}$ and $k_{e\sigma}$) as a two-step model, parallel to the two-step equations in the microscopic phonon–electron interaction model (Anisimòv et al., 1974; Kaganov et al., 1957; Qiu and Tien, 1993).

6.4.2 Equivalence with dual-phase-lagging heat conduction

The two-equation model developed by applying the multiscale theorems can be used to establish equivalence between the dual-phase-lagging and two-phase system heat conduction (Wang and Wei, 2008; Wang et al., 2008a). We first rewrite Equations (991) and (992) in their operator form

$$\begin{bmatrix} \gamma_\beta(\partial/\partial t) - k_\beta\Delta + h & -k_{\beta\sigma}\Delta - ha_v \\ -k_{\beta\sigma}\Delta - ha_v & \gamma_\beta(\partial/\partial t) - k_\beta\Delta + ha_v \end{bmatrix} \begin{bmatrix} \langle T_\beta\rangle^\beta \\ \langle T_\sigma\rangle^\sigma \end{bmatrix} = 0 \tag{1002}$$

We then obtain an uncoupled form by evaluating the operator determinant such that

$$\left[(\gamma_\beta \frac{\partial}{\partial t} - k_\beta\Delta + ha_v)(\gamma_\beta \frac{\partial}{\partial t} - k_\beta\Delta + ha_v) - (k_{\beta\sigma}\Delta - ha_v)^2\right]\langle T_i\rangle^i = 0 \tag{1003}$$

where the index i can take β or σ. Its explicit form reads, after dividing by $ha_v(\gamma_\beta + \gamma_\sigma)$

$$\frac{\partial\langle T_i\rangle^i}{\partial t} + \tau_q \frac{\partial^2\langle T_i\rangle^i}{\partial t^2} = \alpha\Delta\langle T_i\rangle^i + \alpha\tau_T \frac{\partial}{\partial t}(\Delta\langle T_i\rangle^i) + \frac{\alpha}{k}\left[S(\vec{x},t) + \tau_q \frac{\partial S(\vec{x},t)}{\partial t}\right] \tag{1004}$$

where

$$\tau_q = \frac{\gamma_\beta\gamma_\sigma}{ha_v(\gamma_\beta + \gamma_\sigma)}, \qquad \tau_T = \frac{\gamma_\beta k_\sigma + \gamma_\sigma k_\beta}{ha_v(k_\beta + k_\sigma + 2k_{\beta\sigma})},$$

$$k = k_\beta + k_\sigma + 2k_{\beta\sigma}, \qquad \alpha = \frac{k}{\rho c} = \frac{k_\beta + k_\sigma + 2k_{\beta\sigma}}{\gamma_\beta + \gamma_\sigma}, \tag{1005}$$

$$S(\vec{x},t) + \tau_q \frac{\partial S(\vec{x},t)}{\partial t} = \frac{k_{\beta\sigma}^2 - k_\beta k_\sigma}{ha_v}\Delta^2\langle T_i\rangle^i$$

Therefore, $\langle T_\beta\rangle^\beta$ and $\langle T_\sigma\rangle^\sigma$ satisfy *exactly* the same dual-phase-lagging heat conduction equation (Equation (1004); Tzou, 1995, 1997; Wang and Zhou, 2000, 2001; Wang et al., 2008a, b; Xu and Wang, 2005). The dual-phase-lagging heat conduction equation originates from the first law of thermodynamics and the dual-phase-lagging constitutive relation of heat flux density (Tzou, 1995, 1997; Wang and Zhou, 2000, 2001; Wang et al., 2008a, b). It is developed in examining energy transport involving high-rate heating in which the non-equilibrium thermodynamic transition and the microstructural effect become important associated with a shortening of the response time. In addition to its application for ultra fast pulse-laser heating, the dual-phase-lagging heat conduction equation also describes and predicts phenomena such as the propagation of temperature pulses in superfluid liquid helium, non-homogeneous-lagging responses in porous media, thermal lagging in amorphous materials, and the effects of material defects and thermomechanical coupling (Tzou, 1997). Furthermore, the dual-phase-lagging heat conduction equation forms a generalized, unified equation with the wave equation, the potential equation, the classical parabolic heat conduction equation, the hyperbolic heat conduction equation, the energy equation in the phonon scattering model (Guyer and Krumhansi, 1966; Joseph and Preziosi, 1989), and the energy equation in the phonon–electron interaction model (Anisimòv et al., 1974; Kaganov et al., 1957; Qiu and Tien, 1993) as its special cases (Tzou, 1995, 1997; Wang and Zhou, 2000, 2001; Wang et al., 2008a, b). This, with the rapid growth of microscale heat conduction of high-rate heat flux, has attracted the recent research effort on the dual-phase-lagging heat conduction equations (Tzou, 1995, 1997; Wang and Zhou, 2000, 2001; Wang et al., 2008a, b; Xu and Wang, 2005).

Note that Equations (991) and (992) are the mathematical representation of the first law of thermodynamics and the Fourier law of heat conduction for heat conduction processes in two-phase systems at the macroscale. Therefore, we have an *exact* equivalence between dual-phase-lagging heat conduction and Fourier heat conduction in two-phase systems. This is significant because all results in these two fields become mutually applicable. In particular, all analytical methods and results (such as the solution structure theorems) in Xu and Wang (2005), Wang and Zhou (2000, 2001), Tzou (1997), Wang (2000a), Wang et al. (2001, 2008b), Wang and Xu (2002), and Xu and Wang (2002) can be applied to study heat conduction in two-phase systems.

By Equation (1005), we can readily obtain that, in two-phase-system heat conduction

$$\frac{\tau_T}{\tau_q} = 1 + \frac{\gamma_\beta^2 k_\sigma + \gamma_\sigma^2 k_\beta - 2\gamma_\beta \gamma_\sigma k_{\beta\sigma}}{\gamma_\beta \gamma_\sigma (k_\beta + k_\sigma + 2k_{\beta\sigma})} \tag{1006}$$

It can be large, equal, or smaller than 1 depending on the sign of $\gamma_\beta^2 k_\sigma + \gamma_\sigma^2 k_\beta - 2\gamma_\beta \gamma_\sigma k_{\beta\sigma}$. Therefore, by the condition for the existence of thermal waves in Xu and Wang (2002), we may have thermal oscillation and resonance for heat conduction in two-phase systems subject to a lack of local thermal equilibrium. This agrees with the experimental data of casting sand tests in Tzou (1997). Discarding the coupled conductive terms in Equations (991) and (992) assumes $k_{\beta\sigma} = 0$ so that τ_T/τ_q is always larger than 1, which leads to the exclusion of thermal oscillation and resonance (Vadasz, 2005a, c, 2006a, b) and generates an inconsistency between theoretical and experimental results in the literature regarding the possibility of thermal waves and resonance in two-phase-system heat conduction (Tzou, 1997; Vadasz, 2005a, c, 2006a, b). The coupled conductive terms in Equations (991) and (992) are thus responsible for thermal oscillation and resonance in two-phase-system heat conduction that is subject to a lack of local thermal equilibrium.

Although each τ_T and τ_q is ha_v dependent, the ratio τ_T/τ_q is not; this makes its evaluation much simpler as detailed in Vadasz (2005c). The readers are also referred to Vadasz (2005c) for the correlations among physical properties in Equation (1005) and those in Hays-Stang and Haji-Sheikh (1999) and Minkowycz et al. (1999).

6.5 Macroscale phase equations in porous and multiphase systems

There are three approaches for deriving macroscale phase models in porous and multiphase systems: the mixture theory of continuum mechanics, the averaging method and the global balance method for each phase Gray and Hassanizadeh (1998). The multiscale theorems play a significant role for all three approaches. We show this in the following section.

6.5.1 Mixture theory of continuum mechanics

This is a model-downscaling approach for developing the macroscale phase equations in porous and multiphase systems. In this approach, a multiphase system is viewed as a superposition in space of a number of single-phase continua (Bedford and Drumheller, 1983; Bowen, 1982; Dobran, 1984; Marle, 1982; Prvost, 1980; Rajagopal and Tao, 1995; Sampaio and Williams, 1979; Thigpen and Berryman, 1985). No microscale presentation of the system is provided and microscale quantities are not introduced. Phase properties are defined at the macroscale. The global balance equations are formed in terms of macroscale properties such that the integrands are macroscale quantities. These global equations can then be localized to obtain the macroscale point equations.

There are two drawbacks associated with this approach. The first is the lack of connection between microscale and macroscale properties. The second is the difficulty in extension to multiphase systems with distinct properties of interfaces and common curves.

In this approach, a global balance analogous to Equation (10) is written with the integrand properties at the macroscale l_{ma} for a property of the α-phase based on the conservation equations or the balance principles (Eringen and Ingram, 1965)

$$\frac{\mathrm{d}}{\mathrm{d}t}\int_V \rho^\alpha \varepsilon^\alpha \psi^\alpha \mathrm{d}V + \int_S \mathbf{n}^* \cdot [\rho^\alpha \varepsilon^\alpha (\mathbf{v}^\alpha - \mathbf{w}_\mathrm{b})\psi^\alpha - \varepsilon^\alpha \mathbf{i}^\alpha]\mathrm{d}S - \int_V \rho^\alpha \varepsilon^\alpha f^\alpha \mathrm{d}V - \int_V \rho^\alpha \varepsilon^\alpha G^\alpha \mathrm{d}V = 0 \qquad (1007)$$

The main differences between this equation and Equation (10) are the superscript α to indicate the α-phase microscale properties and the volume fraction ε^α of the macroscale point occupied by the α-phase. Note also that the integrand quantities are at the macroscale in Equation (1007) but at the microscale in Equation (10). Therefore, the integrand quantities may require a different interpretation and may represent different physical quantities from those in Equation (10). For example, the generation term G in Equation (10) is zero for the conservation equations of mass, momentum, and energy; whereas G^α in Equation (1007) may be non-zero to account for the interaction among phases present at a given *point*.

To transform Equation (1007) to a differential form, we must use Theorems 1 and 2 (Equations (311) and (320)) to bring the time derivative inside the integral and to convert the boundary integral to a volume integral so that

$$\int_V \left[\frac{\partial(\rho^\alpha \varepsilon^\alpha \psi^\alpha)}{\partial t} + \nabla \cdot (\rho^\alpha \varepsilon^\alpha \mathbf{v}^\alpha \psi^\alpha) - \nabla \cdot (\varepsilon^\alpha \mathbf{i}^\alpha) - \rho^\alpha \varepsilon^\alpha f^\alpha - \rho^\alpha \varepsilon^\alpha G^\alpha\right] \mathrm{d}V = 0 \qquad (1008)$$

Because the size of the volume is arbitrary, by the localization theorem Wang (1997) and Wang and Zhou (2000), the integrand in Equation (1008) must be zero as long as Axiom 1 in Section 1.4 is satisfied so that the macroscale point equation can be obtained as

$$\frac{\partial(\rho^\alpha \varepsilon^\alpha \psi^\alpha)}{\partial t} + \nabla \cdot (\rho^\alpha \varepsilon^\alpha \mathbf{v}^\alpha \psi^\alpha) - \nabla \cdot (\varepsilon^\alpha \mathbf{i}^\alpha) - \rho^\alpha \varepsilon^\alpha f^\alpha - \rho^\alpha \varepsilon^\alpha G^\alpha = 0 \qquad (1009)$$

In principle, this approach can be applied to interfaces and common lines if proper macroscale properties can be identified. However, this is often difficult and may lead to inconsistency. For example, some mixture theories do not consider interfaces, but treat the fluid phases as immiscible (Bowen, 1982; Thigpen and Berryman, 1985). Note also that Equation (1009) is at the macroscale. It does not describe the conservation in as much microscale detail as Equation (14). As a matter of fact, Equation (1009) is in some sense an average representation of Equation (14); the quantities in the two equations must be somehow related. However, mixture theory does not provide a precise correspondence between the terms and quantities in the two equations. It is not clear, for example, how to obtain the macroscale value of ρ^α at a macroscale point from the microscale values of α-phase density in the vicinity of that point.

6.5.2 Averaging method

This approach involves both model-downscaling (Section 1.3.2) and model-upscaling (Section 1.3.1). It first scales down the global balance equations for a single phase continuum (Equation (10)) to microscale point equations (Equation (14)) by applying Theorems 1 and 2 (Equations (311) and (320)). The microscale point equations are assumed to be applicable at each and every point within any particular phase, and are then scaled up to obtain macroscale equations (Equation (9)) using Theorems 39 and 40 (Equations (614) and (627)). The details of these scaling-down and scaling-up are discussed in Section 1.3.

The disadvantage of the averaging approach is its two-step procedure: first scaling-down from the global integral equations to obtain microscale equations and then scaling-up from the microscale equations to obtain the macroscale equations. It is much more desirable if we can obtain the macroscale equations directly by only scaling-down from the global integral equations. Note also that if a microscale quantity such as the heat flux vector is non-local, the global equation containing this quantity may not be localized to the microscale. However, it may still be localized to the macroscale provided that the non-local scale is smaller than the macroscale.

6.5.3 Global balance method for each phase

This approach simply involves a direct scaling-down of the global balance equations to the macroscale, like that in the mixture theory of continuum mechanics. Unlike in mixture theory, global integral balance equations for each phase are formed in terms of microscale properties of that phase such that the integrands are microscale quantities. Then macroscale equations are obtained directly by applying the appropriate multiscale theorems, with precise correspondence between microscale and macroscale quantities (Gray and Hassanizadeh, 1998). Similar procedures can also be used to obtain macroscale equations for interfaces (Section 6.6.1) and for common curves where three phases come together (Section 6.7).

Consider a volume of α-phase δV^{α} that is a portion of a multiphase spherical volume δV (REV). Both δV^{α} and δV are at the macroscale. The boundary of δV^{α} consists of two parts: δS^{α} (portions of the surface that intersect the α-phase and is coincident with the boundary of δV) and $S_{\alpha\beta}$ (portions of the boundary that is an interface between the α-phase and all other phases). The volume δV of the REV is fixed in space and non-deforming so that the normal velocity of δS^{α} vanishes. However, $S_{\alpha\beta}$ may have a normal velocity component because of the possible phase change and deformation. The general conservation equation for the property ψ (Equation (10)) becomes

$$\frac{d}{dt}\int_{\delta V^{\alpha}} \rho\psi dV + \int_{\delta S^{\alpha}} \mathbf{n}^{*} \cdot [\rho\mathbf{v}\psi - \mathbf{i}]dS + \underbrace{\sum}_{\beta\neq\alpha} \int_{S_{\alpha\beta}} \mathbf{n}^{\alpha} \cdot [\rho(\mathbf{v}-\mathbf{w}_{b})\psi - \mathbf{i}]\Big|_{\alpha} dS - \int_{\delta V^{\alpha}} \rho f dV = \int_{\delta V^{\alpha}} \rho G dV \qquad (1010)$$

where $\mathbf{n}^{*}$ is the unit vector normal to $S_{\alpha\beta}$ that is pointing outward from the α-phase.

By Theorems 1 and 39 (Equations (311) and (614)), we have

$$\frac{d}{dt}\int_{V} f dv = \frac{\partial}{\partial t}\int_{V_{\alpha}} f dv \qquad (1011)$$

This is physically grounded because the external boundary of δV is fixed. Note that Equation (1011) holds only for the δV^{α} under consideration, that being the volume of the α-phase within a particular δV located at a particular point in space. While δV^{α} may vary with time and may have different shapes within different δV_{s} at different positions, each δV is fixed in space with its shape independent of position.

By Theorems 40 and 41 (Equations (627) and (634)), we have

$$\int_{\delta S^{\alpha}} \mathbf{n}^{*} \cdot \mathbf{f} ds = \nabla \cdot \int_{\delta V^{\alpha}} \mathbf{f} dv \qquad (1012)$$

Applying Equations (1011) and (1012) into Equation (1010) yields a macroscale balance equation in terms of microscale quantities

$$\frac{\partial}{\partial t}\int_{\delta V_{\alpha}} \rho\psi dV + \nabla \cdot \int_{\delta V_{\alpha}} [\rho\mathbf{v}\psi - \mathbf{i}]dv + \underbrace{\sum}_{\beta\neq\alpha} \int_{S_{\alpha\beta}} \mathbf{n}^{\alpha} \cdot [\rho(\mathbf{v}-\mathbf{w})\psi - \mathbf{i}]\Big|_{\alpha} dS - \int_{\delta V_{\alpha}} \rho f dV = \int_{\delta V_{\alpha}} G dV \qquad (1013)$$

which is the same as Equation (4) and can be written in terms of averaged quantities (see Section 1.3.2 for details; Equation (9))

$$\frac{\partial(\varepsilon^\alpha \langle\rho\rangle^\alpha \bar{\psi}^\alpha)}{\partial t} + \nabla\cdot(\varepsilon^\alpha\langle\rho\rangle^\alpha \bar{\mathbf{v}}^\alpha \bar{\psi}^\alpha)$$
$$-\nabla\cdot\left\{\varepsilon^\alpha\left[\langle\mathbf{i}\rangle^\alpha - \langle\rho\rangle^\alpha\left(\overline{\bar{\mathbf{v}}^\alpha\hat{\psi}}^\alpha + \overline{\hat{\mathbf{v}}^\alpha\bar{\psi}}^\alpha + \overline{\hat{\mathbf{v}}^\alpha\hat{\psi}}^\alpha\right)\right]\right\} - \varepsilon^\alpha\langle\rho\rangle^\alpha\bar{f}^\alpha$$
$$= \varepsilon^\alpha\langle G\rangle^\alpha + \underbrace{\sum}_{\beta\neq\alpha}(\hat{e}^\alpha_{\alpha\beta}\bar{\psi}^\alpha + \hat{I}^\alpha_{\alpha\beta}) \tag{1014}$$

By comparing Equations (1009) and (1014), we may obtain a correspondence between the macroscale quantities in Equation (1009) and the averaged microscale quantities in Equation (1014):

$$\rho^\alpha = \langle\rho\rangle^\alpha \tag{1015}$$

$$\mathbf{v}^\alpha = \bar{\mathbf{v}}^\alpha \tag{1016}$$

$$\psi^\alpha = \bar{\psi}^\alpha \tag{1017}$$

$$f^\alpha = \bar{f}^\alpha \tag{1018}$$

$$\mathbf{i}^\alpha = \langle\mathbf{i}\rangle^\alpha - \langle\rho\rangle^\alpha(\overline{\bar{\mathbf{v}}^\alpha\hat{\psi}}^\alpha + \overline{\hat{\mathbf{v}}^\alpha\bar{\psi}}^\alpha + \overline{\hat{\mathbf{v}}^\alpha\hat{\psi}}^\alpha) \tag{1019}$$

$$\varepsilon^\alpha G^\alpha = \varepsilon^\alpha\langle G\rangle^\alpha + \underbrace{\sum}_{\beta\neq\alpha}(\hat{e}^\alpha_{\alpha\beta}\bar{\psi}^\alpha + \hat{I}^\alpha_{\alpha\beta}) \tag{1020}$$

For $l_{mi} \ll l_{ma}$, the correlation between deviation quantities at the microscale l_{mi} and average quantities at the macroscale l_{ma} is negligible so that Equation (1019) reduces to

$$\mathbf{i}^\alpha = \langle\mathbf{i}\rangle^\alpha - \langle\rho\rangle^\alpha\overline{\hat{\mathbf{v}}^\alpha\hat{\psi}}^\alpha \tag{1021}$$

The second term on the right-hand side of Equation (1021) is an average of a product of spatial deviations, the counterpart of the average of products of temporal deviations that gives rise to Reynolds stresses in turbulence (Davidson, 2004; Piquet, 1999; Pope, 2000; Wang, 1997, 2002). Microscale convection and diffusion at the interface are the source terms for the property under consideration in the α-phase. This leads to the second term on the right-hand side of Equation (1021).

By making an appropriate choice of ψ, $\mathbf{i}$, f, and G, we can obtain the specific macroscale phase equations that are listed in Table 9.

6.6 Macroscale interface equations in porous and multiphase systems

In this section, we show the application of the multiscale theorems for developing macroscale interface equations in porous and multiphase systems by applying the global balance method (Section 6.5.1) to the general conservation equation

Table 9 Macroscale balance equations for α-phase

Conservation law	Equation
Mass	$\frac{\partial(\varepsilon^\alpha\rho^\alpha)}{\partial t}+\nabla\cdot(\varepsilon^\alpha\rho^\alpha\mathbf{v}^\alpha)=\sum_{\beta\neq\alpha}\hat{e}^\alpha_{\alpha\beta}$
Momentum	$\varepsilon^\alpha\rho^\alpha\frac{D^\alpha\mathbf{v}^\alpha}{Dt}-\nabla\cdot(\varepsilon^\alpha\mathbf{t}^\alpha)-\varepsilon^\alpha\rho^\alpha\mathbf{g}^\alpha=\sum_{\beta\neq\alpha}\hat{\mathbf{T}}^\alpha_{\alpha\beta}$ $\frac{D^\alpha}{Dt}=\frac{\partial}{\partial t}+\mathbf{v}^\alpha\cdot\nabla$ $\mathbf{t}^\alpha=\langle\mathbf{t}\rangle^\alpha-\rho^\alpha\overline{\tilde{\mathbf{v}}^\alpha\tilde{\mathbf{v}}^\alpha}$ $\hat{\mathbf{T}}^\alpha_{\alpha\beta}=\frac{\varepsilon^\alpha}{\delta V_\alpha}\int_{S_{\alpha\beta}}\mathbf{n}^\alpha\cdot[\mathbf{t}-\rho(\mathbf{v}-\mathbf{w})\tilde{\mathbf{v}}^\alpha]\vert_\alpha dS$
Angular momentum	$\mathbf{t}^\alpha=\mathbf{t}^{\alpha T}$
Energy	$\varepsilon^\alpha\rho^\alpha\frac{D^\alpha E^\alpha}{Dt}-\nabla\cdot(\varepsilon^\alpha\mathbf{q}^\alpha)-\varepsilon^\alpha\mathbf{t}^\alpha:\nabla\mathbf{v}^\alpha-\varepsilon^\alpha\rho^\alpha h^\alpha$ $=\sum_{\beta\neq\alpha}\hat{\mathbf{Q}}^\alpha_{\alpha\beta}$
Entropy	$\varepsilon^\alpha\rho^\alpha\frac{D^\alpha\eta^\alpha}{Dt}-\nabla\cdot(\varepsilon^\alpha\psi^\alpha)-\varepsilon^\alpha\rho^\alpha b^\alpha$ $=\varepsilon^\alpha\Lambda^\alpha\sum_{\beta\neq\alpha}\hat{\Phi}^\alpha_{\alpha\beta}$ $\eta^\alpha=\bar{\eta}^\alpha,\ b^\alpha=\bar{b}^\alpha,\ \Lambda^\alpha=\langle\Lambda\rangle^\alpha$ $\psi^\alpha=\langle\psi\rangle^\alpha-\rho^\alpha\overline{\tilde{v}^\alpha\tilde{\eta}^\alpha}$ $\hat{\Phi}^\alpha_{\alpha\beta}=\frac{\varepsilon^\alpha}{\delta V_\alpha}\int_{S_{\alpha\beta}}\mathbf{n}^\alpha\cdot[\psi-\rho(\mathbf{v}-\mathbf{w})\tilde{\eta}^\alpha]\vert_\alpha dS$

(Gray and Hassanizadeh, 1998). We also apply the averaging method (Section 6.5.2) to derive the general surface equations of mass and momentum conservation for demonstration of multiscale theorems' applications in obtaining some basic important results.

6.6.1 By the global balance method

Consider an α–β interface in a volume δV (REV). The general conservation equation for its property ψ is

$$\begin{aligned}\frac{\mathrm{d}}{\mathrm{d}t}\int_{S_{\alpha\beta}}(\rho\psi)\Big|_{\alpha\beta}\mathrm{d}S &+ \int_{C_{\alpha\beta}}\boldsymbol{\nu}^{*}\cdot[\rho(\mathbf{v}-\mathbf{w}_{\mathrm{b}})\psi-\mathbf{i}]\Big|_{\alpha\beta}\mathrm{d}C\\ &+\underbrace{\sum}_{\varepsilon\neq\alpha,\beta}\int_{C_{\alpha\beta\varepsilon}}\boldsymbol{\nu}^{\alpha\beta}\cdot[\rho(\mathbf{v}-\mathbf{u}_{\mathrm{b}})\psi-\mathbf{i}]\Big|_{\alpha\beta}\mathrm{d}C\\ &-\underbrace{\sum}_{i=\alpha,\beta}\int_{S_{\alpha\beta}}\mathbf{n}^{i}\cdot[\rho(\mathbf{v}-\mathbf{w})\psi-\mathbf{i}]\Big|_{i}\mathrm{d}S-\int_{S_{\alpha\beta}}(\rho f)\Big|_{\alpha\beta}\mathrm{d}S\\ &=\int_{S_{\alpha\beta}}(\rho G)\Big|_{\alpha\beta}\mathrm{d}S\end{aligned} \tag{1022}$$

where $|_{\alpha\beta}$ indicates a microscale surficial property of the α–β interface, $S_{\alpha\beta}$ is the two-dimensional spatial region occupied by the α–β interface, $C_{\alpha\beta}$ the intersection curve of the α–β interface with the boundary of $\delta V, C_{\alpha\beta\varepsilon}$ the common curve formed at the intersection of $\alpha\beta$, $\alpha\varepsilon$, and $\beta\varepsilon$ interfaces. $\rho|_i$ is the mass density of the α- or β-phase on the side of the interface indicated with the vertical bar $[M/L^3]$, $\rho|_{\alpha\beta}$ the excess mass density of the interface $[M/L^2]$, ψ the property of interest per unit excess mass of the interface. $\mathbf{v}$ is the velocity of the material on the interface, $\mathbf{u}_{\mathrm{b}}$ the velocity on the internal boundary of the interface, i.e. a common curve, $\mathbf{w}$ the velocity on the interface, $\mathbf{w}_{\mathrm{b}}$ the velocity on the boundary edge of the interface on the boundary of δV. $\mathbf{i}$ is a non-convective flux from the interface into an adjacent phase or a non-convective flux from the curve bounding the interface, $\boldsymbol{\nu}^*$ a unit normal vector of $C_{\alpha\beta}$, tangent to $S_{\alpha\beta}$ and oriented outward with respect to the REV, $\boldsymbol{\nu}^{\alpha\beta}$ a unit normal vector of $C_{\alpha\beta\varepsilon}$ and also tangent to $S_{\alpha\beta}$ that is pointing outward, $\mathbf{f}$ the external supply of ψ, G the net production of ψ within the α–β interface, and $\sum_{\varepsilon\neq\alpha,\beta}$ stands for a summation over all common curves bounding the α–β interface.

The first term in Equation (1022) is the rate of ψ accumulation on the interface. The second and third integrals come from the flux out of the interface across the exterior boundary of the volume where it intersects the interface, i.e. on the curve $C_{\alpha\beta}$, and a common curve forming the boundary of the interface in the interior of the volume, i.e. on the curve $C_{\alpha\beta\varepsilon}$, respectively. The fourth integral accounts for the addition of ψ to the interface from the phases on the two sides of the interface. The last integral on the left-hand side of the equation is the external supply term, and the right-hand side of the equation is the net production term.

By Theorems 11 and 42 (Equations (385) and (636)), we have

$$\frac{\mathrm{d}}{\mathrm{d}t}\int_{S_{\alpha\beta}}\mathbf{f}|_{\alpha\beta}\mathrm{d}s-\int_{C_{\alpha\beta}}\boldsymbol{\nu}^{*}\cdot\mathbf{w}f|_{\alpha\beta}\mathrm{d}c=\frac{\partial}{\partial t}\int_{S_{\alpha\beta}}\mathbf{f}|_{\alpha\beta}\mathrm{d}s+\nabla\cdot\int_{S_{\alpha\beta}}\mathbf{n}^{\alpha}\mathbf{n}^{\alpha}(\mathbf{w}f)|_{\alpha\beta}\mathrm{d}s \tag{1023}$$

By Theorems 43 and 45 (Equations (650) and (676)), we have

$$\int_{C_{\alpha\beta}}\boldsymbol{\nu}^{*}\cdot\mathbf{f}|_{\alpha\beta}\mathrm{d}c=\nabla\cdot\int_{S_{\alpha\beta}}\mathbf{f}^{\mathrm{s}}|_{\alpha\beta}\mathrm{d}s \tag{1024}$$

where

$$\mathbf{f}^{s}|_{\alpha\beta} = \mathbf{f}|_{\alpha\beta} - \mathbf{n}^{\alpha}\mathbf{n}^{\alpha}\cdot\mathbf{f}|_{\alpha\beta} \tag{1025}$$

Applying Equations (1023) and (1024) into Equation (1022) yields

$$\begin{aligned}
&\frac{\partial}{\partial t}\int_{S_{\alpha\beta}}(\rho\psi)\Big|_{\alpha\beta}\mathrm{d}s + \nabla\cdot\int_{S_{\alpha\beta}}[\rho\mathbf{v}\psi - \mathbf{i}^{s}]\Big|_{\alpha\beta}\mathrm{d}s \\
&+\underbrace{\sum}_{\varepsilon\neq\alpha,\beta}\int_{C_{\alpha\beta\varepsilon}}\mathbf{v}^{\alpha\beta}\cdot[\rho(\mathbf{v}-\mathbf{u}_{b})\psi - \mathbf{i}]\Big|_{\alpha\beta}\mathrm{d}C \\
&-\underbrace{\sum}_{i=\alpha,\beta}\int_{S_{\alpha\beta}}\mathbf{n}^{i}\cdot[\rho(\mathbf{v}-\mathbf{w})\psi - \mathbf{i}]\Big|_{i}\mathrm{d}S - \int_{S_{\alpha\beta}}(\rho f)\Big|_{\alpha\beta}\mathrm{d}S \\
&= \int_{S_{\alpha\beta}}(\rho G)\Big|_{\alpha\beta}\mathrm{d}S
\end{aligned} \tag{1026}$$

where we used

$$\mathbf{v}|_{\alpha\beta} = \mathbf{v}^{s}|_{\alpha\beta} + \mathbf{n}^{\alpha}\mathbf{n}^{\alpha}\cdot\mathbf{v}|_{\alpha\beta}$$

This is the macroscale balance equation and may be written in terms of average quantities after dividing by the constant averaging volume δV

$$\begin{aligned}
&\frac{\partial(\varepsilon^{\alpha\beta}\langle\rho\psi\rangle^{\alpha\beta})}{\partial t} + \nabla\cdot(\varepsilon^{\alpha\beta}\langle\rho\mathbf{v}\psi\rangle^{\alpha\beta}) - \nabla\cdot(\varepsilon^{\alpha\beta}\langle\mathbf{i}^{s}\rangle^{\alpha\beta}) \\
&+\frac{1}{\delta V}\underbrace{\sum}_{\varepsilon\neq\alpha,\beta}\int_{C_{\alpha\beta\varepsilon}}\mathbf{v}^{\alpha\beta}\cdot[\rho(\mathbf{v}-\mathbf{u})\psi - \mathbf{i}]|_{\alpha\beta}\mathrm{d}c \\
&+\underbrace{\sum}_{i=\alpha,\beta}((\hat{e}^{i}_{\alpha\beta}\psi^{i} + \hat{\mathbf{I}}^{i}_{\alpha\beta}) - \varepsilon^{\alpha\beta}\langle\rho f\rangle^{\alpha\beta} = \varepsilon^{\alpha\beta}\langle G\rangle^{\alpha\beta}
\end{aligned} \tag{1027}$$

where

$$\langle F\rangle^{\alpha\beta} = \frac{1}{S_{\alpha\beta}}\int_{S_{\alpha\beta}}F|_{\alpha\beta}\mathrm{d}s \tag{1028}$$

$$\hat{e}^{i}_{\alpha\beta} = \frac{1}{\delta V}\int_{S_{\alpha\beta}}\mathbf{n}^{i}\cdot[\rho(\mathbf{w}-\mathbf{v})]|_{i}\mathrm{d}S \tag{1029}$$

$$\hat{\mathbf{I}}^{i}_{\alpha\beta} = \frac{1}{V}\int_{S_{\alpha\beta}}\mathbf{n}^{i}\cdot[\mathbf{i} - \rho(\mathbf{v}-\mathbf{w})\hat{\psi}^{i}]|_{i}\mathrm{d}S \tag{1030}$$

and

$$\varepsilon^{\alpha\beta} = \frac{1}{\delta V}\int_{S_{\alpha\beta}}\mathrm{d}s = \frac{S_{\alpha\beta}}{\delta V} \tag{1031}$$

It is also useful to introduce the mass weighted area average and the deviation of a microscale quantity from the macroscale mean value as

$$\bar{F}^{\alpha\beta} = \frac{1}{\langle\rho\rangle^{\alpha\beta}S_{\alpha\beta}}\int_{S_{\alpha\beta}}(\rho F)|_{\alpha\beta}\mathrm{d}s = \frac{\langle\rho F\rangle^{\alpha\beta}}{\langle\rho\rangle^{\alpha\beta}} \tag{1032}$$

and

$$\hat{F}^{\alpha\beta} = F - \bar{F}^{\alpha\beta} \tag{1033}$$

Therefore, Equation (1027) becomes

$$\frac{\partial\left(\varepsilon^{\alpha\beta}\langle\rho\rangle^{\alpha\beta}\bar{\psi}^{\alpha\beta}\right)}{\partial t} + \nabla\cdot\left(\varepsilon^{\alpha\beta}\langle\rho\rangle^{\alpha\beta}\bar{\mathbf{v}}^{\alpha\beta}\bar{\psi}^{\alpha\beta}\right) - \nabla.\left[\varepsilon^{\alpha\beta}\left(\langle\mathbf{i}^{s}\rangle^{\alpha\beta}\right) - \langle\rho\rangle^{\alpha\beta}\left(\overline{\bar{\mathbf{v}}^{\alpha\beta}\hat{\psi}^{\alpha\beta}}^{\alpha\beta}\right)\right]$$
$$+\underbrace{\sum}_{i=\alpha,\beta}\left(\hat{e}^{i}_{\alpha\beta}\psi^{i} + \hat{\mathbf{I}}^{i}_{\alpha\beta}\right) - \varepsilon^{\alpha\beta}\langle\rho\rangle^{\alpha\beta}\bar{f}^{\alpha\beta}$$
$$= \varepsilon^{\alpha\beta}\langle G\rangle^{\alpha\beta} + \underbrace{\sum}_{\varepsilon\neq\alpha,\beta}\left(\hat{e}^{\alpha\beta}_{\alpha\beta\varepsilon}\bar{\psi}^{\alpha\beta} + \hat{\mathbf{I}}^{\alpha\beta}_{\alpha\beta\varepsilon}\right) \tag{1034}$$

where, with $ij = \alpha\beta, \alpha\varepsilon, \beta\varepsilon$,

$$\hat{e}^{ij}_{\alpha\beta\varepsilon} = \frac{1}{\delta V}\int_{S_E}\mathbf{v}^{ij}\cdot[\rho(\mathbf{u}-\mathbf{v})]\big|_{ij}\mathrm{d}C$$

and

$$\hat{\mathbf{I}}^{ij}_{\alpha\beta\varepsilon} = \frac{1}{\delta V}\int_{C_{\alpha\beta\varepsilon}}\mathbf{v}^{ij}\cdot\left[\mathbf{i}-\rho(\mathbf{v}-\mathbf{u})\hat{\psi}^{ij}\right]\Big|_{ij}\mathrm{d}C$$

The quantities $\hat{e}^{ij}_{\alpha\beta\varepsilon}$ and $\hat{\mathbf{I}}^{ij}_{\alpha\beta\varepsilon}$ account for interactions of an ij interface with a bounding common curve.

By defining macroscale interface properties in terms of averages analogous to the definitions in Equations (1015)–(1020) for phase properties, we can rewrite Equation (1034) as

$$\frac{\partial\left(\rho^{\alpha\beta}\varepsilon^{\alpha\beta}\psi^{\alpha\beta}\right)}{\partial t} + \nabla\cdot\left(\rho^{\alpha\beta}\varepsilon^{\alpha\beta}\mathbf{v}^{\alpha\beta}\psi^{\alpha\beta}\right) - \nabla\cdot\left(\varepsilon^{\alpha\beta}\mathbf{i}^{\alpha\beta}\right) - \rho^{\alpha\beta}\varepsilon^{\alpha\beta}f^{\alpha\beta}$$
$$= \rho^{\alpha\beta}\varepsilon^{\alpha\beta}G^{\alpha\beta} - \underbrace{\sum}_{i\neq\alpha,\beta}\left(\hat{e}^{i}_{\alpha\beta}\psi^{i} + \hat{\mathbf{I}}^{i}_{\alpha\beta}\right) + \underbrace{\sum}_{\varepsilon\neq\alpha,\beta}\left(\hat{e}^{\alpha\beta}_{\alpha\beta\varepsilon}\bar{\psi}^{\alpha\beta} + \hat{\mathbf{I}}^{\alpha\beta}_{\alpha\beta\varepsilon}\right) \tag{1035}$$

By making an appropriate choice of ψ, $\mathbf{i}^{s}$, f, and G, we can obtain the specific macroscale interface equations that are listed in Table 10.

6.6.2 By the averaging method

Moving interfaces arise in a variety of natural and industrial processes. Examples are: growth boundaries of crystal, ice and grain, and melting boundaries of ice, snow, and metal. They also appear in the multiphase flows and moving boundary problems. The correct description of their motion is thus of considerable practical importance. Because of the complexities of the problem (Pamplin, 1980), initial attempts to describe physical processes have relied on descriptive models of a more or less intuitive or empirical nature (Kynch, 1952; Mullins, 1956; Mullins, 1960). These models are generally limited in application to particular problems and usually have a narrow range of validity.

Table 10 Macroscale balance equations for $\alpha\beta$-interface

Conservation law	Equation
Mass	$\frac{\partial(\varepsilon^{\alpha\beta}\rho^{\alpha\beta})}{\partial t} + \nabla\cdot(\varepsilon^{\alpha\beta}\rho^{\alpha\beta}\mathbf{v}^{\alpha\beta}) = \sum_{\varepsilon\neq\alpha,\beta}\hat{e}^{\alpha\beta}_{\alpha\beta\varepsilon} - (\hat{e}^{\alpha}_{\alpha\beta} + \hat{e}^{\beta}_{\alpha\beta})$
Momentum	$\varepsilon^{\alpha\beta}\rho^{\alpha\beta}\frac{D^{\alpha\beta}\mathbf{v}^{\alpha\beta}}{Dt} - \nabla\cdot(\varepsilon^{\alpha\beta}\mathbf{S}^{\alpha\beta}) - \varepsilon^{\alpha\beta}\rho^{\alpha\beta}\mathbf{g}^{\alpha\beta}$
	$= \sum_{\varepsilon\neq\alpha,\beta}\hat{\mathbf{T}}^{\alpha\beta}_{\alpha\beta\varepsilon} - \sum_{i=\alpha,\beta}\left(\hat{e}^{i}_{\alpha\beta}\mathbf{v}^{i,\alpha\beta} + \hat{\mathbf{T}}^{i}_{\alpha\beta}\right)$
	$\frac{D^{\alpha\beta}}{Dt} = \frac{\partial}{\partial t} + \mathbf{v}^{\alpha\beta}\cdot\nabla$
	$\mathbf{v}^{i,\alpha\beta} = \mathbf{v}^{i} - \mathbf{v}^{\alpha\beta}$
	$\mathbf{S}^{\alpha\beta} = \langle\mathbf{t}\rangle^{\alpha\beta} - \rho^{\alpha\beta}\overline{\tilde{\mathbf{v}}^{\alpha\beta}\tilde{\mathbf{v}}^{\alpha\beta}}^{\alpha\beta}$
	$\mathbf{t}^{s}\vert_{\alpha\beta} = \mathbf{t}\vert_{\alpha\beta} - \mathbf{n}^{\alpha}\mathbf{n}^{\alpha}\cdot\mathbf{t}\vert_{\alpha\beta}\cdot\mathbf{n}^{\alpha}\mathbf{n}^{\alpha}$
	$\hat{\mathbf{T}}^{ij}_{\alpha\beta\varepsilon} = \frac{1}{\delta V}\int_{C_{\alpha\beta\varepsilon}}\mathbf{v}^{ij}\cdot[\mathbf{t}^{s} - \rho(\mathbf{v}-\mathbf{u})\tilde{\mathbf{v}}^{ij}]\vert_{ij}dC$
	$ij = \alpha\beta,\ \alpha\varepsilon,\ \beta\varepsilon$
Angular momentum	$\mathbf{S}^{\alpha\beta} = \mathbf{S}^{\alpha\beta T}$
Energy	$\varepsilon^{\alpha\beta}\rho^{\alpha\beta}\frac{D^{\alpha\beta}E^{\alpha\beta}}{Dt} - \nabla\cdot(\varepsilon^{\alpha\beta}\mathbf{q}^{\alpha\beta}) - \varepsilon^{\alpha\beta}\mathbf{S}^{\alpha\beta} : \nabla\mathbf{v}^{\alpha\beta} - \varepsilon^{\alpha\beta}\rho^{\alpha\beta}h^{\alpha\beta}$
	$= \sum_{\varepsilon\neq\alpha,\beta}\hat{\mathbf{Q}}^{\alpha\beta}_{\alpha\beta\varepsilon} - v\sum_{i=\alpha,\beta}\{\hat{e}^{i}_{\alpha\beta}[E^{i,\alpha\beta} + (\mathbf{v}^{i,\alpha\beta})^2/2]\}$
	$+\hat{\mathbf{T}}^{i}_{\alpha\beta}\cdot\mathbf{v}^{i,\alpha\beta} + \hat{\mathbf{Q}}^{i}_{\alpha\beta}$

Table 10 (*Continued*)

Conservation law	Equation
	$E^{\alpha\beta} = \overline{E}^{\alpha\beta} + \overline{(\tilde{\mathbf{v}}^{\alpha\beta})^2}^{\alpha\beta}/2$
	$\mathbf{q}^{\alpha\beta} = \langle \mathbf{q} \rangle^{\alpha\beta} + \langle \mathbf{t} \cdot \tilde{\mathbf{v}}^{\alpha\beta} \rangle^{\alpha\beta} - \langle \rho \rangle^{\alpha\beta} \overline{\tilde{\mathbf{v}}^{\alpha\beta}[\tilde{E}^{\alpha\beta} + (\tilde{\mathbf{v}}^{\alpha\beta})^2/2]}^{\alpha\beta}$
	$h^{\alpha\beta} = \overline{h}^{\alpha\beta} + \overline{\mathbf{g} \cdot \tilde{\mathbf{v}}^{\alpha\beta}}^{\alpha\beta}$
	$\hat{\mathbf{Q}}^{ij}_{\alpha\beta\varepsilon} = \frac{1}{\delta V} \int_{C_{\alpha\beta\varepsilon}} \mathbf{v}^{ij} \cdot [\mathbf{q} + \mathbf{t} \cdot \mathbf{v}^{ij} - \rho(\mathbf{v} - \mathbf{u})(\tilde{E}^{ij} + (\tilde{\mathbf{v}}^{ij})^2/2]\|_{ij} dC$
	$E^{i,\alpha\beta} = E^i - E^{\alpha\beta}$
Entropy	$\varepsilon^{\alpha\beta} \rho^{\alpha\beta} \frac{D^{\alpha\beta} \eta^{\alpha\beta}}{Dt} - \nabla \cdot (\varepsilon^{\alpha\beta} \psi^{\alpha\beta}) - \varepsilon^{\alpha\beta} \rho^{\alpha\beta} b^{\alpha\beta}$ $= \varepsilon^{\alpha\beta} \Lambda^{\alpha\beta} + \sum_{\varepsilon \neq \alpha,\beta} \hat{\boldsymbol{\Phi}}^{\alpha\beta}_{\alpha\beta\varepsilon} - \sum_{i=\alpha,\beta} (\hat{e}^i_{\alpha\beta} \eta^{i,\alpha\beta} + \tilde{\boldsymbol{\Phi}}^i_{\alpha\beta})$
	$\eta^{\alpha\beta} = \overline{\eta}^{\alpha\beta}, b^{\alpha\beta} = \overline{b}^{\alpha\beta}, \Lambda^{\alpha\beta} = \langle \Lambda \rangle^{\alpha\beta}$
	$\psi^{\alpha\beta} = \langle \psi^s \rangle^{\alpha\beta} - \rho^{\alpha\beta} \overline{\tilde{v}^{\alpha\beta} \tilde{\eta}^{\alpha\beta}}^{\alpha\beta}$
	$\hat{\boldsymbol{\Phi}}^{ij}_{\alpha\beta\varepsilon} = \frac{1}{\delta V} \int_{C_{\alpha\beta\varepsilon}} \mathbf{v}^{ij} \cdot [\psi^s - \rho(\mathbf{v} - \mathbf{w}) \tilde{\eta}^{ij}]\|_{ij} dC$

The formulation of these models typically begins with some empirical assumptions regarding kinematics of moving interfaces. As well, most of them are kinematic rather dynamic models in nature.

A second possible approach to modeling the moving interfaces can be referred to as molecular models. In this method, the statistical theory is used to obtain a phenomenological description of interfaces from a molecular point of view. This method has been successfully used to model static fluid interfaces (Buff, 1956, 1960; Buff and Saltsburg, 1957a, b; Ono and Kondo, 1960). However, we are not aware of any work on the moving interfaces.

A third approach employs continuum mechanics (Deemer and Slattery, 1978; Iannece, 1994; Scriven, 1960; Slattery, 1964a, b, 1967a). In this approach, a moving interface is viewed as a separate two-dimensional material surface with possible different constitutive equations from those applicable outside the interfacial region. Associated with the surface are the usual thermodynamic quantities which are continuous along the surface but may be discontinuous across the surface. The classical conservation laws of continuum mechanics along with the constitutive equations govern the motion of the interface, resulting in a relatively general dynamic model.

Of the three methods, the continuum mechanical approach appears to provide the best framework to advance the understanding of the moving interfaces. However, works in this group suffer three fundamental defects, resulting in a set of complex equations (Wang, 1997, 1998b). To avoid such defects, a systematic continuum approach was introduced in Wang (1997, 1998b), which leads to two models for the moving interfaces: a mechanical model for the case without thermal action (Wang, 1998b) and a thermodynamic model for the case with thermal action (Wang, 1997). The development of the models, however, relied on extensive knowledge of differential geometry, metric tensors, and rational thermodynamics. The similar mathematical complexity was also encountered in modeling static surfaces (Steigmann and Li, 1995a, b).

Within the framework of the averaging method and multiscale theorems, we may develop a dynamic model of a deforming and arbitrarily shaped surface without such mathematical complexity. Consider now microscale equations of mass and momemtum conservation for a single-phase fluid,

$$\frac{\partial \rho}{\partial t} + \nabla \cdot (\rho \mathbf{v}) = 0 \tag{1036}$$

and

$$\frac{\partial (\rho \mathbf{v})}{\partial t} + \nabla \cdot (\rho \mathbf{v}\mathbf{v}) - \rho \mathbf{g} - \nabla \cdot \mathbf{T} = 0 \tag{1037}$$

where ρ is the fluid density, $\mathbf{v}$ the fluid velocity, $\mathbf{g}$ the gravitational acceleration, and $\mathbf{T}$ the stress tensor. Integrating Equations (1036) and (1037) over a fixed straight-line segment of length ΔL yields

$$\int_{\Delta L} \frac{\partial \rho}{\partial t} \mathrm{d}L + \int_{\Delta L} \nabla \cdot (\rho \mathbf{v}) \mathrm{d}L = 0 \tag{1038}$$

and

$$\int_{\Delta L} \frac{\partial (\rho \mathbf{v})}{\partial t} \mathrm{d}L + \int_{\Delta L} \nabla \cdot (\rho \mathbf{v}\mathbf{v}) \mathrm{d}L - \int_{\Delta L} \rho \mathbf{g} \mathrm{d}L - \int_{\Delta L} \nabla \cdot \mathbf{T} \mathrm{d}L = 0 \tag{1039}$$

Applying Theorems (23) and (25) to them leads to

$$\begin{aligned} &\frac{\partial}{\partial t} \int_{\Delta L} \rho \mathrm{d}L + \nabla^{S} \cdot \int_{\Delta L} \rho \mathbf{v}^{s} \mathrm{d}L + \int_{\Delta L} \nabla \cdot \Lambda\Lambda \cdot \rho \mathbf{v} \mathrm{d}L \\ &+ \left[\frac{\mathbf{n}^{*} \cdot \rho (\mathbf{v} - \mathbf{w})}{\mathbf{e} \cdot \mathbf{n}^{*}} \right] \Big|_{\text{ends}} = 0 \end{aligned} \tag{1040}$$

and

$$\frac{\partial}{\partial t}\int_{\Delta L}\rho\mathbf{v}\mathrm{d}L+\nabla^{S}\cdot\int_{\Delta L}\rho\mathbf{v}^{s}\mathbf{v}\mathrm{d}L+\int_{\Delta L}\nabla\cdot\boldsymbol{\Lambda}\boldsymbol{\Lambda}\cdot\rho\mathbf{v}\mathbf{v}\mathrm{d}L$$
$$-\int_{\Delta L}\rho\mathbf{g}\mathrm{d}L-\nabla^{S}\cdot\int_{\Delta L}(\mathbf{T}-\boldsymbol{\Lambda}\boldsymbol{\Lambda}\cdot\mathbf{T})\mathrm{d}L-\int_{\Delta L}\nabla\cdot\boldsymbol{\Lambda}\boldsymbol{\Lambda}\cdot\mathbf{T}\mathrm{d}L$$
$$+\left\{\frac{\mathbf{n}^{*}\cdot[\rho\mathbf{v}(\mathbf{v}-\mathbf{w})-\mathbf{T}]}{\mathbf{e}\cdot\mathbf{n}^{*}}\right\}\bigg|_{\text{ends}}=0 \tag{1041}$$

where $\mathbf{w}$ is the velocity of the line segment ends and $\mathbf{v}$ the velocity of fluid that is entering or leaving at the segment end. By introducing

$$\langle\rho\rangle=\frac{1}{\Delta L}\int_{\Delta L}\rho\mathrm{d}L,\qquad \bar{\mathbf{v}}=\frac{1}{\langle\rho\rangle\Delta L}\int_{\Delta L}\rho\mathbf{v}\mathrm{d}L,$$
$$\bar{\mathbf{T}}=\frac{1}{\langle\rho\rangle\Delta L}\int_{\Delta L}\rho\mathbf{T}\mathrm{d}L,\qquad m_{s}=\langle\rho\rangle\Delta L, \tag{1042}$$
$$\mathbf{S}=(\langle\mathbf{T}\rangle-\langle\rho\rangle\overline{(\mathbf{v}-\bar{\mathbf{v}})(\mathbf{v}-\bar{\mathbf{v}})})\Delta L$$

we obtain

$$\frac{\partial m_{s}}{\partial t}+\nabla^{S}\cdot(m_{s}\bar{\mathbf{v}}^{s})+\nabla\cdot\boldsymbol{\Lambda}\boldsymbol{\Lambda}\cdot m_{s}\bar{\mathbf{v}}+\left[\frac{\mathbf{n}^{*}\cdot\rho(\mathbf{v}-\mathbf{w})}{\mathbf{e}\cdot\mathbf{n}^{*}}\right]\bigg|_{\text{ends}}=0 \tag{1043}$$

and

$$\frac{\partial(m_{s}\bar{\mathbf{v}})}{\partial t}+\nabla^{S}\cdot(m_{s}\bar{\mathbf{v}}^{s}\bar{\mathbf{v}})+\nabla\cdot\boldsymbol{\Lambda}\boldsymbol{\Lambda}\cdot m_{s}\bar{\mathbf{v}}\bar{\mathbf{v}}-m_{s}\mathbf{g}$$
$$-\nabla^{S}\cdot(\mathbf{S}-\boldsymbol{\Lambda}\boldsymbol{\Lambda}\cdot\mathbf{S})-\nabla\cdot\boldsymbol{\Lambda}\boldsymbol{\Lambda}\cdot\mathbf{S}+\left\{\frac{\mathbf{n}^{*}\cdot[\rho\mathbf{v}(\mathbf{v}-\mathbf{w})-\mathbf{T}]}{\mathbf{e}\cdot\mathbf{n}^{*}}\right\}\bigg|_{\text{ends}}=0 \tag{1044}$$

As $\Delta L\rightarrow 0$, we have

$$\lim_{\Delta L\rightarrow 0}\mathbf{e}=\mathbf{n}^{*},\qquad \lim_{\Delta L\rightarrow 0}\bar{\mathbf{v}}=\mathbf{w},$$
$$\lim_{\Delta L\rightarrow 0}\mathbf{n}^{*}=\pm\boldsymbol{\Lambda},\quad \lim_{\Delta L\rightarrow 0}[\mathbf{f}\cdot\mathbf{n}^{*}]|_{\text{ends}}=\|\mathbf{f}\|\cdot\boldsymbol{\Lambda} \tag{1045}$$

where $\|\ \mathbf{f}\ \|$ stands for the jump in $\mathbf{f}$ when crossing the surface. Therefore, in the limit of $\Delta L\rightarrow 0$, Equations (1043) and (1044) become

$$\frac{\partial m_{s}}{\partial t}+\nabla^{S}\cdot(m_{s}\mathbf{w}^{s})+\nabla\cdot\boldsymbol{\Lambda}\boldsymbol{\Lambda}\cdot m_{s}\mathbf{w}+\big\|\rho(\mathbf{v}-\mathbf{w})\big\|\cdot\boldsymbol{\Lambda}=0 \tag{1046}$$

and

$$\frac{\partial(m_{s}\mathbf{w})}{\partial t}+\nabla^{S}\cdot(m_{s}\mathbf{w}^{s}\mathbf{w})+\nabla\cdot\boldsymbol{\Lambda}\boldsymbol{\Lambda}\cdot m_{s}\mathbf{w}\mathbf{w}-m_{s}\mathbf{g}$$
$$-\nabla^{S}\cdot(\mathbf{S}-\boldsymbol{\Lambda}\boldsymbol{\Lambda}\cdot\mathbf{S})-\nabla\cdot\boldsymbol{\Lambda}\boldsymbol{\Lambda}\cdot\mathbf{S}+\big\|\rho\mathbf{v}(\mathbf{v}-\mathbf{w})-\mathbf{T}\big\|\cdot\boldsymbol{\Lambda}=0 \tag{1047}$$

where we have used the symmetric property of $\mathbf{T}$. Here $\mathbf{w}$ is the fluid velocity in the surface, $\mathbf{w}^{s}$ the surface fluid tangent velocity, and $\mathbf{v}$ the velocity of the bulk fluid adjacent to the surface. Equations (1046) and (1047) form a general dynamic model of deforming and arbitrarily shaped surface with clear physical meanings associated with each terms. The four terms on the left side of Equation (1046) account for, for example, the mass accumulation in the surface, the net convective mass flux within the surface, the mass change due to surface curvature and normal movement and the mass exchange with the fluid on both sides of the surface, respectively.

By applying the multiscale theorems and the averaging method, therefore, we can obtain the surface flow equations (Equations (1046) and (1047)) from the microscale governing equations of a single phase fluid in a three-dimensional space. This is different from the existing continuum mechanical methods (Deemer and Slattery, 1978; Iannece, 1994; Scriven, 1960; Slattery, 1964a, b, 1967; Steigmann and Li, 1995a, b; Wang, 1998). The existing methods develop the dynamic model directly for the surface flow based on the conservation laws without referring to the governing equations for the three-dimensional case Wang, 1997, 1998b; Scriven, 1960). The kinematics of the surface is pre-required to develop such dynamic models. Note that the kinematics of the surface is very complicated because the surface is a two-dimensional non-Euclidean space, can move within a higher-dimension three-dimensional space surrounding it, and from the outset demands a full tensorial treatment. Whereas mathematically rigorous, the methods are not physically satisfying in the sense that the deriving process and final equations depend on extensive knowledge of differential geometry and metric tensors. The application of the multiscale theorems and the averaging method avoids such mathematical difficulty by deriving the surface flow equations from the classical governing equations for flows in a three-dimensional space, being Euclidean and thus always admitting of a Cartesian frame of reference. The multiscale theorems are the key for transforming the problem into a three-dimensional Euclidean space which can be manipulated much easily than the one in a two-dimensional non-Euclidean space.

An interface between two phases of a multiphase system can be viewed as a surface modeled by Equations (1046) and (1047) because for most systems the interface thickness is small compared to the dimensions of the phases themselves. The mass within the interface is normally neglected. For such massless interfaces, Equation (1046) reduces to

$$\left\| \rho(\mathbf{v} - \mathbf{w}) \right\| \cdot \Lambda = 0 \tag{1048}$$

which is the commonly used jump condition of mass across a singular surface (Bowen, 1989; Eringen, 1980; Joseph, 1990; Joseph and Renardy, 1993; Slattery, 1972, 1999; Slattery et al., 2007; Truesdell and Rajagopal, 2000) and states that whatever mass leaves one fluid phase immediately enters the adjacent fluid phase for a massless interface.

For an interface that is massless and also incapable of sustaining any stess, Equation (1047) reduces to

$$\left\| \rho\mathbf{v}(\mathbf{v} - \mathbf{w}) - \mathbf{T} \right\| \cdot \Lambda = 0 \tag{1049}$$

which is the standard jump condition of momentum across a singular surface (Bowen, 1989; Eringen, 1980; Joseph, 1990; Joseph and Renardy, 1993; Slattery, 1972, 1999; Slattery et al., 2007; Truesdell and Rajagopal, 2000). When the interface is considered massless but capable of sustaining a stress, Equation (1047) becomes

$$-\nabla^{S} \cdot (\mathbf{S} - \Lambda\Lambda \cdot \mathbf{S}) - \nabla \cdot \Lambda\Lambda \cdot \mathbf{S} + \left\| \rho\mathbf{v}(\mathbf{v} - \mathbf{w}) - \mathbf{T} \right\| \cdot \Lambda = 0 \tag{1050}$$

When the interface and the two fluids on each side are all Stokesian, we have

$$\begin{aligned}\mathbf{T} &= -p\mathbf{I} + \boldsymbol{\tau} \\ \mathbf{S} &= \sigma(\mathbf{I} - \boldsymbol{\Lambda}\boldsymbol{\Lambda} \cdot \mathbf{I}) + \mathbf{s}\end{aligned} \tag{1051}$$

where p, σ and $\mathbf{s}$ are the pressure, the surface tension, and the dissipative stress in the interface, respectively. $\mathbf{I}$ is the identity tensor. For the case of no mass transfer across the interface such that $(\mathbf{v}-\mathbf{w}) \cdot \boldsymbol{\Lambda} = 0$ on both sides of the interface, Equation (1050) becomes

$$-\nabla^{S}\sigma - \boldsymbol{\Lambda}\sigma(\nabla^{S} \cdot \boldsymbol{\Lambda}) + \|p\|\boldsymbol{\Lambda} - \nabla^{S} \cdot (\mathbf{s} - \boldsymbol{\Lambda}\boldsymbol{\Lambda} \cdot \mathbf{s}) - \nabla \cdot \boldsymbol{\Lambda}\boldsymbol{\Lambda} \cdot \mathbf{S} + \|\tau\|\boldsymbol{\Lambda} = 0 \tag{1052}$$

where we have used Equation (1051). For a static interface, Equation (1052) reduces to

$$-\nabla^{S}\sigma - \boldsymbol{\Lambda}\sigma(\nabla^{S} \cdot \boldsymbol{\Lambda}) + \|p\|\boldsymbol{\Lambda} = 0 \tag{1053}$$

This yields the two well-known results, by taking its dot product with $(\mathbf{I}-\boldsymbol{\Lambda}\boldsymbol{\Lambda} \cdot \mathbf{I})$ and $\boldsymbol{\Lambda}$ respectively,

$$\begin{aligned}\nabla^{S}\sigma &= 0 \\ \sigma(\nabla^{S} \cdot \boldsymbol{\Lambda}) + \|p\| &= 0\end{aligned} \tag{1054}$$

where $\nabla^{S} \cdot \boldsymbol{\Lambda}$ is the inverse of the mean curvature. The former states that the surface gradient of surface tension vanishes for a static interface. The latter is referred to as the Young–Laplace equation in the literature (Bikerman, 1975; Joseph and Renardy, 1993; Slattery et al., 2007) and indicates that the pressure jump across a static interface is equal to the ratio of the surface tension over the mean curvature of the interface.

6.7 Macroscale equations of a common curve in porous and multiphase systems

In this section, we develop the macroscale equations of a common curve in porous and multiphase systems by applying the global balance method and the multiscale theorems (Gray and Hassanizadeh, 1998).

Consider a $\alpha\beta\varepsilon$-common curve in a volume δV (REV). The general conservation equation for its property ψ is

$$\begin{aligned}&\frac{d}{dt}\int_{C_{\alpha\beta\varepsilon}} (\rho\psi)\Big|_{\alpha\beta\varepsilon} dC + \sum_{C_{\alpha\beta\varepsilon\text{ends}}} \boldsymbol{\lambda}^{*} \cdot [\rho(\mathbf{v} - \mathbf{w}_{\mathrm{b}})\psi - \mathbf{i}]\Big|_{\alpha\beta\varepsilon} \\ &+ \underbrace{\sum}_{\gamma\neq\alpha,\beta,\varepsilon} \left\{\boldsymbol{\lambda}^{\alpha\beta\varepsilon} \cdot [\rho(\mathbf{v} - \mathbf{u}_{\mathrm{p}})\psi - \mathbf{i}]\Big|_{\alpha\beta\varepsilon}\right\}\Big|_{P_{\alpha\beta\varepsilon\gamma}} \\ &- \underbrace{\sum}_{ij=\alpha\beta,\alpha\varepsilon,\beta\varepsilon} \int_{C_{\alpha\beta\varepsilon}} \mathbf{v}^{ij} \cdot [\rho(\mathbf{v} - \mathbf{w}_{\mathrm{b}})\psi - \mathbf{i}]\Big|_{ij} dC \\ &- \int_{C_{\alpha\beta\varepsilon}} (\rho f)\Big|_{\alpha\beta\varepsilon} dC = \int_{C_{\alpha\beta\varepsilon}} (\rho G)\Big|_{\alpha\beta\varepsilon} dC\end{aligned} \tag{1055}$$

where $C_{\alpha\beta\varepsilon\text{end}}$ stands for the intersection of the common curve with the boundary of δV, $P_{\alpha\beta\varepsilon\gamma}$ denotes the common point formed at the intersection of the common curves $\alpha\beta\varepsilon, \alpha\beta\gamma, \alpha\varepsilon\gamma$, and $\beta\varepsilon\gamma$. $\rho|_{\alpha\beta\varepsilon}$ is the excess mass density of the common curve $[M/L]$, ψ the property of interest per unit excess mass of the interface, $\psi|_{\alpha\beta\varepsilon}$ the property of interest per unit mass, $\mathbf{v}|_{\alpha\beta\varepsilon}$ the velocity of the material on the common curve, $\mathbf{v}_\mathrm{p}$ the velocity of the common point where different common curves terminate, $\mathbf{w}_\mathrm{b}$ the velocity of the boundary point of a common curve at the exterior boundary of δV. $\mathbf{i}|_{\alpha\beta\varepsilon}$ is a non-convective flux, $\boldsymbol{\lambda}^*$ the unit tangent vector of $C_{\alpha\beta\varepsilon}$ that is oriented outward from a point on the external boundary of δV, $\boldsymbol{\lambda}^{\alpha\beta\varepsilon}$ the unit tangent vector of $C_{\alpha\beta\varepsilon}$ oriented outward from the curve at a common point $P_{\alpha\beta\varepsilon\gamma}$ on the external boundary of δV. $f|_{\alpha\beta\varepsilon}$ is the external supply of $\psi|_{\alpha\beta\varepsilon}$, $G|_{\alpha\beta\varepsilon}$ the net production of $\psi|_{\alpha\beta\varepsilon}$ within the $\alpha\beta\varepsilon$ common curve, and the summation in the second term is over all common points formed at the intersection of $C_{\alpha\beta\varepsilon}$ and all other common curves.

The first term in Equation (1053) is the rate of ψ accumulation in the common curve. The second and third summations come from the flux out of the common curve across the exterior boundary of the volume where it intersects the surface (i.e. at the points where a common curve pierces the shell of the volume) and at common points interior to the volume where four phases coincide (i.e. at points $P_{\alpha\beta\varepsilon\gamma}$), respectively. The fourth term comes from addition of ψ to the common curve from the three interfaces that intersect to form the curve. The last term on the left-hand side of the equation is the external supply term, and the right-hand side of the equation comes from the net production.

By Theorems 35 and 46 (Equations (569) and (678)), we have

$$\frac{\mathrm{d}}{\mathrm{d}t}\int_{C_{\alpha\beta\varepsilon}} f|_{\alpha\beta\varepsilon}\mathrm{d}C - \sum_{C_{\alpha\beta\varepsilon\text{ends}}} \boldsymbol{\lambda}^* \cdot \mathbf{w}_\mathrm{b} f\big|_{\alpha\beta\varepsilon} = \frac{\partial}{\partial t}\int_{C_{\alpha\beta\varepsilon}} f\big|_{\alpha\beta\varepsilon}\mathrm{d}C + \nabla\cdot\int_{C_{\alpha\beta\varepsilon}} (\mathbf{u}^\mathrm{s} f)\big|_{\alpha\beta\varepsilon}\mathrm{d}C \quad (1056)$$

where

$$\mathbf{u}^\mathrm{s}|_{\alpha\beta\varepsilon} = \mathbf{v}|_{\alpha\beta\varepsilon} - \boldsymbol{\lambda}^{\alpha\beta\varepsilon}\boldsymbol{\lambda}^{\alpha\beta\varepsilon}\cdot\mathbf{v}|_{\alpha\beta\varepsilon} \quad (1057)$$

By Theorems 47 and 49 (Equation (692) and (709)), we have

$$\sum_{C_{\alpha\beta\varepsilon\text{ends}}} \boldsymbol{\lambda}^* \cdot \mathbf{f}|_{\alpha\beta\varepsilon} = \nabla\cdot\int_{C_{\alpha\beta\varepsilon}} \mathbf{f}^\mathrm{c}\big|_{\alpha\beta\varepsilon}\mathrm{d}C \quad (1058)$$

where

$$\mathbf{f}^\mathrm{c}\big|_{\alpha\beta\varepsilon} = \boldsymbol{\lambda}^{\alpha\beta\varepsilon}\boldsymbol{\lambda}^{\alpha\beta\varepsilon}\cdot\mathbf{f}|_{\alpha\beta\varepsilon} \quad (1059)$$

Applying Equations (1056) and (1058) into Equation (1055) yields

$$\frac{\partial}{\partial t}\int_{C_{\alpha\beta\varepsilon}} (\rho\psi)\big|_{\alpha\beta\varepsilon}\mathrm{d}C + \nabla.\int_{C_{\alpha\beta\varepsilon}} [\rho\mathbf{v}\psi - \mathbf{i}^\mathrm{c}]\big|_{\alpha\beta\varepsilon}\mathrm{d}C$$
$$+ \underbrace{\sum}_{\gamma\neq\alpha,\beta,\varepsilon} \left\{\boldsymbol{\lambda}^{\alpha\beta\varepsilon}\cdot\left[\rho(\mathbf{v}-\mathbf{u}_\mathrm{p})\psi - \mathbf{i}\right]\big|_{\alpha\beta\varepsilon}\right\}\Big|_{P_{\alpha\beta\varepsilon\gamma}}$$

$$- \underbrace{\sum}_{ij=\alpha\beta,\alpha\varepsilon,\beta\varepsilon} \int_{C_{\alpha\beta\varepsilon}} \mathbf{v}^{ij} \cdot [\rho(\mathbf{v}-\mathbf{u})\psi - \mathbf{i}]\big|_{ij} \mathrm{d}C$$

$$- \int_{C_{\alpha\beta\varepsilon}} (\rho f)\big|_{\alpha\beta\varepsilon} \mathrm{d}C = \int_{C_{\alpha\beta\varepsilon}} (\rho G)\big|_{\alpha\beta\varepsilon} \mathrm{d}C \tag{1060}$$

where we used

$$\mathbf{v}|_{\alpha\beta\varepsilon} = \mathbf{u}^{\mathrm{s}}|_{\alpha\beta\varepsilon} + \mathbf{v}^{\mathrm{c}}|_{\alpha\beta\varepsilon}$$

This is the macroscale balance equation and may be written in terms of average (over the common curve) quantities after dividing by the constant averaging volume δV

$$\frac{\partial\left(\varepsilon^{\alpha\beta\varepsilon}\langle\rho\psi\rangle^{\alpha\beta\varepsilon}\right)}{\partial t} + \nabla\cdot\left(\varepsilon^{\alpha\beta\varepsilon}\langle\rho\mathbf{v}\psi\rangle^{\alpha\beta\varepsilon}\right) - \nabla\cdot\left(\varepsilon^{\alpha\beta\varepsilon}\langle\mathbf{i}^{\mathrm{c}}\rangle^{\alpha\beta\varepsilon}\right)$$

$$+ \frac{1}{\delta V} \underbrace{\sum}_{\gamma\neq\alpha,\beta,\varepsilon} \left\{\boldsymbol{\lambda}^{\alpha\beta\varepsilon}\cdot\left[\rho(\mathbf{v}-\mathbf{u}_{\mathrm{p}})\psi - \mathbf{i}\right]\big|_{\alpha\beta\varepsilon}\right\}\Big|_{P_{\alpha\beta\varepsilon\gamma}}$$

$$+ \underbrace{\sum}_{ij=\alpha\beta,\alpha\varepsilon,\beta\varepsilon} \left(\hat{e}^{ij}_{\alpha\beta\varepsilon}\psi^{ij} + \widehat{\mathbf{I}}^{ij}_{\alpha\beta\varepsilon}\right) - \varepsilon^{\alpha\beta\varepsilon}\langle\rho f\rangle^{\alpha\beta\varepsilon} = \varepsilon^{\alpha\beta\varepsilon}\langle G\rangle^{\alpha\beta\varepsilon} \tag{1061}$$

where

$$\langle F\rangle^{\alpha\beta\varepsilon} = \frac{1}{C_{\alpha\beta\varepsilon}} \int_{C_{\alpha\beta\varepsilon}} F|_{\alpha\beta\varepsilon} \mathrm{d}C \tag{1062}$$

$$\hat{e}^{i}_{\alpha\beta\varepsilon} = \frac{1}{\delta V} \int_{C_{\alpha\beta\varepsilon}} \mathbf{n}^{i}\cdot[\rho(\mathbf{w}-\mathbf{v})]\big|_{i} \mathrm{d}C \tag{1063}$$

$$\widehat{\mathbf{I}}^{i}_{\alpha\beta\varepsilon} = \frac{1}{V} \int_{C_{\alpha\beta\varepsilon}} \mathbf{n}^{i}\cdot\left[\mathbf{i} - \rho(\mathbf{v}-\mathbf{w})\hat{\psi}^{i}\right]_{i}\Big|_{i} \mathrm{d}C \tag{1064}$$

and

$$\varepsilon^{\alpha\beta\varepsilon} = \frac{1}{\delta V} \int_{C_{\alpha\beta\varepsilon}} \mathrm{d}C = \frac{C_{\alpha\beta\varepsilon}}{\delta V} \tag{1065}$$

It is also useful to introduce the mass weighted common curve average and the deviation of a microscale quantity from the macroscale mean as

$$\bar{F}^{\alpha\beta\varepsilon} = \frac{1}{\langle\rho\rangle^{\alpha\beta\varepsilon} C_{\alpha\beta\varepsilon}} \int_{C_{\alpha\beta\varepsilon}} (\rho F)\big|_{\alpha\beta\varepsilon} \mathrm{d}C = \frac{\langle\rho F\rangle^{\alpha\beta\varepsilon}}{\langle\rho\rangle^{\alpha\beta\varepsilon}} \tag{1066}$$

and

$$\hat{F}^{\alpha\beta\varepsilon} = F|_{\alpha\beta\varepsilon} - \bar{F}^{\alpha\beta\varepsilon} \tag{1067}$$

Therefore, Equation (1061) becomes

$$\frac{\partial\left(\varepsilon^{\alpha\beta\varepsilon}\langle\rho\rangle^{\alpha\beta\varepsilon}\bar{\psi}^{\alpha\beta\varepsilon}\right)}{\partial t}+\nabla\cdot\left(\varepsilon^{\alpha\beta\varepsilon}\langle\rho\rangle^{\alpha\beta\varepsilon}\bar{\mathbf{v}}^{\alpha\beta\varepsilon}\bar{\psi}^{\alpha\beta\varepsilon}\right)$$
$$-\nabla\cdot\left[\varepsilon^{\alpha\beta\varepsilon}\left(\left\langle\mathbf{i}^{c}\right\rangle^{\alpha\beta\varepsilon}-\langle\rho\rangle^{\alpha\beta\varepsilon}\left(\overline{\mathbf{v}^{\alpha\beta\varepsilon}\hat{\psi}^{\alpha\beta\varepsilon}}^{\alpha\beta\varepsilon}\right)\right)\right]+\underbrace{\sum}_{ij=\alpha\beta,\alpha\varepsilon,\beta\varepsilon}\left(\hat{e}^{ij}_{\alpha\beta\varepsilon}\psi^{ij}+\hat{\mathbf{I}}^{ij}_{\alpha\beta\varepsilon}\right)$$
$$-\varepsilon^{\alpha\beta\varepsilon}\langle\rho\rangle^{\alpha\beta\varepsilon}\bar{f}^{\alpha\beta\varepsilon}=\varepsilon^{\alpha\beta\varepsilon}\langle G\rangle^{\alpha\beta\varepsilon}+\underbrace{\sum}_{\gamma\neq\alpha,\beta,\varepsilon}\left(\hat{e}^{\alpha\beta\varepsilon}_{\alpha\beta\varepsilon\gamma}\bar{\psi}^{\alpha\beta\varepsilon}+\hat{\mathbf{I}}^{\alpha\beta\varepsilon}_{\alpha\beta\varepsilon\gamma}\right) \tag{1068}$$

where, with $ijk=\alpha\beta\varepsilon,\alpha\beta\gamma,\alpha\varepsilon\gamma,\beta\varepsilon\gamma$,

$$\hat{e}^{ijk}_{\alpha\beta\varepsilon\gamma}=\frac{1}{\delta V}\sum_{\gamma\neq\alpha,\beta,\varepsilon}\{\boldsymbol{\lambda}^{ijk}\cdot[\rho(\mathbf{u}_{\mathrm{p}}-\mathbf{v})]|_{ijk}\}|_{P_{\alpha\beta\varepsilon\gamma}}$$

and

$$\hat{\mathbf{I}}^{ijk}_{\alpha\beta\varepsilon}=\frac{1}{\delta V}\sum_{\gamma\neq\alpha,\beta,\varepsilon}\left\{\boldsymbol{\lambda}^{ijk}\cdot\left[\mathbf{i}-\rho(\mathbf{v}-\mathbf{u}_{\mathrm{p}})\hat{\psi}^{ijk}\right]\Big|_{ijk}\right\}\Bigg|_{P_{\alpha\beta\varepsilon\gamma}}$$

The quantities $\hat{e}^{ijk}_{\alpha\beta\varepsilon\gamma}$ and $\hat{\mathbf{I}}^{ijk}_{\alpha\beta\varepsilon\gamma}$ come from the interaction of the common curve with the points at the end of the curve within the averaging volume δV.

By defining macroscale common curve properties in terms of averages analogous to the definitions in Equations (1015)–(1019) for phase and interface properties, we can rewrite Equation (1068) as

$$\frac{\partial\left(\rho^{\alpha\beta\varepsilon}\varepsilon^{\alpha\beta\varepsilon}\psi^{\alpha\beta\varepsilon}\right)}{\partial t}+\nabla\cdot\left(\rho^{\alpha\beta\varepsilon}\varepsilon^{\alpha\beta\varepsilon}\mathbf{v}^{\alpha\beta\varepsilon}\psi^{\alpha\beta\varepsilon}\right)-\nabla\cdot\left(\varepsilon^{\alpha\beta\varepsilon}\mathbf{i}^{\alpha\beta\varepsilon}\right)-\rho^{\alpha\beta\varepsilon}\varepsilon^{\alpha\beta\varepsilon}f^{\alpha\beta\varepsilon}$$
$$=\rho^{\alpha\beta\varepsilon}\varepsilon^{\alpha\beta\varepsilon}G^{\alpha\beta\varepsilon}-\underbrace{\sum}_{ij=\alpha\beta,\alpha\varepsilon,\beta\varepsilon}\left(\hat{e}^{ij}_{\alpha\beta\varepsilon}\psi^{ij}+\hat{\mathbf{I}}^{ij}_{\alpha\beta\varepsilon}\right)+\underbrace{\sum}_{\gamma\neq\alpha,\beta,\varepsilon}\left(\hat{e}^{\alpha\beta\varepsilon}_{\alpha\beta\varepsilon}\bar{\psi}^{\alpha\beta\varepsilon}+\hat{\mathbf{I}}^{\alpha\beta\varepsilon}_{\alpha\beta\varepsilon}\right) \tag{1069}$$

This equation can be used to obtain the specific macroscale equations of a common curve in Table 11 by making appropriate choices of ψ, $\mathbf{i}^{c}$, f, and G.

6.8 Multiscale deviation theorems

The thermodynamically constrained averaging theory (TCAT) approach has been recently developed in Gray and Miller (2005, 2006) and Miller and Gray (2005) for modeling flow and transport phenomena in multiscale porous-medium systems. The approach provides a systematic framework not only for producing macroscale models of transport phenomena in porous-medium systems that are consistent with microscale representations of transport phenomena and thermodynamic constraints, but also for any development of large scale equations based on physical and thermodynamic processes. Multiscale deviation theorems are routinely used for developing complete, closed-form TCAT models for

Table 11 Macroscale balance equations for $\alpha\beta\varepsilon$ common curve

Conservation law	Equation
Mass	$\frac{\partial(\varepsilon^{\alpha\beta\varepsilon}\rho^{\alpha\beta\varepsilon})}{\partial t} + \nabla\cdot\left(\varepsilon^{\alpha\beta\varepsilon}\rho^{\alpha\beta\varepsilon}\mathbf{v}^{\alpha\beta\varepsilon}\right) = \sum_{\gamma\neq\alpha,\beta,\varepsilon} \hat{e}^{\alpha\beta\varepsilon}_{\alpha\beta\varepsilon\gamma} - \left(\hat{e}^{\alpha\beta}_{\alpha\beta\varepsilon} + \hat{e}^{\alpha\varepsilon}_{\alpha\beta\varepsilon} + \hat{e}^{\beta\varepsilon}_{\alpha\beta\varepsilon}\right)$
Momentum	$\varepsilon^{\alpha\beta\varepsilon}\rho^{\alpha\beta\varepsilon}\frac{D^{\alpha\beta\varepsilon}\mathbf{v}^{\alpha\beta\varepsilon}}{Dt} - \nabla\cdot\left(\varepsilon^{\alpha\beta\varepsilon}\mathbf{C}^{\alpha\beta\varepsilon}\right) - \varepsilon^{\alpha\beta\varepsilon}\rho^{\alpha\beta\varepsilon}\mathbf{g}^{\alpha\beta\varepsilon}$ $= \sum_{\gamma\neq\alpha,\beta,\varepsilon}\hat{\mathbf{T}}^{\alpha\beta\varepsilon}_{\alpha\beta\varepsilon\gamma} - \sum_{ij=\alpha\beta,\alpha\varepsilon,\beta\varepsilon}\left(\hat{e}^{ij}_{\alpha\beta\varepsilon}\mathbf{v}^{ij,\alpha\beta\varepsilon} + \hat{\mathbf{T}}^{ij}_{\alpha\beta\varepsilon}\right)$
	$\frac{D^{\alpha\beta\varepsilon}}{Dt} = \frac{\partial}{\partial t} + \mathbf{v}^{\alpha\beta\varepsilon}\cdot\nabla$
	$\mathbf{v}^{ij,\alpha\beta\varepsilon} = \mathbf{v}^{ij} - \mathbf{v}^{\alpha\beta\varepsilon}$
	$\mathbf{C}^{\alpha\beta\varepsilon} = \left\langle\mathbf{t}^c\right\rangle^{\alpha\beta\varepsilon} - \rho^{\alpha\beta\varepsilon}\overline{\tilde{\mathbf{v}}^{\alpha\beta\varepsilon}\tilde{\mathbf{v}}^{\alpha\beta\varepsilon}}^{\alpha\beta\varepsilon}$
	$\mathbf{t}^c\big\vert_{\alpha\beta\varepsilon} = \boldsymbol{\lambda}^{\alpha\beta\varepsilon}\boldsymbol{\lambda}^{\alpha\beta\varepsilon}\cdot\mathbf{t}\vert_{\alpha\beta\varepsilon}\cdot\boldsymbol{\lambda}^{\alpha\beta\varepsilon}\boldsymbol{\lambda}^{\alpha\beta\varepsilon}$
	$\hat{\mathbf{T}}^{ijk}_{\alpha\beta\varepsilon\gamma} = \frac{1}{\delta V}\sum_{\gamma\neq\alpha,\beta,\varepsilon}\left\{\boldsymbol{\lambda}^{ijk}\cdot\left[\mathbf{t}^c - \rho(\mathbf{v}-\mathbf{u}_p)\tilde{\mathbf{v}}^{ijk}\right]\big\vert_{ijk}\right\}\Big\vert_{P_{\alpha\beta\varepsilon\gamma}}$
	$ijk = \alpha\beta\varepsilon,\ \alpha\beta\gamma,\ \alpha\varepsilon\gamma,\ \beta\varepsilon\gamma$
Angular momentum	$\mathbf{C}^{\alpha\beta\varepsilon} = \mathbf{C}^{\alpha\beta\varepsilon T}$
Energy	$\varepsilon^{\alpha\beta\varepsilon}\rho^{\alpha\beta\varepsilon}\frac{D^{\alpha\beta\varepsilon}E^{\alpha\beta\varepsilon}}{Dt} - \nabla\cdot(\varepsilon^{\alpha\beta\varepsilon}\mathbf{q}^{\alpha\beta\varepsilon}) - \varepsilon^{\alpha\beta\varepsilon}\mathbf{C}^{\alpha\beta\varepsilon}:\nabla\mathbf{v}^{\alpha\beta\varepsilon} - \varepsilon^{\alpha\beta\varepsilon}\rho^{\alpha\beta\varepsilon}h^{\alpha\beta\varepsilon}$ $= \sum_{\gamma\neq\alpha,\beta,\varepsilon}\hat{\mathbf{Q}}^{\alpha\beta\varepsilon}_{\alpha\beta\varepsilon\gamma} - \sum_{ij=\alpha\beta,\alpha\varepsilon,\beta\varepsilon}\left\{\hat{e}^{ij}_{\alpha\beta\varepsilon}\left[E^{ij,\alpha\beta\varepsilon} + (\mathbf{v}^{ij,\alpha\beta\varepsilon})^2/2\right] + \hat{\mathbf{T}}^{ij}_{\alpha\beta\varepsilon}\cdot\mathbf{v}^{ij,\alpha\beta\varepsilon} + \hat{\mathbf{Q}}^{ij}_{\alpha\beta\varepsilon}\right\}$

Table 11 (*Continued*)

Conservation law	Equation
	$E^{\alpha\beta\varepsilon} = \bar{E}^{\alpha\beta\varepsilon} + \overline{(\tilde{\mathbf{v}}^{\alpha\beta\varepsilon})^2}^{\alpha\beta\varepsilon}/2$
	$\mathbf{q}^{\alpha\beta\varepsilon} = \langle \mathbf{q} \rangle^{\alpha\beta\varepsilon} + \langle \mathbf{t}_{\alpha\beta\varepsilon} \cdot \mathbf{v}^{\alpha\beta\varepsilon} - \langle \rho \rangle^{\alpha\beta\varepsilon} \overline{\tilde{\mathbf{v}}^{\alpha\beta\varepsilon} \left[\tilde{E}^{\alpha\beta\varepsilon} + (\tilde{\mathbf{v}}^{\alpha\beta\varepsilon})^2/2 \right]}^{\alpha\beta\varepsilon}$
	$h^{\alpha\beta\varepsilon} = \bar{h}^{\alpha\beta\varepsilon} + \overline{\mathbf{g} \cdot \tilde{\mathbf{v}}^{\alpha\beta\varepsilon}}^{\alpha\beta\varepsilon}$
	$\hat{\mathbf{Q}}^{ijk}_{\alpha\beta\varepsilon\gamma} = \frac{1}{\delta V} \left\{ \boldsymbol{\lambda}^{ijk} \cdot \left[\mathbf{q} + \mathbf{t} \cdot \mathbf{v}^{ijk} - \rho(\mathbf{v} - \mathbf{u}_p)\left(\tilde{E}^{ijk} + (\tilde{\mathbf{v}}^{ijk})^2/2 \right) \right] \Big\|_{ijk} \right\} \Big\|_{P_{\alpha\beta\varepsilon\gamma}}$
	$E^{ij,\alpha\beta\varepsilon} = E^{ij} - E^{\alpha\beta\varepsilon}$
Entropy	$\varepsilon^{\alpha\beta\varepsilon}\rho^{\alpha\beta\varepsilon}\frac{D^{\alpha\beta\varepsilon}\eta^{\alpha\beta\varepsilon}}{Dt} - \nabla \cdot (\varepsilon^{\alpha\beta\varepsilon}\psi^{\alpha\beta\varepsilon}) - \varepsilon^{\alpha\beta\varepsilon}\rho^{\alpha\beta\varepsilon}b^{\alpha\beta\varepsilon} = \varepsilon^{\alpha\beta\varepsilon}\Lambda^{\alpha\beta\varepsilon} + \sum_{\gamma \neq \alpha,\beta,\varepsilon} \tilde{\Phi}^{\alpha\beta\varepsilon}_{\alpha\beta\varepsilon\gamma} - \sum_{ij=\alpha\beta,\alpha\varepsilon,\beta\varepsilon} (\hat{e}^{ij}_{\alpha\beta\varepsilon}\eta^{ij,\alpha\beta\varepsilon} + \tilde{\Phi}^{ij}_{\alpha\beta\varepsilon})$
	$\eta^{\alpha\beta\varepsilon} = \bar{\eta}^{\alpha\beta\varepsilon},\ b^{\alpha\beta\varepsilon} = \bar{b}^{\alpha\beta\varepsilon},\ \Lambda^{\alpha\beta\varepsilon} = \langle \Lambda \rangle^{\alpha\beta\varepsilon}$
	$\psi^{\alpha\beta\varepsilon} = \langle \psi^c \rangle^{\alpha\beta\varepsilon} - \overline{\rho^{\alpha\beta\varepsilon}\tilde{v}^{\alpha\beta\varepsilon}\tilde{\eta}^{\alpha\beta\varepsilon}}^{\alpha\beta\varepsilon}$
	$\hat{\Phi}^{\alpha\beta\varepsilon}_{\alpha\beta\varepsilon\gamma} = \frac{1}{\delta V} \left\{ \boldsymbol{\lambda}^{\alpha\beta\varepsilon} \cdot \left[\psi^c - \rho(\mathbf{v} - \mathbf{u}_p)\tilde{\eta}^{\alpha\beta\varepsilon} \right] \Big\|_{\alpha\beta\varepsilon} \right\} \Big\|_{P_{\alpha\beta\varepsilon\gamma}}$

single- and multiple-fluid-phase porous-medium systems (Gray and Miller, 2006). These multiscale deviation theorems refer to transformations involving the evaluation of integrals of differential operators of the deviation between microscale and macroscale quantities. A set of the multiscale deviation theorems has been recently derived in Miller and Gray (2005). To demonstrate the applicability of the multiscale theorems for developing new theorems, we follow Miller and Gray (2005) in deriving three multiscale deviation theorems for a phase volume, an interface, and a common curve, respectively.

6.8.1 Multiscale deviations for a phase volume

Theorem 72. The volume average of a product of a microscale quantity g_α with a material derivative referenced to the macroscale mass average velocity of an entity ι of the difference between a microscale quantity f_α and its macroscale-weighted average $f^{\bar{\bar{\alpha}}}$ can be expressed as a function of relative velocities at interfacial boundaries and macroscale quantities of the form

$$\frac{1}{V}\int_{\Omega_\alpha} g_\alpha \frac{D^{\bar{\iota}}\left(f_\alpha - f^{\bar{\bar{\alpha}}}\right)}{Dt}\mathrm{d}v = \frac{D^{\bar{\iota}}\left\{\varepsilon^\alpha\left[(g_\alpha f_\alpha)^\alpha - g_\alpha f^{\bar{\bar{\alpha}}}\right]\right\}}{Dt} + \frac{D^{\bar{\iota}}(\varepsilon^\alpha g^\alpha)}{Dt} f^{\bar{\bar{\alpha}}} - \frac{1}{V}\int_{\Omega_\alpha} \frac{D^{\bar{\iota}} g_\alpha}{Dt} f_\alpha \mathrm{d}v + \sum_{\beta\neq\alpha}\frac{1}{V}\int_{\Omega_{\alpha\beta}} \mathbf{n}_\alpha\cdot(\mathbf{v}^\iota - \mathbf{v}_{\alpha\beta}) g_\alpha f_\alpha \mathrm{d}v \tag{1070}$$

Here $\mathbf{v}^{\bar{\iota}}$ is the macroscale mass averaged velocity of entity ι, $\mathbf{n}_\alpha$ the microscale normal vector pointing outward from the α-phase, $\mathbf{n}_\alpha \cdot \mathbf{v}_{\alpha\beta}$ the component of the microscale velocity of the α–β interface that is normal to the interface. V is independent of time and position and is the measure of the averaging domain Ω, $\Omega_\alpha \subset \Omega$ the phase volume of the α-phase, $\Omega_{\alpha\beta}$ the interface between the α- and β-phases, and the material derivative is defined as

$$\frac{D^{\bar{\iota}}}{Dt} = \frac{\partial}{\partial t} + \mathbf{v}^{\bar{\iota}}\cdot\nabla \tag{1071}$$

Proof. Note that

$$\frac{1}{V}\int_{\Omega_\alpha} g_\alpha \frac{D^{\bar{\iota}}\left(f_\alpha - f^{\bar{\bar{\alpha}}}\right)}{Dt}\mathrm{d}v = \frac{1}{V}\int_{\Omega_\alpha} g_\alpha \frac{D^{\bar{\iota}} f_\alpha}{Dt}\mathrm{d}v - \frac{1}{V}\int_{\Omega_\alpha} g_\alpha \frac{D^{\bar{\iota}} f^{\bar{\bar{\alpha}}}}{Dt}\mathrm{d}v \tag{1072}$$

Miller and Gray (2005) apply the chain rule to rearrange Equation (1072) into

$$\frac{1}{V}\int_{\Omega_\alpha} g_\alpha \frac{D^{\bar{\iota}}\left(f_\alpha - f^{\bar{\bar{\alpha}}}\right)}{Dt}\mathrm{d}v = \frac{1}{V}\int_{\Omega_\alpha} \frac{D^{\bar{\iota}}(g_\alpha f_\alpha)}{Dt}\mathrm{d}v - \frac{1}{V}\int_{\Omega_\alpha} \frac{D^{\bar{\iota}}\left(g_\alpha f^{\bar{\bar{\alpha}}}\right)}{Dt}\mathrm{d}v - \frac{1}{V}\int_{\Omega_\alpha} \frac{D^{\bar{\iota}} g_\alpha}{Dt} f_\alpha \mathrm{d}v + \frac{1}{V}\int_{\Omega_\alpha} \frac{D^{\bar{\iota}} g_\alpha}{Dt} f^{\bar{\bar{\alpha}}}\mathrm{d}v \tag{1073}$$

The first term in the right-hand side of Equation (1073) can be written as

$$\frac{1}{V}\int_{\Omega_\alpha} \frac{D^{\bar{\iota}}(g_\alpha f_\alpha)}{Dt}\mathrm{d}v = \frac{1}{V}\int_{\Omega_\alpha} \frac{\partial(g_\alpha f_\alpha)}{\partial t}\mathrm{d}v + \frac{1}{V}\int_{\Omega_\alpha} \mathbf{v}^{\bar{\iota}}\cdot\nabla(g_\alpha f_\alpha)\mathrm{d}v \tag{1074}$$

Application of Theorem 39 (Equation (614)) to the first term on the right-hand side of Equation (1074) yields

$$\frac{1}{V}\int_{\Omega_\alpha} \frac{\partial(g_\alpha f_\alpha)}{Dt}\mathrm{d}v = \frac{\partial}{\partial t}\left(\frac{1}{V}\int_{\Omega_\alpha} g_\alpha f_\alpha \mathrm{d}v\right) - \sum_{\beta\neq\alpha}\frac{1}{V}\int_{\Omega_{\alpha\beta}} \mathbf{n}_\alpha\cdot\mathbf{v}_{\alpha\beta} g_\alpha f_\alpha \mathrm{d}v \tag{1075}$$

Applying Theorem 40 (Equation (626)) to rearrange the second term on the right-hand side of Equation (1074) leads to

$$\frac{1}{V}\int_{\Omega_\alpha} \mathbf{v}^{\bar{\iota}}\cdot\nabla(g_\alpha f_\alpha)\mathrm{d}v = \mathbf{v}^{\bar{\iota}}\cdot\nabla\left(\frac{1}{V}\int_{\Omega_\alpha} g_\alpha f_\alpha \mathrm{d}v\right) + \sum_{\beta\neq\alpha}\mathbf{v}^{\bar{\iota}}\cdot\frac{1}{V}\int_{\Omega_{\alpha\beta}} \mathbf{n}_\alpha g_\alpha f_\alpha \mathrm{d}v \tag{1076}$$

Substituting Equations (1075) and (1076) into Equation (1074) yields

$$\frac{1}{V}\int_{\Omega_\alpha}\frac{D^{\bar{\imath}}(g_\alpha f_\alpha)}{Dt}dv=\frac{\partial}{\partial t}\left(\frac{1}{V}\int_{\Omega_\alpha}g_\alpha f_\alpha dv\right)+\mathbf{v}^{\bar{\imath}}\cdot\nabla\left(\frac{1}{V}\int_{\Omega_\alpha}g_\alpha f_\alpha dv\right)+\sum_{\beta\neq\alpha}\frac{1}{V}\int_{\Omega_{\alpha\beta}}\mathbf{n}_\alpha\cdot\left(\mathbf{v}^{\bar{\imath}}-\mathbf{v}_{\alpha\beta}\right)g_\alpha f_\alpha dv \tag{1077}$$

Expanding the second term in the right-hand side of Equation (1073) gives

$$\frac{1}{V}\int_{\Omega_\alpha}\frac{D^{\bar{\imath}}\left(g_\alpha f^{\bar{\bar{\alpha}}}\right)}{Dt}dv=\frac{1}{V}\int_{\Omega_\alpha}\frac{\partial\left(g_\alpha f^{\bar{\bar{\alpha}}}\right)}{\partial t}dv+\frac{1}{V}\int_{\Omega_\alpha}\mathbf{v}^{\bar{\imath}}\cdot\nabla\left(g_\alpha f^{\bar{\bar{\alpha}}}\right)dv \tag{1078}$$

Applying Theorems 39 and 40 (Equations (614) and (626)) to the right-hand side of Equation (1073) and rearranging leads to

$$\frac{1}{V}\int_{\Omega_\alpha}\frac{D^{\bar{\imath}}\left(g_\alpha f^{\bar{\bar{\alpha}}}\right)}{Dt}dv=\frac{\partial}{\partial t}\left(\frac{1}{V}\int_{\Omega_\alpha}g_\alpha f^{\bar{\bar{\alpha}}}dv\right)+\mathbf{v}^{\bar{\imath}}\cdot\nabla\left(\frac{1}{V}\int_{\Omega_\alpha}g_\alpha f^{\bar{\bar{\alpha}}}dv\right)+\sum_{\beta\neq\alpha}\frac{1}{V}\int_{\Omega_{\alpha\beta}}\mathbf{n}_\alpha\cdot\left(\mathbf{v}^{\bar{\imath}}-\mathbf{v}_{\alpha\beta}\right)g_\alpha f^{\bar{\bar{\alpha}}}dv \tag{1079}$$

or

$$\frac{1}{V}\int_{\Omega_\alpha}\frac{D^{\bar{\imath}}\left(g_\alpha f^{\bar{\bar{\alpha}}}\right)}{Dt}dv=\frac{D^{\bar{\imath}}\left(\varepsilon^\alpha g^\alpha f^{\bar{\bar{\alpha}}}\right)}{Dt}+\sum_{\beta\neq\alpha}\frac{1}{V}\int_{\Omega_{\alpha\beta}}\mathbf{n}_\alpha\cdot\left(\mathbf{v}^{\bar{\imath}}-\mathbf{v}_{\alpha\beta}\right)g_\alpha f^{\bar{\bar{\alpha}}}dv \tag{1080}$$

The third term in the right-hand side of Equation (1073) involves the product of a microscale quantity and the material derivative of a microscale quantity. It cannot be simplified in any obvious manner (Miller and Gray, 2005).

The fourth term in the right-hand side of Equation (1073) can be written as

$$\frac{1}{V}\int_{\Omega_\alpha}\frac{D^{\bar{\imath}}g_\alpha}{Dt}f^{\bar{\bar{\alpha}}}dv=\frac{1}{V}\int_{\Omega_\alpha}\frac{\partial g_\alpha}{\partial t}f^{\bar{\bar{\alpha}}}dv+\frac{1}{V}\int_{\Omega_\alpha}\mathbf{v}^{\bar{\imath}}\cdot\nabla g_\alpha f^{\bar{\bar{\alpha}}}dv \tag{1081}$$

Applying Theorems 39 and 40 (Equations (614) and (626)) to the right-hand side of Equation (1081) yields

$$\frac{1}{V}\int_{\Omega_\alpha}\frac{D^{\bar{\imath}}g_\alpha}{Dt}f^{\bar{\bar{\alpha}}}dv=\frac{\partial}{\partial t}\left(\frac{1}{V}\int_{\Omega_\alpha}g_\alpha dv\right)f^{\bar{\bar{\alpha}}}+\mathbf{v}^{\bar{\imath}}\cdot\nabla\left(\frac{1}{V}\int_{\Omega_\alpha}g_\alpha dv\right)f^{\bar{\bar{\alpha}}}+\sum_{\beta\neq\alpha}\frac{1}{V}\int_{\Omega_{\alpha\beta}}\mathbf{n}_\alpha\cdot\left(\mathbf{v}^{\bar{\imath}}-\mathbf{v}_{\alpha\beta}\right)g_\alpha f^{\bar{\bar{\alpha}}}dv \tag{1082}$$

or

$$\frac{1}{V}\int_{\Omega_\alpha}\frac{D^{\bar{\imath}}g_\alpha}{Dt}f^{\bar{\bar{\alpha}}}dv=\frac{D^{\bar{\imath}}\left(\varepsilon^\alpha g^\alpha\right)}{Dt}f^{\bar{\bar{\alpha}}}+\sum_{\beta\neq\alpha}\frac{1}{V}\int_{\Omega_{\alpha\beta}}\mathbf{n}_\alpha\cdot\left(\mathbf{v}^{\bar{\imath}}-\mathbf{v}_{\alpha\beta}\right)g_\alpha f^{\bar{\bar{\alpha}}}dv \tag{1083}$$

A substitution of Equations (1077), (1080), and (1083) into Equation (1073) yields

$$\frac{1}{V}\int_{\Omega_\alpha} g_\alpha \frac{D^{\bar{\iota}}\left(f_\alpha - f^{\bar{\bar{\alpha}}}\right)}{Dt}\mathrm{d}v = \frac{\partial}{\partial t}\left(\frac{1}{V}\int_{\Omega_\alpha} g_\alpha f_\alpha \mathrm{d}v\right) + \mathbf{v}^{\bar{\iota}}\cdot\nabla\left(\frac{1}{V}\int_{\Omega_{\alpha\beta}} g_\alpha f_\alpha \mathrm{d}v\right)$$
$$+\sum_{\beta\neq\alpha}\frac{1}{V}\int_{\Omega_{\alpha\beta}} \mathbf{n}_\alpha\cdot\left(\mathbf{v}^{\bar{\iota}} - \mathbf{v}_{\alpha\beta}\right) g_\alpha f_\alpha \mathrm{d}v - \frac{D^{\bar{\iota}}\left(\varepsilon^\alpha g^\alpha f^{\bar{\bar{\alpha}}}\right)}{Dt}$$
$$-\frac{1}{V}\int_{\Omega_\alpha}\frac{D^{\bar{\iota}} g_\alpha}{Dt} f_\alpha \mathrm{d}v + \frac{D^{\bar{\iota}}(\varepsilon^\alpha g^\alpha)}{Dt} f^{\bar{\bar{\alpha}}} \tag{1084}$$

Note also that

$$\frac{\partial}{\partial t}\left(\frac{1}{V}\int_{\Omega_\alpha} g_\alpha f_\alpha \mathrm{d}v\right) + \mathbf{v}^{\bar{\iota}}\cdot\nabla\left(\frac{1}{V}\int_{\Omega_\alpha} g_\alpha f_\alpha \mathrm{d}v\right) = \frac{D^{\bar{\iota}}\left[\varepsilon^\alpha \left(g_\alpha f_\alpha\right)^\alpha\right]}{Dt} \tag{1085}$$

Therefore, Equation (1084) becomes

$$\frac{1}{V}\int_{\Omega_\alpha} g_\alpha \frac{D^{\bar{\iota}}\left(f_\alpha - f^{\bar{\bar{\alpha}}}\right)}{Dt}\mathrm{d}v = \frac{D^{\bar{\iota}}\left\{\varepsilon^\alpha\left[\left(g_\alpha f_\alpha\right)^\alpha - g_\alpha f^{\bar{\bar{\alpha}}}\right]\right\}}{Dt} + \frac{D^{\bar{\iota}}(\varepsilon^\alpha g^\alpha)}{Dt} f^{\bar{\bar{\alpha}}} - \frac{1}{V}\int_{\Omega_\alpha}\frac{D^{\bar{\iota}} g_\alpha}{Dt} f_\alpha \mathrm{d}v$$
$$+\sum_{\beta\neq\alpha}\frac{1}{V}\int_{\Omega_{\alpha\beta}} \mathbf{n}_\alpha\cdot\left(\mathbf{v}^{\bar{\iota}} - \mathbf{v}_{\alpha\beta}\right) g_\alpha f_\alpha \mathrm{d}v \tag{1086}$$

which is the same as Equation (1070).

6.8.2 Multiscale deviations for an interface

Theorem 73. The volume average of a product of a microscale quantity $g_{\alpha\beta}$ with a material derivative (referenced to the macroscale mass average velocity of an entity ι) that is restricted to a position on a potentially moving interface $\alpha\beta$ of the difference between a microscale quantity $f_{\alpha\beta}$ and a macroscale quantity $f^{\overline{\overline{\alpha\beta}}}$ can be expressed as a function of relative velocities and macroscale quantities of the form

$$\frac{1}{V}\int_{\Omega_{\alpha\beta}} g_{\alpha\beta}\frac{D'^{\bar{\iota}}\left(f_{\alpha\beta} - f^{\overline{\overline{\alpha\beta}}}\right)}{Dt}\mathrm{d}S$$
$$= \frac{D^{\bar{\iota}}\left\{\varepsilon^{\alpha\beta}\left[\left(g_{\alpha\beta} f_{\alpha\beta}\right)^{\alpha\beta} - g^{\alpha\beta} f^{\overline{\overline{\alpha\beta}}}\right]\right\}}{Dt} + \frac{D^{\bar{\iota}}\left(\varepsilon^{\alpha\beta} g^{\alpha\beta}\right)}{Dt} f^{\overline{\overline{\alpha\beta}}}$$
$$-\nabla\cdot\left(\frac{1}{V}\int_{\Omega_{\alpha\beta}} \mathbf{n}_\alpha\mathbf{n}_\alpha\cdot\left(\mathbf{v}^{\bar{\iota}} - \mathbf{v}_{\alpha\beta}\right) g_{\alpha\beta}\left(f_{\alpha\beta} - f^{\overline{\overline{\alpha\beta}}}\right)\mathrm{d}S\right)$$
$$-\nabla\cdot\left(\frac{1}{V}\int_{\Omega_{\alpha\beta}} \mathbf{n}_\alpha\mathbf{n}_\alpha\cdot\left(\mathbf{v}^{\bar{\iota}} - \mathbf{v}_{\alpha\beta}\right) g_{\alpha\beta}\mathrm{d}S\right) f^{\overline{\overline{\alpha\beta}}}$$
$$+\frac{1}{V}\int_{\Omega_{\alpha\beta}} (\nabla'\cdot\mathbf{n}_\alpha)\mathbf{n}_\alpha\cdot\left(\mathbf{v}^{\bar{\iota}} - \mathbf{v}_{\alpha\beta}\right) g_{\alpha\beta} f_{\alpha\beta}\mathrm{d}S$$

$$+\left(\frac{1}{V}\int_{\Omega_{\alpha\beta}} g_{\alpha\beta}f_{\alpha\beta}\mathbf{n}_\alpha\mathbf{n}_\alpha \mathrm{d}S\right)\otimes \mathbf{S}^{\bar{\bar{\iota}}} - \frac{1}{V}\int_{\Omega_{\alpha\beta}} \frac{D'^{\bar{\iota}} g_{\alpha\beta}}{Dt} f_{\alpha\beta}\mathrm{d}S$$
$$+\sum_{\gamma\neq\alpha,\beta}\frac{1}{V}\int_{\Omega_{\alpha\beta\gamma}} \mathbf{m}_{\alpha\beta}\cdot\left(\mathbf{v}^{\bar{\iota}}-\mathbf{v}_{\alpha\beta\gamma}\right)g_{\alpha\beta}f_{\alpha\beta}\mathrm{d}S \tag{1087}$$

Here $\mathbf{v}^{\bar{\iota}}$ is the macroscale mass averaged velocity of entity ι, $\varepsilon^{\alpha\beta}$ the specific interfacial area of the α–β interface, $\mathbf{n}_\alpha$ the microscale normal vector to the α–β interface that is pointing outward from the α-phase. $\mathbf{n}_\alpha\cdot\mathbf{v}_{\alpha\beta}$ is the component of the microscale velocity of the α–β interface that is normal to the interface, $\mathbf{S}^{\bar{\bar{\iota}}}$ the macroscale rate of strain tensor of entity ι. $\mathbf{m}_{\alpha\beta}$ is a unit vector that is normal to the common curve edge of $\alpha\beta$ and tangent to $\alpha\beta$, V independent of time and position and is the measure of the averaging domain Ω, $\Omega_{\alpha\beta}$ the α–β interface within Ω, $\Omega_{\alpha\beta\gamma}$ (for all γ) is the common curve that bounds $\Omega_{\alpha\beta}$ within Ω, and $D'^{\bar{\iota}}/Dt$ is a material derivative referenced to $\mathbf{v}^{\iota}$ and restricted to the α–β interface defined as

$$\frac{D'^{\bar{\iota}}}{Dt}=\frac{\partial'}{\partial t}+\mathbf{v}^{\bar{\iota}}\cdot\nabla' \tag{1088}$$

Proof. Note that

$$\frac{1}{V}\int_{\Omega_{\alpha\beta}} g_{\alpha\beta}\frac{D'^{\bar{\iota}}\left(f_{\alpha\beta}-\overline{\overline{f^{\alpha\beta}}}\right)}{Dt}\mathrm{d}S$$
$$=\frac{1}{V}\int_{\Omega_{\alpha\beta}} g_{\alpha\beta}\frac{D'^{\bar{\iota}}f_{\alpha\beta}}{Dt}\mathrm{d}S-\frac{1}{V}\int_{\Omega_{\alpha\beta}} g_{\alpha\beta}\frac{D'^{\bar{\iota}}\overline{\overline{f^{\alpha\beta}}}}{Dt}\mathrm{d}S \tag{1089}$$

where

$$\frac{D'^{\bar{\iota}}}{Dt}=\frac{\partial}{\partial t}+\left(\mathbf{v}_{\alpha\beta}-\mathbf{v}^{\bar{\iota}}\right)\cdot\mathbf{n}_\alpha\mathbf{n}_\alpha\cdot\nabla+\mathbf{v}^{\bar{\iota}}\cdot\nabla=\frac{D^{\bar{\iota}}}{Dt}+\left(\mathbf{v}_{\alpha\beta}-\mathbf{v}^{\bar{\iota}}\right)\cdot\mathbf{n}_\alpha\mathbf{n}_\alpha\cdot\nabla \tag{1090}$$

After applying the chain rule, the first term on the right-hand side of Equation (1089) can be written as

$$\frac{1}{V}\int_{\Omega_{\alpha\beta}} g_{\alpha\beta}\frac{D'^{\bar{\iota}}f_{\alpha\beta}}{Dt}\mathrm{d}S=\frac{1}{V}\int_{\Omega_{\alpha\beta}}\frac{D'^{\bar{\iota}}\left(g_{\alpha\beta}f_{\alpha\beta}\right)}{Dt}\mathrm{d}S-\frac{1}{V}\int_{\Omega_{\alpha\beta}}\frac{D'^{\bar{\iota}}g_{\alpha\beta}}{Dt}f_{\alpha\beta}\mathrm{d}S \tag{1091}$$

Note that $\mathbf{v}^{\bar{\iota}}$ is a macroscale quantity so it may be moved outside the integral. After applying Equation (1088) to expand the derivation in the first term on the right-hand side, Equation (1091) becomes

$$\frac{1}{V}\int_{\Omega_{\alpha\beta}} g_{\alpha\beta}\frac{D'^{\bar{\iota}}f_{\alpha\beta}}{Dt}\mathrm{d}S=\frac{1}{V}\int_{\Omega_{\alpha\beta}}\frac{\partial'(g_{\alpha\beta}f_{\alpha\beta})}{\partial t}\mathrm{d}S+\mathbf{v}^{\bar{\iota}}\cdot\frac{1}{V}\int_{\Omega_{\alpha\beta}}\nabla'(g_{\alpha\beta}f_{\alpha\beta})\mathrm{d}S$$
$$-\frac{1}{V}\int_{\Omega_{\alpha\beta}}\frac{D'^{\bar{\iota}}g_{\alpha\beta}}{Dt}f_{\alpha\beta}\mathrm{d}S \tag{1092}$$

Applying Theorem 42 (Equation (636)) to the first integral and Theorem 43 (Equation (649)) to the second integral on the right-hand side of Equation (1092) gives

$$\frac{1}{V}\int_{\Omega_{\alpha\beta}} g_{\alpha\beta}\frac{D'^{\bar{\iota}}f_{\alpha\beta}}{Dt}\mathrm{d}S = \frac{\partial}{\partial t}\left(\frac{1}{V}\int_{\Omega_{\alpha\beta}} g_{\alpha\beta}f_{\alpha\beta}\mathrm{d}S\right) + \nabla\cdot\frac{1}{V}\int_{\Omega_{\alpha\beta}} \mathbf{n}_\alpha\mathbf{n}_\alpha\cdot\mathbf{v}_{\alpha\beta}g_{\alpha\beta}f_{\alpha\beta}\mathrm{d}S$$
$$-\frac{1}{V}\int_{\Omega_{\alpha\beta}}(\nabla'\cdot\mathbf{n}_\alpha)\mathbf{n}_\alpha\cdot\mathbf{v}_{\alpha\beta}g_{\alpha\beta}f_{\alpha\beta}\mathrm{d}S$$
$$-\sum_{\gamma\neq\alpha,\beta}\frac{1}{V}\int_{\Omega_{\alpha\beta\gamma}}\mathbf{m}_{\alpha\beta}\cdot\mathbf{v}_{\alpha\beta\gamma}g_{\alpha\beta}f_{\alpha\beta}\mathrm{d}S$$
$$+\mathbf{v}^{\bar{\iota}}\cdot\frac{1}{V}\int_{\Omega_{\alpha\beta}} g_{\alpha\beta}f_{\alpha\beta}\mathrm{d}S - \nabla\cdot\left(\frac{1}{V}\int_{\Omega_{\alpha\beta}} g_{\alpha\beta}f_{\alpha\beta}\mathbf{n}_\alpha\mathbf{n}_\alpha\mathrm{d}S\right)\cdot\mathbf{v}^{\bar{\iota}}$$
$$+\frac{1}{V}\int_{\Omega_{\alpha\beta}}(\nabla'\cdot\mathbf{n}_\alpha)\mathbf{n}_\alpha\cdot\mathbf{v}^{\bar{\iota}}g_{\alpha\beta}f_{\alpha\beta}\mathrm{d}S$$
$$+\sum_{\gamma\neq\alpha,\beta}\frac{1}{V}\int_{\Omega_{\alpha\beta\gamma}}\mathbf{m}_{\alpha\beta}\cdot\mathbf{v}^{\bar{\iota}}g_{\alpha\beta}f_{\alpha\beta}\mathrm{d}C - \frac{1}{V}\int_{\Omega_{\alpha\beta}}\frac{D'^{\bar{\iota}}g_{\alpha\beta}}{Dt}f_{\alpha\beta}\mathrm{d}S \quad (1093)$$

where

$$\nabla' = \nabla - \mathbf{n}_\alpha\mathbf{n}_\alpha\cdot\nabla$$

Note that $\mathbf{v}^{\bar{\iota}}$ is a macroscale quantity so it may be moved inside the integral. After applying the chain rule, we have

$$\nabla\cdot\left(\frac{1}{V}\int_{\Omega_{\alpha\beta}} g_{\alpha\beta}f_{\alpha\beta}\mathbf{n}_\alpha\mathbf{n}_\alpha\mathrm{d}S\right)\cdot\mathbf{v}^{\bar{\iota}} = \nabla\cdot\left(\frac{1}{V}\int_{\Omega_{\alpha\beta}}\mathbf{n}_\alpha\mathbf{n}_\alpha\cdot\mathbf{v}^{\bar{\iota}}g_{\alpha\beta}g_{\alpha\beta}\mathrm{d}S\right)$$
$$-\left(\frac{1}{V}\int_{\Omega_{\alpha\beta}} g_{\alpha\beta}f_{\alpha\beta}\mathbf{n}_\alpha\mathbf{n}_\alpha\mathrm{d}S\right)\otimes\mathbf{S}^{\bar{\bar{\iota}}} \quad (1094)$$

where

$$\mathbf{S}^{\bar{\bar{\iota}}} = \frac{1}{2}\left[\nabla\mathbf{v}^{\bar{\iota}} + \left(\nabla\mathbf{v}^{\bar{\iota}}\right)^T\right] \quad (1095)$$

Substituting Equation (1094) into Equation (1093) yields

$$\frac{1}{V}\int_{\Omega_{\alpha\beta}} g_{\alpha\beta}\frac{D'^{\bar{\iota}}f_{\alpha\beta}}{Dt}\mathrm{d}S = \frac{D'^{\bar{\iota}}\left[\varepsilon^{\alpha\beta}\left(g_{\alpha\beta}f_{\alpha\beta}\right)^{\alpha\beta}\right]}{Dt} - \nabla\cdot\left(\frac{1}{V}\int_{\Omega_{\alpha\beta}}\mathbf{n}_\alpha\mathbf{n}_\alpha\cdot(\mathbf{v}^{\bar{\iota}}-\mathbf{v}_{\alpha\beta})g_{\alpha\beta}f_{\alpha\beta}\mathrm{d}S\right.$$
$$+\frac{1}{V}\int_{\Omega_{\alpha\beta}}(\nabla'\cdot\mathbf{n}_\alpha)\mathbf{n}_\alpha\cdot(\mathbf{v}^{\bar{\iota}}-\mathbf{v}_{\alpha\beta})g_{\alpha\beta}f_{\alpha\beta}\mathrm{d}S$$
$$+\sum_{\gamma\neq\alpha,\beta}\frac{1}{V}\int_{\Omega_{\alpha\beta\gamma}}\mathbf{m}_{\alpha\beta}\cdot(\mathbf{v}^{\bar{\iota}}-\mathbf{v}_{\alpha\beta\gamma})g_{\alpha\beta}f_{\alpha\beta}\mathrm{d}C$$
$$-\frac{1}{V}\int_{\Omega_{\alpha\beta}}\frac{D'^{\bar{\iota}}g_{\alpha\beta}}{Dt}f_{\alpha\beta}\mathrm{d}S - \left(\frac{1}{V}\int_{\Omega_{\alpha\beta}} g_{\alpha\beta}f_{\alpha\beta}\mathbf{n}_\alpha\mathbf{n}_\alpha\mathrm{d}S\right)\otimes\mathbf{S}^{\bar{\bar{\iota}}} \quad (1096)$$

By applying Equation (1090), we may write the second term on the right-hand side of Equation (1089) as

$$\frac{1}{V}\int_{\Omega_{\alpha\beta}} g_{\alpha\beta}\frac{D'^{\bar{i}}\overline{\overline{f^{\alpha\beta}}}}{Dt}\mathrm{d}S=\frac{1}{V}\int_{\Omega_{\alpha\beta}} g_{\alpha\beta}\frac{D^{\bar{i}}\overline{\overline{f^{\alpha\beta}}}}{Dt}\mathrm{d}S+\frac{1}{V}\int_{\Omega_{\alpha\beta}} g_{\alpha\beta}\left(\mathbf{v}_{\alpha\beta}-\mathbf{v}^{\bar{i}}\right)\cdot\mathbf{n}_\alpha\mathbf{n}_\alpha\cdot\nabla\overline{\overline{f^{\alpha\beta}}}\mathrm{d}S \tag{1097}$$

The material derivative in the first integral and the gradient in the second integral on the right-hand side may be moved outside the integrals because they are at the macroscale. The remaining part of the first integral defines a surficial average so that

$$\frac{1}{V}\int_{\Omega_{\alpha\beta}} g_{\alpha\beta}\frac{D'^{\bar{i}}\overline{\overline{f^{\alpha\beta}}}}{Dt}\mathrm{d}S=\varepsilon^{\alpha\beta}g^{\alpha\beta}\frac{D^{\bar{i}}\overline{\overline{f^{\alpha\beta}}}}{Dt}+\left(\frac{1}{V}\int_{\Omega_{\alpha\beta}} g_{\alpha\beta}\left(\mathbf{v}_{\alpha\beta}-\mathbf{v}^{\bar{i}}\right)\cdot\mathbf{n}_\alpha\mathbf{n}_\alpha\mathrm{d}S\right)\cdot\nabla\overline{\overline{f^{\alpha\beta}}} \tag{1098}$$

Applying the chain rule to both terms yields

$$\begin{aligned}\frac{1}{V}\int_{\Omega_{\alpha\beta}} g_{\alpha\beta}\frac{D'^{\bar{i}}\overline{\overline{f^{\alpha\beta}}}}{Dt}\mathrm{d}S=&\frac{D^{\bar{i}}\left[\varepsilon^{\alpha\beta}\left(g^{\alpha\beta}\overline{\overline{f^{\alpha\beta}}}\right)\right]}{Dt}-\overline{\overline{f^{\alpha\beta}}}\frac{D^{\bar{i}}(\varepsilon^{\alpha\beta}(g^{\alpha\beta}))}{Dt}\\&+\nabla\cdot\left(\frac{1}{V}\int_{\Omega_{\alpha\beta}} g_{\alpha\beta}\overline{\overline{f^{\alpha\beta}}}\mathbf{n}_\alpha\mathbf{n}_\alpha\cdot\left(\mathbf{v}_{\alpha\beta}-\mathbf{v}^{\bar{i}}\right)\mathrm{d}S\right)\\&-\nabla\cdot\left(\frac{1}{V}\int_{\Omega_{\alpha\beta}} g_{\alpha\beta}\mathbf{n}_\alpha\mathbf{n}_\alpha\cdot\left(\mathbf{v}_{\alpha\beta}-\mathbf{v}^{\bar{i}}\right)\mathrm{d}S\right)\overline{\overline{f^{\alpha\beta}}}\end{aligned} \tag{1099}$$

Substituting Equations (1096) and (1099) into Equation (1089) yields

$$\begin{aligned}\frac{1}{V}\int_{\Omega_{\alpha\beta}} g_{\alpha\beta}\frac{D'^{\bar{i}}\left(f_{\alpha\beta}-\overline{\overline{f^{\alpha\beta}}}\right)}{Dt}\mathrm{d}S=&\frac{D^{\bar{i}}\left\{\varepsilon^{\alpha\beta}\left[\left(g_{\alpha\beta}f_{\alpha\beta}\right)^{\alpha\beta}-g^{\alpha\beta}\overline{\overline{f^{\alpha\beta}}}\right]\right\}}{Dt}+\frac{D^{\bar{i}}\left(\varepsilon^{\alpha\beta}g^{\alpha\beta}\right)}{Dt}\overline{\overline{f^{\alpha\beta}}}\\&-\nabla\cdot\left(\frac{1}{V}\int_{\Omega_{\alpha\beta}}\mathbf{n}_\alpha\mathbf{n}_\alpha\cdot\left(\mathbf{v}^{\bar{i}}-\mathbf{v}_{\alpha\beta}\right)g_{\alpha\beta}\left(f_{\alpha\beta}-\overline{\overline{f^{\alpha\beta}}}\right)\mathrm{d}S\right)\\&-\nabla\cdot\left(\frac{1}{V}\int_{\Omega_{\alpha\beta}}\mathbf{n}_\alpha\mathbf{n}_\alpha\cdot\left(\mathbf{v}^{\bar{i}}-\mathbf{v}_{\alpha\beta}\right)g_{\alpha\beta}\mathrm{d}S\right)\overline{\overline{f^{\alpha\beta}}}\\&+\frac{1}{V}\int_{\Omega_{\alpha\beta}}(\nabla'\cdot\mathbf{n}_\alpha)\mathbf{n}_\alpha\cdot\left(\mathbf{v}^{\bar{i}}-\mathbf{v}_{\alpha\beta}\right)g_{\alpha\beta}f_{\alpha\beta}\mathrm{d}S\\&+\left(\frac{1}{V}\int_{\Omega_{\alpha\beta}} g_{\alpha\beta}f_{\alpha\beta}\mathbf{n}_\alpha\mathbf{n}_\alpha\mathrm{d}S\right)\otimes\mathbf{S}^{\bar{i}}-\frac{1}{V}\int_{\Omega_{\alpha\beta}}\frac{D'^{\bar{i}}g_{\alpha\beta}}{Dt}f_{\alpha\beta}\mathrm{d}S\\&+\sum_{\gamma\neq\alpha,\beta}\frac{1}{V}\int_{\Omega_{\alpha\beta\gamma}}\mathbf{m}_{\alpha\beta}\cdot\left(\mathbf{v}^{\bar{i}}-\mathbf{v}_{\alpha\beta\gamma}\right)g_{\alpha\beta}f_{\alpha\beta}\mathrm{d}C\end{aligned} \tag{1100}$$

which completes the proof.

6.8.3 Multiscale deviations for a common curve

Theorem 74. The volume average of a product of a microscale quantity $g_{\alpha\beta\varepsilon}$ with a material derivative referenced to the macroscale mass average velocity of an entity ι restricted to a position on a potentially moving curve $\alpha\beta\varepsilon$ of the difference between a microscale quantity $f_{\alpha\beta\varepsilon}$ and a macroscale quantity $f^{\overline{\overline{\alpha\beta\varepsilon}}}$ can be expressed as a function of relative velocities and macroscale quantities of the form

$$\begin{aligned}
&\frac{1}{V}\int_{\Omega_{\alpha\beta\varepsilon}} g_{\alpha\beta\varepsilon}\frac{D''^{\bar{\iota}}\left(f_{\alpha\beta\varepsilon}-f^{\overline{\overline{\alpha\beta\varepsilon}}}\right)}{Dt}\mathrm{d}C \\
&=\frac{D^{\bar{\iota}}\left\{\varepsilon^{\alpha\beta\varepsilon}\left[\left(g_{\alpha\beta\varepsilon}f_{\alpha\beta\varepsilon}\right)^{\alpha\beta\varepsilon}-g^{\alpha\beta\varepsilon}f^{\overline{\overline{\alpha\beta\varepsilon}}}\right]\right\}}{Dt}+\frac{D^{\bar{\iota}}\left(\varepsilon^{\alpha\beta\varepsilon}g^{\alpha\beta\varepsilon}\right)}{Dt}f^{\overline{\overline{\alpha\beta\varepsilon}}} \\
&-\nabla\cdot\left(\frac{1}{V}\int_{\Omega_{\alpha\beta}}(\mathbf{I}-\mathbf{l}_{\alpha\beta\varepsilon}\mathbf{l}_{\alpha\beta\varepsilon})\cdot\left(\mathbf{v}^{\bar{\iota}}-\mathbf{v}_{\alpha\beta\varepsilon}\right)g_{\alpha\beta\varepsilon}\left(f_{\alpha\beta\varepsilon}-f^{\overline{\overline{\alpha\beta\varepsilon}}}\right)\mathrm{d}C\right) \\
&-\nabla\cdot\left(\frac{1}{V}\int_{\Omega_{\alpha\beta\varepsilon}}(\mathbf{I}-\mathbf{l}_{\alpha\beta\varepsilon}\mathbf{l}_{\alpha\beta\varepsilon})\cdot\left(\mathbf{v}^{\bar{\iota}}-\mathbf{v}_{\alpha\beta\varepsilon}\right)g_{\alpha\beta\varepsilon}\mathrm{d}C\right)f^{\overline{\overline{\alpha\beta\varepsilon}}} \\
&-\frac{1}{V}\int_{\Omega_{\alpha\beta\varepsilon}}(\mathbf{l}_{\alpha\beta\varepsilon}\cdot\nabla''\mathbf{l}_{\alpha\beta\varepsilon}\cdot\left(\mathbf{v}^{\bar{\iota}}-\mathbf{v}_{\alpha\beta\varepsilon}\right)g_{\alpha\beta\varepsilon}f_{\alpha\beta\varepsilon}\mathrm{d}C \\
&+\left(\frac{1}{V}\int_{\Omega_{\alpha\beta\varepsilon}}g_{\alpha\beta\varepsilon}f_{\alpha\beta\varepsilon}(\mathbf{I}-\mathbf{l}_{\alpha\beta\varepsilon}\mathbf{l}_{\alpha\beta\varepsilon})\mathrm{d}C\right)\otimes\mathbf{S}^{\bar{\bar{\iota}}} \\
&-\frac{1}{V}\int_{\Omega_{\alpha\beta}}\frac{D''^{\bar{\iota}}g_{\alpha\beta\varepsilon}}{Dt}f_{\alpha\beta\varepsilon}\mathrm{d}C \\
&+\frac{1}{V}\sum_{\gamma\neq\alpha,\beta,\varepsilon}\left[\mathbf{e}_{\alpha\beta\varepsilon}\cdot\left(\mathbf{v}^{\bar{\iota}}-\mathbf{v}_{\alpha\beta\varepsilon\gamma}\right)g_{\alpha\beta\varepsilon}f_{\alpha\beta\varepsilon}\right]\Big|_{\alpha\beta\varepsilon\gamma}
\end{aligned}\tag{1101}$$

Here $\mathbf{v}^{\bar{\iota}}$ is the macroscale mass averaged velocity of entity $\iota, \varepsilon^{\alpha\beta\varepsilon}$ the specific length of the $\alpha\beta\varepsilon$ common curve, $\mathbf{l}_{\alpha\beta\varepsilon}$ the microscale unit vector tangent to the $\alpha\beta\varepsilon$ common curve, $(\mathbf{I}-\mathbf{l}_{\alpha\beta\varepsilon}\mathbf{l}_{\alpha\beta\varepsilon})\cdot\mathbf{v}_{\alpha\beta\varepsilon}$ the component of the microscale velocity of the $\alpha\beta\varepsilon$ common curve that is normal to the curve. $\mathbf{S}^{\bar{\bar{\iota}}}$ is the macroscale rate of strain tensor of entity ι, $\mathbf{e}_{\alpha\beta\varepsilon}$ a unit vector tangent to the $\alpha\beta\varepsilon$ common curve at its endpoints that is positive outward from the curve, V the independent of time and position and is the measure of the averaging domain Ω, $\Omega_{\alpha\beta\varepsilon}$ the $\alpha\beta\varepsilon$ common curve within Ω, $\Omega_{\alpha\beta\varepsilon\gamma}$ (for all γ) is the set of end points of the $\alpha\beta\varepsilon$ common curve within Ω, and $D''^{\bar{\iota}}/Dt$ is a material derivative referenced to $\mathbf{v}^{\bar{\iota}}$ and restricted to the $\alpha\beta\varepsilon$ common curve defined as

$$\frac{D''^{\bar{\iota}}}{Dt}=\frac{\partial''}{\partial t}+\mathbf{v}^{\bar{\iota}}\cdot\nabla''\tag{1102}$$

Proof. Note that

$$\frac{1}{V}\int_{\Omega_{\alpha\beta\varepsilon}} g_{\alpha\beta\varepsilon}\frac{D''^{\bar{\imath}}\left(f_{\alpha\beta\varepsilon}-\overline{\overline{f^{\alpha\beta\varepsilon}}}\right)}{Dt}\mathrm{d}C=\frac{1}{V}\int_{\Omega_{\alpha\beta\varepsilon}} g_{\alpha\beta\varepsilon}\frac{D''^{\bar{\imath}}f_{\alpha\beta\varepsilon}}{Dt}\mathrm{d}C-\frac{1}{V}\int_{\Omega_{\alpha\beta\varepsilon}} g_{\alpha\beta\varepsilon}\frac{D''^{\bar{\imath}}\overline{\overline{f^{\alpha\beta\varepsilon}}}}{Dt}\mathrm{d}C \tag{1103}$$

where

$$\begin{aligned}\frac{D''^{\bar{\imath}}}{Dt}&=\frac{\partial}{\partial t}+\left(\mathbf{v}_{\alpha\beta\varepsilon}-\mathbf{v}^{\bar{\imath}}\right)\cdot\left(\mathbf{I}-\mathbf{l}_{\alpha\beta\varepsilon}\mathbf{l}_{\alpha\beta\varepsilon}\right)\cdot\nabla+\mathbf{v}^{\bar{\imath}}\cdot\nabla\\&=\frac{D^{\bar{\imath}}}{Dt}+\left(\mathbf{v}_{\alpha\beta}-\mathbf{v}^{\bar{\imath}}\right)\cdot\left(\mathbf{I}-\mathbf{l}_{\alpha\beta\varepsilon}\mathbf{l}_{\alpha\beta\varepsilon}\right)\cdot\nabla\end{aligned} \tag{1104}$$

After applying the chain rule, the first term on the right-hand side of Equation (1103) can be written as

$$\frac{1}{V}\int_{\Omega_{\alpha\beta\varepsilon}} g_{\alpha\beta}\frac{D''^{\bar{\imath}}f_{\alpha\beta\varepsilon}}{Dt}\mathrm{d}C=\frac{1}{V}\int_{\Omega_{\alpha\beta\varepsilon}}\frac{D''^{\bar{\imath}}(g_{\alpha\beta\varepsilon}f_{\alpha\beta\varepsilon}}{Dt}\mathrm{d}C-\frac{1}{V}\int_{\Omega_{\alpha\beta\varepsilon}}\frac{D''^{\bar{\imath}}g_{\alpha\beta\varepsilon}}{Dt}f_{\alpha\beta\varepsilon}\mathrm{d}C \tag{1105}$$

Note that $\mathbf{v}^{\bar{\imath}}$ is a macroscale quantity so that it may be moved outside the integral. After applying Equation (1102) to expand the derivation in the first term on the right-hand side, Equation (1105) becomes

$$\frac{1}{V}\int_{\Omega_{\alpha\beta\varepsilon}} g_{\alpha\beta\varepsilon}\frac{D''^{\bar{\imath}}f_{\alpha\beta\varepsilon}}{Dt}\mathrm{d}C=\frac{1}{V}\int_{\Omega_{\alpha\beta\varepsilon}}\frac{\partial''\left(g_{\alpha\beta\varepsilon}f_{\alpha\beta\varepsilon}\right)}{\partial t}\mathrm{d}C+\mathbf{v}^{\bar{\imath}}\cdot\frac{1}{V}\int_{\Omega_{\alpha\beta\varepsilon}}\nabla'\left(g_{\alpha\beta\varepsilon}f_{\alpha\beta\varepsilon}\right)\mathrm{d}C-\frac{1}{V}\int_{\Omega_{\alpha\beta\varepsilon}}\frac{D''^{\bar{\imath}}g_{\alpha\beta\varepsilon}}{Dt}f_{\alpha\beta\varepsilon}\mathrm{d}C \tag{1106}$$

Applying Theorems 46 and 47 (Equations (678) and (691)) to the first and second integrals on the right-hand side of Equation (1106), respectively, yields

$$\begin{aligned}&\frac{1}{V}\int_{\Omega_{\alpha\beta\varepsilon}} g_{\alpha\beta\varepsilon}\frac{D''^{\bar{\imath}}f_{\alpha\beta\varepsilon}}{Dt}\mathrm{d}C\\&=\frac{\partial}{\partial t}\left(\frac{1}{V}\int_{\Omega_{\alpha\beta\varepsilon}} g_{\alpha\beta\varepsilon}f_{\alpha\beta\varepsilon}\mathrm{d}C\right)+\nabla\cdot\frac{1}{V}\int_{\Omega_{\alpha\beta\varepsilon}}\left(\mathbf{I}-\mathbf{l}_{\alpha\beta\varepsilon}\mathbf{l}_{\alpha\beta\varepsilon}\right)\cdot\mathbf{v}_{\alpha\beta\varepsilon}g_{\alpha\beta\varepsilon}f_{\alpha\beta\varepsilon}\mathrm{d}C\\&\quad-\frac{1}{V}\int_{\Omega_{\alpha\beta\varepsilon}}\left(\mathbf{l}_{\alpha\beta\varepsilon}\cdot\nabla''\mathbf{l}_{\alpha\beta\varepsilon}\right)\cdot\mathbf{v}_{\alpha\beta\varepsilon}g_{\alpha\beta\varepsilon}f_{\alpha\beta\varepsilon}\mathrm{d}C-\frac{1}{V}\sum_{\gamma\neq\alpha,\beta,\varepsilon}\left(\left(\mathbf{e}_{\alpha\beta\varepsilon}\cdot\mathbf{v}_{\alpha\beta\varepsilon\gamma}\right)g_{\alpha\beta\varepsilon}f_{\alpha\beta\varepsilon}\right)\Big|_{\alpha\beta\varepsilon\gamma}\\&\quad+\mathbf{v}^{\bar{\imath}}\cdot\nabla\left(\frac{1}{V}\int_{\Omega_{\alpha\beta\varepsilon}} g_{\alpha\beta\varepsilon}f_{\alpha\beta\varepsilon}\mathrm{d}C\right)-\nabla\cdot\left(\frac{1}{V}\int_{\Omega_{\alpha\beta\varepsilon}} g_{\alpha\beta\varepsilon}f_{\alpha\beta\varepsilon}\left(\mathbf{I}-\mathbf{l}_{\alpha\beta\varepsilon}\mathbf{l}_{\alpha\beta\varepsilon}\right)\mathrm{d}C\right)\cdot\mathbf{v}^{\bar{\imath}}\\&\quad-\left(\frac{1}{V}\int_{\Omega_{\alpha\beta\varepsilon}} g_{\alpha\beta\varepsilon}f_{\alpha\beta\varepsilon}\left(\mathbf{l}_{\alpha\beta\varepsilon}\cdot\nabla''\mathbf{l}_{\alpha\beta\varepsilon}\right)\mathrm{d}C\right)\cdot\mathbf{v}^{\bar{\imath}}+\frac{1}{V}\sum_{\varepsilon\neq\alpha,\beta}\left(\mathbf{e}_{\alpha\beta\varepsilon}g_{\alpha\beta\varepsilon}f_{\alpha\beta\varepsilon}\right)\Big|_{\alpha\beta\varepsilon\gamma}\cdot\mathbf{v}^{\bar{\imath}}\\&\quad-\frac{1}{V}\int_{\Omega_{\alpha\beta\varepsilon}}\frac{D''^{\bar{\imath}}g_{\alpha\beta\varepsilon}}{Dt}f_{\alpha\beta\varepsilon}\mathrm{d}C\end{aligned} \tag{1107}$$

Note that $\mathbf{v}^{\bar{\imath}}$ is a macro scale quantity so that it may be moved inside the integral. After applying the chain rule, we have

$$\nabla\cdot\left(\frac{1}{V}\int_{\Omega_{\alpha\beta\varepsilon}} g_{\alpha\beta\varepsilon}f_{\alpha\beta\varepsilon}(\mathbf{I}-\mathbf{l}_{\alpha\beta\varepsilon}\mathbf{l}_{\alpha\beta\varepsilon})\mathrm{d}C\right)\cdot\mathbf{v}^{\bar{\imath}}=\nabla\cdot\left(\frac{1}{V}\int_{\Omega_{\alpha\beta\varepsilon}} g_{\alpha\beta\varepsilon}f_{\alpha\beta\varepsilon}(\mathbf{I}-\mathbf{l}_{\alpha\beta\varepsilon}\mathbf{l}_{\alpha\beta\varepsilon})\cdot\mathbf{v}^{\bar{\imath}}\mathrm{d}C\right)$$
$$-\left(\frac{1}{V}\int_{\Omega_{\alpha\beta\varepsilon}} g_{\alpha\beta\varepsilon}f_{\alpha\beta\varepsilon}(\mathbf{I}-\mathbf{l}_{\alpha\beta\varepsilon}\mathbf{l}_{\alpha\beta\varepsilon})\mathrm{d}C\right)\otimes\mathbf{S}^{\bar{\bar{\imath}}} \quad (1108)$$

where

$$\mathbf{S}^{\bar{\bar{\imath}}}=\frac{1}{2}\left[\nabla\mathbf{v}^{\bar{\imath}}+\left(\nabla\mathbf{v}^{\bar{\imath}}\right)^{T}\right] \quad (1109)$$

Substituting Equation (1108) into Equation (1107) yields

$$\frac{1}{V}\int_{\Omega_{\alpha\beta\varepsilon}} g_{\alpha\beta\varepsilon}\frac{D''^{\bar{\imath}}f_{\alpha\beta\varepsilon}}{Dt}\mathrm{d}C=\frac{D^{\bar{\imath}}\left[\varepsilon^{\alpha\beta\varepsilon}\left(g_{\alpha\beta\varepsilon}f_{\alpha\beta\varepsilon}\right)^{\alpha\beta\varepsilon}\right]}{Dt}$$
$$-\nabla\cdot\left(\frac{1}{V}\int_{\Omega_{\alpha\beta\varepsilon}}(\mathbf{I}-\mathbf{l}_{\alpha\beta\varepsilon}\mathbf{l}_{\alpha\beta\varepsilon})\cdot(\mathbf{v}^{\bar{\imath}}-\mathbf{v}_{\alpha\beta\varepsilon})g_{\alpha\beta\varepsilon}f_{\alpha\beta\varepsilon}\mathrm{d}C\right.$$
$$+\frac{1}{V}\int_{\Omega_{\alpha\beta\varepsilon}}(\mathbf{l}_{\alpha\beta\varepsilon}.\nabla''\mathbf{l}_{\alpha\beta\varepsilon})\cdot(\mathbf{v}^{\bar{\imath}}-\mathbf{v}_{\alpha\beta\varepsilon})g_{\alpha\beta\varepsilon}f_{\alpha\beta\varepsilon}\mathrm{d}C$$
$$+\frac{1}{V}\sum_{\gamma\neq\alpha,\beta,\varepsilon}\left[\mathbf{e}_{\alpha\beta\varepsilon}\cdot(\mathbf{v}^{\bar{\imath}}-\mathbf{v}_{\alpha\beta\varepsilon\gamma})g_{\alpha\beta\varepsilon}f_{\alpha\beta\varepsilon}\right]\Big|_{\alpha\beta\varepsilon\gamma}\mathrm{d}C$$
$$-\frac{1}{V}\int_{\Omega_{\alpha\beta\varepsilon}}\frac{D''^{\bar{\imath}}g_{\alpha\beta\varepsilon}}{Dt}f_{\alpha\beta\varepsilon}\mathrm{d}C-\left(\frac{1}{V}\int_{\Omega_{\alpha\beta\varepsilon}} g_{\alpha\beta\varepsilon}f_{\alpha\beta\varepsilon}(\mathbf{I}-\mathbf{l}_{\alpha\beta\varepsilon}\mathbf{l}_{\alpha\beta\varepsilon})\mathrm{d}C\right)\otimes\mathbf{S}^{\bar{\bar{\imath}}} \quad (1110)$$

By applying Equation (1104), we may write the second term on the right-hand side of Equation (1103) as

$$\frac{1}{V}\int_{\Omega_{\alpha\beta\varepsilon}} g_{\alpha\beta\varepsilon}\frac{D''^{\bar{\imath}}f^{\overline{\overline{\alpha\beta\varepsilon}}}}{Dt}\mathrm{d}C=\frac{1}{V}\int_{\Omega_{\alpha\beta\varepsilon}} g_{\alpha\beta\varepsilon}\frac{D^{\bar{\imath}}f^{\overline{\overline{\alpha\beta\varepsilon}}}}{Dt}\mathrm{d}C$$
$$+\frac{1}{V}\int_{\Omega_{\alpha\beta\varepsilon}} g_{\alpha\beta\varepsilon}(\mathbf{v}_{\alpha\beta\varepsilon}-\mathbf{v}^{\bar{\imath}})\cdot(\mathbf{I}-\mathbf{l}_{\alpha\beta\varepsilon}\mathbf{l}_{\alpha\beta\varepsilon})\cdot\nabla f^{\overline{\overline{\alpha\beta\varepsilon}}}\mathrm{d}C \quad (1111)$$

The material derivative in the first integral and the gradient in the second integral on the right-hand side may be moved outside the integrals because they are at the macro scale. The remaining part of the first integral defines an average over the common curve so that

$$\frac{1}{V}\int_{\Omega_{\alpha\beta\varepsilon}} g_{\alpha\beta\varepsilon}\frac{D''^{\bar{\imath}}f^{\overline{\overline{\alpha\beta\varepsilon}}}}{Dt}\mathrm{d}C=\varepsilon^{\alpha\beta\varepsilon}g^{\alpha\beta\varepsilon}\frac{D^{\bar{\imath}}f^{\overline{\overline{\alpha\beta\varepsilon}}}}{Dt}$$
$$+\left(\frac{1}{V}\int_{\Omega_{\alpha\beta\varepsilon}} g_{\alpha\beta\varepsilon}(\mathbf{v}_{\alpha\beta\varepsilon}-\mathbf{v}^{\bar{\imath}})\cdot(\mathbf{I}-\mathbf{l}_{\alpha\beta\varepsilon}\mathbf{l}_{\alpha\beta\varepsilon})\mathrm{d}C\right)\cdot\nabla f^{\overline{\overline{\alpha\beta\varepsilon}}} \quad (1112)$$

Applying the chain rule to both terms yields

$$\frac{1}{V}\int_{\Omega_{\alpha\beta\varepsilon}} g_{\alpha\beta\varepsilon}\frac{D''^{\bar{\imath}} \overline{\overline{f^{\alpha\beta\varepsilon}}}}{Dt}\mathrm{d}C = \frac{D^{\bar{\imath}}\left[\varepsilon^{\alpha\beta\varepsilon}\left(g^{\alpha\beta\varepsilon}\overline{\overline{f^{\alpha\beta\varepsilon}}}\right)\right]}{Dt} - \overline{\overline{f^{\alpha\beta\varepsilon}}}\frac{D^{\bar{\imath}}\left(\varepsilon^{\alpha\beta\varepsilon}g^{\alpha\beta\varepsilon}\right)}{Dt}$$
$$+\nabla\cdot\left(\frac{1}{V}\int_{\Omega_{\alpha\beta\varepsilon}} g_{\alpha\beta\varepsilon}\overline{\overline{f^{\alpha\beta\varepsilon}}}\left(\mathbf{I}-\mathbf{l}_{\alpha\beta\varepsilon}\mathbf{l}_{\alpha\beta\varepsilon}\right)\cdot\left(\mathbf{v}_{\alpha\beta\varepsilon}-\mathbf{v}^{\bar{\imath}}\right)\mathrm{d}C\right)$$
$$-\nabla\cdot\left(\frac{1}{V}\int_{\Omega_{\alpha\beta\varepsilon}} g_{\alpha\beta\varepsilon}\left(\mathbf{I}-\mathbf{l}_{\alpha\beta\varepsilon}\mathbf{l}_{\alpha\beta\varepsilon}\right)\cdot\left(\mathbf{v}_{\alpha\beta\varepsilon}-\mathbf{v}^{\bar{\imath}}\right)\mathrm{d}C\right)\overline{\overline{f^{\alpha\beta\varepsilon}}} \tag{1113}$$

Substituting Equations (1110) and (1113) into Equation (1103) yields

$$\frac{1}{V}\int_{\Omega_{\alpha\beta\varepsilon}} g_{\alpha\beta\varepsilon}\frac{D''^{\bar{\imath}}\left(f_{\alpha\beta\varepsilon}-\overline{\overline{f^{\alpha\beta\varepsilon}}}\right)}{Dt}\mathrm{d}C$$
$$=\frac{D^{\bar{\imath}}\left\{\varepsilon^{\alpha\beta\varepsilon}\left[\left(g_{\alpha\beta\varepsilon}f_{\alpha\beta\varepsilon}\right)^{\alpha\beta\varepsilon}-g^{\alpha\beta\varepsilon}\overline{\overline{f^{\alpha\beta\varepsilon}}}\right]\right\}}{Dt}+\frac{D^{\bar{\imath}}\left(\varepsilon^{\alpha\beta\varepsilon}g^{\alpha\beta\varepsilon}\right)}{Dt}\overline{\overline{f^{\alpha\beta\varepsilon}}}$$
$$-\nabla\cdot\left(\frac{1}{V}\int_{\Omega_{\alpha\beta}}\left(\mathbf{I}-\mathbf{l}_{\alpha\beta\varepsilon}\mathbf{l}_{\alpha\beta\varepsilon}\right)\cdot\left(\mathbf{v}^{\bar{\imath}}-\mathbf{v}_{\alpha\beta\varepsilon}\right)g_{\alpha\beta\varepsilon}\left(f_{\alpha\beta\varepsilon}-\overline{\overline{f^{\alpha\beta\varepsilon}}}\right)\mathrm{d}C\right)$$
$$-\nabla\cdot\left(\frac{1}{V}\int_{\Omega_{\alpha\beta\varepsilon}}\left(\mathbf{I}-\mathbf{l}_{\alpha\beta\varepsilon}\mathbf{l}_{\alpha\beta\varepsilon}\right)\cdot\left(\mathbf{v}^{\bar{\imath}}-\mathbf{v}_{\alpha\beta\varepsilon}\right)g_{\alpha\beta\varepsilon}\mathrm{d}C\right)\overline{\overline{f^{\alpha\beta\varepsilon}}}$$
$$-\frac{1}{V}\int_{\Omega_{\alpha\beta\varepsilon}}\left(\mathbf{l}_{\alpha\beta\varepsilon}\cdot\nabla''\mathbf{l}_{\alpha\beta\varepsilon}\cdot\left(\mathbf{v}^{\bar{\imath}}-\mathbf{v}_{\alpha\beta\varepsilon}\right)g_{\alpha\beta\varepsilon}f_{\alpha\beta\varepsilon}\mathrm{d}C\right.$$
$$+\left(\frac{1}{V}\int_{\Omega_{\alpha\beta\varepsilon}} g_{\alpha\beta\varepsilon}f_{\alpha\beta\varepsilon}\left(\mathbf{I}-\mathbf{l}_{\alpha\beta\varepsilon}\mathbf{l}_{\alpha\beta\varepsilon}\right)\right)\otimes\mathbf{S}^{\bar{\imath}}$$
$$-\frac{1}{V}\int_{\Omega_{\alpha\beta\varepsilon}}\frac{D''^{\bar{\imath}}g_{\alpha\beta\varepsilon}}{Dt}f_{\alpha\beta\varepsilon}\mathrm{d}C+\frac{1}{V}\sum_{\gamma\neq\alpha,\beta,\varepsilon}\left[\mathbf{e}_{\alpha\beta\varepsilon}\cdot\left(\mathbf{v}^{\bar{\imath}}-\mathbf{v}_{\alpha\beta\varepsilon\gamma}\right)g_{\alpha\beta\varepsilon}f_{\alpha\beta\varepsilon}\right]\Big|_{\alpha\beta\varepsilon\gamma} \tag{1114}$$

which completes the proof.

7. CONCLUDING REMARKS

Predicting and analyzing multiscale phenomena are research challenges of rare potential but daunting difficulty. The potential comes from both scientific and practical opportunities in nearly all fields of science and technology. The difficulty reflects the issues related to scales and scaling. In the last decade, much effort has been made in various fields on *process scales* (properties of process scales; the intrinsic relationship between time and space scales; scale properties and ways to characterize various multiscale processes; preferred scales; the existence of a separation of scales), *observation scales* (appropriate observation scales and choice of scales within logistical, technological and perceptual constraints; properties of observation scales), *modeling scales* (integration of processes operating at diverse scales; ways to suppress irrelevant details; finding relevant scales for modeling; ways to develop predictive and investigative models) and *interplay* among process, observation, and modeling scales. Research work on scaling has mainly focused on developing technologies of upscaling (including model-upscaling), downscaling (including model-downscaling), and multiscale analysis (including multiscale modeling).

As it stands, multiscale science is still in its infancy despite the significant progress which has been made in the last decade. Because of the complexities of problems, there appear no universal systematic approaches available in multiscale science. Available approaches are more or less haphazard and *ad hoc* with applications usually limited to particular fields or even particular multiscale problems. Development of the multiscale theorems represents an important advance towards a universal systematic approach for scaling.

The multiscale theorems transform derivatives of a function from one scale to another, a fundamental and essential task in multiscale science. The derivatives can be spatial or temporal. The former includes the gradient, divergence, and curl. Each of these three spatial operators can be spatial, surficial, or curvilinear. The latter refers to the partial derivative with respect to time of a function defined in space, on a surface or along a curve. The multiscale theorems developed in the present work consist of 38 integration theorems and 33 averaging theorems. The integration theorems change any or all spatial scales of a derivative from the microscale (or any continuum scale) to the megascale by integration. The integration domain may translate and deform with time, and can be a volume, a surface, or a curve. The averaging theorems change any or all spatial scales of a derivative from microscale to macroscale by averaging. The averaging volume is located in space and integration is performed over volumes, surfaces, or curves contained within the averaging volume. Whereas the 71 theorems in the present work form a set of generalized and universal integration theorems and are likely to be adequate for many applications involving the dynamics of multiscale systems in space, on surfaces and along curves, applications may arise where variations of these theorems are required. These variations may be developed either by following the same approach as that used in developing the 71 theorems or by combining some of them and making use of some standard relations from calculus. In either case, these theorems will facilitate analyzing problems systematically and with rigor based on fundamental laws.

As with all mathematical theorems, the tools used to prove each theorem and the logic behind the proof itself also hold significant value. The advantage of the approach used in the present work is that theorems for complicated geometries, such as interfaces and common curves in a multiphase system, become tractable. In fact, the mechanics of the proofs of each theorem are the same regardless of the complexity of the geometry or the dimensionality of the derivative operator, although the derivations become lengthier with system complexity.

Indicator functions are essential for the approach. These functions can be used to identify a region of interest by taking the value of one in the interior and zero exterior to the region. By using indicator functions, curvilinear, surficial, and volumetric integrals are transformed to integrals overall space with the indicator functions and/or their derivatives appearing in the integrand. The spatial and temporal dependence of integration domains is, thus, moved into the integrand. All subsequent manipulations involve the integrand with minimal concern for the limits of integration. The differentiation and integration properties of the indicator functions provide a simplified route to scale changes and are very useful in many fields, especially those requiring inter-conversion of integrals over curves, surfaces, and volumes. In a very real sense, indicator functions are mathematical catalysts: they facilitate the derivations but they do not appear in the end product. As with chemistry wherein a reaction would not take place without the presence of a catalyst, the derivations here rely on the catalytic activity of the indicator functions.

The multiscale theorems are developed by averaging over an integration domain or over volumes, surfaces, or curves contained within an averaging volume. The averaging process has taken into account the geometrical structures of and interactions between various entities within the integration domain. They are thus endowed with important information regarding the geometrical and topological structures of and interactions among various entities such as phases, interfaces, common curves, and common points. Therefore, the 71 multiscale theorems provide not only simple tools to model multiscale phenomena at various scales and to interchange among these scales, but also the critical information for resolving the closure problems routinely encountered in multiscale science. Applications of these theorems to transport-phenomena modeling and scaling result in the macroscale equations of turbulence eddies of various geometrical and topological structures, the macroscale equations of transient heat conduction in two-phase systems, the macroscale equations for phases, interfaces, and common curves in porous and multiphase systems, and the multiscale deviation theorems for the TCAT approach of modeling flow and transport phenomena in multiscale porous-medium systems. All these macroscale equations and multiscale deviation theorems contain terms regarding the interactions among various entities that are of considerable significance for developing closure. Applications of the multi-scale theorems also lead to the scale-by-scale energy budget equations in real space for turbulent flows and the energy flux rates among various groups of eddies. Furthermore, the application of multiscale theorems to two-phase-system heat conduction builds the intrinsic equivalence between dual-phase-lagging heat conduction and Fourier heat conduction in two-phase systems that are subject to a lack of local thermal equilibrium. This result reconciles the inconsistency between theoretical and experimental results in the literature regarding the possibility of thermal waves and resonance in two-phase-system heat conduction, uncovers the mechanism responsible for such waves and resonance, and enables us to apply the methods and results in one field to another.

NOMENCLATURE

A	planar surface and its area (m^2)
a_v	interfacial area per unit volume (m^{-1})
$A_{\beta\sigma}$	area of the β–σ interface (m^2)
b	external entropy source (J/(kg s K))
c	specific heat (J/(kg K))
C	curve bounding a surface and its length (m)
$C_{\alpha\beta}$	the curve of intersection of $S_{\alpha\beta}$ surface with the external surface of the REV and its length (m)
$C_{\alpha\beta\varepsilon}$	the boundary curve of $S_{\alpha\beta}$ within the REV where the α-and β-phases meet other phases and its length (m)
e	specific internal energy (J/kg)
$\mathbf{e}$	unit tangent vector (–)
$\mathbf{e}_{\alpha\beta\varepsilon}$	unit vector tangent to the $\alpha\beta\varepsilon$ common curve and oriented positive outward at the endpoints (–)
E	kinetic energy (J/kg)
$\mathcal{E}$	set of all types of entities (–)
$\mathbf{f}$	scalar function (–)
$\mathbf{f}$	vector function (–)
$\mathbf{f}^c$	vector function tangent to C or $C_{\alpha\beta\varepsilon}$ (–)
$\mathbf{f}^s$	vector function normal to C or tangent to S (–)
g	external supply of momentum (gravity) (m/s^2)
h	external supply of energy (J/(kg s))
h	film heat transfer coefficient ($W/(m^2 K)$)
$\mathbf{I}$	identity tensor (–)
$\mathcal{I}$	index set of all types of entities (–)
k	wave number (–)
k_β	effective thermal conductivity of the β-phases (W/(m K))
k_σ	effective thermal conductivity of the σ-phases (W/(m K))
$k_{\beta\sigma}$	cross-effective thermal conductivity between the β- and σ-phases (W/(m K))
$k_{e\beta}$	equivalent effective thermal conductivity of the β-phases (W/(m K))
$k_{e\sigma}$	equivalent effective thermal conductivity of the σ-phases (W/(m K))
$\mathbf{K}_{\text{eff}}$	one-equation model effective thermal conductivity tensor (W/(m K))
$\mathbf{K}_{\beta\beta}$	two-equation model effective thermal conductivity tensor associated with $\nabla \langle T_\beta \rangle^\beta$ in the β-phase equation (W/(m K))
$\mathbf{K}_{\beta\sigma}$	two-equation model effective thermal conductivity tensor associated with $\nabla \langle T_\sigma \rangle^\sigma$ in the β-phase equation (W/(m K))
$\mathbf{K}_{\sigma\sigma}$	two-equation model effective thermal conductivity tensor associated with $\nabla \langle T_\sigma \rangle^\sigma$ in the σ-phase equation (W/(m K))
$\mathbf{K}_{\sigma\beta}$	two-equation model effective thermal conductivity tensor associated with $\nabla \langle T_\beta \rangle^\beta$ in the σ-phase equation (W/(m K))
l	length scale (m)
$\mathbf{l}_{\alpha\beta\varepsilon}$	unit vector tangent to the $\alpha\beta\varepsilon$ common curve (–)
L	straight-line segment of integration and its length (m)

L_T	characteristic length of $\nabla\langle T\rangle$ (m)
L_{T1}	characteristic length of $\nabla\nabla\langle T\rangle$ (m)
$\mathbf{m}_{\alpha\beta}$	unit vector orthogonal to $\mathbf{n}_\alpha$ and outward normal from the σ–β interface along the edge of the interface (–)
$\mathbf{n}$	outward unit normal vector (–)
$\mathbf{n}_{\beta\alpha}$	$= -\mathbf{n}_{\sigma\beta}$ outward unit normal vector pointing from the β-phase toward the σ-phase (–)
n_C	number of common curves (–)
n_I	number of interfaces (–)
n_P	number of phases (–)
n_{Pt}	number of common points (–)
$\mathbf{n}_\alpha$	outward normal vector from the α-phase (–)
$\mathbf{N}$	unit normal vector of planar surface (–)
p	pressure (N/m^2)
$\mathcal{P}$	set of system properties (–)
$\mathbf{q}$	heat flux density vector (W/m^2)
$\mathbf{r}$	position vector (m)
r	microscale spatial moment arm (m)
Re	Reynolds number (–)
S	surface and its area (m^2)
S_α	the portion of the REV occupied by the α-phase and its volume (m^3)
$S_{\alpha\beta}$	the interface within the REV between the α-phase and all other phases and its area (m^2)
$\mathbf{S}$	rate of strain tensor (s^{-1})
t	time (s)
$\mathbf{t}$	stress tensor (N/m^2)
T	temperature (K)
$\mathbf{u}$	velocity (m/s)
U	characteristic velocity (m/s)
$\mathbf{v}$	velocity (m/s)
V	volume measure, volume of integration (m^3)
V_α	the portion of the REV occupied by the α-phase and its volume (m^3)
$\mathbf{w}$	microscale velocity vector of an interface whose tangential components may differ from the angential velocity of the material in the interface (m/s)
$\mathbf{w}^c$	tangent velocity of C (m/s)
$\mathbf{w}^s$	normal velocity of C (m/s)
$\mathbf{x}$	position vector (m)

GREEKS

α	thermal diffusivity (m^2/s)
δl	change in length scale (m)
ε	precision estimate for property (–)

ε	volume fraction (–)
ε_t	turbulence dissipation rate (J/(kg s))
γ	indicator function used to identify a volume in space (–)
γ^{s}	indicator function used to identify a surface (–)
γ^{c}	indicator function used to identify a curve (–)
γ_β	β-phase effective thermal capacity (J/(m^3 K))
γ_σ	σ-phase effective thermal capacity (J/(m^3 K))
Γ	boundary (–)
η	Kolmogorov length scale (m)
η	entropy (J/(kg K))
$\boldsymbol{\lambda}$	unit vector tangent to one of the surficial coordinates in S (–)
Λ	net entropy production (J/(kg s K))
μ	dynamic viscosity (kg/(m s))
ν	kinematic viscosity (m^2/s)
$\boldsymbol{\nu}$	unit vector tangent to another of the surficial coordinates in S (–)
Ω	domain (–)
ϕ	entropy flux vector (J/(m^2 s K))
ρ	density (kg/m^3)
$\boldsymbol{\sigma}$	stress tensor (N/m^2)
τ_q	phase-lag of heat flux vector (s)
τ_T	phase-lag of temperature gradient (s)
φ	porosity (–)

OTHER MATHEMATICAL SYMBOLS

$\langle\ \rangle$	averaging operator (–)
$D^{\bar{I}}/Dt$	material derivative (s^{-1})
$D'^{\bar{I}}/Dt$	material derivative restricted to an interface (s^{-1})
$D''^{\bar{I}}/Dt$	material derivative restricted to a common curve (s^{-1})
$\partial'/\partial t$	partial derivative of a point on a potentially moving interface (s^{-1})
$\partial''/\partial t$	partial derivative of a point on a potentially moving common curve (s^{-1})
∇	gradient operator (m^{-1})
∇'	microscale surficial del operator on an interface (m^{-1})
∇''	microscale curvilinear del operator on a common curve (m^{-1})
∇^{c}	curvilineal operator (m^{-1})
∇^{S}	surficial operator (m^{-1})
Δ	Laplacian (m^{-2})

SUBSCRIPTS AND SUPERSCRIPTS

-	mass average qualifier (–)
=	specifically defined average (–)
c	curve (–)
e	external boundary qualifier (–)

I	interface qualifier (–)
i	general index qualifier (–)
i	internal qualifier (–)
j	general index qualifier (–)
k	general index qualifier (–)
ma	macroscale qualifier (–)
me	megascale qualifier (–)
mi	microscale qualifier (–)
mo	molecular scale qualifier (–)
P	phase qualifier (–)
Pt	point qualifier (–)
s	surface (–)
T	transpose operator (–)
α	α-phase (–)
$\alpha\beta$	(or other pair of Greek letters) entity qualifier for an interface between the α- and β-phases (or other pair of phases) (–)
$\alpha\beta\varepsilon$	entity qualifier for a common curve that is on the boundary of the α-, β- and ε-phases (–)
$\alpha\beta\varepsilon\gamma$	entity qualifier for a common point where the boundaries of the α-, β-, ε- and γ-phases are in contact (–)
β	β-phase (–)
γ	γ-phase (–)
σ	σ-phase (–)
ε	ε-phase (–)
ι	general entity qualifier which could indicate a phase, interface, or common curve (–)

ABBREVIATIONS

LES	large eddy simulations (–)
n	non-wetting phase (–)
ns	non-wetting-solid phase interface (–)
RAS	Reynolds average simulations (1)
REV	representative elementary volume (–)
RHS	right-hand side (–)
s	solid phase (–)
SGS	subgrid scale (–)
TCAT	thermodynamically constrained averaging theory (–)
w	wetting phase (–)
wn	wetting–non-wetting phase interface (–)
ws	wetting-solid phase interface (–)
wns	wetting–non-wetting-solid phase common curve (–)
1D	one-dimensional (–)
2D	two-dimensional (–)
3D	three-dimensional (–)

ACKNOWLEDGEMENTS

We are indebted to G. B. Marin and two anonymous referees for their critical review and constructive comments/suggestions on the original manuscript. We owe much to our many colleagues in the fields of multiscale phenomena whose insights fill many of the pages of this work. We benefited immensely from the stimulating work by J. L. Auriault, V. Balakotaiah, G. S. Beavers, A. Bejan, F. B. Busse, G. Blöschl, R. M. Bowen, J. H. Cushman, D. A. Drew, B. Engquist, J. Glimm, W. G. Gray, S. M. Hassanizadeh, T. Y. Hou, T. J. R. Hughes, D. D. Joseph, M. Lesieur, S. Levin, J. H. Li, J. L. Lumley, S. B. Pope, M. Quintard, K. R. Rajagopal, L. E. Scriven, M. Sivapalan, J. C. Slattery, D. J. Steigmann, D. Y. Tzou, P. Vadasz, D. G. Vlachos, E. Weinan, and S. Whitaker. We are very grateful to K. Senechal who has provided us with valuable comments and suggestions for improving our manuscript, and to Y. X. Zhang who made the technical drawings of Figures 1, 2, and 5. The support of our research program by the Research Grants Council of Hong Kong SAR (GRF717508) and the CRCG of the University of Hong Kong is also greatly appreciated.

REFERENCES

Aanonsen, S. I., and Eydinov, D. *Comput. Geosci.* **10**, 97–117 (2006).
Aarnes, J. E., Kippe, V., and Lie, K. A. *Adv. Water Resour.* **28**, 257–271 (2005).
Aarnes, J. E., Krogstad, S., and Lie, K. A. *Multiscale Model. Simul.* **5**, 337–363 (2006).
Abdallah, N. B., Arnold, A., Degond, P., Gamba, I. M., Glassey, R. T., Levermore, C. D., and Ringhofer, C., "Dispersive Transport Equations and Multiscale Models". Springer, New York (2004).
Abdullaev, F. K., and Kraenkel, R. A. *Phy. Rev. A* **62**, (2000).
Abgrall, R., and Perrier, V. *Multiscale Model. Simul.* **5**, 84–115 (2006).
Abildtrup, J., Audsley, E., Fekete-Farkas, M., Giupponi, C., Gylling, M., Rosato, P., and Rounsevell, M. *Environ. Sci. Policy* **9**, 101–115 (2006).
Achanta, S., and Cushman, J. H. *J. Colloid Interface Sci.* **168**, 266–268 (1994).
Achanta, S., Cushman, J. H., and Okos, M. R. *Int. J. Eng. Sci.* **32**, 1717–1738 (1994).
Acharya, A. *Comput. Meth. Appl. Mech. Eng.* **194**, 3067–3089 (2005).
Adams, K. L., King, J. R., and Tew, R. H. *J. Eng. Math.* **45**, 197–226 (2003).
Ahn, K. H., Lookman, T., and Bishop, A. R. *Nature* **428**, 401–404 (2004).
Alarcon, T., Byrne, H. M., and Maini, P. K. *Prog. Biophys. Mol. Biol.* **85**, 451–472 (2004).
Alber, M., Chen, N., Glimm, T., and Lushnikov, P. M. *Phy. Rev. E* **73**, (2006).
Aldama, A. A., "Filtering Techniques for Turbulent Flow Simulation". Springer, Berlin (1990).
Aleman-Flores, M., and Alvarez-Leon, L. *Lect. Notes Comput. Sci.* **2695**, 479–493 (2003).
Al-Ghoul, M., Boon, J. P., and Coveney, P. V. *Philos. Trans. R. Soc. Lond. A: Math. Phys. Eng. Sci.* **362**, 1551–1552 (2004).
Alkire, R. *J. Electroanal. Chem.* **559**, 3–12 (2003).
Alkire, R. C., and Braatz, R. D. *Aiche J.* **50**, 2000–2007 (2004).
Allen, M. P. *Comput. Phys. Commun.* **169**, 433–437 (2005).
Anderson, T. B., and Jackson, R. *Ind. Eng. Chem. Fundam.* **6**, 527–539 (1967).
Anderson, M. C., Kustas, W. P., and Norman, J. M. *Agron. J.* **95**, 1408–1423 (2003).
Anderson, M. C., Norman, J. M., Mecikalski, J. R., Torn, R. D., Kustas, W. P., and Basara, J. B. *J. Hydrometeor.* **5**, 343–363 (2004).
Anderson, M. C., Neale, C. M. U., Li, F., Norman, J. M., Kustas, W. P., Jayanthi, H., and Chavez, J. *Remote Sens. Environ.* **92**, 447–464 (2004a).
Anderson, M. C., Norman, J. M., Mecikalski, J. R., Torn, R. D., Kustas, W. P., and Basara, J. B. *J. Hydrometeor.* **5**, 343–363 (2004b).
Anisimòv, S. I., Kapeliovich, B. L., and Perelman, T. L. *Sov. Phys. JETP* **39**, 375–377 (1974).
Antic, S., Laprise, R., Denis, B., and de Elia, R. *Clim. Dyn.* **23**, 473–493 (2004).
Antic, S., Laprise, R., Denis, B., and de Elia, R. *Clim. Dyn.* **26**, 305–325 (2006).
Antonic, N., Van Duijn, C. J., Jger, W., and Mikelic, A. (Eds.), "Multiscale Problems in Science and Technology: Challenges to Mathematical Analysis and Perspectives". Springer, Berlin (2002).

Antoniou, I., and Lumer, G., "Generalized Functions, Operator Theory, and Dynamical Systems". CRC Press, Boca Raton (1999).
Antropov, V. P., and Belashchenko, K. D. *J. Appl. Phys.* **93**, 6438–6443 (2003).
Araujo, M. B., Thuiller, W., Williams, P. H., and Reginster, I. *Global Ecol. Biogeogr.* **14**, 17–30 (2005).
Arbogast, T. *Siam J. Numer. Anal.* **42**, 576–598 (2004).
Arbogast, T., and Boyd, K. J. *Siam J. Numer. Anal.* **44**, 1150–1171 (2006).
Aris, R., "Vectors, Tensors, and the Basic Equations of Fluid Mechanics". Prentice-Hall, Englewood Cliffs (1962).
Artus, V., and Noetinger, B. *Oil Gas Sci. Technol. Rev. Inst. Francais Petrole* **59**, 185–195 (2004).
Artus, V., Noetinger, B., and Ricard, L. *Transp. Porous Media* **56**, 283–303 (2004).
Ates, H., Bahar, A., El-Abd, S., Charfeddine, M., Kelkar, M., and Datta-Gulpta, A. *Spe Reserv. Eval. Eng.* **8**, 22–32 (2005).
Attinger, S., and Koumoutsakos, P. Eds., "Multiscale Modelling and Simulation". Springer, Berlin (2004).
Auclair, F., Estournel, C., Marsaleix, P., and Pairaud, I. *Geophys. Res. Lett.* **33** (2006).
Auriault, J. L. *Int. J. Eng. Sci.* **18**, 775–785 (1980).
Auriault, J. L. *Int. J. Eng. Sci.* **29**, 785–795 (1991).
Auriault, J. L., Borne, L., and Chambon, R. *J. Acoust. Soc. Am.* **77**, 1641–1650 (1985).
Babadagli, T. *Math. Geol.* **38**, 33–50 (2006).
Bacchi, B., and Ranzi, R. *Hydrol. Earth Syst. Sci.* **7**, 784–798 (2003).
Bachmat, Y. *Isr. J. Technol.* **10**, 391–403 (1972).
Bacon, D. J., and Osetsky, Y. N. *Mater. Sci. Eng. A.Struct. Mater.* **365**, 46–56 (2004).
Badaroglu, M., Wambacq, P., Van der Plas, G., Donnay, S., Gielen, G. G. E., and De Man, H. J. *IEEE Trans. Circuits Syst. I Regular Papers* **53**, 296–305 (2006).
Badas, M. G., Deidda, R., and Piga, E. *Nat. Hazards Earth Syst. Sci.* **6**, 427–437 (2006).
Banks, H. T., and Pinter, G. A. *Multiscale Model. Simul.* **3**, 395–412 (2005).
Baraka-Lokmane, S., and Liedl, R. *Int. J. Geophy.* **166**, 1440–1453 (2006).
Bar-Joseph, Z., El-Yaniv, R., Lischinski, D., and Werman, M. *IEEE Trans. Vis. Comput. Graph.* **7**, 120–135 (2001).
Bardossy, A., Bogardi, I., and Matyasovszky, I. *Theor. Appl. Climatol.* **82**, 119–129 (2005).
Barth, T. J., Chan, T., and Haimes, R. (Eds.), "Multiscale and Multiresolution Methods: Theory and Applications". Springer, Berlin (2002).
Barthel, R., Rojanschi, V., Wolf, J., and Braun, J. *Phys. Chem. Earth* **30**, 372–382 (2005).
Basquet, R., Jeannin, L., Lange, A., and Braun, J. *Spe Reserv. Eval. Eng.* **7**, 378–384 (2004).
Bassi, A. L., Bottani, C. E., Casari, C., and Beghi, M. *Appl. Sur. Sci.* **226**, 271–281 (2004).
Batchelor, G. K., "An Introduction to Fluid Mechanics". Cambridge University Press, Cambridge (1967).
Becker, A., Blöschl, G., and Hall, A. *J. Hydrol.* **217**, 169–335 (1999).
Bedford, A., and Drumheller, D. S. *Int. J. Eng. Sci.* **21**, 863–960 (1983).
Bejan, A., "Convection Heat Transfer". 3rd Ed. Wiley, New York (2004).
Bejan, A., Dincer, I., Lorente, A., Miguel, A. F., and Reis, A. H., "Porous and Complex Flow Structures in Modern Technologies". Springer, New York (2004).
Belashchenko, K. D., and Antropov, V. P. *J. Magn. Magn. Mater.* **253**, L87–L95 (2002a).
Belashchenko, K. D., and Antropov, V. P. *Phys. Rev. B* **66**, (2002b).
Benassi, A., Cohen, S., and Istas, J. *Bernoulli* **8**, 97–115 (2002).
Bennethum, L. S., and Cushman, J. H. *Int. J. Eng. Sci.* **34**, 125–145 (1996a).
Bennethum, L. S., and Cushman, J. H. *Int. J. Eng. Sci.* **34**, 147–169 (1996b).
Bent, J., Hutchings, L. R., Richards, R. W., Gough, T., Spares, R., Coates, P. D., Grillo, I., Harlen, O. G., Read, D. J., Graham, R. S., Likhtman, A. E., Groves, D. J., Nicholson, T. M., and McLeish, T. C. B. *Science* **301**, 1691–1695 (2003).
Berentsen, C. W. J., Verlaan, M. L., and van Kruijsdijk, C. P. J. W. *Phys. Rev. E* **71**, (2005).
Bergant, K., and Kajfez-Bogataj, L. *Theor. Appl. Climatol.* **81**, 11–23 (2005).
Berkowitz, B., Cortis, A., Dentz, M., and Scher, H. *Rev. Geophy.* **44**, (2006).
Berryman, J. G. *J. Eng. Mech.-asce* **131**, 928–936 (2005).

Bezzo, F., Macchietto, S., and Pantelides, C. C. *Comput. Chem. Eng.* **28**, 501–511 (2004).
Bierkens, M. F. P., Finke, P. A., and Willigen, P. (Eds.), "Upscaling and Down-scaling Methods for Environmental Research". Kluwer Academic Publishers, Dordrecht (2000).
Bikerman, J. J. *Centaurus* **19**, 182–206 (1975).
Billock, V. A., de Guzman, G. C., and Kelso, J. A. S. *Physica D* **148**, 136–146 (2001).
Bird, R. B. *AICHE J.* **50**, 273–287 (2004).
Blöschl, G. *Hydrol. Processes* **13**, 2149–2175 (1999).
Blöschl, G., Statistical upscaling and downscaling in hydrology, *in* "Encyclopedia of Hydrological Sciences" (M. G. Anderson Ed.), pp. 135–154. Wiley, Chichester (2005).
Blöschl, G., and Sivapalan, M. *Hydrol. Processes* **9**, 251–290 (1995).
Blöschl, G. *Hydrol. Processes* **15**, 709–711 (2001).
Blöschl, G., "Scale and Scaling in Hydrology: A Framework for Thinking and Analysis". Wiley, Chichester (2004).
Blöschl, G., Grayson, R. B., and Sivapalan, M. *Hydrol. Processes* **9**, 313–330 (1995).
Blöschl, G., Sivapalan, M., Gupta, V. K., and Beven, K. *Water Resour. Res.* **33**, 2881–2999 (1997).
Blum, P., Mackay, R., Riley, M. S., and Knight, J. L. *Int. J. Rock Mech. Min. Sci.* **42**, 781–792 (2005).
Boghosian, B. M., Yepez, J., Coveney, P. V., and Wager, A. *Proc. R. Soc. Lond. A: Math. Phys. Eng. Sci.* **457**, 717–766 (2001).
Boghosian, B. M., Love, P. J., Coveney, P. V., Karlin, I., Succi, S., and Yepez, J. *Phys. Rev. E* **68**, 025103 (R) (2003).
Bonilla, L. L., Sanchez, A. L., and Carretero, M. *Siam J. Appl. Math.* **61**, 528–550 (2000).
Bonn, D., Couder, Y., Van Dam, P. H. J., and Douady, S. *Phys. Rev. E* **47**, R28–R31 (1993).
Borges, A. V. *Estuaries* **28**, 3–27 (2005).
Borri-Brunetto, M., Carpinteri, A., and Chiaia, B. *Rock Mech. Rock Eng.* **37**, 117–126 (2004).
Borucki, L. Taking on the multiscale challenge, in "Dispersive Transport Equations and Multiscale Models" (N. B. Abdallah, A. Arnold, P. Degond, I. M. Gamba, R. T. Glassey, C. D. Levermore and C. Ringhofer Eds.), pp. 25–35. Springer, Berlin (2003).
Bourgeat, A., Gipouloux, O., and Marusic-Paloka, E. *Math. Meth. Appl. Sci.* **27**, 381–403 (2004).
Bowen, R. M. *Int. J. Eng. Sci.* **20**, 697–735 (1982).
Bowen, R. M., "Introduction to Continuum Mechanics for Engineers". Plenum Press, New York (1989).
Brachet, M. E. *C. R. Acade. Sci. Paris, Série II* **311**, 775 (1990).
Brachet, M. E. *Fluid Dyn. Res.* **8**, 1–8 (1991).
Bramble, J. H., Cohen, A., and Dahmen, W. (Eds.), "Multiscale Problems and Methods in Numerical Simulations". Springer, Berlin (2003).
Brandt, A., Multiscale scientific computation: Review (2001), *in* "Multiscale and Multiresolution Methods: Theory and Applications" (T. J. Barth, T. Chan, and R. Haimes Eds.), pp. 3–95. Springer, Berlin (2002).
Brandt, A. *Comput. Phys. Commun.* **169**, 438–441 (2005).
Braun, C., Helmig, R., and Manthey, S. *J. Contam. Hydrol.* **76**, 47–85 (2005).
Breakspear, M., and Stam, C. J. *Philos. Trans. R. Soc. B Biol. Sci.* **360**, 1051–1074 (2005).
Briere, J. *J. Nerv. Ment. Dis.* **194**, 78–82 (2006).
Brueckner, S. A., Serugendo, G. D. M., Karageorgos, A., and Nagpal, R. (Eds.), "Engineering Self-Organising Systems: Methodologies and Applications". Springer, New York (2005).
Bruinink, C. M., Peter, M., Maury, P. A., De Boer, M., Kuipers, L., Huskens, J., and Reinhoudt, D. N. *Adv. Funct. Mater.* **16**, 1555–1565 (2006).
Brun, M., Demadrille, R., Rannou, P., Pron, A., Travers, J. P., and Grevin, B. *Adv. Mater.* **16**, 2087 (2004).
Buff, F. P. *J. Chem. Phys.* **25**, 146–153 (1956).
Buff. F. P. Encyclopedia of physics, *in* "Structure of liquids" (by S. Flügge, Ed.), vol. 10, p. 281. Springer, Berlin (1960).
Buff, F. P., and Saltsburg, H. *J. Chem. Phys.* **26**, 1526–1533 (1957a).
Buff, F. P., and Saltsburg, H. *J. Chem. Phys.* **26**, 23–31 (1957b).
Buiron, N., Hirsinger, L., and Billardon, R. *J. Phys. IV* **9**, 187–196 (1999).

Buldum, A., Busuladzic, I., Clemons, C. B., Dill, L. H., Kreider, K. L., Young, G. W., Evans, E. A., Zhang, G., Hariharan, S. I., and Kiefer, W. *J. Appl. Phys.* **98** (2005a).
Buldum, A., Clemons, C. B., Dill, L. H., Kreider, K. L., Young, G. W., Zheng, X., Evans, E. A., Zhang, G., and Hariharan, S. I. *J. Appl. Phys.* **98**, (2005b).
Burger, G., and Chen, Y. *J. Hydrol.* **311**, 299–317 (2005).
Busch, A., Gensterblum, Y., Krooss, B. M., and Littke, R. *Int. J. Coal Geol.* **60**, 151–168 (2004).
Busuioc, A., Giorgi, F., Bi, X., and Ionita, M. *Theor. Appl. Climatol.* **86**, 101–123 (2006).
Byrne, H. M., Owen, M. R., Alarcon, T., Murphy, J., and Maini, P. K. *Math. Mod. Meth. Appl. Sci.* **16**, 1219–1241 (2006).
Byun, K. T., Kim, K. Y., and Kwak, H. Y. *J. Korean Phys. Soc.* **47**, 1010–1022 (2005).
Cardells-Tormo, F., and Arnabat-Benedicto, J. *IEEE Trans. Circuits Syst. II-Express Briefs* **53**, 522–526 (2006).
Cassiraga, E. F., Fernandez-Garcia, D., and Gomez-Hernandez, J. J. *Int. J. Rock Mech. Min. Sci.* **42**, 756–764 (2005).
Cavallotti, C., Di Stanislao, M., Moscatelli, D., and Veneroni, A. *Electrochim. Acta* **50**, 4566–4575 (2005).
Chai, L. H. *Prog.Chem.* **17**, 186–191 (2005).
Chalon, F., Mainguy, M., Longuemare, P., and Lemonnier, P. *Int. J. Numer. Anal. Meth. Geomech.* **28**, 1105–1119 (2004).
Charlton, R., Fealy, R., Moore, S., Sweeney, J., and Murphy, C. *Clim. Change* **74**, 475–491 (2006).
Charpentier, J. C., and McKenna, T. F. *Chem. Eng. Sci.* **59**, 1617–1640 (2004).
Chastanet, J., Royer, P., and Auriault, J. L. *Transp. Porous Media* **56**, 171–198 (2004).
Chatterjee, A., Snyder, M. A., and Vlachos, D. G. *Chem. Eng. Sci.* **59**(22-23), 5559–5567 (2004).
Chaturvedi, R., Huang, C., Kazmierczak, B., Schneider, T., Izaguirre, J. A., Glimm, T., Hentschel, H. G. E., Glazier, J. A., Newman, S. A., and Alber, M. S. *J. R. Soc. Interface* **2**, 237–253 (2005).
Chave, J., and Levin, S. *Environ. Resour. Econ.* **26**, 527–557 (2003).
Chawathe, A., and Taggart, I. *Spe Reserv. Eval. Eng.* **7**, 285–296 (2004).
Chen, Z. M. *J. Comput. Math.* **24**, 393–400 (2006).
Chen, Y. G., and Durlofsky, L. J. *Multiscale Model.Simul.* **5**, 445–475 (2006a).
Chen, Y. G., and Durlofsky, L. J. *Trans. Porous Media* **62**, 157–185 (2006b).
Chen, Z. M., and Yue, X. Y. *Multiscale Model. Simul.* **1**, 260–303 (2003).
Chen, L. Y., Goldenfeld, N., and Oono, Y. *Phys. Rev. E* **54**, 376–394 (1996).
Chen, Z. M., Deng, W. B., and Ye, H. A. *Commun. Math. Sci.* **3**, 493–515 (2005a).
Chen, Z. M., Deng, W. B., and Ye, Y. H. *Discrete Cont. Dyn. Syst.* **13**, 941–960 (2005b).
Chen, D. L., Achberger, C., Rälsänen, J., and Hellström, C. *Adv. Atmos. Sci.* **23**, 54–60 (2006).
Cheng, K. C., and Wang, L. Q. *J. Flow Visu. Image Process.* **3**, 237–246 (1996).
Chiaravalloti, F., Milovanov, A. V., and Zimbardo, G. *Phys. Scr.* **122**, 79–88 (2006).
Churchill, S. W. *Adv. Heat Transf.* **34**, 255–361 (2001).
Ciofalo, M. *Adv. Heat Transf.* **25**, 321–419 (1994).
Coleman, B. D., and Noll, W. *Arch. Rat. Mech. Anal.* **13**, 168–178 (1963).
Collis, S. S., Joslin, R. D., Seifert, A., and Theofilis, V. *Prog. Aerospace Sci.* **40**, 237–289 (2004).
Corwin, D. L., Hopmans, J., and de Rooij, G. H. *Vadose Zone J.* **5**, 129–139 (2006).
Costanza, R. *Futures* **35**, 651–671 (2003).
Coulibaly, P. *Geophys. Res. Lett.* **31** (2004).
Coulibaly, P., Dibike, Y. B., and Anctil, F. *J. Hydrometeor.* **6**, 483–496 (2005).
Coveney, P. V. *Mesoscale Phenomena in Fluid Syst. ACS Symp.* **861**, 206–226 (2003a).
Coveney, P. V. *Philos. Trans. R. Soc. Lond. A: Math. Phys. Eng. Sci.* **361**, 1057–1079 (2003b).
Coveney, P. V., and Fowler, P. W. *J. R. Soc. Interface* **2**, 267–280 (2005).
Coveney, P. V., and Wattis, J. A. D. *Mol. Phys.* **104**, 177–185 (2006).
Crawley, M. J., and Harral, J. E. *Science* **291**, 864–868 (2001).
Crow, W. T., yu, D., and Famiglietti, J. S. *Adv. Water Resour.* **28**, 1–14 (2005).
Curtarolo, S., and Ceder, G. *Phys. Rev. Lett.* **88**, (2002).
Cushman, J. H. (Ed.), "Dynamics of Fluids in Hierarchical Porous Media". Academic Press, London (1990).

Cushman, J. H., "The Physics of Fluids in Hierarchical Porous Media: Angstroms to Miles". Kluwer Academic Publishers, Dordrecht (1997).
Cushman, J. H., Bennethum, L. S., and Hu, B. X. *Adv. Water Resour.* **25**, 1043–1067 (2002).
Dadvar, M., and Sahimi, M. *Chem. Eng. Sci.* **58**, 4935–4951 (2003).
Dahmen, W., Kurdila, A. J., and Oswald, P. (Eds.), "Multiscale Wavelet Methods for Partial Differential Equations". Academic Press, San Diego (1997).
Daly, E., Porporato, A., and Rodriguez-Iturbe, I. *J. Hydrometeor.* **5**, 546–558 (2004a).
Daly, E., Porporato, A., and Rodriguez-Iturbe, I. *J. Hydrometeor.* **5**, 559–566 (2004b).
Daniel, L., Hubert, O., and Billardon, R. *Int. J. Appl. Electromagn. Mech.* **19**, 293–297 (2004).
Darrah, P. R., Jones, D. L., Kirk, G. J. D., and Roose, T. *Eur. J. Soil Sci.* **57**, 13–25 (2006).
Das, D., and Hassanizadeh, S. *Transp. Porous Media* **58**, 1–3 (2005).
Das, D. B., Hassanizadeh, S. M., Rotter, B. E., and Ataie-Ashtiani, B. *Transp. Porous Media* **56**, 329–350 (2004).
DasGupta, D., Basu, S., and Chakraborty, S. *Phys. Lett. A* **348**, 386–396 (2006).
Davidson, P. A., "Turbulence: An Introduction for Scientists and Engineers". Oxford University Press, Oxford (2004).
Davis, J. A., Curtis, G. P., and Kent, D. B. *Abstr. Pap. Am. Chem. Soc.* **231**, (2006).
DeCoursey, D. G. Hydrological, climatological and ecological systems scaling: a review of selected literature and comments, USDA-ARS-NPA Internal Progress Report, Great Plains Systems Research Unit, Fort Collins, Co, USA (1996).
De Fabritiis, G. D., Coveney, P. V., and Flekkoy, E. G. *Philos. Trans. R. Soc. Lond. A: Math. Phys. Eng. Sci.* **360**, 317–331 (2002).
De la Rubia, T. D., Zbib, H. M., Khraishi, T. A., Wirth, B. D., Victoria, M., and Caturla, M. J. *Nature* **406**, 871–874 (2000).
De Rooy, W. C., and Kok, K. *Weather Forecast.* **19**, 485–495 (2004).
Dean, D. W., and Russell, T. F. *Adv. Water Resour.* **27**, 445–464 (2004).
Deemer, A. R., and Slattery, J. C. *Int. J. Multiphase Flow* **4**, 171–192 (1978).
Defranoux, N. A., Stokes, C. L., Young, D. L., and Kahn, A. J. *J. Bone Miner. Res.* **20**, 1079–1084 (2005).
Deidda, R., Badas, M. G., and Piga, E. *Water Resour. Res.* **40** (2004).
Deidda, R., Badas, M. G., and Piga, E. *J. Hydrol.* **322**, 2–13 (2006).
Demidov, A. S., "Generalized Functions in Mathematical Physics: Main Ideas and Concepts". Nova Science, Huntington (2001).
Diaz-Nieto, J., and Wilby, R. L. *Clim. Change* **69**, 245–268 (2005).
Dibike, Y. B., and Coulibaly, P. *J. Hydrol.* **307**, 145–163 (2005).
Dibike, Y. B., and Coulibaly, P. *Neural Netw.* **19**, 135–144 (2006).
Diego, B., David, L., Girard-Reydet, E., Lucas, J. M., and Denizart, O. *Poly. Int.* **53**, 515–522 (2004).
Diez, E., Primo, C., Garcia-Moya, J. A., Gutierrez, J. M., and Orfila, B. *Tellus A Dyn. Meteorol. Oceanogr.* **57**, 409–423 (2005).
Ding, Y. *J. Petrol. Sci. Eng.* **43**, 87–97 (2004).
Dittmann, R., Richter, J., Vital, A., Piazza, D., Aneziris, C., and Graule, T. *Adv. Eng. Mater.* **7**, 354–360 (2005).
Dobran, F. *Int. J. Multiphase Flow* **10**, 273–305 (1984).
Dobrovitski, V. V., Katsnelson, M. I., and Harmon, B. N. *J. Magn. Magn. Mater.* **221**, L235–L242 (2000).
Dobrovitski, V. V., Katsnelson, M. I., and Harmon, B. N. *J. Appl. Phys.* **93**, 6432–6437 (2003).
Dokholyan, N. V. *Curr. Opin. Struct. Biol.* **16**, 79–85 (2006).
Dollet, A., de Persis, S., Pons, M., and Matecki, M. *Surf. Coat. Tech.* **177**, 382–388 (2004).
Donoho, D. L., and Huo, X. M. *IEEE Trans. Inform. Theor.* **47**, 2845–2862 (2001).
Dormieux, L., and Ulm, F. J. Eds., "Applied Micromechanics of Porous Materials". Springer, New York (2005).
Douady, S., and Couder, Y., On the dynamical structures observed in 3D turbulence, *in* "Turbulence in Spatially Extended Systems" (R. Benzi, C. Basdevant, and S. Cilberto Eds.), pp. 3–17. Nova Science, New York (1993).
Douady, S., Couder, Y., and Brachet, M. E. *Phys. Rev. lett.* **67**, 983–986 (1991).
Dowell, E. H., and Tang, D. *J. Appl. Mech. Trans. Aame* **70**, 328–338 (2003).

Drew, D. A. *Stud. Appl. Math.* **50**, 133 (1971).
Drew, D. A., and Passman, S. L., "Theory of Multicomponent Fluids". Springer, New York (1999).
Drews, T. O., Webb, E. G., Ma, D. L., Alameda, J., Braatz, R. D., and Alkire, R. C. *Aiche J.* **50**, 226–240 (2004).
Drolon, H., Druaux, F., and Faure, A. *Pattern Recognit. Lett.* **21**, 473–482 (2000).
Drolon, H., Hoyez, B., Druaux, F., and Faure, A. *Math. Geol.* **35**, 805–817 (2003).
Droujinine, A. *J. Geophys. Eng.* **3**, 59–81 (2006).
Dumoulin, S., Busso, E. P., O'Dowd, N. P., and Allen, D. *Philos. Mag.* **83**, 3895–3916 (2003).
Dunbabin, V. M., McDermott, S., and Bengough, A. G. *Plant Soil* **283**, 57–72 (2006).
Eberhard, J. *J. Phys. A Math. Gen.* **37**, 9587–9602 (2004).
Eberhard, J. *Multiscale Model.Simul.* **3**, 957–976 (2005a).
Eberhard, J. P. *Phys. Rev. E* **72** (2005b).
Eberhard, J., Attinger, S., and Wittum, G. *Multiscale Model.Simul.* **2**, 269–301 (2004).
Eberhard, J. P., Efendiev, Y., Ewing, R., and Cunningham, A. *Int. J. Multiscale Comput. Eng.* **3**, 499–516 (2005).
Efendiev, Y., and Pankov, A. *Siam J. Appl. Math.* **65**, 43–68 (2004a).
Efendiev, Y., and Pankov, A. *Multiscale Model. Simul.* **2**, 237–268 (2004b).
Efendiev, Y., Datta-Gupta, A., Osaka, I., and Mallick, B. *Adv. Water Resour.* **28**, 303–314 (2005).
Egermann, P., and Lenormand, R. *Petrophysics* **46**, 335–345 (2005).
Ehleringer, J. R., and Field, C. B., "Scaling Physiological Processes: Leaf to Globe". Academic Press, San Diego (1992).
Eidsvik, K. J. *Wind Energy* **8**, 237–249 (2005).
Engquist, B., and Runborg, O. Eds., "Multiscale Methods in Science and Engineering". Springer, Berlin (2005).
Enke, W., Deutschlander, T., Schneider, F., and Kuchler, W. *Meteorol. Z.* **14**, 247–257 (2005a).
Enke, W., Schneider, F., and Deutschlander, T. *Theor. Appl. Climatol.* **82**, 51–63 (2005b).
Eringen, A. C., "Mechanics of Continua". Krieger, New York (1980).
Eringen, A. C., and Ingram, J. D. *Int. J. Eng. Sci.* **3**, 197–212 (1965).
Fabry, F., Flamant, G., and Fulcheri, L. *Chem. Eng. Sci.* **56**, 2123–2132 (2001).
Farassat, F. *J. Sound Vib.* **55**, 165–193 (1977).
Farina, A., "Principles and Methods in Landscape Ecology". Chapman & Hall, London (1998).
Fauchais, P., and Vardelle, A. *Int. J. Thermal Sci.* **39**, 852–870 (2000).
Feddersen, H., and Andersen, U. *Tellus A Dyn. Meteorol. Oceanogr.* **57**, 398–408 (2005).
Federov, A. G. *Int. J. Multiscale Comput. Eng.* **3**, 1–3 (2005).
Feng, X., Fryxell, G. E., Wang, L. Q., Kim, A. Y., Liu, J., and Kemner, K. M. *Science* **276**, 923–926 (1997).
Ferreira, G. C., "Introduction to the Theory of Distributions". Addison Wesley, London (1997).
Field, C. B., and Ehleringer, J. R., Introduction: Questions of scale, *in* "Scaling Physiological Processes: Leaf to Globe" (J. R. Ehleringer, and C. B. Field Eds.), pp. 1–4. Academic Press, San Diego (1993).
Filippova, O., Succi, S., Mazzocco, F., Arrighetti, C., Bella, G., and Hanel, D. *J. Comput. Phys.* **170**, 812–829 (2001).
Fischer, M., Dewitte, B., and Maitrepierre, L. *Geophys. Res. Lett.* **31** (2004).
Fish, J., and Yu, Q. *Int. J. Numer. Meth. Eng.* **52**, 159–191 (2001).
Flad, H. J., Hackbusch, W., Luo, H. J., and Kolb, D. *Phy. Rev. B* **71** (2005).
Flekkoy, E. G., Coveney, P. V., and Fabritiis, G. D. *Phys. Rev. E* **62**, 2140–2157 (2000).
Flodin, E. A., Durlofsky, L. J., and Aydin, A. *Petrol. Geosci.* **10**, 173–181 (2004).
Fouque, J. P., Papanicolaou, G., Sircar, R., and Solna, K. *Multiscale Model. Simul.* **2**, 22–42 (2003).
Fox-Rabinovitz, M., Cote, J., Dugas, B., Deque, M., and McGregor, J. L. *J. Geophys. Res. Atmos.* **111**, (2006).
Frantziskonis, G. *Probabilistic Eng. Mech.* **17**, 359–367 (2002a).
Frantziskonis, G. *Probabilist. Eng. Mech.* **17**, 349–357 (2002b).
Fredberg, J. J., and Kamm, R. D. *Annu. Rev. Phys.* **68**, 507–541 (2006).
Freeden, W., and Michel, V., "Multiscale Potential Theory: With Applications to Geoscience". Birkhuser, Boston (2004).

Freeden, W., and Michel, V. *Int. J. Wavelets Multires. Inform. Process.* **3**, 523–558 (2005).
Frisch, U., "Turbulence: The Legacy of A. N. Kolmogorov". Cambridge University Press, Cambridge (1995).
Fung, K. T., and Siu, W. C. *IEEE Trans. Image Process.* **15**, 394–403 (2006).
Gaffin, S. R., Rosenzweig, C., Xing, X. S., and Yetman, G. *Global Environ. Change Hum. Policy Dimens.* **14**, 105–123 (2004).
Gallegher, R., and Appenzeller, T., (Eds.). Special issue on complex systems, *Science* **284**, 79–109 (1999).
Galves, A., and Jonalasinio, G. *Ann. Inst. Henri Poincare-Phys. Theor.* **55**, 590 (1991).
Gangopadhyay, S., Clark, M., and Rajagopalan, B. *Water Resour. Res.* **41**, (2005).
Gasda, S. E., and Celia, M. A. *Adv. Water Resour.* **28**, 493–506 (2005).
Gaslikova, L., and Weisse, R. *Ocean Dyn.* **56**, 26–35 (2006).
Gatski, T. B., Hussaini, M. Y., and Lumley, J. L., "Simulation and Modeling of Turbulent Flows". Oxford University Press, Oxford (1996).
Gatto, A., Feigl, T., Kaiser, N., Garzella, D., De Ninno, G., Couprie, M. E., Marsi, M., Trovo, M., Walker, R., Grewe, M., Wille, K., Paoloni, S., Reita, V., Roger, J. P., Boccara, C., Torchio, P., Albrand, G., and Amra, C. *Nucl. Instrum. Methods Phys. Res A* **483**, 172–176 (2002).
Geers, M. G. D., Kouznetsova, V., and Brekelmans, W. A. M. *J. Electr. Packaging* **127**, 255–261 (2005).
Geindreau, C., Sawicki, E., Auriault, J. L., and Royer, P. *Int. J. Numer. Anal. Meth. Geomech.* **28**, 229–249 (2004).
Gelhar, L. W. *Water Resour. Res.* **22**, 135S–145S (1986).
Geraerts, M., Michiels, M., Baekelandt, V., Debyser, Z., and Gijsbers, R. *J. Gene Med.* **7**, 1299–1310 (2005).
Gerde, E., and Marder, M. *Nature* **413**, 285–288 (2001).
Gerritsen, M. G., and Durlofsky, L. J. *Annu. Rev. Fluid Mech.* **37**, 211–238 (2005).
Ghan, S. J., and Shippert, T. *J. Clim.* **19**, 1589–1604 (2006).
Ghan, S. J., Shippert, T., and Fox, J. *J. Clim.* **19**, 429–445 (2006).
Ghoniem, N. M., Busso, E. P., Kioussis, N., and Huang, H. C. *Philos. Mag.* **83**, 3475–3528 (2003).
Ghosh, S., and Mujumdar, P. P. *Curr. Sci.* **90**, 396–404 (2006).
Giardino, M., Giordan, D., and Ambrogio, S. *Nat. Haz. Earth Syst. Sci.* **4**, 197–211 (2004).
Gibson, C. C., Ostrom, E., and Ahn, T. K. *Ecol. Econ.* **32**, 217–239 (2000).
Gibson, C. A., Meyer, J. L., Poff, N. L., Hay, L. E., and Georgakakos, A. *River Res. Appl.* **21**, 849–864 (2005).
Gill, S. P. A., Jia, Z., Leimkuhler, B., and Cocks, A. C. F. *Phys. Rev. B* **73**, (2006).
Givon, D., Kupferman, R., and Stuart, A. *Nonlinearity* **17**, R55–R127 (2004).
Glatzmaier, G. C., and Ramirez, W. F. *Chem. Eng. Sci.* **43**, 3157–3169 (1988).
Glimm, J. and Sharp, D. H. Multiscale science, SIAM News 30, October (1997).
Golledge, R. G., and Stimson, R. J., "Spatial Behavior: A Geographic Perspective". Guilford Press, New York (1997).
Goyeau, B., Benihaddadene, T., Gobin, D., and Quintard, M. *Transp. Porous Media* **28**, 19–50 (1997).
Granbakken, D., Haarberg, T., Rollheim, M., Ostvold, T., Read, P., and Schmidt, T. *Acta Chem. Scand.* **45**, 892–901 (1991).
Gray, W. G. *Adv. Water Resour.* **22**, 521–547 (1999).
Gray, W. G., and Hassanizadeh, S. M. *Int. J. Multiphase Flow* **15**, 81–95 (1989).
Gray, W. G., and Hassanizadeh, S. M. *Water Resour. Res.* **27**, 1847–1854 (1991a).
Gray, W. G., and Hassanizadeh, S. M. *Water Resour. Res.* **27**, 1855–1863 (1991b).
Gray, W. G., and Hassanizadeh, M. *Adv. Water Resour.* **21**, 261–281 (1998).
Gray, W. G., and Lee, P. C. Y. *Int. J. Multiphase Flow* **3**, 333–340 (1977).
Gray, W. G., and Miller, C. T. *Adv. Water Resour.* **28**, 161–180 (2005).
Gray, W. G., and Miller, C. T. *Adv. Water Resour.* **29**, 1745–1765 (2006).
Gray, W. G., Leijnse, A., Kolar, R. L., and Blain, C. A., "Mathematical Tools for Changing Spatial Scales in the Analysis of Physical Systems". CRC Press, Boca Raton (1993).
Grayson, R. B., and Blöschl, G. Eds., "Spatial Patterns in Catchment Hydrology: Observations and Modelling". Cambridge University Press, Cambridge (2000).

Grayson, R. B., Moore, I. D., and McMahon, T. A. *Water Resour. Res.* **26**, 2659–2666 (1992).
Grayson, R. B., Western, A. W., Chiew, F. H. S., and Blöschl, G. *Water Resour. Res.* **33**, 2897–2908 (1997).
Grigoriev, M. M., and Dargush, G. F. *Comput. Meth. Appl. Mech. Eng.* **192**, 4281–4298 (2003).
Groenenberg, J. E., Bonten, L. T., Romkens, F. A. M., and Rietra, R. P. J. J. *Abstr. Pap. Am. Chem. Soc.* **227**, u1114–u1114 (2004).
Grohens, Y., Castelein, G., Carriere, P., Spevacek, J., and Schultz, J. *Langmuir* **17**, 86–94 (2001).
Gueguen, Y., Le Ravalec, M., and Ricard, L. *Pure Appl. Geophys.* **163**, 1175–1192 (2006).
Guo, W. L., and Tang, C. *Int. J. Multiscale Comput. Eng.* **4**, 115–125 (2006).
Gupta, V. K., and Waymire, E. C. *J. Appl. Meteorol.* **32**, 251–267 (1993).
Gupta, V. K., Rodrguez-Iturbe, I., and Wood, E. F. (Eds.), "Scale Problems in Hydrology". D. Reidel, Dordrecht (1986).
Gusev, A. A. *Izv.Phys. Solid Earth* **41**, 798–812 (2005).
Gutierrez, J. M., Cano, R., Cofino, A. S., and Sordo, C. *Tellus A Dyn. Meteorol. Oceanogr.* **57**, 435–447 (2005).
Gutkowski, W., and Kowalewski, T. A., "Mechanics of the 21st Century". Springer, Dordrecht (2005).
Guyer, R. A., and Krumhansi, J. A. *Phys. Rev.* **148**, 766–778 (1966).
Haarberg, T., Jakobsen, J., and Ostvold, T. *Acta Chem. Scand.* **44**, 907–915 (1990).
Hadsell, F. A., "Tensors of Geophysics: Vol. 2, Generalized Functions and Curvilinear Coordinates".Society of Exploration Geophysicists, Tulsa (1999).
Hall, O., Hay, G. J., Bouchard, A., and Marceau, D. J. *Landscape Ecol.* **19**, 59–76 (2004).
Hanssen-Bauer, I., Achberger, C., Benestad, R. E., Chen, D., and Forland, E. J. *Clim. Res.* **29**, 255–268 (2005).
Harari, I. *Comput. Meth. Appl. Mech. Eng.* **195**, 1594–1607 (2006).
Haro, M. L., Rio, J. A., and Whitaker, S. *Transp. Porous Media* **25**, 167–192 (1996).
Harpham, C., and Wilby, R. L. *J. Hydrol.* **312**(1-4), 235–255 (2005).
Hassanizadeh, S. M., and Gray, W. G. *Adv. Water Resour.* **2**, 131–144 (1979a).
Hassanizadeh, M., and Gray, W. G. *Adv. Water Resour.* **2**, 191–208 (1979b).
Hassanizadeh, M., and Gray, W. G. *Adv. Water Resour.* **3**, 25–40 (1980).
Hassanizadeh, S. M., and Gray, W. G. *Adv. Water Resour.* **13**, 169–186 (1990).
Hauke, G. *Comput. Meth. Appl. Mech. Eng.* **191**, 2925–2947 (2002).
Hauke, G., and Doweidar, M. H. *Comput. Meth. Appl. Mech. Eng.* **194**, 691–725 (2005a).
Hauke, G., and Doweidar, M. H. *Comput. Meth. Appl. Mech. Eng.* **194**, 45–81 (2005b).
Hauke, G., and Doweidar, M. H. *Comput. Meth. Appl. Mech. Eng.* **195**, 6158–6176 (2006).
Hauke, G., and Garcia-Olivares, A. *Comput. Meth. Appl. Mech. Eng.* **190**, 6847–6865 (2001).
Hayes, R. L., Fago, M., Ortiz, M., and Carter, E. A. *Multiscale Model. Simul.* **4**, 359–389 (2005).
Haylock, M. R., Cawley, G. C., Harpham, C., Wilby, R. L., and Goodess, C. M. *Int. J. Climatol.* **26**, 1397–1415 (2006).
Hays-Stang, K. J., and Haji-Sheikh, A. *Int. J. Heat Mass Transf.* **42**, 455–465 (1999).
Held, R., Attinger, S., and Kinzelbach, W. *Water Resour. Res.* **41**, (2005).
Hermes, L., and Buhmann, J. M. *IEEE Trans. Image Process.* **12**, 1243–1258 (2003).
Hewitson, B. C., and Crane, R. G. *Int. J. Climatol.* **26**, 1315–1337 (2006).
Hidy, G. M. *Energy Fuels* **16**, 270–281 (2002).
Hilfer, R., and Helmig, R. *Adv. Water Resour.* **27**, 1033–1040 (2004).
Hoffman, M. B., and Coveney, P. V. *Mol. Simul.* **27**, 157–168 (2001).
Hong, F., and Root, D. D. *Drug Discov. Today* **11**, 640–645 (2006).
Hong, Y., Lin, Y. Y., Ling, Y., Lai, L. Z., Feng, J., Lai, Y. F., Qiao, B. W., Tang, T. G., Cai, B. C., and Chen, B. M. *Integr. Ferroelectr.* **78**, 153–163 (2006).
Hontans, T., and Terpolilli, P. *Comput. Geosci.* **9**, 219–245 (2005).
Hoofman, R. J. O. M., Verheijden, G. J. A. M., Michelon, J., Iacopi, F., Travaly, Y., Baklanov, M. R., Tokei, Z., and Beyer, G. P. *Microelectron. Eng.* **80**, 337–344 (2005).
Hosokawa, I., and Yamamoto, K. *J. Phys. Soc. Jpn.* **59**, 401–404 (1990).
Hou, T. Y. *Int. J. Numer. Meth. Fluids* **47**, 707–719 (2005).
Hou, T. Y. *Int. J. Numer. Meth. Fluids* **47**, 707–719 (2005).

Hubert, P., Schertzer, D., Tchiguirinskaia, I., Bendjoudi, H., Lovejoy, S., Hallegatte, S., and Larcheveque, M. *Houille Blan.-Rev. Int. Eau* **4**, 31–33 (2002).
Hughes, T. J. R. *Comput. Meth. Appl. Mech. Eng.* **127**, 387–401 (1995).
Hughes, T. J. R., Feijóo, G. R., Mazzei, L., and Quincy, J. B. *Comput. Meth. Appl. Mech. Eng.* **166**, 3–24 (1998).
Hughes, T. J. R., Mazzei, L., and Jansen, K. E. *Comput. Visual Sci.* **3**, 47–59 (2000).
Hughes, T. J. R., Oberai, A. A., and Mazzei, L. *Phys. Fluids* **13**, 1784–1799 (2001).
Hughes, T. J. R., Calo, V. M., and Scovazzi, G., Variational and multiscale methods in turbulence, *in* "Mechanics of the 21st Century" (W. Gutkowski, and T. A. Kowalewski Eds.), pp. 153–163. Springer, Dordrecht (2005).
Hui, M. H., Zhou, D., Wen, X. H., and Durlofsky, L. J. *Spe Reserv. Eval. Eng.* **8**, 189–195 (2005).
Humby, S. J., Biggs, M. J., and Tuzun, U. *Chem. Eng. Sci.* **57**, 1955–1968 (2002).
Huth, R. *J. Clim.* **17**, 640–652 (2004).
Huth, R. *Int. J. Climatol.* **25**, 243–250 (2005).
Iannece, D. *Int. J. Eng. Sci.* **32**, 1801–1809 (1994).
Idris, Z., Orgeas, L., Geindreau, C., Bloch, J. F., and Auriault, J. L. *Model. Simul. Mater. Sci. Eng.* **12**, 995–1015 (2004).
Illman, W. A. *Geophys. Res. Lett.* **31**, (2004).
Imkeller, P., and Von Storch, J. S. Eds., "Stochastic Climate Models". Birkhuser Verlag, Basel (2001).
Ingram, G. D., Cameron, I. T., and Hangos, K. M. *Chem. Eng. Sci.* **59**, 2171–2187 (2004).
Ishii, M., "Thermo-Fluid Dynamic Theory of Two-Phase Flow". Eyrolles, Paris (1975).
Israeli, N., and Goldenfeld, N. *Phys. Rev. Lett.* **92**, 074105 (2004).
Jaubert, G., and Stein, J. *Quar. J. R. Meteor. Soc.* **129**, 755–776 (2003).
Javaux, M., and Vanclooster, M. *J. Hydrol.* **327**, 376–388 (2006).
Jenny, P., Lee, S. H., and Tchelepi, H. A. *Multiscale Model. Simul.* **3**, 50–64 (2004).
Jewitt, G. P. W., and Grgens, A. H. M., Scale and model interfaces in the contex of integrated water resources management for the rivers of the Kruger national park, Report 627/1/00. Water Research Commission, Pretoria, South Africa (2000).
Joseph, D. D., "Fluid Dynamics of Viscoelastic Liquids". Springer, New York (1990).
Joseph, D. D., and Preziosi, L. *Rev.Mod. Phys.* **61**, 41–73 (1989).
Joseph, D. D., and Renardy, Y. Y., "Fundamentals of Two-Fluid Dynamics". Springer, New York (1993).
Juanes, R., and Patzek, T. W. *J. Hydraulic Res.* **42**, 131–140 (2004).
Jungen, A., Pfenninger, M., Tonteling, M., Stampfer, C., and Hierold, C. *J. Micromech. Microeng.* **16**, 1633–1638 (2006).
Kadowaki, H., and Liu, W. K. *Cmes-Comput. Model. Eng. Sci.* **7**, 269–282 (2005).
Kafer, J., Hogeweg, P., and Maree, A. F. M. *Plos Comput. Biol.* **2**, 518–529 (2006).
Kaganov, M. I., Lifshitz, I. M., and Tanatarov, M. V. *Sov. Phys. JETP* **4**, 173–178 (1957).
Kaipio, J., "Statistical and Computational Inverse Problems". Springer, New York (2005).
Kalma, J. D., and Sivapalan, M. Eds., "Scale Issues in Hydrological Modeling". Wiley, Chichester (1995).
Kaminski, M. M., "Computational Mechanics of Composite Materials: Sensitivity, Randomness, and Multiscale Behaviour". Springer, London (2005).
Kang, Q. J., Zhang, D. X., and Chen, S. Y. *Phys. Rev. E* **66**, 056307 (2002).
Kanwal, R. P., "Generalized Functions: Theory and Applications". Birkhuser, Boston (2004).
Karniadakis, G., Beskok, A., and Aluru, N., "Microflows and Nanoflows: Fundamentals and Simulation". Springer, New York (2005).
Karsch, F., Monien, B., and Satz, H. (Eds.), "Multiscale Phenomena and their Simulation". World Scientific, Singapore (1997).
Karssenberg, D. *Adv. Water Resour.* **29**, 735–759 (2006).
Katsoulakis, M. A., Majda, A. J., and Sopasakis, A. *Commun. Math. Sci.* **3**, 453–478 (2005).
Kaviany, M., "Principles of Heat Transfer in Porous Media". Springer, Berlin (1995).
Kerr, R. *J. Fluid Mech.* **153**, 31 (1985).
Kettle, H., and Thompson, R. *Clim. Res.* **26**, 97–112 (2004).
Kfoury, M., Ababou, R., Noetinger, B., and Quintard, M. *C. R. Mecha.* **332**, 679–686 (2004).

Kfoury, M., Ababou, R., Noetinger, B., and Quintard, M. *J. Appl. Mech. Trans. Asme* **73**, 41–46 (2006).
Khan, M. S., Coulibaly, P., and Dibike, Y. *Hydrol. Processes* **20**, 3085–3104 (2006a).
Khan, M. S., Coulibaly, P., and Dibike, Y. *J. Hydrol.* **319**, 357–382 (2006b).
Khurram, R. A., and Masud, A. *Comput. Mech.* **38**, 403–416 (2006).
Kim, S., Kavvas, M. L., and Yoon, J. *J. Hydrol. Eng.* **10**, 151–159 (2005).
Kim, J., Guo, Q., Baldocchi, D. D., Leclerc, M., Xu, L., and Schmid, H. P. *Agric. Forest Meteor.* **136**, 132–146 (2006).
Kinnmark, I. P. E., and Gray, W. G. *Adv. Water Resour.* **7**, 113–115 (1984).
Klein, R. *Esaim-Math. Model. Numer. Anal.* **39**, 537–559 (2005).
Knight, M., Thomas, D. S. G., and Wiggs, G. F. S. *Geomorphology* **59**, 197–213 (2004).
Knudby, C., Carrera, J., Bumgardner, J. D., and Fogg, G. E. *Adv. Water Resour.* **29**, 590–604 (2006).
Kok, S., Bharathi, M. S., Beaudoin, A. J., Fressengeas, C., Ananthakrishna, G., Kubin, L. P., and Lebyodkin, M. *Acta Materialia* **51**, 3651–3662 (2003).
Kolaczyk, E. D., and Huang, H. Y. *Geogr. Anal.* **33**, 95–118 (2001).
Konstandopoulos, A. G., Kostoglou, M., and Vlachos, N. *Int. J. Veh. Design* **41**, 256–284 (2006).
Korostyshevskaya, O., and Minkoff, S. E. *Siam J. Numer. Anal.* **44**, 586–612 (2006).
Koumoutsakos, P. *Annu. Rev. Fluid Mech.* **37**, 457–487 (2005).
Krause, R., and Rank, E. *Comput. Meth. Appl. Mech. Eng.* **192**, 3959–3983 (2003).
Kretzschmar, M., and Consolini, G. *Adv. Space Res.* **37**, 552–558 (2006).
Krumhansl, J. A. *Mater. Sci. Forum* **327-3**, 1–7 (2000).
Kruzik, M., and Prohl, A. *Siam Rev.* **48**, 439–483 (2006).
Kunstmann, H., Schneider, K., Forkel, R., and Knoche, R. *Hydrol. Earth Syst. Sci.* **8**, 1030–1044 (2004).
Kynch, G. J. *Trans. Faraday Soc.* **48**, 166–176 (1952).
Ladevze, P., A bridge between the micro- and mesomechanics of laminates: fantasy or reality, *in* "Mechanics of the 21st Century" (W. Gutkowski, and T. A. Kowalewski Eds.), pp. 187–201. Springer, Berlin (2005).
Lam, R., and Vlachos, D. G. *Phy. Rev. B* **64**, (2001).
Lambert, R. K., Castile, R. G., and Tepper, R. S. *J. Appl. Physiol.* **96**, 688–692 (2004).
Lanfredi, M., Lasaponara, R., Simoniello, T., Cuomo, V., and Macchiato, M. *Geophys. Res. Lett.* **30**, (2003).
Larachi, F. *Topics Catal.* **33**, 109–134 (2005).
Larachi, F., and Desvigne, D. *Chem. Eng. Sci.* **61**, 1627–1657 (2006).
Lartigue-Korinek, S., Carry, C., and Priester, L. *J. Eur. Ceramic Soc.* **22**, 1525–1541 (2002).
Lasseux, D., Quintard, M., and Whitaker, S. *Transp. Porous Media* **24**, 107–137 (1996).
Lassila, D. H., McElfresh, M. W., Rudd, R. E., Lightstone, F. C., Balhorn, R. L., Vitalis, E. A., White, D. A., Lee, C. L., Darnell, I. M., and Becker, R. C. Multiscale modeling of nano-scale phenomena: towards a multiphysics simulation capability for design and optimization of sensor systems, A Multidirectorate White Paper, Lawrence Livermore National Laboratory, UCRL-TR-201247 (2003).
Lechelle, J., Bleuet, P., Martin, P., Girard, E., Bruguier, F., Martinez, M. A., Somogyi, A., Simionovici, A., Ripert, M., Valdivieso, F., and Goeuriot, P. *IEEE Trans. Nucl. Sci.* **51**, 1657–1661 (2004).
Lee, P. D., Chirazi, A., Atwood, R. C., and Wang, W. *Mater. Sci. Eng.A Struct. Mater.* **365**, 57–65 (2004).
Lemaire, T., Naili, S., and Remond, A. *Biomech. Model. Mechanobiol.* **5**, 39–52 (2006).
Lesieur, M., Mtais, O., and Comte, P., "Large-Eddy Simulations of Turbulence". Cambridge University Press, Cambridge (2005).
Leveque, D., Schieffer, A., Mavel, A., and Maire, J. F. *Composites Sci. Technol.* **65**, 395–401 (2005).
Levin, S. A., The problem of relevant detail, *in* "Differential Equations: Models in Biology, Ecology and Epidemiology" (S. Busenberg, and M. Martelli Eds.), pp. 9–15. Springer, Berlin (1991).
Levin, S. A. *Ecology* **73**, 1943–1967 (1992).
Levin, S. A., Concepts of scale at the local level, *in* "Scaling Physiological Processes: Leaf to Globe" (J. R. Ehleringer, and C. B. Field Eds.), pp. 7–19. Academic Press, San Diego (1993).
Levin, S. A. *Ecosystems* **3**, 498–506 (2000).
Levin, S. A. *Bull. Am. Math. Soc.* **40**, 3–19 (2002).

Levin, S. A., and Pacala, S. W., Theories of simplification and scaling of spatially distributed processes, *in* "Spatial Ecology: The Role of Space in Population Dynamics and Interspecific Interactions" (D. Tilman, and P. Kareiva Eds.), pp. 271–296. Princeton University Press, Princeton (1997).
Levin, S. A., Grenfell, B., Hastings, A., and Perelson, A. S. *Science* **275**, 334–343 (1997).
Levitas, V. I. *Phy. Rev. B* **70**, (2004).
Lewandowska, J., Szymkiewicz, A., and Auriault, J. L. *Adv. Water Resour.* **28**, 1159–1170 (2005).
Li, J. H., and Kwauk, M. S. *Chem. Eng. Sci.* **58**, 521–535 (2003).
Li, J. H., and Kwauk, M. S. *Chem. Eng. Sci.* **59**, 1611–1612 (2004).
Li, C. S., Mosier, A., Wassmann, R., Cai, Z. C., Zheng, X. H., Huang, Y., Tsuruta, H., Boonjawat, J., and Lantin, R. *Global Biogeochem. Cycle* **18**, Art. No. Gb1043 (2004a).
Li, J. H., Zhang, J. Y., Ge, W., and Liu, X. H. *Chem. Eng. Sci.* **59**, 1687–1700 (2004b).
Li, S. F., Liu, X. H., and Gupta, A. *Int. J. Numer. Meth. Eng.* **62**, 1264–1294 (2005).
Li, L., Peters, C. A., and Celia, M. A. *Adv. Water Resour.* **29**, 1351–1370 (2006).
Liang, X. Z., Pan, J. P., Zhu, J. H., Kunkel, K. E., Wang, J. X. L., and Dai, A. G. *J. Geophys. Res. Atmos.* **111**, (2006).
Liao, X. X., Wang, L. Q., and Yu, P., "Stability of Dynamical Systems". Elsevier, London (2007).
Likos, W. J., and Lu, N. *Clays Clay Miner.* **54**, 515–528 (2006).
Lindeberg, T. *J. Appl. Stat.* **21**, 224–270 (1994a).
Lindeberg, T., "Scale-Space Theory in Computer Vision". Kluwer Academic Publishers, Dordrecht (1994b).
Lindeberg, T. *Int. J. Comput. Vision* **30**, 77–116 (1998).
Lindeberg, T. Methods for automatic scale selection, in "Handbook on Computer Vision and Applications" (B. Jahne, H. Haussecker, and P. Geissler Eds.), Vol. 2, pp. 239–274. Academic Press, Boston (1999).
Lindeberg, T., and Romeny, B. T. H., Linear scale-space: I. basic theory, II. early visual operations, *in* "Geometry-Driven Diffusion in Computer Vision" (B. T. H. Romeny Ed.), pp. 1–77. Kluwer Academic Publishers, Dordrecht (1994).
Liou, W. W., and Fang, Y. C., "Microfluid Mechanics: Principles and Modeling". McGraw-Hill, New York (2006).
Lipowsky, R., and Klumpp, S. *Phys. A Stat. Mech. Appl.* **352**, 53–112 (2005).
Liu, Y., Wang, Z. W., and Tay, J. H. *Biotechnol. Adv.* **23**, 335–344 (2005).
Lock, P. A., Jing, X. D., and Zimmerman, R. W. *J. Hydraulic Res.* **42**, 3–8 (2004).
Love, P. J., Nekovee, M., Coveney, P. V., Chin, J., Gonzalez-Segredo, N., and Martin, J. M. R. *Comput. Phys. Commun.* **153**, 340–358 (2003).
Lovell, C., Mandondo, A., and Moriarty, P. *Conserv. Ecol.* **5**, (2002).
Lpinoux, J., Mazire, D., Pontikis, V., and Saada, G. (Eds.), "Multiscale Phenomena in Plasticity: From Experiments to Phenomenology, Modelling and Materials Engineering". Kluwer Academic Publishers, Dordrecht (2000).
Lu, Y. H., and Fu, B. J. *Acta Ecol. Sin.* **21**, 2096–2105 (2001).
Ludwig, D., and Walters, C. J. *Can. J. Fish Aquat. Sci.* **42**, 1066–1072 (1985).
Ludwig, R., Taschner, S., and Mauser, W. *Hydrol. Earth Syst. Sci.* **7**, 833–847 (2003).
Lui, A. T. Y. *J. Atmos. Sol.-Terr. Phys.* **64**, 125–143 (2002).
Ma, J. S., Couples, G. D., and Harris, S. D. *Water Resour. Res.* **42**, (2006).
Mack, G. *Commun. Math. Phys.* **219**, 141–178 (2001).
MacLachlan, S. P., and Moulton, J. D. *Water Resour. Res.* **42**, (2006).
Madec, R., Devincre, B., Kubin, L., Hoc, T., and Rodney, D. *Science* **301**(September), 1879–1882 (2003).
Magesa, S. M., Lengeler, C., DeSavigny, D., Miller, J. E., Njau, R. J. A., Kramer, K., Kitua, A., and Mwita, A. *Malaria J.* **4**, (2005).
Mander, U., Muller, F., and Wrbka, T. *Ecol. Indicators* **5**, 267–272 (2005).
Mander, U., Muller, F., and Wrbka, T. *Ecol. Indicators* **5**, 267–272 (2005).
Marin, G. B. (Ed.), Multiscale Analysis, Advances in Chemical Engineering, Vol. 30. Elsevier, Amsterdam (2005).
Marle, C. M. *Int. J. Eng. Sci.* **20**, 643–662 (1982).
Maroudas, D., and Gungor, M. R. *Comput. Mater. Sci.* **23**, 242–249 (2002).

Maroudas, D., Zepeda-Ruiz, L. A., Pelzel, R. I., Nosho, B. Z., and Weinberg, W. H. *Comput. Mater. Sci.* **23**, 250–259 (2002).
Masi, M. *J. Phys. IV* **11**, 117–128 (2001).
Masi, M., Di Stanislao, M., and Veneroni, A. *Progr. Cryst. Growth Character. Mater.* **47**, 239–270 (2003).
Mateos, J., Vasallo, B. G., Pardo, D., and Gonzalez, T. *Appl. Phys. Lett.* **86**, (2005).
Matulla, C. *Meteorol. Z.* **14**, 31–45 (2005).
McLeary, E. E., Jansen, J. C., and Kapteijn, F. *Microporous Mesoporous Mater.* **90**, 198–220 (2006).
Meakin, P., "Fractals, Scaling and Growth Far from Equilibrium". Cambridge University Press, Cambridge (1998).
Medjdoub, F., Zaknoune, M., Wallart, X., Gaquiere, C., Dessenne, F., Thobel, J. L., and Theron, D. *IEEE Trans. Electron Devices* **52**, 2136–2143 (2005).
Mehrotra, R., and Sharma, A. *J. Geophys. Res. Atmos.* **110**, (2005).
Mehrotra, R., and Sharma, A. *J. Geophys. Res. Atmos.* **111**, (2006a).
Mehrotra, R., and Sharma, A. *Adv. Water Resour.* **29**, 987–999 (2006b).
Mehrotra, R., and Sharma, A. *New J. Phys.* **8**, (2006c).
Mehrotra, R., Sharma, A., and Cordery, I. *J. Geophys. Res. Atmos.* **109**, (2004).
Mei, C. C., and Auriault, J. L. *Proc. R. Soc. Lond. A* **426**, 391–423 (1989).
Mei, C. C., and Auriault, J. L. *J. Fluid Mech.* **222**, 647–663 (1991).
Meier, H. E. M. *Clim. Dyn.* **27**, 39–68 (2006).
Meier, H. E. M., Kjellstrom, E., and Graham, L. P. *Geophys. Res. Lett.* **33**, (2006).
Merlin, O., Chehbouni, A., Kerr, Y. H., and Goodrich, D. C. *Remote Sens. Environ.* **101**, 379–389 (2006).
Mezghani, M., and Roggero, F. *Spe J.* **9**, 79–87 (2004).
Mezzacappa, A. *Annu. Rev. Nucl. Particle Sci.* **55**, 467–515 (2005).
Mezzenga, R., Folmer, B. M., and Hughes, E. *Langmuir* **20**, 3574–3582 (2004).
Miehle, P., Livesley, S. J., Li, C. S., Feikema, P. M., Adams, M. A., and Arndt, S. K. *Global Change Biol.* **12**, 1421–1434 (2006).
Mika, J., Molnar, J., and Tar, K. *Phys. Chem. Earth* **30**, 135–141 (2005).
Miksovsky, J., and Raidl, A. *Nonlinear Processes Geophy.* **12**, 979–991 (2005).
Miller, C. T., and Gray, W. G. *Adv. Water Resour.* **28**, 181–202 (2005).
Miller, R. E., Shilkrot, L. E., and Curtin, W. A. *Acta Mater.* **52**, 271–284 (2004).
Minkowycz, W. J., Haji-Sheikh, A., and Vafai, K. *Int. J. Heat Mass Transf.* **42**, 3373–3385 (1999).
Mizuseki, H., Hongo, K., Kawazoe, Y., and Wille, L. T. *Comput. Mater. Sci.* **24**, 88–92 (2002).
Moffatt, H. K., Some remarks on topological fluid mechanics, *in* "An Introduction to the Geometry and Topology of Fluid Flows" (R. Ricca Ed.), pp. 3–10. Kluwer, Amsterdam (2001).
Moffatt, H. K., The topology of turbulence, *in* "New Treads in Turbulence" (M. Lesieur, A. Yaglom, and F. David Eds.), pp. 319–340. Springer, Berlin (2001).
Moffatt, H. K., The topology of scalar fields in 2D and 3D turbulence, *in* "Proceedings of IUTAM Symposiumof Geometry and Statistics of Turbulence" (T. Kambe, T. Nakano, and T. Miyauchi Eds.), pp. 13–22. Kluwer, Amsterdam (2002).
Moin, P., and Apte, S. V. *Aiaa J.* **44**, 698–708 (2006).
Moore, M. (Ed.), "Spatial Statistics; Methodological Aspects and Applications". Springer, New York (2001).
Moriondo, M., and Bindi, M. *Clim. Res.* **30**, 149–160 (2006).
Moseler, M., Gumbsch, P., Casiraghi, C., Ferrari, A. C., and Robertson, J. *Science* **309**, 1545–1548 (2005).
Mosler, J. *Comput. Struct.* **83**, 369–382 (2005).
Mounier, F., Echevin, V., Mortier, L., and Crepon, M. *Progr. Oceanogr.* **66**, 251–269 (2005).
Mueller, G., and Jochen, F. *J. Rare Earths* **24**, 200–207 (2006).
Muller, R., Jonge, S., Myny, K., Wouters, D. J., Genoe, J., and Heremans, P. *Solid-State Electron.* **50**, 601–605 (2006).
Mullins, W. W. *J. Appl. Phys.* **27**, 900–904 (1956).
Mullins, W. W. *Trans. Met. Soc. AIME* **218**, 354–361 (1960).
Murad, M. A., and Cushman, J. H. *Int. J. Eng. Sci.* **34**, 313–336 (1996).
Murad, M. A., and Cushman, J. H. *Int. J. Eng. Sci.* **38**, 517–564 (2000).
Murad, M. A., Bennethum, L. S., and Cushman, J. H. *Transp. Porous Media* **19**, 93–122 (1995).

Myers, J. D., Rahn, L., Leahy, D., Pancerella, C. M., von Laszewski, G., Ruscic, B., and Green, W. H. *Abstr.Papers Am. Chem. Soc.* **227**, (2004).
Myers, J. D., Allison, T. C., Bittner, S., Didier, B., Frenklach, M., Green, W. H., Ho, Y. L., Hewson, J., Koegler, W., Lansing, C., Leahy, D., Lee, M., McCoy, R., Minkoff, M., Nijsure, S., Von Laszewski, G., Montoya, D., Oluwole, L., Pancerella, C., Pinzon, R., Pitz, W., Rahn, L. A., Ruscic, B., Schuchardt, K., Stephan, E., Wagner, A., Windus, T., and Yang, C. *Cluster Computing: J. Networks Software Tools Appl.* **8**, 243–253 (2005).
Narayanan, B., Pryamitsyn, V. A., and Ganesan, V. *Macromolecules* **37**, 10180–10194 (2004).
Nardin, A., and Schrefler, B. A. *Comput. Mech.* **36**, 343–359 (2005).
Nash, D. A., and Ragsdale, D. J. *IEEE Trans. Syst. Man Cybern. A Syst. Hum.* **31**, 327–331 (2001).
Neale, K. W., Inal, K., and Wu, P. D. *Int. J. Mech. Sci.* **45**, 1671–1686 (2003).
Neuweiler, I., and Cirpka, O. A. *Water Resour. Res.* **41**, (2005).
Nicot, F. *Eur. J. Mech. A Solids* **22**, 325–340 (2003).
Nicot, F. *Int. J. Solids Struct.* **41**, 3317–3337 (2004).
Niedda, M. *Water Resour. Res.* **40**, (2004).
Nield, D. A., and Bejan, A., "Convection in Porous Media". 3rd Ed. Springer, New York (2006).
Nieminen, R. M. *Curr. Opin. Solid State Mater. Sci.* **4**, 493–498 (1999).
Nigmatulin, R. I. *Int. J. Multiphase Flow* **5**, 353–385 (1979).
Niu, Y. Y., and Lin, Y. M. *Numer. Heat Transf. Part A –Appl.* **50**, 545–560 (2006).
Noetinger, B., and Gallouet, T. *Oil Gas Sci. Technol. Revue de 1 Institut Francais du Petrole* **59**, 117–118 (2004).
Noetinger, B., and Zargar, G. *Oil Gas Sci. Technol. Rev. Inst. Francaisu Petrole* **59**, 119–139 (2004).
Noetinger, B., Artus, V., and Ricard, L. *Transp. Porous Media* **56**, 305–328 (2004).
Noetinger, B., Artus, V., and Zargar, G. *Hydrogeol. J.* **13**, 184–201 (2005).
Nordbotten, J. M., Rodriguez-Iturbe, I., and Celia, M. A. *Proc. R. Soc. A Math. Phys. Eng. Sci.* **462**, 2359–2371 (2006).
Novak, M. M. (Ed.), "Thinking in Patterns: Fractals and Related Phenomena in Nature". World Scientific, Singapore (2004).
Novikov, A. *J. Comput. Phys.* **195**, 341–354 (2004).
Nuttinck, S. *IEEE Trans. Electron Devices* **53**, 1193–1199 (2006).
Nuttinck, S., Scholten, A. J., Tiemeijer, L. F., Cubaynes, F., Dachs, C., Detcheverry, C., and Hijzen, E. A. *IEEE Trans. Electron Devices* **53**, 153–157 (2006).
O'Connell, P. E., A historical perspective, *in* "Recent Advances in the Modeling of Hydrologic Systems" (D. S. Bowles, and P. E. O'Connell Eds.), pp. 3–30. Kluwer, Dordrecht (1991).
Ogunlana, D. O., and Mohanty, K. K. *J. Petrol. Sci. Eng.* **46**, 1–21 (2005).
Oja, T., Alamets, K., and Parnamets, H. *Ecol. Indicators* **5**, 314–332 (2005a).
Oja, T., Alamets, K., and Parnamets, H. *Ecol. Indicators* **5**, 314–321 (2005b).
Olsson, J., Uvo, C. B., Jinno, K., Kawamura, A., Nishiyama, K., Koreeda, N., Nakashima, T., and Morita, O. *J. Hydrol. Eng.* **9**, 1–12 (2004).
Ono, S., and Kondo, S. Encyclopedia of physics, *in* "Structure of Liquids" (S. Flügge, Ed.), Vol. 10, p. 134. Springer, Berlin (1960).
Orgeas, L., Idris, Z., Geindreau, C., Bloch, J. F., and Auriault, J. L. *Hemical Eng. Sci.* **61**, 4490–4502 (2006).
Osetsky, Y. N., and Bacon, D. J. *Nucl. Instrum.Meth. Phys. Res. B* **202**, 31–43 (2003).
Padmanabhan, G., and Rao, A. R. *Wat. Resour. Res.* **25**, 1519–1533 (1988).
Paeth, H., Born, K., Podzun, R., and Jacob, D. *Meteorol. Z.* **14**, 349–367 (2005).
Pai, R. A., Humayun, R., Schulberg, M. T., Sengupta, A., Sun, J. N., and Watkins, J. J. *Science* **303**, 507–510 (2004).
Painter, S., and Cvetkovic, V. *Water Resour. Res.* **41**, (2005).
Pamplin, B. R., "Crystal Growth". Pergamon, Oxford (1980).
Pardo-Iguzquiza, E., Chica-Olmo, M., and Atkinson, P. M. *Remote Sens. Environ.* **102**, 86–98 (2006).
Park, M., and Cushman, J. H. *J. Comput. Phys.* **217**, 159–165 (2006).
Park, E., and Parker, J. C. *Adv. Water Resour.* **28**, 1280–1291 (2005).
Pasini, A., Lore, M., and Ameli, F. *Ecol. Model.* **191**, 58–67 (2006).

Pathak, H., Li, C., and Wassmann, R. *Biogeosciences* **2**, 113–123 (2005).
Patil, G. P., Balbus, J., Biging, G., Jaja, J., Myers, W. L., and Taillie, C. *Environ. Ecol. Stat* **11**, 113–138 (2004).
Paulson, S. *Human Org.* **62**, 242–254 (2003).
Paulson, S., Gezon, L. L., and Watts, M. *Human Org.* **62**, 205–217 (2003).
Pavan, V., Marchesi, S., Morgillo, A., Cacciamani, C., and Doblas-Reyes, F. J. *Tellus A Dyn. Meteorol. Oceanogr.* **57**, 424–434 (2005).
Pennock, D., Farrell, R., Desjardins, R., Pattey, E., and MacPherson, J. I. *Can. J. Soil Sci.* **85**, 113–125 (2005).
Perfect, E., Gentry, R. W., Sukop, M. C., and Lawson, J. E. *Geoderma* **134**, 240–252 (2006).
Perotto, S. *Esaim Math. Model. Numer. Anal.* **40**(3), 469–499 (2006).
Perry, R. I., and Ommer, R. E. *Fish. Oceangr.* **12**, 513–522 (2003).
Pettorelli, N., Mysterud, A., Yoccoz, N. G., Langvatn, R., and Stenseth, N. C. *Proc. R. Soc. Lond. B Biol. Sci.* **272**, 2357–2364 (2005).
Petts, G. E., Nestler, J., and Kennedy, R. *Hydrobiologia* **565**, 277–288 (2006).
Pickup, G. E., Stephen, K. D., Ma, J., Zhang, P., and Clark, J. D. *Transp. Porous Media* **58**, 191–216 (2005).
Pingault, M., Bruno, E., and Pellerin, D. *IEEE Trans. Image Process.* **12**, 1416–1426 (2003).
Piquet, J., "Turbulent Flows: Models and Physics". Springer, Berlin (1999).
Pont, V., Fontan, J., and Lopez, A. *Atmos. Res.* **66**, 83–105 (2003).
Ponziani, D., Pirozzoli, S., and Grasso, F. *Int. J. Numer. Meth. Fluids* **42**, 953–977 (2003).
Pope, S. B., "Turbulent Flows". Cambridge University Press, Cambridge (2000).
Pozdnyakova, L., Gimenez, D., and Oudemans, P. V. *Agron. J.* **97**, 49–57 (2005).
Pradhan, N. R., Tachikawa, Y., and Takara, K. *Hydrol. Processes* **20**, 1385–1405 (2006).
Prevost, M., Lepage, F., Durlofsky, L. J., and Mallet, J. L. *Petrol. Geosci.* **11**, 339–345 (2005).
Pruhs, S., Dinter, C., Blume, T., Schutz, A., Harre, M., and Neh, H. *Org. Process Res. Dev.* **10**, 441–445 (2006).
Prvost, J. *Int. J. Eng. Sci.* **18**, 787–800 (1980).
Pryor, S. C., Schoof, J. T., and Barthelmie, R. J. *J. Geophys. Res. Atmos.* **110**, (2005a).
Pryor, S. C., Schoof, J. T., and Barthelmie, R. J. *Clim. Res.* **29**, 183–198 (2005b).
Qi, D. S., and Hesketh, T. *Petrol. Sci. Technol.* **22**, 1595–1624 (2004a).
Qi, D. S., and Hesketh, T. *Petrol. Sci. Technol.* **22**, 1625–1640 (2004b).
Qi, D. S., and Hesketh, T. *Petrol. Sci. Technol.* **23**, 1291–1302 (2005a).
Qi, D. S., and Hesketh, T. *Petrol. Sci. Technol.* **23**, 827–842 (2005b).
Qin, K. R., Jiang, Z. L., Sun, H., Gong, K. Q., and Liu, Z. R. *Acta Mech. Sin.* **22**, 76–83 (2006).
Qiu, T. Q., and Tien, C. L. *J. Heat Transf.* **115**, 835–841 (1993).
Quarteroni, A., and Veneziani, A. *Multiscale Model. Simul.* **1**, 173–195 (2003).
Quintard, M., and Whitaker, S. *Adv. Heat Transf.* **23**, 369–464 (1993).
Quintard, M., and Whitaker, S. *Transp. Porous Media* **14**, 179–206 (1994a).
Quintard, M., and Whitaker, S. *Transp. Porous Media* **15**, 31–49 (1994b).
Quintard, M., and Whitaker, S. *Adv. Water Resour.* **19**, 29–47 (1996).
Quintard, M., Bletzacker, L., Chenu, D., and Whitaker, S. *Chem. Eng. Sci.* **61**, 2643–2669 (2006).
Raderschall, N. *Meteorol. Appl.* **11**, 311–318 (2004).
Rafii-Tabar, H., Shodja, H. M., Darabi, M., and Dahi, A. *Mech. Mat.* **38**, 243–252 (2006).
Raimondeau, S., and Vlachos, D. G. *Chem. Eng. J.* **90**, 3–23 (2002).
Rajagopal, K. R., and Tao, L., "Mechanics of Mixtures". World Scientific, Singapore (1995).
Rebora, N., Ferraris, L., von Hardenberg, J., and Provenzale, A. *J. Hydrometeor.* **7**, 724–738 (2006a).
Rebora, N., Ferraris, L., von Hardenberg, J., and Provenzale, A. *Nat. Hazards Earth Syst. Sci.* **6**, 611–619 (2006b).
Ren, X. F., and Malik, J. *Lect. Notes Comput. Sci.* **2350**, 312–327 (2002).
Ren, X. B., and Otsuka, K. *MRS Bull.* **27**, 115–120 (2002).
Ren, W. Q., and Weinan, E. *J. Comput. Phys.* **204**, 1–26 (2005).
Ren, L. Q., Sinton, D., and Li, D. Q. *J. Micromech. Microeng.* **13**, 739–747 (2003).
Rengel, R., Pardo, D., and Martin, M. J. *Nanotechnology* **15**, S276–S282 (2004).
Riitters, K. H. *Ecol. Indicators* **5**, 273–279 (2005).

Robinson, A. R., and Brink, K. H. Eds., "The Global Coastal Ocean: Multiscale Interdisciplinary Processes". Harvard University Press, Boston (2005).
Robinson, P. A., Rennie, C. J., Rowe, D. L., and O'Connor, S. C. *Hum. Brain Mapp.* **23**, 53–72 (2004).
Robinson, P. A., Rennie, C. J., Rowe, D. L., O'Connor, S. C., and Gordon, E. *Philos. Trans. R. Soc. B-Biol. Sci.* **360**, 1043–1050 (2005).
Rodgers, P., Soulsby, C., and Waldron, S. *Hydrol. Processes* **19**, 2291–2307 (2005).
Ronayne, M. J., and Gorelick, S. M. *Phys. Rev. E* **73**, (2006).
Rouch, P., and Ladeveze, P. *Comput. Meth. Appl. Mech. Eng.* **192**, 3301–3315 (2003).
Rundle, J. B., Turcotte, D. L., Shcherbakov, R., Klein, W., and Sammis, C. *Rev. Geophys.* **41**, (2003).
Rusli, E., Drews, T. O., and Braatz, R. D. *Chem. Eng. Sci.* **59**, 5607–5613 (2004).
Russo, D. *Water Resour. Res.* **39**, (2003).
Saedi, A. *J. Electroanal. Chem.* **592**, 95–102 (2006).
Sagar, B. S. D., and Rao, C. B. *Int. J. Pattern Recognit. Artif. Intell.* **17**, 163–165 (2003).
Sagaut, P., "Large Eddy Simulation for Incompressible Flows: An Introduction". 3rd Ed. Springer, Berlin (2006a).
Sagaut, P., "Large Eddy Simulation for Incompressible Flows: An Introduction". Springer, Berlin (2006b).
Sagaut, P., Deck, S., and Terracol, M. Multiscale and Multiresolution Approaches in Turbulence, World Scientific Imperial Col. Pr., London (2006).
Sakiyama, Y., Takagi, S., and Matsumoto, Y. *Phys. Fluids* **16**, 1620–1629 (2004).
Sakiyama, Y., Iga, Y., Yamaguchi, H., Takagi, S., and Matsumoto, Y. *Surf. Coat. Technol.* **200**, 3385–3388 (2006).
Salathe, E. P. *Int. J. Climatol.* **25**, 419–436 (2005).
Salis, H., Sotiropoulos, V., and Kaznessis, Y. N. *BMC Bioinform.* **7**, (2006).
Samantray, A. K., Dashti, Q. M., Ma, E. D. C., and Kumar, P. S. *Spe Reserv. Eval. Eng.* **9**, 15–23 (2006).
Sampaio, R., and Williams, W. O. *J. Mecanique* **18**, 19–45 (1979).
Sanyal, D., Ramachandrarao, P., and Gupta, O. P. *Comput. Mater. Sci.* **37**, 166–177 (2006).
Schiehlen, W., and Seifried, R. *Multibody Syst. Dyn.* **12**, 1–16 (2004).
Schlecht, E., and Hiernaux, P. *Nutr. Cycl. Agroecosyst.* **70**(3), 303–319 (2004).
Schmidli, J., Frei, C., and Vidale, P. L. *Int. J. Climatol.* **26**, 679–689 (2006).
Schneider, D. C. *Ecosystems* **5**, 736–748 (2002).
Schulze, R. *Agric. Ecosyst. Environ.* **82**, 185–212 (2000).
Scriven, L. E. *Chem. Eng. Sci.* **12**, 98–108 (1960).
Seeboonruang, U., and Ginn, T. R. *J. Contam. Hydrol.* **84**, 127–154 (2006a).
Seeboonruang, U., and Ginn, T. R. *J. Contam. Hydrol.* **84**, 155–177 (2006b).
Segall, D. E., Li, C., and Xu, G. *Philos. Mag.* **86**, 5083–5101 (2006).
She, Z. S., Jackson, E., and Orszag, S. A. *Proc. R. Soc. Lond. A* **434**, 101–124 (1991).
Shehadeh, M. A., Zbib, H. M., and de la Rubia, T. D. *Int. J. Plasticity* **21**, 2369–2390 (2005).
Shilkrot, L. E., Miller, R. E., and Curtin, W. A. *J. Mech. Phys. Solids* **52**, 755–787 (2004).
Shilov, G. E., "Generalized Functions and Partial Differential Equations". Gordon and Breach, New York (1968).
Shin, D. W., Bellow, J. G., Larow, T. E., Cocke, S., and O'Brien, J. J. *J. Appl. Meteorol. Climatol.* **45**, 686–701 (2006).
Siegert, F., Ruecker, G., Hinrichs, A., and Hoffmann, A. A. *Nature* **414**, 437–440 (2001).
Sih, G. C. *Theor. Appl. Fract. Mech.* **37**, 335–369 (2001).
Sinha, S., and Goodson, K. E. *Int. J. Multiscale Comput. Eng.* **3**, 107–133 (2005).
Sitnov, M. L., Sharma, A. S., Papadopoulos, K., and Vassiliadis, D. *Phys. Rev. E* **65**, (2002).
Sivapalan, M., Zhang, L., Vertessy, R., and Blöschl, G. *Hydrol. Processes* **17**, 2099–2326 (2003).
Skar, J., and Coveney, P. V. *Philos. Trans. R. Soc. Lond. A: Math. Phys. Eng. Sci.* **361**, 1047 (2003a).
Skar, J., and Coveney, P. V. *Philos. Trans. R. Soc. Lond. A: Math. Phys. Eng. Sci.* **361**, 1313–1317 (2003b).
Skar, J. *Philos. Trans. R. Soc. Lond. A: Math. Phys. Eng. Sci.* **361**, 1049–1056 (2003).
Skoien, J. O., Blöschl, G., and Western, A. W. *Water Resour. Res.* **39**, 1304 (2003).
Slattery, J. C. *Chem. Eng. Sci.* **19**, 379–385 (1964a).
Slattery, J. C. *Chem. Eng. Sci.* **19**, 453–455 (1964b).

Slattery, J. C. *I & EC Fundam.* **6**, 108–115 (1967a).
Slattery, J. C. *AICHE J.* **13**, 1066 (1967b).
Slattery, J. C., "Momentum, Energy and Mass Transfer in Continua". McGraw-Hill, New York (1972).
Slattery, J. C., "Advanced Transport Phenomena". Cambridge University Press, New York (1999).
Slattery, J. C., Sagis, L., and Oh, E. S., "Interfacial Transport Phenomena". Springer, New York (2007).
Solecki, W. D., and Oliveri, C. *J. Environ. Manage.* **72**, 105–115 (2004).
Sornette, D., and Zhou, W. X. *Int. J. Forecast.* **22**, 153–168 (2006).
Soulard, O., Sabel'nikov, V., and Gorokhovski, M. *Int. J. Heat Fluid Flow* **25**, 875–883 (2004).
Sposito, G. (Ed.), "Scale Dependence and Scale Invariance in Hydrology". Cambridge University Press, Cambridge (1998).
Steigmann, D. J., and Li, D. Q. *Proc. R. Soc. Lond. A* **449**, 223–231 (1995a).
Steigmann, D. J., and Li, D. Q. *IMA J. Appl. Math.* **55**, 1–17 (1995b).
Stone, H. A. *Phys. Fluids A* **2**, 111–112 (1990).
Stull, R. B., "An Introduction to Boundary Layer Meteorology". Kluwer Academic, Dordrech (1988).
Succi, S., Filippova, O., Smith, G., and Kaxiras, E. *Comput. Sci. Eng.* **2001**, 26–37 (2001).
Sudderth, E. B., Wainwright, M. J., and Willsky, A. S. *IEEE Trans. Signal Process.* **52**, 3136–3150 (2004).
Sun, Q. Y., Vrieling, E. G., van Santen, R. A., and Sommerdijk, N. A. J. M. *Curr. Opin. Solid State Mater. Sci.* **8**, 111–120 (2004).
Sun, L. Q., Moncunill, D. F., Li, H. L., Moura, A. D., Filho, F. D. A. D. S., and Zebiak, S. E. *J. Clim.* **19**, 1990–2007 (2006).
Takeuchi, K. *Hydrol. Processes* **18**, 2967–2976 (2004).
Temizel, A., and Vlachos, T. *J. Electr. Imaging* **14**, (2005).
Teppola, P., and Minkkinen, P. *J. Chemometr.* **14**, 383–399 (2000).
Tewfik, A. H., Kim, M., and Deriche, M., Multiscale signal processing techniques: A review, *in* "Signal Processing and Its Applications" (N. K. Bose, and C. R. Rao Eds.), pp. 619–882. Amsterdam, North-Holland (1993).
Thigpen, L., and Berryman, J. G. *Int. J. Eng. Sci.* **23**, 1203–1214 (1985).
Tian, Z. J., Xu, Y. P., and Lin, L. W. *Chem. Eng. Sci.* **59**, 1745–1753 (2004).
Tidriri, M. D. *Nonlinear Anal. Theor. Meth. Appl.* **47**, 4995–5008 (2001).
Timbal, B. *Clim. Res.* **26**, 233–249 (2004).
Toniolo, H., Parker, G., Voller, V., and Beaubouef, R. T. *J. Sediment. Res.* **76**, 798–818 (2006).
Tononi, G. *Sleep Med.* **6**(Suppl.), S8–S12 (2005).
Toomanian, N., Gieske, A. S. M., and Akbary, M. *Int. J. Remote Sens.* **25**, 4945–4960 (2004).
Tretter, T., and Jones, M. G. *Sci. Teacher* **70**, 22–25 (2003).
Truesdell, C., and Rajagopal, K. R., "An Introduction to the Mechanics of Fluids". Birkhäuser, Boston (2000).
Truskey, G. A., Yuan, F., and Katz, D. F., "Transport Phenomena in Biological Systems". Pearson Prentice Hall, Upper Saddle River (2004).
Tung, W. W., Moncrieff, M. W., and Gao, J. B. *J. Clim.* **17**, 2736–2751 (2004).
Tureyen, O. I., and Caers, J. *Comput. Geosci.* **9**, 75–98 (2005).
Tworek, S. *Ornis Fenn.* **80**, 49–62 (2003).
Tzou, D. Y. *J.Heat Transf.* **117**, 8–16 (1995).
Tzou, D. Y., "Macro- to Microscale Heat Transfer: the Lagging Behavior". Taylor & Francis, Washington (1997).
Ukhorskiy, A. Y., Sitnov, M. I., Sharma, A. S., and Papadopoulos, K. *Ann. Geophys.* **21**, 1913–1929 (2003).
Uva, G., and Salerno, G. *Int. J. Solids Struct.* **43**, 3739–3769 (2006).
Vadasz, P. *Int. J. Heat Mass Transf.* **48**, 2822–2828 (2005a).
Vadasz, P. *Transp. Porous Media* **59**, 341–355 (2005b).
Vadasz, P. *J. Heat Transf.* **127**, 307–314 (2005c).
Vadasz, P. *J. Heat Transf.* **128**, 465–477 (2006a).
Vadasz, P. *Int. J. Heat Mass Transf.* **49**, 4886–4892 (2006b).
Valenza, M., Hoffmann, A., Sodini, D., Laigle, A., Martinez, F., and Rigaud, D. *IEE Proc. Circuits Devices Syst.* **151**, 102–110 (2004).
Valluzzi, R., Probst, W., Jacksch, H., Zellmann, E., and Kaplan, D. L. *Soft Mater.* **1**, 245–262 (2003).

Van den Akker, H. E. A. *Adv. Chem. Eng.* **31**, 151–229 (2006).
Van Dommelen, J. A. W., Schrauwen, B. A. G., Van Breemen, L. C. A., and Govaert, L. E. *J. Poly. Sci. Part B: Poly. Phys.* **42**, 2983–2994 (2004).
Van Gardingen, P. R., Foody, G. M., and Curran, P. J. (Eds.), "Scaling-Up: From Cell to Landscape". Cambridge University Press, Cambridge (1997).
Varshney, A., and Armaou, A. *Chem. Eng. Sci.* **60**, 6780–6794 (2005).
Vdovina, T., Minkoff, S. E., and Korostyshevskaya, O. *Multiscale Model. Simul.* **4**, 1305–1338 (2005).
Veneziano, D., Furcolo, P., and Iacobellis, V. *J. Hydrol.* **322**, 105–119 (2006).
Verburg, P. H., van Bodegom, P. M., van der Gon, H. A. C. D., Bergsma, A., and van Breemen, N. *Plant Ecol.* **182**, 89–106 (2006a).
Verburg, P. H., Schulp, C. J. E., Witte, N., and Veldkamp, A. *Agric. Ecosyst. Environ.* **114**(1), 39–56 (2006b).
Vidakovic, B. D., Katul, G. G., and Albertson, J. D. *J. Geophy. Res. Atmos.* **105**, 27049–27058 (2000).
Vignon, I. E., and Taylor, C. A. *Wave Motion* **39**, 361–374 (2004).
Vincent, A., and Meneguzzi, M. *J. Fluid Mech.* **225**, 1–25 (1991).
Vinogradov, A., and Hashimoto, S. *Mater. Trans. JIM* **42**, 74–84 (2001).
Vitale, M., Gerosa, G., Ballarin-Denti, A., and Manes, F. *Atmos. Environ.* **39**, 3267–3278 (2005).
Vlachos, D. G. *Adv. Chem. Engng.* **30**, 1–61 (2005).
Vladimirov, V. S., "Equations of Mathematical Physics". M. Dekker, New York (1971).
Vladimirov, V. S., "Methods of the Theory of Generalized Functions". Taylor & Francis, New York (2002).
Voigt, A. (Ed.), "Multiscale Modeling in Epitaxial Growth". Birkhuser, Boston (2005).
Voyiadjis, G. Z., Abu Al-Rub, R. K., and Palazotto, A. N. *Arch. Mech.* **55**, 39–89 (2003).
Vvedensky, D. D. *J. Phys. Condens. Mat.* **16**, R1537–R1576 (2004).
Waldrop, M. M., "Complexity: The Emerging Science at The Edge of Order and Chaos". Simon & Schuster, New York (1992).
Walker, D. D., Gylling, B., and Selroos, J. O. *Ground Water* **43**, 40–51 (2005).
Wang, L. Q. *Int. J. Heat Mass Transf.* **37**, 2627–2634 (1994).
Wang, L. Q. *J. Fluid Mech.* **352**, 341–358 (1997a).
Wang, L. Q. *Phys. Rev. E* **55**, 1732–1738 (1997b).
Wang, L. Q. *Appl. Phys. Lett.* **73**, 1329–1330 (1998a).
Wang, L. Q. *Surf. Rev. Lett.* **5**, 1015–1022 (1998b).
Wang, L. Q. *Int. J. Non-linear Mech.* **34**, 35–50 (1999a).
Wang, L. Q. *Progr. Theor. Phys.* **101**, 541–557 (1999b).
Wang, L. Q. *Int. J. Heat Mass Transf.* **43**, 365–373 (2000a).
Wang, L. Q. *Transp. Porous Med.* **39**, 1–24 (2000b).
Wang, L. Q. *Trend Heat Mass Moment. Transf.* **8**, 153–195 (2002).
Wang, L. Q. *Chaos Solitons Fractals* **34**, 368–375 (2007).
Wang, L. Q., and Cheng, K. C. *Phys. Rev. E* **51**, 1155–1161 (1995).
Wang, L. Q., and Cheng, K. C. *Phys. Fluids* **8**, 1553–1573 (1996).
Wang, L. Q., and Cheng, K. C. *Int. J. Rotat. Machin.* **3**, 215–231 (1997).
Wang, L. Q., and Wei, X. H. *Int. J. Heat Mass Transf.* **51**, 1751–1756 (2008).
Wang, L. Q., and Xu, M. T. *Int. J. Heat Mass Transf.* **45**, 1165–1171 (2002).
Wang, L. Q., and Yang, T. L. *Adv. Heat Transf.* **38**, 203–255 (2004).
Wang, L. Q., and Zhou, X. S., "Dual-Phase-Lagging Heat-Conduction Equations". Shandong University Press, Jinan (2000).
Wang, L. Q., and Zhou, X. S., "Dual-Phase-Lagging Heat-Conduction Equations: Problems and Solutions". Shandong University Press, Jinan (2001).
Wang, L. Q., Xu, M. T., and Zhou, X. S. *Int. J. Heat Mass Transf.* **44**, 1659–1669 (2001).
Wang, Y. Q., Leung, L. R., McGregor, J. L., Lee, D. K., Wang, W. C., Ding, Y. H., and Kimura, F. *J. Meteorol. Soc. Jpn.* **82**, 1599–1628 (2004).
Wang, L. Q., Xu, M. T., and Wei, X. H., Dual-phase-lagging and porous-medium heat conduction processes, *in* "Emerging Topics in Heat and Mass Transfer in Porous Media-from Bioengineering and Microelectronics to Nanotechnology" (P. Vadasz Ed.), pp. 1–37. Springer, Berlin (2008a).

Wang, L. Q., Zhou, X. S., and Wei, X. H., "Heat Conduction: Mathematical Models and Analytical Solutions". Springer, Berlin (2008b).
Ward, A. L., Zhang, Z. F., and Gee, G. W. *Adv. Water Resour.* **29**, 268–280 (2006).
Watkins, N. W., Freeman, M. P., Chapman, S. C., and Dendy, R. O. *J. Atmos. Sol.-Ter. Phys.* **63**, 1435–1445 (2001).
Weinan, E., Engquist, B., and Huang, Z. Y. *Phys. Rev. B* **67**, 092101 (2003).
Western, A. W., and Blöschl, G. *J. Hydrol.* **217**, 203–224 (1999).
Western, A. W., Grayson, R. B., and Blöschl, G. (Ed.), Scaling of soil moisture: a hydrologic perspective. *Annu. Rev. Earth Planet. Sci.* **30**, 149–180 (2002).
Wett, B. *Water Sci. Technol.* **53**(12), 121–128 (2006).
Whitaker, S. *AICHE J.* **13**, 420 (1967).
Whitaker, S. *Ind. Eng.. Chem.* **61**, 14–28 (1969).
Whitaker, S. *Transp. Porous Media* **1**, 3–25 (1986).
Whitaker, S. *Transp. Porous Media* **25**, 27–61 (1996).
Whitaker, S., "The Method of Volume Averaging". Kluwer Academic, Dordrecht (1999).
Wilby, R. L. *Hydrol. Processes* **19**, 3201–3219 (2005).
Wong, M. T. F., and Asseng, S. *Plant Soil* **283**, 203–215 (2006).
Wood, E. F., Sivapalan, M., Beven, K., and Band, L. *J. Hydrol.* **102**, 29–47 (1988).
Wood, A. W., Leung, L. R., Sridhar, V., and Lettenmaier, D. P. *Clim. Change* **62**, 189–216 (2004).
Woth, K., Weisse, R., and von Storch, H. *Ocean Dyn.* **56**, 3–15 (2006).
Xiao, S. P., and Yang, W. X. *Comput. Mater. Sci.* **37**, 374–379 (2006).
Xiao, C. W., Janssens, I. A., Yuste, J. C., and Ceulemans, R. *Trees Struct. Func.* **20**, 304–310 (2006).
Xie, M., Agus, S. S., Schanz, T., and Kolditz, O. *Int. J. Numer. Analy. Meth. Geomech.* **28**, 1479–1502 (2004).
Xoplaki, E., Gonzalez-Rouco, J. F., Luterbacher, J., and Wanner, H. *Clim. Dyn.* **23**, 63–78 (2004).
Xu, Y., and Subramaniam, S. *Phys. Fluids* **18**, (2006).
Xu, M. T., and Wang, L. Q. *Int. J. Heat Mass Transf.* **45**, 1055–1061 (2002).
Xu, M. T., and Wang, L. Q. *Int. J. Heat Mass Transf.* **48**, 5616–5624 (2005).
Xu, C. Y., Widen, E., and Halldin, S. *Adv. Atmos. Sci.* **22**(6), 789–797 (2005).
Yang, T. L., and Wang, L. Q. *Comput. Mech.* **26**, 520–527 (2000).
Yang, T. L., and Wang, L. Q. *Comput. Mech.* **29**, 520–531 (2002).
Yang, T. L., and Wang, L. Q. *Int. J. Heat Mass Transf.* **46**, 613–629 (2003).
Yang, C., Chandler, R. E., Isham, V. S., Annoni, C., and Wheater, H. S. *J. Hydrol.* **302**, 239–254 (2005).
Yong, Z. *Chinese Sci. Bull.* **49**, 2415–2423 (2004).
Yoon, S., and MacGregor, J. F. *Aiche J.* **50**, 2891–2903 (2004).
Yu, H. L., and Christakos, G. *Siam J. Appl. Math.* **66**, 433–446 (2006).
Zagar, N., Zagar, M., Cedilnik, J., Gregoric, G., and Rakovec, J. *Tellus A Dyn. Meteorol. Oceanogr.* **58**, 445–455 (2006).
Zbib, H. M., and de la Rubia, T. D. *Int. J. Plasticity* **18**, 1133–1163 (2002).
Zbib, H. M., de la Rubia, T. D., and Bulatov, V. *J. Eng. Mater. Technol. Trans. ASME* **124**, 78–87 (2002).
Zhang, X. C. *Agric. Forest Meteorol.* **135**, 215–229 (2005).
Zhang, Z. F., Ward, A. L., and Gee, G. W. *Water Resour. Res.* **40**, (2004a).
Zhang, Z. F., Ward, A. L., and Gee, G. W. *J. Hydraulic Res.* **42**, 93–103 (2004b).
Zhang, L. F., Aoki, J., and Thomas, B. G. *Metal. Mater. Trans. B Process Metal. Mater. Process. Sci.* **37**, 361–379 (2006).
Zhu, J. T., and Mohanty, B. P. *Vadose Zone J.* **3**, 1464–1470 (2004).
Zhu, J., Morgan, C. L. S., Norman, J. M., Yue, W., and Lowery, B. *Geoderma* **118**, 321–334 (2004a).
Zhu, J. T., Mohanty, B. P., Warrick, A. W., and van Genuchten, M. T. *Vadose Zone J.* **3**, 527–533 (2004b).
Zhu, J. T., Zhao, J. C., Liao, Y. G., and Jiang, W. *J. Poly. Sci. Part B: Poly. Phys.* **43**, 2874–2884 (2005).
Zlokarnik, M., "Scale-Up in Chemical Engineering". Wiley-VCH, Weinheim (2006).

SUBJECT INDEX

CONTENTS OF VOLUMES IN THIS SERIAL

Volume 6 (1966)

Volume 7 (1968)

Volume 8 (1970)

Volume 9 (1974)

Volume 10 (1978)

Volume 11 (1981)

Volume 20 (1994)

Volume 21 (1995)

Volume 22 (1995)

Volume 23 (1996)

Volume 29 (2004)

Volume 30 (2005)

Volume 31 (2006)

Volume 32 (2007)

Volume 33 (2007)

Volume 34 (2008)

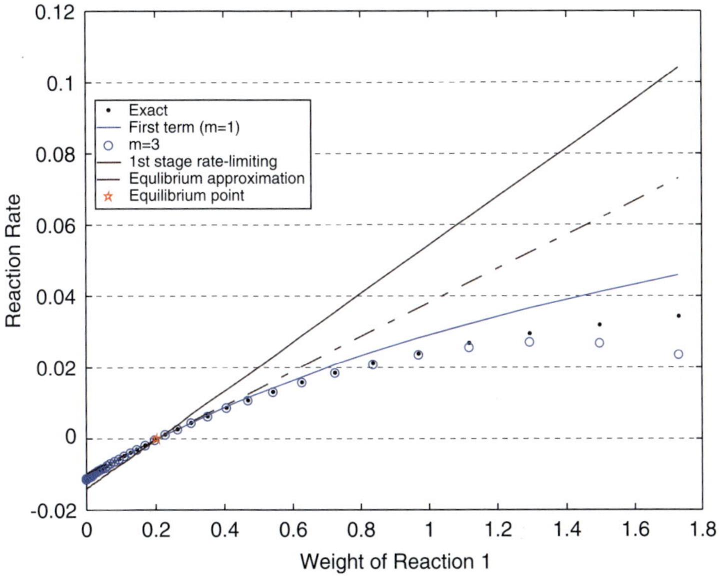

Plate 1 Overall reaction rate and its approximations: LH mechanism. Parameters: $r_1 = 0.1$, $f_2 = 14$, $r_2 = 10$, $f_3 = 1$ and $r_3 = 2$ (for Black and White version, see page 85).

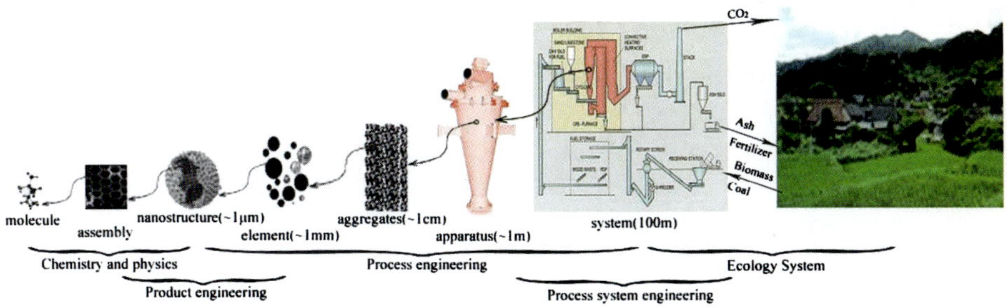

Plate 2 Multiscale chemical processes (after Li et al., 2004b) (for Black and White version, see page 182).

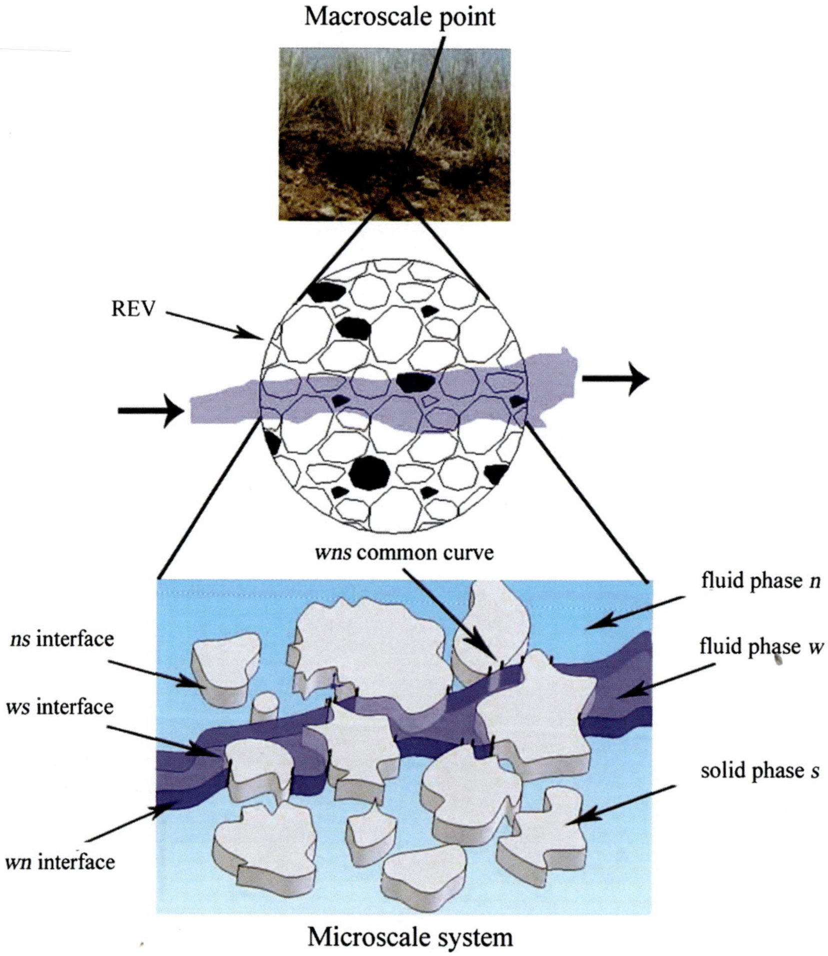

Plate 3 Three-phase microscale system (s, solid phase; w, wetting fluid phase; n, non-wetting fluid phase; after Gray, 1999) (for Black and White version, see page 204).